Lecture Notes in Computer Science 2542

Edited by G. Goos, J. Hartmanis, and J. van Leeuwen

Springer
Berlin
Heidelberg
New York
Barcelona
Hong Kong
London
Milan
Paris
Tokyo

Ivan Dimov Ivan Lirkov
Svetozar Margenov Zahari Zlatev (Eds.)

Numerical Methods and Applications

5th International Conference, NMA 2002
Borovets, Bulgaria, August 20-24, 2002
Revised Papers

Series Editors

Gerhard Goos, Karlsruhe University, Germany
Juris Hartmanis, Cornell University, NY, USA
Jan van Leeuwen, Utrecht University, The Netherlands

Volume Editors

Ivan Dimov
Ivan Lirkov
Svetozar Margenov
Bulgarian Academy of Sciences, Central Laboratory for Parallel Processing
Acad. G. Bonchev, bl. 25A, 1113 Sofia, Bulgaria
E-mail: ivdimov@bas.bg, {ivan,margenov}@parallel.bas.bg

Zahari Zlatev
National Environmental Research Institute
Frederiksborgvej 399, P.O. Box 358, 4000 Roskilde, Denmark
E-mail: zz@dmu.dk

Cataloging-in-Publication Data applied for

A catalog record for this book is available from the Library of Congress.

Bibliographic information published by Die Deutsche Bibliothek.
Die Deutsche Bibliothek lists this publication in the Deutsche Nationalbibliografie; detailed bibliographic data is available in the Internet at <http://dnb.ddb.de>.

CR Subject Classification (1998): G.1, F.2.1, G.4, J.2, J.6

ISSN 0302-9743
ISBN 3-540-00608-7 Springer-Verlag Berlin Heidelberg New York

Springer-Verlag Berlin Heidelberg New York
a member of BertelsmannSpringer Science+Business Media GmbH

http://www.springer.de

Printed in Germany

Typesetting: Camera-ready by author, data conversion by Boller Mediendesign
Printed on acid-free paper SPIN: 10871615 06/3142 5 4 3 2 1 0

Preface

This volume of the Springer Series "Lecture Notes in Computer Science" contains refereed papers which were presented at the Fifth International Conference on Numerical Methods and Applications, NMA 2002, held in Borovets, Bulgaria, during 20–24 August 2002. The NMA 2002 Conference was organized by the Central Laboratory for Parallel Processing at the Bulgarian Academy of Sciences in cooperation with SIAM (Society for Industrial and Applied Mathematics) and GAMM (Gesellschaft für Angewandte Mathematik und Mechanik). Co-organizers of this traditional scientific meeting were the Institute of Mathematics and Informatics at the Bulgarian Academy of Sciences and the Faculty of Mathematics and Informatics at the University of Sofia. The conference was devoted to the 70th anniversary of the distinguished Bulgarian mathematician academician Blagovest Sendov — the founder of the Bulgarian school in numerical analysis, the teacher of dozens of Ph.D. students, the author of more than 270 papers, textbooks, and monographs. He is the scholar who has been at the center of the mathematical life in Bulgaria for about 50 years. Over the years he headed the Bulgarian Academy of Sciences, the University of Sofia, the Faculty of Mathematics and Informatics of the University of Sofia, the Central Laboratory for Parallel Processing of the Bulgarian Academy of Sciences, the Department of Mathematical Modeling of the Institute of Mathematics and Informatics, the Union of Bulgarian Mathematicians, etc. He represented with distinction Bulgarian science in such world-recognized organizations as IFIP (International Federation for Information Processing), the International Association of Universities, the International Council of Scientific Unions, the Balkan Mathematical Union, etc. He has been the Editor-in-Chief of the journal "Mathematica Balkanica" for many years. Without his enormous pioneering work in the field of numerical analysis, this book would not have been completed and many of its papers would not have been written. His efforts and moral support have played an important role in the developments in numerical analysis and related fields in the last several decades in Bulgaria.

The conference follows the traditions of the first four conferences on Numerical Methods and Applications held in Sofia in 1984, 1988, 1994, and 1998 in providing possibilities for exchanging ideas between scientists who develop and study numerical methods and researchers and engineers who use them for solving real-life problems.

The subjects of the conference ranged from basic research to applications in physics, mechanics, engineering, environmental sciences, and other areas. Most of the presentations covered numerical methods based on the use of finite difference, finite element, and finite volume methods, boundary element methods, Monte Carlo methods, multigrid and domain decomposition, numerical linear algebra, parallel algorithms, numerical methods for nonlinear problems, computational mechanics, large-scale modeling, and engineering applications. Five special minisymposia were also held at the conference:

- Monte Carlo and Quasi-Monte Carlo methods;
- numerical analysis of problems with blow-up solutions;
- robust iterative solution methods and applications;
- control and uncertain systems; and
- numerical methods for sensor data processing.

The selected papers in this volume contain a series of new mathematical tools, which will be useful for a wide community of specialists working on the development of efficient numerical methods and their application in science and engineering. The presented results will also be of special interest to university lecturers in the areas of numerical analysis and its applications in the solution of scientific and engineering problems.

The recent results in numerical analysis together with the increasing speed and expanded storage capacity of modern computers have greatly improved the ability to solve real-life computational problems. Many scientists recognize the impact of the new generation of computers today. Moreover, the new computers and the new computational techniques stimulate new research and solutions in the field of numerical analysis. On the other hand, the progress in numerical analysis is stimulating the creation of new and more efficient computational tools.

The Fifth International Conference on Numerical Methods and Applications and the present volume in particular are the outcome of the joint efforts of many colleagues from various institutions and organizations. First of all, we would like to thank all the members of the Program Committee for their valuable contribution towards forming the "scientific face" of the conference, as well as for their help in reviewing and editing the contributed papers. We would like to specially thank the organizers of the minisymposia: Aneta Karaivanova, Todor Gurov, Stefka Dimova, Rita Meyer-Spasche, Oleg Iliev, Maya Neytcheva, Svetoslav Markov, Mikhail Krastanov, Vladimir Veliov, Tzvetan Semerdjiev, and Herman Bruyninckx. The organizers of the minisymposia significantly contributed to the improved scientific level and the success of the conference.

Special support for the conference from the Information Society Centre of Excellence for Education Science and Technology BIS-21 through a grant from the European Commission (contract number ICA1-CT-2000-70016) is gratefully acknowledged.

Details regarding the conference are available on its home page: http://www.bas.bg/clpp/nma02.html.

Sofia, December 2002

Ivan Dimov
Ivan Lirkov
Svetozar Margenov
Zahari Zlatev

Table of Contents

I Invited Papers

II Monte Carlo and Quasi-Monte Carlo Methods

III Robust Iterative Solution Methods and Applications

IV Control and Uncertain Systems

V Numerical Methods for Sensor Data Processing

VI Contributed Papers

Part I

Invited Papers

Forgivable Variational Crimes

Robert Beauwens

Service de Métrologie Nucléaire
Université Libre de Bruxelles, C.P. 165/84
50, Av. F.D. Roosevelt, B - 1050 Brussels

Abstract. It is common use to call variatonal crimes those applications of the variational method or more generally, of the Galerkin method, where not all the assumptions needed to validate the method are exactly satisfied. This covers a.o. the use of trial functions for second order elliptic problems that are only approximately continuous along element boundaries or the use of quadrature formulas to only approximately compute the entries of the stiffness matrix or the replacement of the exact domain boundary by an approximate one, to cite the main examples. Classical techniques based on bounding the so-called consistency error terms have been used to analyse and most often absolve such crimes.
In the present contribution, we propose an alternate approach for a restricted class of variational crimes such that there is no approximation on the representation of the RHS of the exact equation and such that a generalized variational principle can be introduced in such way that both the exact and the approximate problems appear as Galerkin approximations of this generalized problem. This reduces their analysis to successive applications of the variational method itself and produces essentially the same error bounds as perfectly legal applications of the variational method, whence our suggestion to consider such crimes as forgivable.
Examples of applications including and generalizing the PCD method presented elsewhere in this conference are considered by way of illustration.

1 Introduction

The main purpose of the present contribution is to draw attention on a particular kind of variational crime whose analysis may be reduced to successive applications of the Ritz-Galerkin procedure. In such case, a priori error bounds can be derived from interpolation (or any other approximation) theory in the corresponding spaces, on the basis of Cea's lemma only as in the case of conformal finite elements. For that reason, we suggest to consider such variational crimes as *forgivable.*

We wish to proceed with a concrete example, namely the PCD discretization method introduced elsewhere [1] in this conference. It will lead us to an alternate introduction of that method presenting it as a Ritz-Galerkin approximation of a generalized variational problem, at least inasmuch as its RHS is represented exactly.

I. Dimov et al. (Eds.): NMA 2002, LNCS 2542, pp. 3–11, 2003.

We shall also take this opportunity to show possible ways to handle more general triangular and quadrangular meshes, hereby extending its range of applicability.

2 The Structure of PCD Discreti ations

The main feature of PCD discretizations consists in representing the unknown distribution and its derivatives by piecewise constant distributions but on distinct meshes. For instance in the case of a second order boundary value problem, where we need to represent distributions v and their first derivatives $\partial_i v$, we introduce piecewise constant distributions $v_h, \partial_{hi} v_h$ each defined on a specific mesh. In the case of a regular rectangular mesh (i. e. without mesh refinement) in 2 dimensions, a possible way to proceed consists in endowing each rectangular mesh element with the three submeshes shown on Fig. 1 together with the difference quotients shown on the same figure to relate the values of $\partial_{hi} v_h$ to those of v_h and adding the condition that v_h must be continuous across element boundaries.

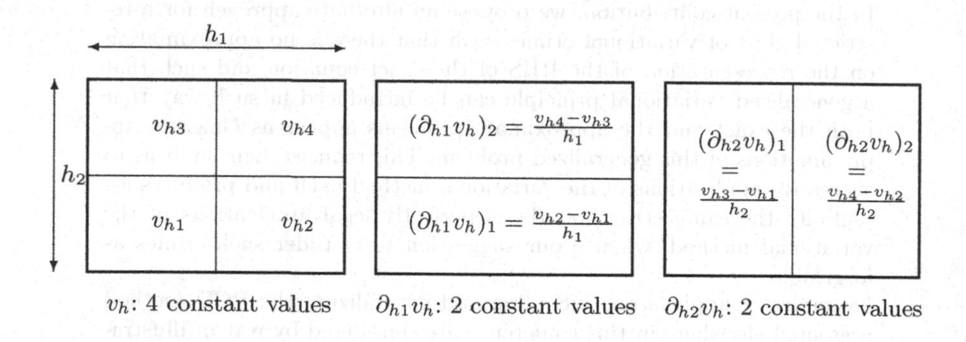

Fig. 1. A regular rectangular element

To handle such a discretization, we need to introduce a space structure such as depicted on Fig. 2. To fix ideas, we consider the case of a second order elliptic boundary value problem defined on some rectangular domain Ω and discretized with the rectangular PCD elements of Fig. 1. Here and in the following,

- H denotes the Sobolev space used in the variational formulation of the given boundary value problem; for the present example, it is some space between $H_0^1(\Omega)$ and $H^1(\Omega)$, with $H^1(\Omega)$ scalar product;
- H_h denotes the space of piecewise constant distributions v_h endowed with an appropriate scalar product $(u_h, v_h)_h$; for the present example, we use $(u_h, v_h)_h = (u_h, v_h) + \sum_i (\partial_{hi} u_h, \partial_{hi} v_h)$; we shall also use the notation H_{h0}, to denote the same space but with the $L^2(\Omega)$ scalar product;
- $E = (L^2(\Omega))^3$;

- F denotes the subspace of E made up of $U = (u_0, u_1, u_2)$ such that $u_i = \partial_i u_0$, $i = 1, 2$ (together with the requirement that u_0 has zero trace on appropriate parts of $\partial\Omega$, which makes sense since $U \in F \Rightarrow u_0 \in H^1(\Omega)$);
- F_h denotes the subspace of E made up of $U_h = (u_{h0}, u_{h1}, u_{h2})$ such that $u_{hi} = \partial_{hi} u_{h0}$, $i = 1, 2$ (together with the requirement that u_{h0} has zero trace on appropriate parts of $\partial\Omega$).

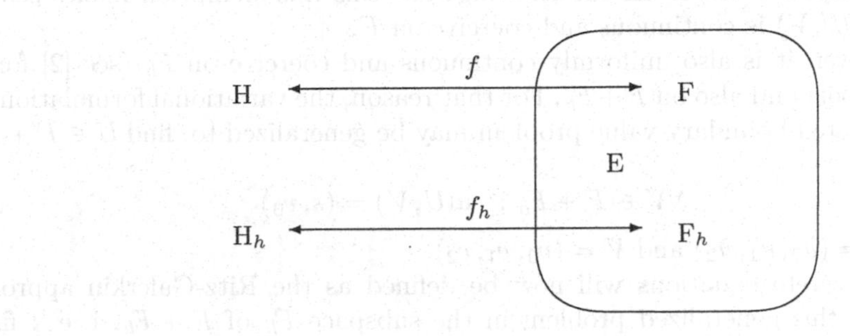

Fig. 2. Discretization structure

We need such a structure because we cannot directly compare elements of H and H_h while we can measure the distance between their images in E through the one-to-one correspondances f and f_h

- $f : H \rightarrow F : u \rightarrow U = (u, \partial_1 u, \partial_2 u)$
- $f_h : H_h \rightarrow F_h : u_h \rightarrow U_h = (u, \partial_{h1} u_h, \partial_{h2} u_h)$

3 Discrete E uations

Let us now consider the continuous and coercive bilinear form $a(u, v)$ on $H \times H$ associated with the variational formulation of a second order elliptic boundary value problem defined on some domain Ω in R^2. Typically such $a(u, v)$ has the form

$$a(u, v) = \int_{\Omega} (p(x)\partial_1 u(x)\partial_1 v(x) + p(x)\partial_2 u(x)\partial_2 v(x) + q(x)u(x)v(x))dx \qquad (1)$$

with possibly additional boundary terms which we don't need to include here as they don't modify the principles of the method. For the continuity of $a(u, v)$, we assume that p and $q \in L^\infty(\Omega)$ and, for its coerciveness, we assume that $p(x) \geq p_0 > 0$ and $q(x) \geq 0$ on Ω. Given some RHS $s \in L^2(\Omega)$, the associated boundary value problem may be defined as: find $u \in H$ such that

$$\forall v \in H : \quad a(u, v) = (s, v) \qquad (2)$$

Such a form may now be redefined on the subspace F of E through

$$a(U,V) = \int_\Omega (p(x)u_1(x)v_1(x) + p(x)u_2(x)v_2(x) + q(x)u_0(x)v_0(x))dx \quad (3)$$

where $U = (u_0, u_1, u_2)$, $V = (v_0, v_1, v_2)$ and $U, V \in F$. The last assumption means that $u_i = \partial_i u_0$ and $v_i = \partial_i v_0$, $i = 1, 2$ and thus implies, as already stressed, that u_0 and v_0 have square summable derivatives on Ω, whence also square summable traces on $\partial\Omega$ showing that this new definition makes sense. Clearly, $a(U, V)$ is continuous and coercive on F.

Moreover, it is also uniformly continuous and coercive on F_h (see [2] for a formal proof) and also on $F + F_h$. For that reason, the variational formulation of the associated boundary value problem may be generalized to: find $\tilde{U} \in F + F_h$ such that

$$\forall V \in F + F_h : \quad a(\tilde{U}, V) = (s, v_0) \quad (4)$$

where $\tilde{U} = (\tilde{u}_0, \tilde{u}_1, \tilde{u}_2)$ and $V = (v_0, v_1, v_2)$.

The discrete equations will now be defined as the Ritz-Galerkin approximation of this generalized problem in the subspace F_h of $F + F_h$, i. e. : find $U_h = (u_h, \partial_{h1} u_h, \partial_{h2} u_h)$ such that

$$\forall V_h = (v_h, \partial_{h1} v_h, \partial_{h2} v_h) \in F_h : \quad a(U_h, V_h) = (s, v_h) \quad (5)$$

We further notice that the exact given boundary value problem, namely: find $U \in F$ such that

$$\forall V \in F : \quad a(U, V) = (s, v_0) \quad (6)$$

appears now as the Ritz-Galerkin approximation of the generalized problem (4) in the subspace F of $F + F_h$.

4 Convergence Analysis

Summarizing, we have the following situation:

- a generalized problem: eq. (4),
- the exact problem: eq. (6),
- an approximate problem: eq. (5).

Both the exact and approximate problems are Ritz-Galerkin approximations of the generalized problem, in F and F_h respectively. Therefore U is the a-orthogonal projection of $\tilde{U}$ on F and U_h is the a-orthogonal projection of $\tilde{U}$ on F_h and we have that

$$\|U - \tilde{U}\|_E \leq C \min_{V \in F} \|V - \tilde{U}\|_E \quad (7)$$

$$\|U_h - \tilde{U}\|_E \leq C \min_{V \in F_h} \|V - \tilde{U}\|_E \quad (8)$$

whence also

$$\|U - U_h\|_E \leq C \left(\min_{V \in F} \|V - \tilde{U}\|_E + \min_{V \in F_h} \|V - \tilde{U}\|_E\right) . \quad (9)$$

But $\tilde{U} = \tilde{U}_F + \tilde{U}_{F_h}$ with $\tilde{U}_F \in F$ and $\tilde{U}_{F_h} \in F_h$ whence

$$\min_{V \in F} \|V - \tilde{U}\|_E = \min_{V \in F} \|V - \tilde{U}_{F_h}\|_E \tag{10}$$

since $\tilde{U}_F$ can be represented exactly in this case and

$$\min_{V \in F} \|V - \tilde{U}\|_E = \min_{V \in F_h} \|V - \tilde{U}_F\|_E \tag{11}$$

since $\tilde{U}_{F_h}$ can now be represented exactly. Therefore

$$\|U - U_h\|_E \le C \, (\min_{V \in F} \|V - \tilde{U}_{F_h}\|_E + \min_{V \in F_h} \|V - \tilde{U}_F\|_E) \, . \tag{12}$$

In particular, we may use for the first term any V for which $\tilde{U}_{F_h}$ is an interpolant and for the second term the interpolant of $\tilde{U}_F$ in F_h (at least in those cases where $\tilde{U}_F$ is sufficiently regular). Of course, one may also use any other approximation of $U \in F$ by some appropriate element $U_h \in F_h$, for example its L^2-orthogonal projection rather than interpolation.

This however is not the whole story because we have disregarded the RHS, assuming that it was exactly represented in all cases. And while this is true for the exact and generalized problems, it may fail for the approximate one. To be true in the latter case, the source itself has to be exact on H_{h0}. Otherwise, we need an approximation s_h to s. This raises a new source of error, namely

$$(s, v) - (s_h, v) \, . \tag{13}$$

Letting Π_h be any type of approximation of v in H_{h0}, we may for example choose

$$(s_h, v) = (s, \Pi_h v) \, . \tag{14}$$

Writing u for the exact solution with RHS s and u_R^h for the exact solution with RHS s_h (not to be confused with u_h which lies in H_h while $u_R^h \in H$), we get the following expressions for the error

$$\begin{aligned}(s, v) - (s_h, v) &= (s - s_h, v) \\ &= (p(x)(\partial_1 u - \partial_1 u_R^h), \partial_1 v) + (p(x)(\partial_2 u - \partial_2 u_R^h), \partial_2 v) \\ &\quad + (q(x)(u - u_R^h), v)\end{aligned} \tag{15}$$

and

$$\begin{aligned}(s, v) - (s_h, v) &= (s, v) - (s, v_h) \; = \; (s, v - v_h) \\ &= (p(x)\partial_1 u, \partial_1 v - \partial_1 v_h) + (p(x)\partial_2 u, \partial_2 v - \partial_2 v_h) \\ &\quad + (q(x)u, v - v_h)\end{aligned} \tag{16}$$

where, in the last line, it should be noticed that $\partial_i v_h$ reduces to Dirac distributions along the edges of the regions where v_h is constant, weighted by the corresponding jumps of v_h (This slight abuse of notation can easily be justified through careful cell by cell integration by parts). The last expression may be used to bound this error.

Of course, when $s = s_h$, we have $u_R^h = u$ and this error term cancels. For our crime to forgivable, we do assume here that $s = s_h$.

5 ther PCD Elements

5.1 Triangular Elements

Another possible choice for regular rectangular elements is shown on Fig. 3; as in the case of Fig. 1, we require that v_h be continuous across element boundaries. All what we did with the former element can also be done with this new one. It is further clear that this element can also be obtained by assembling two rectangular triangular elements of the type shown on Fig. 4 which were introduced by Tahiri [2] to better accomodate arbitrary shapes,

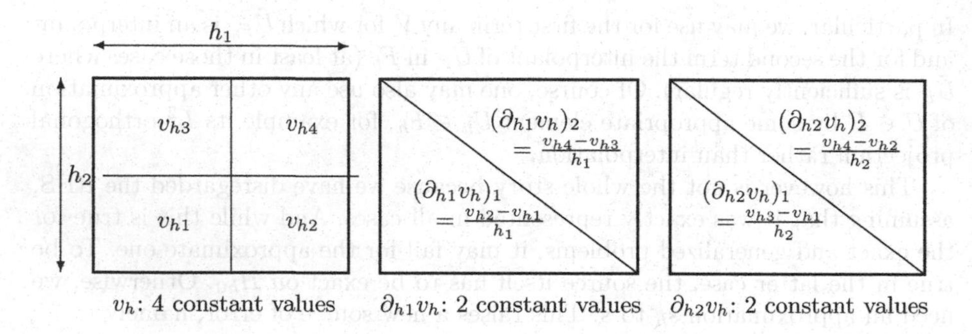

Fig. 3. Another regular rectangular element

The situation in this regard is similar to that of conformal finite elements with the element of Fig. 1 playing here the role of the BLQ conformal element, the element of Fig. 4 playing that of the PLT conformal element and the element of Fig. 3 playing that of the rectangular conformal element obtained by assembling two PLT elements. In both cases, we obtain rectangular elements that are comparable but not identical.

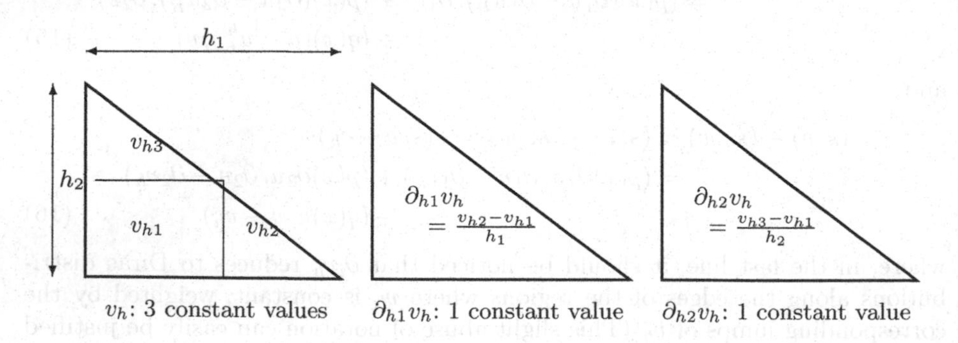

Fig. 4. A rectangular triangular element

These comparisons open the way to design more general elements. A first step in this direction is the triangular element shown on Fig. 5. For a general triangle

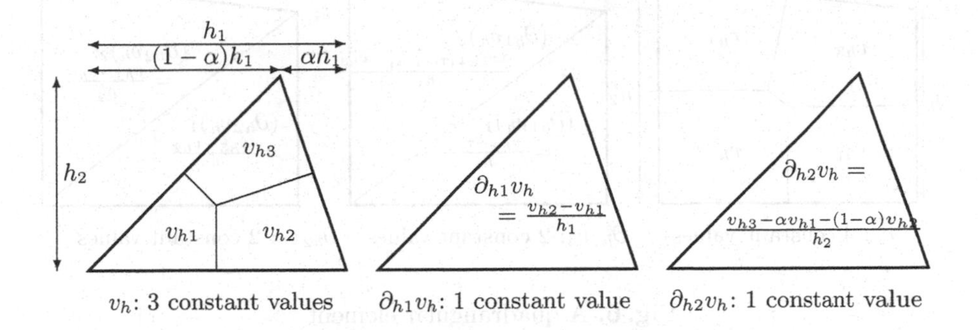

Fig. 5. A triangular element

$P_1P_2P_3$ with vertices P_i of coordinates (x_i, y_i), $i = 1, 2, 3$, H_{h0} submesh determined by the mediatrices of its edges with value v_{hi} on the mesh cell around P_i, $i = 1, 2, 3$, the formulas to compute the derivatives (constant on the triangle) are easily derived in the same way, leading to

$$\partial_{h1} v_h = \frac{\sum_{i=1}^{3} \begin{vmatrix} v_i & y_i \\ v_{i+1} & y_{i+1} \end{vmatrix}}{\sum_{i=1}^{3} \begin{vmatrix} x_i & y_i \\ x_{i+1} & y_{i+1} \end{vmatrix}}, \quad \partial_{h2} v_h = \frac{\sum_{i=1}^{3} \begin{vmatrix} x_i & v_i \\ x_{i+1} & v_{i+1} \end{vmatrix}}{\sum_{i=1}^{3} \begin{vmatrix} x_i & y_i \\ x_{i+1} & y_{i+1} \end{vmatrix}} \tag{17}$$

where index values are understood modulo 3.

The interpolation theory of distributions in F by distriburions in F_h developed in [2] in the case of rectangular PCD elements, relies on the use of the mean value theorem but is otherwise essentially independent of the geometrical shape of the elements. Therefore, in order to extend it to the case of general triangular PCD elements, the interpolations implied by the formulas (17) must represent true interpolations, not extrapolations. For that reason, one should require that all interior angles of all triangles be no larger than 90°, a restriction comparable to those used to get bounded form factors in the conformal finite element theory.

5.2 Quadrangular Elements

It becomes now straightforward to generate more general quadrangular elements since it is sufficient to assemble more general triangular elements. An example is shown on Fig. 6, where $\alpha = (h'_2 - h_2)/h'_2$. Of course, one calls for non rectangular triangles only when necessary. Thus, in the case of a trapezoidal element as considered here, it is sufficient to assemble a rectangular triangle and a triangle of the type shown on Fig. 5.

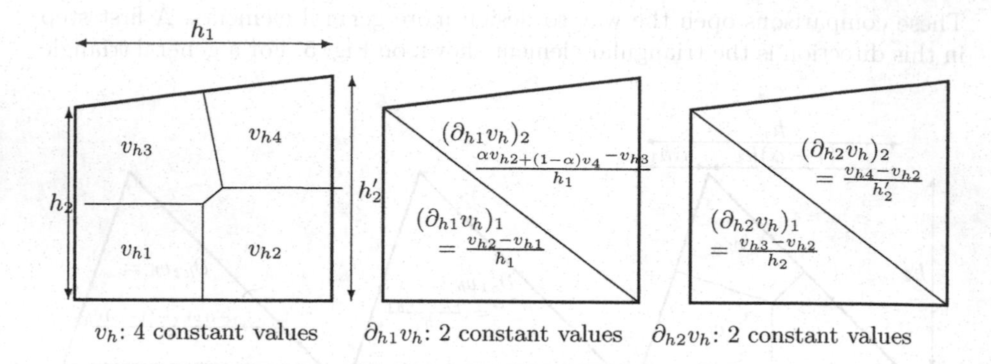

Fig. 6. A quadrangular element

In the same case, one might prefer using another quadrangular element, based on generalizing the rectangular element of Fig. 1, and this concern leads to the PCD element shown on Fig. 7.

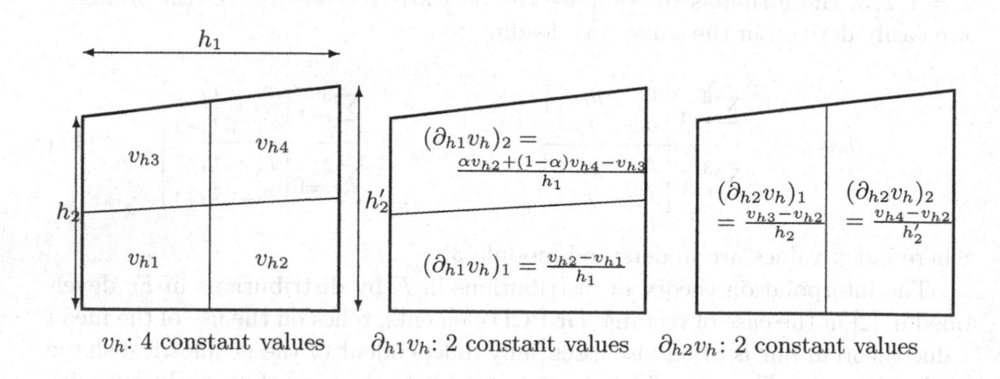

Fig. 7. Another quadrangular element

It should now be clear that a wide variety of PCD elements can be developed showing that the PCD method has indeed a wide scope of application and can in particular accomodate arbitrary polygonal domains.

5.3 A Compact Discretization Method

In the case of conformal finite elements, there is no specific advantage connected with using rectangular rather than arbitrary (well shaped) triangles for the boundary value problem associated with the general bilinear form (1) because the mass matrix is not diagonal. The situation is in this respect quite different in the case of PCD elements because they have a diagonal mass matrix. However, the more general PCD elements introduced in the last two sections introduce additional links due to interpolation and this induces a less sparse stiffness matrices.

A possible solution to recover compact discrete stencils is proposed in Tahiri [2]. It consists in combining rectangular triangular elements with a mesh refinement technique that avoids the use of slave nodes. The technique only rests on the appropriate choice of the submeshes used to represent the derivatives and the associated difference quotient formulas, which should not introduce undesired links between nodes and then, letting the variational principle determine by itself the best interpolation. We refer to [2] for a complete description.

6 Concluding Remarks

By way of conclusion, we may stress that the convergence of PCD discretization methods depends only on approximation results such as interpolation theory of distributions in F by distributions in F_h inasmuch as the source term is exactly represented and that the same conclusion holds, mutatis mutandis, whenever the discretization method is a forgivable variational crime.

A contrario, this result also shows that, when such error behaviour is not observed, it is entirely due to inadequate representation of the source. In many practical applications, the source is actually fairly well represented by piecewise constant distributions and the error behaviour will thus be close to the behaviour with exact source representation.

References

1. Tahiri, A.: The PCD method, these proceedings.
2. Tahiri, A.: A compact discretization method with local mesh refinement. PhD thesis, ULB, CP 165/84, Brussels, Belgium. In preparation (2002).

An Introduction to the Theory of Plausible and Paradoxical Reasoning

Jean Dezert

Onera, DTIM/IED,
29 Avenue de la Division Leclerc, 92320 Châtillon, France
Jean.Dezert@onera.fr

Abstract. This paper presents the basic mathematical settings of a new theory of plausible and paradoxical reasoning and describes a rule of combination of sources of information in a very general framework where information can be both uncertain and paradoxical. Within this framework, the rule of combination which takes into account explicitly both conjunctions and disjunctions of assertions in the fusion process, appears to be more simple and general than the Dempster's rule of combination. Through two simple examples, we show the strong ability of this new theory to solve practical but difficult problems where the Dempster-Shafer theory usually fails.

1 Introduction

The processing of uncertain information has always been a hot topic of research since mainly the 18th century. Up to middle of the 20th century, most theoretical advances have been devoted to the theory of probabilities. With the development of computer science, the last half of the 20th century has became very prolific for the development new original theories dealing with uncertainty and imprecise information. Mainly three major theories are available now as alternative of the theory of probabilities for the automatic plausible reasoning in expert systems: the fuzzy set theory developed by L. Zadeh in sixties (1965), the Shafer's theory of evidence in seventies (1976) and the theory of possibilities by D. Dubois and H. Prade in eighties (1985) and, very recently, the *avant-gardiste* neutrosophy unifying theory by F. Smarandache (2000). This paper is a brief introduction of the new theory of plausible and paradoxical reasoning developped by the author which can be interpreted as a generalization of the theory of evidence. Due to space limitation, only a very short presentation of the Dempster-Shafer theory will be presented in the next section to help to set up the foundations of our new theory in section 3. The full presentation of this theory is presented in [5]. A discussion on the justification of the new rule of combination of uncertain and paradoxical sources of evidences will appear also in section 3. Two simple illustrative examples of the power and usefulness of this new theory will also presented at the end of this paper. The mathematical foundations of this new theory can be found in [5].

I. Dimov et al. (Eds.): NMA 2002, LNCS 2542, pp. 12–23, 2003.

2 The Dem ster-Shafer Theory of Evidence

The Dempster-Shafer theory of evidence (DST) is usually considered as a generalization of the bayesian theory of subjective probability [10] and offers a simple and direct representation of ignorance [15]. The DST has shown its compatibility with the classical probability theory, with boolean logic and has a feasible computational complexity for problems of small dimension. It is a powerful theoretical tool which can be applied for the representation of incomplete knowledge, belief updating, and for combination of evidence through the Demspter's rule of combination.

2.1 Basic Belief Masses

Let $\Theta = \{\theta_i, i = 1, \ldots, n\}$ be a finite discrete set of *exhaustive* and *exclusive* elements (hypotheses) called elementary elements. Θ is called the frame of discernment of hypotheses or universe of discourse. The cardinality (number of elementary elements) of Θ is denoted $|\Theta|$. The power set $\mathcal{P}(\Theta)$ of Θ which is the set of all subsets of Θ is usually noted $\mathcal{P}(\Theta) = 2^\Theta$ because its cardinality is exactly $2^{|\Theta|}$. Any element of 2^Θ is then a composite event (disjunction) of the frame of discernment.

Definition 1. *The DST starts by defining a map associated to a body of evidence $\mathcal{B}$ (source of information), called basic belief assignment (bba)*[1] *or information granule $m(.) : 2^\Theta \rightarrow [0, 1]$ such that*

$$m(\emptyset) = 0 \quad \text{and} \quad \sum_{A \in 2^\Theta} m(A) \equiv \sum_{A \subseteq \Theta} m(A) = 1 \tag{1}$$

$m(A)$ corresponds to the measure of the partial belief that is committed *exactly* to A (degree of truth supported exactly by A) by the body of evidence $\mathcal{B}$ but not the total belief committed to A. All subsets A for which $m(A) > 0$ are called focal elements of m. The set of all focal elements of $m(.)$ is called the core $\mathcal{K}(m)$ of m. Note that $m(A_1)$ and $m(A_2)$ can both be 0 even if $m(A_1 \cup A_2) \neq 0$. Even more peculiar, note that $A \subset B \nRightarrow m(A) < m(B)$ (i.e. $m(.)$ is not monotone to inclusion). Hence, the bba $m(.)$ is in general different from a probability distribution $p(.)$.

2.2 Belief and Plausibility Functions

Definition 2. *To measure the total belief committed to $A \in 2^\Theta$, Glenn Shafer has defined the belief (credibility) function Bel(.) : $2^\Theta \rightarrow [0, 1]$ associated with bba $m(.)$ as*

$$Bel(A) = \sum_{B \subseteq A} m(B) \tag{2}$$

[1] This terminology suggested by Philippe Smets to the author appears to be less confusing than the basic probability assignment terminology (bpa) originally adopted by Glenn Shafer

It can been shown [10] that a belief function Bel(.) can be characterized without reference to the information granule $m(.)$ and that from any given belief function Bel(.), one can always associate an unique information granule $m(.)$ from the Möbius inversion formula.

Definition 3. *The plausibility $Pl(A)$ of any assertion $A \subset 2^\Theta$, which measures the total belief mass that can move into A (interpreted sometimes as the upper probability of A), is defined by*

$$Pl(A) \triangleq 1 - Bel(A^c) = \sum_{B \subseteq \Theta} m(B) - \sum_{B \subseteq A^c} m(B) = \sum_{B \cap A \neq \emptyset} m(B) \quad (3)$$

Bel(A) summarizes all our reasons to believe in A and Pl(A) expresses how much we could believe in A. Let now $(\Theta, m(.))$ be a source of information, then it is always possible to build the following *pignistic* probability [3,16] (bayesian belief function) by choosing $\forall \theta_i \in \Theta, P\{\theta_i\} = \sum_{B \subseteq \Theta | \theta_i \in B} \frac{1}{|B|} m(B)$. One always gets

$$\forall A \subseteq \Theta, \qquad \text{Bel}(A) \leq \quad [P(A) = \sum_{\theta_i \in A} P\{\theta_i\}] \quad \leq \text{Pl}(A) \quad (4)$$

2.3 The Dempster's Rule of Combination

G. Shafer has proposed the Dempster's rule of combination ($\oplus$ operator), to combine two so-called distinct bodies of evidences $\mathcal{B}_1$ and $\mathcal{B}_2$ over the same frame of discernment Θ. The global belief function Bel(.) = $\text{Bel}_1(.) \oplus \text{Bel}_2(.)$ is obtained from the combination of the information granules $m_1(.)$ and $m_2(.)$ relative to $\mathcal{B}_1$ and $\mathcal{B}_2$, as follows: $m(\emptyset) = 0$ and for any $C \neq \emptyset$ and $C \subseteq \Theta$,

$$m(C) \triangleq [m_1 \oplus m_2](C) = \frac{\sum_{A \cap B = C} m_1(A) m_2(B)}{1 - \sum_{A \cap B = \emptyset} m_1(A) m_2(B)} \quad (5)$$

$m(.)$ is a proper bba if $K \triangleq= 1 - k \equiv 1 - \sum_{A \cap B = \emptyset} m_1(A) m_2(B) \neq 0$. The quantity k is called the *weight of conflict* between the bodies of evidences. When $K = 0$ (i.e. $k = 1$), $m(.)$ *does not exist* and the bodies of evidences $\mathcal{B}_1$ and $\mathcal{B}_2$ are said to be totally contradictory. Such case arises whenever the cores of $\text{Bel}_1(.)$ and $\text{Bel}_2(.)$ are disjoint. The same problem of existence has already been pointed out in the presentation of the optimal Bayesian fusion rule in [4].

The Dempster's rule of combination proposed by G. Shafer in [10] has been strongly criticized by the disparagers of the DST in the past decades because it had not been completly well justified by the author in his book, even if this has been corrected later in [11]. The DS rule is now accepted since the axiomatic of the transferable belief model (TBM) developed by Smets in [13,7,8,14,15] from an idea initiated by Cheng and Kashyap in [1]. The Dempster's and the optimal bayesian fusion rules [4] coincide exactly when $m_1(.)$ and $m_2(.)$ become bayesian basic probability assignments and if we accept the principle of indifference within the optimal Bayesian fusion rule. Many numerical examples of the Dempster's rule of combination can be found in [10]. What is more interesting now, is to focuse our attention on the following disturbing example.

Example 1. A simple but disturbing example

In 1982, Lofti Zadeh has given to Philippe Smets during a dinner at Acapulco, the following example of a use of the Dempster's rule which shows an unexpected result drawn from the DST. Two doctors examine a patient and agree that it suffers from either meningitis (M), concussion (C) or brain tumor (T). Thus $\Theta = \{M, C, T\}$. Assume that the doctors agree in their low expectation of a tumor, but disagree in likely cause and provide the following diagnosis

$$m_1(M) = 0.99 \qquad m_1(T) = 0.01 \qquad \text{and} \qquad m_2(C) = 0.99 \qquad m_2(T) = 0.01$$

The DS rule yields the unexpected result $m(T) = \frac{0.0001}{1-0.0099-0.0099-0.9801} = 1$ which means that the patient suffers with certainty from brain tumor !!!. This unexpected result arises from the fact that the two doctors agree that patient does not suffer from tumor but are in almost full contradiction for the other causes of the disease. This very simple but practical example shows the limitations of practical use of the DST for automated reasoning. Some extreme caution on the degree of conflict of the sources must always be taken before taking a final decision based on the Dempster's rule of combination.

3 A New Theory for Plausible and Paradoxical Reasoning

3.1 Presentation

As seen in previous example, the use of the DST must be done only with extreme caution if one has to take a final and important decision from the result of the Dempter's rule of combination. In most of practical applications based on the DST, some ad-hoc or heuristic recipes must always be added to the fusion process to correctly manage or reduce the possibility of high degree of conflict between sources. Otherwise, the fusion results lead to a very dangerous conclusions (or cannot provide a reliable results at all). Even if nowadays, the DST provides fruithfull results in many applications, we strongly argue that this theory is still too limited because it is based on the two following restrictive constraints as already reported in literature

C1- The DST considers a discrete and finite frame of discernment based on a set of exhaustive and exclusive elementary elements.

C2- The bodies of evidence are assumed independent (each source of information does not take into account the knowledge of other sources) and provide a belief function on the power set 2^{Θ}.

These two constraints do not allow us to deal with the more general and practical problems involving uncertain reasoning and the fusion of uncertain, imprecise and paradoxical sources of information. The constraint $C1$ is very strong actually since it does not allow paradoxes between elements of the frame of discernment Θ. The DST accepts as foundation the commonly adopted principle of the third exclude. Even if at first glance, it makes sense in the traditional classical thought, we present here a new theory which does not accept this principle of the third exclude and accepts and deals with paradoxes.

The constraint C1 assumes that each elementary hypothesis of Θ is finely and precisely well defined and we are able to discriminate between all elementary hypotheses without ambiguity and difficulty. We argue that this constraint is too limited and that it is not always possible in practice to choose and define Θ satisfying C1 even for some very simple problems where each elementary hypothesis corresponds to a vague concept or attributes. In such cases, the elementary elements of Θ cannot be precisely separated without ambiguity such that no refinement of Θ satisfying the first constraint is possible. Our second remark concerns the universal nature of the frame of discernment. It is clear that, in general, the *same* Θ is interpreted differently by the bodies of evidence or experts. Some subjectivity on the information provided by a source of information is almost unavoidable, otherwise this would assume, as within the DST, that all bodies of evidence have an objective/universal (possibly uncertain) interpretation or measure of the phenomena under consideration. This corresponds to the C2 constraint. This vision seems to be too excessive because usually independent bodies of evidence provide their beliefs about some hypotheses only with respect to their own worlds of knowledge and experience without reference to the (inaccessible) absolute truth of the space of possibilities. Therefore, C2 is, in many cases, also a too strong hypothesis to accept as foundations for a general theory of probable and paradoxical reasoning. A general theory has to include the possibility to deal with evidences arising from different sources of information which don't have access to absolute interpretation of the elements Θ under consideration. This yields to accept the paradoxical information as the basis for a new general theory of probable reasoning. Actually, the paradoxical information arising from the fusion of several bodies of evidence is very informative and can be used to help us to take legitimous final decision as it will be seen. Our new theory can be interpreted as a general and direct extension of probability theory and the Dempster-Shafer theory in the following sense. Let $\Theta = \{\theta_1, \theta_2\}$ be the simpliest frame of discernment involving only two elementary hypotheses (with no more additional assumptions on θ_1 and θ_2), then

- the probability theory deals with basic probability assignments $m(.) \in [0,1]$ such that $m(\theta_1) + m(\theta_2) = 1$
- the Dempster-Shafer theory deals with bba $m(.) \in [0,1]$ such that $m(\theta_1) + m(\theta_2) + m(\theta_1 \cup \theta_2) = 1$
- our general theory deals with new bba $m(.) \in [0,1]$ such that

$$m(\theta_1) + m(\theta_2) + m(\theta_1 \cup \theta_2) + m(\theta_1 \cap \theta_2) = 1$$

3.2 Notion of Hyper-Power Set

Let $\Theta = \{\theta_1, \ldots, \theta_n\}$ be a set of n elementary elements considered as exhaustive which cannot be precisely defined and separated so that no refinement of Θ in a new larger set Θ_{ref} of disjoint elementary hypotheses is possible and let's consider the classical set operators $\cup$ (disjunction) and $\cap$ (conjunction). The exhaustivity assumption about Θ is not a strong constraint since when $\theta_i, i = 1, n$ does not constitute an exhaustive set of elementary possibilities, we can always add an extra element θ_0 such that $\theta_i, i = 0, n$ describes now an exhaustive set. We will assume therefore, from now on and in the following, that Θ characterizes an exhaustive frame of discernment. Θ will be called a *general* frame of discernment in the sequel to emphaze the fact that Θ does not satisfy the Dempster-Shafer C1 constraint.

Definition 4. *The classical power set* $\mathcal{P}(\Theta) = 2^\Theta$ *has been defined as the set of all proper subsets of* Θ *when all elements* θ_i *are disjoint. We extend here this notion and define now the hyper-power set* D^Θ *as the set of all composite possibilities build from* Θ *with* $\cup$ *and* $\cap$ *operators such that* $\forall A \in D^\Theta, B \in D^\Theta, (A \cup B) \in D^\Theta$ *and* $(A \cap B) \in D^\Theta$.

The cardinality of D^Θ is majored by 2^{2^n} when $\text{Card}(\Theta) = \mid \Theta \mid = n$. The generation of hyper-power set D^Θ corresponds to the famous Dedekind's problem on enumerating the set of monotone Boolean functions [2]. The choice of letter D in our notation D^Θ to represent the hyper-power set of Θ is in honour of the great mathematician R. Dedekind. The general solution of the Dedekind's problem (for $n > 10$) has not been found yet although this problem is more than one century old .... We just know that the cardinality numbers of D^Θ follow the Dedekind's numbers (minus one) when $\text{Card}(\Theta) = n$ increases, i.e. $\mid D^\Theta \mid = 1, 2, 5, 19, 167, 7580, 7828353, \ldots$ when $\text{Card}(\Theta) = n = 0, 1, 2, 3, 4, 5, 6, \ldots$ Obviously, one would always have $D^\Theta \subset 2^{\Theta_{ref}}$ if the refined power set $2^{\Theta_{ref}}$ could be defined and accessible which is unfortunately not possible in general as already argued.

Example 2.

1. for $\Theta = \{\}$ (empty set), $D^\Theta = \{\emptyset\}$ and $\mid D^\Theta \mid = 1$
2. for $\Theta = \{\theta_1\}$, $D^\Theta = \{\emptyset, \theta_1\}$ and $\mid D^\Theta \mid = 2$
3. for $\Theta = \{\theta_1, \theta_2\}$, $D^\Theta = \{\emptyset, \theta_1, \theta_2, \theta_1 \cup \theta_2, \theta_1 \cap \theta_2\}$ and $\mid D^\Theta \mid = 5$
4. for $\Theta = \{\theta_1, \theta_2, \theta_3\}$,

$$\begin{aligned} D^\Theta = \{&\emptyset, \theta_1, \theta_2, \theta_3, \\ &\theta_1 \cup \theta_2, \theta_1 \cup \theta_3, \theta_2 \cup \theta_3, \theta_1 \cap \theta_2, \theta_1 \cap \theta_3, \theta_2 \cap \theta_3, \theta_1 \cup \theta_2 \cup \theta_3, \theta_1 \cap \theta_2 \cap \theta_3, \\ &(\theta_1 \cup \theta_2) \cap \theta_3, (\theta_1 \cup \theta_3) \cap \theta_2, (\theta_2 \cup \theta_3) \cap \theta_1, (\theta_1 \cap \theta_2) \cup \theta_3, (\theta_1 \cap \theta_3) \cup \theta_2, (\theta_2 \cap \theta_3) \cup \theta_1, \\ &(\theta_1 \cup \theta_2) \cap (\theta_1 \cup \theta_3) \cap (\theta_2 \cup \theta_3)\} \end{aligned}$$

and $\mid D^\Theta \mid = 19$

3.3 The General Basic Belief Masses $m(.)$

Definition 5. *Let* Θ *be a general frame of discernment of the problem under consideration. We define a map* $m(.) : D^\Theta \rightarrow [0, 1]$ *associated to a given body of evidence* $\mathcal{B}$ *which can support paradoxical information, as follows*

$$m(\emptyset) = 0 \qquad \textit{and} \qquad \sum_{A \in D^\Theta} m(A) = 1 \tag{6}$$

The quantity $m(A)$ is called A's *general basic belief number* (gbba) or the general basic belief mass for A. As in the DST, all subsets $A \in D^\Theta$ for which $m(A) > 0$ are called focal elements of $m(.)$ and the set of all focal elements of $m(.)$ is also called the core $\mathcal{K}(m)$ of m.

Definition 6. *The belief and plausibility functions are defined in the same way as in the DST, i.e.*

$$Bel(A) = \sum_{B \in D^\Theta, B \subseteq A} m(B) \quad and \quad Pl(A) = \sum_{B \in D^\Theta, B \cap A \neq \emptyset} m(B) \tag{7}$$

Note that, we don't define here explicitly the complementary A^c of a proposition A since $m(A^c)$ cannot be precisely evaluated from $\cup$ and $\cap$ operators on D^Θ since we include the possibility to deal with a complete paradoxical source of information such that $\forall A \in D^\Theta, \forall B \in D^\Theta, m(A \cap B) > 0$. These definitions are compatible with the DST definitions when the sources of information become uncertain but rational (they do not support paradoxical information). We still have $\forall A \in D^\Theta, \mathrm{Bel}(A) \leq \mathrm{Pl}(A)$.

3.4 Construction of Pignistic Probabilities from gbba $m(.)$

The construction of a pignistic probability measure from the general basic belief masses $m(.)$ over D^Θ with $|\Theta| = n$ is still possible and is given by the general expression of the form

$$\forall i = 1, \ldots, n \qquad P\{\theta_i\} = \sum_{A \in D^\Theta} \alpha_{\theta_i}(A) m(A) \tag{8}$$

where $\alpha_{\theta_i}(A) \in [0,1]$ are weighting coefficients which depend on the inclusion or non-inclusion of θ_i with respect to proposition A. No general analytic expression for $\alpha_{\theta_i}(A)$ has been derived yet even if $\alpha_{\theta_i}(A)$ can be obtained explicitly for simple examples. When general bba $m(.)$ reduces to classical bba (i.e. the DS bba without paradoxe), then $\alpha_{\theta_i}(A) = \frac{1}{|A|}$ when $\theta_i \subseteq A$. We present here an example of a pignistic probabilities reconstruction from a general and non degenerated bba $m(.)$ (i.e. $\nexists A \in D^\Theta$ with $A \neq \emptyset$ such that $m(A) = 0$) over D^Θ.

Example 3. If $\Theta = \{\theta_1, \theta_2, \theta_3\}$ then $P\{\theta_1\}$ equals

$$\begin{aligned}
& m(\theta_1) + \frac{1}{2} m(\theta_1 \cup \theta_2) + \frac{1}{2} m(\theta_1 \cup \theta_3) + \frac{1}{2} m(\theta_1 \cap \theta_2) + \frac{1}{2} m(\theta_1 \cap \theta_3) \\
& + \frac{1}{3} m(\theta_1 \cup \theta_2 \cup \theta_3) + \frac{1}{3} m(\theta_1 \cap \theta_2 \cap \theta_3) \\
& + \frac{1/2 + 1/3}{3} m((\theta_1 \cup \theta_2) \cap \theta_3) + \frac{1/2 + 1/3}{3} m((\theta_1 \cup \theta_3) \cap \theta_2) \\
& + \frac{1/2 + 1/2 + 1/3}{3} m((\theta_2 \cup \theta_3) \cap \theta_1) + \frac{1/2 + 1/2 + 1/3}{5} m((\theta_1 \cap \theta_2) \cup \theta_3) \\
& + \frac{1/2 + 1/2 + 1/3}{5} m((\theta_1 \cap \theta_3) \cup \theta_2) + \frac{1 + 1/2 + 1/2 + 1/3}{5} m((\theta_2 \cap \theta_3) \cup \theta_1) \\
& + \frac{1/2 + 1/2 + 1/3}{4} m((\theta_1 \cup \theta_2) \cap (\theta_1 \cup \theta_2) \cap (\theta_2 \cup \theta_3))
\end{aligned}$$

Same kind of expressions can be derived for $P\{\theta_2\}$ and $P\{\theta_2\}$. The evaluation of weighting coefficients $\alpha_{\theta_i}(A)$ has been obtained from the geometrical interpretation of the relative contribution of the distinct parts of A with proposition θ_i under consideration. For example, consider $A = (\theta_1 \cap \theta_2) \cup \theta_3$ which corresponds to the area $a_1 \cup a_2 \cup a_3 \cup a_4 \cup a_5$ on the following Venn diagram.

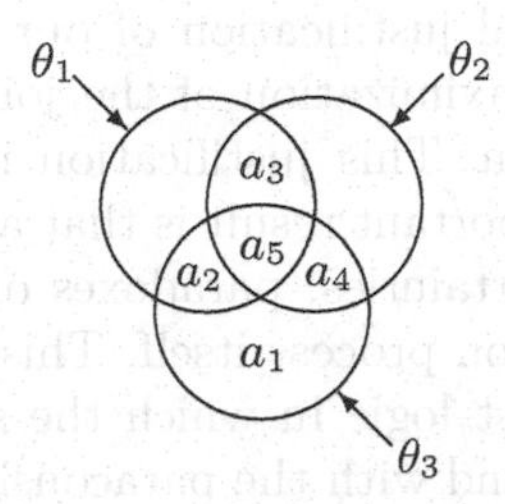

Fig. 1. Representation of $A = (\theta_1 \cap \theta_2) \cup \theta_3 \equiv a_1 \cup a_2 \cup a_3 \cup a_4 \cup a_5$

a_1 which is shared only by θ_3 will contribute to θ_3 with weight 1; a_2 which is shared by θ_1 and θ_3 will contribute to θ_3 with weight 1/2; a_3 which is not shared by θ_3 will contribute to θ_3 with weight 0; a_4 which is shared by θ_2 and θ_3 will contribute to θ_3 with weight 1/2; a_5 which is shared by both θ_1,θ_2 and θ_3 will contribute to θ_3 with weight 1/3. Since moreover, one must have $\forall A \in D^\Theta$ with $m(A) \neq 0$, $\sum_{i=1}^{n} \alpha_{\theta_i}(A)m(A) = m(A)$, it is necessary to normalize $\alpha_{\theta_i}(A)$. Therefore $\alpha_{\theta_1}(A)$, $\alpha_{\theta_2}(A)$ and $\alpha_{\theta_3}(A)$ will be given by

$$\alpha_{\theta_1}(A) = \alpha_{\theta_2}(A) = \frac{1/2 + 1/2 + 1/3}{5} \qquad \alpha_{\theta_3}(A) = \frac{1 + 1/2 + 1/2 + 1/3}{5}$$

All $\alpha_{\theta_i}(A), \forall A \in D^\Theta$ entering in derivation of the *pignistic* probabilities $P\{\theta_i\}$ can be obtained using similar process.

3.5 General Rule of Combination of Paradoxical Sources of Evidence

Let's consider now two distinct (but potentially paradoxical) bodies of evidences $\mathcal{B}_1$ and $\mathcal{B}_2$ over the same frame of discernment Θ with belief functions $\text{Bel}_1(.)$ and $\text{Bel}_2(.)$ associated with information granules $m_1(.)$ and $m_2(.)$.

Definition 7. *The combined global belief function Bel(.) = Bel₁(.) ⊕ Bel₂(.) is obtained through the combination of the granules* $m_1(.)$ *and* $m_2(.)$ *by the simple rule*

$$\forall C \in D^\Theta, \qquad m(C) \triangleq [m_1 \oplus m_2](C) = \sum_{A,B \in D^\Theta, A \cap B = C} m_1(A)m_2(B) \qquad (9)$$

Since D^Θ is closed under $\cup$ and $\cap$ operators, this new rule of combination guarantees that $m(.) : D^\Theta \rightarrow [0,1]$ is a proper general information granule statisfying (6). The global belief function Bel(.) is then obtained from the granule $m(.)$ through (7). This rule of combination is commutative and associative and can always be used for fusion of rational or paradoxical sources of information. Obviously, the decision process will have to be made with more caution to take the final decision based on the general granule $m(.)$ when internal paradoxical conflicts arise. The theoretical justification of our rule of combination can be obtained as in [17] by the maximization of the joint entropy of the two paradoxical sources of information. This justification is reported in details in the companion paper [5]. The important result is that any fusion of sources of information generates either uncertainties, paradoxes or more generally both. This is intrinsic to the general fusion process itself. This general fusion rule can also be used within the intuitionist logic in which the sum of bba is allowed to be less than one ($\sum m(A) < 1$) and with the paraconsistent logic in which the sum of bba is allowed to be greater than one ($\sum m(A) > 1$) as well. In such cases, the fusion result does not provide in general $\sum m(A) = 1$. In practice, for the sake of fair comparison between several alternatives or choices, it is better and simplier to deal with normalized bba to take a final important decision for the problem under consideration. A nice property of the new rule of combination of non-normalized bba is its invariance to the pre- or post-normalization process.

3.6 Zadeh's Example Revisited

Let's take back the disturbing Zadeh's example given in section 2.4. Two doctors examine a patient and agree that it suffers from either meningitis (M), concussion (C) or brain tumor (T). Thus $\Theta = \{M, C, T\}$. Assume that the two doctors agree in their low expectation of a tumor, but disagree in likely cause and provide the following diagnosis

$$m_1(M) = 0.99 \qquad m_1(T) = 0.01$$

and $\forall A \in D^\Theta, A \neq T, A \neq M, m_1(A) = 0$

$$m_2(C) = 0.99 \qquad m_2(T) = 0.01$$

and $\forall A \in D^\Theta, A \neq T, A \neq C, m_2(A) = 0$.
The new general rule of combination (9), yields the following combined information granule

$$m(M \cap C) = 0.9801 \qquad m(M \cap T) = 0.0099$$

$$m(C \cap T) = 0.0099 \qquad m(T) = 0.0001$$

From this granule, one gets

$$\begin{aligned}
\text{Bel}(M) &= m(M \cap C) + m(M \cap T) = 0.99 \\
\text{Bel}(C) &= m(M \cap C) + m(T \cap C) = 0.99 \\
\text{Bel}(T) &= m(T) + m(M \cap T) + m(C \cap T) = 0.0199
\end{aligned}$$

If both doctors can be considered as equally reliable, the combined information granule $m(.)$ mainly focuses weight of evidence on the paradoxical proposition $M \cap C$ which means that patient suffers both meningitis and concussion but almost surely not from brain tumor. This conclusion is coherent with the common sense actually. Then, no therapy for brain tumor (like heavy and ever risky brain surgical intervention) will be chosen in such case. This really helps to take important decision to save the life of the patient in this example. A deeper medical examination adapted to both meningitis and concussion will almost surely be done before applying the best therapy for the patient. Just remember that in this case, the DST had concluded that the patient had brain tumor with certainty ....

3.7 Mahler's Example Revisited

Let's consider now the following example excerpt from the R. Mahler's paper [9]. We consider that our classification knowledge base consists of the three (imaginary) new and rare diseases corresponding to following frame of discernment

$$\Theta = \{\theta_1 = \textit{kotosis}, \theta_2 = \textit{phlegaria}, \theta_3 = \textit{pinpox}\}$$

We assume that the three diseases are equally likely to occur in the patient population but there is some evidence that *phlegaria* and *pinpox* are the same disease and there is also a small possibility that *kotosis* and *phlegaria* might be the same disease. Finally, there is a small possibility that all three diseases are the same. This information can be expressed by assigning a priori bba as follows

$$m_0(\theta_1) = 0.2 \qquad m_0(\theta_2) = 0.2 \qquad m_0(\theta_3) = 0.2$$
$$m_0(\theta_2 \cap \theta_3) = 0.2 \; m_0(\theta_1 \cap \theta_2) = 0.1 \; m_0(\theta_1 \cap \theta_2 \cap \theta_3) = 0.1$$

Let Bel(.) the prior belief measure corresponding to this prior bba $m(.)$. Now assume that Doctor D_1 and Doctor D_2 examine a patient and deliver diagnoses with following reports:

- Report for D_1: $m_1(\theta_1 \cup \theta_2 \cup \theta_3) = 0.05 \quad m_1(\theta_2 \cup \theta_3) = 0.95$
- Report for D_2: $m_2(\theta_1 \cup \theta_2 \cup \theta_3) = 0.20 \quad m_2(\theta_2) = 0.80$

The combination of the evidences provided by the two doctors $m' = m_1 \oplus m_2$ obtained by the general rule of combination (9) yields the following bba $m'(.)$

$$m'(\theta_2) = 0.8 \qquad m'(\theta_2 \cup \theta_3) = 0.19 \qquad m'(\theta_1 \cup \theta_2 \cup \theta_3) = 0.01$$

The combination of bba $m'(.)$ with prior evidence $m_0(.)$ yields the final bba $m = m_0 \oplus m' = m_0 \oplus [m_1 \oplus m_2]$ with

$$m(\theta_1) = 0.002 \qquad m(\theta_2) = 0.200 \qquad m(\theta_3) = 0.040$$
$$m(\theta_1 \cap \theta_2) = 0.260 \qquad m(\theta_2 \cap \theta_3) = 0.360 \; m(\theta_1 \cap \theta_2 \cap \theta_3) = 0.100$$
$$m(\theta_1 \cap (\theta_2 \cup \theta_3)) = 0.038$$

Therefore the final belief function given by (7) is

$$\begin{aligned}
\text{Bel}(\theta_1) &= 0.002 + 0.260 + 0.100 + 0.038 = 0.400\\
\text{Bel}(\theta_2) &= 0.200 + 0.260 + 0.360 + 0.100 = 0.920\\
\text{Bel}(\theta_3) &= 0.040 + 0.360 + 0.100 = 0.500\\
\text{Bel}(\theta_1 \cap \theta_2) &= 0.260 + 0.100 = 0.360\\
\text{Bel}(\theta_2 \cap \theta_3) &= 0.360 + 0.100 = 0.460\\
\text{Bel}(\theta_1 \cap (\theta_2 \cup \theta_3)) &= 0.038 + 0.100 = 0.138\\
\text{Bel}(\theta_1 \cap \theta_2 \cap \theta_3) &= 0.100
\end{aligned}$$

Thus, on the basis of all the evidences one has, we are able to conclude with high a degree of belief that the patient has phlegaria which is coherent with the Mahler's conclusion based on his Conditioned Dempster-Shafer theory developed from his conditional event algebra although a totally new and simpliest approach has been adopted here.

4 Conclusion

In this paper, the foundations for a new theory of paradoxical and plausible reasoning have been shortly presented. This theory takes into account in the combination process itself the possibility for both uncertain and paradoxical information. The basis for the development of this theory is to work with the hyper-power set of the frame of discernment relative to the problem under consideration rather than its classical power set since, in general, the frame of discernment cannot be fully described in terms of an exhaustive and exclusive list of disjoint elementary hypotheses. In this general framework, no refinement is possible to apply directly the classical Dempster-Shafer theory of evidence. In our new theory, the rule of combination is justified from the maximum entropy principle and there is no mathematical impossibility to combine sources of evidence even if they appear at first glance in contradiction (in the Shafer's sense) since the paradox between sources is fully taken into account in our formalism. We have shown that in general, the combination of evidence yields unavoidable paradoxes and, through two simple, but illustrative, examples, that conclusion drawn from this new theory, provides a result which agrees with the human reasoning and becomes useful to take a decision on complex problems where DST usually fails. A complete presentation of this new theory is available in [5].

References

1. Cheng Y., Kashyap R.L., "Study of the Different Methods for Combining Evidence", Proceedings of SPIE on Applications of Artificial Intelligence, Vol. 635, pp. 384-393, 1986.
2. Dedekind R.,"Über Zerlegungen von Zhalen durch ihre grössten gemeinsammen Teiler", In Gesammelte Werke, Bd. 1., pp. 103-148, 1897.
3. Dezert J., "Autonomous Navigation with Uncertain Reference Points using the PDAF", Chapter 9 in Multitarget-Multisensor Tracking: Applications and Advances, Vol. 2, pp. 271-324,Y. Bar-Shalom Editor, Artech House, 1991.

4. Dezert J., "Optimal Bayesian Fusion of Multiple Unreliable Classifiers", Proceedings of 4th Intern. Conf. on Information Fusion (Fusion 2001), Montréal, Aug. 7-10, 2001.
5. Dezert J., "Foundations for a new theory of plausible and paradoxical reasoning", To appear in next issue of the International Journal of Information & Security, edited by Prof. Tzv. Semerdjiev, CLPP, Bulgarian Academy of Sciences, 40 pages, 2002.
6. Dubois D., Prade H.,"Théories des Possibilités. Application à la Représentation des Connaissances en Informatique", Editions Masson, Paris, 1985.
7. Dubois D., Garbolino P., Kyburg H.E., Prade H., Smets, Ph.,"Quantified Uncertainty", J. Applied Non-Classical Logics, Vol.1, pp. 105-197, 1991.
8. Klawonn F., Smets Ph., "The dynamic of belief in the transferable belief model and specialization- generalization matrices", in Uncertainty in Artificial Intelligence 92, pp 130-137, Dubois D. and Wellman M. P. and D'Ambrosio B. and Smet, Ph. Editors, Morgan Kaufman, San Mateo, Ca, 1992.
9. Mahler R., "Combining Ambiguous Evidence with Respect to Abiguous a priori Knowledge, I: Boolean Logic", IEEE Trans. on SMC, Part 1: Systems and Humans, Vol. 26, No. 1, pp. 27-41, 1996.
10. Shafer G.,"A Mathematical Theory of Evidence", Princeton University Press, Princeton, New Jersey, 1976.
11. Shafer G.,Tversky A., "Languages and designs for probability", Cognitive Sc., Vol.9, pp. 309-339, 1985.
12. Smarandache F.,"An Unifying Field in Logics: Neutrosophic Logic", (Second Edition), American Research Press, Rehoboth, 2000 (ISBN 1-879585-76-6).
13. Smets Ph.,"The Combination of Evidence in the Transferable Belief Model", IEEE Trans. on PAMI, Vol. 12, no. 5, 1990.
14. Smets Ph.,"The alpha-junctions: combination operators applicable to belief function", Qualitative and quantitative practical reasoning, Springer, Gabbay D.M and Kruse R. and Nonnengart A. and Ohlbach H.J. Editors, pp. 131-153, 1997.
15. Smets Ph.,"The transferable belief model for quantified belief representation", Handbook of Defeasible Reasoning and Uncertainty Management Systems, D. M. Gabbay and Ph. Smets (Editors), Vol. 1, Kluwer, Doordrecht, The Netherlands, 1998.
16. Smets Ph., "Data Fusion in the Transferable Belief Model", Proceedings of 3rd Int. Conf. on Inf. Fusion (Fusion 2000), pp. PS-21–PS33, Paris, July 10-13, 2000 (http://www.onera.fr/fusion2000)
17. Sun H., he K., Zhang B.,"The Performance of Fusion Judgment on Dempster-Shafer Rule", Chinese Journal of Electronics, Vol. 8, no. 1, Jan. 1999.
18. Zadeh L.A.,"A Theory of Approximate Reasoning" Machine Intelligence, J. Hayes, D. Michie and L. Mikulich Eds, Vol. 9, pp. 149-194, 1979.

Comparison of Ten Methods for the Solution of Large and Sparse Linear Algebraic Systems

Kyle A. Gallivan[1], Ahmed H. Sameh[2], and Zahari Zlatev[3]

[1] CSIT, Florida State University, Tallahassee, Florida, USA
gallivan@csit.fsu.edu
[2] CS Dept., Purdue University, West Lafayette, Indiana, USA
sameh@cs.purdue.edu
[3] National Environmental Research Institute
Frederiksborgvej 399, P. O. Box 358, DK-4000 Roskilde, Denmark
zz@dmu.dk

Abstract. The treatment of systems of linear algebraic equations is very often the most time-consuming part when large-scale applications arising in different fields of science and engineering are to be handled on computers. These systems can be very large, but in the most of the cases they are sparse (i.e. many of the elements in their coefficient matrices are zeros). Therefore, it is very important to select fast, robust and sufficiently accurate methods for the solution of large and sparse systems of linear algebraic equations. Tests with ten well-known methods have been carried out. Most of the methods are preconditioned conjugate gradient-type methods. Two important issues are mainly discussed: (i) the problem of finding automatically a good preconditioner and (ii) the development of robust and reliable stopping criteria. Numerical examples, which illustrate the efficiency of the developed algorithms for finding the preconditioner and for stopping the iterations when the required accuracy is achieved, are presented. The performance of the different methods for solving systems of linear algebraic equations is compared. Several conclusions are drawn, the main of them being the fact that it is necessary to include several different methods for the solution of large and sparse systems of linear algebraic equations in software designed to be used in the treatment of large-scale scientific and engineering problems.

1 Efficient Solvers of Sparse Linear Algebraic Equations

The solution of systems of linear algebraic equations is normally an essential part of the computational work when large mathematical models arising in different fields of science and engineering are to be handled numerically. Even when low accuracy of the solution is required, one must ensure that the imposed accuracy requirements are satisfied. Two examples will be given in order to demonstrate the importance of this requirement.

In many cases it is desirable to establish how the results vary when some key parameter is varied. This leads to many runs with different values of the key parameter in order to determine the dependence of the model results on the

I. Dimov et al. (Eds.): NMA 2002, LNCS 2542, pp. 24–35, 2003.

selected parameter. This relationship can be determined only if the errors arising from the solution of the systems of linear algebraic equations are sufficiently small and do not interfere with the changes caused by the variation of the parameter chosen.

If stiff systems of ordinary differential equations (ODEs) are solved by implicit integration methods, then normally the errors of the method for solving ODEs must be controlled. The control will be efficient only if the errors arising in the solution of the systems of linear algebraic equations do not interfere with the errors caused by the method for solving stiff ODEs.

Many more examples can be given, but these two examples are showing that it is necessary to check the accuracy of the solution obtained by using the selected algorithm for solving systems of linear algebraic equations. This is especially true when iterative methods are used. In the latter case, one controls as a rule not the accuracy of the solution of the system of linear algebraic equations, but other quantities as, for example, the smallness of some norm of the residual vector. It will be shown that such a control is not always providing reliable results.

Ten different methods are selected (including a direct algorithm based on the Gaussian elimination, the simple iterative refinement algorithm, pure conjugate gradient-type methods and preconditioned conjugate gradient-type methods). The automatic computation of preconditioners and the choice of stopping criteria for the iterative methods are discussed. It is shown that the traditionally used stopping criteria (based on continuing the computations until some norm of the residual vector or the scaled residual vector becomes smaller than a prescribed in advance quantity) are in many cases not sufficient to ensure robust and reliable control of the error arising in the solution of the system of linear algebraic equations which is solved. Other stopping criteria are suggested. These criteria are based on two important principles: (a) an attempt to evaluate the rate of convergence and to use it in the stopping criteria and (b) a check of the variability of some important parameters, which are calculated and used at each iteration of the iterative process. Stopping the iterative process is discouraged when these parameters (i) vary too much from one iteration to another, (ii) are increasing too quickly or (iii) are decreasing too quickly. It is shown that the iterative process produces more reliable results when such stopping criteria are implemented (i) for all selected methods and, if preconditioning is used, (ii) for the cases where the accuracy of the preconditioners is varied in a wide range.

2 Selection of Solvers

Ten numerical methods for solving systems of linear algebraic equations are used in this paper. The methods are listed in Table 1.

The method for calculating directly the solution (DIR) is based on the use of sparse matrix technique for calculation of an LU factorization of the coefficient matrix of the system of linear algebraic equations. This method is also used to calculate approximate LU factorizations, which can be used as preconditioners in the iterative methods. A special parameter RELTOL (relative drop tolerance),

Table 1. The solvers for systems of linear algebraic equations that are used in this paper.

No.	Method	Abbreviation	Reference
1	Direct solution	DIR	[10,12]
2	Iterative Refinement	IR	[11,12]
3	Pure Modified Orthomin	PURE	[6,12]
4	Preconditioned Modified Orthomin	ORTH	[6,12]
5	Conjugate Gradients Squared	CGS	[5]
6	Bi-Conjugate Gradients STAB	BiCGSTAB	[7]
7	Transpose-Free Quasi Minimum Residual	TFQMR	[2]
8	Generalized Minimum Residual	GMRES	[4]
9	Eirola-Nevanlinna Method	EN	[1,8]
10	Block Eirola-Nevanlinna Method	BEN	[9]

$0 \leq RELTOL < 1$, is used to calculate an approximate LU factorization. If the absolute value of an element a_{ij} of the original coefficient matrix is smaller than the product of $RELTOL$ and the absolute value of the diagonal element a_{ii}, then a_{ij} is dropped (replaced by zero). The same principle of dropping is used in the course of the Gaussian elimination. The method is described in detail in [12]. If $RELTOL$ is set equal to zero, then the direct solution is calculated. If $RELTOL$ becomes larger, then in general the obtained LU factorization is becoming less accurate, but sparser. This means that the number of iterations needed to obtain the required accuracy of the solution of the system of linear algebraic equations will tend to become greater for larger values of $RELTOL$, but the computational work per iteration will in general become smaller. If $RELTOL$ is very large, then the rate of convergence can be very slow or the preconditioned iterative method may even not converge. This shows why it is very important to select robust and reliable stopping criteria when preconditioned iterative methods are used. However, it should be emphasized here that robust and reliable stopping criteria are also needed when pure iterative methods are applied.

Preconditioning is always used when the system of linear algebraic equations is solved by last six methods in Table 1. The modified Orthomin algorithm ([12]) is used both as a pure iterative method (PURE) and as a preconditioned iterative method (ORTH). If an approximate LU factorization is used in IR, then this method can be considered as a preconditioned iterative method.

3 Stopping Criteria

Let x_i and $r_i = b - Ax_i$ be the approximation of the exact solution x of the system of linear algebraic equations $Ax = b$ obtained after i iterations of the chosen iterative method and the corresponding residual vector. We shall assume here that all components of x are of the same order of magnitude. Let $ACCUR$ be the desired accuracy of the approximate solution. One of the following stopping criteria is as a rule selected when an iterative method is used (see, for example,

[2,4,7]): stop the iterative process if (i) $\|x_i - x_{i-1}\| \leq ACCUR$, (ii) $\|r_i\| \leq ACCUR$, (iii) $\|x_i - x_{i-1}\|/\|x_i\| \leq ACCUR$ or (iv) $\|r_i\|/\|r_0\| \leq ACCUR$.

It is intuitively clear that if the iterative process is slowly convergent, then (i) can be satisfied even if the exact error $\|x_i - x\|$ is much larger than $ACCUR$.

Assume that $x_i = x + \epsilon_i$, where ϵ_i is the exact error after i iterations. Then $r_i = b - Ax_i = b - A(x+\epsilon_i) = -A\epsilon_i$. From this relationship we can conclude that the fact that (ii) is satisfied does not always mean that the required accuracy is achieved and the fact that (ii) is not satisfied does not necessarily mean that $\|x_i - x\| > ACCUR$. Indeed, consider the scalar case (the number of equations is equal to one) and assume that $ACCUR = 10^{-4}$, $\epsilon = 10^{-4}$ and $A = 1$. Then (ii) is telling us that the required accuracy has been achieved and this is true. Replace now $A = 1$ with $A = 100$. Now (ii) is telling us that the required accuracy has not been achieved, while the opposite is true. Finally, replace $\epsilon = 10^{-4}$ with $\epsilon = 10^{+4}$ and $A = 1$ with $A = 10^{-10}$. In this case (ii) is telling us that the required accuracy has been achieved, which is certainly not true. This example show us that in the scalar case, the check (ii) will give good results if the problem is scaled so that $A = 1$. It is not immediately clear what kind of scaling should be used in the multi-dimensional case. However, (ii) will give good results if a very good preconditioner has been selected. To show that this is true, assume that the preconditioner C is equal to A^{-1}. Let $ACCUR = 10^{-4}$ and $\|\epsilon\| = 10^{-4}$ Then the norm of the quasi-residual vector $\|CA\epsilon_i\| = \|\epsilon_i\| \leq ACCUR$. Thus, very good preconditioning makes (ii) reliable (let us emphasize, however, that this is true only under the assumption made in the beginning of this section that all components of the solution vector x are of the same order of magnitude).

It is clear that if the first approximation x_0 is very bad and if the rate of convergence is very slow, then (iii) will probably give very bad results.

In some evolutionary problems, where many time steps are used, one normally uses as a first approximation x_0 the accepted solution at the previous time-step. If the process approaches the steady state case, then x_0 chosen in this way can be very accurate. If this happens, then $\|r_0\|$ will become very small and, thus, the check (iv) can cause severe problems.

The above analysis shows that the simple stopping criteria (i) - (iv) may cause problems in some situations. We shall illustrate by numerical examples in Section 5 that this actually happens when the problem is difficult for the iterative methods.

Some more reliable stopping criteria can be derived as follows. In many iterative methods the key operation is the calculation of a new approximation $x_{i+1} = x_i + \alpha_i p_i$. If the process converges then we have that $x_i \to x$ as $i \to \infty$ implies $x = x_0 + \sum_{i=0}^{\infty} \alpha_i p_i$, where p_i is some vector calculated during iteration i. The convergence of the infinite sum implies that $\alpha_i p_i \to 0$ as $i \to \infty$. The reverse relations do not hold, but it is nevertheless reasonable to expect that if (a) $\alpha_i p_i \to 0$ as $i \to \infty$ where (b) α_i does not vary too much and is of order $O(1)$, then $x_i \to x$ as $i \to \infty$.

The restrictions (a) and (b) indicate that it is reasonable to require that the following conditions are satisfied: (A) α_i does not decrease too quickly, (B) α_i

does not increase too quickly and (C) α_i does not oscillate too much. If any of the requirements (A) - (C) is not satisfied, then the stopping criteria are becoming more stringent (which essentially leads to reducing the value of $ACCUR$).

Assume again that the iterative process is convergent. Then the following inequality holds:

$$\|x - x_i\| \le |\alpha_i| \, \|p_i\| \left(1 + \sum_{j=i+1}^{\infty} \frac{|\alpha_j| \|p_j\|}{|\alpha_i| \|p_i\|} \right) \quad (1)$$

Assume that

$$\frac{|\alpha_j| \, \|p_j\|}{|\alpha_i| \|p_i\|} = (RATE)^{j-i} < 1 \quad for \quad \forall \, j > i. \quad (2)$$

Then (1) reduces to

$$\|x - x_i\| \le |\alpha_i| \, \|p_i\| \left(\sum_{j=0}^{\infty} (RATE)^j \right) = \frac{|\alpha_i| \, \|p_i\|}{1 - RATE}. \quad (3)$$

In general, $RATE$ varies from one iteration to another. Denote by $RATE_i$ the value of $RATE$ at iteration i. If $RATE_i$ does not vary too much during several iterations and if all three conditions (A) -(C) are satisfied, then (3) can be used with $RATE$ replaced by some appropriate parameter (as for example, the mean value of $RATE_i$ during the last four iterations; [3]).

If the requirement that $RATE_i$ does not vary too much is not satisfied, then some extra requirements (leading again to the use of a reduced value of $ACCUR$) are to be imposed.

The particular rules (based on the above principles) for the Modified Orthomin Method are described in [3]. Similar rules were developed for the other methods. In some of the methods updating is made not by the formula $x_{i+1} = x_i + \alpha_i \, p_i$, but by a formula which contains some linear combination of vectors instead of the term $\alpha_i \, p_i$. However, the rules sketched above can easily be extended to cover these cases too.

It should be mentioned that relative stopping criteria are used in the experiments which will be presented in Section 5. These criteria are obtained by dividing (3) by $\|x_i\|$. We shall use criteria based on norms of the involved vectors. One of the additional benefits of using stopping criteria based on (3) is the fact that these can easily be modified for component-wise checks. A relative component-wise criterion derived from (3) for the component k of the error can be written in the following way:

$$\frac{|x_k - x_{ki}|}{|x_{ki}|} \le \frac{|\alpha_i| \, |p_{ki}|}{(1 - RATE) \, |x_{ki}|}. \quad (4)$$

Component-wise stopping criteria are useful in the cases where the components of the solution vector are varying in a wide range (this is the case for some problems arising in atmospheric chemistry). These criteria require some more careful work in the cases where (A) - (C) are not satisfied and/or when the values of $RATE_i$ vary considerably from one iteration to another. This is why only relative stopping criteria based on (3) will be used in Section 5.

4 Test-Matrices

In order to ensure systematic investigation of the abilities of the discussed in the previous section stopping criteria to produce reliable results we need test-matrices which have the following properties: (i) matrices of different orders can be created, (ii) the pattern of the non-zero elements can be varied, (iii) the condition number of the matrices can be varied. A matrix generator which creates matrices that satisfy these three conditions has been developed. The matrices depend on four parameters: (a) the order N of the matrix, (b) a parameter C by which the location of some of the non-zero elements can be varied and (c) parameters δ and γ by which the size of certain elements can be varied. N and C are integers, while δ and γ are reals. All four parameters can be freely varied. The non-zero elements of a matrix created by this matrix generator are obtained by using the following rules: (A) all elements on the main diagonal are equal to 4, (B) the elements on the diagonal with first positions (1,2) and (1,C+1) are equal to $-1+\delta$, (C) the elements on the diagonals with first position (2,1) and (C+1,1) are equal to $-1-\delta$ and (D) the elements on the diagonal with first position (1,C+2) are equal to γ. This means that (i) symmetric and positive definite matrices (similar to the matrices obtained by the discretization of the Laplacian operator by the 5-point rule) are created when $\delta = \gamma = 0$, (ii) the matrices are symmetric in structure when $\delta \neq 0$ and $\gamma = 0$, (iii) general matrices are produced when $\gamma \neq 0$. The condition numbers of the created matrices are normally increased when the order is increased and/or the values of the parameters δ and γ are increased. Thus, the requirements stated in the beginning of this section are satisfied. As an illustrations the matrix created with $N = 16$, $C = 4$, $\delta = 3$ and $\gamma = 5$ is given in Fig. 1. The matrices generated in this way are in fact Toeplitz matrices. Thus, these matrices may also be useful when Toeplitz solvers are tested.

5 Numerical Results

Matrices have been created by selecting five values of δ ($\delta = 0,\ 2,\ 4,\ 8,\ 16$). For each value of δ, six values of γ were chosen with $\gamma = 0(1)5$. Thus, 30 pairs (δ, γ) were used in the experiments. For every pair (δ, γ), 20 matrices were created with $C = 10(10)200$ and $N = C^2$. Every matrix was factorized by using 10 values of the relative drop-tolerance $RELTOL$, $RELTOL = 0$ and $RELTOL = 2^{-k}$ where $k = 9(-1)1$. The LU factorizations calculated by the sparse solver from [12] were used as preconditioners of the iterative methods. An experiment with matrices produced by a fixed pair (δ, γ) leads to the solution of 2000 systems of linear algebraic equations (200 per method) in the range from $N = 100$ to

4	2	0	0	2	5	0	0	0	0	0	0	0	0	0	0
-4	4	2	0	0	2	5	0	0	0	0	0	0	0	0	0
0	-4	4	2	0	0	2	5	0	0	0	0	0	0	0	0
0	0	-4	4	2	0	0	2	5	0	0	0	0	0	0	0
-4	0	0	-4	4	2	0	0	2	5	0	0	0	0	0	0
0	-4	0	0	-4	4	2	0	0	2	5	0	0	0	0	0
0	0	-4	0	0	-4	4	2	0	0	2	5	0	0	0	0
0	0	0	-4	0	0	-4	4	2	0	0	2	5	0	0	0
0	0	0	0	-4	0	0	-4	4	2	0	0	2	5	0	0
0	0	0	0	0	-4	0	0	-4	4	2	0	0	2	5	0
0	0	0	0	0	0	-4	0	0	-4	4	2	0	0	2	5
0	0	0	0	0	0	0	-4	0	0	-4	4	2	0	0	2
0	0	0	0	0	0	0	0	-4	0	0	-4	4	2	0	0
0	0	0	0	0	0	0	0	0	-4	0	0	-4	4	2	0
0	0	0	0	0	0	0	0	0	0	-4	0	0	-4	4	2
0	0	0	0	0	0	0	0	0	0	0	-4	0	0	-4	4

Fig. 1. A matrix created with $N = 16$, $C = 4$, $\delta = 3$ and $\gamma = 5$.

$N = 40000$. The right-hand-side b for each system of linear algebraic equations is created with a solution x all components of which are equal to 1.

An accuracy requirement is defined by setting $ACCUR = 10^{-10}$, i.e. it is desirable to obtain a solution the relative error of which is less than 10^{-10} ($\|x - x_i\|/\|x\| \leq 10^{-10}$) when iterative methods are used. The code tries to evaluate the relative error and to stop the calculations when the evaluated relative error becomes less than $ACCUR$. The run is counted as successful every time when this happens. The maximal number of successful runs for a given pair (δ, γ) is 200 for the preconditioned iterative methods (20 matrices, the corresponding system being solved by using 10 preconditioners obtained with the 10 values of $RELTOL$) and 20 for DIR and PURE.

The evaluated by the code error can be less than $ACCUR$, while for the exact error this relation is not satisfied. If this happens, then the code is "lying". It is also necessary to count how many times the code is "lying". The run is accepted as successful when the exact error of the solution obtained by the code is less than $10 * ACCUR$, which means that the accuracy requirement is slightly relaxed when the number of accepted runs is counted. Thus, it is declared that the code is "lying" when it is telling us that the accuracy requirement is satisfied, but the exact error is greater than $10 * ACCUR$.

During the iterarive process, the requirements for stopping the iterations are increased when certain conditions are not satisfied (when some appropriate parameters of the method, $RATE$, α and others, are varying too much; see more details in Section 3). Because of these increased requirements, which are necessary in the efforts to make the stopping criteria more reliable and more robust, there is a danger that the code will tell us that the computed solution is not sufficiently accurate (the evaluated error is greater than $ACCUR$), while the exact error satisfies the relaxed requirement $\|x - x_i\|/\|x\| \leq 10 * ACCUR$.

Therefore, it is also necessary to count the cases in which the code cannot find out that the accuracy requirement has been satisfied.

Some results are given in Table 2. The conclusions made in this section as well as those made in the following section are based on results obtained in all runs discussed above. Mainly the ability of the stopping criteria to terminate successfully the iterative process is discussed, but in the end of the section some computing times are also given.

One should expect that the iterative methods will perform best when the pair ($\delta = 0,\ \gamma = 0$) is selected, because in this case the matrices are symmetric and positive definite. It is seen from Table 2 that this is so.

Table 2. Numerical results obtained when matrices obtained by using three pairs (δ, γ) are used. "SU" refers to the number of successful runs. "WR" refers to number of cases where the code is reporting successful runs, but the exact error does not satisfy the relaxed accuracy requirement. "UN" refers to number of cases in which the code is not able to detect the fact that the accuracy requirement is satisfied. If DIR is used, then the run is successful only when $RELTOL = 0$. If PURE is used the solution does not depend on the value of $RELTOL$ ($RELTOL$ is only used to calculate a preconditioner).

	$\delta = 0,\ \gamma = 0$			$\delta = 4,\ \gamma = 0$			$\delta = 4,\ \gamma = 4$		
Method	SU	WR	UN	SU	WR	UN	SU	WR	UN
DIR	20	0	0	20	0	0	20	0	0
IR	173	0	1	154	0	2	159	0	0
PURE	16	0	0	19	0	1	17	0	1
ORTH	194	0	0	170	0	0	177	0	0
CGS	174	0	0	163	1	0	144	0	0
BiCGSTAB	200	1	0	177	2	0	176	0	0
TFQMR	175	3	5	167	5	3	177	2	3
GMRES	194	0	2	173	0	8	181	0	0
EN	174	0	6	158	0	0	152	0	1
BEN	188	0	6	159	0	0	182	0	2

It was also expected that the results obtained by using the pair ($\delta = 4,\ \gamma = 0$) will be better than the results obtained by using the third pair, ($\delta = 4,\ \gamma = 4$), because in the former case the matrices are symmetric in structure and the number of non-zero elements is smaller. However, the results for the third pair are often better.

The number of cases where the code is telling us that the accuracy required is achieved, while the exact error is greater than $10 * ACCUR$ is not very high (see the results in the columns under "WR" in Table 2). Also the cases where the fact that the accuracy required has been achieved remains undetected are low (see the results in the columns under "UN" in Table 2). Nevertheless, the results indicate that some more work on the stopping criteria is desirable.

It is also necessary to test the performance of the traditionally used stopping criteria, i.e. the stopping criteria,which are based on one of the following four tests (without any attempt to evaluate the convergence rate and/or the reliability of any of the important iteration parameters): (i) $\|r_i\| < ACCUR$, (ii) $\|r_i\|/\|r_0\| < ACCUR$, (iii) $\|x_i - x_{i-1}\| < ACCUR$ and (iv) $\|x_i - x_{i-1}\|/\|x_i\| < ACCUR$. Some results are given in Table 3. It should immediately be mentioned, that it was extremely difficult to run the experiment when the second of these criteria is used. This was especially true when the preconditioner was a very good approximation of A^{-1}, which is, for example, true when the preconditioner is obtained by using $RELTOL = 0$ (see also the discussion in Section 3). Therefore, we replaced this criterion with (ii.a) $\|r_i\|/\max(\|r_0\|, 1.0) < ACCUR$. This means that a pure residual test is carried out when $\|r_0\| \leq 1.0$, while a relative residual test is used in the stopping criteria when $\|r_0\| < 1.0$. The results in Table 3 were obtained by using the mixed criterion (ii.a). Moreover, only the results obtained by using stopping criteria based on (i), (ii.a) and (iii) are given in Table 3. The results obtained by using a stopping criterion based on (iv) are very similar to those obtained by using (iii), because all components of the solution vectors are equal to one in the experiments.

Table 3. Numerical results obtained when three traditionally used stopping criteria are used in the solution of 20 problems with matrices obtained with the parameters $\delta = 0$ and $\gamma = 0$ and with 10 values of the relative drop tolerance for each matrix. The quantities used in the stopping criteria are given in the first row of the table. "SU" refers to the number of successful runs. "WR" refers to number of cases where the code is reporting successful runs, but the exact error does not satisfy the relaxed accuracy requirement. "UN" refers to number of cases in which the code is not able to detect the fact that the accuracy requirement is satisfied. If DIR is used, then the run is successful only when $RELTOL = 0$. If PURE is used the solution does not depend on the value of $RELTOL$ ($RELTOL$ is only used to calculate a preconditioner).

	$\|r_i\|$			$\|r_i\|/\|r_0\|$			$\|x_i - x_{i-1}\|$		
Method	SU	WR	UN	SU	WR	UN	SU	WR	UN
DIR	20	0	0	20	0	0	20	0	0
IR	180	150	0	176	138	0	180	144	0
PURE	17	15	0	16	4	1	16	7	1
ORTH	194	97	0	194	13	2	194	21	0
CGS	178	37	0	178	22	0	178	0	0
BiCGSTAB	200	100	0	200	18	0	200	41	0
TFQMR	192	78	0	191	32	1	200	36	0
GMRES	200	116	0	200	34	0	200	116	0
EN	184	141	0	180	9	0	184	14	0
BEN	198	160	0	194	24	0	196	30	0

The comparison of the results given in Table 3 with the corresponding results in Table 2 (the results in columns 2-4) shows that (i) the numbers of "successful runs" (i.e. the cases where the code tells us that the problem is successfully solved) is increased when the traditionally used stopping criteria is used, however, (ii) the comparison of the calculated solution with the exact one reveals the fact that this is very often not true. On the other hand, there are only a few cases where the code cannot detect that the desired accuracy is achieved.

The comparison of the results in column 2-4 in Table 2 with the results in Table 3 indicates that one must be careful when the stopping criteria are selected for the iterative methods that are to be used in the treatment of large scientific and engineering problems. In order to achieve better reliability and robustness of the iterative methods, it is desirable to apply more advanced stopping criteria in which attempts are made (i) to evaluate the convergence rate and ii) to prevent stopping the iterative process when some iteration parameters vary in a wide range. Stopping criteria based on these principles have been applied in an attempt to improve the performance of the fourth traditionally used stopping criterion (that based on checking $\|x_i - x_{i-1}\|/\|x_i\|$). It is clear that similar principles might be applied in the efforts to improve the performance of the other three stopping criteria.

The results in Table 2 indicate that both DIR and PURE seem to work rather well. Therefore it is also important to compare the computing times (especially in the important case where the matrix is large). The results in Table 4 show that great savings in computing time can be achieved when the preconditioned methods are used with a well-chosen drop-tolerance. The computing times are reduced in some cases by a factor more than 100 when preconditioned methods are used instead of the pure iterative method. Sometimes the preconditioned iterative methods converge and give good results while the pure iterative method is not convergent. There are also some problems. If only one system is to be solved, then it is difficult to find the best value of $RELTOL$. However, in practice one has normally to solve a long sequence of problems with similar matrices. In such a case, it will be profitable to spent some extra time in the beginning of the process in order to find a good value of $RELTOL$. After that one should expect to solve the remaining systems efficiently by using the obtained good value of the drop-tolerance.

6 Concluding Remarks

The results shown in this paper show that the preconditioned methods can be a very useful tool when large scale problems have to be handled (see the results in Table 4). However, several requirements have to be satisfied in order to achieve efficiency. First, one should be able to improve the preconditioner when the iterative process is slowly convergent or even not convergent at all. In our case this can be achieved by reducing the value of the relative drop-tolerance $RELTOL$.

Table 4. Best computing times for $N = 40000$ together with the value of the *RELTOL* for which they were obtained. "N.C" refers to the case where the method does not converge. The runs have been carried out on a SUN workstation.

	$\delta = 0,\ \gamma = 0$		$\delta = 4,\ \gamma = 0$		$\delta = 4,\ \gamma = 4$	
Method	Time	RELTOL	Time	RELTOL	Time	RELTOL
DIR	45.3	0	46.9	0	409.9	0
IR	45.4	0	2.9	2^{-5}	39.6	2^{-8}
PURE	222.7	-	301.5	-	N. C.	-
ORTH	19.6	2^{-9}	2.9	2^{-5}	36.8	2^{-7}
CGS	12.3	2^{-9}	2.7	2^{-5}	38.0	2^{-8}
BiCGSTAB	11.2	2^{-8}	2.7	2^{-5}	37.5	2^{-8}
TFQMR	13.7	2^{-8}	3.0	2^{-5}	37.5	2^{-8}
GMRES	15.0	2^{-8}	3.0	2^{-6}	37.7	2^{-8}
EN	13.3	2^{-8}	3.1	2^{-5}	37.5	2^{-8}
BEN	13.6	2^{-8}	2.8	2^{-5}	37.2	2^{-8}

It is also highly desirable to achieve the required accuracy. More advanced stopping criteria were developed and used for this purpose. The results indicate that the solution process becomes more reliable when these stopping criteria are properly implemented in the codes.

Many hundreds of runs have been used in the experiments. Nevertheless, more experiments might give additional useful information, which can be used to improve further the efficiency. The development of a benchmark of matrices, where matrices of different orders, with different structure and with different properties can be easily generated and used to check systematically the efficiency of the numerical algorithms for solving systems of linear algebraic equations, is a challenging tasks. Such a benchmark will be very useful both for developers of new algorithms and for users in their search for an appropriate algorithm. Some first steps in this directions were made by introducing the set of matrices that were discussed in Section 4.

The efficiency of the results obtained by the different algorithms is normally depending on the matrices which are solved. The efficiency does not automatically mean a small CPU time. Other parameters may also be important for a particular user. For example, some users might be interested in reduction of the storage requirements. This indicates that several different algorithms must be included in a good package for solving systems of linear algebraic equations.

References

1. Eirola, T. and Nevanlinna, O.: Accelerating with rank-one updates. Lin. Alg. Appl., Vol. 121 (1989) 511–520.
2. Freund, R. W.: A transpose-free quasi-minimal residual algorithm for non-Hermitian linear systems. SIAM J. Sci. Stat. Comput., Vol. 14 (1993) 470–482.
3. Gallivan, K. A., Sameh, A. H., and Zlatev, Z.: A parallel hybrid sparse linear system solver. Computing Systems in Engineering, Vol. 1 (1990) 183–195.

4. Saad, Y. and Schultz, M. H.: GMRES: a generalized minimal residual algorithm for solving nonsymmetric linear systems. SIAM J. Sci. Stat. Comput., Vol. 7 (1986) 856–869.
5. Sonneveld, P.: CGS, a fast Lanczos-type solver for nonsymmetric linear systems. SIAM J. Sci. Stat. Comput., Vol. 10 (1989) 36–52.
6. Vinsome, P.: ORTHOMIN, an iterative method for solving sparse sets of simultaneous linear equations. Proc. Fourth Sympos. on Reservoir Simulation, Society of Petr. Eng. of AIME (1976)
7. van der Vorst, H. A.: BI-CGSTAB: A fast and smoothly converging variant of BI-CG for the solution of nonsymmetric linear systems. SIAM J. Sci. Stat. Comput., Vol. 13 (1992) 631–644.
8. Vuik, C. and van der Vorst, H. A.: A comparison of some GMRES-like methods. Lin. Alg. Appl., Vol. 160 (1992) 131–160.
9. Yang, U. Meier and Gallivan, K.: A Family of Preconditioned Iterative Solvers for Sparse Linear Systems. Applied Numerical Mathematics, Vol. 30 (1999) 155–173.
10. Zlatev, Z.: On some pivotal strategies in Gaussian elimination by sparse technique. SIAM J. Numer. Anal., Vol. 17 (1980) 18–30.
11. Zlatev, Z.: Use of iterative refinement in the solution of sparse linear systems. SIAM J. Numer. Anal., Vol. 19 (1982) 381–399.
12. Zlatev, Z.: Computational methods for general sparse matrices. Mathematics and Its Applications, Vol. 65, Kluwer Academic Publishers, Dordrecht-Boston-London (1991)

Variable-Coefficient Difference Schemes for Quasilinear Evolution Problems

Rita Meyer-Spasche

Max-Planck-Institut für Plasmaphysik, EURATOM-Association,
D-85748 Garching, Germany
meyer-spasche@ipp-garching.mpg.de

Abstract. We review several integration methods leading to variable-coefficient schemes and/or to exact schemes for ODEs (functional fitting; Principle of Coherence). Conditions for obtaining coefficients that are independent of the time t and of the time step τ are investigated. It is shown that some of the discussed schemes lead to efficient difference schemes for problems from applications, in particular for highly oscillatory ODEs and for parabolic equations with blow-up solutions.

1 Introduction

Time discretization for the numerical solution of initial value problems means that we approximate a continuous dynamical system by a family of discrete dynamical systems. We introduce the additional parameter $\Delta t =: \tau$ and require $\mathcal{O}(\tau^s)$-convergence for $\tau \to 0$, $s \geq 1$. If the dynamics of the discrete and continuous systems are very different for larger τ, the step-size τ must be small for satisfactory results. If the dynamics of the systems are very similar, τ may be larger, computations are more efficient. In the ideal case, the step-size of the computations is determined by the solution to be computed: by its structure and by the accuracy required. In many applications, for instance in equilibrium computations and turbulence computations in plasma physics, the bounds for the step-size have to be determined by properties of the numerical method instead.

Let us look at a very simple example:

$$\dot{u} = u^2, \qquad u(0) = u_0 > 0\ , \tag{1}$$

with solution

$$u(t) = \frac{u_0}{1 - tu_0}\ . \tag{2}$$

This solution ceases to exist when the denominator vanishes, i.e. at its blow-up time $T = 1/u_0$. We take $u_0 = U_0$ and compute discrete solutions $\{U_n\}_{n=1}^N$.

If we discretize eq. (1) with the explicit forward Euler scheme, we obtain

$$\frac{U_{n+1} - U_n}{\tau} = U_n^2\ , \quad \text{or} \quad U_{n+1} = U_n + \tau U_n^2\ , \tag{3}$$

and the iterates exist for all times, independent of the value of U_0. Moreover, the step-size τ must be small enough to prohibit instability of the scheme.

I. Dimov et al. (Eds.): NMA 2002, LNCS 2542, pp. 36–47, 2003.

If we discretize eq. (1) with the implicit backward Euler scheme, we obtain

$$\frac{U_{n+1}-U_n}{\tau}=U_{n+1}^2\,,\quad \text{or}\quad U_{n+1}=\frac{1}{2\tau}(1-\sqrt{1-4\tau U_n})\,, \tag{4}$$

and the iterates exist for all times, provided the step-size τ is small enough. Moreover, the step-size τ must be small enough to ensure uniqueness and to prohibit superstability of the scheme.

If we discretize eq. (1) with the 'nonstandard scheme'

$$\frac{U_{n+1}-U_n}{\tau}=U_nU_{n+1}\,,\quad \text{or}\quad U_{n+1}=\frac{U_n}{1-\tau U_n}, \tag{5}$$

we find that this scheme is exact, i.e. for any step-size τ it reproduces the solution (2) without discretization error, as long as $n\cdot\tau<T=1/u_0$.

Given any individual differential equation, how to find an optimal scheme for it? How should nonlinear terms in differential equations be discretized? Attempts are made for developing a theory of nonstandard schemes 'optimal for individual differential equations' [10]. In the case of eq. (1) and $f(u)=u^2$, however, we notice that

$$\frac{U_{n+1}-U_n}{\tau}=U_nU_{n+1}=f(U_n)+f'(U_n)\frac{U_{n+1}-U_n}{2}\,, \tag{6}$$

and this is a linearly implicit *standard* scheme, a so-called Rosenbrock-Wanner scheme. Such schemes are standard in the numerical treatment of differential-algebraic and of stiff differential equations [5, vol.II]. Also other 'nonstandard' schemes found in the literature turned out to be standard [9]. In the following we shall discuss on which functions given, well-known schemes are exact (on which $f(u)$, on which $u(t)$?) and we shall discuss several methods for finding schemes which are exact on given s-dimensional function spaces. These investigations mostly give results for simple quasilinear equations. Exact schemes for simpler equations can lead to new efficient schemes for more involved equations (Denk-Bulirsch schemes, LeRoux schemes, Kojouharov-Chen schemes [10, chap.2]).

2 Exact Schemes

We start by considering non-autonomous systems

$$\dot{u}=f(t,u)\,,\quad u(0)=u_0\,, \tag{7}$$

with smooth functions $f:(t_1,t_2)\times I\!R^q\to I\!R^q$, $q\geq 1$. We assume $[0,T]\subset(t_1,t_2)\subset I\!R$ and consider one-step schemes

$$U_{n+1}=\mathcal{A}(f)(U_n,\tau),\qquad U_0=u_0, \tag{8}$$

for the numerical solution of system (7) on the interval $[0,T]$. Here $\tau=\Delta t$ is the time step, U_n is an approximation to the exact solution $u(t_n)$ at time

$t_n = n\tau$, $n = 0, 1, 2, \ldots$, and $\mathcal{A}(f)$ denotes the evolution map given by the numerical scheme. Note that for implicit methods, the map $\mathcal{A}(f)$ requires a non-linear solve. In this text we assume that the explicit form (8) can always be obtained uniquely and exact. We also allow numerical methods for which the evolution map $\mathcal{A}(f)$ involves derivatives of f with respect to u.

We define the *truncation error* $T(u_0, \tau, f)$ of scheme (8) by

$$T(u_0, \tau, f) := \frac{1}{\tau}(u(\tau) - U_1) = \frac{1}{\tau}(u(\tau) - \mathcal{A}(f)(u_0, \tau)). \tag{9}$$

Scheme (8) is
* *of order m on eq. (7)*, if m is the largest integer such that

$$\lim_{\tau \to 0} \frac{||T(u_0, \tau, f)||}{\tau^m} < \infty \tag{10}$$

for all smooth f and arbitrary u_0;
* *exact on the solution $u(t; u_0)$ of eq. (7) for given f*, if $T(u_0, \tau, f)$ vanishes for arbitrary step size $\tau \leq \tau_0 \in [0, T]$;
* *exact on eq. (7) for given f*, if $T(u_0, \tau, f)$ vanishes for arbitrary initial value $u_0 \in I\!R^p$ and arbitrary step size $\tau \leq \tau_0 \in [0, T]$.

2.1 Error Expansions for Constant-Coefficient Schemes

We now confine to autonomous scalar initial value problems

$$\dot{u} = f(u)\ , \quad u(0) = u_0 \in I\!R\ . \tag{11}$$

For the analysis we expand $T(u_0, \tau, f)$ in a Taylor series in τ,

$$T(u_0, \tau, f) = \sum_{j=0}^{\infty} B_j(f) \tau^j . \tag{12}$$

Scheme (8) is *of order m* on eq. (11), if $B_j(f) = 0$ for all $j < m$, arbitrary u_0 and all smooth functions f. Scheme (8) is *exact* on eq. (11) *for a given function f*, if $B_j(f) = 0$ for arbitrary $u_0 \in I\!R$ and for all $j \geq 0$.

Lemma 1. *The trapezoidal rule*

$$\frac{U_{n+1} - U_n}{\tau} = \frac{f(U_{n+1}) + f(U_n)}{2} \tag{13}$$

is exact on equation (11) for those functions $f : I\!R \to I\!R$ satisfying

$$f'' f + (f')^2 = 0\ , \tag{14}$$

i.e. for $f(u) = \pm\sqrt{a_1 u + a_2}$, a_1, a_2 constant, $a_1 u + a_2 \geq 0$. It follows that it is exact for solutions $u(t)$ satisfying $u(t) \in span\{1, t, t^2\}$.

The proof was already given in [8] and [3]. We thus only sketch it here. For the truncation error we obtain the expansion (12) with $B_j(f) = 0$ for $j < 2$ and

$$B_j(f) = \left(\frac{1}{(j+1)!} - \frac{1}{2j!}\right) \left.\frac{d^j f(u(t))}{dt^j}\right|_{t=0} \quad \text{for } j \geq 2. \tag{15}$$

For general smooth f we thus obtain $B_0 = B_1 = 0$. The method is second order in general. For those f which satisfy eq. (14), we obtain $B_2(f) = 0$, and moreover $B_j(f) = 0$ for all j. Thus the trapezoidal rule is exact for those f satisfying eq. (14). We then get by integration $f(u) = \pm(a_1 u + a_2)^{1/2}$, a_1, a_2 constants. From the differential equation we obtain that the second derivative of $f(u(t))$ w.r.t. t is equal to the third derivative of $u(t)$. So both vanish and the solution space is $U = \operatorname{span}\{1, t, t^2\}$. ◇ In a similar way the following lemma was proved as well.

Lemma 2. *The implicit midpoint rule*

$$\frac{U_{n+1} - U_n}{\tau} = f(\frac{U_{n+1} + U_n}{2}) \tag{16}$$

is exact on equation (11) for those functions $f : \mathbb{R} \to \mathbb{R}$ satisfying

$$f'' f - 2(f')^2 = 0\,, \tag{17}$$

i.e. for $f(u) = a_2(a_3 - a_1 u)^{-1}$, a_1, a_2, a_3 constants, $a_3 - a_1 u(t) \neq 0$ for all t. Solutions u of eq. (11) then satisfy

$$u(t) = a_2 t + u_0 \qquad \text{for } a_1 = 0,\ a_3 = 1, \qquad \text{or} \tag{18}$$

$$u(t) = \frac{a_3 \pm \sqrt{(a_3 - a_1 u_0)^2 - 2a_1 t}}{a_1} \qquad \text{for } a_1 \neq 0,\ a_3 - a_1 u_0 \neq 0,\ a_2 = 1. \tag{19}$$

The solutions form a nonlinear manifold, not a linear space. In particular, $u(t) = \sqrt{t}$ does not belong to the manifold for $t_0 = 0$. Also, exact schemes are not unique: Already in [3] it was shown that both the second-order Taylor method $U_{n+1} = U_n + \tau f(U_n) + \tau^2 f'(U_n) f(U_n)/2$ and the trapezoidal rule (13) are exact on the same set of differential equations (11). They are clearly different difference schemes, one explicit, one implicit, and their error expansions are different. All B_j, $j \geq 2$, are different from those in (15), but vanish for the same fs. This non-uniqueness should not surprise: difference schemes are equations to be satisfied by the approximate solutions of the differential equations under consideration. There is not *the* exact scheme, there might be many of them, differing by terms which vanish for those differential equations on which they are exact. What we have to require, of course, is the unique solvability of the difference scheme for given initial value and sufficiently small step size τ.

The trapezoidal rule and the implicit midpoint rule both are Runge-Kutta methods. We now look at exact schemes within the framework of RK methods.

2.2 Constant-Coefficient RK Methods as Exact Schemes

We start by collecting some facts on RK methods for use lateron (see [2, p.58] and [5, vol.I, p.211f] for more details). We consider system (7) again, and we will always assume $U_0 = u_0$ for the discrete iteration. Let $b_i, a_{ij} (i, j = 1, \ldots, s)$ be real numbers and let

$$c_i = \sum_{j=1}^{s} a_{ij} , \qquad i = 1, \ldots, s . \tag{20}$$

The method defined by

$$Y_i = U_n + \tau \sum_{j=1}^{s} a_{ij} f(t_n + c_j\tau, Y_j) , \qquad i = 1, \ldots, s , \tag{21}$$

$$U_{n+1} = U_n + \tau \sum_{i=1}^{s} b_i f(t_n + c_i\tau, Y_i) . \tag{22}$$

is called an *s-stage Runge-Kutta scheme (RK scheme)*. There is a certain redundancy: formally different schemes can define the same numerical integration method, even when they have different stage numbers s_1 and s_2. In the following we consider *Runge-Kutta methods*, assuming that the resulting numerical integration method is at least of first order, and that the scheme representing it has minimal stage number s and satisfies $c_i \neq c_j$ for $i \neq j$.

If an s-stage RK method satisfies the simplifying conditions

$$B(\xi) : \qquad \sum_{i=1}^{s} b_i c_i^{k-1} = 1/k , \quad 1 \leq k \leq \xi, \tag{23}$$

and is used for integrating

$$\dot{y} = f(t) , \qquad y(t_n) = 0 , \tag{24}$$

on the interval (t_n, t_{n+1}), then eq. (22) is an integration method of order ξ. Equation (22) is then exact on $f \in \text{span}\{1, t, \ldots, t^{\xi-1}\}$. Trivial consequence: all consistent RK schemes are exact on $\dot{u} = const$.

If an RK method satisfies the simplifying conditions

$$C(\xi) : \qquad \sum_{j=1}^{s} a_{ij} c_j^{k-1} = \frac{1}{k} c_i^k , \quad 1 \leq k \leq \xi, \quad 1 \leq i \leq s, \tag{25}$$

then the stage equations (21) define integration methods of order ξ for eq. (24) on the intervals $(t_n, t_n + c_i\tau)$. Thus they are exact there for $f \in \text{span}\{1, t, \ldots, t^{\xi-1}\}$. Note that the validity of $C(1)$ is ensured by eq. (20).

If an s-stage RK method satisfies $B(s)$ and $C(s)$, it is called a *collocation method* (after Burrage 1978).

Apply a collocation method (21), (22) to the differential equation (7). The *collocation polynomial*, i.e. the polynomial $p(t)$ interpolating the numerical solution of (21), (22) in the points t_n and $t_{ni} := t_n + c_i\tau$, $i = 1, \ldots, s$, then has degree s. Its derivative $\dot{p}$ has degree $s - 1$ and is integrated exactly by the stage equations (21) on the intervals $[t_n, t_{ni}]$. We thus obtain

$$p(t_n + c_i\tau) = p(t_n) + \tau \sum_{j=1}^{s} a_{ij}\, \dot{p}(t_n + c_j\tau)\,, \qquad i = 1, \ldots, s, \tag{26}$$

$$p(t_n + \tau) = p(t_n) + \tau \sum_{i=1}^{s} b_i\, \dot{p}(t_n + c_i\tau)\,. \tag{27}$$

Put $p(t_n) = u_n$, where $u_n = u(t_n)$ and $u(t)$ is the solution of eq. (7) to be approximated. Then the collocation polynomial satisfies eq. (7) at the internal abscissas $t_{ni} = t_n + c_i\tau$, $i = 1, \ldots, s$.

Given a positive integer s and numbers $c_1, \ldots, c_s \in I\!R$, $c_i \neq c_j$ for $i \neq j$, $0 \leq c_i \leq 1$, $i = 1, \ldots, s$. The collocation method satisfying

$$\begin{aligned} p(t_n) &= U_n \\ \dot{p}(t_n + c_i\tau) &= f(t_n + c_i\tau, p(t_n + c_i\tau))\,, \quad i = 1, \ldots, s, \\ p(t_n + \tau) &=: U_{n+1} \end{aligned} \tag{28}$$

is equivalent to the s-stage RK method (21), (22) with coefficients

$$a_{ij} := \int_0^{c_i} \ell_j(t)dt\,, \qquad b_j := \int_0^1 \ell_j(t)dt, \qquad i, j = 1, \ldots, s\,, \tag{29}$$

where the $\ell_j(t)$ are the Lagrange polynomials $\ell_j(t) = \prod_{m \neq j}(t - c_m)(c_j - c_m)^{-1}$. Note that $c_i \neq c_j$ for $i \neq j$ is essential here. ◇

Consider the μ-dependent family of RK schemes

$$\begin{aligned} Y_1 &= U_n, \\ Y_2 &= U_n + \tau(\mu f(t_n, Y_1) + (1 - \mu) f(t_n + \tau, Y_2))\,, \\ U_{n+1} &= U_n + \tau(f(t_n, Y_1) + f(t_n + \tau, Y_2))/2\,. \end{aligned} \tag{30}$$

For $\mu = 1/2$ the scheme represents the trapezoidal rule. It is a second-order 2-stage method and it satisfies the simplifying conditions $B(2)$ and $C(2)$. It thus is a collocation method and integrates eq. (24) exactly for $f \in \text{span}\{1, t\}$. It is also exact on autonomous eqs. (11) if f satisfies eq. (14).

The implicit midpoint rule can also be written as a Runge-Kutta method,

$$\begin{aligned} Y_1 &= U_n + \tau f(t_n + \tau/2, Y_1)/2\,, \\ U_{n+1} &= U_n + \tau f(t_n + \tau/2, Y_1)\,. \end{aligned} \tag{31}$$

It is a second-order 1-stage method satisfying $B(2)$ and $C(1)$. It thus is a collocation method and integrates eq. (24) exactly for $f \in \text{span}\{1, t\}$. It is also exact on autonomous eqs. (11) if f satisfies eq. (17).

The question thus arises if every RK method is exact on some nontrivial differential equation. The answer is *no*. Consider again scheme (30), but now choose $\mu = 2/3$. For $\mu = 2/3$ and autonomous f we obtain $B_0 = B_1 = 0$ and

$$B_2(f) = (\frac{1}{2} \cdot 0 \cdot 1 + \frac{1}{2} \cdot \frac{1}{3} \cdot 1 - \frac{1}{6}) f f'^2 + (\frac{1}{2} \cdot \frac{1}{2} \cdot 1 - \frac{1}{6}) f^2 f'' = \frac{1}{12} f^2 f'' . \quad (32)$$

Hence the scheme is second order in general. $B_2(f)$ vanishes if either $f = 0$ or $f'' = 0$. This means $f(u) = a_1 u + a_2$ for arbitrary constants a_1 and a_2. Thus for the differential equation

$$\dot{u} = a_1 u + a_2 \quad (33)$$

the scheme is at least third order. With $f'' = 0$ we find $B_3(f) \not\equiv 0$. Hence B_3 does not vanish when B_2 does. Thus the 2nd order scheme (30) with $\mu = 2/3$ is only third order for eq. (33) and not exact. The vanishing of the first non-zero term in the error expansion by particular choice of the r.h.s. function f does not guarantee exactness as one might have hoped for.

Different approaches to be discussed next allow to find exact schemes for equation (33) and for linear systems of type (33). It turns out, however, that the coefficients of the schemes must be allowed to depend on the step-size.

2.3 Functional Fitting: Variable-Coefficient RK Schemes

In recent years much research has been performed for finding efficient numerical methods for systems (7) with oscillatory solutions. If a good estimate of the frequency is known in advance, exact integration of the linear part of system (7) leads to very useful integration methods. Here we take a closer look at two different methods in this family: the application of the Principle of Coherence by Denk and the *functional-fitting RK methods* as introduced by Ozawa.

Now we allow variable coefficients in the RK schemes, i.e. we consider schemes whose coefficients a_{ij}, b_i depend on the independent variable t and on the step size τ. Ozawa [11] proved the following results:

Lemma 3. *Let $\{c_i\}_{i=1}^s \in \mathbb{R}$ be given, $c_i \neq c_j$ for $i \neq j$. Let $\{v_m(t)\}_{m=1}^s$ be admissible, smooth linearly independent functions defined on an open interval containing $[0, T]$. Then the linear system*

$$\begin{aligned} v_m(t + c_i\tau) &= v_m(t) + \tau \textstyle\sum_{j=1}^s a_{ij}(t,\tau) \dot{v}_m(t + c_j\tau) \\ v_m(t + \tau) &= v_m(t) + \tau \textstyle\sum_{i=1}^s b_i(t,\tau) \dot{v}_m(t + c_i\tau) \end{aligned} \quad (34)$$

is uniquely solvable for $a_{ij}(t,\tau)$ and $b_i(t,\tau)$, with $t, t+\tau \in [0,T]$ fixed, $0 < \tau < \tau_0$, τ_0 small enough.

For the definitions of 'admissible' and 'smooth enough' and for the proof see [11] or [3]. The constant $u(t) \equiv c$ is not admissible. Nevertheless it will be included in any linear space spanned by functions satisfying system (34): it satisfies (34) for given a_{ij}, b_i for *any* admisssible set of functions $\{v_m\}_{m=1}^s$. Now we consider the performance of a scheme obtained via Lemma 3.

Lemma 4. *Let the coefficients of the variable-coefficient s-stage RK scheme*

$$\begin{aligned} Y_i &= U_n + \tau \textstyle\sum_{j=1}^{s} a_{ij}(t_n,\tau) f(t_n + c_j\tau, Y_j) \\ U_{n+1} &= U_n + \tau \textstyle\sum_{i=1}^{s} b_i(t_n,\tau) f(t_n + c_i\tau, Y_i) \\ i &= 1,\ldots,s \qquad t_n = t_0 + n\tau, \qquad U_0 = u_0, \end{aligned} \tag{35}$$

be obtained according to Lemma 3. Then the order of the scheme is p, where p depends on the choice of the $\{c_i\}$ and satisfies $s \le p \le 2s$.

The s-stage RK scheme obtained with the linearly independent functions $\{v_m\}$ is exact on the solution $u(t)$ whenever $u(t) \in U = \text{span}\{1, v_1, ...v_s\}$. If all solutions of (7) happen to belong to U, the scheme is exact on (7), because we can first construct the linear combination of the basis functions and afterwards we replace $\dot{u}$ by $f(t,u)$. It is thus of interest to use functional fitting RK-schemes whenever there is some knowledge about the solution in advance. If the scheme is not exact, the remaining part of the solution is captured by the order of the RK-scheme; and the constants in the error expansion will be small as long as the scheme is in a small neighborhood of an exact scheme.

In the *proof* [11] it is shown that the first terms $a_{ij}^{(0)}$ and $b_i^{(0)}$ in the power expansions of $a_{ij}(t,\tau)$ and $b_i(t,\tau)$ with respect to τ satisfy the simplifying conditions $B(s)$ and $C(s)$ and depend only on $\{c_i\}_{i=1}^{s}$, but not on the generating functions $\{v_m\}_{m=1}^{s}$. They thus agree with the coefficients of the corresponding collocation scheme. When we apply Lemma 3, the resulting scheme might have constant or variable coefficients, see the following examples.

For constant-coefficient schemes for non-autonomous differential equations, it is a *convention* to satisfy eq. (20) when designing new schemes. Because condition (20) implies that $t_n + c_i\tau = U_n + c_i\tau$ for $u(t) = t$ [2, p. 56]. In the case of Lemma 3, condition (20) is satisfied if $u(t) = t$ is one of the generating functions.

Example 1: We chose $s = 2$, $v_1(t) = t$, $v_2(t) = t^2$ and use c_1, c_2 as parameters. Solving system (34) we find the coefficients of the collocation schemes obtained for $\{c_1, c_2\}$ via eq. (29), because of the remark on $a_{ij}^{(0)}$, $b_i^{(0)}$. For varying c_1, c_2 with $c_1 \neq c_2$ this is a 2–parameter family of RK schemes. For $c_1 = 0$, $c_2 = 1$ we obtain the trapezoidal rule, as expected. Though all these schemes are equivalent when exact, they differ in their order and numerical performance when not exact.

Example 2: With $s = 2$, $c_1 = 0$, $c_2 = 1$ and $v_1(t) = t$, $v_2(t) = \exp \lambda t$, $1/\lambda \notin [0,T]$, we obtain the coefficients

$$\begin{aligned} a_{21}(\tau) &= \frac{1 - (1-\lambda\tau)\exp\lambda\tau}{\tau\lambda(\exp\lambda\tau - 1)}\,, \qquad a_{22}(\tau) = \frac{-1 - \lambda\tau + \exp\lambda\tau}{\tau\lambda(\exp\lambda\tau - 1)}\,, \\ a_{11} &= 0\,, \quad a_{12} = 0\,, \qquad\qquad b_1(\tau) = a_{21}(\tau)\,, \quad b_2(\tau) = a_{22}(\tau)\,. \end{aligned} \tag{36}$$

These coefficients have an apparent singularity in the limit $\tau \to 0$. The limiting values of $a_{21}(\tau), a_{22}(\tau), b_1(\tau)$, and $b_2(\tau)$ computed by L'Hopital's rule all are equal to $1/2$. For $\tau \to 0$ the scheme becomes the related collocation scheme.

Example 3: With $s = 2$, $v_1(t) = t$, $v_2(t) = (t+1)^{-1}$, and pairwise differing c_is as parameters, we obtain with $d(c) := (t + 1 + c\tau)^{-1}$, $\quad d(c)^2 = d(c) \cdot d(c)$,

$$
\begin{aligned}
a_{i1}(t,\tau) &= \frac{d(c_i) - d(0) + \tau c_i d(c_2)^2}{(d(c_2)^2 - d(c_1)^2)\tau}, \quad a_{i2} = c_i - a_{i1}, \quad i = 1,2 \\
b_1(t,\tau) &= \frac{d(1) - d(0) + \tau d(c_2)^2}{(d(c_2)^2 - d(c_1)^2)\tau}, \qquad b_2 = 1 - b_1 .
\end{aligned}
\tag{37}
$$

This example shows that quite simple functions v_m can lead to complicated coefficients which depend on t and τ. RK schemes with variable coefficients are nonstandard. Time-dependent coefficients are very unusual and would be inconvenient in computations. Below we will show how to avoid them. Coefficients depending on the step-size τ are not quite that unusual: such coefficients are always obtained in the context of exponential fitting [12] and of evaluating the Principle of Coherence [1] and seem to be unavoidable in certain situations. We now ask for general conditions such that the coefficients are constant, i.e. independent of t and/or τ.

Lemma 5. *Let the assumptions of Lemma 3 be satisfied. Assume that the coefficients computed according to Lemma 3 are independent of t. Then the linear space $\tilde{U} := span\{1, v_1(t), \ldots, v_s(t)\}$ is closed with respect to differentiation.*

Proof: Computing the time derivative of every equation of system (34) and using that the time-derivatives of all coefficients vanish, we obtain

$$
\begin{aligned}
\dot{v}_m(t + c_i\tau) &= \dot{v}_m(t) + \tau \textstyle\sum_{j=1}^{s} a_{ij}(t,\tau)\ddot{v}_m(t + c_j\tau), \\
\dot{v}_m(t + \tau) &= \dot{v}_m(t) + \tau \textstyle\sum_{i=1}^{s} b_i(t,\tau)\ddot{v}_m(t + c_i\tau).
\end{aligned}
\tag{38}
$$

Time-independent coefficients have to satisfy both this system and system (34). Since system (34) already determines the coefficients uniquely, system (38) cannot add new conditions but must consist of linear combinations of equations contained in system (34), possibly adding the trivial equation for the constant. This means that $\text{span}\{\dot{v}_1, \ldots, \dot{v}_s\}$ is a subset of $\text{span}\{1, v_1, \ldots, v_s\}$. Thus the linear space $\tilde{U}$ is closed with respect to differentiation. ◇

Examples: Each of $\text{span}\{e^{\omega t}\}$, $\text{span}\{e^{-\omega t}\}$ and $\text{span}\{\sin \omega t, \cos \omega t\}$ is closed with respect to differentiation. The linear spaces $\text{span}\{t^s, t^{s-1}, \ldots, t, 1\}$, $s \in I\!N$, and $\text{span}\{1, t, e^{\omega t}\}$ are also closed with respect to differentiation. This agrees with the time-independence of the coefficients of Example 1 and of (36). There is no finite-dimensional linear space containing $\text{span}\{(t + c)^{-1}\}$, $c \in I\!R$, and closed with respect to differentiation. This is in agreement with the time-dependence of coefficients (37). There is no finite-dimensional linear space containing $\text{span}\{(1+t)^{1/2}\}$ and closed with respect to differentiation. We thus get (quite complicated) coefficients which depend both on t and τ. As we have seen earlier, however, there *is* a 1-stage constant-coefficient RK scheme exact on $u(t) = (t + 1)^{1/2}$, *when applied to differential equation (11) with (17)*: the implicit midpoint rule (put $u_0 = 1$, $a_1 = -2$, $a_3 = 0$ in eq. (19)). The nonlinearities f play an important role for the results of section 2.1.

For finding conditions for the existence of coefficients which are independent of the step-size τ, we could procede same way as for finding time-independent coefficients. Computing the τ-derivative of every equation of system (34) and using their τ-independence, however, leads to conditions that are too complicated to be of practical value.

2.4 The Principle of Coherence

Exact schemes for linear differential equations and for linear systems have been derived by many authors, sometimes by using the known continuous solution. A straight-forward procedure for deriving equations to be satisfied by the coefficients of exact schemes is the Principle of Coherence introduced by Hersch in 1958. The basic idea of the Principle of Coherence was formulated by Hersch [4] as *Successive approximations should not contradict each other.* We explain this by the following example: Using central finite differences at t with step-size τ on $\ddot{z}(t) + \lambda^2 z(t) = 0\,,\ \ \lambda > 0$, we obtain $z(t-\tau) - (2 - \lambda^2\tau^2)z(t) + z(t+\tau) = 0$. We write instead $z(t-\tau) - A(\tau)z(t) + z(t+\tau) = 0$, where the coefficient $A(\tau)$ is to be determined. With step-size 2τ we obtain similarly $z(t-2\tau) - A(2\tau)z(t) + z(t+2\tau) = 0$. By linear combination of three difference equations with step size τ centered at $t-\tau$, t and $t+\tau$ we obtain on the other hand $z(t-2\tau) - (A(\tau)^2 - 2)z(t) + z(t+2\tau) = 0$. For a coherent numerical approximation of the differential equation, the two schemes involving the points $t-2\tau, t$ and $t+2\tau$ must coincide. This means $A(2\tau) = A(\tau)^2 - 2$. This is satisfied for $A(\tau) = 2\cos\kappa\tau,\ \kappa \in \mathbb{R}$. Moreover, the resulting difference scheme has to be consistent with the differential equation. Thus we obtain $\kappa = \lambda$. The coherent scheme therefore is $z(t-\tau) - 2\cos(\lambda\tau)z(t) + z(t+\tau) = 0$. Comparison shows that this scheme is exact on the differential equation, while the second-order central difference formula uses the first two terms of the Taylor expansion of $A(\tau)$.

3 Applications

We conclude by showing that exact schemes are of practical relevance in scientific computing. We give two examples where exact schemes for simpler equations led to efficient schemes for more complicated equations.

3.1 Denk-Bulirsch Schemes

Denk used the Principle of Coherence explained in subsection 2.4 for designing a numerical integration method which is capable of simulating highly oscillatory circuits efficiently and reliably. He treated first order systems

$$\dot{u} + Au = f(t,u)\,, \quad A \in \mathbb{R}^{q\times q}\,. \tag{39}$$

Applying the Principle of Coherence with the step-sizes τ and 2τ to $\dot{z} + Az = 0$ leads to the exact scheme $Z(t_0+\tau) = \exp(-A\tau)Z(t_0)$. For approximation of eq. (39), Denk combined this with a multistep approach (explicit or implicit). Explicit:

$$U_m - \exp(-A\tau)U_{m-1} = \tau \sum_{i=0}^{m} \beta_i(\tau) f(t_{i-1}, U_{i-1}) , \tag{40}$$

where $U_m = U(t + m\tau)$, $t_m = t + m\tau$, and $\beta_i(\tau)$ are *matrix coefficients* of the multistep scheme. In the case $m = 1$ this leads to

$$\begin{aligned} \beta_1(\tau) &= -\left(I - (I - \exp(-A\tau))(A\tau)^{-1}\right)(A\tau)^{-1} , \\ \beta_0(\tau) &= (I - \exp(-A\tau))\,(A\tau)^{-1} - \beta_1 . \end{aligned} \tag{41}$$

Note that the coefficients $\beta_i(\tau)$ have an apparent singularity for $\tau \to 0$, similar to the coefficients given in (36). This seems to be unavoidable.

The resulting scheme is consistent of order m, $A(0)$-stable and convergent. It is A-stable without order restrictions. This is no contradiction to Dahlquist's order barriers because those barriers were shown to hold for constant-coefficient schemes. In numerical experiments, some of them in a time interval corresponding to about 250 000 oscillations, the code HERSCH using these ideas for equations more general than eq. (39) proved to be more efficient than the codes LSODE, DOPRI5 and RADAU5 on highly oscillatory problems [1].

3.2 LeRoux Schemes

Consider the parabolic problem

$$\begin{aligned} v_t - \Delta v^m &= \alpha v^m && \text{for } x \in \Omega \subset I\!R^d ,\ t > 0 , \\ v(x,t) &= 0 && \text{for } x \in \partial\Omega ,\ t > 0, \\ v(x,0) &= v_0(x) > 0 && \text{for } x \in \Omega , \end{aligned} \tag{42}$$

where Ω is a smooth bounded domain, $\alpha \geq 0$ real and $m > 1$ an integer. Let (λ_1, u_1) be the principal eigenpair of $-\Delta u = \lambda u$, $u|_{\partial\Omega} = 0$, satisfying $u_1(x) > 0$ in Ω and $||u_1||_{L^1(\Omega)} = 1$. Let v_1 be a real smooth function satisfying $v_1^m(x) = u_1(x)$ in Ω. Then θv_1, $\theta > 0$, is a steady-state solution of (42) for $\alpha = \lambda_1$. A steady-state solution for all α is $v(x,t) \equiv 0$. The time-dependent solutions of (42) for given initial function $v_0(x) > 0$ were investigated by Sacks and others, and the results are [6]: If $0 \leq \alpha < \lambda_1$ and $v_0 \in L^q(\Omega)$, $q > 1$, problem (42) has a solution existing for all times and decaying to zero for $t \to \infty$. If $\alpha = \lambda_1$ and $v_0 \in L^q(\Omega)$, $q > 1$, problem (42) has a solution existing for all times and tending to $\theta v_1 = \theta u_1^{1/m}$ for $t \to \infty$. The factor θ depends on the initial function v_0. If $\alpha > \lambda_1$ there exists $T > 0$ such that problem (42) blows up at $t = T$. It has for given v_0 a unique weak solution v on $[0, T]$ with $\lim_{t \to T_-} ||v(\cdot, t)||_{L^\infty(\Omega)} = +\infty$. The only nonnegative solution of problem (42) which exists for all times is $v(x,t) \equiv 0$.

To construct a numerical scheme whose solution has similar properties as the solution of the continuous problem, Le Roux [6] used the exact scheme (5) for

$$d(w^m)/dt = \beta(w^m)^2 , \qquad w(0) = w_0 > 0 ,\ \beta \in I\!R , \tag{43}$$

to derive the approximate semi-discrete scheme

$$(m-1)^{-1}(V_n^{1-m}V_{n+1}^m - V_{n+1}) - \tau \Delta V_{n+1}^m = \alpha\tau V_{n+1}^m \tag{44}$$

for eq. (42). Here τ is the time step and $V_n = V(x, n\tau)$ approximates $v(x,t)$ at $t = n\tau$. Note that solving eq. (44) for V_{n+1} with given V_n means solving a quasilinear elliptic boundary value problem with $V_{n+1|\partial\Omega} = 0$, and this has to be done at each time step. This can be reformulated as $\tau\Delta U_{n+1} = f(U_{n+1}; U_n)$, which is a standard quasilinear elliptic problem for U_{n+1}. Le Roux [6] proved that, under mild assumptions and for sufficiently small $\tau < \tau_0$, scheme (44) produces qualitatively correct numerical solutions which satisfy known estimates for the exact solutions. In further work this scheme and its mathematical analysis were extended to more general cases, see her review in JCAM *97*, 121–136. Such schemes were successfully applied in plasma physics for diffusion equations with slow diffusion modelling the density of particles, and with fast diffusion for the temperature of the particles [7].

Acknowledgement The author thanks Prof. Dr. Dr. h.c.mult R. Bulirsch and Profs. G. Vanden Berghe, M. Van Daele and S. Dimova for valuable discussions. She thanks Prof. S. Dimova for her hospitality at the Bulgarian Academy of Sciences in Sofia. She is grateful to the organizers of the meeting in Borovets.

References

1. G. Denk (1993): A new numerical method for the integration of highly oscillatory second-order ordinary differential equations. Appl. Num. Math. **13**, 57–67; see also Report TUM-M9413, 35 pages
2. K. Dekker, J.G. Verwer (1984): *Stability of Runge-Kutta Methods for Stiff Nonlinear Differential Equations*, Amsterdam: North Holland Publ.
3. M.J. Gander, R. Meyer-Spasche: 'An Introduction to Numerical Integrators Preserving Physical Properties', chap.5 in [10]
4. J. Hersch (1958): Contribution à la méthode des équations aux différences. Z. Angew. Math. Phys. **2**, 129–180
5. E. Hairer, S.P. Nørsett, G. Wanner: *Solving Ordinary Differential Equations*, vol. I, 2nd ed., 1992; vol. II, 1991, Springer Verlag
6. M.N. Le Roux (1994): Semidiscretization in time of nonlinear parabolic equations with blowup of the solution, SIAM J. Numer. Anal. **31**, 170–195
7. M.N. Le Roux et al. (1992): Simulation of a coupled dynamic system of temperature and density in a fusion plasma, Phys. Scripta **46**, 457–462
8. R. Meyer-Spasche: 'Difference schemes of optimum degree of implicitness for a family of simple ODEs with blow-up solutions' JCAM **97** (1998), 137–152
9. R. Meyer-Spasche, D. Düchs (1997): A general method for obtaining unconventional and nonstandard difference schemes. Dyn.Cont.Discr.Imp.Sys. **3**, 453–467
10. R.E. Mickens (ed.): *'Applications of Nonstandard Finite Difference Schemes'*, World Scientific, Singapore, 2000
11. K. Ozawa (2001): 'Functional fitting Runge-Kutta method with variable coefficients', Jap J Ind Appl Math **18** 107–130
12. G. Vanden Berghe, L. Gr. Ixaru, H. De Meyer (2001): Frequency determination and step-length control for exponentially-fitted Runge-Kutta methods, JCAM **132**, 95–105

Additive Schemes for Systems of Time-Dependent Equations of Mathematical Physics

Alexander Samarskii and Petr Vabishchevich

Institute for Mathematical Modeling, RAS
4-A Miusskaya Square, 125047 Moscow, Russia
vab@imamod.ru

Abstract. Additive difference schemes are derived via a representation of an operator of a time-dependent problem as a sum of operators with a more simple structure. In doing so, transition to a new time level is performed as a solution of a sequence of more simple problems. Such schemes in various variants are employed for approximate solving complicated time-dependent problems for PDEs. In the present work construction of additive schemes is carried out for systems of parabolic and hyperbolic equations of second order. As examples there are considered dynamic problems of the elasticity theory for materials with variable properties, dynamics problems for an incompressible fluid with a variable viscosity, general 3D problems of magnetic field diffusion.

1 Introduction

Nowadays, various classes of additive difference schemes are developed by splitting the problem operator into certain components [1,2,3,4,5]. Classical examples of additive difference schemes are so called economical difference schemes for multi-dimensional problems of mathematical physics. The simplest examples of these economical schemes are the well-known alternating direction schemes, locally-one-dimensional schemes etc. In this case we have additive difference schemes where the splitting is done with respect to the separate directions.

Very often the components of the splitting operator are connected with description of differed components of the process under investigation. For instance, in continuum mechanics the process of transport of a substance can be decomposed into transport due to medium motion and transport by means of diffusion. To highlight such a peculiarity of the problem, this decomposition is referred to as splitting with respect to physical processes.

Domain decomposition methods are in common use in constructing computational algorithms for the solution of time-dependent problems for PDEs on modern parallel computers [5,6]. In this case, a particular processor solves boundary value problem in a separate subdomain. The corresponding additive difference schemes are referred to as regional-additive ones.

Numerical solution of systems of time-dependent PDEs is often required in applications of mathematics to scientific, industrial and environmental problems.

I. Dimov et al. (Eds.): NMA 2002, LNCS 2542, pp. 48–60, 2003.

As a rule, the unknown variables are included into several equations, what makes impossible to find one component independently from others. In this case, the operator corresponding to the system can be split into more simple operators, connected with separate components of the solution. Thus, the transition to a new time level can be based on solving consecutively separate problems for each of the component of the solution. Possibilities for constructing such additive schemes are considered in the present work. Hydrodynamics problems for an incompressible fluid with a variable viscosity and for diffusion of a magnetic field are considered as typical examples.

2 Problem Formulation

Let us consider real grid functions y from a finite-dimensional real Hilbert space H. In this space we have the dot product and norm $(\cdot,\cdot)$, $\|y\| = \sqrt{(y,y)}$, respectively. For an operator $D = D^* > 0$ we introduce H_D as space H with dot product $(y,w)_D = (Dy,w)$ and norm $\|y\|_D = \sqrt{(Dy,y)}$.

We will search solution $u(t) \in H$ of the Cauchy problem for the following evolutionary equation of first order

$$\frac{du}{dt} + Au = f(t), \quad 0 < t \le T, \tag{1}$$

$$u(0) = u_0. \tag{2}$$

Let us consider the simplest case of positive self-adjoint steady-state operator A, i.e. $A \neq A(t) = A^* > 0$.

In a similar way, the Cauchy problem for the evolutionary equation of second order is formulated. In this case we search $u(t) \in H$ from the equation

$$\frac{d^2u}{dt^2} + Au = f(t), \quad 0 < t \le T, \tag{3}$$

$$u(0) = u_0, \tag{4}$$

$$\frac{du}{dt}(0) = u_1. \tag{5}$$

The following estimate, concerning stability of the solution with respect to the initial data and to the RHS, is fulfilled for the above problem (1),(2) (it will further serve us as a checkpoint in constructing difference schemes).

Lemma 1. *The following estimate holds for problem* (1), (2)

$$\|u(t)\| \le \|u_0\| + \int_0^t \|f(s)\| ds.$$

In a similar way, it is possible to formulate a statement concerning stability of the solution of problem(3) - (5) with respect to the initial data and to the right hand side.

Lemma 2. *The following estimate holds for problem* (3) - (5)

$$\|u(t)\|_* \le \|u_0\|_A + \|u_1\| + \int_0^t \|f(s)\| ds,$$

where

$$\|u(t)\|_*^2 \equiv \|u\|_A^2 + \left\| \frac{du}{dt} \right\|^2 .$$

Our aim is to construct additive schemes for problems (1), (2) and (3) - (5). Assume that the operator A has an additive representation

$$A = \sum_{\alpha=1}^{p} A^{(\alpha)}, \quad A^{(\alpha)} \neq A^{(\alpha)}(t) = (A^{(\alpha)})^* \ge 0, \quad \alpha = 1, 2, \dots, p. \tag{6}$$

Additive difference schemes are derived on the basis of representation (6), where a transition from one time-level t^n, to the next one $t^{n+1} = t^n + \tau$ (here $\tau > 0$ — time-increment), is connected with solving problems for the separate operators $A^{(\alpha)}$, $\alpha = 1, 2, \dots, p$. Thus, the initial problem is decomposed into p more simple subproblems.

3 Some Classes of Additive Schemes

3.1 Schemes of Component-wise Splitting

Additive difference schemes for problems with splitting into three and more operators (which are pair-wise non-commutative ones), are constructed on the basis of the concept of the additive approximation. Schemes with component-wise splitting (locally one-dimensional schemes) serve as a prototype in this case. The following difference schemes are employed for problem (1), (2), (6)

$$\frac{y^{n+\alpha/p} - y^{n+(\alpha-1)/p}}{\tau} + A^{(\alpha)}(\sigma_\alpha y^{n+\alpha/p} + (1-\sigma_\alpha) y^{n+(\alpha-1)/p}) = f_\alpha^n, \tag{7}$$

$\alpha = 1, 2, ..., p, \quad n = 0, 1, ...,$ where $f^n = \sum_{\alpha=1}^{p} f_\alpha^n$.

For $\sigma_\alpha \ge 0.5$, the component-wise splitting scheme (7) is unconditionally stable. Let us derive the corresponding a priori estimate for stability with respect to the initial data and to the right hand side. A special representation is used for the right hand sides f_α^n, $\alpha = 1, 2, ..., p$:

$$f_\alpha^n = \overset{\circ}{f_\alpha^n} + \overset{*}{f_\alpha^n}, \quad \alpha = 1, 2, ..., p, \quad \sum_{\alpha=1}^{p} \overset{\circ}{f_\alpha^n} = 0. \tag{8}$$

Such a form of the right hand side is essential in the consideration of the problem for the error of the additive scheme. The following statement holds for the component-wise splitting scheme.

Theorem 1. *For $0.5 \le \sigma_\alpha \le 2$, $\alpha = 1,2,...,p$ and $\tau > 0$, the following a priori estimate holds for the solution of the Cauchy problem for difference equations (6), (8)*

$$\|y^{n+1}\| \le \|u_0\| + \sum_{k=0}^{n} \tau \sum_{\alpha=1}^{p} \left(\| \overset{*}{f}{}_\alpha^k \| + \tau \|A^{(\alpha)} \sum_{\beta=\alpha}^{p} \overset{\circ}{f}{}_\beta^k \| \right).$$

3.2 Additively-Averaged Schemes

Additively-averaged schemes, based on component-wise splitting can be used successfully on modern parallel computers. In this case the transition to a new time level is performed as follows:

$$\frac{y_\alpha^{n+1} - y^n}{p\tau} + A^{(\alpha)}(\sigma_\alpha y_\alpha^{n+1} + (1-\sigma_\alpha)y^n) = f_\alpha^n, \tag{9}$$

$$\alpha = 1,2,\ldots,p, \quad n = 0,1,\ldots, \quad y^{n+1} = \frac{1}{p}\sum_{\alpha=1}^{p} y_\alpha^{n+1}.$$

The stability conditions for these schemes are the same as for standard scheme of component-wise splitting. Similarly to Theorem 1, the following statement can be proved.

Theorem 2. *For $\sigma_\alpha \ge 0.5$, $\alpha = 1,2,...,p$ and any $\tau > 0$, the following a priori estimate holds for the solution of (6), (9)*

$$\|y^{n+1}\| \le \|u_0\| + \sum_{k=0}^{n} \tau \sum_{\alpha=1}^{p} \left(\| \overset{*}{f}{}_\alpha^k \| + p\tau\sigma_\alpha \|A^{(\alpha)} \overset{\circ}{f}{}_\alpha^k \| \right).$$

A potential advantage of the additively-averaged scheme (9) is connected with the fact that it is possible to perform parallel evaluation of grid functions y_α^{n+1}, $\alpha = 1,2,...,p$.

3.3 Regularized Additive Schemes

It is convenient to construct additive schemes on the basis of the regularization principle for the difference schemes. An example is the additive scheme constructed by perturbing each separate operator in the additive representation (6):

$$\frac{y^{n+1} - y^n}{\tau} + \sum_{\alpha=1}^{p} (E + \sigma_\alpha \tau A^{(\alpha)})^{-1} A^{(\alpha)} y^n = f^n, \quad n = 0,1,\ldots. \tag{10}$$

Theorem 3. *For $\sigma_\alpha \ge p/2$, $\alpha = 1,2,...,p$ and any $\tau > 0$, the following a priori estimate holds for the solution of equation (6), (10)*

$$\|y^{n+1}\| \le \|u_0\| + \sum_{k=0}^{n} \tau \|f^k\|.$$

The considered regularized scheme (10) is closely connected with the above considered additively-averaged scheme. To illustrate this fact, let us introduce fictitious grid unknowns y_α^{n+1}, $\alpha = 1, 2, ..., p$. These functions have no independent sense and are used for auxiliary purposes. Let us rewrite the scheme (10) as follows:

$$\frac{y_\alpha^{n+1} - y^n}{p\tau} + (E + \sigma\tau A^{(\alpha)})^{-1} A^{(\alpha)} y^n = f_\alpha^n,$$

$$\alpha = 1, 2, \ldots, p, \quad n = 0, 1, \ldots, \; y^{n+1} = \frac{1}{p} \sum_{\alpha=1}^{p} y_\alpha^{n+1}.$$

Thus, we again obtain an additively-averaged scheme, but in this case it is constructed without using the concept of the additive approximation. This scheme differs from the early presented scheme (9) in approximation of the right hand sides.

3.4 Second-Order Equation

Certain problems arise in constructing operator-splitting schemes for evolutionary equations of second order. Let us discuss now some classes of regularized additive schemes (so called full approximation schemes) for evolutionary equations of second order. The multiplicative regularization for the problem (3) - (5) results in the scheme

$$\frac{y^{n+1} - 2y^n + y^{n-1}}{\tau^2} + \sum_{\alpha=1}^{p} (E + \sigma\tau^2 A^{(\alpha)})^{-1} A^{(\alpha)} y^n = f^n, \quad n = 1, 2, \ldots. \quad (11)$$

Theorem 4. *The additive difference scheme (11) for the problem (3) – (5) is unconditionally stable for* $\sigma_\alpha \geq p/4$, $\alpha = 1, 2, \ldots, p$.

The scheme (11) can be implemented as follows:

$$(E + \sigma\tau^2 A^{(\alpha)}) \frac{y_\alpha^{n+1} - 2y^n + y^{n-1}}{p\tau^2} + A^{(\alpha)} y^n = \frac{1}{p}(E + \sigma\tau^2 A^{(\alpha)}) f^n,$$

$$\alpha = 1, 2, \ldots, p, \quad n = 1, 2, \ldots, \; y^{n+1} = \frac{1}{p} \sum_{\alpha=1}^{p} y_\alpha^{n+1}.$$

Thus we obtain a special additively-averaged scheme.

4 Alternating Triangle Method

The alternating triangle method has been developed in [7] for solving the Cauchy problem for linear ODEs with a symmetric matrix. It is based on splitting of the equation matrix into two triangular matrices. A general description of the alternating triangle method is presented below along with the discussion of possibilities for constructing new classes of additive alternating triangle schemes. Namely, this class of additive schemes is convenient for solving boundary value problems for systems of time-dependent equations of mathematical physics.

4.1 General Description of the Alternating Triangle Method

Let us consider Cauchy problem (1), (2) with a time-independent self-adjoint positive definite operator A. The alternating triangle method is defined by the two-component additive splitting

$$A = A^{(1)} + A^{(2)} > 0, \quad (A^{(1)})^* = A^{(2)}. \tag{12}$$

Let (1), (2) be the operator formulation of the Cauchy problem for the system of linear ODEs of first order

$$\frac{du_i(t)}{dt} + \sum_{j=1}^{m} a_{ij}u_j(t) = f_i(t), \quad t > 0, \tag{13}$$

$$u_i(0) = u_i^0, \quad i = 1, 2, \ldots, m. \tag{14}$$

Here $u = \{u_1, u_2, \ldots u_m\}$ stands for the vector of unknowns, $f = \{f_1, f_2, \ldots f_m\}$ — the specified vector of right hand sides, and $A = \{a_{ij}\}$ — the symmetric real matrix with elements $a_{ij} = a_{ji}$, $i, j = 1, 2, \ldots, m$.

Under the above conditions, the matrix A of the problem (13), (14) is considered as a self-adjoint linear operator in a finite-dimensional Hilbert (Euclidean) space $H = l_2$ with dot product $(y, v) = \sum_{i=1}^{m} y_i v_i$ and norm $\|y\| = \sqrt{(y, y)}$.

For the elements of the matrices

$$A^{(\alpha)} = \{a_{ij}^{(\alpha)}\}, \quad \alpha = 1, 2,$$

in correspondence with the decomposition (12), we have

$$a_{ij}^{(1)} = \begin{cases} a_{ij}, & i < j, \\ \frac{1}{2}\, a_{ii}, & i = j, \\ 0, & i > j, \end{cases} \qquad a_{ij}^{(2)} = \begin{cases} 0, & i < j, \\ \frac{1}{2}\, a_{ii}, & i = j, \\ a_{ij}, & i > j. \end{cases}$$

Thus, the matrix A is decomposed into two triangular matrices.

The standard variant of the alternating triangle method employs the alternating direction scheme by Peaceman-Rachford [8] for solving the problem (1), (2), (12)

$$\frac{y^{n+1/2} - y^n}{0.5\tau} + A^{(1)} y^{n+1/2} + A^{(2)} y^n = f^n, \tag{15}$$

$$\frac{y^{n+1} - y^{n+1/2}}{0.5\tau} + A^{(1)} y^{n+1/2} + A^{(2)} y^{n+1} = f^n. \tag{16}$$

Implementation of the above additive schemes is connected with the consecutive inversion of the upper and the lower triangular matrices, what explains the name "the alternating triangle method" for these additive schemes.

4.2 Equations of First Order

Additive schemes of the alternating triangle method can be investigated in the most complete way using results of the general theory of stability for operator-difference schemes [2,9]. Let us rewrite the two-level factorized scheme of the alternating triangle method in the canonical form

$$B\frac{y^{n+1}-y^n}{\tau}+Ay^n=f^n, \quad t_n\in\omega_\tau \tag{17}$$

$$B=(E+\sigma\tau A^{(1)})(E+\sigma\tau A^{(2)}). \tag{18}$$

The scheme (17), (18) is equivalent to the scheme (15), (16) when the weight parameter σ equals one half: $\sigma=0.5$.

Theorem 5. *Factorized scheme of alternating triangle method (12), (17), (18) is unconditionally stable for $\sigma\geq 0.5$ in H_A. The following a priori estimate holds*

$$\|y^{n+1}\|_A^2\leq\|y^0\|_A^2+\frac{1}{2}\sum_{k=0}^{n}\tau\|f^k\|^2.$$

The alternating triangle additive scheme for three-level operator-difference schemes is constructed in a similar way. Schemes from this class are also constructed for problems with non-selfadjoint operators in the case of subordination of the skew-symmetric part of the operator [5].

4.3 Equation of Second Order

In constructing additive schemes for the Cauchy problem for evolutionary equation of second order (3) - (5), the emphasize is on the additive alternating triangle schemes.

For three-level schemes

$$D\frac{y_{n+1}-2y_n+y_{n-1}}{\tau^2}+Ay_n=\varphi_n, \quad n=1,2,\dots \tag{19}$$

with splitting (12), we define a factorized operator

$$D=(E+\sigma\tau^2A^{(1)})(E+\sigma\tau^2A^{(2)}). \tag{20}$$

Theorem 6. *Factorized scheme of alternating triangle method (12), (19), (20) is unconditionally stable for $\sigma\geq 0.25$.*

This statement follows from the estimate

$$D=D^*\geq E+\sigma\tau^2A$$

and from the general results of the stability theory for three-level operator-difference schemes [2].

5 Hydrodynamics Problems for an Incompressible Fluid with Variable Viscosity

In modeling flows of an inhomogeneous fluid, in particular, in solving hydrodynamics problems with free boundaries, it is necessary to consider problems with a variable viscosity. The primary peculiarities of such problems are connected with the fact, that equations for the particular components of the velocity are strongly coupled (through the principal derivatives). There is no such a coupling in the problems with a constant viscosity. Let us recall, that in the constant viscosity case, the (linearized) system of momentum equations can be naturally decomposed and it is possible to solve independently equations for the velocity components. The emphasis in the case of variable viscosity is on constructing special additive schemes. They are based on splitting with respect to the physical processes and take into account the peculiarities of the problems with variable viscosity [10].

5.1 Problem Formulation

Let ϱ be the density, p — the pressure, $\mathbf{u}$ — the velocity and μ — the viscosity of an incompressible fluid. The momentum equation has the form

$$\varrho \left(\frac{\partial \mathbf{u}}{\partial t} + (\mathbf{u} \cdot \mathrm{grad})\mathbf{u} \right) + \mathrm{grad} p = \mathrm{Div} \sigma + \varrho \mathbf{f}, \tag{21}$$

where σ stands for the viscous stresses tensor and $\mathbf{f}$ stands for the volumetric force (e.g., the buoyant force in free convection problems).

In the case of a Newtonian fluid, the viscous stresses tensor has the following coordinate-wise representation:

$$\sigma_{ij} = \mu \left(\frac{\partial u_i}{\partial x_j} + \frac{\partial u_j}{\partial x_i} \right).$$

In addition, for an incompressible fluid the continuity equation has the form

$$\mathrm{div} \mathbf{u} = 0. \tag{22}$$

For simplicity, let us consider motion of a fluid with a constant density. Thus, instead of (21) we employ the equation

$$\frac{\partial \mathbf{u}}{\partial t} + (\mathbf{u} \cdot \mathrm{grad})\mathbf{u} + \mathrm{grad} p' = \mathrm{Div} \sigma' + \varrho \mathbf{f}', \tag{23}$$

where the pressure, the tensor of viscous stresses and the force are normalized by the density in such a way that, for instance,

$$\sigma_{ij} = \nu \left(\frac{\partial u_i}{\partial x_j} + \frac{\partial u_j}{\partial x_i} \right),$$

where $\nu = \mu / \varrho$ is the kinematic viscosity.

5.2 The 2D Problem

The detailed consideration will be conducted for planar-parallel flows. Let $\mathbf{x} = (x_1, x_2)$, $\mathbf{v} = (v_1, v_2)$, then equations (22), (23) in the coordinate-wise representation take the following form

$$\frac{\partial v_1}{\partial t} + v_1 \frac{\partial v_1}{\partial x_1} + v_2 \frac{\partial v_1}{\partial x_2} + \frac{\partial p'}{\partial x_1} =$$

$$= \frac{\partial}{\partial x_1}\left(2\nu \frac{\partial v_1}{\partial x_1}\right) + \frac{\partial}{\partial x_2}\left(\nu \frac{\partial v_1}{\partial x_2}\right) + \frac{\partial}{\partial x_2}\left(\nu \frac{\partial v_2}{\partial x_1}\right) + f_1'(\mathbf{x}, t), \quad (24)$$

$$\frac{\partial v_2}{\partial t} + v_1 \frac{\partial v_2}{\partial x_1} + v_2 \frac{\partial v_2}{\partial x_2} + \frac{\partial p'}{\partial x_2} =$$

$$= \frac{\partial}{\partial x_1}\left(\nu \frac{\partial v_1}{\partial x_2}\right) + \frac{\partial}{\partial x_1}\left(\nu \frac{\partial v_2}{\partial x_1}\right) + \frac{\partial}{\partial x_2}\left(2\nu \frac{\partial v_2}{\partial x_2}\right) + f_2'(\mathbf{x}, t), \quad (25)$$

$$\frac{\partial v_1}{\partial x_1} + \frac{\partial v_2}{\partial x_2} = 0. \quad (26)$$

Using the representation (24) - (26), it is more easy to see the coupling between components v_α, $\alpha = 1, 2$ in the right hand side of the momentum equation.

5.3 Operator of Viscous Stresses

Let us define the standard Hilbert space $\mathcal{H} = L_2(\Omega)$ with the dot product and norm

$$(u, v) = \int_\Omega u(\mathbf{x}) v(\mathbf{x}) d\mathbf{x}, \quad \|u\| = (u, u)^{1/2}.$$

For the 2D vectors $\mathbf{u}$, $\mathbf{v}$ we introduce the Hilbert space $\mathcal{H}_2 = \mathcal{H} \oplus \mathcal{H}$ with dot product given by

$$(\mathbf{u}, \mathbf{v}) = (u_1, v_1) + (u_2, v_2).$$

On the set of vector functions $\mathbf{v}$ equal to zero on $\partial\Omega$, we define the operator $\mathcal{N}$,

$$\mathcal{N}\mathbf{v} = -\operatorname{Div} \sigma.$$

In accordance with (24), (25), the operator $\mathcal{N}$ is nothing but the matrix

$$\mathcal{N} = \begin{pmatrix} \mathcal{N}_{11} & \mathcal{N}_{12} \\ \mathcal{N}_{21} & \mathcal{N}_{22} \end{pmatrix},$$

where

$$\mathcal{N}_{11} v_1 = -\frac{\partial}{\partial x_1}\left(2\nu \frac{\partial v_1}{\partial x_1}\right) + \frac{\partial}{\partial x_2}\left(\nu \frac{\partial v_1}{\partial x_2}\right),$$

$$\mathcal{N}_{12} v_2 = -\frac{\partial}{\partial x_2}\left(\nu \frac{\partial v_2}{\partial x_1}\right),$$

$$\mathcal{N}_{21} v_1 = -\frac{\partial}{\partial x_1}\left(\nu \frac{\partial v_1}{\partial x_2}\right),$$

$$\mathcal{N}_{22} v_2 = -\frac{\partial}{\partial x_1}\left(\nu \frac{\partial v_2}{\partial x_1}\right) + \frac{\partial}{\partial x_2}\left(2\nu \frac{\partial v_2}{\partial x_2}\right).$$

It is easy to see that

$$(\mathcal{N}\mathbf{v}, \mathbf{u}) = (\mathbf{v}, \mathcal{N}\mathbf{u}),$$

i.e. operator $\mathcal{N} = \mathcal{N}^* > 0$ in $\mathcal{H}_2$.

In constructing unconditionally stable difference schemes, it is necessary to derive schemes where terms with viscous stresses are implicitly discretized. On the other hand, the fully implicit treatment of this operator is not convenient due to the above mentioned coupling of the velocity components. So, it is necessary to select an operator that is close to $\mathcal{N}$ but is more convenient in computational implementation. It is natural to take the following operator

$$\mathcal{D} = \begin{pmatrix} \mathcal{D}_{11} & 0 \\ 0 & \mathcal{D}_{22} \end{pmatrix},$$

where

$$\mathcal{D}_{11} v_1 = -\frac{\partial}{\partial x_1}\left(\nu \frac{\partial v_1}{\partial x_1}\right) + \frac{\partial}{\partial x_2}\left(\nu \frac{\partial v_1}{\partial x_2}\right),$$

$$\mathcal{D}_{22} v_2 = -\frac{\partial}{\partial x_1}\left(\nu \frac{\partial v_2}{\partial x_1}\right) + \frac{\partial}{\partial x_2}\left(\nu \frac{\partial v_2}{\partial x_2}\right).$$

Thus

$$\mathcal{D}\mathbf{v} = -\operatorname{div}(\nu \operatorname{grad} \mathbf{v}).$$

It is easy to show that operators $\mathcal{N}$ and $\mathcal{D}$ are energy equivalent. Namely, the following estimate holds

$$\nu(\mathcal{D}\mathbf{v}, \mathbf{v}) \le (\mathcal{N}\mathbf{v}, \mathbf{v}) \le 2\nu(\mathcal{D}\mathbf{v}, \mathbf{v}).$$

The facts that operator $\mathcal{N}$ is self-adjoint and positive definite in space $\mathcal{H}_2$ and, moreover, it is energy equivalent to the Laplace operator demonstrate its most important properties. Just these peculiarities will be preserved in constructing of the discrete analogs of the differential operators.

For of the numerical implementation we will employ representation of the operator matrix $\mathcal{N}$ as a sum of two operators:

$$\mathcal{N} = \mathcal{N}^{(1)} + \mathcal{N}^{(2)}, \quad (\mathcal{N}^{(1)})^* = \mathcal{N}^{(2)}. \tag{27}$$

Taking into account the expression for the operator $\mathcal{N}$, we have

$$\mathcal{N}^{(1)} = \begin{pmatrix} \frac{1}{2}\mathcal{N}_{11} & 0 \\ \mathcal{N}_{21} & \frac{1}{2}\mathcal{N}_{22} \end{pmatrix}, \quad \mathcal{N}^{(2)} = \begin{pmatrix} \frac{1}{2}\mathcal{N}_{11} & \mathcal{N}_{12} \\ 0 & \frac{1}{2}\mathcal{N}_{22} \end{pmatrix}.$$

In using additive schemes calculations are connected with inversion of operators $E + \sigma\tau\mathcal{N}^{(\alpha)}$, $\alpha = 1, 2$ (without convective transport). Thus, we have the decomposed system of equations with separate solution of elliptic boundary value problems for particular components of the velocity.

In fact, we have two closely related possibilities. The first possibility is connected with constructing difference schemes on the basis of regularizer $\mathcal{D}$, the second one - on using triangular splitting (27). Viscous terms are split in both cases.

5.4 Additive Difference Schemes

Differential equations (22), (23) can be written as a single equation

$$\frac{d\mathbf{v}}{dt} + \mathcal{V}\mathbf{v} + \mathcal{P}\mathbf{v} + \mathcal{N}\mathbf{v} = \mathbf{f}'$$

for solenoidal functions $\mathbf{v} \in \mathcal{H}_2$ (functions that satisfy condition (22)). Here $\mathcal{V}$ is the convective transport operator, whereas $\mathcal{P}$ is the operator connected with the pressure. The boundary and the initial conditions are treated in the standard way.

Let us construct the simplest difference schemes for equation (27), based on the uniform integration in time with increment $\tau > 0$. As the basis we consider the scheme where the convective transport is taken from the previous time-level. Let us derive regularized additive schemes. Unfortunately, the traditional schemes of ADI-type are not possible here.

Using the regularizer $\mathcal{D}$, we can employ the scheme

$$\frac{\mathbf{v}^{n+1} - \mathbf{v}^n}{\tau} + \mathcal{V}^n\mathbf{v}^n + (E + \sigma_1\tau\mathcal{D})^{-1}\mathcal{N}(E + \sigma_1\tau\mathcal{D})^{-1}\mathbf{v}^n + \\ + (E + \sigma_2\tau\mathcal{P})^{-1}\mathcal{P}\mathbf{v}^n = \mathbf{f}'_n. \qquad (28)$$

The proposed scheme can be implemented in various ways. For example, it is possible to introduce some auxiliary function $\mathbf{v}^{n+1/2}$, which is to be determined from the equation

$$\frac{\mathbf{v}^{n+1/2} - \mathbf{v}^n}{\tau} + \mathcal{V}^n\mathbf{v}^n + (E + \sigma_1\tau\mathcal{D})^{-1}\mathcal{N}(E + \sigma_1\tau\mathcal{D})^{-1}\mathbf{v}^n = \mathbf{f}'_n.$$

In such a way we calculate the transport due to the convection, due to viscosity and due to volumetric forces.

The second stage is the calculation of the transport due to the pressure:

$$\frac{\mathbf{v}^{n+1} - \mathbf{v}^{n+1/2}}{\tau} + (E + \sigma_2\tau\mathcal{P})^{-1}\mathcal{P}\mathbf{v}^n = 0.$$

In fact, schemes of such a type are in common use. The primary peculiarity is in calculation of the first stage. This complicated construction is necessary for deriving a stable scheme. Implementation consists in double solution of elliptic problem $E+\sigma_1\tau\mathcal{D}$. Such an increasing of the computational costs is not essential, compare to the advantages of the scheme.

In a similar way we can design schemes with triangular decomposition (27). For instance, an analog of scheme (28) will be the following one

$$\frac{\mathbf{v}_{n+1}-\mathbf{v}_n}{\tau} + (E+\sigma_1\tau\mathcal{N}^{(1)})^{-1}(\mathcal{N}+\mathcal{V}_n)\,(E+\sigma_1\tau\mathcal{N}^{(2)})^{-1}\mathbf{v}_n + \\ + (E+\sigma_2\tau\mathcal{P})^{-1}\mathcal{P}\mathbf{v}_n = \mathbf{f}_n. \quad (29)$$

The advantage of scheme (29) compare to scheme (28) consists in the fact, that only one elliptic problem for evaluation of the velocity components is solved here at each time-level (but this elliptic problem is slightly more complicated).

6 enerali ations

In a similar way we can construct additive operator-difference schemes for solving the system of equation for the diffusion of a magnetic field. Induction of a magnetic field is governed by the equation

$$\frac{\partial \mathbf{B}}{\partial t} - \mathrm{rot}\left(\frac{1}{\sigma}\,\mathrm{rot}\mathbf{B}\right) = 0. \quad (30)$$

In the general 3D case it is no possible to formulate boundary value problems for particular components of vector **B**. For equation (30) we can design additive alternating triangular difference schemes.

Evolutionary equations of second order arise in considering dynamic problems of the elasticity theory. Some possibilities of designing additive operator-difference schemes for this case are discussed in the work [11].

Acknowledgments

This research was supported by Russian Foundation for Basic Research under grant No. 02-01-00555.

References

1. Yanenko, N. N.: The Method of Fractional Steps. Springer–Verlag, New York (1967)
2. Samarskii, A. A.: The Theory of Difference Schemes. Marcell Dekker (2001)
3. Marchuk, G. I.: Splitting and alternating direction methods. In: Ciarlet,P. G., Lions, J.-L. (eds): Handbook of Numerical Analysis, Vol. 1. North–Holland, Amsterdam (1990) 197–462

4. Samarskii, A. A., Vabishchevich, P. N.: Computational Heat Transfer, Vol.1,2. Wiley, Chichester (1995)
5. Samarskii, A. A., Vabishchevich, P. N.: Additive Schemes for Problems of Mathematical Physics. Nauka, Moscow (1999) (in Russian)
6. Samarskii, A. A., Vabishchevich, P. N., Matus, P. P.: Difference Schemes with Operator Factors. Minsk (1998) (in Russian)
7. Samarskii, A. A.: An economical algorithm for numerical solution of systems of differential and algebraic equations. Zhurnal Vychislitel'noi Matematiki i Matematicheskoi Fiziki. 4 (1964) 580–585 (in Russian)
8. Peaceman, D. W.,Rachford, H. H.: The numerical solution of parabolic and elliptic differential equations. J. SIAM. 3 (1955) 28–41
9. Samarskii, A. A., Gulin, A. V.: Stability of Difference Schemes. Nauka, Moscow (1973) (in Russian)
10. Vabishchevich, P. N., Samarskii, A. A.: Solution of problems of incompressible fluid dynamics with variable viscosity. Zhurnal Vychislitel'noi Matematiki i Matematicheskoi Fiziki. 40 (2000) 1813–1822 (in Russian)
11. Lisbona, F. L., Vabishchevich, P. N.: Operator-splitting schemes for solving elasticity problems. Comput. Methods in Applied Math. 1 (2001) 188–198

Geometry of Polynomials and Numerical Analysis

Blagovest Sendov

Central Laboratory for Parallel Processing, Bulgarian Academy of Sciences
"Acad. G. Bonchev" street, Block 25A, 1113 Sofia, Bulgaria
bsendov@argo.bas.bg,
http://WWW.copern.bas.bg/~bsendov/

Abstract. The purpose of this paper is to demonstrate the potential of a fruitful collaboration between Numerical Analysis and Geometry of Polynomials. This is natural as the polynomials are still a very important instrument in Numerical Analysis, regardless many new instruments as splines, wavelets and others. On the other hand, the Numerical Analysis through computers is a powerful instrument for experimentation in almost every mathematical discipline.

Key words: *Geometry of polynomials, Gauss–Lucas Theorem, zeros of polynomials, critical points, Smale's Conjecture, Sendov's Conjecture.*
Mathematics Subject Classification: 30C10

1 Introduction

The Geometry of Polynomials studies the geometrical relation between the zeros of an algebraic polynomial and the zeros of its derivatives as points in the complex plan $\mathcal{C}$. The term was coined in the nineteenth century by Félix Lucas in his paper *Gëométrie des polynômes* [21], see also [22]. In this paper F. Lucas gave a formal proof of the Gauss mechanical conclusion that:

Gauss–Lucas Theorem. *The convex hull of the zeros of a polynomial p contains all zeros of its derivative p'.*

This is one of the first and basic theorems in the Geometry of Polynomials, a lively subject through the twentieth century. The Gauss–Lucas Theorem may be considered as an analog of the Roll Theorem for real polynomials and real zeros.

The purpose of this paper is to demonstrate the potential of a fruitful collaboration between mathematicians working in both fields, Numerical Analysis and Geometry of Polynomials. This is natural as the polynomials are still a very important instrument in Numerical Analysis, regardless many new instruments as splines, wavelets and others. On the other hand, the Numerical Analysis through computers is a powerful instrument for experimentation in almost every mathematical discipline.

I. Dimov et al. (Eds.): NMA 2002, LNCS 2542, pp. 61–69, 2003.

1.1 Geometry of Polynomials in Numerical Analysis

First we consider a problem in Geometry of Polynomials, the famous Mean Value Conjecture of the Fields medalist Stephen Smale from 1981 [31], formulated in connection with a numerical method for successive approximation of the zeros of a polynomial p. For this method it was desirable to find the smallest positive constant c such that

$$|p'(z)| \geq \frac{1}{c}\left|\frac{p(\zeta)-p(z)}{\zeta - z}\right| \tag{1}$$

for at least one zero of p' (critical point of p) and all $z \in \mathcal{C}$. It is sufficient to establish (1) only for $z = 0$ and all polynomials p such that $p(0) = 0$ and $p'(0) \neq 0$. Concerning the smallest constant c, S. Smale formulated the following conjecture, known as *Smale's mean value conjecture.*

Conjecture 1 *Let p be a polynomial of degree n such that $p(0) = 0$ and $p'(0) \neq 0$. Then*

$$\min\left\{\left|\frac{p(\zeta)}{\zeta p'(0)}\right| : p'(\zeta) = 0\right\} \leq c, \tag{2}$$

where $c = 1$ or possibly $c = (n-1)/n$.

G. Schmeisser [29] formulated Conjecture 1 as a problem from the Geometry of Polynomials as follows. Consider an arbitrary polynomial $p(z) = a_n \prod_{k=1}^{n}(z - z_n)$ of degree n having z_j as a simple zero. Then

$$g(z) = p(z + z_j) \tag{3}$$

is a polynomial which satisfies the hypothesis of Conjecture 1. Conversely, by (3), every polynomial g satisfying the hypothesis of Conjecture 1 can be associated with a polynomial p that has z_j as a simple zero. Clearly, the critical points of g are obtained from those of p by subtracting z_j. Hence, according to Conjecture 1 for g, there should exist a critical point ζ of p such that

$$\left|\frac{g(\zeta - z_j)}{(\zeta - z_j)g'(0)}\right| \leq c.$$

In terms of the zeros of p, we have inequality

$$\left(\prod_{k=1,\, k\neq j}^{n} |\zeta - z_j|\right) \Big/ \left(\prod_{k=1,\, k\neq j}^{n} |z_j - z_j|\right) \leq c,$$

and also

$$\left(\prod_{k=1,\, k\neq j}^{n} |\zeta - z_j|\right)^{1/(n-1)} \leq c^{1/(n-1)} \left(\prod_{k=1,\, k\neq j}^{n} |z_k - z_j|\right)^{1/(n-1)}. \tag{4}$$

For $c < 1$, this means that, for each simple zero z_j of p, there exists a critical point ζ of p which is, in geometric mean, closer to the remaining zeros than z_j.

1.2 Numerical Analysis in Geometry of Polynomials

As a second example we consider one of our conjectures, which is more than 40 years old, but was published for the first time by W. K Hayman [16], for more information see [30].

Conjecture 2 *Let*

$$p = (z - z_1)(z - z_2)\cdots(z - z_n) \quad \text{and} \quad p'(z) = n(z - \zeta_1)(z - \zeta_2)\cdots(z - \zeta_{n-1}),$$

then for every $k = 1, 2, \ldots, n$ *there exists at least one* ζ_j*, such that*

$$|z_k - \zeta_j| \le \max\{|z_l| : l = 1, 2, \ldots, n\}.$$

We shall formulate Conjecture 2 in another way. The operator of differentiation $\mathcal{D} = \partial/\partial z$ may be considered as a mapping of the set $A = \{z_1, z_2, \ldots, z_n\}$ of n points (the zeros of the polynomial p) from the complex plane $\mathcal{C}$ in to the set $A' = \{\zeta_1, \zeta_2, \ldots, \zeta_{n-1}\}$ of $n-1$ points (the zeros of its derivative). By applying a linear transformation, it suffices to prove Conjecture 2 only for the polynomials p from **the set $\mathcal{P}_n$ of all monic polynomials** $p(z) = (z-z_1)(z-z_2)\cdots(z-z_n)$**, for which the unit disk $D = D(0,1)$ is the smallest disc containing all zeros of p.**

Define the deviation $\rho(A; B)$ of a finite set of points $A \subset \mathcal{C}$ from an other such set of points B as follows:

$$\rho(A; B) = \max\{\min\{|a - b| : b \in B\} : a \in A\}. \tag{5}$$

The substance of Conjecture 2 is to determine **how large** may be the deviation $\rho(A(p); A(p'))$. In this notation, Conjecture 2 may be reformulated as follows.

Conjecture 2 *If* $p \in \mathcal{P}_n$*, then* $\rho(A(p); A(p')) \le 1$.

From the Gauss–Lucas Theorem it follows that $\rho(A(p); A(p') \le 2$ for every $p \in \mathcal{P}_n$. B. Bojanov, Q. I. Rahman and J. Szynal [3] proved that

$$\kappa_{n,1} = \sup\{\rho(A(p); A(p')); \ p \in \mathcal{P}_n\} \le 1.08006\ldots \quad \text{and} \quad \lim_{n\to\infty} \kappa_{n,1} = 1.$$

A problem, which is in a sense inverse to Conjecture 2, was formulated and solved by A. Aziz [1].

Theorem 1. (A. Aziz). *If* $p \in \mathcal{P}_n$*, then* $\rho(A(p'); A(p)) \le 1$.

According to Conjecture 2, if z_k is a zero of $p \in \mathcal{P}_n$ then the closed disk $D(z_k, 1) = \{z : |z - z_k| \le 1\}$ with center z_k and radius 1 contains at list one critical point of p. A conjecture, stronger than Conjecture 2, announced in A. G. Goodman, Q. I. Rahman and J. Ratti [15] and independently in G. Schmeisser [28], is the following:

Conjecture 3 *If* $p \in \mathcal{P}_n$*, then for every zero* z_k *of* p *the disk* $D(z_k/2, 1-|z_k|/2)$ *contains at least one zero of* $p'(z)$.

Obviously, Conjecture 2 follows from Conjecture 3, as the disk $D(z_k/2, 1-|z_k|/2)$ lies in the disk $D(z_k, 1)$. Conjecture 3 is true if z_k is a peripheral zero, i. e., if $|z_k| = 1$, see [15] and [28], hence Conjecture 2 is also true for peripheral zeros. Conjecture 3 was proved in 1971 by G. GACS [14] for $2 \leq n \leq 5$. Next progress was stopped for two decades. Then, the Numerical Analysis and computers were used in the investigations. M. J. MILLER [24], using computers, constructed counterexamples for Conjecture 3 for polynomials of degree 6, 8, 10 and 12. Counterexamples for polynomials of degree 7, 9 and 11 were found by S. KUMAR AND B. G. SHENOY [19]. Now it remains for the specialists in Geometry of Polynomials to prove that Conjecture 3 is not true for $n > 12$ and for specialists in Numerical Analysis to increase the limit 12.

In Section 2 we consider Smale's Conjecture and prove it for the simplest cases. In Section 3 some conjectures connected with Conjecture 2 are considered, which may be treated by Numerical Analysis and computers.

2 The Smale's Con ecture

Let p be a polynomial of degree n such that $p(0) = 0$, $p'(0) \neq 0$ and let

$$\rho(p) = \min \left\{ \left| \frac{p(\zeta)}{\zeta p'(0)} \right| : \zeta \in A(p') \right\}.$$

It is strait forward that if $q(z) = ap(bz)$, $ab \neq 0$, then $\rho(q) = \rho(p)$, therefore it is sufficient to consider only polynomials $p(z)$ from the set

$$\mathcal{Q}_n = \{z^n + a_{n-1}z^{n-1} + \cdots + a_2 z^2 + nz : (a_2, a_3, \ldots, a_{n-1}) \in \mathcal{C}^{n-2}\}. \quad (6)$$

Conjecture 1 is proved only for $n = 2, 3, 4$ and for the following special cases. Denote by $\mathcal{R}_n$ the set of all polynomials $p \in \mathcal{Q}_n$ with equal modules of all their critical points. The next theorem is due to D. Tischler [32].

Theorem 2. *Conjecture 1 is true for every polynomial $p \in \mathcal{R}_n$.*

By Theorem 2, to prove Conjecture 1 it is sufficient to prove that $\rho(p)$ has its maximal value for $p \in \mathcal{R}_n$. For a weaker conclusion of Conjecture 1, one may allow c to be larger than 1. S. Smale verified (2) for $c = 4$. Recently, A. F. Beardon, D. Minda and T. W. Ng [2] proved (2) for

$$c = \alpha_n = 4^{(n-2)/(n-1)}$$

and G. Schmeisser [29] proved very simply that (2) is true for

$$c = \beta_n = \frac{2^n - (n+1)}{n(n-1)}.$$

Asymptotically β_n is much bigger than α_n, but for $2 \leq n \leq 7$ we have $\beta_n < \alpha_n$.

Let $p \in \mathcal{Q}_n, p'(z) = n(z-\zeta_1)(z-\zeta_2)\cdots(z-\zeta_{n-1}),\ \zeta_j \neq 0$ for $j = 1, 2, \ldots, n-1$ and denote

$$\eta_{k,j} = \frac{\zeta_k}{\zeta_j}, \quad \text{for} \quad j = 1, 2, \ldots, k-1 \quad \text{and} \quad \eta_{k,j} = \frac{\zeta_k}{\zeta_{j+1}}, \quad \text{for} \quad j = k, k+1, \ldots, n-2.$$

It is clear that there is a one to one correspondence between the set $\mathcal{Q}_n$ and the collection of all (non ordered) sets $\{\eta_{k,1}, \eta_{k,2}, \ldots, \eta_{k,n-2}\}$ of $n-2$ complex numbers. A polynomial $p \in \mathcal{Q}_n$ is represented by its critical points

$$p(z) = z^n - \frac{n}{n-1}\sigma_{n-1,1}z^{n-1} + \cdots + (-1)^{n-1}n\sigma_{n-1,n-1}z,$$

where

$$\sigma_{n-1,1} = \sum_{j=1}^{n-1}\zeta_j, \quad \sigma_{n-1,2} = \sum_{j=1}^{n-1}\sum_{l=j+1}^{n-1}\zeta_j\zeta_l, \quad \ldots, \quad \sigma_{n-1,n-1} = \prod_{j=1}^{n-1}\zeta_j$$

are the elementary symmetric functions of the variables $\zeta_1, \ldots, \zeta_{n-1}$. Then

$$\frac{p(\zeta_k)}{\zeta_k p'(0)} = \frac{\zeta_k^{n-1} - (n/(n-1))\sigma_{n-1,1}\zeta_k^{n-2} + \cdots + (-1)^{n-1}n\sigma_{n-1,n-1}}{n(-1)^{n-1}\sigma_{n-1,n-1}} =$$

$$1 - \frac{\sigma_{n-1,n-2}\zeta_k}{2\sigma_{n-1,n-1}} + \frac{\sigma_{n-1,n-3}\zeta_k^2}{3\sigma_{n-1,n-1}} + \cdots + \frac{(-1)^{n-1}\zeta_k^{n-1}}{n\sigma_{n-1,n-1}}$$

or

$$\frac{p(\zeta_k)}{\zeta_k p'(0)} = \frac{1}{1\cdot 2} - \frac{1}{2\cdot 3}\sigma_{n-2,1} + \frac{1}{3\cdot 4}\sigma_{n-2,2} + \cdots + \frac{(-1)^{n-2}}{(n-1)n}\sigma_{n-2,n-2}, \tag{7}$$

where $\sigma_{n-2,k},\ k = 1, 2, \ldots, n-2$ are the elementary symmetric functions of $\eta_{k,1}, \eta_{k,2}, \ldots, \eta_{k,n-2}$.

The assertion of Conjecture 1 is trivially satisfied for $n = 2$, as from (7), for the single zero ζ_1 of p' it follows

$$\frac{p(\zeta_1)}{\zeta_1 p'(0)} = \frac{1}{1\cdot 2} = 1 - \frac{1}{2}.$$

For $n = 3$, from (7) we have

$$\frac{p(\zeta_k)}{\zeta_k p'(0)} = \frac{1}{1\cdot 2} - \frac{1}{2\cdot 3}\sigma_{1,1} = \frac{1}{2} - \frac{1}{6}\eta_{k,1}, \quad k = 1.2$$

and $\eta = \eta_{1,1} = \zeta_1/\zeta_2 = 1/\eta_{2,1}$. Then

$$\rho(p) = \min\left\{\left|\frac{p(\zeta_1)}{\zeta_1 p'(0)}\right|, \left|\frac{p(\zeta_2)}{\zeta_2 p'(0)}\right|\right\} = \min\left\{\left|\frac{1}{2} - \frac{\eta}{6}\right|, \left|\frac{1}{2} - \frac{1}{6\eta}\right|\right\}$$

$$\leq \min\left\{\frac{1}{2} + \frac{1}{6}|\eta|, \frac{1}{2} + \frac{1}{6}|\eta|^{-1}\right\}.$$

It is clear that

$$\sup\left\{\min\left\{\frac{1}{2}+\frac{1}{6}|\eta|,\frac{1}{2}+\frac{1}{6}|\eta|^{-1}\right\}:\ \eta\in\mathcal{C}\right\}=\frac{2}{3}=1-\frac{1}{3}$$

is obtained for $|\eta|=1$. This is a proof of Conjecture 1 for $n=3$. The proof for $n=4$ is much more complicated. By elementary calculations we find that for $n=4$

$$\rho(p)\le\sup\{\varphi(\eta_1,\eta_2):\ \eta_1,\eta_2\in\mathcal{C}\},$$

where

$$\varphi(\eta_1,\eta_2)=\min\left\{\left|\frac{1}{2}-\frac{1}{6}(\eta_1+\eta_2)+\frac{1}{12}\eta_1\eta_2\right|,\right.$$

$$\left.\left|\frac{1}{2}-\frac{1}{6}(1+\eta_2)\frac{1}{\eta_1}+\frac{1}{12}\eta_1\eta_2\frac{1}{\eta_1^3}\right|,\left|\frac{1}{2}-\frac{1}{6}(\eta_1+1)\frac{1}{\eta_2}+\frac{1}{12}\eta_1\eta_2\frac{1}{\eta_2^3}\right|\right\}.$$

We claim that the supremum is obtained for $\eta_1=e^{2\pi i/3},\eta_2=e^{4\pi i/3}$ and

$$\sup\{\varphi(\eta_1,\eta_2):\ \eta_1,\eta_2\in\mathcal{C}\}$$

$$=\varphi(e^{2\pi i/3},e^{4\pi i/3})=\frac{1}{2}+\frac{1}{6}+\frac{1}{12}=\frac{3}{4}=1-\frac{1}{4}.$$

It is interesting that

$$\frac{1}{1\cdot 2}+\frac{1}{2\cdot 3}+\frac{1}{3\cdot 4}+\cdots+\frac{1}{(n-1)\cdot n}=1-\frac{1}{n}.$$

3 Con ectures around Con ecture 2

Conjecture 2 is trivial for polynomials of degree 2. After it appeared in Hayman's book [16], a number of proofs have been published for polynomials of degree 3, see [6,26,28,13,5] and of degree 4, see [26,28,23,17,12,5]. The proof for $n\le 5$ was given by A. Meir and A. Sharma [23] in 1969, see also [20,5]. More than 20 years later a proof for $n\le 6$ was published by J. Brown [8] in 1991, see also [4,5]. The case $n=7$ was proved first by J. Borcea [5] in 1996 and by J. Brown [9] in 1997. In a recent paper J. E. Brown and G. Xiang (1999) [11] proved Conjecture 2 for $n\le 8$. The proof is very elaborated. It is based on obtaining good upper and lower estimates on the product of the module of the critical points of p. The method of proof in [11] could be probably extended to $n=9$ but this is becoming too laborious.

Conjecture 2 is proved also for every polynomial with 3,4,5,6,7 and 8 distinct zeros, see [27,7,20,11]. G. L. Cohen and G. H. Smith [12] proved that Conjecture 2 is true for polynomials of degree n with m distinct zeros, if $n\ge 2^{m-1}$. The general case is still open. Interestingly enough, it is not proved even for polynomials with real coefficients and only real critical points, see [10].

We generalized Conjecture 2, see [30], as follows:

Conjecture 4 *If $p \in \mathcal{P}_n$ and $n \geq s+1$, then*

$$\rho(A(p); A(p^{(s)})) \leq \frac{2s}{s+1}.$$

For $s = 1$, Conjecture 4 is Conjecture 2, and for $s = n-1$ it is simple to prove it, since $p^{(n-1)}(z)$ has only one zero. Conjecture 4 has been proved also for $s = n-2$ and $n \geq 3$, for $s = n-3$ and $n \geq 4$, for $s = n-4$ and $n \geq 6$, see [30].

For every natural $n \geq 2$, the set $\mathcal{P}_n$ is compact. Therefore, for every $n \geq s+1$ and $s = 1, 2, 3 \ldots$, there exists[1] a polynomial $p_{n,s} \in \mathcal{P}_n$, such that

$$\rho_{n,s} = \rho(A(p_{n,s}); A(p_{n,s}^{(s)})) = \sup\{\rho(A(p); A(p^{(s)})) : p \in \mathcal{P}_n\}. \tag{8}$$

The polynomial $p_{n,s}$ is called **extreme for** $\rho(A(p); A(p^{(s)}))$ **in** $\mathcal{P}_n$ and a zero $z_1 \in A(p_{n,s})$ is called **extreme zero of** $p_{n,s}$, if $\rho(z_1; A(p_{n,s}^{(s)})) = \rho_{n,s}$.

In 1972 D. Phelps and R. S. Rodriguez [25] conjectured that, *if a polynomial p is extreme for $\rho(A(p); A(p'))$ in $\mathcal{P}_n$, then $p'(z) = nz^{n-1}$*. We generalize this conjecture as follows.

Conjecture 5 *If a polynomial p is extreme for $\rho(A(p); A(p^{(s)}))$ in $\mathcal{P}_n$, then*

$$p^{(s)}(z) = \frac{n!}{(n-s)!}(z - \lambda_{n,s})^{n-s},$$

where $\lambda_{n,s}$ is a constant.

We proved in [30] that an extreme polynomial for $\rho(A(p); A(p''))$ in the set of polynomials $p \in \mathcal{P}_4$ with real coefficients, is the polynomial

$$p(z) = (z-1)^2\left(z^2 + \frac{2}{3}z + 1\right) \quad \text{with} \quad p''(z) = 12\left(z - \frac{1}{3}\right)^2. \tag{9}$$

Until now, we fail to prove that the polynomial (9) is extreme for $\rho(A(p); A(p''))$ in $\mathcal{P}_4$ and that $\rho_{4,2} = 2/\sqrt{3}$. It is possible to find a counterexample by a computer. Observe that, according to Conjecture 4, we have $\rho_{4,2} \leq 4/3 = \rho_{3,2}$.

References

1. Aziz, A.: *On the zeros of a polynomial and its derivative*, Bull. Austr. Math. Soc. **31**, # 4 (1985), 245–255.
2. Beardon, A. F., D. Minda and T. W. Ng: *Smale's mean value conjecture and the hyperbolic metric*, Mathematische Annalen (to appear).

[1] Not unique in general.

3. Bojanov, B. D., Q. I. Rahman and J. Szynal: *On a conjecture of Sendov about the critical points of a polynomial*, Math. Z. **190** (1985), 281–285.
4. Borcea, J.: *On the Sendov conjecture for polynomials with at most six distinct zeros*, J. Math. Anal. Appl. **200**, #1 (1996), 182–206.
5. Borcea, J.: *The Sendov conjecture for polynomials with at most seven distinct roots*, Analysis **16** (1996), 137–159.
6. Brannan, D. A.: *On a conjecture of Ilieff*, Math. Proc. Cambridge Phil. Soc. **64** (1968), 83–85.
7. Brown, J. E.: *On the Ilieff–Sendov conjecture*, Pacific J. Math. **135** (1988), 223–232.
8. Brown, J. E.: *On the Sendov conjecture for sixth degree polynomials*, Proc. Amer. Math. Soc. **113**, # 4 (1991), 939–946.
9. Brown, J. E.: *A proof of the Sendov conjecture for polynomials of degree seven.* Complex Variables Theory Appl. **33**, # 1-4 (1997), 75–95.
10. Brown, J. E.: *On the Sendov's Conjecture for polynomials with real critical points*, Contemporary Mathematics **252** (1999), 49–62.
11. Brown, J. E. and G. Xiang: *Proof of the Sendov conjecture for polynomials of degree at most eight*, J. Math. Anal. Appl. **232**, # 2 (1999), 272–292.
12. Cohen, G. L. and G. H. Smith: *A proof of Iliev's conjecture for polynomials with four zeros*, Elemente d. Math. **43** (1988), 18–21.
13. Cohen, G. L. and G. H. Smith: *A simple verification of Iliev's conjecture for polynomials with three zeros*, Amer. Math. Mountly **95** (1988), 734–737.
14. Gacs, G.: *On polynomials whose zeros are in the unit disk*, J. Math. Anal. Appl. **36** (1971), 627–637.
15. Goodman, A. G., Q. I. Rahman and J. Ratti: *On the zeros of a polynomial and its derivative*, Proc. Amer. Math. Soc. **21** (1969), 273–274.
16. Hayman, W. K.: *Research Problems in Function Theory*, Althlone Press, London, 1967.
17. Joyal, A.: *On the zeros of a polynomial and its derivative*, J. Math. Anal. Appl. **25** (1969), 315–317.
18. Kakeya, S.: *On zeros of polynomial and its derivative*, Tôhoku Math. J. **11** (1917), 5–16.
19. Kumar, S. and B. G. Shenoy: *On some counterexamples for a conjecture in geometry of polynomials*, Zeszyty Nauk. Politech. Rzeszowskiej Mat. Fiz. # 12 (1991), 47 - 51.
20. Kumar, S. and B. G. Shenoy: *On the Sendov–Ilieff conjecture for polynomials with at most five zeros*, J. Math. Anal. Appl. **171** (1992), 595–600.
21. Lucas, F.: *Géométrie des polynômes*, J. École Polytech. (1) **46** (1879), 1–33.
22. Marden, M.: *Geometry of Polynomials*, 2nd edition, Math. Surveys Monographs, **3**, Amer. Math. Soc., Providence, RI, 1966.
23. Meir, A. and A. Sharma: *On Ilieff's conjecture*, Pacific J. Math. **31** (1969), 459–467.
24. Miller, M. J.: *Maximal polynomials and the Ilieff–Sendov conjecture*, Trans. Amer. Math. Soc. **321** (1990), 285–303.
25. Phelps, D. and R. S. Rodriguez: *Some properties of extremal polynomials for the Ilieff conjecture*, Kodai Math. Sem. Rep. **24** (1972), 172–175.
26. Rubinstein, Z.: *On a problem of Ilieff*, Pacific J. Math. **26** (1968), 159–161.
27. Saff, E. B. and J. B. Twomey: *A note on the location of critical points of polynomials*, Proc. Amer. Math. Soc. **27**, # 2 (1971), 303–308.
28. Schmeisser, G.: *Bemerkungen zu einer Vermutung von Ilieff*, Math. Z. **111** (1969), 121–125.

29. SCHMEISSER, G.: *The Conjectures of Sendov and Smale*, Approximation Theory (B. Bojanov, Ed.), DARBA, Sofia, 2002, 353 - 369.
30. SENDOV, BL.: *Hausdorff Geometry of Polynomials*, East Journal on Approximation, **7** # 2 (2001), 1–56.
31. SMALE, S.: *The fundamental theorem of algebra and complexity theory*, Bull. Amer. Math. Soc. **4** (1981), 1–36.
32. TISCHLER, D: *Critical points and values of complex ploynomials*, J. of Complexity **5** (1989), 438–456.

Part II

Monte Carlo and Quasi-Monte Carlo Methods

Hybrid Monte Carlo Methods for Matrix Computation

Vassil Alexandrov and Bo Liu

Department of Computer Science
The University of Reading
Whiteknights P.O. Box 225
RG6 6AY, United Kingdom
{v.n.alexandrov, b.liu}@rdg.ac.uk

Abstract. In this paper we consider hybrid (fast stochastic approximation and deterministic refinement) algorithms for Matrix Inversion (MI) and Solving Systems of Linear Equations (SLAE). Monte Carlo methods are used for the stochastic approximation, since it is known that they are very efficient in finding a quick rough approximation of the element or a row of the inverse matrix or finding a component of the solution vector. We show how the stochastic approximation of the MI can be combined with a deterministic refinement procedure to obtain MI with the required precision and further solve the SLAE using MI. We employ a splitting $A = D - C$ of a given non-singular matrix A, where D is a diagonal dominant matrix and matrix C is a diagonal matrix. In our algorithm for solving SLAE and MI different choices of D can be considered in order to control the norm of matrix $T = D^{-1}C$, of the resulting SLAE and to minimize the number of the Markov Chains required to reach given precision. Experimental results with dense and sparse matrices are presented.
Keywords: Monte Carlo Method, Markov Chain, Matrix Inversion, Solution of sytem of Linear Equations, Matrix Decomposition, Diagonal Dominant Matrices, SPAI.

1 Introduction

The problem of inverting a real $n \times n$ matrix (MI) and solving system of linear algebraic equations (SLAE) is of an unquestionable importance in many scientific and engineering applications: e.g. communication, stochastic modelling, and many physical problems involving partial differential equations. For example, the direct parallel methods of solution for systems with dense matrices require $O(n^3/p)$ steps when the usual elimination schemes (e.g. non-pivoting Gaussian elimination, Gauss-Jordan methods) are employed [4]. Consequently the computation time for very large problems or real time problems can be prohibitive and prevents the use of many established algorithms.

It is known that Monte Carlo methods give statistical estimation of the components of the inverse matrix or elements of the solution vector by performing

I. Dimov et al. (Eds.): NMA 2002, LNCS 2542, pp. 73–82, 2003.

random sampling of a certain random variable, whose mathematical expectation is the desired solution. We concentrate on Monte Carlo methods for MI and solving SLAEs, since, firstly, only $O(NL)$ steps are required to find an element of the inverse matrix where N is the number of chains and L is an estimate of the chain length in the stochastic process, which are independent of matrix size n and secondly, the process for stochastic methods is inherently parallel.

Several authors have proposed different coarse grained Monte Carlo parallel algorithms for MI and SLAE [7,8,9,10,11]. In this paper, we investigate how Monte Carlo can be used for diagonally dominant and some general matrices via a general splitting and how efficient hybrid (stochastic/deterministic) parallel algorithms can be derived for obtaining an accurate inversion of a given non-singular matrix A. We employ either uniform Monte Carlo (UM) or almost optimal Monte Carlo (MAO) methods [7,8,9,10,11]. The relevant experiments with dense and sparse matrices are carried out.

Note that the algorithms are built under the requirement $\|T\| < 1$. Therefore to develop efficient methods we need to be able to solve problems with matrix norms greater than one. Thus we developed a spectrum of algorithms for MI and solving SLAEs ranging from special cases to the general case. Parallel MC methods for SLAEs based on Monte Carlo Jacobi iteration have been presented by Dimov [11]. Parallel Monte Carlo methods using minimum Makrov Chains and minimum communications between master/slave processors are presented in [5,1]. Most of the above approaches are based on the idea of balancing the stochastic and systematic errors [11]. In this paper we go a step further and have designed hybrid algorithms for MI and solving SLAEs by combining two ideas: iterative Monte Carlo methods based on the Jacobi iteration and deterministic procedures for improving the accuracy of the MI or the solution vector of SLAEs.

The generic Monte Carlo ideas are presented in Section 2, the main algorithms are described in Section 3 and the parallel approach and some numerical experiments are presented in Section 4 and 5 respectively.

2 Monte Carlo and Matrix Computation

Assume that the system of linear algebraic equations (SLAE) is presented in the form:

$$Ax = b \tag{1}$$

where A is a real square $n \times n$ matrix, $x = (x_1, x_2, ..., x_n)^t$ is a $1 \times n$ solution vector and $b = (b_1, b_2, ..., b_n)^t$.

Assume the general case $\|A\| > 1$. We consider the splitting $A = D - C$, where off-diagonal elements of D are the same as those of A, and the diagonal elements of D are defined as $d_{ii} = a_{ii} + \gamma_i ||A||$, choosing in most cases $\gamma_i > 1, i = 1, 2, ..., n$. We further consider $D = B - B_1$ where B is the diagonal matrix of D, e.g. $b_{ii} = d_{ii}$, $i = 1, 2, ..., n$. As shown in [1] we could transform the system (1) to

$$x = Tx + f \tag{2}$$

where $T = D^{-1}C$ and $f = D^{-1}b$. The multipliers γ_i are chosen so that, if it is possible, they reduce the norm of T to be less than 1 and reduce the number of Markov chains required to reach a given precision. We consider two possibilities, first, finding the solution of $x = Tx + f$ using Monte Carlo (MC) method if $\|T\| < 1$ or finding D^{-1} using MC and after that finding A^{-1}. Then, if required, obtaining the solution vector is found by $x = A^{-1}b$.

Consider first the stochastic approach. Assume that $\|T\| < 1$ and that the system is transformed to its iterative form (2). Consider the Markov chain given by:

$$s_0 \to s_1 \to \cdots \to s_k, \tag{3}$$

where the $s_i, i = 1, 2, \cdots, k$, belongs to the state space $S = \{1, 2, \cdots, n\}$. Then for $\alpha, \beta \in S, p_0(\alpha) = p(s_0 = \alpha)$ is the probability that the Markov chain starts at state α and $p(s_{j+1} = \beta | s_j = \alpha) = p_{\alpha\beta}$ is the transition probability from state α to state β. The set of all probabilities $p_{\alpha\beta}$ defines a transition probability matrix $P = \{p_{\alpha\beta}\}_{\alpha,\beta=1}^{n}$ [3,9,10]. We say that the distribution $(p_1, \cdots, p_n)^t$ is acceptable for a given vector g, and that the distribution $p_{\alpha\beta}$ is acceptable for matrix T, if $p_\alpha > 0$ when $g_\alpha \neq 0$, and $p_\alpha \geq 0$, when $g_\alpha = 0$, and $p_{\alpha\beta} > 0$ when $T_{\alpha\beta} \neq 0$, and $p_{\alpha\beta} \geq 0$ when $T_{\alpha\beta} = 0$ respectively. We assume $\sum_{\beta=1}^{n} p_{\alpha\beta} = 1$, for all $\alpha = 1, 2, \cdots, n$. Generally, we define

$$W_0 = 1, W_j = W_{j-1} \frac{T_{s_{j-1}s_j}}{p_{s_{j-1}s_{j-1}}} \tag{4}$$

for $j = 1, 2, \cdots, n$.

Consider now the random variable $\theta[g] = \frac{g_{s_0}}{p_{s_0}} \sum_{i=1}^{\infty} W_i f_{s_i}$. We use the following notation for the partial sum:

$$\theta_i[g] = \frac{g_{s_0}}{p_{s_0}} \sum_{j=0}^{i} W_j f_{s_j}. \tag{5}$$

Under condition $\|T\| < 1$, the corresponding Neumann series converges for any given f, and $E\theta_i[g]$ tends to (g, x) as $i \to \infty$. Thus, $\theta_i[g]$ can be considered as an estimate of (g, x) for i sufficiently large. To find an arbitrary component of the solution, for example, the rth component of x, we should choose, $g = e(r) = \underbrace{(0, ..., 1}_{r}, 0, ..., 0)$ such that

$$e(r)_\alpha = \delta_{r\alpha} = \begin{cases} 1 & if \quad r = \alpha \\ 0 & otherwise \end{cases} \tag{6}$$

It follows that

$$(g, x) = \sum_{\alpha=1}^{n} e(r)_\alpha x_\alpha = x_r. \tag{7}$$

The corresponding Monte Carlo method is given by:

$$x_r = \hat{\Theta} = \frac{1}{N} \sum_{s=1}^{N} \theta_i[e(r)]_s,$$

where N is the number of chains and $\theta_i[e(r)]_s$ is the approximate value of x_r in the sth chain. It means that using Monte Carlo method, we can estimate only one, few or all elements of the solution vector. We consider Monte Carlo with uniform transition probability (UM) $p_{\alpha\beta} = \frac{1}{n}$ and Almost optimal Monte Carlo method (MAO) with $p_{\alpha\beta} = \frac{|T_{\alpha\beta}|}{\sum_{\beta=1}^{n} |T_{\alpha\beta}|}$, where $\alpha, \beta = 1, 2, \ldots, n$. Monte Carlo MI is obtained in a similar way [3].

To find the inverse $A^{-1} = C = \{c_{rr'}\}_{r,r'=1}^{n}$ of some matrix A, we must first compute the elements of matrix $M = I - A$, where I is the identity matrix. Clearly, the inverse matrix is given by

$$C = \sum_{i=0}^{\infty} M^i, \tag{8}$$

which converges if $\|M\| < 1$.

To estimate the element $c_{rr'}$ of the inverse matrix C, we let the vector f be the following unit vector

$$f_{r'} = e(r'). \tag{9}$$

We then can use the following Monte Carlo method for calculating elements of the inverse matrix C:

$$c_{rr'} \approx \frac{1}{N} \sum_{s=1}^{N} \left[\sum_{(j|s_j=r')} W_j \right], \tag{10}$$

where $(j|s_j = r')$ means that only

$$W_j = \frac{M_{rs_1} M_{s_1 s_2} \ldots M_{s_{j-1} s_j}}{p_{rs_1} p_{s_1 s_2} \ldots p_{s_{j-1} p_j}} \tag{11}$$

for which $s_j = r'$ are included in the sum (10).

Since W_j is included only into the corresponding sum for $r' = 1, 2, \ldots, n$, then the same set of N chains can be used to compute a single row of the inverse matrix, which is one of the inherent properties of MC making them suitable for parallelization.

The *probable error* of the method, is defined as $r_N = 0.6745\sqrt{D\theta/N}$, where $P\{|\bar{\theta} - E(\theta)| < r_N\} \approx 1/2 \approx P\{|\bar{\theta} - E(\theta)| > r_N\}$, if we have N independent realizations of random variable (r.v.) θ with mathematical expectation $E\theta$ and average $\bar{\theta}$ [6].

3 The Hybrid MC Algorithm

The basic idea is to use MC to find the approximate inverse of matrix D, refine the inverse and find A^{-1}. We can then find the solution vector through A^{-1}.

Algorithm: Finding A^{-1}.

1. **Initial data:** Input matrix A, parameters γ and ϵ.
2. **Preprocessing:**
   2.1 **Split** $A = D - (D - A)$, where D is a diagonally dominant matrix.
   2.2 **Set** $D = B - B_1$ where B is a diagonal matrix $b_{ii} = d_{ii}$ $i = 1, 2, ..., n$.
   2.3 **Compute** the matrix $T = B^{-1}B_1$.
   2.4 **Compute** $||T||$, the Number of Markov Chains $N = (\frac{0.6745}{\epsilon}.\frac{1}{(1-||T||)})^2$.
3. **For** i=1 to n;
   3.1 **For** j=1 to j=N;
   Markov Chain Monte Carlo Computation:
   3.1.1 **Set** $t_k = 0$(stopping rule), $W_0 = 1$, $SUM[i] = 0$ and $Point = i$.
   3.1.2 **Generate** an uniformly distributed random number *nextpoint*.
   3.1.3 **If** $T[point][netxpoint]! = 0$.
   LOOP
   3.1.3.1 **Compute** $W_j = W_{j-1}\frac{T[point][netxpoint]}{P[point][netxpoint]}$.
   3.1.3.2 **Set** $Point = nextpoint$ and $SUM[i] = SUM[i] + W_j$.
   3.1.3.3 **If** $|W_j| < \gamma$, $t_k = t_k + 1$
   3.1.3.4 **If** $t_k \geq n$, end LOOP.
   3.1.4 **End If**
   3.1.5 **Else** go to step 3.1.2.
   3.2 **End of loop j.**
   3.3 **Compute** the average of results.
4. **End of loop i.**
5. **Obtain** The matrix $V = (I - T)^{-1}$.
6. **Therefore** $D^{-1} = VB^{-1}$.
7. **Compute** the MC inversion $D^{-1} = B(I - T)^{-1}$.
8. **Set** $D_0 = D^{-1}$ (approximate inversion) and $R_0 = I - DD_0$.
9. **use filter procedure** $R_i = I - DD_i$, $D_i = D_{i-1}(I + R_{i-1})$, $i = 1, 2, ..., m$, where $m \leq k$.
10. **Consider the accurate inversion of D** by step 9 given by $D_0 = D_k$.
11. **Compute** $S = D - A$ where S can be any matrix with all non-zero elements in diagonal and all of its off-diagonal elements are zero.
12. **Main function** for obtaining the inversion of A based on D^{-1} step 9:
    12.1 **Compute** the matrices $S_i, i = 1, 2, ..., k$, where each S_i has just one element of matrix S.
    12.2 **Set** $A_0 = D_0$ and $A_k = A + S$
    12.3 **Apply** $A_k^{-1} = A_{k+1}^{-1} + \frac{A_{k+1}^{-1}S_{i+1}A_{k+1}^{-1}}{1-trace(A_{k+1}^{-1}S_{i+1})}$, $i = k-1, k-2, ..., 1, 0$.
13. **Print**the inversion of matrix A.
14. **End** of algorithm.

According to the general definition of a regular splitting [2], if A, M and N are three given matrices satisfying $A = M - N$, then the pair of matrices M, N are called regular splitting of A, if M is nonsingular and M^{-1} and N are non-negative. Therefore, let A be a nonsingular matrix. If we find a regular splitting of A such as $A = D - C$, then the SLAE $x^{(k+1)} = Tx^{(k)} + f$,where $T = D^{-1}C$, and $f = D^{-1}b$ converges to the unique solution x^* if and only if $\|T\| < 1$ [2].

The efficiency of inverting diagonally dominant matrices is an important part of the process enabling MC to be applied to diagonally dominant and some general matrices. The basic MC algorithms, covering the inversion of diagonally dominant matrices are presented in [12]. Here we outline only the hybrid MC algorithm for inverting an arbitrary non-singular matrix. Note that in some cases to obtain a very accurate inversion of matrix D (and matrix A)some filter procedures can be applied.

4 Parallel Implementation

We have implemented the algorithms proposed on a cluster of workstations, a Silicon Graphics ONYX2 machine and IBM SP3 machine under PVM. We have applied master/slave approach.

Inherently, Monte Carlo methods for solving SLAE allow us to have minimal communication, i.e. to partition the matrix A, pass the non-zero elements of the dense (sparse) matrix to every processor, to run the algorithm in parallel on each processor computing $\lceil n/p \rceil$ rows (components) of MI or the solution vector and to collect the results from slaves at the end without any communication between sending non-zero elements of A and receiving partitions of A^{-1} or x. The splitting procedure and refinement are also parallelised and integrated in the parallel implementation. Even in the case, when we compute only k components $(1 \leq k \leq n)$ of the MI (solution vector) we can divide evenly the number of chains among the processors, e.g. distributing $\lceil kN/p \rceil$ chains on each processor. The only communication is at the beginning and at the end of the algorithm execution which allows us to obtain very high efficiency of parallel implementation.

In this way we can obtain for the parallel time complexity of the MC procedure of the algorithm $O(nNL/p)$ where N denotes the number of Markov Chains. According to central limit theorem for the given error ϵ we have $N \geq \left(\frac{0.6745}{\epsilon \times (1 - \|T\|)}\right)^2$, L denotes the length of the Markov chains and L$\leq \left(\frac{\log(\gamma)}{\log\|T\|}\right)$, where ϵ, γ show the accuracy of Monte Carlo approximation [3]. Parameters ϵ, γ are used for the stochastic and systematic error. The absolute error of the solution for matrix inversion is $\left\| I - \hat{A}^{-1}A \right\|$, where A is the matrix whose inversion has to be found, and $\hat{A}^{-1}$ is the approximate MI. The computational time is shown in seconds.

5 Experimental Results

The algorithms run on a 12 processor (four 195 MHZ, eight 400MHZ) $ONYX2$ Silicon Graphics machine with 5120 Mbytes main memory and a 32 processor IBM SP3 machine. We have carried test with low precision $10^{-1} - 10^{-2}$ and higher precision $10^{-5} - 10^{-6}$ [12] in order to investigate the balance between stochastic and deterministic components of the algorithms based on the principle of balancing of errors (e.g. keeping the stochastic and systematic error of the same order) [7].

Here we consider the comparison of the hybrid Monte Carlo algorithm, which is written in C using PVM and the SPAI algorithm which is written in C using MPI. To test the two approaches we consider matrices from the Matrix Market (http://math.nist.gov/MatrixMarket/) and some general matrices randomly generated by the generator we provide. The generator generates random numbers for the matrices with predefined sparsity. We run both algorithms with the corresponding parameters. The computing time of the Monte Carlo calculation and the SPAI preconditioner are shown separately. The full time required to solve the SLAEs for hybrid algorithms and SPAI with BICGSTAB is also shown. Some of the test matrices used in this experiment are given below.

- **BWM200.MTX** size n=200, with nz=796 nonzero elements, from Matrix Market web site;
- **CAVITY03.MTX**size n=327, with nz=7327 nonzero elements, from Matrix Market web site;
- **GEN200_DENSE.MC** size n=200, dense, from the generator;
- **GEN200_60SPAR.MC** size n=200, with nz=16000 nonzero elements, from the generator;
- **GEN800_DENSE.MC** size n=800, dense, from the generator;
- **GEN800_70SPAR.MC** size n=800, with nz=192000 nonzero elements, from the generator;
- **GEN2000_99SPAR.MC** size n=2000, with nz=40000 nonzero elements, from the generator;
- **GEN500_90SPAR.MC** size n=500, with nz=25000 nonzero elements, from the generator;
- **GEN1000_95SPAR.MC** size n=1000, with nz=50000 nonzero elements, from the generator;

The default parameters we set up in Monte Carlo methods are

1. $\epsilon = 0.05$ denotes a given stochastic error;
2. $\delta = 0.01$ denotes the accuracy of Monte Carlo approximation;
3. *step* = 1 denotes how many steps are spent on the refinement function, in each single step of using the refinement function two matrix-by-matrix multiplications are computed.

We apply the appropriate parameters in SPAI and show the best performance with the best combination of these parameters. R-MC denotes the residual computing time of Monte Carlo approach which includes the time for the refinement procedure, the retrieval procedure and obtaining the approximate solution vector. MC denotes the time required for Monte Carlo algorithm only. Therefore, TOTAL-MC is the time required for the MC and R-MC. R-SPAI denotes the residual computing time of SPAI, which includes the time of BICGSTAB for obtaining the approximate solution vector (the block procedure and scalar-matrix procedure while using block algorithm). SPAI denotes the time required by the SPAI preconditioner. TOTAL-SPAI is the total time of SPAI and R-SPAI. ERROR denotes the absolute error of the approximate solution vector given below.

$$ERROR = \frac{||x_{exact} - x_{approximate}||_2}{||x_{exact}||_2} \tag{12}$$

Table 1. BWM200.MTX Solution with parameters: ERROR 10^{-4} and P=10 processors. SPAI is tested with different values of ϵ, and both blocked and non-blocked.

Para	BL	SPAI	R-SPAI	TOTAL-SPAI	MC	R-MC	TOTAL-MC
0.5	1×1	0.9303448	85.6012096	86.5315544			
	3×3	0.9759936	75.2054016	76.1813952			
0.6	1×1	0.8856872	220.1436288	221.0293160	2.186874	3.164163	5.351037
	3×3	0.9230960	150.3301808	151.2532768			
0.7	1×1	1.3930344	351.6003544	352.9933888			
	3×3	1.1148200	160.7620560	161.8768760			

Table 2. GEN200_DENSE.MC Solution with parameters: ERROR = 10^{-4} and P=10 processors. SPAI is tested with different values of ϵ, and both blocked and non-blocked.

Para	BL	SPAI	R-SPAI	TOTAL-SPAI	MC	R-MC	TOTAL-MC
0.5	1×1	2.9910024	nan	nan			
	3×3	1.1972320	nan	nan			
0.6	1×1	3.1077280	nan	nan	0.9338300	2.346462	3.280292
	3×3	1.2565296	nan	nan			
0.7	1×1	2.6192136	nan	nan			
	3×3	0.9718288	nan	nan			

Table 3. CAVITY03.MTX ERROR=10^{-4} and P=2, 5, 8, 12 processors respectively. SPAI is run with the default parameter of iterations in BICGSTAB, 500, and *nan* means we can not obtain the results under the setting, or it is converging slowly.

P#	MC	TOTAL-MC	SPAI	TOTAL-SPAI
2	76.20574	85.813108	0.2917584	nan
5	36.16877	60.095253	0.5471512	nan
8	28.29067	46.830696	0.3889352	nan
12	21.07290	33.248477	1.3224760	nan

Table 4. GEN800_70SPAR.MC ERROR=10^{-4} and P=2, 5, 8, 12 processors respectively.

P#	MC	TOTAL-MC	SPAI	TOTAL-SPAI
2	156.1388	586.202544	130.6839168	nan
5	47.70701	385.626890	67.6120952	nan
8	36.74342	304.847875	24.0445288	nan
12	26.53539	294.639845	23.1435808	nan

Table 5. ERROR=10^{-4} and P=10 processors

Matrix	MC	TATOL-MC	SPAI	TOTAL-SPAI
GEN200_60SPAR.MC	1.047589	4.720657	1.1548728	nan
GEN500_90SPAR.MC	7.730741	84.94246	4.1712408	nan
GEN800_DENSE.MC	24.322690	300.3137	29.0005580	nan
GEN1000_95SPAR.MC	46.274798	746.9745	9.1504488	nan
GEN2000_99SPAR.MC	302.626325	4649.0219003	6.2901104	nan

The examples show that for some cases MC converges much faster than SPAI. In case of some general matrices it can be seen that SPAI is converging very slowly. There are also cases where SPAI is performing much better than MC. Further experiments are required to carry out detailed analysis.

The latest results from the experiments with the Danish Air Pollution model show the superiority of Monte Carlo method for large tri-diagonal systems. The table below shows that Monte Carlo performs better than LU decomposition with the growth of the problem size.

32×32×32	Error	Starting Computation(Sec)	Advection	Total Time
LU-Decomposition	2.50E-4	0.01	31.06	31.07
Monte Carlo	8.52E-3	3.52	29.23	32.75

96×96×96	Error	Starting Computation(Sec)	Advection	Total Time
LU-Decomposition	3.4E-4	0.0	227.57	227.57
Monte Carlo	3.5E-3	63.8	88.8	152.6

6 Conclusion

In this paper we have introduced a hybrid Monte Carlo/deterministic algorithms for Matrix Inversion and finding a solution to SLAEs for any non-singular matrix. Further experiments are required to determine the optimal number of chains required for Monte Carlo procedures and how best to tailor together Monte Carlo and deterministic refinement procedures. We have shown also that the algorithms run efficiently in parallel and the results confirmed their efficiency for MI with dense and sparse matrices. Once good enough inverse is found, we can find the solution vector of SLAE. The accuracy of results are comparable with other known computational algorithms, with the Monte Carlo approach being somewhat better for some general cases and SPAI approach performing better in case of very sparse matrices. We can also conclude that both techniques offer excellent potential for use on high performance computers.

References

1. Fathi Vajargah, B., Liu, B., and Alexandrov, V.: On the preconditioned Monte Carlo methods for solving linear systems. MCM 2001, Salzburg, Austria (presented).
2. Ortega, J.: Numerical Analysis. SIAM edition, USA (1990)
3. Alexandrov V.N.: Efficient parallel Monte Carlo Methods for Matrix Computation. Mathematics and computers in Simulation, Elsevier, Netherlands, Vol. 47 (1998) 113–122.
4. Golub, G.H., Ch., F., Van Loan, *Matrix Computations,* The Johns Hopkins Univ. Press, Baltimore and London, (1996)
5. Taft, K. and Fathi Vajargah, B.: Monte Carlo Method for Solving Systems of Linear Algebraic Equations with Minimum Markov Chains. International Conference PDPTA'2000 Las Vegas (2000)
6. Sobol, I.M.: Monte Carlo Numerical Methods. Moscow, Nauka (1973) (in Russian)
7. Dimov, I., Alexandrov, V.N., and Karaivanova, A.: Resolvent Monte Carlo Methods for Linear Algebra Problems. Mathematics and Computers in Simulation, Vol. 155 (2001) 25–36.
8. Fathi Vajargah, B. and Alexandrov, V.N.: Coarse Grained Parallel Monte Carlo Algorithms for Solving Systems of Linear Equations with Minimum Communication. In: Proc. of PDPTA, June 2001, Las Vegas (2001) 2240–2245.
9. Alexandrov, V.N. and Karaivanova, A.: Parallel Monte Carlo Algorithms for Sparse SLAE using MPI. Lecture Notes in Computer Science, Vol. 1697, Springer-Verlag (1999) 283–290.
10. Alexandrov, V.N., Rau-Chaplin, A., Dehne, F., and Taft, K.: Efficient Coarse Grain Monte Carlo Algorithms for matrix computation using PVM. Lecture Notes in Computer Science, Vol. 1497, Springer-Verlag (1998) 323–330.
11. Dimov, I.T., Dimov, T.T., Gurov, T.V.: A new iterative Monte Carlo Approach for Inverse Matrix Problem. J. of Computational and Applied Mathematics, 92 (1998) 15–35.
12. Fathi Vajargah, B., Bo Liu, and Alexandrov, V.N.: Mixed Monte Carlo Parallel Algorithms for Matrix Computation. Lecture Notes in Computer Science, Vol. 2330, Springer-Verlag (2002) 609–618.

A New Efficient Algorithm for Generating the Scrambled Sobol' Sequence*

Emanouil I. Atanassov

Central Laboratory for Parallel Processing
Bulgarian Academy of Sciences, Sofia, BULGARIA
emanouil@parallel.bas.bg

Abstract. The Sobol' sequence is the most widely deployed low-discrepancy sequence, and is used for calculating multi-dimensional integrals and in quasi-Monte Carlo simulation. Owen first proposed the idea of scrambling this sequence in a manner that maintained its low discrepancy. One of his motivations was to use these scrambled sequences to provide quasi-Monte Carlo methods with simple error estimates like those in normal Monte Carlo. In fact, it is now common for the Sobol' sequence as well as (t, m, s)-nets and (t, s)-sequences to be used with scrambling. However, many have pointed out that scrambling is often difficult to implement and time consuming. In this paper we describe a new generation algorithm that allows consecutive terms of the scrambled Sobol' sequence to be obtained with essentially only two operations per coordinate: one floating point addition and one bit-wise *xor* operation. Note: this omits operations that are needed only once per tuple. This scrambling is achieved at no additional computational cost over that of unscrambled generation as it is accomplished totally in the initialization. In addition, the terms of the sequence are obtained in their normal order, without the usual permutation introduced by Gray code ordering used to minimize the cost of computing the next Sobol' element. This algorithm is relatively simple and is quite suitable for parallelization and vectorization. An example implementation of the algorithm, written in pseudo-code is presented. Possible improvements of the algorithm are discussed along with the presentation of some timing results.

1 Introduction

The Sobol' sequence is perhaps the most generally used low-discrepancy sequence. It was introduced by Sobol' in [2] to speed the convergence of multidimensional Monte Carlo integration. To help describe the generation of the Sobol' sequence we require the following definition:

---

* Supported by the project of European Commission - BIS 21 under contract ICA1-CT-2000-70016 and by the Ministry of Education and Science of Bulgaria under contracts NSF I-811/98 and NSF MM-902/99

I. Dimov et al. (Eds.): NMA 2002, LNCS 2542, pp. 83–90, 2003.

Definition 1. *Let $A_1, \ldots, A_s$ be infinite matrices*

$$A_k = \left\{ a_{ij}^{(k)} \right\}, \quad i, j = 0, 1, \ldots,$$

with $a_{ij}^{(k)} \in \{0,1\}$, such that $a_{ii}^{(k)} = 1$ for all i and k, $a_{ij}^{(k)} = 0$ if $i < j$. The $\tau^{(1)}, \ldots, \tau^{(s)}$ are sequences of permutations of the set $\{0,1\}$. Each non-negative integer n may be represented in the binary number system as

$$n = \sum_{i=0}^{r} b_i 2^i.$$

Then the nth term of the low-discrepancy sequence σ is defined by

$$x_n^{(k)} = \sum_{j=0}^{r} 2^{-j-1} \tau_j^{(k)} \left(\bigoplus_{i=0}^{j} b_i a_{ij}^{(k)} \right),$$

where by "$\bigoplus$" we denote the operation "bit-wise addition modulo 2."

When only the first m rows of the matrices A_k are used, and n is in the interval $[0, 2^m - 1]$, then we can generate at most 2^m terms of the low-discrepancy sequence, and so we speak of a net or point set instead of a sequence. In this paper we do not deal with the "best" choice for the matrices $A_1, \ldots, A_s$: we consider them as given.

The above definition is general, covering most digital (t, m, s)-nets in base 2. The Sobol' sequence is a (t, s)-sequence in base 2 and is a particular case of the above definition with the matrices A_k defined as in [2].

Owen, [4], first proposed the idea of scrambling low-discrepancy sequences in a manner that maintains their equidistribution. A major motivation was to use these scrambled sequences to provide quasi-Monte Carlo methods with simple statistical error estimates like those in normal Monte Carlo. The purpose of this paper is to describe a new, fast, generation algorithm for the Sobol' sequence that efficiently implements Owen's idea. Our algorithm is equally applicable to all sequences or nets which satisfy Definition 1.

A natural approach for generating terms of the sequence 1 is to use information available when calculating an element in the sequence in the generation of the subsequent term. However, many known algorithms seeking efficiency, like [3], permute the sequence by making use of so-called Gray-code ordering. When generating $N = 2^m$ elements in this manner, one obtains a permutation of the original sequence. However, when N is not a power-of-two, a different sequence is obtained with Gray-code ordering. In our algorithm, we generate the sequence in its natural order. In addition, our algorithm allows consecutive terms of the scrambled Sobol' sequence to be obtained with essentially only two operations per coordinate: one floating point addition and one bit-wise *xor* operation. Note: this accounting omits operations that are needed only once per tuple. This scrambling is achieved at no additional computational cost over that of unscrambled generation, as it is accomplished totally in the initialization.

When quasi-Monte Carlo calculations are performed in parallel with n processors there are essentially two approaches that are used. Suppose that nN terms of the sequence have to be generated. In the first (blocking) approach the processor P_i computes and uses the terms in the sequence with indices from $(i-1)N$ to iN. With the second approach (leap-frogging), processor P_i computes and uses the terms of the sequence with indices $2^k j+i-1, \quad j=0,\dots,N-1$. Our algorithm is applicable in both cases with the caveat that leap-frogging works for us only when n is a power-of-two. The practical difference from the sequential version is only in the parameters passed to the initialization routine. In fact, after initialization of a parallel version of our algorithm, no further interprocessor communication is required.

The plan of the paper is as follows. In §2 we describe some properties of sequences satisfying the above definition, and important points from the computer representation of floating point numbers, which are the basis of our algorithm. In §3, the algorithm is presented in pseudo-code. In §4, some issues related to the hardware-dependent optimization are discussed, and timing results comparing the reference (unoptimized) implementation to various optimized versions are given.

2 Mathematical Foundations of the Algorithm

In this section we elaborate on some properties of sequences or nets from Definition 1.

Lemma 1. *Let σ be a sequence or net satisfying Definition 1. Define the related sequences $\mu^{(1)},\dots,\mu^{(s)}$ by*

$$\mu_j^{(i)} = \tau_j^{(i)}(0).$$

If $n \geq 0$ has its binary representation as

$$n = \sum_{i=0}^{r} b_i 2^i,$$

then the nth term of the sequence σ can be equivalently obtained via

$$x_n^{(k)} = \sum_{j=0}^{r} 2^{-j-1}\left(\mu_j^{(k)} \bigoplus \left(\bigoplus_{i=0}^{j+1} b_i a_{i+1j}^{(k)}\right)\right).$$

The Lemma follows directly from the equality $\pi(x) = \pi(0) \bigoplus x$ for any permutation, π, of the set $\{0,1\}$. The lemma permits us to reduce the information needed to implement our algorithm since for each sequence of permutations we need only store their values at 0. In addition, we use this to reduce the scrambling operation to *xor* operations in initialization.

This next Lemma explains how we obtain consecutive terms of the sequence, σ.

Lemma 2. *Let σ be a sequence or net satisfying Definition 1, and let the non-negative integers n, p, m be given. Suppose that we desire the first p binary digits of elements in σ with indices of the form $2^m j + n < 2^p$; this implicitly defines a set of compatible j's. Thus the only numbers we need compute are*

$$y_j^{(k)} = \lfloor 2^p x_{2^m j+n}^{(k)} \rfloor.$$

The integers $\left\{ v_r^{(k)} \right\}_{r=0}^{\infty}$, which we call "twisted direction numbers," are defined by

$$v_r^{(k)} = \sum_{t=0}^{p-1} 2^{p-1-t} \bigoplus_{j=m}^{p-1} a_{tj}^{(k)}.$$

Suppose that the largest power-of-two that divides $2^m(j+1)+n$ is l, i. e. $2^m(j+1)+n = 2^l(2K+1)$. Then the following equality holds

$$y_{j+1}^{(k)} = y_j^{(k)} \bigoplus v_l^{(k)}.$$

From this Lemma it is obvious how one could obtain the next set of numbers $y_j^{(k)}$, $k = 1, \ldots, s$ from previous values, with only one *xor* operation for each coordinate. Note that generation of y_{j+1} follows from the y_j through a single *xor* operations with the appropriate twisted direction numbers. Since scrambling commutes with this operation, if the scrambling is applied once, for example to y_0, then it need not be applied again to obtain a scrambled output.

It is obvious that by multiplying the numbers $y_j^{(k)}$ by 2^{-p}, one could obtain p binary digits of the terms of the sequence σ. However, this implies the need to convert $y_j^{(k)}$ from integer to floating point and to perform a floating-point multiply by 2^{-p}. However, for reasons of efficiency we use another approach, which works on computers that adhere to the IEEE floating-point standard (see [6]). Such IEEE-compliant machine use 32 bits for single precision and 64 bits for double precision. With our method we replace the need for conversion to floating-point and a floating-point multiplication with a single floating-point addition. In `C` we can implicitly do the conversion with a type cast on the respective pointers, while in FORTRAN the same can be achieved by using the `EQUIVALENCE` statement.

To more readily describe our procedure we require the following.

Definition 2. *Let y be a positive real number, such that $y = (1+x)2^r$, where $0 \le x < 1$. The binary single-precision floating-point representation of y is $d_0 \ldots d_{31}$, here d_0 is the most-significant bit. Also $d_0 = 0$; $d_1 \ldots d_8$ is the binary representation of $r + 127$, where r is the floating-point exponent; and $d_9 \ldots d_{31}$ is the binary representation of the integer part of $x2^{23}$, the mantissa. Recall that an IEEE single-precision mantissa is 23 bits. Here we suppose that $-126 < r < 127$. The analogous double-precision floating-point representation of y is $d_0 \ldots d_{63}$, where $d_0 = 0$; $d_1 \ldots d_{11}$ is the bits of $r + 1023$, where r is the double-precision exponent; and $d_12 \ldots d_{63}$ is the binary representation of the integer part of $x2^{52}$, the double-precision mantissa. In this case we need $-1022 < r < 1023$. Zero is represented by a sequence of zeroes.*

Here numbers are rounded towards zero, and thus the terms in the Sobol' sequence will also be rounded towards zero. This is usually not a problem, since rounding will be at most 2^{-23} and 2^{-52} in absolute magnitude respectively.

Now that we understand floating-point representations we describe a method which avoids the multiplication and conversion from integer to floating point.

Lemma 3. *Let x be a real in the interval $[0, 1)$, and let y be the integer part of $x2^{23}$. Suppose y is stored as a 32-bit integer. If one xor's* 001111111 *to the nine most-significant bits of y, what remains in memory is the floating-point representation of $1 + x$. Again this representation is rounded towards zero.*

Note: a similar statement for double precision can be stated where 64-bit integers are used, 23 is replaced by 52, and the *xor*ed most significant bits are 001111111111. Thus, we can operate in place on the integer y with integer operations and at the end subtract one to obtain x.

3 Description of the Algorithm

One main aspect of the algorithm is to store terms of the sequence with one added to them in a floating-point array, y, to take advantage of the optimizations described above. Also, instead of working with the matrices A_i, we store the "twisted direction numbers" in a table with columns indexed by dimension. These twisted direction numbers are computed once for all in initialization. In order to obtain the coordinates of the next element of the sequence, one first determines which row of the table is needed and performs an operation between y, thought of as an integer array, and the appropriate row, store the result, and then subtract 1 to correct for the representation.

We have developed two versions of the algorithm, depending on the required precision: single or double. Because of the properties of the (unscrambled) Sobol' sequence, it's first 2^{23} terms can be represented exactly in single precision assuming that the machine has 32-bit IEEE compliant floating-point numbers. Our single-precision code correctly computes elements in the Sobol' sequence with indices up to 2^{23}. In `C`, here the corresponding variable types are `float` and `int`.

In the double-precision code, which requires IEEE compliant double precision and 64-bit integers, the `C` data type for integers is a `long long int`. Note that the type `long long int` is non-standard, and some architectures provide 64-bit integers in `C` with type `long int`. However, since we only need to perform **xor** operations on these integers, in the rare case when such 64-bit integers are not available, we can also implement the 64-bit integers ourselves with arrays of 32-bit integers. Note: our code does not require special treatment of little or big endian architectures.

In the sequel we describe the algorithm in more detail. Note that in practice, multiplication by powers of two is replaced by integer left-shift operations. What follows is a step-by-step description of both the single- and double-precision algorithm. We use a fairly verbose "pseudo-code" to describe it.

1. Input initial data:
   - if the precision is single, set the number of bits b to 32, and the maximal power of two p to 23, otherwise set b to 64 and p to 52;
   - dimension s;
   - direction vectors $\{a_{ij}\}$, $i = 0, p$, $j = 1, \dots, s$ representing the matrices $A_1, \dots, A_d$ (always $a_{ij} < 2^{i+1}$);
   - scrambling terms $d_1, \dots, d_s$ - arbitrary integers less than 2^p, if all of them are equal to zero, then no scrambling is used;
   - index of the first term to be generated - n;
   - scaling factor m, so the program should generate elements with indices $2^m j + n, \quad j = 0, 1, \dots$.
2. Allocate memory for $s * l$ b-bit integers (or floating point numbers in the respective precision) $y_1, \dots, y_s$..
3. Preprocessing: calculate the twisted direction numbers v_{ij}, $i = 0, \dots, p - 1$, $j = 0, \dots, s$:
   - for all j from 1 to s do
   - for $i = 0$ to $p - 1$ do
   - if i=0, then $v_{ij} = a_{ij} 2^{p-m}$, else

$$v_{ij} = v_{i-1j} \mathbf{xor}(a_{i+m,j} * (2^{p-i-m}));$$

4. Calculate the coordinates of the n-th term of the Sobol' sequence (with the scrambling applied) using any known algorithm (this operation is performed only once). Add +1 to all of them and store the results as floating point numbers in the respective precision in the array y.
5. Set the counter N to $\left\lfloor \frac{n}{2^m} \right\rfloor$.
6. Generate the next point of the sequence:
   - When a new point is required, the user supplies a buffer x with enough space to hold the result.
   - The array y is considered as holding floating point numbers in the respective precision, and the result of subtracting 1. from all of them is placed in the array x.
   - Add 1 to the counter N;
   - Determine the first nonzero binary digit k of N so that $N = (2M + 1)2^k$ (on the average this is achieved in 2 iterations);
   - consider the array y as an array of b-bit integers and updated it by using the kth row of twisted direction numbers:
     for $i = 1$ to d do
   - $y_i = y_i xor v_{ki}$.
   - return the control to the user. When a new point is needed, go to 6.

4 Discussion of Algorithmic Improvements and Some Timings

The above implementations of the algorithm should be regarded only as an outline. However, a "reference" implementation of both the single- and double-precision versions of the algorithm, written in C, is available on the web at the following URL: `http://parmac1.bas.bg/emanouil/sobol.html`. In practice, several well known optimization techniques may be used to improve performance.

When the dimension is sufficiently large, most of the computational time is spent in (integer) *xor* and floating-point addition operations applied to whole vectors. Therefore if the processor has special vector instructions for these operations, they can be utilized to speed up the computation. For instance, one can use the MMX or SSE1/SSE2 instruction sets for the Pentium class processors and AltiVec instructions for the Power PC G4. Subroutines for performing such operations on large vectors of data are usually included in libraries for digital signal processing (see e. g. [1]).

In the following table we present some timings of our algorithm on an AMD Athlon processor running at 950MHz. We compare the speed of the reference implementation of the algorithm mentioned above with a hand-tuned code using MMX instructions. We measure the CPU time required for generating 1,000,000 terms of the sequence in 360 dimensions as this dimension is natural in many problems in financial mathematics (see e.g. [5]). By BS and BD we denote the unoptimized (basic) code, in single and in double precision respectively, and by VS and VD we denote the single- and double-precision versions of the vectorized MMX code. The speed is computed by dividing the time required to generate the millions of coordinates (in this case it is always 360,000,000) by the CPU time. It can be seen that the improvement from using vector instructions is quite

Table 1. Timing results of the generation of 10^6 points of the scrambled Sobol' sequence in 360 dimensions with the Basic and Vectorized version of the algorithm in single and double precision.

Implementation	BS	BD	VS	VD
CPU time (seconds)	1.87	5.65	1.26	1.50
Speed (million of coordinates per second)	192	53	285	240

substantial. However, there is still room for improvement in the reference code. For example, one could manually unroll some of the crucial loops to obtain a substantial speedup.

In these measurements we only generated elements of the scrambled Sobol' sequence, without performing any calculations with them. In order to gauge the generation speed relative the rest of a quasi-Monte Carlo computation, we also summed up all the coordinates computed. It is remarkable to note that the time required for summing up all the coordinates of all the generated points was an additional 1.46 seconds in single precision and 1.72 seconds in double precision. From this, we may safely conclude that using our algorithm for generating the Sobol' sequence in any quasi-Monte Carlo application would require only a small fraction of the total CPU time.

Also note that parallelization and scrambling can be achieved simply by changing some parameters in the "Preprocessing" part of the code, without affecting the computational work required in the main generation function. Some timings of a parallel and vector implementation of this algorithm were described in [7]. Since no interprocessor communication is needed after initialization, one can see that the parallel efficiency must be close to 100%.

Looking at the times in Table 1, we may fairly conclude that with this new algorithm the Sobol' sequence can be generated with speeds comparable or superior to those of pseudorandom number generators.

Acknowledgments

The author would like to thank Prof. Michael Mascagni for his considerable editorial input.

References

1. Intel Signal Processing Library: `http://developer.intel.com/software/products/perflib/spl/index.htm`.
2. Sobol', I. M.: The distribution of points in a cube and approximate evaluation of integrals. Zh. Vychisl. Mat. Mat. Fiz. 7 (1967) 784–802 (in Russian).
3. Sobol', I. M.: Multi-dimensional Quadrature Formulae and Haar Functions. Nauka, Moscow (1969)
4. Owen, A.: Monte Carlo Extension of Quasi-Monte Carlo. In: Medieiros, D. J., Watson, E. F., Manivannan, M., and Carson, J. (eds.): Proceedings of WSC'98 (1998) 571–577.
5. Paskov, S. and Traub, J.: Faster Valuation of Financial Derivatives. Journal of Portfolio Management, 22:1, Fall (1995) 113–120.
6. ANSI/IEEE Standard 754-1985: Standard for Binary Floating Point Arithmetic.
7. Atanassov, E. I.: Measuring the Performance of a Power PC Cluster. Sloot, Peter M. A., Kenneth Tan, C. J., Dongarra, Jack, and Hoekstra, Alfons G. (eds.): Proceedings of the 2002 International Conference on Computational Science (ICCS 2002). Lecture Notes in Computer Science, Vol. 2329, Springer-Verlag, Berlin (2002) 628–634.

Generating and Testing the Modified Halton Sequences*

Emanouil I. Atanassov and Mariya K. Durchova

Central Laboratory for Parallel Processing - BAS, Sofia
emanouil@parallel.bas.bg, mabs@parallel.bas.bg

Abstract. The Halton sequences are one of the most popular low-discrepancy sequences, used for calculating multi-dimensional integrals or in quasi-Monte Carlo simulations. Various techniques for their randomization exist. One of the authors proved that for one such modification an estimate of the discrepancy with a very small constant before the leading term can be proved. In this paper we describe an efficient algorithm for generating these sequences on computers and show timing results, demonstrating the efficiency of the algorithm. We also compare the integration error of these sequences with that of the classical Halton sequences on families of functions widely used for such benchmarking purposes. The results demonstrate that the modified Halton sequences can be used successfully in quasi-Monte Carlo methods.

1 Introduction

The Halton sequences were introduced by Halton in [4]. Faure in [3] considers a wider class of sequences, and shows how sequences with small discrepancy can be found in this class, when the dimension is one. He calls them generalized Van der Corput sequences. We give here the definitions of the generalized Van der Corput sequences and of the Halton sequences.

Definition 1. *Let $p \geq 2$ be a fixed integer, and let $\tau = \{\tau_j\}_{j=0}^{\infty}$ be a sequence of permutations of the numbers $\{0, \dots, p-1\}$. The terms of the corresponding generalized Van der Corput sequence are obtained by representing $n \geq 0$ as $n = \sum_{j=0}^{\infty} a_j p^j$, $\quad a_j \in \{0, \dots, p-1\}$, and putting $x_n = \sum_{j=0}^{\infty} \tau_j(a_j) p^{-j-1}$. The one-dimensional Van der Corput - Halton sequence in base p is obtained by setting $\tau_j(i) = i$.*

Definition 2. *Let $p_1, \dots, p_s$ be pairwise relatively prime integers, $p_i \geq 2$. The Halton sequence $\sigma(p_1, \dots, p_s) = \left\{\left(x_n^{(1)}, \dots, x_n^{(s)}\right)\right\}_{n=0}^{\infty}$ is constructed by setting each sequence $\left\{x_n^{(i)}\right\}_{n=0}^{\infty}$ to be a Van der Corput - Halton sequence in base p_i.*

* Supported by the project of European Commission — BIS 21 under contract ICA1-CT-2000-70016 and by the Ministry of Education and Science of Bulgaria under contract NSF I-1201/02 and NSF MM-902/99

I. Dimov et al. (Eds.): NMA 2002, LNCS 2542, pp. 91–98, 2003.

In the applications usually the numbers $p_1, \ldots, p_s$ are taken to be the first s primes.

E. Atanassov in his paper [1] introduced a modification of the Halton sequences in multiple dimensions, using Faure's definition, and proved that for these sequences an estimate of their discrepancy with a very small leading term can be obtained (see Theorem 2.1 and Theorem 2.2). His construction is based on the existence of some numbers, called "admissible".

Definition 3. *Let $p_1, \ldots p_s$ be distinct primes. The integers $k_1, \ldots k_s$ are called "admissible" for them, if $p_i \nmid k_i$ and for each set of integers $m_1, \ldots, m_s, p_i \nmid m_i$, there exists a set of integers $\alpha_1, \ldots, \alpha_s$, satisfying the congruences*

$$k_i^{\alpha_i} \prod_{1 \le j \le s, j \ne i} p_j^{\alpha_j} \equiv m_i \,(\mathrm{mod}\ p_i)\,, \quad i = 1, \ldots, s. \tag{1}$$

In that paper one can see a constructive proof of the existence of such "admissible" numbers for each set of distinct primes. Now we give the definition of the modified Halton sequences.

Definition 4. *Let $p_1, \ldots, p_s$ be distinct primes, and let the integers $k_1, \ldots, k_s$ be "admissible" for them. The modified Halton sequence $\sigma(p_1, \ldots, p_s; k_1, \ldots, k_s) = \left\{\left(x_n^{(1)}, \ldots, x_n^{(s)}\right)\right\} n = 0^\infty$ is constructed by setting each sequence $\left\{x_n^{(i)}\right\}_{n=0}^{\infty}$ to be a generalized Van der Corput-Halton sequence in base p_i (see Definition 1), with the sequence of permutations $\tau^{(i)}$ defined by taking $\tau_j^{(i)}(t)$ to be the remainder of tk_i^j modulo p_i, $\tau_j^{(i)}(t) \in \{0, \ldots, p_i - 1\}$.*

These results are mostly theoretical. In practice three important questions arise:

- Is it possible to add scrambling.
- How to choose the modifiers $k_1, \ldots, k_s$.
- What is the price of using these modifiers in CPU time and is it worth it.

In this paper we try to answer these questions. The first of them is relatively easy. The simple solution is to change the formula for the permutations to

$$\tau_j^{(i)}(t) \equiv tk_i^{(j+1)} + b_j^{(i)} (\mathrm{mod}\ p_i),$$

where the integers $b_j^{(i)}$ are chosen independently in the interval $[0, p_i - 1]$. The power has been changed from j to $j+1$, because it is intuitively better, and the theoretical estimates are satisfied for these sequences too, because if for any integers $m_1, \ldots, m_s$ the congruences

$$k_i^{\alpha_i} \prod_{1 \le j \le s, j \ne i} p_j^{\alpha_j} \equiv m_i \,(\mathrm{mod}\ p_i)\,, \quad i = 1, \ldots, s \tag{2}$$

have a solution, then the same is true for the congruences

$$k_i^{\alpha_i + 1} \prod_{1 \le j \le s, j \ne i} p_j^{\alpha_j} + b_j^i \equiv m_i \,(\mathrm{mod}\ p_i)\,, \quad i = 1, \ldots, s. \tag{3}$$

This observation permits the same proof to be carried out for these version of the modified Halton sequences.

In the next section we describe an efficient algorithm for generating the modified Halton sequences. Since the Halton sequences have simpler definition, a simplified version of these algorithm can be used for their generation too. In the third section we compare the speed of these two versions of our algorithm with that of two known algorithms for generating the Halton sequences. In the last section we use the Halton and the modified Halton sequences for calculating some multi-dimensional integrals. One can see how when the modified Halton sequences are used, the integration error is definitely smaller. These results are encouraging, since a small increase in the computational time permits us to obtain a large decrease in the integration error.

2 Algorithm Description

Various algorithms are known for the generation of the Halton sequences, see e.g. Lecot [5], Bratley and Fox [2]. Lecot's algorithm requires only one floating point operation for generating one coordinate of the sequence. However, his algorithm stores all the terms of the sequence in RAM, which is impossible when the dimension or the number of terms of the sequence are large. Since access to RAM is much slower than access to cache memory, the algorithm runs at memory speed, i.e. the speed is limited by the memory bandwidth. In our testing configuration this algorithm actually run slower than the others.

The algorithm of Bratley and Fox runs much faster, but has the disadvantage that it doesn't generate the terms of the sequence exactly, but with a small error, which increases with the number of terms to be generated, and when this number is relatively large, the program refuses to run, since the error becomes noticeable. Therefore the user must perform some error analysis before using this algorithm.

We designed a simple algorithm for generating the terms of the modified Halton sequence, which requires a small amount of memory and generates the terms of the sequence with maximal error less than 10^{-14} when 10^6 terms are generated.

The terms of the sequence are generated consecutively, so some information from each invocation of the generation function is kept and used in the next invocation. In C such information can be saved using the keyword `static` (the development of a thread-safe version of the algorithm is also possible, but essentially trivial).

All the algorithms were implemented in double precision.

For the modified Halton sequences we require not only the primes, but also the modifiers as initial data. In [1] the existence of "admissible" numbers for each set of primes is proven, and the proof is constructive. The proof is by induction on the dimension, and the induction step involves calculating $s-1$ determinants. However, it is easy to see that the same result can be achieved by a diagonalization procedure similar to Gaussian elimination. Since usually the first s primes are used for the $s-$ dimensional Halton sequence, we fixed the maximal dimension to be 1000, found "admissible" numbers for the first 1000 primes, and the procedure guarantees that the first s of them are "admissible" for the set of the first s primes (something which is not true in general). These calculations were performed only once and took several seconds of CPU time. Then the primes and the modifiers were hard-coded into the generation program.

The data that should be kept for the next invocation of the function is organized with the idea to use efficiently the cache memory. Now we are going to explain how the ith coordinate of the sequence is generated. Let the representation of the index N in a p_i-adic number system be the following:

$$N = \sum_{r=0}^{63} a_r p_i^r.$$

Let j be any number between 0 and 63. For any i and j we make sure when the generation function is invoked, the following data is available:

1. the remainder term $c_j^{(i)} = \mathrm{mod}\,(a_j k_i^{j+1} + b_j^{(i)}, p_i)$;
2. the prime number p_i;
3. the number $d_j^{(i)} = \mathrm{mod}\,(k_i^{j+1}, p_i)$;
4. the scrambling term $b_j^{(i)}$;
5. the partial sum

$$\Sigma_j^{(i)} = \sum_{r=0}^{j-1} (0 + b_r^{(i)}) p_i^{-r-1} + \sum_{r=j}^{63} \left(\mathrm{mod}\,(a_r k_i^{r+1} + b_r^{(i)}, p_i) \right) p_i^{-r-1};$$

6. the number $A_j^{(i)} = c_j^{(i)} p_i^{-j-1}$;
7. the number $B_j^{(i)} = p_i^{-j}$.

The first four quantities are kept as 16 bit short integers, the next 3 as double precision numbers. When j is zero, the partial sum is actually the ith coordinate of the modified Halton sequence. Therefore we must only explain how this data is transformed in each invocation of the function. In order to obtain the ith coordinate, we perform the following operations:

1. store the partial sum $\Sigma_0^{(i)}$ in the buffer supplied by the user;
2. set $j = 0$;
3. add $A_j^{(i)}$ to $\Sigma_j^{(i)}$;
4. add $d_j^{(i)}$ to $c_j^{(i)}$;
5. if $c_j^{(i)} = p_i$ then set $c_j^{(i)} = c_j^{(i)} - p_i$ and $\Sigma_j^{(i)} = \Sigma_j^{(i)} - B_j^{(i)}$;
6. if $c_j^{(i)}$ is equal to the scrambling term $b_j^{(i)}$, then increase j and go to 3, otherwise set all the partial sums $\Sigma_k^{(i)}$, $k = 0, \ldots, j-1$ to be equal to $\Sigma_j^{(i)}$.

Observe that when $c_j^{(i)}$ becomes equal to $b_j^{(i)}$, the digit $a_j^{(i)}$ is necessarily zero, which justifies the algorithm.

Therefore 32 bytes are needed for each p_i-adic digit - 4 16-bit short integers and 3 double precision floating point numbers.

One should choose some reasonable way to order all these numbers. Probably the optimal ordering is machine-dependent, so we do not discuss this issue here.

Let us consider what changes have to be made in order to generate the original Halton sequence. In this case, the remainder terms $c_j^{(i)}$ are in fact the corresponding digits. We do not need the scrambling terms, since they are all zeroes. The numbers $d_j^{(i)}$ are all ones, and therefore are redundant too. The terms $A_j^{(i)}$ are equal to p_i^{-j-1} and the terms $B_j^{(i)}$ are not needed. In each invocation of the generation function for each i we perform the following operations:

1. store the partial sum $\Sigma_0^{(i)}$ in the buffer supplied by the user;
2. set $j = 0$;
3. add $A_j^{(i)}$ to $\Sigma_j^{(i)}$;
4. if $c_j^{(i)} = p_i$ then set $c_j^{(i)} = 0$;
5. if $c_j^{(i)} = 0$ then increase j and go to 3, otherwise set all the partial sums $\Sigma_k^{(i)}, \quad k = 0, \ldots, j-1$ to be equal to $\Sigma_j^{(i)}$.

This algorithm requires less memory and involves less computation, that is why in the next section we shall see that it runs significantly faster.

3 Timings

In this section we compare the two algorithms introduced above with other two algorithms for generation the Halton sequence — the algorithm of C. Lecot and the algorithm of Bratley and Fox. Our algorithms are called in the sequel A.M and A.H, the algorithm of Lecot is called C.L and the algorithm of Bratley and Fox is called B.F. The tests are performed on a PC running Linux with CPU AMD Athlon at 950Mhz. The same options were used for the `gcc` and the `g77` compilers

```
-O3 -funroll-all-loops
```

in all the tests presented here. We measured the time used for the generation of the terms of the sequence, when no operations are performed with the results. The CPU time needed for the generation of 10^6 terms in 10 and 20 dimensions is measured and the results are shown in Table 1.

Table 1. CPU time required for generating 1000000 terms of the sequences

Dim	A.H	A.M	B.F	C.L
10	0.24s	0.37s	0.26s	2.88s
20	0.34s	0.59s	0.46s	5.58s

It is remarkable that both our algorithms outperform the algorithm of Lecot, which uses only one floating point operation for generating one coordinate. The reason is that his algorithm is memory intensive and generates too many cache misses. The difference in speed between our algorithms and the algorithm of Bratley and Fox is also very small.

While our algorithm for generating the modified Halton sequences A.M is inferior in speed to A.H and B.F, it still runs relatively fast, and can be used in quasi-Monte Carlo calculations, since the other calculations will take much more time than the generation part. In the next section we demonstrate how with this algorithm we obtain more accurate results for a small increase in the total computational time.

4 Numerical Results

In this section we compare the modified sequences with the original Halton sequences for the approximate calculation of several integrals, widely used for benchmarking low-discrepancy sequences. The integrals are

$$I_k = \int_{E^s} f_k(x)dx, \qquad k = 1, 2, 3,$$

and they are approximated by the formula

$$I_k \approx \frac{1}{N} \sum_{j=0}^{N-1} f_k(x_j),$$

where the sequence x_j is the Halton or the modified Halton sequence (we always used the first s prime numbers). The functions $f_k, \quad k = 1, 2, 3$ are defined by

$$f_1(x) = \prod_{i=1}^{s} |4x_i - 2|,$$

$$f_2(x) = \prod_{i=1}^{s} (x_i^3 + \frac{3}{4}),$$

$$f_3(x) = \sum_{i=1}^{s} \prod_{j=1}^{i} (-1)^j x_j.$$

The integrals I_1 and I_2 are equal to 1 and the exact value of I_3 is

$$I_3 = \frac{1}{3}\left(1 - \frac{(-1)^s}{2^s}\right).$$

The numerical results from calculating these integrals with the Halton sequences (denoted by "H") and the modified Halton sequences (denoted by "M") are shown in Tables 2, 3, 4. In Fig.1 we show a plot of the results for I_1 in 10 dimensions. One can see how the error decreases as some power of the number of points N, but the constants in the case of the modified sequences are much smaller.

Table 2. Absolute errors when the integral I_1 is approximated using N terms of the Halton and Modified Halton sequences.

Dim	5		10		20	
N	H	M	H	M	H	M
10^3	$9.28\ 10^{-3}$	$3.13\ 10^{-4}$	$5.00\ 10^{-2}$	$1.04\ 10^{-3}$	$2.09\ 10^{-1}$	$1.49\ 10^{-3}$
10^4	$1.31\ 10^{-3}$	$5.66\ 10^{-5}$	$8.45\ 10^{-3}$	$8.95\ 10^{-5}$	$5.15\ 10^{-2}$	$3.59\ 10^{-4}$
10^5	$1.58\ 10^{-4}$	$2.41\ 10^{-7}$	$1.21\ 10^{-3}$	$2.56\ 10^{-6}$	$1.05\ 10^{-2}$	$1.69\ 10^{-4}$
10^6	$1.87\ 10^{-5}$	$1.23\ 10^{-7}$	$1.32\ 10^{-4}$	$1.88\ 10^{-6}$	$1.63\ 10^{-3}$	$4.25\ 10^{-5}$
10^7	$1.69\ 10^{-6}$	$3.74\ 10^{-8}$	$1.58\ 10^{-5}$	$2.40\ 10^{-7}$	$1.65\ 10^{-4}$	$1.24\ 10^{-5}$

Table 3. Absolute errors when the integral I_2 is approximated using N terms of the Halton and Modified Halton sequences.

Dim	5		10		20	
N	H	M	H	M	H	M
10^3	$5.75\ 10^{-3}$	$5.13\ 10^{-4}$	$4.56\ 10^{-1}$	$1.78\ 10^{-2}$	$5.48\ 10^{2}$	$1.46\ 10^{-2}$
10^4	$1.56\ 10^{-3}$	$1.74\ 10^{-4}$	$3.78\ 10^{-2}$	$1.06\ 10^{-3}$	$5.46\ 10^{1}$	$7.24\ 10^{-3}$
10^5	$3.71\ 10^{-5}$	$4.69\ 10^{-6}$	$1.79\ 10^{-3}$	$2.75\ 10^{-4}$	$5.39\ 10^{0}$	$7.06\ 10^{-4}$
10^6	$1.28\ 10^{-5}$	$5.75\ 10^{-7}$	$6.44\ 10^{-4}$	$2.15\ 10^{-5}$	$5.37\ 10^{-1}$	$9.11\ 10^{-4}$
10^7	$8.82\ 10^{-7}$	$3.97\ 10^{-8}$	$7.01\ 10^{-6}$	$1.41\ 10^{-5}$	$5.04\ 10^{-2}$	$4.89\ 10^{-4}$

Table 4. Absolute errors when the integral I_3 is approximated using N terms of the Halton and Modified Halton sequences.

Dim	5		10		20	
N	H	M	H	M	H	M
10^3	$1.01\ 10^{-2}$	$3.55\ 10^{-5}$	$3.92\ 10^{-3}$	$6.29\ 10^{-5}$	$1.03\ 10^{-4}$	$6.30\ 10^{-5}$
10^4	$4.33\ 10^{-3}$	$1.05\ 10^{-5}$	$2.76\ 10^{-3}$	$1.18\ 10^{-5}$	$7.13\ 10^{-4}$	$5.01\ 10^{-6}$
10^5	$9.39\ 10^{-4}$	$3.95\ 10^{-7}$	$3.32\ 10^{-3}$	$8.92\ 10^{-7}$	$2.12\ 10^{-4}$	$6.20\ 10^{-7}$
10^6	$5.32\ 10^{-4}$	$2.64\ 10^{-10}$	$2.20\ 10^{-3}$	$9.55\ 10^{-8}$	$3.16\ 10^{-5}$	$3.79\ 10^{-8}$
10^7	$1.99\ 10^{-3}$	$5.41\ 10^{-10}$	$1.18\ 10^{-3}$	$4.34\ 10^{-9}$	$4.53\ 10^{-4}$	$4.08\ 10^{-8}$

References

1. Atanassov, E.I.: On the discrepancy of the Halton sequences. Mathematica Balkanika (accepted for publication).
2. Bratley, P., Fox, B.: Algorithm 659: Implementing Sobol's quasirandom sequence generator. ACM Transactions on Mathematical Software (TOMS), Vol. 14, 1, ACM Press, New York, NY, USA (1988) 88–100.
3. Faure, H.: Discrépance de suites associées à un système de numéracion (en dimension un). H. Bull. Soc. Math. France, 109 (1981) 143–182.
4. Halton, J.H.: On the efficiency of certain quasi-random sequences of points in evaluating multi-dimensional integrals. Numer. math. 2 (1960) 84–90.

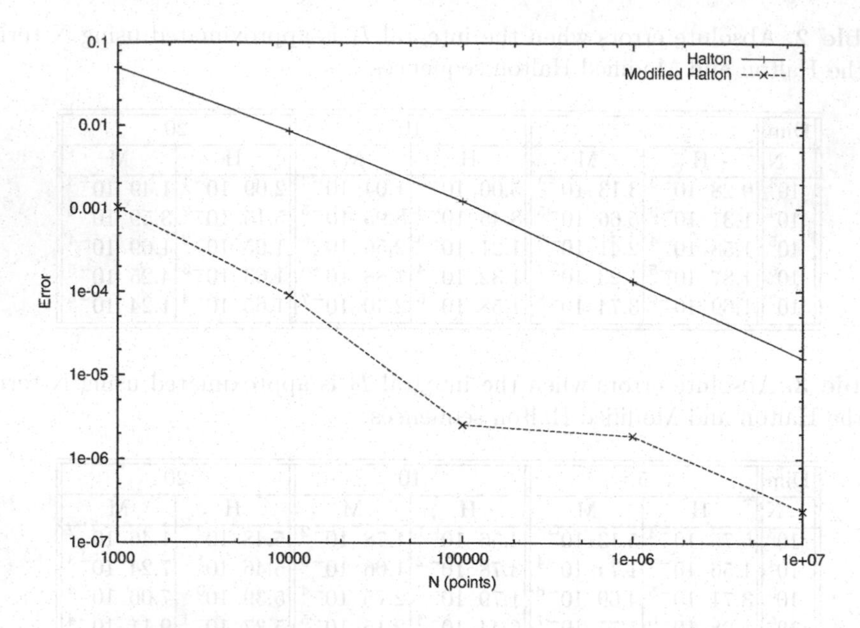

Fig. 1. Absolute error for the integral I_1 in 10 dimensions in log-log scale.

5. Lecot, C.: An algorithm for generating low discrepancy sequences on vector computers. Parallel Computing, 11 (1989) 113–116.
6. Ökten, G., Srinivasan, A.: Parallel Quasi-Monte Carlo Methods on a Heterogeneous Cluster. In: Fang, K. T., Hickernell, F. J., and Niederreiter, H. (eds.): Monte Carlo and Quasi-Monte Carlo Methods 2000. Springer-Verlag, Berlin (2002) 406–421,
7. Schmidt, W. Ch., Uhl, A.: Tehniques for parallel quasi-Monte Carlo integration with digital sequences and associated problems. Mathematics and Computers in Simulation, Vol. 55, 1-3, North-Holland (2001) 249–257.

Parallel Importance Separation and Adaptive Monte Carlo Algorithms for Multiple Integrals*

Ivan Dimov, Aneta Karaivanova, Rayna Georgieva, and Sofiya Ivanovska

CLPP - Bulgarian Academy of Sciences
Acad. G. Bonchev St., Bl.25A, 1113 Sofia, Bulgaria
ivdimov@bas.bg, anet@copern.bas.bg,
rayna@copern.bas.bg, sofia@copern.bas.bg

Abstract. Monte Carlo Method (MCM) is the only viable method for many high-dimensional problems since its convergence is independent of the dimension. In this paper we develop an adaptive Monte Carlo method based on the ideas and results of the *importance separation*, a method that combines the idea of separation of the domain into uniformly small subdomains with the Kahn approach of importance sampling. We analyze the error and compare the results with crude Monte Carlo and *importance sampling* which is the most widely used variance reduction Monte Carlo method. We also propose efficient parallelizations of the importance separation method and the studied adaptive Monte Carlo method. Numerical tests implemented on PowerPC cluster using MPI are provided.

1 Introduction

Multidimensional numerical quadratures are of great importance in many practical areas, ranging from atomic physics to finance. The crude Monte Carlo method has rate of convergence $O(N^{-1/2})$ which is independent of the dimension of the integral, and that is why Monte Carlo integration is the only practical method for many high-dimensional problems. Much of the efforts to improve Monte Carlo are in construction of variance reduction methods which speed up the computation.

Importance sampling is probably the most widely used Monte Carlo variance reduction method, [5]. One use of importance sampling is to emphasize rare but important events, i.e., small regions of space in which the integrand is large. One of the difficulties in this method is that sampling from the importance density is required, but this can be performed using acceptance-rejection.

It is also known that importance sampling can greatly increase the variance in some cases, [11]. In Hesterberg (1995, [8]) a method of defensive importance sampling is presented; when combined with suitable control variates, defensive importance sampling produces a variance that is never worse than the crude

* Supported by Center of Excellence BIS-21 Grant ICA1-2000-70016 and by the Ministry of Education and Science of Bulgaria under Grants I-1201/02 and MM-902/99

I. Dimov et al. (Eds.): NMA 2002, LNCS 2542, pp. 99–107, 2003.

Monte Carlo variance, providing some insurance against the worst effects of importance sampling. Defensive importance sampling can however be much worse than the original importance sampling.

Owen and Zhow (1999) recommend an importance sampling from a mixture of m sampling densities with m control variates, one for each mixture component. In [11] it is shown that this method is never much worse than pure importance sampling from any single component of the mixture.

Another method, multiple importance sampling, similar to defensive importance sampling, is presented in Veach&Guibas (1995, [13]) and Veach (1997, [14]). It is standard practice to weight observations in inverse proportion to their sampling probability. Multiple importance sampling can break that rule, and do so in a way that still results in an unbiased estimate of the integral. The idea is that in some parts of the sample space, the integrand may be roughly proportional to one of sampling densities while other densities are appropriate to other parts of the space. The goal is to place greater weight on those locally most appropriate densities.

In [9,10] a method called importance separation that combines ideas from importance sampling and stratification is presented and studied. This method has the best possible rate of convergence for certain class of functions but its disadvantage is the increased computational complexity.

Another group of algorithms, widely used for numerical calculation of multidimensional integrals, are the adaptive algorithms. Most of the adaptive algorithms use a sequence of increasingly finer subdivisions of the original region, chosen to concentrate integrand evaluations on subregions with difficulties. Two main types of subdivision strategies are in common use: local and global subdivision. The main disadvantage of local subdivision strategy is that it needs a local absolute accuracy requirement which will be met after the achievement of the global accuracy requirement. The main advantage of the local subdivision strategy is that it allows a very simple subregion management (there is no need to store inactive subregions). Globally adaptive algorithms usually require more working storage than locally adaptive routines, and accessing the region collection is slower. These algorithms try to minimize the global error as fast as possible, independent of the specified accuracy requirement. For example, see [2], where an improved adaptive algorithm for the approximate calculation of multiple integrals is presented - this algorithm is similar to a globally adaptive algorithm for single integrands first described by van Dooren and de Ridder [6]. The modifications are imposed by that the new algorithm applies to a vector of integrands.

The adaptive algorithms proved to be very efficient but they do not have the inherent parallel properties of crude Monte Carlo. In recent years, two approaches to parallel adaptive integration have emerged, for comparison see Bull&Freeman (1998, [4]). One is based on adapting the ideas of sequential globally adaptive algorithms to the parallel context by selecting a number of subdomains of the integration domain according to the associated error estimate, see, for example, Bull&Freeman (1994, [3]). The other approach proceeds by imposing an initial static partitioning of the domain and treats the resulting problems as independent. This approach needs a mechanism for detecting load imbalance and for redistributing work to other processors, see, for example, [7]. Let us mention that a central feature of the parallel adaptive algorithms is the list containing the subintervals and corresponding error estimates. Fundamentally different parallel algorithms result depending on whether the list is maintained as a single shared data structure accessible to all processors, or else as the union of nonoverlapping sublists, each private to a processor.

In this paper, we use the ideas of importance separation to create an adaptive algorithm for integration. We describe parallelization of these algorithms, study their parallel properties and compare them with importance sampling.

2 Importance Separation and Adaptive Monte Carlo Method

Consider the problem of approximate calculation of the multiple integral

$$I = \int_G f(x)p(x)\,dx, \quad G \equiv [0;1]^d \tag{1}$$

where $f(x)$ is an integrable function for any $x \in G \subset \mathbb{R}^d$ and $p(x) \geq 0$ is a probability density function, such that $\int_G p(x)\,dx = 1$.

The Monte Carlo quadrature formula is based on the probabilistic interpretation of an integral. If $\{x_n\}$ is a sequence in G sampled with density $p(x)$, then the Monte Carlo approximation to the integral is, [12],

$$I \approx I_N = \frac{1}{N}\sum_{n=1}^{N} f(x_n)$$

with the integration error $\varepsilon_N = |I - I_N| \approx \sqrt{\frac{Var(f)}{N}}$.

2.1 Importance Separation

Here we briefly present a Monte Carlo method called importance separation first described and studied in [9,10]. This method combines the ideas of stratification and importance sampling and has the best rate of convergence (see [1]) for the class of functions with bounded derivatives.

The method of importance separation uses a special partion of the domain and computes the given integral as a sum of the integrals on the subdomains. First, let us describe the method in the one-dimensional case - when the domain is an interval, say $[0,1]$ and $f(x) \in \mathbb{C}_{[0,1]}$. Partition [0,1] into M subintervals in the following way: $x_0 = 0;\ \ x_M = 1;\ \ G_i \equiv [x_{i-1}, x_i];$

$$x_i = \frac{C_i}{f(x_{i-1})(M-i+1)}, \quad i = 1, ..., M-1 \tag{2}$$

where

$$C_i = 1/2[f(x_{i-1}) + f(1)](1 - x_{i-1}),\ i = 1, \dots, M - 1.$$

Obviously, $I = \int_0^1 f(x)p(x)\,dx = \sum_{i=1}^{M} \int_{x_{i-1}}^{x_i} f(x)p(x)dx$. If $f(x) \in H(1, L)_{[0,1]}$, there exist constants $L_i \left(L \geq \max_i L_i\right)$, such that

$$L_i \geq \left|\frac{\partial f}{\partial x}\right| \quad \text{for any } x \in G_i. \tag{3}$$

Moreover, for the above scheme there exist constants c_{1_i} and c_{2_i} such that

$$p_i = \int_{G_i} p(x)\,dx \leq c_{1_i}/M,\ i = 1, \dots, M \tag{4}$$

$$\sup_{x_{1_i}, x_{2_i} \in G_i} |x_{1_i} - x_{2_i}| \leq c_{2_i}/M,\ i = 1, \dots, M. \tag{5}$$

The following theorem, proved in [9], gives the rate of convergence:

Theorem Let $f(x) \in H(1, L)_{[0,1]}$ and $M = N$. Then using the importance separation (3)-(5) of G we have the following Monte Carlo integration error:

$$\varepsilon_N \approx \sqrt{2}[1/N \sum_{j=1}^{N} (L_j c_{1_j} c_{2_j})^2]^{1/2} N^{-3/2}.$$

Now consider the multidimensional case. For an importance separation with analogous properties (for each coordinate we apply the already described one-dimensional scheme (2) in the same manner), we have the following integration error ($M = N$):

$$\varepsilon_N \approx \sqrt{2}d \left[\frac{1}{N} \sum_{i=1}^{N} (L_i c_{1_i} c_{2_i})^2\right]^{1/2} N^{-1/2-1/d}.$$

The disadvantage of the above described methods is the increased computational complexity. The accuracy is improved (in fact, importance separation gives the theoretically optimal accuracy, [10]) but the price is increased number of additional computations which makes these methods impractical for large d.

2.2 Adaptive Monte Carlo Method

Based on advantages and disadvantages of importance separation we develop an adaptive approach for calculation of the desired scalar variable I. Our adaptive method does not use any a priori information about the smoothness of the integrand, but it uses a posteriori information for the variance. The idea of the method consists in the following: the domain of integration G is separated into subdomains with identical volume. The interval [0;1] on every dimension coordinate is partitioned into M subintervals, i.e.

$$G = \sum_j G_j, \quad j = 1, M^d.$$

Denote by p_j and I_{G_j} the following expressions:

$$p_j = \int_{G_j} p(x)\,dx \quad \text{and} \quad I_{G_j} = \int_{G_j} f(x)p(x)\,dx.$$

Consider now a random point $\xi^{(j)} \in G_j$ with a density function $p(x)/p_j$ and in this case

$$I_{G_j} = \mathrm{E}\left[\frac{p_j}{N}\sum_{i=1}^{N} f(\xi_i^{(j)})\right] = \mathrm{E}\theta_N.$$

The algorithm starts with a relatively small number M which is given as input data. For every subdomain the integral I_{G_j} and the variance are evaluated. Then the variance is compared with a preliminary given value. The obtained information is used for the next refinement of the domain and for increasing the density of the random points. In order to choose the first and the next dimension coordinate on which an additive division is made, we use random numbers. To avoid the irregular separation on different coordinates a given coordinate recurs only even all other coordinates have been already chosen. In the end an approximation for the integral $I = \sum_j I_{G_j}$ is obtained. The algorithm is described below.

Algorithm

1. **Input data**: *number of points N, number of subintervals M on every dimension coordinate, constant ε (estimation for the variance), constant δ (stop criterion; estimation for the length of subintervals on every coordinate).*
2. **For** $j = 1, M^d$
   - 2.1 **Calculate** *the approximation of I_{G_j} and the variance D_{G_j} in subdomain G_j based on N independent realizations of random variable θ_N*
   - 2.2 **If** ($\mathrm{D}_{G_j} \geq \varepsilon$) **then**
     - 2.2.1 **Choose** *the axis direction on which the partition will perform*
     - 2.2.2 **Divide** *the current domain into two (G_{j_1}, G_{j_2}) along the chosen direction*
     - 2.2.3 **If** *the length of obtained subinterval is less than δ* **then go to** *step 2.2.1* **else** $j = j_1$ *(G_{j_1} is the current domain) and* **go to** *step 2.1*
   - 2.3 **Elseif** ($\mathrm{D}_{G_j} < \varepsilon$), *but an approximation of $I_{G_{j_2}}$ has not been calculated yet* **then** $j = j_2$ *(G_{j_2} is the current domain along the corresponding direction) and* **go to** *step 2.1*
   - 2.4 **Elseif** ($\mathrm{D}_{G_j} < \varepsilon$), *but there are subdomains along the other axis directions* **then go to** *step 2.3*
   - 2.5 **Else** *Accumulation in the approximation I_N of I*

3 Parallel Implementation

In this section we present the parallel importance separation and parallel adaptive Monte Carlo algorithms for evaluation of multiple integrals. The crude

Monte Carlo possesses inherent parallelism which is based on the possibility to calculate simultaneously realizations of the random variable on different processors. For our two algorithms (importance separation and simple adaptive) we have additional work: partitioning of the domain and assigning the correspondent portions of work to the available processors. This has to be done very carefully in order to have good load balancing. We consider a multiprocessor configuration with p nodes.

N uniformly distributed random points $x_i \in [0;1]^d, \; i = 1, \ldots, N$ are used to obtain an approximation with given accuracy of the integral (1). For generation of $d-$dimensional random point we need d random numbers.To estimate the performance of the parallel algorithms we use:

$\mathbf{ET_p(A)}$ mathematical expectation of time, required for a set of p processing elements to solve the problem using algorithm A

$\mathbf{S_p(A)} = \frac{\mathbf{ET_1(A)}}{\mathbf{ET_p(A)}}$ speed-up

$\mathbf{E_p(A)} = \frac{\mathbf{S_p(A)}}{\mathbf{p}}$ parallel efficiency.

In parallel version of the adaptive algorithm the processors are separated as "master" and "slaves". The "master" calculates the variance in every subdomain G_j, analyzes it and sends information about the new partition — the limits of new subdomains - to the "slaves".

4 Numerical Experiments

We now present the numerical results for the accuracy and the convergence of crude, importance sampling, adaptive, importance separation MCM for numerical integration. The numerical tests are implemented on a cluster of 4 two-processor computers Power Macintosh using MPI. The results are presented as a function of the sample size, N. For each case the error is computed with respect to the exact solution. A lot of numerical tests were performed. Here we present the results of evaluating the 5-dimensional integral over $I^5 = [0;1]^5$ of the function $f(x) = \exp\left(\sum_{i=1}^{5} a_i x_i^2 \frac{2+\sin(\sum_{j=1, j\neq i}^{5} x_j)}{2}\right)$ using the positive definite importance function (density) $h(x) = \frac{1}{\eta}\exp\left(\sum_{i=1}^{5} a_i x_i^2\right)$ where $\mathbf{a} = (1, 0.5, 0.2, 0.2, 0.2)$ and $\eta = \int_{I^5} exp(\sum_{i=1}^{5} a_i x_i^2)\, dx$ so that h is normalised. We denote by I_N an estimation of the integral using one sample of N random points.

We compare the results (accuracy and CPU time in seconds) for four methods: crude Monte Carlo, importance sampling, importance separation (IMS) and the proposed adaptive Monte Carlo (AMC). The results shown in the Table 1 and Figure 1 illustrate the superior behaviour of the considered in this paper methods - for twice more computational time (6 seconds), IMS gives approximately 10 times better accuracy than crude Monte Carlo and importance sampling. Let us mention also that the results obtained with the adaptive method and with importance separation are very similar when the sample size increases. A

Table 1. Comparison between Crude MCM, Importance sampling, Importance separation and Adaptive MCM. Input data: $M = 6, \varepsilon = 0.6, \delta = 0.1e-06$ (calculations are implemented on one processor).

N	Crude MCM		Imp. sampling		Imp. separation		Adaptive MCM	
	$\|I - I_N\|$	T_1	$\|I - I_N\|$	T_1	$\|I - I_N\|$	T_1	$\|I - I_N\|$	T_1
100	0.009532	0.001	0.081854	0.008	0.000316	6	0.001102	20
500	0.092960	0.004	0.007102	0.036	0.000003	31	0.000246	121
2500	0.009027	0.020	0.006381	0.175	0.000068	152	0.000131	587
10000	0.006611	0.076	0.004673	0.697	0.000061	610	0.000036	2390
50000	0.008443	0.386	0.003212	3.489	0.000021	3047	0.000009	12093

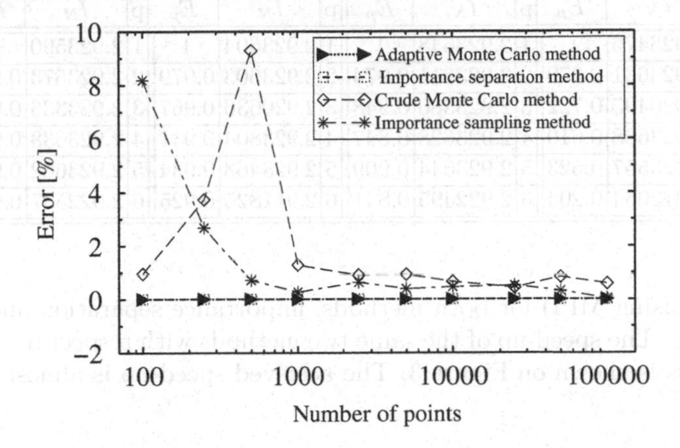

Fig. 1. Comparison of the accuracy of Crude MCM, Importance sampling, Importance separation and Adaptive MCM.

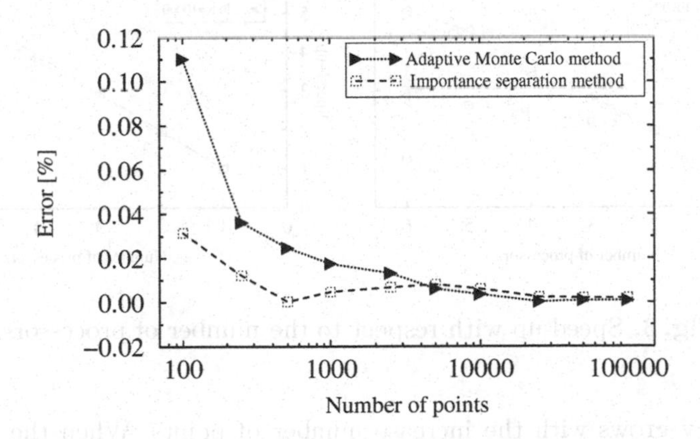

Fig. 2. Comparison of the accuracy of Importance separation and Adaptive MCM.

comparison of these two methods is given in Figure 2. It can be seen that importance separation method needs less points than adaptive method to achieve the desired accuracy. The reason for these results is the additional partitioning of the domain for adaptive method using only a posteriori information about the variance. The Table 2 presents the achieved efficiency of the parallel imple-

Table 2. Implementation of Adaptive MCM and Importance separation using MPI ($I = 2.923651$).

Adaptive MCM						Importance separation					
$N = 1000$			$N = 10000$			$N = 1000$			$N = 10000$		
p	I_N	E_p	p	I_N	E_p	p	I_N	E_p	p	I_N	E_p
1	2.923476	1	1	2.923648	1	1	2.923604	1	1	2.923590	1
2	2.923631	0.970	2	2.923611	0.974	2	2.923603	0.979	2	2.923573	0.985
3	2.920495	0.722	3	2.923380	0.948	3	2.920636	0.967	3	2.923336	0.983
4	2.923608	0.610	4	2.923628	0.847	4	2.923804	0.941	4	2.923638	0.980
5	2.923557	0.523	5	2.923644	0.909	5	2.923463	0.934	5	2.923602	0.979
6	2.912064	0.204	6	2.922495	0.841	6	2.911825	0.925	6	2.922537	0.977

mentation (using MPI) for both methods, importance separation and adaptive Monte Carlo. The speed-up of the same two methods with respect to the number of processors is shown on Figure 3. The achieved speed-up is almost linear and

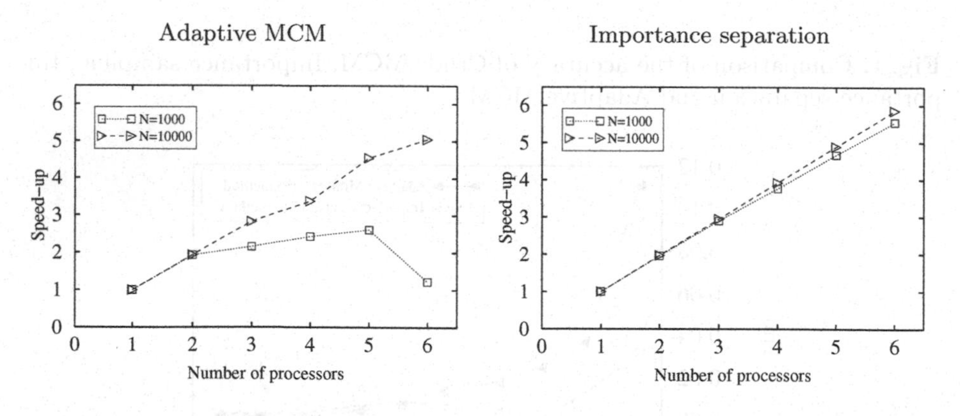

Fig. 3. Speed-up with respect to the number of processors.

the efficiency grows with the increase number of points. When the number of processors is large, but CPU time (even on one processor) is small, the efficiency is not very good because most of CPU time is used for communications between processors. An additional explanation of the efficiency of the adaptive method is that the "master" carries out some preliminary calculations (computing of the limits of two new subregions obtained after division).

5 Conclusion and Future Directions

In this paper we present and study the parallel properties of importance separation and an adaptive Monte Carlo algorithms. We compare the results with crude Monte Carlo and with importance sampling. Our algorithms show some advantages.

While the results of applying our methods for approximate calculation of integrals are very satisfying, we would like to extend these methods for solving integral equations. We already started working on these algorithms.

References

1. Bahvalov, N. S.: On the optimal estimations of convergence of the quadrature processes and integration methods. Numerical Methods for Solving Differential and Integral Equations, Nauka, Moscow (1964) 5–63 (in Russian).
2. Berntsen, J., Espelid, T. O., and Genz, A.: An adaptive algorithm for the approximate calculation of multiple integrals. ACM Trans. Math. Softw., 17 (1991) 437–451.
3. Bull, J. M. and Freeman, T. L.: Parallel globally adaptive quadrature on the KSR-1. Advances in Comp. Mathematics, 2 (1994) 357–373.
4. Bull, J. M. and Freeman, T. L.: Parallel algorithms for multi-dimensional integration. Parallel and Distributed Computing Practices, 1(1) (1998) 89–102.
5. Caflisch, R. E.: Monte Carlo and quasi-Monte Carlo methods. Acta Numerica, 7 (1998) 1–49.
6. van Dooren, P. and de Ridder, L.: An adaptive algorithm for numerical integration over an N-dimensional cube. Journal of Computational and Applied Mathematics, 2 (1976) 207–217.
7. Freeman, T. L. and Bull, J. M.: A comparison of parallel adaptive algorithms for multi-dimensional integration. Proceedings of 8th SIAM Conference on Parallel Processing for Scientific Computing (1997)
8. Hesterberg, T.: Weighted average importance sampling and defensive mixture distributions. Technometrics, 37(2) (1995) 185–194.
9. Karaivanova, A.: Adaptive Monte Carlo methods for numerical integration. Mathematica Balkanica, 11 (1997) 391–406.
10. Karaivanova, A. and Dimov, I.: Error analysis of an adaptive Monte Carlo method for numerical integration. Mathematics and Computers in Simulation, 47 (1998) 201–213.
11. Owen, A. and Zhou, Y.: Safe and effective importance sampling. Technical report, Stanford University, Statistics Department (1999)
12. Sobol', I. M.: Monte Carlo Numerical Methods. Nauka, Moscow (1973) (in Russian).
13. Veach, E. and Guibas, L. J.: Optimally combining sampling techniques for Monte Carlo rendering. Computer Graphics Proceedings, Annual Conference Series, ACM SIGGRAPH '95 (1995) 419–428.
14. Veach, E.: Robust Monte Carlo Methods for Light Transport Simulation. Ph.D. dissertation, Stanford Universty (1997)

Monte Carlo and Quasi-Monte Carlo Algorithms for the Barker-Ferry Equation with Low Complexity*

T.V. Gurov[1], P.A. Whitlock[2], and I.T. Dimov[1]

[1] CLPP - BAS, Acad. G. Bonchev st, bl. 25 A,
1113 Sofia, Bulgaria,
gurov@copern.bas.bg, ivdimov@bas.bg

[2] Dep. of Comp. and Inf. Sci., Brooklyn College - CUNY
2900 Bedford Ave, Brooklyn, NY 11210, USA
whitlock@sci.brooklyn.cuny.edu

Abstract. In this paper we study the possibility to use the Sobol' and Halton quasi-random number sequences (QRNs) in solving the Barker-Ferry (B-F) equation which accounts for the quantum character of the electron-phonon interaction in semiconductors. The quasi-Monte Carlo (QMC) solutions obtained by QRNs are compared with the Monte Carlo (MC) solutions in case when the scalable parallel random number generator (SPRNG) library is used for producing the pseudo-random number sequences (PRNs).
In order to solve the B-F equation by a MC method, a transition density with a new sampling approach is suggested in the Markov chain.

1 Introduction

The B-F equation [1] describes a femtosecond relaxation process of optically excited electrons which interact with phonons in an one-band semiconductor [2]. We consider an one time-dimension integral form of this quantum kinetic equation [3]:

$$f(\mathbf{k},t) = \int_0^t dt'' \int d^3\mathbf{k}' \{\mathcal{K}(\mathbf{k}',\mathbf{k},t-t'')f(\mathbf{k}',t'') - \mathcal{K}(\mathbf{k},\mathbf{k}',t-t'')f(\mathbf{k},t'')\} + \phi(\mathbf{k}), \quad (1)$$

with a kernel

$$\mathcal{K}(\mathbf{k}',\mathbf{k},t-t'') = \frac{e^2\omega_{\mathbf{q}}}{2\pi^2\hbar}\left|\frac{1}{\epsilon_\infty} - \frac{1}{\epsilon_s}\right|\frac{1}{(\mathbf{k}'-\mathbf{k})^2} \times \quad (2)$$

$$\times\left\{\frac{(n_{\mathbf{q}}+1)\Gamma_{\mathbf{k}',\mathbf{k}}}{\Omega^2_{\mathbf{k}',\mathbf{k}}+\Gamma^2_{\mathbf{k}',\mathbf{k}}}[1+G(\mathbf{k}',\mathbf{k},t,t'')] + \frac{n_{\mathbf{q}}\Gamma_{\mathbf{k},\mathbf{k}'}}{\Omega^2_{\mathbf{k},\mathbf{k}'}+\Gamma^2_{\mathbf{k},\mathbf{k}'}}[1+G(\mathbf{k},\mathbf{k}',t,t'')]\right\},$$

* Supported by the Center of Excellence BIS-21 grant ICA1-2000-70016 and by the NSF of Bulgaria under grants MM-902/99 and I-1201/02.

I. Dimov et al. (Eds.): NMA 2002, LNCS 2542, pp. 108–116, 2003.

$$G(\mathbf{k}', \mathbf{k}, t, t'') = \left(\frac{\Omega_{\mathbf{k}',\mathbf{k}}}{\Gamma_{\mathbf{k}',\mathbf{k}}} \sin(\Omega_{\mathbf{k}',\mathbf{k}}(t - t'')) - \cos(\Omega_{\mathbf{k}',\mathbf{k}}(t - t'')) \right) \exp(-\Gamma_{\mathbf{k}',\mathbf{k}}(t - t'')),$$

where

$\mathbf{k}$ and t are the momentum and the evolution time, respectively;

$f(\mathbf{k}, t)$ is the distribution function;

$\phi(\mathbf{k})$ is the initial electron distribution function;

$n_{\mathbf{q}} = 1/(exp(\hbar\omega_{\mathbf{q}}/KT) - 1)$ is the Bose function, where K is the Boltzmann constant and T is the temperature of the crystal, corresponds to an equilibrium distributed phonon bath;

$\Gamma_{\mathbf{k}',\mathbf{k}} = \Gamma_{\mathbf{k}'} + \Gamma_{\mathbf{k}}$ is related to the finite carrier lifetime for the scattering process:

$$\Gamma_{\mathbf{k}} = \int d^3\mathbf{k}' \frac{e^2\omega_{\mathbf{q}}}{4\pi} \left| \frac{1}{\epsilon_\infty} - \frac{1}{\epsilon_s} \right| \sum_{\pm} \frac{1}{(\mathbf{k}' - \mathbf{k})^2} \delta(\varepsilon(\mathbf{k}') - \varepsilon(\mathbf{k}) \pm \hbar\omega_{\mathbf{q}}) \left(n_{\mathbf{q}} + \frac{1}{2} \pm \frac{1}{2} \right);$$

ϵ_∞ and ϵ_s are the optical and static dielectric constants;

$\Omega_{\mathbf{k}',\mathbf{k}} = (\varepsilon(\mathbf{k}') - \varepsilon(\mathbf{k}) - \hbar\omega_{\mathbf{q}})/\hbar$, where $\omega_{\mathbf{q}}$ is the phonon frequency, $\hbar\omega_{\mathbf{q}}$ is the phonon energy which generally depends on $\mathbf{q} = \mathbf{k}' - \mathbf{k}$, and $\varepsilon(\mathbf{k}) = (\hbar^2\mathbf{k}^2)/(2m)$ is the electron energy.

Note the kernel (2) can be decomposed into a time-independent part and a part which depends explicitly on the time. Consider the problem for evaluating the following functional

$$J_g(f) \equiv (g, f) = \int_0^{\mathcal{T}} \int_G g(\mathbf{k}, t) f(\mathbf{k}, t) d^3\mathbf{k} dt,$$

by a MC method. Here we specify that the wave vector $\mathbf{k}$ belongs to a finite domain G which is sphere with radius Q and $t \in (0, \mathcal{T})$. The case, when $g(\mathbf{k}, t) = \delta(\mathbf{k} - \mathbf{k_0})\delta(t - t_0)$, is of special interest, because we are interested in calculating the value of f at a fixed point $(\mathbf{k_0}, t_0)$. Now Eq.(1) can be written in the following form:

$$f(\mathbf{k}, t) = \int_0^t dt'' \int_G d^3\mathbf{k}' \{K_1(\mathbf{k}, \mathbf{k}', t, t'') f(\mathbf{k}', t'') + K_2(\mathbf{k}, \mathbf{k}', t, t'') f(\mathbf{k}, t'')\} + \phi(\mathbf{k}), \tag{3}$$

where $K_1(\mathbf{k}, \mathbf{k}', t, t'') = \mathcal{K}(\mathbf{k}', \mathbf{k}, t - t'')$ and $K_2(\mathbf{k}, \mathbf{k}', t, t'') = -\mathcal{K}(\mathbf{k}, \mathbf{k}', t - t'')$. We note that the Neumann series of the integral equation (3) converges [3] and the solution can be evaluated by a MC estimator.

2 Monte Carlo and Quasi-Monte Carlo Algorithms

The biased MC estimator for the solution of Eq.(3) at the fixed point $(\mathbf{k_0}, t_0)$ using backward time evolution of the numerical trajectories [4] has the following form:

$$\xi_n[\mathbf{k_0}, t_0] = \phi(\mathbf{k_0}) + \sum_{j=1}^{n} W_j^\alpha \phi_\alpha(\mathbf{k}_j), \tag{4}$$

where

$$\phi(\mathbf{k}_j^\alpha) = \begin{cases} \phi(\mathbf{k}_j), & \text{if } \alpha = 1 \\ \phi(\mathbf{k}_{j-1}), & \text{if } \alpha = 2, \end{cases}$$

$$W_j^\alpha = W_{j-1}^\alpha \frac{K_\alpha(\mathbf{k}_{j-1}, \mathbf{k}_j, t_{j-1}, t_j)}{p_\alpha p(\mathbf{k}_{j-1}, \mathbf{k}_j, t_{j-1}, t_j)}, \quad W_0^\alpha = 1, \ \alpha = 1, 2, \ j = 1, \ldots, n.$$

The probabilities $p_\alpha, (\alpha = 1, 2)$ are chosen to be proportional to the absolute value of the kernels. Every point $(\mathbf{k}_j, t_j) \in G \times (0, t_{j-1})$ in the Markov chain $(\mathbf{k}_0, t_0) \to \ldots \to (\mathbf{k}_j, t_j) \to \ldots \to (\mathbf{k}_n, t_n)$, $j = 1, 2, \ldots, n$ is sampled using a transition density function $p(\mathbf{k}, \mathbf{k}', t, t'')$ which is tolerant[1] to both kernels in Eq.(3). The Markov chain terminates in time $t_n < \varepsilon_1$, where ε_1 is a fixed small positive number called a truncation parameter.

Here we suggest the following transition density function: $p(\mathbf{k}, \mathbf{k}', t, t'') = p(\mathbf{k}'/\mathbf{k})p(t, t'')$, where $p(t, t'') = 1/t$. In spherical coordinates (r', θ', φ') with a center $\mathbf{k}$, the function $p(\mathbf{k}'/\mathbf{k})$ is chosen by the following way: $p(\mathbf{k}'/\mathbf{k}) = (4\pi)^{-1}(r')^{-2}l(\mathbf{w})^{-1}$, where $\mathbf{w} = (\mathbf{k}' - \mathbf{k})/r'$, $r' = |\mathbf{k}' - \mathbf{k}|$ and $l(\mathbf{w})$ is distance in the direction of the unit vector $\mathbf{w}$ from $\mathbf{k}$ to the boundary of the domain G. This function satisfies the condition for a transition density. Indeed,

$$\int_G p(\mathbf{k}'/\mathbf{k}) d^3\mathbf{k}' = \oint (4\pi)^{-1} d\mathbf{w} \int_0^{l(\mathbf{w})} r'^{2-2} l(\mathbf{w})^{-1} dr' = 1.$$

Using the spherical symmetry of the task we suppose that $\mathbf{k} = (0, 0, k)$. Thus $\mathbf{k}'$ can be found by the following steps:

1. **Sample** a random unit vector $\mathbf{w} = (\sin\theta', 0, \cos\theta')$ in the plane $\varphi' = 0$ as $\mu = \cos\theta' = 2\beta_1 - 1$ and β_1 is an uniformly distributed number in (0, 1);
2. **Calculate** $l(\mathbf{w}) = -\mu k + (Q^2 - k^2(1-\mu^2))^{\frac{1}{2}}$, where Q is the radius of G;
3. **Sample** $r' = l(\mathbf{w})\beta_2$, where β_2 is an uniformly distributed number in (0, 1);
4. **Calculate** $\mathbf{k}' = \mathbf{k} + r'\mathbf{w}$ and $k' = (r'^2 + k^2 + 2kr'\mu)^{\frac{1}{2}}$.

To complete one transition $(\mathbf{k}, t) \to (\mathbf{k}', t'')$ in the Markov chain we take again an uniformly distributed number $\beta \in (0,\ 1)$. The new time $t'' \in (0,\ t)$ is defined by the equality $t'' = t\beta$.

The solution of Eq.(3) at the fixed point $(\mathbf{k}_0, t_0)$ is evaluated by N independent samples of the estimator (4), i.e.

$$\frac{1}{N}\sum_{i=1}^{N}(\xi_n[\mathbf{k}_0, t_0])_i \xrightarrow{P} J_\delta(f_n) \approx J_\delta(f),$$

where $\xrightarrow{P}$ means stochastic convergence as $N \to \infty$; f_n is the iterative solution obtained by the Neumann series of Eq.(3), and n is the number of iterations.

To solve the above problem we consider two cases for producing uniformly distributed numbers.

[1] $r(x)$ is tolerant of $g(x)$ if $r(x) > 0$ when $g(x) \neq 0$ and $r(x) \geq 0$ when $g(x) = 0$.

Case 1. We use PRNs obtained by the SPRNG library [5,6]. In this case the algorithm is called the MC-SPRNG algorithm. The well known "law of three sigmas" gives the rate of convergence [7] that depends on the variance, $Var(\xi_n[\mathbf{k}_0, t_0])$, and on N,i.e.

$$P\left(\left| \frac{1}{N} \sum_{i=1}^{N} (\xi_n[\mathbf{k}_0, t_0])_i - J_\delta(f_n) \right| < 3 \frac{Var(\xi_n[\mathbf{k}_0, t_0])^{1/2}}{N^{1/2}} \right) \approx 0.997 .$$

Thus, as N increases, the statistical error decreases as $O(N^{-1/2})$.

Case 2. The uniformly distributed numbers that are necessary in the calculation of every transition in the Markov chain are obtained from the Sobol' and Halton QRNs [8,9]. In this case, we obtain two QMC algorithms called the QMC-S algorithm and the QMC-H algorithm, respectively.

We note QRNs are constructed to minimize a measure of their deviation from uniformity of a sequence of real numbers. This measure is called discrepancy. In particular, the discrepancy of s points $x_1, \ldots, x_s \in [0,\, 1]^d, d \geq 1$, is defined by

$$D_s^{(d)} = \sup_E \left| \frac{A(E;s)}{s} - \lambda(E) \right|,$$

where the supremum is taken over all the subsets of $[0,\, 1]^d$ of the form $E = [0, u_1) \times \ldots \times [0, u_d), 0 \leq u_j \leq 1, 1 \leq j \leq d$, λ denotes the Lebesgue measure, and $A(E;s)$ denotes the number of the x_j that are contained in E [10].

A sequence $x_1, x_2, \ldots$ of points in $[0,\, 1]^d$ is a low discrepancy sequence iff

$$D_s^{(d)} \leq c(d) \frac{(\log s)^d}{s}, \forall s > 1,$$

where the constant $c(d)$ depends only on the dimension d [10]. The Sobol' and Halton sequences are low discrepancy sequences [7,11].

Suppose that number of the transitions, n, in the Markov chain is fixed. To model every transition we need three numbers in $[0, 1]$. Therefore, using (3n)-dimensional Sobol' or Halton sequences and applying the Koksma-Hlawka inequality [12], we have the following error bound:

$$\left| J_\delta(f_n) - \frac{1}{N} \sum_{i=1}^{N} (\xi_n[\mathbf{k}_0, t_0])_i \right| \leq C_1(K_\alpha(\mathbf{k}, \mathbf{k}', t, t''), \phi(\mathbf{k})) D_N^{(3n)}, \qquad (5)$$

where the constant $C_1(.,.)$ depends on the kernels of Eq.(3) and on the initial condition, and $D_N^{(3n)}$ has order $O((\log^{3n} N)/N)$. For n fixed and N large, the error $(\log^{3n} N)/N$ is better than the MC error $N^{-1/2}$. But for N fixed and n large, the $(\log^{3n} N)/N$ factor looks ominous. Therefore, it can supposed that QMC algorithms should not be used for high-dimensional problems.

We mention that in the iterative MC algorithms (as the MC-SPRNG algorithm), N is connected with the stochastic error while the parameter n is connected with the systematic error [4]. Thus, we can say the dimension of

QRNs is connected with the systematic error when they are used for estimating of iterative solutions.

The computational complexity of the MC-SPRNG algorithm can be measured by the quantity $F_{mc} = Nnt_{mc}$. Here n is the average number of transitions in the Markov chain and t_{mc} is the mean time for modeling one transition when the SPRNG library is used. When we use the QMC-S and QMC-H algorithms the computational complexity is measured by the quantities $F_S = Nnt_S$ and $F_H = Nnt_H$, respectively. Here t_S (t_H) is the mean time for modeling one transition in the Markov chain in case of the Sobol' (Halton) $3n$-dimensional points used, and n is fixed.

Results for the computational cost of the above algorithms and the accuracy of the MC and QMC solutions are compared with the best MC algorithm called OTDIMC algorithm that is suggested in [4].

3 Numerical Results

The results discussed in the following have been obtained for finite temperature. Material parameters for $GaAs$ have been used: the electron effective mass is 0.063, the optimal phonon energy is $36meV$, the static and optical dielectric constants are $\varepsilon_s = 10.92$ and $\varepsilon_\infty = 12.9$. The initial condition at $t = 0$ is given by a function which is Gaussian in energy, ($\phi(k) = exp(-(b_1k^2 - b_2)^2)$ $b_1 = 96$ and $b_2 = 24$), scaled in a way to ensure, that the peak value is equal to unity. The solution $f(\mathbf{k}, t)$ is estimated in 65 points of the simulation domain G between 0 and $Q = 66 \times 10^7/m$. The quantity presented on the y-axes in all figures is $|\mathbf{k}| * f(\mathbf{k}, t)$, i.e. it is proportional to the distribution function multiplied by the density of states. It is given in arbitrary units. The quantity $\mathbf{k}^2$, given on the x-axes in units of $10^{14}/m^2$, is proportional to the electron energy.

All the algorithms were implemented in C and compiled with the "cc" compiler. The numerical tests were performed on a PowerPC (G4 w/AltiVec) 450 MHz, running YDL 2.0, using the PRNs and QRNs under consideration.

The results for the computational cost (CPU time for all 65 points) of the MC and QMC algorithms are shown in Table 1. Here, N is the number of random walks need to obtain approximately smooth solutions using the different MC algorithms and σ_N is the average estimate of the standard deviation, $(Var(\xi_{l_\varepsilon}[\kappa_0, \tau_0]))^{1/2}$ for all 65 points. We see the MC-SPRNG, QMC-S, and QMC-H algorithms are faster than the OTDIMC algorithm with 10%, 15%, and 30%, respectively. Therefore, the presented algorithms have lower computational complexity. Comparison of the electron energy distribution, which is obtained by all algorithms, is shown on Figures 1-3. The solution of Eq.(3) is estimated at different evolution times as the data for N and t are taken from Table 1. Here $\varepsilon_1 = 0.0001$ for the MC-SPRNG and OTDIMC algorithms, and $n = 16$ for the QMC-S and QMC-H algorithms. We see (on Figure 1) that the MC-SPRNG and OTDIMC solutions approximately coincide and are smooth. Therefore, the use of the MC-SPRNG algorithm is correct. The results on Figures 2 and 3 show noise in the QMC solutions when the evolution time increases. This result can

Table 1. Comparison of the computational complexity of the MC-SPRNG, QMC-S, and QMC-H algorithms with OTDIMC algorithm. The lattice temperature is $-273.15°C$.

	t	N	n	CPU time	σ_N
	$150fs$	100 000	15.56	21m18.32s	0.99
OTDIMC	$200fs$	1 mln	15.95	217m12.02s	2.59
algorithm	$250fs$	3 mln	16.29	658m42.59s	6.75
	$300fs$	15 mln	16.58	3380m29.85s	21.51
	$150fs$	100 000	15.55	18m48.57s	0.97
MC-SPRNG	$200fs$	1 mln	15.99	195m53.01s	2.76
algorithm	$250fs$	3 mln	16.34	596m02.49s	7.75
	$300fs$	15 mln	16.65	3016m14.57s	23.36
	$150fs$	100 000	16	18m13.53s	-
QMC-S	$200fs$	1 mln	16	187m29.23s	-
algorithm	$250fs$	3 mln	16	574m35.40s	-
	$300fs$	15 mln	16	2911m58.48s	-
	$150fs$	100 000	16	15m57.20s	-
QMC-H	$200fs$	1 mln	16	163m48.09s	-
algorithm	$250fs$	3 mln	16	503m13.18s	-
	$300fs$	15 mln	16	2549m33.72s	-

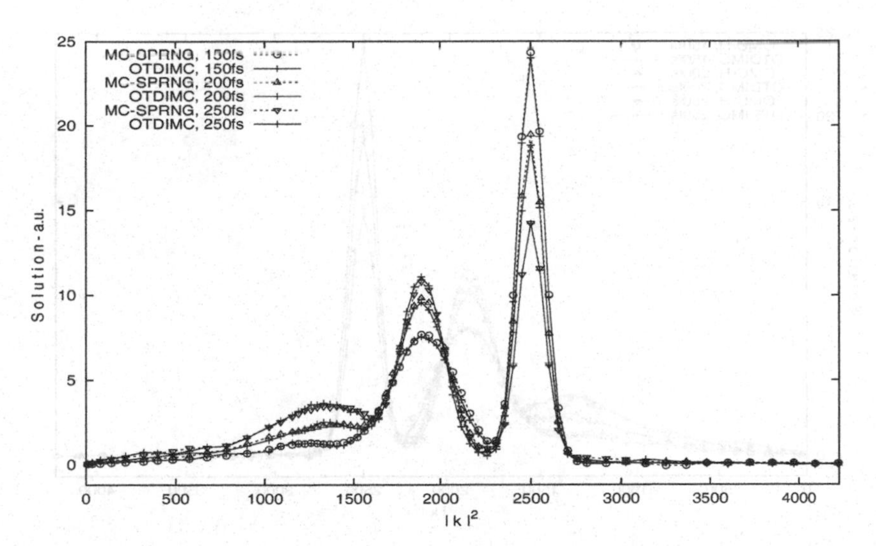

Fig. 1. Comparison of the electron energy distribution $\mathbf{k} * f(\mathbf{k}, t)$ versus $|\mathbf{k}|^2$ obtained by MC-SPRNG and OTDIMC algorithms. $\varepsilon_1 = 0.0001$.

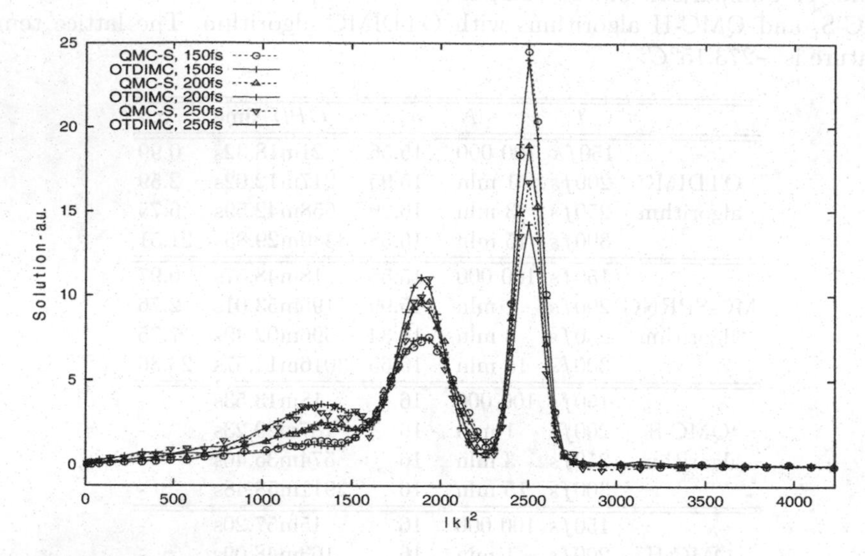

Fig. 2. Comparison of the electron energy distribution $\mathbf{k} * f(\mathbf{k}, t)$ versus $|\mathbf{k}|^2$ obtained by QMC-S and OTDIMC algorithms.

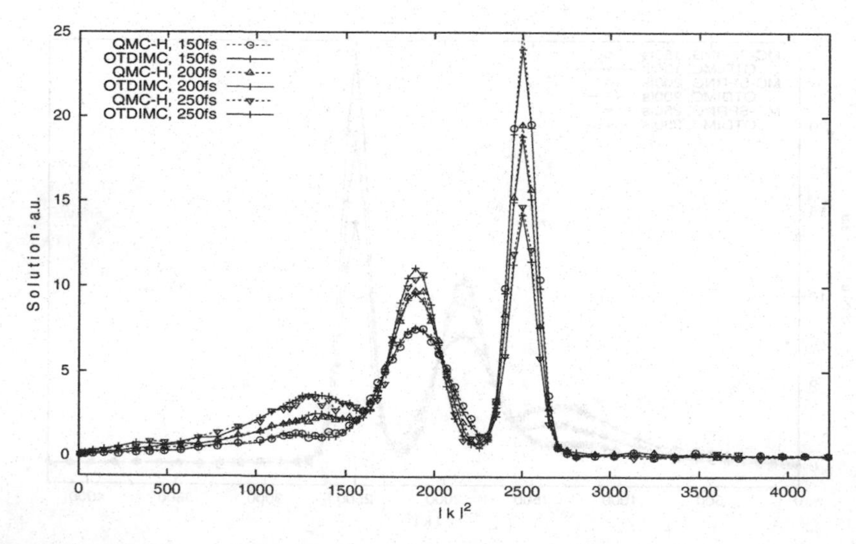

Fig. 3. Comparison of the electron energy distribution $\mathbf{k} * f(\mathbf{k}, t)$ versus $|\mathbf{k}|^2$ obtained by QMC-H and OTDIMC algorithms.

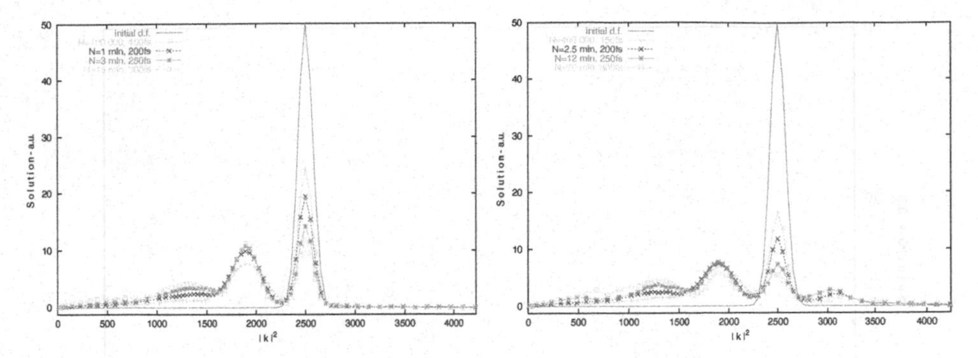

Fig. 4. The electron energy distribution $\mathbf{k} * f(\mathbf{k}, t)$ versus $|\mathbf{k}|^2$ obtained by MC-SPRNG algorithms at different evolution times. The lattice temperature is $T = -273.15°C$ on the left graphics and $T = 18°C$ on the right graphics.

be explained by either the discrepancy increases with increasing the evolution time or there isn't balance between the systematic error and the error from Eq.(5). Therefore, the presented QMC algorithms for solving Eq.(3) are under future investigation. We note that the standard deviation in the cases when MC algorithms are used, increases with increasing the evolution time (see Table 1). Figures 4 and 5 show the electron energy distribution at evolution times (up $300fs$) and at different lattice temperatures. The relaxation leads to a time-dependent broadening of the replicas. The solution in the classically forbidden region, on the right of the initial condition, demonstrates enhancement of the electron population with the growth of the lattice temperature (see Figure 5).

In this paper, we have presented MC and QMC algorithms using a new transition density in the Markov chain that solves the B-F equation. The algorithms have lower complexity when compared with the fast algorithm from [4]. However, noise appeared in the QMC solutions as the evolution time increased. Therefore, an open problem is how to improve the accuracy of the QMC algorithms, QMC-S and QMC-H, while keeping their low complexity. The new MC algorithm, MC-SPRNG, was successfully used to solve the B-F equation at several different evolution times and lattice temperatures.

References

1. Barker, J., Ferry, D.: Self-scattering path-variable formulation of high field time-dependent quantum kinetic equations for semiconductor transport in the finite-collision-duration regime. Physical Review Letters. series 42, 26 (1979) 1779–1781.
2. Schilp, J., Kuhn, T., Mahler, G.: Electron-phonon quantum kinetics in pulse-excited semiconductors: Memory and renormalization effects. Physical Review B., series 47, 8 (1994) 5435–5447.
3. Gurov, T.V., Whitlock, P.A.: An efficient backward Monte Carlo estimator for solving of a quantum kinetic equation with memory kernel. Math. and Comp. in Simulation., series 60 (2002) 85–105.

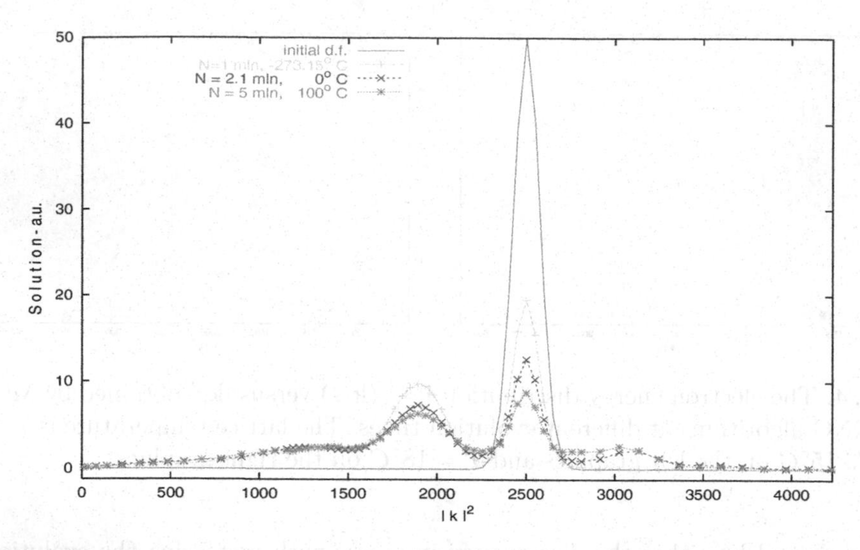

Fig. 5. The electron energy distribution $\mathbf{k} * f(\mathbf{k}, t)$ versus $|\mathbf{k}|^2$ obtained by MC-SPRNG algorithms at different lattice temperatures. The evolution time is $t = 200fs$.

4. Gurov, T.V., Whitlock, P.A.: Statistical Algorithms for Simulation of Electron Quantum Kinetics in Semiconductors - Part I. In: Margenov, S., Waśniewski, J., Yalamov, P. (eds.): Large-Scale Scientific Computing, Lecture Notes in Computer Science, Vol. 2179, Springer-Verlag (2001) 149–156.
5. Mascagni, M.: SPRNG: A Scalable Library for Pseudorandom Number Generation. In: Iliev, O., Kaschiev, M., Margenov, S., Sendov, Bl., Vassilevski, P. (eds.): Recent Advances in Numerical Methods and Applications II. World Scientific, Singapore (1999) 284–295.
6. Scalable Parallel Random Number Generators Library for Parallel Monte Carlo Computations, SPRNG 1.0 and SPRNG 2.0. http://sprng.cs.fsu.edu.
7. Sobol', I.M.: Monte Carlo numerical methods. Nauka, Moscow (1973) (in Russian).
8. Atanassov, E.I.: A New Efficient Algorithm for Generating the Scrambled Sobol' Sequence. In: Dimov, I., Lirkov, I., Margenov, S., Zlatev, Z. (eds.): Numerical methods and applications. Lecture Notes in Computer Science, Vol. 2542, Springer-Verlag (2003) 83–90.
9. Atanassov, E.I., Durtchova, M.K.: Generating and Testing the Modified Halton Sequences. In: Dimov, I., Lirkov, I., Margenov, S., Zlatev, Z. (eds.): Numerical methods and applications. Lecture Notes in Computer Science, Vol. 2542, Springer-Verlag (2003) 91–98.
10. Papageorgiou, A., Traub, J.F.: Faster Evaluation of Multidimensional Integrals. Computers in Physics, November (1997) 574–578.
11. Halton, J.H.: On the efficiency of certain quasi-random sequences of points in evaluating multi-dimensional integrals. Numer. Math., series 2 (1960) 84–90.
12. Karaivanova, A., Georgieva, R.: Solving systems of LAEs using quasirandom numbers. In: Margenov, S., Waśniewski, J., Yalamov, P. (eds.): Large-Scale Scientific Computing. Lecture Notes in Computer Science, Vol. 2179, Springer-Verlag (2001) 166–174.

On Some Weight Monte Carlo Methods for Investigating the Asymptotical Behavior of Solution of the Transfer Equation

Galiya Z. Lotova

Institute of Computational Mathematics and Mathematical Geophysics,
Prosp. Lavrentieva 6, Novosibirsk, 630090, Russia
lot@osmf.sscc.ru

Abstract. In this paper we consider local Monte Carlo estimates for solving transfer equation with a nonstationary particle source. The comparison of local and double-local estimates is presented.

We consider the nonstationary integro-differential transfer equation [1]:

$$\frac{1}{v}\cdot\frac{\partial\Phi(r,\omega,t)}{\partial t}+\omega\,\nabla\Phi(r,\omega,t)+\sigma(r)\Phi(r,\omega,t)=$$

$$=\sigma_s(r)\int\limits_{\Omega} w_s(\omega,\omega';r)\,\Phi(r,\omega',t)\,d\omega'+f(r,\omega,t),$$

where

r is the vector of coordinates;
ω is the unit vector of velocity direction;
Ω is the space of unit vectors ω;
t is the time;
v is the absolute value of velocity;
$\sigma(r)=\sigma_s(r)+\sigma_c(r)$ is the extinction coefficient;
$\sigma_c(r)$, $\sigma_s(r)$ are coefficients of absorption and scattering, respectively;
$\Phi(r,\omega,t)$ is radiation intensity;
$w_s(\omega,\omega';r)$ is the indicatrix of scattering;
$f(r,\ \omega,t)$ is the distribution density of a particle source.

Note that parameters of the medium do not depends on time, and the radiation source is nonstationary.

We calculate the functionals

$$J=\int\limits_{D}\int\limits_{\Omega}\int\limits_{T}\Phi(r,\omega,t)h(r,\omega,t)\,dr\,d\omega\,dt$$

by a Monte Carlo method. Here D is a subset of coordinate space, $r\in D$, $t\in T$, h is the weight function of the form $h(r,\omega,t)=\delta(t-t^*)\delta(r-r^*)\Delta_{\Omega_0}(\omega)$, where

I. Dimov et al. (Eds.): NMA 2002, LNCS 2542, pp. 117–122, 2003.

t^* is a given time, r^* is a given point of space, $\Delta_{\Omega_0}(\omega)$ is the indicator function of the domain Ω_0.

These functionals are usually calculated by local and double-local estimators [1]. The local estimator can be represented by the formulae

$$J = \mathbf{E}\xi_1, \qquad \xi_1 = \sum_{n=0}^{N} Q_n h_1(x_n)\,\delta\left(t_n + \frac{|r_n - r^*|}{v} - t\right), \tag{1}$$

where $x_n = (r_n, \omega_n, t_n)$; r_n, ω_n and t_n are the point, velocity direction and time in the nth state of the Markov chain,

$$h_1(x_n) = \frac{\sigma_s(r_n)}{\sigma(r_n)}\,\frac{w_s(\omega_n, \omega_n^*; r_n)}{|r^* - r_n|^2}\sigma(r^*)\exp(-\tau_{op}(\ell, r_n, \omega_n^*))\Delta_{\Omega_0}(\omega_n^*),$$

$\ell = |r^* - r_n|$, $\omega_n^* = (r^* - r_n)/|r^* - r_n|$, $\tau_{op}(\ell, r_n, \omega) = \int\limits_0^{\ell} \sigma(r_n + s\omega)\,ds$ is the optical length of the particle path, and Q_n are special weights [1].

Let us consider the double-local estimator of the Monte Carlo method:

$$J = \mathbf{E}\xi_2, \qquad \xi_2 = \sum_{n=0}^{N} Q_n h_2(x_n)\,\delta\left(t_n + \frac{|r_n - r'|}{v} + \frac{|r' - r^*|}{v} - t\right), \tag{2}$$

$$h_2(x_n) = \frac{\sigma_s(r_n)}{\sigma(r_n)}\,\frac{w_s(\omega_n, \omega'; r_n)}{|r' - r_n|^2}\,w_s(\omega', \omega^*; r')\sigma(r^*)\exp(-\tau_{op}(\ell, r_n, \omega')).$$

Here $\omega' = (r_n - r')/|r_n - r'|$. The intermediate point r' is constructed by the following way. The direction ω^* from the point r^* and the randomized free path of the length ℓ^* are simulated. Hence $r' = r^* - \omega^*\ell^*$. Since estimators (1) and (2) contain delta-functions of time, they are not usable.

Let us apply a new modification of Monte Carlo estimators for nonstationary problems [2]. The corresponding estimators become

$$J(t) = \mathbf{E}\xi_t, \qquad \xi_t = \sum_{n=0}^{N} Q_n h_{1,2}(r_n, \omega_n) f(r_0, \omega_0, t - t_n)/f_0(r_0, \omega_0). \tag{3}$$

Here $h_{1,2}$ is equal to h_1 or h_2, $f(r, \omega, t)$ is the distribution density of particle source in the initial medium, $f_0(r, \omega)$ is the distribution density of the first point (r_0, ω_0) of trajectory in the model medium. Note that for this estimator a sufficiently smooth distribution density is needed [2].

It is known that for a semi-infinite homogeneous medium with scattering the function $J(t)$ has the following asymptotic with respect to time:

$$J(t) \sim At^{-\beta}\exp(-\sigma_c t) = \tilde{J}(t),$$

where A and β are constants. The calculation of the functional for large values of t is time-consuming. Hence, the coefficient β and time t_a from which the intensity is close to the asymptotic are usually obtained by Monte Carlo methods and then the second part of the function is replaced by the corresponding asymptotic.

1. Let us compare double-local and local estimators with modification by calculating the reflected light intensity in a simplified system "atmosphere-ocean". This system is determined as follows. The light scattering medium ("ocean") fills the half-space $z \leq 0$. The upper half-space $z > 0$ is vacuum.

Momentary (with the time dependence $\delta(t)$) source of the radiation is located at the point with coordinates $(0, 0, H)$ and emitted particles propagate in the direction $\omega = (0, 0, -1)$ (towards the negative semi-axis z). The detector registers the time distribution of radiation intensity at the same point $(0, 0, H)$. The cosine of scattering angle is simulated by the formulae that corresponds to well-known Henyey-Greenstein indicatrix [3]:

$$w_s(\omega', \omega; r) = \frac{1}{2} \frac{1 - \mu_0^2}{(1 + \mu_0^2 - 2\mu_0\mu)^{3/2}}, \quad \mu = (\omega', \omega), \quad \mu_0 = 0.9.$$

The calculations was made for the model with parameters $\sigma = 0.175\, m^{-1}$, $\sigma_s = 0.175\, m^{-1}$, $H = 5\, m$; the angle of the source was $\gamma = 0.001\, rad$, and $v = 0.225\, m/nsec$. Here $\sigma_c = 0$. Hence it is needed to finish the trajectory when the condition $L_n + \ell > vT_0$ for some T_0 holds, where L_n is the length of the particle path from the entrance to the "ocean".

Instead of the "pulsed" source, for estimating the point of transition to asymptotic we use the artificial source $f(r, v, t) = v^2 t \exp(-vt)$. It is clear that this method gives an overestimated (but practically satisfactory) value of the replacement point and does not change the type of the asymptotic.

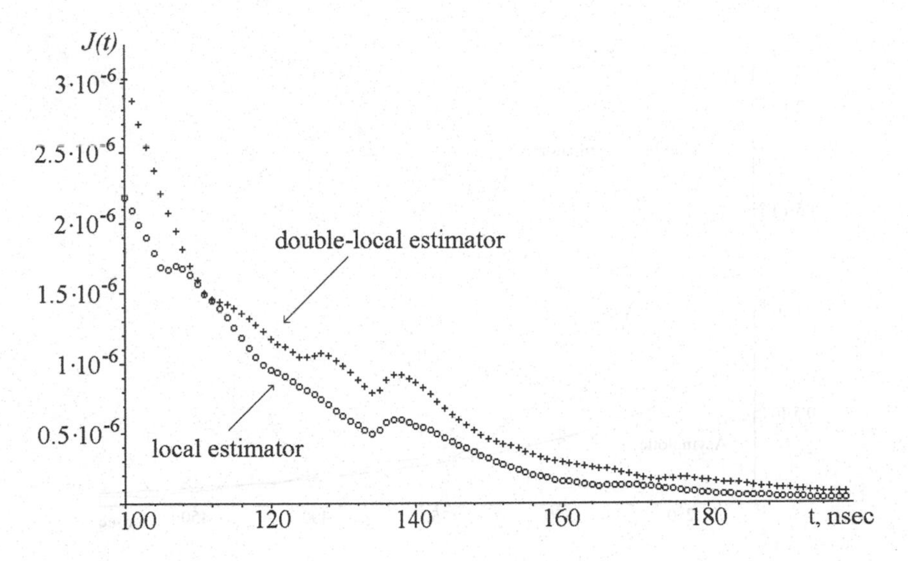

Fig. 1. Graphs of the intensity obtained by local ("circles") and double-local ("crosses") estimators.

The graphs of intensity are shown in Fig.1. They are obtained by local ("circles") and double-local ("crosses") estimators. The results are obtained with a moderate number of trajectories $M = 10^4$ and with the detector angle $\gamma_{pr} = 0.1 rad$. Here the double-local estimator gives a better result. The difference between the graphs increases when the angle decreases. The difference vanishes (up to statistical errors) when the number of trajectories grows. Our investigations show that the double-local estimator should be used when the number of trajectories is small and the detector angle is narrow. The smoother graphs can be obtained by choosing the optimal artificial source.

2. The double-local estimator described above has an infinite variance. Now we propose a combined estimator with a finite variance. The medium is divided into two parts. All vectors $r - r^*$ in the first part have the angle with axis z less than $\gamma_{pr} + \Delta\gamma$ for a small value $\Delta\gamma$. If the previous point is in the first part, then we calculate the local estimator. In the second part, we calculate the double-local estimator. Our investigations show that the value J_1 obtained in the first part is greater than the value J_2 obtained in the second one. The value $J_1/(J_1 + J_2)$ increases when the angle γ_{pr} decreases. The combined estimator contains advantages of both local estimators and can be used for calculating the time t_a.

It is known [1] that for intensity the coefficient β is equal to 2.5. The intensity obtained by the combined estimator and corresponding asymptotic are shown in Fig.2. Here we use 10^8 trajectories. It is seen that the time t_a is close to 350.

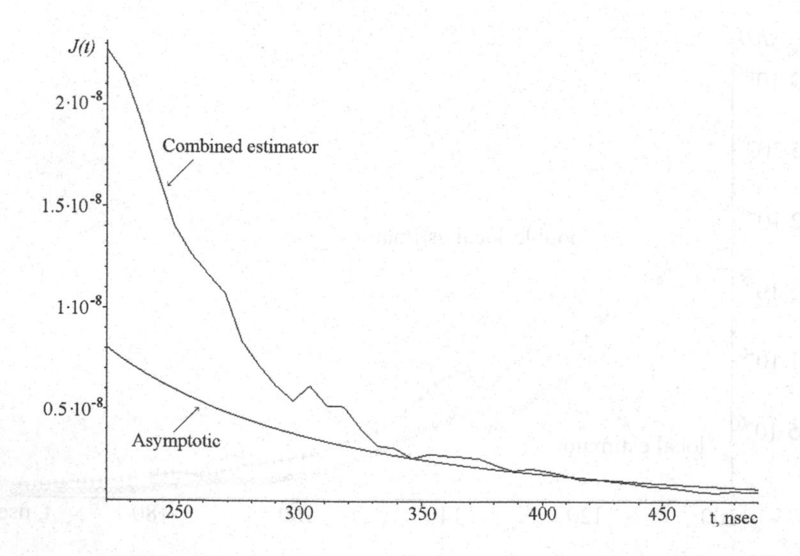

Fig. 2. The intensity $J(t)$ obtained by the combined estimator and the asymptotic for $\beta = 2.5$.

3. Let us use the modified estimator (3) to find derivatives of intensity with respect to time. The corresponding estimator takes the form

$$J^{(m)}(t) = \mathbf{E}\xi_t^{(m)}, \qquad \xi_t^{(m)} = \sum_{n=0}^{N} Q_n h_{1,2}(r_n, \omega_n) f^{(m)}(r_0, \omega_0, t - t_n)/f_0(r_0, \omega_0).$$

It is clear that

$$\frac{J'(t)}{J(t)} \approx -\frac{\beta}{t} - \sigma_c.$$

If we obtain two values of this quotient for different values of time then we can find the coefficient β. We obtain $\beta \approx 4$ for our example.

The intensity (for $2 \cdot 10^9$ trajectories) and the asymptotic are shown in Fig.3.

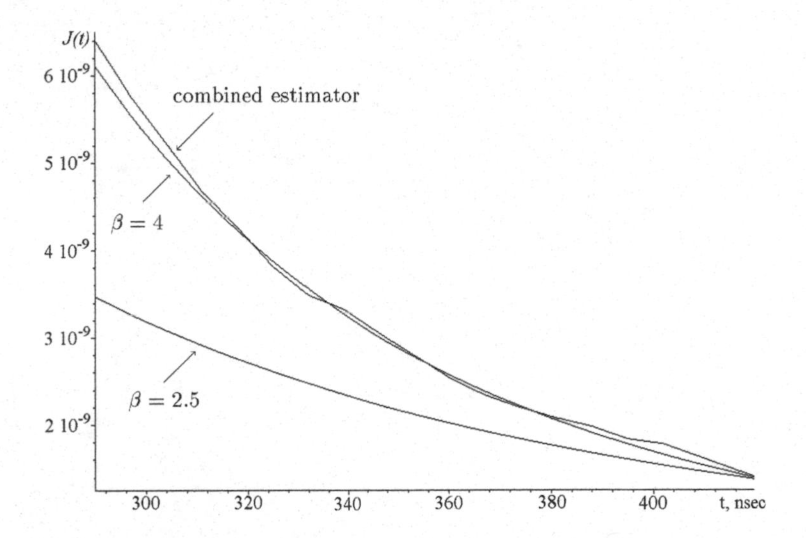

Fig. 3. The intensity $J(t)$ obtained by the combined estimator, the asymptotic for $\beta = 4$ and asymptotic for $\beta = 2.5$.

The modified estimator (3) allows us to find the intensity and its derivatives with respect to time by the same trajectories and calculate the coefficient of asymptotic.

The work was supported by INTAS (project 01- 0239) and Integration grant of SB RAS-2000 (project 43).

References

1. Marchuk, G.I., Mikhailov, G.A., Nazaraliev, M.A., et al.: The Monte Carlo Method in Atmospheric Optics. Springer, Berlin Heidelberg (1980)

2. Lotova, G.Z., Mikhailov, G.A.: New Monte Carlo methods for the solution of non-stationary problems in the radiation transfer theory. Russ. J. Numer. Anal. Math. Modelling, Vol. 15 **3-4** (2000) 285–295
3. Mikhailov, G.A., Voytishek, A.V.: Numerical constructing of special non-Gaussian fields in solving problems of the radiation transfer theory for stochastically inhomogeneous media. Russ. J. Numer. Anal. Math. Modelling, Vol.10 **3** (1995) 213–232

A Monte Carlo Approach for Finding More than One Eigenpair*

Michael Mascagni[1] and Aneta Karaivanova[1,2]

[1] Department of Computer Science and School of Computational Science and Information Technology,
Florida State University, Tallahassee, FL 32306, USA,
mascagni@cs.fsu.edu, http://www.cs.fsu.edu/~mascagni

[2] Bulgarian Academy of Sciences, Central Laboratory for Parallel Processing,
Acad G. Bonchev St., bl. 25A, 1113 Sofia, Bulgaria,
aneta@csit.fsu.edu, http://parallel.bas.bg/~anet/

Abstract. The Monte Carlo method has been successfully used for computing the extreme (largest and smallest in magnitude) eigenvalues of matrices. In this paper we study computing eigenvectors as well with the Monte Carlo approach. We propose and study a Monte Carlo method based on applying the ergodic theorem and compare the results with those produced by a Monte Carlo version of the power method. We also study the problem of computing more than one eigenpair combining our Monte Carlo method and deflation techniques.

1 Introduction

Many important problems in computational physics and chemistry can be reduced to the computation of dominant eigenvalues of matrices of high or infinite order (for example, quantum mechanical Hamiltonians, Markov matrices and transfer matrices, [11]). The analogy of the time-evolution operator in quantum mechanics, on the one hand, and the transfer matrix and the Markov matrix in statistical mechanics, on the other, allows these two fields to share numerical techniques. Specifically, a transfer matrix, G, of a statistical-mechanical system in d dimensions often can be interpreted as the evolution operator in discrete, imaginary time, t, of a quantum-mechanical analog in $d-1$ dimensions. That is, $G \approx exp(-tH)$, where H is the Hamiltonian of a system in $d-1$ dimensions, the quantum mechanical analog of the statistical-mechanical system. From this point of view, the computation of the partition function in statistical mechanics, and of the ground-state energy in quantum mechanics are essentially the same problems: finding the largest eigenvalue of G and $exp(-tH)$, respectively. Another issue is the computation of the relaxation time of a system with stochastic dynamics. This problem is equivalent to the computation of the second largest eigenvalue of the Markov matrix.

* Supported, in part, by the U.S. Army Research Office under Contract # DAAD19-01-1-0675

I. Dimov et al. (Eds.): NMA 2002, LNCS 2542, pp. 123–131, 2003.

Another important problem is estimating the Fiedler vector, [13], which is the eigenvector corresponding to the second smallest eigenvalue of a matrix. This task is the most computationally intensive component of several applications, such as graph partitioning, graph coloring, envelope reduction, and seriation. Some difficulties in applying Monte Carlo approach to this problem have been reported in [13]

These numerous examples which require the calculation of more than just the dominant eigenpair have motivated us to study the problem of finding one or more eigenvectors of a matrix via Monte Carlo. The problem of using Monte Carlo and quasi-Monte Carlo methods for finding an extremal eigenvalue has been extensively studied, for example, [7,3,4,8,9]. In this paper, we study the problem how to accurately find both the dominant eigenvector and the second largest eigenpair using deflation techniques coupled with Monte Carlo or quasi-Monte Carlo power iterations. Due to space considerations we do not describe here the quasirandom sequences used as this can be found in many articles and books, for example, [2,10].

2 The Problem and the Method of Solution

Consider the eigenvalue problem for determining complex number-vector pairs, (λ, x), for which the following matrix equation

$$Ax = \lambda x,$$

has a non-trivial solution. One (extremal) solution depends upon the convergence, for almost all choices of initial values, x_0, on the sequence

$$x^{(k)} = Ax^{(k-1)}/\lambda_k,$$

where λ_k is chosen for normalization, i. e., so that $||x^{(k)}|| = 1$ in some vector norm, [1]. Then the λ_k converge to the dominant (largest) eigenvalue, λ_1, of A, and $x^{(k)}$ converges to the corresponding eigenvector. We suppose that A is $n \times n$ and that its n eigenvalues are ordered as follows $|\lambda_1| > |\lambda_2| \geq \ldots \geq |\lambda_{n-1}| \geq |\lambda_n|$. Choosing the initial vector, f, and a vector h (both of dimension n), the construction of the desired Monte Carlo Method (MCM) begins with defining a Markov chain $k_0 \to k_1 \to \ldots \to k_i$, of the natural numbers, $k_j \in \{1, 2, \ldots, n\}$ for $j = 1, \ldots, i$. The k_j's defining the Markov chain can be thought of as a random walk on the n dimensions in the vector space we find ourselves. We then define an initial density vector, $p = \{p_\alpha\}_{\alpha=1}^n$, to be permissible to the vector h and a transition density matrix, $P = \{p_{\alpha\beta}\}_{\alpha\beta=1}^n$, to be permissible to A, [3]. We then define the following random variable on the given Markov chain:

$$W_0 = \frac{h_{k_0}}{p_{k_0}}, \quad W_j = W_{j-1}\frac{a_{k_{j-1}k_j}}{p_{k_{j-1}k_j}}, \quad j = 1, \ldots, i. \tag{1}$$

The Monte Carlo method for estimating the eigenvector that corresponds to the dominant eigenvalue is based on the following expected value identity[12,3]:

$$(h, A^i f) = E[W_i f_{k_i}], \quad i = 1, 2, \ldots.$$

By setting $h = e(r) = (0, \ldots, 0, \underbrace{1}_{r}, 0, \ldots, 0)^T$, the rth canonical unit vector, for $r = 1, \ldots, n$ the above random variable has expected value equal to the r-th component of the dominant eigenvector. The Monte Carlo estimate for the dominant eigenvalue is based on the fact that [14,3]:

$$\lambda_{max} \approx \frac{E[W_i f_{k_i}]}{E[W_{i-1} f_{k_{i-1}}]}. \tag{2}$$

2.1 Convergence

The fact that the system of the eigenvectors $\{v^{(1)}, v^{(2)}, \ldots, v^{(n)}\}$ is linearly independent and spans $\mathbb{R}^n$ implies that any vector, $x^{(0)}$, can be presented as a linear combination of them as follows

$$x^{(0)} = \sum_{j=1}^{n} \alpha_j v^{(j)}.$$

If we repeatedly apply A to $x^{(0)}$ we get

$$A^k x^{(0)} = \lambda_1^k \sum_{j=1}^{n} \alpha_j (\frac{\lambda_j}{\lambda_1})^k v^{(j)}.$$

Since $|\lambda_1| > |\lambda_j|$ for all $j = 2, 3, \ldots, n$, we have $\lim_{k\to\infty} (\lambda_j/\lambda_1)^k = 0$, and so

$$\lim A^k x^{(0)} = \lim_{k\to\infty} \lambda_1^k \alpha_1 v^{(1)}. \tag{3}$$

Obviously, this sequence converges to zero if $|\lambda_1| < 1$, and diverges if $|\lambda_1| \geq 1$, provided, of course, that $\alpha_1 \neq 1$. Advantage can be made of the relationship (3) by scaling the powers of $A^k x^{(0)}$ appropriately to ensure that the limit in (3) is finite and nonzero. The convergence of this process, and therefore of the Monte Carlo process as well, is most rapid if $x^{(0)}$ is close to $v^{(1)}$. We may be able to enhance the probability of this occurring by replacing x^k by $x^{(k)}/\sum_1^n x_i^{(k)}$ after performing our sampling. This, in effect, gives new weights to the same set of Markov-chain paths, which we hope will give better estimates for the eigenpair.

The power method has rate of convergence of $O((\frac{\lambda_2}{\lambda_1})^k)$. When we use the power method with Monte Carlo iterations we have additional error from the approximate, probabilistic, computation of matrix powers. This error is stochastic, and thus the uncertainty in this average taken from N samples is $O(N^{-1/2})$, by virtue of the central limit theorem. One generic approach to improving the convergence of MCMs has been the use of highly uniform, quasirandom, numbers (QRNs) in place of the usual pseudorandom numbers (PRNs). While PRNs are constructed to mimic the behavior of truly random numbers, QRNs are constructed to be distributed as evenly as mathematically possible. Quasi-MCMs use quasirandom sequences, which are deterministic, and may have correlations between points, and they were designed primarily for integration. For example, with QRNs, the convergence of numerical integration can sometimes be improved to as much as $O(N^{-1})$!

Since we wish to discuss both MCMs and and quasi-MCMs for these problems, it is timely to recall some pertinent estimates. Using MCMs, [3,4], we have

$$|h^T A^i f - \frac{1}{N}\sum_{s=1}^{N}(\theta)_s| \approx Var(\theta)^{1/2}N^{-1/2},$$

where $Var(\theta) = \{(E[\theta])^2 - E[\theta^2]\}$. In addition, we have

$$E[\theta] = E[\frac{h_{k_0}}{p_{k_0}} W_i f_{k_i}] = \sum_{k_0=1}^{n} \frac{h_{k_0}}{p_{k_0}} p_{k_0} \sum_{k_1=1}^{n} \dots \sum_{k_i=1}^{n} \frac{a_{k_0 k_1} \dots a_{k_{i-1} k_i}}{p_{k_0 k_1} \dots p_{k_{i-1} k_i}} p_{k_0 k_1} \dots p_{k_{i-1} k_i}.$$

Using quasi-MCMs, [8,9], we obtain:

$$|h_N^T A_N^i f_N - \frac{1}{N}\sum_{s=1}^{N} h(x_1^{(s)}) a(x_1^{(s)}, x_2^{(s)}) \dots a(x_i^{(s)}, x_{i+1}^{(s)}) f(x_{i+1}^{(s)})| \leq |h|^T |A|^i |f| D_N^*, \tag{4}$$

where the $i+1$-dimensional quasirandom sequence $\{(x_1^{(s)}, x_2^{(s)}, \dots, x_{i+1}^{(s)})\}$, has a star discrepancy of D_N^*.

Let us compare MCM and QMCM errors for computing $(h, A^m f)$: Both are products of two factors (first depends on A, second - on the sequence). The order is $N^{-1/2}$ for MCM and $(log^{m+1}N)N^{-1}$ for QMCM. Moreover, the MCM error is a probabilistic error bound while the QMCM error is the worst-case bound (inequality). In the same time, computational complexity for MCM and QMCM is the same: $O((m+1)N)$, where N is the number of chains, $m+1$ is the length of a single Markov chain.

Then, the convergence rates for power method with MCM or QMCM iterations are:

$$O(\|\frac{\lambda_2}{\lambda_3}\|^m + \sigma N^{-1/2})$$

$$O(\|\frac{\lambda_2}{\lambda_3}\|^m + (log^m N)N^{-1})$$

2.2 The Complex Conjugate Eigenvalue Pair Case

The above approach can be as is used when the dominant eigenvalue is real, a singleton, and separated from the second eigenvalue. However, the method can also be extended to estimate also a complex conjugate pair of eigenvalues, $\lambda_1 = a + bi$ and $\lambda_2 = a - bi$, and their corresponding eigenvectors. These eigenvectors are approximately equal to

$$bx^{k-1} \pm i(ax^{k-1} - x^k),$$

provided the power iterations, $x^k = Ax^{k-1}$, are real. Unfortunately, the eigenvector estimate is inaccurate when the conjugate pair is close to the real axis, i. e., when b is small.[1]
What is the Monte Carlo procedure in this case? It can be shown ([6]) that λ_1 and λ_2 are the roots of the quadratic equation $\lambda^2 + c_1\lambda + c_2 = 0$. In addition, it can be shown that the coefficients, c_2, c_1, and 1 have the property that for any three consecutive iterations x_k, x_{k+1} and x_{k+2} in the power method, they are the coefficients of an approximate linear relation, i. e. , $x_{k+2} + c_1 x_{k+1} + c_2 x_k \approx 0$.

Let $h \in \mathbb{R}^{1\times n}$ be an n-dimensional vector. One can prove that the above system is equivalent to the following one, [4]:

$$d_1(h, Ax_k) + d_2(h, x_k) = -(h, A^2 x_k).$$

Suppose, we have the values of four MC iterations:

$$(h, x_{k-3});\ (h, Ax_{k-3}) = (h, x_{k-2});$$

$$(h, A^2 x_{k-3}) = (h, x_{k-1});\ (h, A^3 x_{k-3}) = (h, x_k).$$

One can prove that the dominant complex conjugate eigenvalues, λ_1 and λ_2, of the matrix A are the solution of the following quadratic equation: $\lambda^2 + d_1\lambda + d_2 = 0$, where

$$d_1 = \frac{(h, x_{k-1})(h, x_{k-2}) - (h, x_k)(h, x_{k-3})}{(h, x_{k-1})(h, x_{k-3}) - (h, x_{k-2})^2} \tag{5}$$

and

$$d_2 = \frac{(h, x_k)(h, x_{k-2}) - (h, x_{k-1})^2}{(h, x_{k-1})(h, x_{k-3}) - (h, x_{k-2})^2}. \tag{6}$$

These expressions give us a Monte Carlo procedure to estimate the complex conjugate eigenvalues and their corresponding eigenvectors.

2.3 A Fast and Rough Estimate of the Eigenvector

By using the ergodic theorem, [5], we can estimate the first eigenvector very fast. In this case we use a single, very long, Markov chain:

$$EigV = lim_{k\to\infty} A^k x^{(0)}.$$

[1] It is obvious, that as b goes to zero, $\lambda_1 = a + bi$ and $\lambda_2 = a - bi$ both approach a and the two eigenvalues coalesce into a single eigenvalue of multiplicity 2.

This approach can be used only when the matrix is a stochastic matrix with principal eigenvalue equal to 1. Such a matrix describes a Markov chain where all states communicate.[2] Our test matrices are not stochastic, but still a rough estimate for the first eigenvector can be computing using the matrix $\frac{1}{\lambda_{max}}A$. The numerical results are given in the Table 2.

2.4 Numerical Tests

We performed many numerical tests using PRNs, and Soboĺ, Halton and Faure quasirandom sequences. Some of the results for the L_2-norm of the error: $err = \sqrt{\sum_{i=1}^{n}(x^k(i) - EigV(i))^2}$, where $EigV$ is the exact eigenvector and x^k is the our approximation after k iterations, are presented in the Tables 1 and 2.

Table 1. Accuracy in computing the first eigenvector of sparse matrices of order n using N_w random or quasirandom walks of length k

n	N_w	k	URAND	SOBOĹ	FAURE	HALTON
128	1280	5	0.282e-01	0.642e-02	0.632e-02	0.496e-02
1024	10240	6	0.375e-02	0.958e-03	0.837e-03	0.778e-03
2000	20000	6	0.158e-02	0.153e-02	0.158e-02	0.156e-02

Table 2. Accuracy in computing the first eigenvector of sparse matrices of order n using one long random or quasirandom walk

n	URAND	SOBOĹ	FAURE	HALTON
128	0.52	0.22	1.32	0.41
1024	0.67	0.14	1.22	0.17
2000	0.86	0.003	1.32	0.003

For our numerical tests we use randomly generated sparse matrices of order n = 128, 1024 and 2000. With $N_w = 10n$ walks we achieve sufficiently good accuracy in estimating the eigenvector using the power method with MCM or quasi-MCM iterations. For large n, this makes the power MCM and the power quasi-MCMs computationally very efficient. We should recall that the computational complexity of both methods is kN_w, where k is the length of the Markov chains (the power in the power method) and N_w is the number of the walks. It

[2] For example, if the matrix has a block structure, it does not satisfy these communication requirements.

should also be noted that the results using QRNs are better than with PRNs, as we hoped; in order to compare the Monte Carlo and the quasi-Monte Carlo approach, we implemented them using the same number and length of walks for both of the methods, and the quasi-Monte Carlo gives better accuracy.

Let us also note, that when we compute the dominant eigenvalue we need to perform many more walks to achieve an accuracy similar to that in case we compute the dominant eigenvector (see the results given in the Table 3).

The Table 2 shows the results using the ergodic theorem - they confirm the theoretical assumption for giving us a rough approximation.

Table 3. Accuracy in computing the dominant eigenvalue of a sparse matrix of order n using N_w random or quasirandom walks of length k

n	N_w	k	URAND	SOBOĹ	FAURE	HALTON
128	12800	5	0.42*e*-01	0.36*e*-02	0.14*e*-01	0.18*e*-01
1024	20480	6	0.37*e*-02	0.23*e*-02	0.57*e*-02	0.43*e*-02
2000	50000	6	0.17	0.15	0.14	0.17

3 Finding More Eigenpairs

Once having the eigenpair, (λ_1, v_1), we deflate the computed eigenvalue from A by constructing a smaller matrix, B. The deflated matrix, B, has one less row and column than A, and the eigenvalues of B are the same as those of A, except that the previously computed λ_1 is missing from B's spectrum.

The deflated matrix is constructed using a Householder matrix, $H = I - 2ww^T$, where w is an n-dimensional vector with components, [6],

$$w(1) = \frac{1}{\sqrt{2s(s+|v_1(1)|)}}(v_1(1) + sign(v_1(1))s),$$

and

$$w(i) = \frac{1}{\sqrt{2s(s+|v_1(1)|)}}v_1(i), \ i = 2, \ldots, n,$$

where $s = (\sum_{i=1}^{n} v_1(i)^2)^{1/2}$. Then ensures that the matrix HAH has the form

$$HAH = \begin{pmatrix} \lambda_1 & b^T \\ 0 & B \end{pmatrix},$$

with eigenpairs (λ, Hv), where λ is an eigenvalue of A and v is the corresponding eigenvector of A. The characteristic polynomial for HAH is: $\det(HAH - \lambda I) = (\lambda_1 - \lambda)\det(B - \lambda I)$. One zero of the characteristic polynomial for HAH is

$\lambda = \lambda_1$, while the remaining zeros are the roots of the equation $det(B - \lambda I) = 0$, i. e., the eigenvalues of B. Hence, the eigenvalues of B are the same as the eigenvalues of A except that λ_1 is missing. Now we can apply power iterations to estimate λ_2 using B.

Our goal is to study the applicability of MCMs and quasi-MCMs for finding the second eigenpair. We need to understand how the use of the approximate eigenvector (estimated by MC or by QMC iterations) will "spoil" the deflated matrix, and how spectrum of B will be changed. Generally speaking, the answer depends on what type of matrix A is, and on how accurate its first eigenvector is computed. If we perform exactly the above procedure, we will obtain the deflated matrix with the accuracy of the approximate eigenvector. Then we again apply the power method with MC or quasi-MC iterations. It is important to note that this second iteration has a rate of convergence depending additionally on $\frac{\lambda_2}{\lambda_3}$.

3.1 Numerical Tests

In our numerical tests, we use a matrix with size $n = 128$ which produced estimates for the first eigenpair with the worst accuracy among the test matrices. We obtained the deflated matrix using the eigenvector computed using PRNs and the Soboĺ sequence. The "approximate" deflated matrix has the same spectral radius as the exactly deflated matrix, and the eigenvalues very close to the exact ones as well. The expected rate of convergence for the power method applied to the deflated matrix for estimating λ_2 is $\frac{\lambda_3}{\lambda_2} = \frac{13}{16}$. The results using PRNs and Soboĺ sequence are presented in the Table 4. We have tabulated the relative error for computing λ_2 and the L_2-error in the eigenvector. Again, the quasi-MCM shows slightly better results than the MCM, but the best results are obtained when we combine results using both types (pseudorandom and quasirandom) of sequences.

4 Conclusions and Future Plans

We have shown that Monte Carlo approach can be used for finding more than one eigenpair of a matrix. We tested the method by finding the first- and second-largest eigenvalues and their eigenvectors using PRNs and quasirandom sequences. In all of our numerical experiments, the quasi-MCM gives better results than the MCM. Given these results, we now have to study the use of Monte

Table 4. The error in computing the second eigenpair of a matrix of order 128 using PRNs, Soboĺ, and both sequences

	Error(λ_2)	Error($v^{(2)}$)
PRNs	0.22	0.3571
Soboĺ	0.022	0.3555
PRNs+Sob	0.007	0.3482

Carlo on the analogous problem at the "other end of the spectrum," i. e., for estimating the second smallest eigenpair based on prior knowledge of the smallest eigenpair. This problem can be solved in a similar way; however, using the resolvent Monte Carlo method on a deflated matrix. However, it is well known that resolvent-type algorithms have more restrictions on applicability and have worse rate of convergence than the power-type algorithms studies here, so a careful study should be undertaken.

References

1. Burden, R. and Faires, J. D.: Numerical Anaysis. PWS Publ. Company, Boston (1997)
2. Caflisch, R. E.: Monte Carlo and quasi-Monte Carlo methods. Acta Numerica, 7 (1998) 1–49.
3. Dimov, I. and Karaivanova, A.: Parallel computations of eigenvalues based on a Monte Carlo approach. Journal of Monte Carlo Methods and Applications, Vol. 4,.1 (1998) 33–52.
4. Dimov, I. and Karaivanova, A.: A Power Method with Monte Carlo iterations. In: Iliev, O., Kaschiev, M., Margenov, S., Sendov, Bl., Vassilevski, P. (eds.): Recent Advances in Numerical Methods and Applications II. World Scientific, Singapore (1999) 239–247.
5. Feller, W.: An introduction to probability theory and its applications. Vol. 1, Wiley, New York (1968)
6. Hager, William W.: Applied Numerical Linear Algebra. Prentice Hall, London (1989)
7. Hammersley, J. M. and Handscomb, D. C.: Monte Carlo Methods. Methuen, London (1964)
8. Mascagni, M. and Karaivanova, A.: Matrix Computations Using Quasirandom Sequences. In: L.Vulkov, Waśniewski, J., Yalamov, P. (eds.): Numerical Analysis and its Applications. Lecture Notes in Computer Science, Vol. 1988, Springer-Verlag (2001) 552–559.
9. Mascagni, M. and Karaivanova, A.: A Parallel Quasi-Monte Carlo Method for Computing Extremal Eigenvalues. Monte Carlo and Quasi-Monte Carlo Methods 2000, Lecture Notes in Statistics, Springer-Verlag (2002) 369–380.
10. Niederreiter, H.: Random number generation and quasi-Monte Carlo methods. SIAM, Philadelphia (1992)
11. Nightingale, M. P., Umrigar, C. J.: Monte Carlo Eigenvalue Methods in Quantum Mechanics and Statistical Mechanics. In: Ferguson, D. M., Siepmann, J. Ilja, and Truhlar, D. G. (eds.): Monte Carlo Methods in Chemistry. Advances in Chemical Physics, Vol. 105, Wiley, NY, in press.
12. Soboĺ, I. M.: Monte Carlo numerical methods. Nauka, Moscow (1973)
13. Srinivasan, A., Mascagni, M.: Monte Carlo Techniques for Estimating the Fiedler Vector in Graph Applications. Lecture Notes in Computer Sccience, Vol. 2330, 2, Springer-Verlag (2002) 635–645.
14. Vladimirov, V. S.: On the application of the Monte Carlo method to the finding of the least eigenvalue, and the corresponding eigenfunction of a linear integral equation. in Russian: Teoriya Veroyatnostej i Yeye Primenenie, 1, No 1 (1956) 113–130.

A New Class of Grid-Free Monte Carlo Algorithms for Elliptic Boundary Value Problems*

R.J. Papancheva, I.T. Dimov, and T.V. Gurov

Central Laboratory for Parallel Processing, Bulgarian Academy of Sciences,
Acad. G. Bonchev St., bl. 25A, 1113 Sofia, Bulgaria.
rumi@cantor.bas.bg, http://www.bas.bg/dpa

Abstract. In this paper we consider the following mathematical model: an elliptic boundary value problem, where the partial differential equation contains advection, diffusion, and deposition parts. A Monte Carlo (MC) method to solve this equation uses a local integral representation by the Green's function and a random process called "Walks on Balls"(WOB). A new class of grid free MC algorithms for solving the above elliptic boundary value problem is suggested and studied. We prove that the integral transformation kernel can be taken as a transition density function in the Markov chain in the case when the deposition part is equal to zero. An acceptance-rejection (AR) and an inverse-transformation methods are used to sample the next point in the Markov chain. An estimate for the efficiency of the AR method is obtained.

1 Formulation of the Problem

Consider the functional

$$J(u) \equiv (g, u) = \int_{\Omega} g(x)u(x)dx, \tag{1}$$

where $\Omega \subset R^3$ and $x = (x_1, x_2, x_3) \in \Omega$ is a point in the Euclidean space R^3. The functions $u(x)$ and $g(x)$ belong to the Banach space X and to the adjoint space X^*, respectively, and $u(x)$ is a unique solution of the following Fredholm integral equation:

$$u(x) = \int_{\Omega} k(x, y)u(y)dy + f(x). \tag{2}$$

The main task is to calculate the functional (1), where $u(x)$ is the solution of the following boundary value problem:

$$Mu = -\Phi(x), \qquad x \in \Omega, \quad \Omega \subset R^3, \tag{3}$$

* Supported by Center of Excellence BIS-21 grant ICA1-2000-70016 and by the NSF of Bulgaria under Grants I-1201/02 and MM-902/99

I. Dimov et al. (Eds.): NMA 2002, LNCS 2542, pp. 132–139, 2003.

$$u = \psi(x), \qquad x \in \partial\Omega, \tag{4}$$

where the operator M is defined by: $M = \sum_{i=1}^{3}\left(\frac{\partial^2}{\partial x_i^2} + b_i(x)\frac{\partial}{\partial x_i}\right) + c(x)$.

As shown in [1,7], if the coefficients of the operator M satisfy the conditions: $b_i(x),\, c(x) \in C^{(0,\lambda)}(\overline{\Omega})$, $c(x) \le 0$, $\Phi \in C^{(0,\lambda)}(\Omega)\bigcap C(\overline{\Omega})$ and $\psi \in C(\partial\Omega)$ in the closed domain $\overline{\Omega} \in A^{(1,\lambda)}$, then the problem (3) - (4) has an unique solution $u(x) \in C^2(\Omega)\bigcap C(\overline{\Omega})$. A definition for the class $A^{(k,\lambda)}$ can be found in [6].

We denote by $B(x)$ the maximal ball inside the domain Ω with radius $R(x)$ and center in the point x, i.e.:

$$B(x) = B_{R(x)} = \{y : r = |y - x| \le R(x)\}. \tag{5}$$

Levy's function for the problem (3)-(4) [8] is:

$$L_p(y,x) = \mu_p(R)\int_r^R \left(\frac{1}{r} - \frac{1}{\rho}\right) p(\rho)d\rho, \quad r \le R, \tag{6}$$

where $p(r)$ is a density function and the following notation is used:

$$r = |x-y| = \left(\sum_{i=1}^{3}(x_i - y_i)^2\right)^{1/2}, \quad \mu_p(R) = (4\pi q_p(R))^{-1}, \quad q_p(R) = \int_0^R p(\rho)d\rho.$$

The components of the vector-function $\mathbf{b}(x)$ are assumed to satisfy the conditions: $b_i(x) \in C^{(1)}(\Omega), i = 1,2,3$ and $div\mathbf{b}(x) = 0$. Then a local integral representation of the solution by the use of the Green's function approach exists for standard domains, lying inside Ω. In addition, taking in account that $\frac{\partial L_p(y,x)}{\partial y_i} = \mu_p(R)\frac{x_i - y_i}{r^3}\int_r^R p(\rho)d\rho$, one can see that the Levy's function satisfies the conditions:

$$\frac{\partial L_p(y,x)}{\partial y_i} = L_p(y,x) = 0 \quad \text{for any} \quad \mathrm{y} \in \partial\Omega. \tag{7}$$

Finally, the solution of the problem (3)-(4) can be written in the following integral form [4,8]:

$$u(x) = \int_{B(x)} M_y^* L_p(y,x)u(y)dy + \int_{B(x)} L_p(y,x)\Phi(y)dy, \tag{8}$$

where $M^* = \sum_{i=1}^{3}\left(\frac{\partial^2}{\partial x_i^2} - b_i(x)\frac{\partial}{\partial x_i}\right) + c(x)$ is the adjoint operator to M and

$$M_y^* L_p(y,x) = \mu_p(R)\frac{p(r)}{r^2} - \mu_p(R)c(y)\int_r^R \frac{p(\rho)}{\rho}d\rho$$

$$+\frac{\mu_p(R)}{r^2}\left[c(y)r + \sum_{i=1}^{3} b_i(y)\frac{y_i - x_i}{r}\right]\int_r^R p(\rho)d\rho. \tag{9}$$

The MC method, that solves our problem uses the local integral representation (8) and a WOB random process.

2 Monte Carlo Algorithms

The Monte Carlo estimator with mathematical expectation equal to $J(u)$ is

$$\Theta[g] = \frac{g(\xi_0)}{\pi(\xi_0)} \sum_{j=0}^{\infty} Q_j f(\xi_i), \tag{10}$$

where $Q_0 = 1, Q_j = Q_{j-1} \frac{k(\xi_{j-1},\xi_j)}{p(\xi_{j-1},\xi_j)}, j = 1,2,3,...$, and $\xi_0, \xi_1, ...$ is a Markov chain in Ω with initial density function $\pi(x)$ and transition densities $p(x,y)$, which are tolerant to $g(x)$ and $k(x,y)$, respectively (see [2,5,9]).

To ensure the convergence of the process, we introduce an ε - strip of the boundary. The process starts at point $\xi_0 = x \in \Omega$, which is chosen correspondingly with the initial density function $\pi(x)$. The next random point is determined by a transition density function $p(x,y)$. This process terminates when the point falls into the ε - strip of the boundary.

The kernel, $k(x,y) = M_y^* L_p(y,x)$, of the integral equation (8) can be used as a transition density in the Markov chain when it is non-negative. This condition is satisfied in the case when the density function $p(r) = e^{-kr}$ and $k = b^* + \overline{R}c^*$, where $b^* = \max_{x\in\Omega} \mid \mathbf{b}(x) \mid$, $c^* = \max_{x\in\Omega} \mid c(x) \mid$, and $\overline{R}$ is the radius of the maximal ball lying inside Ω [3,4].

Here we propose and study new MC algorithms, where the density function $p(r)$ depends only on the advection part $\mathbf{b}(x)$ of the elliptic equation.

The following assertion holds:

Theorem 1. *If for the function $p(r) \geq 0$ the inequality*

$$p(r) \geq b^* \int_r^R p(\rho)d\rho \tag{11}$$

is true, then the function

$$p(x,y) = \frac{\mu_p(R)}{r^2}\left[p(r) + \sum_{i=1}^{3} b_i(y)\frac{y_i - x_i}{r}\int_r^R p(\rho)d\rho\right] \tag{12}$$

can be used as a transition density for Markov process.

Proof. The function (12) is obtained from the kernel $k(x,y)$ when $c(y) \equiv 0$. It is already proved in [4] that $\int_{B(x)} p(x,y)dy = 1$. The remaining task is to find out when $p(x,y) \geq 0$.

By using spherical coordinates:

$$y_1 - x_1 = r\sin\theta\sin\varphi, \quad y_2 - x_2 = r\sin\theta\cos\varphi, \quad y_3 - x_3 = r\cos\theta,$$

and replacing $\omega_i = (y_i - x_i)/r$, the following equations hold:

$$\sum_{i=1}^{3} b_i(y)\frac{y_i - x_i}{r} = \sum_{i=1}^{3} b_i(x + r\mathbf{w})\omega_i = (\mathbf{b}, \mathbf{w}). \tag{13}$$

From the inequalities $|(\mathbf{b}, \mathbf{w})| \le |\mathbf{b}||\mathbf{w}| = |\mathbf{b}| \le b^*$ we obtain the estimate $\sum_{i=1}^{3} b_i(y)(y_i - x_i)/r \ge -b^*$, and finally:

$$p(x, y) \ge \frac{\mu_p(R)}{r^2}\left[p(r) - b^*\int_r^R p(\rho)d\rho\right]. \tag{14}$$

The proof is completed as (14) leads to (11). ◇

This theorem gives us the base for new class MC algorithms, where the choice of $p(r)$ depends only on b^* and does not depend on c^* and $\overline{R}$.

The function $p(x, y)$ can be written in spherical coordinates as:

$$p(r, \mathbf{w}) = \frac{\sin\theta}{4\pi}\frac{p(r)}{q_p(R)}\widetilde{p}(\mathbf{w}|r), \quad \widetilde{p}(\mathbf{w}|r) = 1 + \frac{|\mathbf{b}(x + r\mathbf{w})|\cos(\mathbf{b}, \mathbf{w})}{p(r)}\int_r^R p(\rho)d\rho.$$

Now, the next random point y in the Markov chain depends on the direction $\mathbf{w}$ and on the jump r that is made to an internal point into the maximal ball. First we sample the random jump r with density function $p(r)/q_p(r)$ using an inverse-transformation method. To obtain the random direction $\mathbf{w}$ with density function $p(\mathbf{w}|r) = \frac{\sin\theta}{4\pi}\widetilde{p}(\mathbf{w}|r)$ an acceptance-rejection (AR) method is used.

Since $\widetilde{p}(\mathbf{w}|r) \le \left[1 + \frac{b^*}{p(r)}\int_r^R p(\rho)d\rho\right] = h(r)$, the function $h(r)$ can be accepted as a majorant for the AR method.

Here, the algorithm for one random walk is described:

1. **Calculate** the radius $R(x)$.

2. **Sample** the jump r with density $\frac{p(r)}{q_p(R)}$.

3. **Calculate** the function $h(r)$.

4. **Compute** the independent realizations $\mathbf{w_j}$ of a unit isotropic vector in $\mathbf{R}^3$.

5. **Compute** the independent realizations γ_j of a uniformly distributed random variable in the interval $[0, 1]$.

6. **Repeat** the steps 4 and 5 until define the parameter j_0 from the condition: $j_0 = \min\{j : h(r)\gamma_j \le \widetilde{p}(\mathbf{w_j}|r)\}$. The random vector $\mathbf{w_{j_0}}$ has the density $p(\mathbf{w}|r)$.

7. **Calculate** the next random point y by formula $y = x + r\mathbf{w_{j_0}}$.

8. **Stop** the algorithm when the random process reaches the ε -strip of the boundary. If $y \notin \partial\Omega_\varepsilon$ then the algorithm has to be repeated for $x = y$.

We consider and analyze three possible alternatives for the density function $p(r)$:

$\bullet p(r) = e^{-b^* r}$ $\bullet p(r) = const.$ $\bullet p(r) = e^{b^* r}$

In the first case, when $p(r) = e^{-b^* r}$ we have:

$$b^* \int_r^R p(\rho) d\rho = e^{-b^* r} - e^{-b^* R} \le e^{-b^* r}.$$

Therefore the inequality (11) is always true. The following assertion holds:

Lemma 1. *If $p(r) = e^{-b^* r}$ then the function $p(x, y)$ can be used as a transition density.*

The advantage of such kind of choice is that it does not depend on any additional requirements to the parameters of the problem.

In the second case, when $p(r) = const.$, the inequality (11) is equivalent to: $b^*(R - r) \le 1$. From $0 \le r \le R$ follows: $b^*(R - r) \le b^* R \le b^* \overline{R}$. Thus we can formulate the conclusion:

Lemma 2. *If the parameters of the problem satisfy the inequality*

$$b^* \overline{R} \le 1 \tag{15}$$

and the density function $p(r)$ is chosen to be $p(r) = const.$, then the function $p(x, y)$ can be used as a transition density.

Because of the computational simplicity, one can expect that the use of a constant density leads to decreasing of the computational cost of the algorithm in comparison with the algorithms, using exponential densities.

In the third case, when $p(r) = e^{b^* r}$, the inequality (11) is true when: $b^*(R - r) \le \ln 2$. Finally, we reached to the assertion:

Lemma 3. *If the parameters of the problem satisfy the inequality*

$$b^* \overline{R} \le \ln 2 \tag{16}$$

and the density function $p(r)$ is chosen to be $p(r) = e^{b^* r}$, then the function $p(x, y)$ can be used as a transition density function in Markov process.

Since the function $e^{b^* r}$ is strictly increasing, the choice of a bigger jump r in the ball in the Markov chain is more probable. One can expect that the average moves (El_ε) in the WOB are less and the process is more efficient.

3 Estimates for the Efficiency of the AR Method

Let us estimate the efficiency of the AR method for sampling the vector $\mathbf{w}$ with density $p(\mathbf{w}|r)$.

In the case when $p(r) = e^{-b^* r}$ we can bound the function $h(r)$ more precisely:

$$h(r) = 2 - \frac{e^{b^* r}}{e^{b^* R}} \leq 2 - \frac{1}{e^{b^* R}} = H.$$

Now we can use the constant H as a majorant function for the AR algorithm.
The following formula gives us the efficiency of the AR method:

$$Eff_T = \int_0^{2\pi}\int_0^{\pi} p(\mathbf{w}|r) d\theta d\varphi \Big/ \int_0^{2\pi}\int_0^{\pi} \frac{sin(\theta)}{4\pi} H d\theta d\varphi$$

The majorant functions (H) and the theoretical estimates (Eff_T) for the AR efficiency in the all cases for $p(r)$ are given in Table 1. Taking in account that

Table 1. The majorant functions and the estimates for the AR efficiency.

$p(r)$	H	Eff_T
$e^{-b^* r}$	$2 - 1/e^{b^* R}$	$(2 - 1/e^{b^* R})^{-1}$
$const$	$1 + b^* R$	$(1 + b^* R)^{-1}$
$e^{b^* r}$	$e^{b^* R}$	$(e^{b^* R})^{-1}$

$0 < \frac{1}{e^{b^* R}} < 1$ and inequalities (15) and (16), for all three choices for the $p(r)$, the following estimate for the efficiency of the AR method is obtained:

$$Eff_T \geq 1/2.$$

4 Numerical Tests

As an example the following boundary value problem was solved using the MC algorithms under consideration:

$$\sum_{i=1}^{3} \left(\frac{\partial^2 u}{\partial x_i^2} + b_i(x) \frac{\partial u}{\partial x_i} \right) + c(x) u = 0 \ \text{ in } \ \Omega = [0,1]^3,$$

$$u(x_1, x_2, x_3) = e^{a_1 x_1 + a_2 x_2 + a_3 x_3}, \qquad x \in \partial\Omega_\varepsilon.$$

In our tests, we have: $b_1(x) = a_2 a_3 (x_2 - x_3)$, $b_2(x) = a_3 a_1 (x_3 - x_1)$, $b_3(x) = a_1 a_2 (x_1 - x_2)$, and $c(x) = -(a_1^2 + a_2^2 + a_3^2)$, where a_1, a_2, a_3 are the parameters.

It is easy to see that $div\mathbf{b}(x) = 0$. This condition guarantees the possibility to use the local integral representation by Green's function.

A large number of experiments are done that investigate the computational cost of the presented MC algorithms and the efficiency of the AR method for different values of the coefficients $a_i, i = 1, 2, 3$. Some of them that estimate

Table 2. $u(x)$=1.454991, a_1=a_2=a_3=0.25, N=5000, b^*=0.108, $\overline{R}$=0.5

$p(r)$	Eff_T	$\varepsilon = 0.01$		$\varepsilon = 0.05$	
		Eff_P	err	Eff_P	err
e^{-b^*r}	0.949949	0.987592	-0.003216	0.979795	-0.001380
$const$	0.948653	0.987223	0.001937	0.979358	0.000372
e^{b^*r}	0.947312	0.987142	0.003435	0.979248	0.003498

the solution at a point with coordinates (0.5, 0.5, 0.5) for two ε-strips, $\varepsilon = 0.01$ and $\varepsilon = 0.05$, are presented in Table 2. Here, $u(x) = 1.454991$ is the exact solution; err is the relative error; Eff_T is the value of the theoretical AR efficiency and Eff_P is the AR efficiency from our experiments. We see that the numerical tests for the AR efficiency confirm the theoretical results.

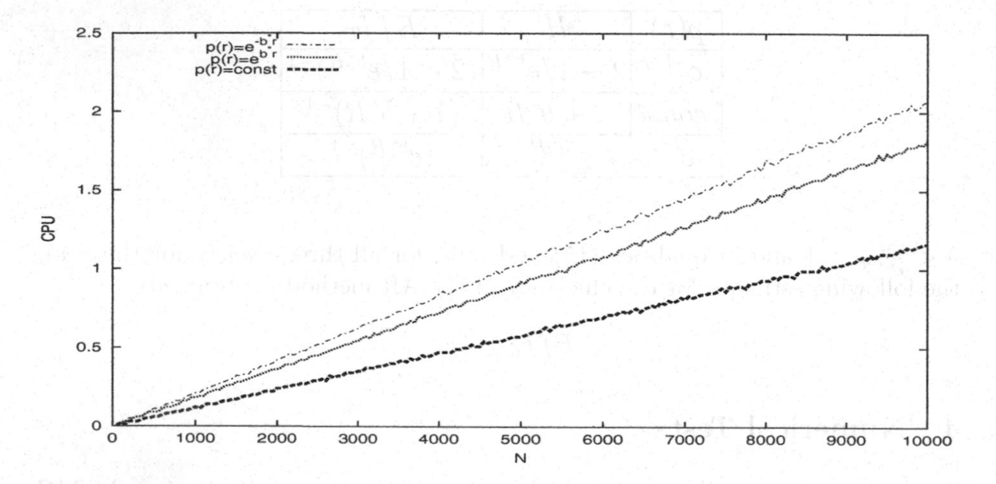

Fig. 1. The CPU times for all algorithms when $\varepsilon = 0.01$, $b^* = 1.732051$.

Figure 1 shows CPU times of the MC algorithms using the three different density functions. In the case, when $p(r) = const.$ the computational cost is less than other two cases for $p(r)$. When the function $p(r) = e^{b^*r}$, the CPU time is less than the case when $p(r) = e^{-b^*r}$. This can be explained with results for the average moves in the WOB that are presented in Table 3. Thus, the numerical results show that it is better to use both constant density and exponential density with positive degree if the parameters of the boundary value problem allow.

In conclusion a new class of grid-free MC algorithms have been studied for solving the elliptic boundary value problem under consideration. The density function $p(\rho)$ which is used in the definition of the Levy's function 6 have be chosen to depend only on the advection. This choice allows the integral transfor-

Table 3. The average number of moves in the WOB. $u(x) = 1.454991$, $\varepsilon = 0.01$

$p(r)$	El_ε		
	$a_i = 0.25,\ b^* = 0.108$	$a_i = 0.5,\ b^* = 0.433$	$a_i = -1,\ b^* = 1.732$
$e^{-b^* r}$	36.319736	36.745987	38.757385
const	36.115093	36.178699	36.148346
$e^{b^* r}$	35.952934	35.590324	33.949802

mation kernel be non-negative in the case when deposition is zero. Thus, it has used as a transition density at the Markov chain. An estimate for the efficiency of the applied AR method have been obtained. This estimate has the same rate as the estimate of Ermakov, Nekrutkin and Sipin [6]. The difference is that their estimate is obtained when $p(\rho)$ is taken to be an exponential density with negative degree. Also, it depends on the advection, on the deposition, and on the radius of the maximal ball lying inside the domain Ω. Therefore, we solve the problem in more common case without some limitations as dependence on the deposition and on the radius of the maximal ball.

References

1. Bitzadze, A.: Equations of the Mathematical Physics. Nauka, Moscow (1982)
2. Curtiss, J.: Monte Carlo methods for the iteration of the linear operators. J. Math. Phys., Vol. 32, 4 (1954), 209–232.
3. Gurov, T., Withlock, P., Dimov, I.: A grid free Monte Carlo algorithm for solving elliptic boundary value problems. In: L.Vulkov, Waśniewski, J., Yalamov, P. (eds.): Numerical Analysis and its Applications. Lecture notes in Comp. Sci., Vol. 1988. Springer-Verlag (2001) 359–367.
4. Dimov, I., Gurov, T.: Estimates of the computational complexity of iterative Monte Carlo algorithm based on Green's function approach. Mathematics and Computers in Simulation, Vol. 47 (2-5) (1988) 183–199.
5. Ermakov, S., Mikhailov, G.: Statistical Simulation. Nauka, Moscow (1982)
6. Ermakov, S., Nekrutkin, V., Sipin, A.: Random Processes for Solving Classical Equations of the Mathematical Physics. Nauka, Moscow (1984)
7. Mikhailov, V.: Partial Differential Equations. Nauka, Moscow (1983)
8. Miranda, C.: Equasioni alle dirivate parziali di tipo ellipttico. Springer-Verlag, Berlin (1955)
9. Sobol, I.: Monte Carlo Numerical Methods. Nauka, Moscow (1973)

Monte-Carlo Integration Using Cryptographically Secure Pseudo-random Generator

Hiroshi Sugita

Faculty of Mathematics, Kyushu University, Fukuoka 812-0851, Japan,
sugita@math.kyushu-u.ac.jp

Abstract. The drastic reduction of randomness by random Weyl sampling enables us to use cryptographically secure pseudo-random generators for Monte-Carlo integration as well as for parallel Monte-Carlo integration.

1 Introduction

In this paper, we look at Monte-Carlo (MC for short) methods from the cryptographic point of view. Then we show that, theoretically speaking, cryptographically secure pseudo-random generation (cf. [1,2]) could be an ideal basic ingredient for MC methods.

But there is a big problem in practice, i.e., the speed. It has been believed that cryptographically secure pseudo-random generators are so slow that we cannot use them in MC methods. However, in this paper, we assert that it is certainly possible to use them, if we consider only *MC integration.*

In [5,6,7], random Weyl sampling(RWS for short) and its variant were introduced to reduce randomness of MC integration drastically. For example, let a random variable W be a function of 500 tosses of a coin. To integrate W numerically, the usual MC integration (i.e., i.i.d.-sampling) with 10^7 samples requires $500 \times 10^7 = 5 \times 10^9$ random bits, while RWS with the same sample size requires only $2 \times \lceil 500 + \log_2 10^7 \rceil = 1048$ random bits(Example 2 of [5]). Nevertheless, both methods have an identical mean square error.

This drastic reduction of randomness enables us to neglect the speed of pseudo-random generation in the total running time. Therefore we can use a cryptographically secure pseudo-random generator for RWS, and as a result, we obtain a very reliable and fast MC integration method.

The cryptographic point of view also provides a new parallel scheme for MC integration, of course, using a cryptographically secure pseudo-random generator.

I. Dimov et al. (Eds.): NMA 2002, LNCS 2542, pp. 140–146, 2003.

2 Formulation of Monte-Carlo Integration

2.1 Basic Probability Space

To describe the mathematical structure of MC methods systematically, we adopt the Lebesgue probability space $([0,1), \mathcal{B}, P)$, $\mathcal{B}$ being the Borel field of $[0,1)$, as our basic probability space.

We use the following notations: For $m \in \mathbf{N}$,

$$D_m := \{i/2^m\,;\, i = 0, \ldots, 2^m - 1\} \tag{1}$$

$$\lfloor x \rfloor_m := \lfloor 2^m x \rfloor / 2^m \,(\mathrm{mod}1) \in D_m, \quad x \in [0,1). \tag{2}$$

Accordingly, we define an increasing sequence of sub σ-fields $\{\mathcal{B}_m\}_{m=1}^{\infty}$ of $\mathcal{B}$ by

$$\mathcal{B}_m := \sigma\{[a,b)\,;\, a, b \in D_m, a < b\} = \sigma(d_1, \ldots, d_m), \quad m \in \mathbf{N},$$

where, the function $d_i(x)$ denotes the i-th digit of $x \in [0,1)$ in its dyadic expansion, i.e.,

$$d_1(x) := \begin{cases} 0 & (0 \le x < 1/2) \\ 1 & (1/2 \le x < 1) \end{cases}, \quad d_i(x) := d_1(2^{i-1}x \,(\mathrm{mod}\,1)), \quad x \in [0,1).$$

It is well-known that the random variables $\{d_i\}_{i=1}^{\infty}$ defined on $([0,1), \mathcal{B}, P)$ are a realization of the coin tossing process.

Let W be a random variable that we wish to integrate numerically. In this paper, we always assume, except the final section, that W is a function of m tosses of a coin, i.e., $W : \{0,1\}^m \to \mathbf{R}$. If we define a $\mathcal{B}_m$-measurable function $\widetilde{W}$ by

$$\widetilde{W}(x) := W(d_1(x), \ldots, d_m(x)), \quad x \in [0,1), \tag{3}$$

then W and $\widetilde{W}$ are identically distributed. In this sense, we will always identify W with $\widetilde{W}$.

2.2 I.i.d.-sampling

Let us assume that $W : [0,1) \to \mathbf{R}$ is $\mathcal{B}_m$-measurable. For $N \in \mathbf{N}$, define a $\mathcal{B}_{Nm}$-measurable function F_1 by

$$F_1(x) := \frac{1}{N} \sum_{i=1}^{N} W\left(X_i^{\langle m \rangle}(x)\right), \quad x \in [0,1), \tag{4}$$

where

$$X_i^{\langle m \rangle}(x) := \left\lfloor 2^{(i-1)m} x \right\rfloor_m, \quad x = \sum_{i=1}^{\infty} 2^{-(i-1)m} X_i^{\langle m \rangle}(x). \tag{5}$$

Note that the sequence $\{X_i^{\langle m \rangle}\}_{i=1}^{\infty}$ is D_m-valued uniform i.i.d. random variables under P. If N is large, the distribution of $F_1(x)$ under P is concentrated near around $\mathbf{E}[W] := \int_0^1 W(x)dx$ (the law of large numbers). More precisely,

$$\mathbf{E}[F_1] = \mathbf{E}[W]$$

and the mean square error

$$\mathbf{V}[F_1] := \mathbf{E}\left[|F_1 - \mathbf{E}[F_1]|^2\right] = \frac{\mathbf{V}[W]}{N}.$$

F_1 is the sample mean of i.i.d.-sampling for W with sample size N.

2.3 Monte-Carlo Integration from Cryptographic Point of View

We here consider how MC integration looks from the cryptographic point of view. A basic recognition in the theory of cryptography is that *pseudo-random numbers are random variables whose randomness comes from a random choice of the seed.*

Theoretically, an MC method assumes an ideal source of randomness. However, in practice, we use a pseudo-random generator instead of a purely random source. Then, what is the random element "ω" in practical MC methods? According to the above recognition, it is the seed of the pseudo-random generator.

A function $g_n : \{0,1\}^n \to \{0,1\}^{\ell(n)}$ is called a *pseudo-random generator*, if $n < \ell(n)$. We take a random seed $\omega \in \{0,1\}^n$ which distributes uniformly in $\{0,1\}^n$, and plug it into g_n. Then we use the output $g_n(\omega) \in \{0,1\}^{\ell(n)}$, which is called *pseudo-random bits* (or *numbers*), instead of the coin tossing process. Since $n < \ell(n)$, the pseudo-random bits cannot be the true coin tossing process.

According to our formulation, using the identification formula (3), we regard g_n as a $\mathcal{B}_n$-measurable function $g_n : [0,1) \longrightarrow D_{\ell(n)}$.

Suppose W is $\mathcal{B}_m$-measurable and that $Nm \le \ell(n)$. Then what we get by i.i.d.-sampling for W using the pseudo-random generator g_n is $F_1(g_n(x))$, where $x \in [0,1)$ is a random element governed by P. Therefore what is crucial is whether the distribution of $F_1(g_n(x))$ is sufficiently close to that of $F_1(x)$. If it is, the i.i.d.-sampling is successful.

3 Cryptographically Secure Pseudo-random Generation

In this section, we introduce the notion of cryptographically secure pseudo-random generation. For details, see [1,2], for example.

A *pseudo-random generator* in the theory of cryptography is a sequence of functions $\{g_n : [0,1) \to D_{\ell(n)},\ \mathcal{B}_n\text{-measurable}\}_n$ with $n < \ell(n)$, rather than a single function g_n. We assume $\{g_n\}_n$ is *feasible*, i.e., $\ell(n)$ and the time complexity $T_g(n)$ of g_n are at most of polynomial order in n. We consider a sequence of functions for a statistical test:

$$A_n : [0,1) \times [0,1) \longrightarrow \{0,1\}, \quad \mathcal{B}_{\ell(n)} \times \mathcal{B}_{s(n)}\text{-measurable},$$

where $s(n)$ is at most of polynomial order in n, too. Now, define

$$\delta_{g,A}(n) := \left|P^2\left(A_n(g_n(x), y) = 1\right) - P^2\left(A_n(x,y) = 1\right)\right|$$
$$S_{g,A}(n) := T_A(n)/\delta_{g,A}(n),$$

where P^2 is the 2-dimensional Lebesgue probability measure, under which (x, y) distributes uniformly in $[0, 1) \times [0, 1)$. If there is no $\{A_n\}_n$ such that $S_{g,A}(n)$ is at most of polynomial order, then $\{g_n\}_n$ is called *cryptographically secure.* This means that $\delta_{g,A}(n)$ decreases very rapidly if $T_A(n)$ is at most of polynomial order. In other words, it is practically impossible to distinguish the distribution of $g_n(x)$ from the uniform distribution over $D_{\ell(n)}$ by any feasible $\{A_n\}_n$.

Remark 1. There is a serious unsolved problem: We do not know whether a cryptographically secure pseudo-random generator exists or not. If it exists, then we can show $\mathbf{P} \neq \mathbf{NP}$, which is a famous unsolved conjecture in computer science (cf. Exercise 8 of [1]). However, there are many pseudo-random generators which are conjectured to be cryptographically secure(cf. Chapter 12 of [2]). In this paper, we assume the existence.

Let $\{g_n : [0, 1) \to D_{\ell(n)}, \mathcal{B}_n\text{-measurable}\}_n$ be a cryptographically secure pseudo-random generator. In the sequel, we will simply say " $g_n : [0, 1) \to D_L$ is a cryptographically secure pseudo-random generator " to mean that n is chosen so that $n < L \leq \ell(n)$, g_n is $\mathcal{B}_n$-measurable, and that it is practically impossible to distinguish the distribution of $\lfloor g_n(x) \rfloor_L$ from the uniform distribution over D_L.

If we use a cryptographically secure pseudo-random generator in MC methods, we can say the following: Let $Y := F(x)$ be a $\mathcal{B}_L$-measurable random variable, where F should be feasible, and let $g_n : [0, 1) \to D_L$ be a cryptographically secure pseudo-random generator. Suppose that we use $Y' := F(g_n(x))$ instead of Y in an MC computation. Then the distributions of Y and Y' cannot be distinguished in practice. (If they could be, then g_n would not be cryptographically secure !) This means that Y' can completely pretend to be Y in the MC computation.

4 Reduction of Randomness by Random eyl Sampling

Although, as we have seen in the previous section, a cryptographically secure pseudo-random generator could be an ideal basic ingredient for MC methods, it has been widely believed that it is too slow for practical MC methods.

However, if we consider only *MC integration*, it is certainly possible to use it. The idea is simple: Reduce randomness by using *pairwise independent* samples instead of i.i.d. samples.

Theorem 1. *Let $j, m \in \mathbf{N}$. Define a D_m-valued $\mathcal{B}_{2(m+j)}$-measurable random variable $\widehat{X}_i^{(m,j)}$ for each $1 \leq i \leq 2^j$ by*

$$\widehat{X}_i^{(m,j)}(x) := \left\lfloor \lfloor x \rfloor_{m+j} + i \times \lfloor 2^{m+j} x \rfloor_{m+j} \right\rfloor_m, \quad x \in [0, 1). \tag{6}$$

Then under the Lebesgue probability measure P, each $\widehat{X}_i^{(m,j)}$ distributes uniformly in D_m and the sequence $\{\widehat{X}_i^{(m,j)}\}_{i=1}^{2^j}$ is pairwise independent.

The proof can be found in Example 2 of [5]. It is clear that the sample mean

$$F_2(x) := \frac{1}{N}\sum_{i=1}^{N} W\left(\widehat{X}_i^{(m,j)}(x)\right), \quad N \le 2^j, \tag{7}$$

satisfies that

$$\mathbf{E}[F_2] = \mathbf{E}[F_1] = \mathbf{E}[W], \qquad \mathbf{V}[F_2] = \mathbf{V}[F_1] = \frac{\mathbf{V}[W]}{N}. \tag{8}$$

This sampling method is called *random Weyl sampling*(RWS for short, [5,7][1]). In practice, we use a pseudo-random generator $g_n : [0,1) \to D_{2(m+j)}$ to compute F_2, i.e., what we practically compute is $F_2(g_n(x))$.

Note that F_2 is $\mathcal{B}_{2(m+j)}$-measurable, while F_1, the sample mean of i.i.d.-sampling (4), is $\mathcal{B}_{Nm}$-measurable. Since we are interested in the case $N \le 2^j \ll 2^m$, this means that F_2 requires much less randomness than F_1, so that in computing F_2, the necessary time for generating the pseudo-random bits is negligible in the total running time. Thus, we can use a slow but precise pseudo-random generator, typically, a cryptographically secure one, in RWS.

5 Parallel Random eyl Sampling

From the cryptographic point of view, what we get as a result of MC integration is a random variable whose randomness comes from the random seed of the pseudo-random generator. This recognition also provides a new idea for parallel MC integration. In this section, we introduce a parallel version of RWS.

Suppose that our target random variable W is again $\mathcal{B}_m$-measurable. Recall that, in RWS, we practically compute $F_2(g_n(x))$ to simulate $F_2(x)$ by means of a pseudo-random generator $g_n : [0,1) \to D_{2(m+j)}$. If g_n is cryptographically secure, the distribution of $F_2(g_n(x))$ cannot be practically distinguished from that of $F_2(x)$.

Let K be the number of processors. Our idea of parallelization is as follows: *Apply i.i.d.-sampling to $F_2(g_n(x))$ with K independent copies of it, by having each copy computed by a distinct processor.*

To do so, let

$$g_n : [0,1) \to D_{\max(2(m+j),Kn)}$$

be again a cryptographically secure pseudo-random generator. Define a sequence of random variables $\{y_k(x)\}_{k=1}^K$ by[2]

$$y_k(x) := X_k^{\langle n\rangle}(g_n(x)) \in D_n, \quad x \in [0,1), \quad k = 1, \ldots, K.$$

[1] In [5,7], this method is called *discrete* random Weyl sampling, but here, since we only deal with discrete cases, we omit the term "discrete".

[2] $X_\bullet^{\langle \# \rangle}$ has been defined by (5).

Since g_n is cryptographically secure, $\{y_k(x)\}_{k=1}^K$ looks like a D_n-valued uniform i.i.d. sequence. We then apply i.i.d.-sampling to $F_2(g_n(x))$ by computing

$$F_3(x) := \frac{1}{K}\sum_{k=1}^{K} F_2\left(g_n(y_k(x))\right), \quad x \in [0,1),$$

where, for each $k = 1, \ldots, K$, after $y_k(x)$ is given, the k-th processor computes $F_2\left(g_n(y_k(x))\right)$ completely separately from the other processors. It is clear that the distribution of F_3 is concentrated around $\mathbf{E}[W]$ with variance $\mathbf{V}[F_3]$ being close to $\mathbf{V}[W]/KN$.

Remark 2. To reduce randomness thoroughly, we can again apply RWS to the random variable $F_2(g_n(x))$: Take j' so that $2^{j'} \geq K$, and define a sequence of random variables $\{\hat{y}_k(x)\}_{k=1}^K$ by[3]

$$\hat{y}_k(x) := \widehat{X}_k^{(n,j')}(g_n(x)) \in D_n, \quad x \in [0,1), \quad k = 1, \ldots, K.$$

Then we define

$$F_4(x) := \frac{1}{K}\sum_{k=1}^{K} F_2\left(g_n\left(\hat{y}_k(x)\right)\right), \quad x \in [0,1),$$

which is the goal. To compute F_4, we need only $\max(2(m+j), 2(n+j'))$ pseudo-random bits, i.e., it is enough to assume

$$g_n : [0,1) \longrightarrow D_{\max(2(m+j),2(n+j'))}.$$

6 Dynamic Random eyl Sampling

In RWS, our target random variable W must be $\mathcal{B}_m$-measurable for a fixed $m \in \mathbf{N}$. But we can remove this restriction in the following way.

Let $\tau : [0,1) \to \mathbf{N}$ be a $\{\mathcal{B}_m\}_m$*-stopping time*, i.e.,

$$\{x \in [0,1) \,;\, \tau(x) \leq m\} \in \mathcal{B}_m, \qquad m \in \mathbf{N}.$$

A random variable W on $([0,1), \mathcal{B}, P)$ is said to be $\mathcal{B}_\tau$*-measurable*, if

$$W(x) = W\left(\lfloor x \rfloor_{\tau(x)}\right), \qquad x \in [0,1).$$

Let $j, K \in \mathbf{N}$ and let $\{(x_l, \alpha_l)\}_{l\in\mathbf{N}}$ be an i.i.d.-sequence of $D_{K+j} \times D_{K+j}$-valued uniformly distributed random variables. Define a sequence of $[0,1)$-valued random variables $\{\mathbf{x}_n\}_{n=1}^{2^j}$ by

$$\mathbf{x}_n := \sum_{l=1}^{\lceil \tau(\mathbf{x}_n)/K \rceil} 2^{-(l-1)K} \lfloor x_l + \nu_{n,l}\alpha_l \rfloor_K$$

$$\nu_{n,l} := \#\{\, 1 \leq i \leq n \,;\, \tau(\mathbf{x}_i) > (l-1)K \,\}.$$

Note that since τ is a stopping time, $\nu_{n,l}$ and $\mathbf{x}_n$ are well-defined.

Then we have the following theorem([6]).

[3] $\widehat{X}_\bullet^{(\#,\flat)}$ has been defined by (6).

Theorem 2. *If a random variable W on $([0,1),\mathcal{B},P)$ is $\mathcal{B}_\tau$-measurable, then $\{W(\mathbf{x}_n)\}_{n=1}^{2^j}$ is a sequence of pairwise independent copies of W.*

Based on Theorem 2, the sampling method using $\{W(\mathbf{x}_n)\}_{n=1}^{2^j}$ is called *dynamic random Weyl sampling*(DRWS for short). A parallel version of DRWS is easily obtained in the similar way as in the previous section.

C and C++ libraries for DRWS as well as its parallel version are open to the public([4]), in which a slow but very precise pseudo-random generator [3] is used.

References

1. Luby, M.: Pseudorandomness and Cryptographic Applications. Princeton Univ. Press (1996)
2. Stinton, D.R.: Cryptography — Theory and Practice. CRC Press (1995)
3. Sugita, H.: Pseudo-random number generator by means of irrational rotation. Monte Carlo Methods and Applications, VSP, 1-1 (1995) 35–57.
4. Sugita, H.: The Random sampler. C and C++ libraries for pseudo-random generation and dynamic random Weyl sampling, available at `http://idisk.mac.com/hiroshi_sugita/Public/imath/mathematics.html`.
5. Sugita, H.: Robust numerical integration and pairwise independent random variables. Jour. Comput. Appl. Math., 139 (2001) 1–8.
6. Sugita, H.: Dynamic random Weyl sampling for drastic reduction of randomness in Monte-Carlo integration. (a special issue for IMACS Seminar on Monte Carlo Methods MCM2001, Salzburg) Mathematics and Computers in Simulation, to appear.
7. Sugita, H. and Takanobu, S.: Random Weyl sampling for robust numerical integration of complicated functions. Monte Carlo Methods and Appl., 6-1 (1999) 27–48.

Part III

Robust Iterative Solution Methods and Applications

Galerkin and Control Volume Finite Element Methods for Large Scale Air Pollution Simulations: A Comparison

Anton Antonov

Wolfram Research Inc.,
100 Trade Center Drive,
Champaign, IL 61820-7237, USA
antonov@wolfram.com

Abstract. The methods compared are implemented within the object-oriented framework of the Danish Eulerian Model for large scale air pollution simulations. Their mathematical formulations are explained in the paper. The results of the simulations are shown, and the corresponding phase and group velocities are discussed.

1 Introduction

In this paper we compare two methods for simulation of incompressible advection dominated flow used in conjunction with a numerical ODE solver for the simulation of the chemical reactions in the large scale air pollution model called the Danish Eulerian Model (DEM)[4,6,8]. All important chemical species (sulfur pollutants, nitrogen pollutants, ammonia-ammonium, ozone, as well as many radicals and hydrocarbons) can be studied by DEM; its space domain contains the whole of Europe (the covered area is $4800km \times 4800km.$). The methods are implemented within the object-oriented version of DEM, the Object-Oriented Danish Eulerian Model (OODEM)[2].

The first method is a Galerkin Finite Element Method (GFEM) with linear basis functions; the second method is a Control Volume Finite Element Method (CVFEM) with linear basis functions. Both methods are over triangular grids; (see Figure 1). They both are described in [5] by Gresho and Sani. In Section 2.2.6 of [5] is given a review of the control volume methods development (and literature). The CVFEM described in [5] is, as the authors state, "a fully legitimate (no cheating) *alternate finite element* technique" [5, Sec 2.2.6]. The most attractive property of the CVFEM is that it has the property of "local conservation" i.e. the conservation law is applied at control volume level. In this article, our goal is to judge what advantages or disadvantages CVFEM has compared with GFEM when they are applied in a large scale air pollution model like DEM. In Section 3 we will describe both methods, and compare their phase and group velocities. Tests and experiments with the methods are presented in Section 4. The mathematical background for DEM is given in Section 2.

I. Dimov et al. (Eds.): NMA 2002, LNCS 2542, pp. 149–157, 2003.

Another reason to make these comparisons is that it was relatively easy to do them. The OODEM is more than just an implementation of a mathematical model. It is a software framework, that makes the usual process of design, implementation, and testing of new methods 20 to 100 times faster[1,2,3].

2 Mathematical Background of the Danish Eulerian Model

Temporal and spatial variations of the concentrations and/or the depositions of various harmful air pollutants can be studied [8] by solving the system (1) of partial differential equations (PDE's):

$$\begin{aligned}\frac{\partial c_s}{\partial t} &= -\frac{\partial(uc_s)}{\partial x} - \frac{\partial(vc_s)}{\partial y} - \frac{\partial(wc_s)}{\partial z} \\ &+\frac{\partial}{\partial x}(K_x\frac{\partial c_s}{\partial x}) + \frac{\partial}{\partial y}(K_y\frac{\partial c_s}{\partial y}) + \frac{\partial}{\partial z}(K_z\frac{\partial c_s}{\partial z}) \\ &+E_s + Q_s(c_1, c_2, ..., c_q) - (k_{1s} + k_{2s})c_s, \\ &s = 1, 2, ..., q.\end{aligned} \tag{1}$$

The different quantities that are involved in this mathematical model have the following meaning: (i) the concentrations are denoted by c_s; (ii) u, v and w are wind velocities; (iii) K_x, K_y and K_z are diffusion coefficients; (iv) the emission sources in the space domain are described by the functions E_s; (v) κ_{1s} and κ_{2s} are deposition coefficients; (vi) the chemical reactions used in the model are described by the non-linear functions $Q_s(c_1, c_2, \ldots, c_q)$. The number of equations q is equal to the number of species that are included in the model.

In the two dimensional DEM the terms $-\frac{\partial(wc_s)}{\partial z}$ are $\frac{\partial}{\partial z}(K_z\frac{\partial c_s}{\partial z})$ are skipped; see [8]. The methods and the tests in this article are for the two-dimensional DEM model.

Remark The functions c_s, u, v, K_x, K_y are space and time dependent, e.g. $c_s = c_s(x, y, t)$.

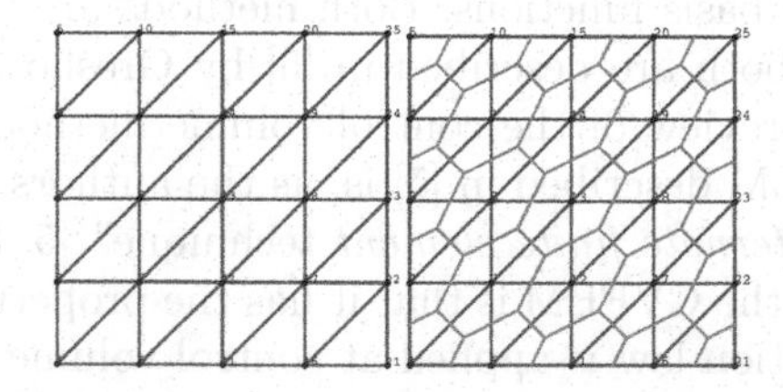

Fig. 1. Parts of the GFEM(left) and CVFEM(right) grids on which the tests and experiments were made. The test/experiment grids have 96×96 nodes.

3 Description of the Methods Theoretical Comparison

After the appliaction of a spliting procedure to (1) [8], the two-dimensional advection-diffusion submodel of DEM is described with the equation:

$$\frac{\partial c}{\partial t} = -\frac{\partial (uc)}{\partial x} - \frac{\partial (vc)}{\partial y} + K_x \frac{\partial^2 c}{\partial x^2} + K_y \frac{\partial^2 c}{\partial y^2}. \tag{2}$$

The species index s is omitted, since the equations are identical for all s. Also, in DEM the diffusion coefficients K_x and K_y are constants.

We will consider the numerical solution of (2) over the two dimensional area Ω with boundary Γ. All space dependent functions considered will have domain that coincides with Ω. With $Supp(f)$ we will denote the support of the function f. When Ω is covered with a finite element grid, the set of the grid elements will be denoted with E, the set of the grid nodes with I. Let $\bigcup_{j=1}^{|E|} e_j = E$, and $\{1, ..., |I|\} = I$. With $\varphi_j = \varphi_j(x,y)$ and $\psi_j = \psi_j(x,y)$ we will denote the basis and test function at node j respectively. With E_j we will denote the minimal set of elements that contains $Supp(\varphi_j)$. The support of any basis or test function is a proper subset of Ω.

3.1 Galerkin Finite Element Method

In GFEM, from (2) we make the week formulation by multiplying (2) by a test function φ_j and integrating over the φ_j's domain, that coincides with Ω for every j:

$$\int_\Omega \frac{\partial c}{\partial t}\varphi_j = \int_\Omega \left(-\frac{\partial (uc)}{\partial x} - \frac{\partial (vc)}{\partial y} + K_x \frac{\partial^2 c}{\partial x^2} + K_y \frac{\partial^2 c}{\partial y^2} \right) \varphi_j,\ j = 1, \ldots, |I| \,.$$

Next we substitute $c(x,y,t)$ with the expansion $c(x,y,t) = \sum_{i=1}^{|I|} g_i(t)\varphi_i(x,y)$:

$$\begin{aligned} &\sum_{i=1}^{|I|} \frac{\partial g_i}{\partial t} \int_\Omega \varphi_i \varphi_j = \\ &\sum_{i=1}^{|I|} g_i \int_\Omega \left(-\frac{\partial (u\varphi_i)}{\partial x}\varphi_j - \frac{\partial (v\varphi_i)}{\partial y}\varphi_j + K_x \frac{\partial \varphi_i}{\partial x}\frac{\partial \varphi_j}{\partial x} + K_y \frac{\partial \varphi_i}{\partial y}\frac{\partial \varphi_j}{\partial y} \right), \\ &j = 1, \ldots, |I| \,. \end{aligned}$$

If $I_j = \{i : Supp(\varphi_i) \cap Supp(\varphi_j) \neq \emptyset\}$ we can rewrite the last equation as

$$\begin{aligned} &\sum_{i \in I_j} \frac{\partial g_i}{\partial t} \int_{E_j} \varphi_i \varphi_j = \\ &\sum_{i \in I_j} g_i \int_{E_j} \left(-\frac{\partial (u\varphi_i)}{\partial x}\varphi_j - \frac{\partial (v\varphi_i)}{\partial y}\varphi_j + K_x \frac{\partial \varphi_i}{\partial x}\frac{\partial \varphi_j}{\partial x} + K_y \frac{\partial \varphi_i}{\partial y}\frac{\partial \varphi_j}{\partial y} \right), \quad (3) \\ &j = 1, \ldots, |I| \,. \end{aligned}$$

If we define the numbers

$$p_{ji} = \int_{E_j} \varphi_i \varphi_j,$$

$$a_{ji} = \int_{E_j} \left(-\frac{\partial(u\varphi_i)}{\partial x}\varphi_j - \frac{\partial(v\varphi_i)}{\partial y}\varphi_j + K_x \frac{\partial\varphi_i}{\partial x}\frac{\partial\varphi_j}{\partial x} + K_y \frac{\partial\varphi_i}{\partial y}\frac{\partial\varphi_j}{\partial y} \right),$$

then (3) can be written as

$$\sum_{i\in I_j} \frac{\partial g_i}{\partial t} p_{ji} = \sum_{i\in I_j} g_i a_{ji},\ j = 1, ..., |I|,$$

which is equivalent to

$$P\frac{\partial \vec{g}(t)}{\partial t} = A\vec{g}(t), \tag{4}$$

where $P = \{p_{ji}\}$, $A = \{a_{ji}\}$, $i, j = 1, ..., dim(S)$, $\vec{g}(t) \in \mathbf{R}^{|I|}$. So the problem to solve approximately (2) is reduced to the problem of solving the system of ODE's (4).

3.2 Control Volume Finite Element Method

In CVFEM we partition Ω into non-overlapping regions in a manner shown with the dashed contours on Figure 1. Each of these regions surrounds exactly one node. So, if a region surrounds the node j, it will be denoted as Ω_j. To each Ω_j we will assign a corresponding test function ψ_j that is 1 within Ω_j and 0 outside of Ω_j. The boundary of Ω_j will be denoted as Γ_j .

In CVFEM, we make the week formulation by multiplying (2) by a test function ψ_j and integrating over the ψ_j's domain, that coincides with Ω for every j:

$$\int_\Omega \frac{\partial c}{\partial t}\psi_j = \int_\Omega \left(-\frac{\partial(uc)}{\partial x} - \frac{\partial(vc)}{\partial y} + K_x\frac{\partial^2 c}{\partial x^2} + K_y\frac{\partial^2 c}{\partial y^2} \right)\psi_j,\ j = 1, \ldots, |I|.$$

But owing to the definition of the test functions, and using the divergence theorem, the above equation is equivalent to

$$\int_{\Omega_j} \frac{\partial c}{\partial t} = \int_{\Gamma_j} \vec{n} \cdot \left(-\vec{W}c + \nabla(\vec{k}c) \right),\ j = 1, \ldots, |I|,$$

where $\vec{W} = \begin{bmatrix} u \\ v \end{bmatrix}$, $\vec{k} = \begin{bmatrix} K_x \\ K_y \end{bmatrix}$. Next, as in GFEM, we substitute $c(x, y, t)$ with the expansion $c(x, y, t) = \sum_{i=1}^{|I|} g_i(t)\varphi_i(x, y)$:

$$\sum_{i=1}^{|I|} \frac{\partial g_i}{\partial t} \int_{\Omega_j} \varphi_i = \sum_{i=1}^{|I|} g_i \int_{\Gamma_j} \vec{n} \cdot \left(-\vec{W}\varphi_i + \nabla(\vec{k}\varphi_i) \right),\ j = 1, \ldots, |I|.$$

If $I_j = \{i : Supp(\varphi_i) \cap Supp(\psi_j) \neq \emptyset\} = \{i : Supp(\varphi_i) \cap \Omega_j \neq \emptyset\}$ we can rewrite the last equation as

$$\sum_{i \in I_j} \frac{\partial g_i}{\partial t} \int_{\Omega_j} \varphi_i = \sum_{i \in I_j} g_i \int_{\Gamma_j} \vec{n} \cdot \left(-\vec{W}\varphi_i + \nabla(\vec{k}\,\varphi_i) \right), \; j = 1, \ldots, |I|. \tag{5}$$

Implementational Remarks

- The computation of the right hand side integral of (5) can be done in two ways. Let $E_j = \{k : e_k \in E \bigwedge e_k \cap \Omega_j \neq \emptyset\}$. It is natural to assume that the wind field $\vec{W} = \vec{W}(x, y, t)$ is given at the nodes of the grid. We can assume further that the wind is constant within the elements. Let $\vec{W}_i$ denote the wind at node i, and let $\vec{W}_{e_i}$ denotes the constant wind within the element $e_i \in E$. Then the right hand side integral of (5),

  $$\int_{\Gamma_j} \vec{n} \cdot \left(-\vec{W}\varphi_i + \nabla(\vec{k}\,\varphi_i) \right),$$

  can be computed either with the formula

  $$\int_{\Gamma_j} \vec{n} \cdot \left(-\vec{W}_j\varphi_i + \nabla(\vec{k}\,\varphi_i) \right), \tag{6}$$

  or with the formula

  $$\sum_{k \in E_j} \int_{\Gamma_j \cap e_k} \vec{n} \cdot \left(-\vec{W}_{e_k}\varphi_i + \nabla(\vec{k}\,\varphi_i) \right). \tag{7}$$

- The dashed contours of the CVFEM grid on Figure 1 are drawn between the midpoints of the triangles.

3.3 Phase and Group Velocities

If we consider the methods with constant wind field on infinite regular grids for the model equation

$$\frac{\partial c}{\partial t} = -W\cos\phi \frac{\partial c}{\partial x} - W\sin\phi \frac{\partial (vc)}{\partial y}, \tag{8}$$

we can derive their phase and group velocities. This was done using a Fourier analysis technique described by Vichnevetsky in [7]; a description how this technique was applied to GFEM in OODEM can be found in [2]. With this technique we find the numerical velocity W^* that corresponds to W when solving (8). The corresponding GFEM and CVFEM phase and group velocities for the grids on Figure 1 are shown on Figure 2.

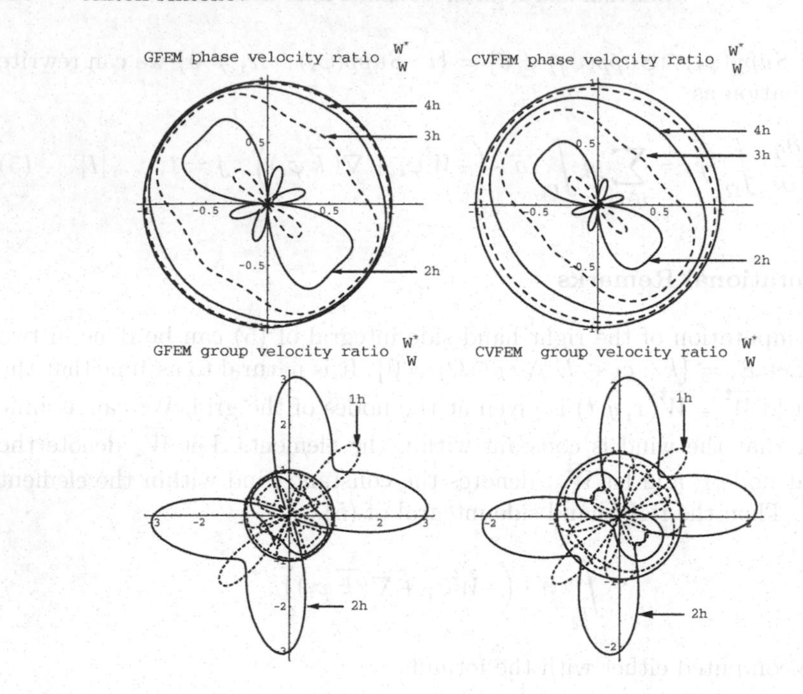

Fig. 2. Phase and group velocities for GFEM and CVFEM for the grids on Figure 1. The velocities are computed for the Currant number $W\Delta t/h = 0.2$ for the wave lengths $\lambda = 1h, 2h, 3h, 4h, 6h, 12h$.

We can make the following observations:

- we can see the famous anisotropy property of the triangle meshes: short-length waves propagate faster in the direction perpendicular to the triangles hypotenuses;
- GFEM has more isotropical phase velocities;
- for both methods the negative group velocities of the $2\Delta x$ wave (see [5, Sec. 2.6] or [2, Ch. 3]) are very anisotropical, and they have similar shapes; (these are the "butterfly" contours;)
- the amplitude of the negative group velocity ratio of the CVFEM $2\Delta x$ wave is 2, and that of GFEM is 3. This is in agreement with the observation in [5, p. 149] that "the *higher* the phase accuracy is for long waves [...], the *larger* is the negative short-wave group velocity";

4 Tests and Experiments Practical Comparisons

4.1 Rotational Tests

For both methods we can do the pure-advection rotational test [8] of a cone concentration distribution. The results of this test, over the 96×96 grids shown on Figure 1, are shown on Figure 3. The CVFEM cone is more "dumped", which fits the prediction by the phase velocity graphics on Figure 2. Both formulas (6) and (7) were used for the CVFEM scheme. The results are almost identical: the cone heights are 83.9175 and 83.9173 for (6) and (7) respectively.

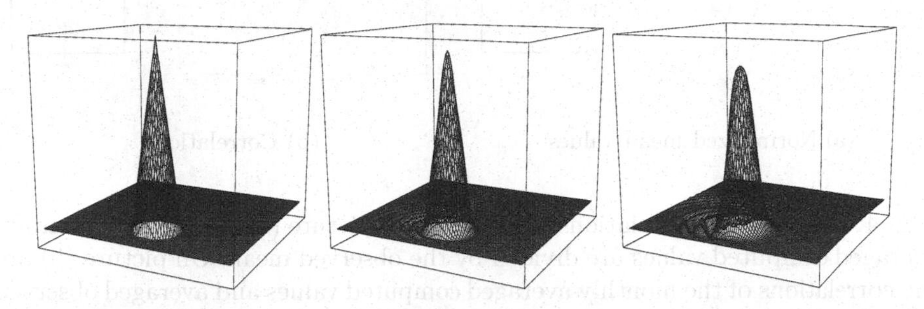

Fig. 3. Rotational tests initial condition (left) and results after one rotation for GFEM (middle) and CVFEM (right). The height of the initial condition cone is 100, the height of the GFEM cone is 92.33, the height of the CVFEM cone is 83.92.

4.2 Experiments with Real Data

Three simulations were made with the following methods/implementations:

1. GFEM
2. CVFEM with formula (6)
3. CVFEM with formula (7)

All simulations were made for July 1995 with the data set used in DEM at the National Environmental Research Institute (NERI) of Denmark. The 96×96 grids used are shown on Figure 1. The quality of the simulations can be judged with scatter plots. The scatter plots are made with the measurements of pollutants for the given period (July'95) and the simulation results. The correspondence between the measurements and the simulation results are summarized on Figure 4. We can see from these figures that CVFEM with formula (7) produces results very close to those by the GFEM method. We did some additional visualizations that suggest to "rule out" CVFEM with formula (6) – it does not produce expected results. Further experiments/comparisons should be done.

5 Conclusions and Future Plans

The main goal of the article was to compare a theoretically correct CVFEM – as described in [5] – with GFEM for a time dependent simulation of an advection dominated flow. The presented theoretical results were obtained with simplifying assumptions, and the presented experimental results are done for just one month. From the theoretical and test results we can conclude that (i) CVFEM

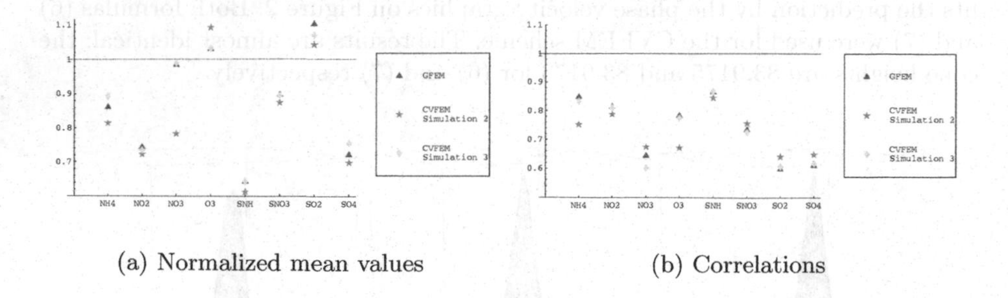

(a) Normalized mean values (b) Correlations

Fig. 4. Results from simulations 1, 2, and 3. On picture (a) the mean of monthly averaged computed values are divided by the observed mean. On picture (b) are the correlations of the monthly averaged computed values and averaged observed values.

is inferior to GFEM within the framework of the Danish Eulerian Model. From the experimental result we can conclude that (ii) if GFEM or CVFEM wins over the other, it will be within "a narrow" margin. (It is predicted in [5] that GFEM would win within "a narrow" margin.) More experiments should be done. These additional experiments should be done for a large number of months in different years. It will be nice if the experiments are repeated with different chemical schemes/solvers.

Other extensions of the work presented can be similar comparisons with regular rectangular grids and comparisons with grids with local refinements. (The later might include tests how CVFEM propagates waves through inner boundaries of local refinements; see [2] about GFEM.) It will be also interesting to see how the theoretically obtained phase velocities fit the rotational tests results.

References

1. Antonov, Anton: Object-Oriented Framework for Large Scale Air Pollution Modeling. In: Margenov, S., Wasniewski, J., and Yalamov, P. (eds.): Large-Scale Scientific Computing. Lecture Notes in Computer Science, Vol. 2179, Springer-Verlag (2001) 247–254.
2. Antonov, Anton: Object-Oriented Framework for Large Scale Air Pollution Models. PhD thesis, The Technical University of Denmark, April 2001. http://www.imm.dtu.dk/~uniaaa/OODEM/ThesisComplete.ps.gz.
3. Antonov, Anton: Framework for Parallel Simulations in Air Pollution Modeling with Local Refinements. In: Welders, P., Ecer, A., Periaux, J., Satofuka, N., and Fox, P. (eds.): Parallel Computational Fluid Dynamics, Practice and Theory. North-Holland (2002).
4. Bastrup-Birk, A., Brandt, J., Uria, I., and Zlatev, Z.: Studying Cumulative Ozone Exposures in Europe During a 7-year Period. Journal of Geophysical Research, 102 (1997) 23917–23935.

5. Gresho, P. M. and Sani, R. L.: Incompressible Flow and Finite Element Method. John Willey and Sons (1998)
6. Skjøth, C. Ambelas, Bastrup-Birk, A., Brandt, J., and Zlatev, Z.: Studying Variations of Pollution Levels in a Given Region of Europe during a Long Time-Period. Systems Analysis Modelling Simulation, 37 (2000) 297–311.
7. Vichnevetsky, R. and Bowles, J. B.: Fourier Analysis of Numerical Approximations of Hyperbolic Equations. SIAM (1982)
8. Zlatev, Z.: Computer Treatment of Large Air Pollution Models. Kluwer (1995)

Robust Preconditioners for Saddle Point Problems

Owe Axelsson[1] and Maya Neytcheva[2]

[1] Department of Mathematics, University of Nijmegen, Toernooiveld 1
6525 ED Nijmegen, The Netherlands,
axelsson@sci.kun.nl

[2] Department of Scientific Computing, Uppsala University
Box 337, 75105 Uppsala, Sweden,
maya@tdb.uu.se

Abstract. We survey preconditioning methods for matrices on saddle point form, as typically arising in constrained optimization problems. Special consideration is given to indefinite matrix preconditioners and a preconditioner which results in a symmetric positive definite matrix, which latter may enable the use of the standard conjugate gradient (CG) method. These methods result in eigenvalues with positive real parts and small or zero imaginary parts. The behaviour of some of these techniques is illustrated on solving a regularized Stokes problem.

1 Introduction

Constrained optimization problems involve at least two variables, one of which is the so-called Lagrange multiplier. In the context of mixed variable or constrained finite element methods, the Lagrange multiplier acts as a natural physical variable. The arising systems are normally symmetric and indefinite, but have a block matrix structure which can be utilized. Various types of preconditioners for such systems have been proposed through the years. They include block diagonal preconditioners and block incomplete factorization preconditioners where the iteration matrix is symmetrizable but still indefinite, requiring some generalized iterative solution algorithm such as a minimum residual conjugate gradient method. Reduced matrix approaches, where a Schur complement system for one of the variables is solved, have also been proposed. The latter system is normally symmetric and positive definite and a symmetric positive definite preconditioner can be chosen. If one uses inner iterations, perturbations occur so one may be forced to use a (truncated) version of a generalized conjugate gradient method, such as GCG or GMRES.

Finally, indefinite preconditioners have been proposed. For the simplest type of those, the corresponding preconditioned matrix has only positive eigenvalues but has, in general, a deficient eigenvector space, i.e., the set of eigenvectors do not span the whole space. This may cause problems in the convergence of the iterative solution method. These problems, however, do not arise when one solves a stabilized saddle point system.

In the sequel, the above methods are surveyed.

I. Dimov et al. (Eds.): NMA 2002, LNCS 2542, pp. 158–166, 2003.

2 Solving Indefinite Matrix Problems by Iteration; Preconditioners for Indefinite Matrix Problems

Consider the solution of linear systems with indefinite matrices of the form

$$\mathcal{A}\begin{bmatrix}\mathbf{u}\\ \mathbf{x}\end{bmatrix} = \begin{bmatrix}M & B^T\\ B & -C\end{bmatrix}\begin{bmatrix}\mathbf{u}\\ \mathbf{x}\end{bmatrix} = \begin{bmatrix}\mathbf{a}\\ \mathbf{b}\end{bmatrix}, \tag{1}$$

where B has full row rank ($= m$) and M, of order ($n \times n$), is symmetric. The block matrix M can be indefinite in general, but we assume that M is positive definite on $ker(B)$. The matrix C is symmetric positive semidefinite and is frequently a zero matrix.

Solving indefinite matrix problems by iteration requires special care, because the iterative method may diverge, or a breakdown (division by zero), or a near breakdown (division by a number which is small) can occur.

Next we consider various approaches to solve (1).

2.1 Schur Complement Approaches

A straightforward approach to solve (1) is via its reduced system, involving the Schur complement $S = -(C+BM^{-1}B^T)$. We then eliminate the first component ($\mathbf{u}$) to get

$$(C + BM^{-1}B^T)\mathbf{x} = BM^{-1}\mathbf{a} - \mathbf{b}, \tag{2}$$

which is solved first, followed by $M\mathbf{u} = \mathbf{a} - B^T\mathbf{x}$. Here $-S$ is symmetric and also positive definite since B has full rank. Clearly, the matrix S is not formed explicitly since only the action of it is needed in each iteration step. However, each action requires the solution of the "inner" system with the matrix M which, in general, for large scale problems is solved by iteration as well. Hence, a coupled inner-outer iteration method must be used. The major expense in such an iterative scheme is normally related to the actions of M^{-1}, and the accuracy with which the systems with M are solved plays thereby an important role.

When M and/or $BM^{-1}B^T$ are ill-conditioned, it can be efficient to use a proper reformulation of the problem, as discussed in detail in, i.e., [7].

2.2 Block-Triangular Preconditioners

We consider next preconditioners for the whole, unreduced, system (1) with $C \equiv 0$.

Let D_1, D_2 be symmetric positive preconditioners to M and $BD_1^{-1}B^T$, respectively. Then $\mathcal{D} = \begin{bmatrix}D_1 & 0\\ B & -D_2\end{bmatrix}$ can act as a simple, but sometimes still efficient preconditioner to $\mathcal{A}$. The matrix $\mathcal{D}$ corresponds to the matrix splitting $\begin{bmatrix}D_1 & 0\\ B & -D_2\end{bmatrix} = \begin{bmatrix}M & B^T\\ B & 0\end{bmatrix} - \begin{bmatrix}M - D_1 & B^T\\ 0 & D_2\end{bmatrix}$. For the analysis of the preconditioner

$\mathcal{D}$ we must analyse the eigenvalues (λ) of $\mathcal{D}^{-1}\mathcal{A}$, which equal $\lambda = 1 + \delta$, where δ denotes the eigenvalues of

$$\begin{bmatrix} D_1 & 0 \\ B & -D_2 \end{bmatrix}^{-1} \begin{bmatrix} M - D_1 & B^T \\ 0 & D_2 \end{bmatrix} = \begin{bmatrix} D_1^{-1}(M - D_1) & D_1^{-1}B^T \\ D_2^{-1}BD_1^{-1}(M - D_1) & D_2^{-1}BD_1^{-1}B^T - I_2 \end{bmatrix}.$$

Letting $\widetilde{M} = D_1^{-\frac{1}{2}}MD_1^{-\frac{1}{2}}$, $\widetilde{B} = D_2^{-\frac{1}{2}}BD_1^{-\frac{1}{2}}$, $\widetilde{\mathbf{u}} = D_1^{\frac{1}{2}}\mathbf{u}$ and $\widetilde{\mathbf{x}} = D_2^{\frac{1}{2}}\mathbf{x}$, one can show (cf. [7]) that the eigenvalues of $\mathcal{D}^{-1}\mathcal{A}$ satisfy $\lambda = 1 + \frac{1}{2}(a + b^2 - 1) \pm \frac{1}{2}\sqrt{(a + b^2 - 1)^2 + 4a}$, where $a = \widetilde{\mathbf{u}}^*(\widetilde{M} - I_1)\widetilde{\mathbf{u}}/\|\widetilde{\mathbf{u}}\|^2$, $b = \|\widetilde{B}\widetilde{\mathbf{u}}\|/\|\widetilde{\mathbf{u}}\|$, $\widetilde{\mathbf{u}} = D_1^{\frac{1}{2}}\mathbf{u}$ and $[\mathbf{u}, \mathbf{x}]$ is an eigenvector of $\mathcal{D}^{-1}\mathcal{A}$.

It follows that if $D_1 = M$ then $a = 0$ and $\delta = \begin{cases} b^2 - 1 \\ 0 \end{cases}$. Hence, the eigenvalues of $\mathcal{D}^{-1}\mathcal{A}$ are real and equal the unit number (with multiplicity at least n) and the eigenvalues of $\widetilde{B}\widetilde{B}^T$, i.e., of $D_2^{-1}BD_1^{-1}B^T$, respectively, which latter are positive. By choosing D_2 sufficiently close to $BD_1^{-1}B^T$ we can hence cluster the eigenvalues around the unit number. This holds also if $|a|$ is small. However, if b is small (which occurs for a nearly rank-deficient matrix B) then δ can take values close to a and -1, which means that the preconditioned matrix is nearly singular. Further, it can be seen that if $-(b+1)^2 < a < -(b-1)^2$, then the eigenvalues are complex.

It can be hence concluded that the above preconditioning method does not have a robust behavior.

Remark 1. A symmetrized form of a block-triangular preconditioner

In [10], a clever one-sided symmetrization of a block Gauss-Seidel preconditioned matrix was proposed. There, the method was formulated as a block-triangular method, which was subsequently symmetrized by choosing a proper inner product (cf. also [14] for a similar approach). The advantages with the above method are that, contrary to the standard block-triangular preconditioner, one can use a classical preconditioned CG method for symmetric positive definite problems and that, contrary to the Schur complement method, no inner iterations are required. The rate of convergence depends on the condition number of both matrices $M_0^{-\frac{1}{2}}MM_0^{-\frac{1}{2}} - I_1$ where M_0 is a s.p.d. preconditioner to M, which has been scaled to satisfy $\mathbf{u}^T M_0 \mathbf{u} \le \alpha_1 \mathbf{u}^T M \mathbf{u}$ for all $\mathbf{u} \in \mathbb{R}^n$ and $0 < \alpha_1 < 1$, and $C + BM^{-1}B^T$. It requires therefore a particularly accurate preconditioner M_0 and it may be less robust when B is nearly rank-deficient.

Remark 2. Preconditioners of the form of $\mathcal{D}$ are also suitable for cases when M itself is nonsymmetric, as for the linearized Navier-Stokes equations, see e.g. [12,9,11]. As pointed out in [11], if D_2 takes the form $D_2 = (BB^T)(BMB^T)^{-1}(BB^T)$, then under a certain additional condition, all eigenvalues of $\mathcal{D}^{-1}\mathcal{A}$ equal the unit number and no Jordan block has order higher than two.

Preconditioners of the form $(BB^T)(BMB^T)^{-1}(BB^T)$ to $BM^{-1}B^T$ have been considered previously in [5] for nonlinear diffusion problems, where it enabled for a cheap update of the nonlinear matrix M.

2.3 A Preconditioner on Stabilized Saddle Point Form

We shall now analyse preconditioners of the same, indefinite, form as the given matrix. For the analysis of them we shall use the next lemma (see [7]).

Lemma 1. *Let B, C, E be real matrices of order $n\times m$, $m\times m$ and $n\times n$ respectively, where B has full rank ($=m$), C is positive definite and E is symmetric. Then the eigenvalues of the generalized eigenvalue problem*

$$\gamma \begin{bmatrix} I & B^T \\ B & -C \end{bmatrix} \begin{bmatrix} \mathbf{u} \\ \mathbf{x} \end{bmatrix} = \begin{bmatrix} E & 0 \\ 0 & 0 \end{bmatrix} \begin{bmatrix} \mathbf{u} \\ \mathbf{x} \end{bmatrix}, \quad |\mathbf{u}| + |\mathbf{x}| \neq \mathbf{0} \tag{3}$$

where $\mathbf{u} \in \mathbb{C}^n$ and $\mathbf{x} \in \mathbb{C}^m$, satisfy

(a) $\gamma = \frac{\mathbf{u}^* E\mathbf{u}}{\mathbf{u}^*(I+B^TC^{-1}B)\mathbf{u}} \neq \mathbf{0}$, if $E\mathbf{u} \neq \mathbf{0}$ and $\gamma = 0$, if and only if $E\mathbf{u} = \mathbf{0}, \mathbf{x} \neq \mathbf{0}$,
(b) the dimension of the eigenvector space corresponding to the zero eigenvalue is $m+q$, where $q = dim\{ker(E)\}$,
(c) the nonzero eigenvalues are contained in the interval $\min\{0, \lambda_{min}(E)\} \leq \gamma \leq \lambda_{max}(E)$ and there are $n-q$ eigenvectors to them.

Consider now the generalized eigenvalue problem (3), where $C = 0$. Here there holds that $\gamma(\mathbf{u}+B^T\mathbf{x}) = E\mathbf{u}$ and $\gamma B\mathbf{u} = \mathbf{0}$. Thus, in this case, at least one of $\gamma = 0$ or $B\mathbf{u} = \mathbf{0}$ must hold. If $B\mathbf{u} = \mathbf{0}$ but $\gamma \neq 0$ then $\gamma BB^T\mathbf{x} = BE\mathbf{u}$ (i.e., there holds here $E\mathbf{u} \neq \mathbf{0}$) and $\gamma\mathbf{u}^*\mathbf{u} = \mathbf{u}^*E\mathbf{u}$, i.e., $\mathbf{x} = \frac{1}{\gamma}(BB^T)^{-1}BE\mathbf{u}$, where $\gamma = \frac{\mathbf{u}^*E\mathbf{u}}{\mathbf{u}^*\mathbf{u}}$. Further, $\gamma\mathbf{u} + B^T(BB^T)^{-1}BE\mathbf{u} = E\mathbf{u}$ or $\gamma\mathbf{u} = (I - B^T(BB^T)^{-1}B)$ $E\mathbf{u}$. Hence, for any $\mathbf{u} \in ker(B), \mathbf{u} \neq \mathbf{0}$ and eigenvalue $\gamma \neq 0$, there holds that $[\mathbf{u}, \frac{1}{\gamma}(BB^T)^{-1}BE\mathbf{u}]^t, \gamma = \mathbf{u}^*E\mathbf{u}/\mathbf{u}^*\mathbf{u}$ is an eigenvector for this γ.

If $\gamma = 0$, then it must hold $E\mathbf{u} = \mathbf{0}$. If $E\mathbf{u} = \mathbf{0}$ then $\gamma = 0$ is an eigenvalue for such a vector $\mathbf{u}$ and any $\mathbf{x}$. Hence, there holds that $[0, \mathbf{x}^{(i)}]^T, i = 1, 2, \cdots, m$ and $[\mathbf{u}^{(j)}, \mathbf{0}]^T, j = 1, 2, \cdots, q$ are eigenvectors for $\gamma = 0$, where $\{\mathbf{x}^{(i)}\}_1^m$ span $\mathbb{R}^m$ and $E\mathbf{u}^{(j)} = \mathbf{0}, \mathbf{u}^{(j)} \neq \mathbf{0}$. Hence, there are $m+q$ linearly independent eigenvectors for $\gamma = 0$.

For some of the eigenvectors $\mathbf{u}^{(j)}$ it may hold $\mathbf{u}^{(j)} \in ker(B)$. If $ker(E) \cap ker(B) = 0$ then the algebraic multiplicity of $\gamma = 0$ is $n + m - (n-m) = 2m$. In this case the index of eigenvector deficiency is $2m - (m+q) = m - q$.

If, however, $ker(E) \cap ker(B) \neq 0$, then the algebraic multiplicity is increased and the index of eigenvector deficiency is correspondingly increased.

Consider now the matrix $\mathcal{A} = \begin{bmatrix} M & B^T \\ B & -C \end{bmatrix}$. Let D be a symmetric and positive definite preconditioner to M and let $\begin{bmatrix} D & B^T \\ B & -C \end{bmatrix}$ be a preconditioner to $\mathcal{A}$. For the generalized eigenvalue problem $\lambda \begin{bmatrix} D & B^T \\ B & -C \end{bmatrix} \begin{bmatrix} \mathbf{u} \\ \mathbf{x} \end{bmatrix} = \mathcal{A} \begin{bmatrix} \mathbf{u} \\ \mathbf{x} \end{bmatrix}$ we have $\gamma \begin{bmatrix} D & B^T \\ B & -C \end{bmatrix}$ $\begin{bmatrix} \mathbf{u} \\ \mathbf{x} \end{bmatrix} = \begin{bmatrix} M-D & 0 \\ 0 & 0 \end{bmatrix} \begin{bmatrix} \mathbf{u} \\ \mathbf{x} \end{bmatrix}$, where $\gamma = \lambda - 1$, or $\gamma \begin{bmatrix} I & \widetilde{B}^T \\ \widetilde{B} & -C \end{bmatrix} \begin{bmatrix} \widetilde{\mathbf{u}} \\ \mathbf{x} \end{bmatrix} = \begin{bmatrix} \widetilde{E} & 0 \\ 0 & 0 \end{bmatrix} \begin{bmatrix} \widetilde{\mathbf{u}} \\ \mathbf{x} \end{bmatrix}$, where $\widetilde{B} = BD^{-\frac{1}{2}}$, $\widetilde{E} = D^{-\frac{1}{2}}MD^{-\frac{1}{2}} - I$ and $\widetilde{\mathbf{u}} = D^{-\frac{1}{2}}\mathbf{u}$.

This problem has the same form as the generalized eigenvalue problem in Lemma 1. Hence, the previous analysis of the eigenvalues and eigenvectors is applicable. If C is positive definite, no eigenvalue deficiency occurs.

In this paper we propose the following preconditioner on regularized form,

$$\begin{bmatrix} D_1 & B^T \\ B & -C \end{bmatrix}, \tag{4}$$

where D_1 is a preconditioner to M. The systems with this preconditioner can be solved via the Schur complement $C + BD_1^{-1}B^T$.

A preconditioner for the Schur complement system must then be found. In applications such as for Stokes problem, where D_1 is a sufficiently accurate preconditioner to M, it can be chosen as a diagonal matrix D_2. The computational effort in solving systems with the preconditioner (using the conjugate gradient method) is therefore not big. Furthermore, as we have already commented on, the number of outer iterations if using a GCG method will be few.

Remark 3. Indefinite preconditioners for the saddle point problem (3) have been proposed previously in [4] and [13]. However, it was not pointed out that the eigenvector space for the unit eigenvalue, $\lambda = 1$, is deficient in general. This was, however, done in [15]. The eigenvector deficiency may cause problems in the iterative solution algorithm. For instance, the rate of convergence of minimal residual iterative methods are based on the expansion of the initial residual using the eigenvectors as basis vectors. This approach is inapplicable when the space is deficient (unless one is able to work in the orthogonal complement to the nullspace). Alternatively one can use estimates, based on the Jordan canonical form (see [7]) but this shows also a possible long delay in convergence. Other preconditioners have been proposed in [15] where one may avoid the eigenvalue deficiency problem. As we have seen, there is no eigenvalue deficiency if one uses the regularized form of the preconditioner.

2.4 An Indefinite Preconditioner on Factorized Form

Following [4], let now the preconditioner be given on factorized form

$$\mathcal{A}_0 = \begin{bmatrix} D & B^T \\ B & -R \end{bmatrix} = \begin{bmatrix} D & 0 \\ B & N \end{bmatrix} \begin{bmatrix} I_1 & D^{-1}B^T \\ 0 & -N^T \end{bmatrix}, \tag{5}$$

where $R = NN^T - BD^{-1}B^T$. Here D is a preconditioner to M and NN^T is a preconditioner to $C + BD^{-1}B^T$. We will use $\mathcal{A}_0$ as a preconditioner to $\mathcal{A} = \begin{bmatrix} M & B^T \\ B & -C \end{bmatrix}$. Depending on the choice of N and D, this preconditioner can be positive definite or indefinite.

For its analysis, we consider then the generalized eigenvalue problem
$\lambda \mathcal{A}_0 \begin{bmatrix} \mathbf{u} \\ \mathbf{x} \end{bmatrix} = \mathcal{A} \begin{bmatrix} \mathbf{u} \\ \mathbf{x} \end{bmatrix}$ or $\gamma \begin{bmatrix} D & B^T \\ B & -R \end{bmatrix} \begin{bmatrix} \mathbf{u} \\ \mathbf{x} \end{bmatrix} = \begin{bmatrix} M - D & 0 \\ 0 & R - C \end{bmatrix} \begin{bmatrix} \mathbf{u} \\ \mathbf{x} \end{bmatrix}$, where $\gamma = \lambda - 1$.

We find then

$$\gamma\begin{bmatrix}\mathbf{u}\\ \mathbf{x}\end{bmatrix}=\begin{bmatrix}D^{-1}-D^{-1}B^TS^{-1}BD^{-1} & D^{-1}B^TS^{-1}\\ S^{-1}BD^{-1} & -S^{-1}\end{bmatrix}\begin{bmatrix}M-D & 0\\ 0 & R-C\end{bmatrix}\begin{bmatrix}\mathbf{u}\\ \mathbf{x}\end{bmatrix},$$

where $S=R+BD^{-1}B^T=NN^T$, or

$$\gamma\begin{bmatrix}\widetilde{\mathbf{u}}\\ \widetilde{\mathbf{x}}\end{bmatrix}=\begin{bmatrix}I_1-\widetilde{B}^T\widetilde{B} & \widetilde{B}^T\\ \widetilde{B} & -I_2\end{bmatrix}\begin{bmatrix}\widetilde{M}-I_1 & 0\\ 0 & \widetilde{R}-\widetilde{C}\end{bmatrix}\begin{bmatrix}\widetilde{\mathbf{u}}\\ \widetilde{\mathbf{x}}\end{bmatrix}, \tag{6}$$

where $\widetilde{\mathbf{u}}=D^{\frac{1}{2}}\mathbf{u}$, $\widetilde{\mathbf{x}}=N^T\mathbf{x}$, $\widetilde{B}=N^{-1}BD^{-\frac{1}{2}}$, $\widetilde{M}=D^{-\frac{1}{2}}MD^{-\frac{1}{2}}$, $\widetilde{R}=N^{-1}RN^{-T}=I_2-\widetilde{B}\widetilde{B}^T$, $\widetilde{C}=N^{-1}CN^{-T}$.

To analyse this further we make first the assumption that $\widetilde{M}\geq I_1$ and $\widetilde{R}\geq\widetilde{C}$, i.e., $I_2\geq\widetilde{C}+\widetilde{B}\widetilde{B}^T$. This holds if D and NN^T are proper preconditioners to M and $C+BD^{-1}B^T$, respectively.

It follows then from (6), that if $\widehat{\mathbf{u}}=(\widetilde{M}-I_1)^{\frac{1}{2}}\widetilde{\mathbf{u}}$ and $\widehat{\mathbf{x}}=(\widetilde{R}-\widetilde{C})^{\frac{1}{2}}\widetilde{\mathbf{x}}$,

$$\gamma\begin{bmatrix}\widehat{\mathbf{u}}\\ \widehat{\mathbf{x}}\end{bmatrix}=\begin{bmatrix}(\widetilde{M}-I_1)^{\frac{1}{2}}(I_1-\widetilde{B}^T\widetilde{B})(\widetilde{M}-I_1)^{\frac{1}{2}} & (\widetilde{M}-I_1)^{\frac{1}{2}}\widetilde{B}^T(\widetilde{R}-\widetilde{C})^{\frac{1}{2}}\\ (\widetilde{R}-\widetilde{C})^{\frac{1}{2}}\widetilde{B}(\widetilde{M}-I_1)^{\frac{1}{2}} & -(\widetilde{R}-\widetilde{C})\end{bmatrix}\begin{bmatrix}\widehat{\mathbf{u}}\\ \widehat{\mathbf{x}}\end{bmatrix}. \tag{7}$$

This matrix is symmetric but indefinite. Its eigenvalues are hence real and the absolute values of them can be controlled by choosing sufficiently accurate preconditioners D to M and NN^T to $C+BD^{-1}B^T$. In this way we may get $\gamma>-1$, so $\lambda=\gamma+1>0$. On the other hand, if $\widetilde{R}\leq\widetilde{C}$, then (6) can be transformed to

$$\gamma\begin{bmatrix}\widehat{\mathbf{u}}\\ \widehat{\mathbf{x}}\end{bmatrix}=\begin{bmatrix}(\widetilde{M}-I_1)^{\frac{1}{2}}(I_1-\widetilde{B}^T\widetilde{B})(\widetilde{M}-I_1)^{\frac{1}{2}} & -(\widetilde{M}-I_1)^{\frac{1}{2}}\widetilde{B}^T(\widetilde{C}-\widetilde{R})^{\frac{1}{2}}\\ (\widetilde{C}-\widetilde{R})^{\frac{1}{2}}\widetilde{B}(\widetilde{M}-I_1)^{\frac{1}{2}} & (\widetilde{C}-\widetilde{R})\end{bmatrix}\begin{bmatrix}\widehat{\mathbf{u}}\\ \widehat{\mathbf{x}}\end{bmatrix},$$

where now $\widehat{\mathbf{x}}=(\widetilde{C}-\widetilde{R})^{\frac{1}{2}}\mathbf{x}$. In this case, the preconditioned matrix, after transformation, is nonsymmetric, but with positive definite symmetric part. Hence the eigenvalues may be complex but there holds that $Re(\lambda)\geq 1$. Further, as before, the eigenvalues cluster around the unit number when D is sufficiently close to M and NN^T is sufficiently close to $C+BD^{-1}B^T$.

Finally, if $\widetilde{M}-I_1$ and/or $\widetilde{R}-\widetilde{C}$ are indefinite, then the eigenvalues may be complex but we can estimate the absolute value of λ by simple norm inequalities and clustering around the unit number occurs as before.

2.5 A Scaling of $\mathcal{A}$

We present now a method which can be seen as an algebraic version of the method proposed in [10], see Remark 1.

Consider a scaled matrix

$$\widehat{\mathcal{A}}=\begin{bmatrix}D_1^{-1} & 0\\ 0 & D_2^{-1}\end{bmatrix}\begin{bmatrix}A_{11} & A_{12}\\ A_{21} & -A_{22}\end{bmatrix}\begin{bmatrix}D_1^{-T} & 0\\ 0 & D_2^{-T}\end{bmatrix}=\begin{bmatrix}\widehat{A}_{11} & \widehat{A}_{12}\\ \widehat{A}_{21} & -\widehat{A}_{22}\end{bmatrix},$$

where $\widehat{A}_{ij}=D_i^{-1}A_{ij}D_j^{-T}, i,j=1,2.$ and where we assume that $\mathbf{x}_1^T\widehat{A}_{11}\mathbf{x}_1\geq\alpha\mathbf{x}_1^T\mathbf{x}_1$ for some $\alpha>1$.

We consider the one-sided (left) preconditioned matrix

$$B = \begin{bmatrix} \widehat{A}_{11} - I_1 & 0 \\ \widehat{A}_{21} & -I_2 \end{bmatrix} \begin{bmatrix} \widehat{A}_{11} & \widehat{A}_{12} \\ \widehat{A}_{21} & -\widehat{A}_{22} \end{bmatrix} = \begin{bmatrix} (\widehat{A}_{11} - I_1)\widehat{A}_{11} & (\widehat{A}_{11} - I_1)\widehat{A}_{12} \\ \widehat{A}_{21}(\widehat{A}_{11} - I_1) & \widehat{A}_{22} + \widehat{A}_{12}\widehat{A}_{12} \end{bmatrix} \quad (8)$$

This matrix is symmetric and also positive definite, because $(\widehat{A}_{11} - I_1)\widehat{A}_{11} > 0$ and its Schur complement S_B turns out to be equal to the Schur complement of $\widehat{A}$, $S_{\widehat{A}} = \widehat{A}_{22} + \widehat{A}_{21}\widehat{A}_{11}^{-1}\widehat{A}_{12}$. Further, $S_{\widehat{A}} = D_2^{-1}(A_{22} + A_{21}A_{11}^{-1}A_{12})D_2^{-T}$. Thus, $D_2D_2^T$ should be a preconditioner to $A_{22} + A_{21}A_{11}^{-1}A_{12}$, or rather to $A_{22} + A_{21}(D_1D_1^T)^{-1}A_{12}(= C + BM_0^{-1}B^T)$.

Finally, B can be scaled itself:

$$\widehat{B} = \begin{bmatrix} \sigma I_1 & 0 \\ 0 & I_2 \end{bmatrix} B \begin{bmatrix} \sigma I_1 & 0 \\ 0 & I_2 \end{bmatrix} = \begin{bmatrix} \sigma^2(\widehat{A}_{11} - I_1)\widehat{A}_{11} & \sigma(\widehat{A}_{11} - I_1)\widehat{A}_{12} \\ \sigma\widehat{A}_{21}(\widehat{A}_{11} - I_1) & \widehat{A}_{22} + \widehat{A}_{21}\widehat{A}_{12} \end{bmatrix},$$

where $\sigma = 1/(\sqrt{\alpha}\sqrt{\alpha - 1})$. The matrix $\widehat{B}$ has a condition number close to unity if $\widehat{A} \approx \alpha I_1$, $\alpha \approx 1$ and $\widehat{A}_{22} + \widehat{A}_{21}\widehat{A}_{12} \approx I_2$. (Note that the off-diagonal blocks of $\widehat{B}$ are of order σ^{-1}.)

We can use an unpreconditioned CG method to solve the corresponding system with $\widehat{B}$ or use the preconditioner $\begin{bmatrix} A_0 & 0 \\ 0 & B_0 \end{bmatrix}$ to

$$\begin{bmatrix} \sigma^2 A_{11}(A_0^{-1}A_{11} - I_1) & \sigma(A_{11}A_0^{-1} - I_1)A_{12} \\ \sigma A_{21}(A_0^{-1}A_{11} - I_1) & A_{22} + A_{12}A_0^{-1}A_{12} \end{bmatrix},$$

where $A_0 = D_1D_1^T$ and $B_0 = D_2D_2^T$. It is then not necessary to have A_0 and B_0 on such factorized forms. For each action of this preconditioner and the matrix, three actions (i.e., solutions of corresponding systems) of A_0^{-1}, two actions of A_{11} and one action of $A_{ii}, i = 1, 2$ and of B_0^{-1} are required.

3 Numerical Tests

The behaviour of the proposed indefinite system preconditioners is illustrated on the stationary Stokes problem, described in detail in [3],

$$\begin{aligned} -\Delta\mathbf{u} + \nabla p &= \mathbf{f} \quad \text{in } \Omega \\ \operatorname{div}\mathbf{u} &= 0 \quad \text{in } \Omega \\ \mathbf{u} &= \mathbf{g} \quad \text{on } \partial\Omega \end{aligned} \quad (9)$$

where $\Omega \subset \mathbb{R}^2$ is a bounded domain with boundary $\partial\Omega$ and $\int_{\partial\Omega} \mathbf{gn} = \mathbf{0}$. Let $\Omega = (0,1)^2$ and $\mathbf{f}$ and the boundary conditions are computed so that the exact solution of (9) is $u(x,y) = x^3 + x^2 - 2xy + x$, $v(x,y) = -3x^2y + y^2 - 2xy - y$ and $p(x,y) = x^2 + y^2$.

We use a regularized method, namely, we solve a system with a matrix $\mathcal{A} = \begin{bmatrix} M & B^T \\ B & -\sigma C \end{bmatrix}$. The choice of the regularization parameter σ is broadly discussed in [3].

The problem is solved using the preconditioned GCG-MR method (cf. [2]) and a preconditioner $\widetilde{\mathcal{A}}$ to $\mathcal{A}$ of the form (4). D_1 in our experiments is the AMLI-preconditioner constructed for the diagonal blocks of M. During each GCG-MR iteration, systems with the preconditioner are solved via its Schur complement $S_{\widetilde{\mathcal{A}}} = -(\sigma C + B\,\mathrm{AMLI}[M]\,B^T)$. Relative stopping criteria are used for both GCG-MR and for the CG methods, namely, $\|\mathbf{r}^{(k)}\|/\|\mathbf{r}^{(0)}\|$ is checked to be less than 10^{-6} and 10^{-4}, respectively. The results of this experiment are shown in Table 1. The systems with $S_{\widetilde{\mathcal{A}}}$ can be solved by an unpreconditioned CG method since for the particular regularized formulation its condition number is independent of the discretization parameter h (column 4 in Table 1). One can further improve the total cost of the method by using a preconditioned CG to solve $S_{\widetilde{\mathcal{A}}}$ (column 5 in Table 1). Finally, the factorized form (5) of the preconditioner on indefinite form is used, where the NN^T is constructed from $\sigma C + BD_0^{-1}B^T$, where D_0 is a diagonal approximation of AMLI[M]. Since in this version no coupled iterations occur, it outperforms the Schur complement based solver. The numerical tests are performed in `Matlab`.

Table 1. Problem (9)

		Method (4)			Method (5)
		GCG-MR	Aver.	Aver.	GCG-MR
h	*size*($\mathcal{A}$)	iter.	CG iter.	PCG iter.	iter.
0.0667	768 (3 × 256)	7	24	14	15
0.0323	3072 (3 × 1024)	7	25	16	18
0.0159	12288 (3 × 4096)	8	25	18	19
0.0039	196608 (3 × 65536)	8	25	18	20

References

1. Axelsson O.: Preconditioning of indefinite problems by regularization. *SIAM Journal on Numerical Analysis*, 16 (1979), 58-69.
2. Axelsson O.: *Iterative Solution Methods*, Cambridge University Press, Cambridge, 1994.
3. Axelsson O., Barker V.A., Neytcheva M., Polman B.: Solving the Stokes problem on a massively parallel computer. *Mathematical Modelling and Analysis*, 4 (2000), 1-22.

4. Axelsson O., Gustafsson I.: An iterative solver for a mixed variable variational formulation of the (first) biharmonic problem. *Computer Methods in Applied Mechanics and Engineering*, 20 (1979), 9-16.
5. Axelsson O., Gustafsson I.: An efficient finite element method for nonlinear diffusion problems. *Bulletin Greek Mathematical Society*, 22 (1991), 45-61.
6. Axelsson O, Makarov M. On a generalized conjugate orthogonal residual method. *Numerical Linear Algebra with Applications*, 1995; 2:467-480.
7. Axelsson O., Neytcheva M.: Preconditioning methods for linear systems arising in constrained optimization problems. Submitted to *Numerical Linear Algebra with Applications*, January 2002.
8. Axelsson O., Vassilevski P.S.: Variable-step multilevel preconditioning methods. I. Selfadjoint and positive definite elliptic problems. *Numer. Linear Algebra Appl.*, 1 (1994), 75-101.
9. Braess D.: *Finite Elements. Theory, fast solvers, and applications in solid mechanics.* Cambridge University Press, Cambridge, 2001. (Second edition)
10. Bramble JH, Pasciak JE.: A preconditioning technique for indefinite systems resulting from mixed approximations of elliptic problems. *Mathematics of Computation*, 1988; 50:1-17.
11. Elman H.C.: Preconditioning for the steady-state Navier-Stokes equations with low viscosity. *SIAM Journal on Scientific Computing*, 20 (1999), 1299-1316.
12. H.C. Elman and D. Silvester, Fast nonsymmetric iterations and preconditioning for Navier-Stokes equations. *SIAM Journal on Scientific Computing*, 17 (1996), 33-46.
13. Ewing R.E., Lazarov R., Lu P., Vassilevski P.: Preconditioning indefinite systems arising from mixed finite element discretization of second order elliptic problems. In Axelsson O., Kolotilina L. (eds.): *Lecture Notes in Mathematics*, 1457, Springer-Verlag, Berlin, 1990.
14. Klawonn A.: Block-triangular preconditioners for saddle point problems with a penalty term. *SIAM Journal on Scientific Computing*, 1998; 19:172-184.
15. Lukšan L., Vlček J.: Indefinitely preconditioned inexact Newton method for large sparse equality constrained non-linear programming problems. *Numerical Linear Algebra with Applications*, 5(1998), 219-247.

Solving Eigenproblems: From Arnoldi via Jacobi-Davidson to the Riccati Method

Jan H. Brandts

Korteweg-de Vries Institute, Faculty of Science, University of Amsterdam,
Plantage Muidergracht 24, 1018 TV Amsterdam, Netherlands,
brandts@science.uva.nl

Abstract. The formulation of eigenproblems as generalized algebraic Riccati equations removes the non-uniqueness problem of eigenvectors. This basic idea gave birth to the Jacobi-Davidson (JD) method of Sleijpen and Van der Vorst (1996). JD converges quadratically when the current iterate is close enough to the solution that one targets for. Unfortunately, it may take quite some effort to get close enough to this solution. In this paper we present a remedy for this. Instead of linearizing the Riccati equation (which is done in JD) and replacing the linearization by a low-dimensional linear system, we propose to replace the Riccati equation by a low-dimensional Riccati equation and to solve it exactly. The performance of the resulting *Riccati algorithm* compares extremely favorable to JD while the extra costs per iteration compared to JD are in fact negligible.

Keywords: Arnoldi, Riccati, Jacobi-Davidson, Krylov subspace.

1 Introduction

The standard eigenvalue problem $Ax = \lambda x$ for possibly non-Hermitian matrices A is one of the basic building blocks in computational sciences. Solution methods, which are necessarily iterative, range from the QR algorithm (if all eigenvalues are wanted) via the Arnoldi [1] method to Jacobi-Davidson [4] for large problems from which only few eigenpairs are needed. As a matter of fact, both the Arnoldi method and the Jacobi-Davidson method can be derived as Ritz-Galerkin projection methods that use subspaces of growing dimension, and in which the expansion of the subspace is governed by adding approximations of solution(s) of a generalized algebraic Riccati equation. We will show this in Section 2, and then discuss in Section 3 a third, very natural method based on the same idea. This method was briefly introduced in [2]. In Section 4, convincing numerical evidence of the success of the new approach is given, using matrices from the Matrix Market test-collection.

2 Projection on Expanding Subspaces

A straightforward tool to tackle the eigenvalue problem $Ax = \lambda x$ in $\mathcal{R}^n$ is to project it on a k dimensional subspace $\mathcal{V}$ of $\mathcal{R}^n$ with $k \ll n$. By this we mean

I. Dimov et al. (Eds.): NMA 2002, LNCS 2542, pp. 167–173, 2003.

the following. Assume that V is an $n \times k$ matrix of which the columns span $\mathcal{V}$. Approximations of eigenpairs of A can be found in $\mathcal{V}$ by computing vectors $v \in \mathcal{V}$ such that, instead of Av itself, the *orthogonal projection* $P_{\mathcal{V}}Av$ of Av onto $\mathcal{V}$ is a (scalar) multiple of v. The condition $v \in \mathcal{V}$ can be expressed as $\exists y \in \mathcal{R}^k : v = Vy$, whereas the condition $w \perp \mathcal{V}$ translates to $V^*w = 0$, i.e., w is orthogonal to each of the columns of V and hence to their span $\mathcal{V}$. So, the problem to solve, usually called the *projected problem*, becomes

$$\text{find } y \in \mathcal{R}^k \text{ and } \mu \in \mathcal{R} \text{ such that } V^*(AVy - \mu Vy) = 0. \tag{1}$$

Note that both V^*AV and V^*V are small $k \times k$ matrices, and that when the columns of V are an *orthonormal basis* for $\mathcal{V}$, then V^*V is the identity matrix, hence (1) a small standard eigenvalue problem, which can be solved routinely by for instance the QR algorithm. The couples (μ, Vy) can be interpreted as *approximate eigenpairs* for A, and are usually called *Ritz-pairs*. If the *residuals* $r = AVy - \mu Vy$, which can easily be computed *a posteriori*, are not small enough according to the wishes of the user, expansion of the subspace, followed by solving a new projected problem, may lead to improvement. The following pseudo-code presents a general framework for this approach. In each execution of the while-loop, the subspace dimension increases by one, in some direction q, to be specified below.

input: A, V, ε; matrix, first subspace, the desired residual reduction
$W = AV$;
$M = V^*W$; *initial projected matrix; note that* $M = V^*AV$
$r = s =$ *residual of projected problem*;
while $\|r\| > \varepsilon\|s\|$
  q $=$ *new expansion vector for* $\mathcal{V}$;
  v $=$ *such that span* $(V|v)$ *equals span* $(V|q)$, *and* $(V|v)$ *is orthonormal*;
  w $= Av$;
  $M = \begin{pmatrix} M & V^*w \\ v^*W & v^*w \end{pmatrix}$ *efficient implementation of projection* $M = (V|v)^*(W|w)$ *using previous* M;
  V $= (V|v)$ *expansion of the subspace* $\mathcal{V}$;
  W $= (W|w)$ *expansion of the subspace* $\mathcal{W}$;
  r $=$ *residual of the new projected problem*;
end (while)

We will now discuss expansion strategies, i.e., how to choose q such that projection on the expanded space may yield better approximate eigenpairs. For this, suppose that for the eigenvalue problem $Ax = \lambda x$ we have an approximate eigenvector v with $\|v\| = 1$ that was obtained by projection of the problem on a subspace $\mathcal{V}$. Consider the affine variety $v^\perp = \{x + v | x^*v = 0\}$. Generally, there will be n points v_j in $v^\perp$ corresponding to eigenvectors of the matrix A, which are the intersections of lines (eigenvector directions) through the origin, with the affine variety $v^\perp$. Each of those points, obviously, can be written as $v_j = v + q_j$, with $q_j^*v = 0$. Writing $\mu = v^*Av$ and $r = Av - \mu v$, it is not hard to show that the vectors q_j are the roots of the following generalized algebraic Riccati equation in q,

$$q^*v = 0 \quad \text{and} \quad (I - vv^*)Aq - q\mu = q(v^*A)q - r. \tag{2}$$

If we intend to use approximations to solutions q_j of this equation to expand the subspace $\mathcal{V}$, we should find a balance between the effort spent on computing such approximations, and what we get back in terms of improved approximations from the bigger space. Note that one of the crudest (hence cheapest) approximations would result from replacing A by the identity I, which gives $\hat{q} = -(1-\mu)^{-1}r$ as approximate root q. The resulting algorithm is in fact the Arnoldi method. Note that the orthogonalization step in the while loop above becomes then superfluous, making the method even less numerically expensive.

The Jacobi-Davidson method [4] results, when the Riccati equation (2) is linearized around $q = 0$. The linearized equation

$$\hat{q}^*v = 0 \quad \text{and} \quad (I - vv^*)A\hat{q} - \hat{q}\mu = -r \tag{3}$$

is, in turn, usually only solved approximately. The approximate solution of (3) is then used to expand the subspace $\mathcal{V}$, and a new approximate eigenvector v is extracted from the expanded space by projection, and the process repeated. Equation (3) can be approximated by projection as well, say on a ℓ-dimensional subspace $\mathcal{U}$ with and $r \in \mathcal{U}$ and $\mathcal{U} \perp v$. Note that the latter requirement assures that an approximation in $v^\perp$ is found. If U is a matrix whose orthonormal columns span $\mathcal{U}$, then the projected equation would be $U^*AU\hat{z} - \hat{z}\mu = -U^*r$. For $\ell = 1$, this gives the Arnoldi method again. Using higher values for ℓ results in a structurally different method, whereas solving (3) exactly is the "full" Jacobi-Davidson method of which can be proved that asymptotically, it converges quadratically. Therefore, much attention has been paid to finding good preconditioners to solve (3) in high precision with little effort.

Perhaps the most important observation is, that due to the linearization, the effort in solving (3) accurately is only worthwhile if the quadratic term in (2) could really be neglected, which is only the case if there is a solution q of (2) with $\|q\|$ small enough, i.e., close enough to zero. Thus, v needs to be a rather good approximation of an eigenvector. It has been quantified in [3] that this is the case if $\sigma^2 - 12\|r\|\|v^*A\| > 0$, where σ is the smallest singular value of A projected on $v^\perp$. It has moreover been observed in experiments that this condition seems necessary, and also that it can be very restrictive. This explains why the Jacobi-Davidson method shows an initial phase of no structural residual reduction, before it plunges into the quadratically convergence region. Especially if the start-vector for Jacobi-Davidson is chosen badly (for instance, randomly), this initial phase can be very long and hence very expensive.

3 Curing the Stagnation The Riccati Method

The phase in the Jacobi-Davidson method in which the current eigenvector approximation is still outside the quadratic convergence region can be significantly reduced, as was firstly observed in Section 4.2 of [2]. There we proposed to project (2) directly on an ℓ-dimensional subspace $\mathcal{U}$, and to observe that if ℓ is moderate, we can refrain from linearizing the resulting ℓ-dimensional nonlinear equation, and compute *all* its roots instead.

Consider the Jacobi-Davidson method. Suppose that in solving (3), projection on a subspace $\mathcal{U}$ with $r \in \mathcal{U}$ and $\mathcal{U} \perp v$ is used, as suggested below equation (3). As before, let U contain an orthonormal basis for $\mathcal{U}$. Then the projected linearized equation reads as

$$U^*AU\hat{z} - \mu\hat{z} = -U^*r, \tag{4}$$

after which $U\hat{z}$ is the approximation of $\hat{q}$ with which $\mathcal{V}$ is going to be expanded. Alternatively, we could also have used the subspace $\mathcal{U}$ to approximate the Riccati equation (2) directly, without linearization, yielding the ℓ-dimensional projected Riccati equation

$$U^*AUz - \mu z = z(v^*AU)z - U^*r. \tag{5}$$

In fact, (4) is the linearization around $z = 0$ of (5), so it seems as if no progress has been made, apart from the fact that linearization and projection commute. There is, however, an important difference, which is that there is no need anymore to linearize the low-dimensional Riccati equation. It can be solved exactly by realizing that it is equivalent to the $(\ell+1) \times (\ell+1)$ eigenvalue problem

$$\left(\begin{array}{c|c} \mu & v^*AU \\ \hline U^*r & U^*AU \end{array}\right) \begin{pmatrix} 1 \\ z \end{pmatrix} = (\mu + v^*AUz) \begin{pmatrix} 1 \\ z \end{pmatrix}. \tag{6}$$

The gain in comparison to Jacobi-Davidson is, that instead of obtaining an approximation of a unique correction $\hat{q}$ of (3), we get a small number $\ell+1$ approximations $\tilde{q}_j$ of the solutions q of (2). This gives extra freedom in deciding how to expand the space $\mathcal{V}$, for example, by computing the Ritz values corresponding to the vectors $v + \tilde{q}_j, j \in \{1, \ldots, \ell+1\}$ and using the $\tilde{q}_j$ that gives a Ritz value closest to a given target to expand $\mathcal{V}$. We will call the resulting method the *Riccati method* if $\mathcal{U}$ is chosen as the ℓ-dimensional Krylov subspace for r and $(I - vv^*)A$. Note that the presence of the projection $(I - vv^*)$ assures that this Krylov subspace $\mathcal{U}$ will be orthogonal to v. The following remarks are of interest:

• For moderately small $\ell \ll n$, the costs for solving the eigenvalue problem (6) are negligible compared to the ℓ matrix-vector multiplications with A that are needed to construct the projected matrix U^*AU, which is needed in both JD (4) and Riccati (6). This shows that the Riccati method is only slightly more expensive per iteration than JD. By one iteration we mean expansion of $\mathcal{V}$ with a new vector q.

• If v is very close to an eigenvector, then there exists a solution q of (2) with small norm. If there exists a solution z of (5) such that Uz is very close to q, then the solution $\hat{z}$ of (4) will yield an accurate approximation $U\hat{z}$ of q, since the quadratic term in z will then be negligible. Hence, JD would give a good approximation of this eigenvector. In the Riccati method, we would have the *option* to expand the space in (almost) the same way as in JD, resulting in similarly good approximations. In this case, the advantage of Riccati is, that if we are not interested in this particular eigenvector but in another one, we can refrain from expansion with the JD correction and point towards a different direction. This shows that it is unlikely that JD will outperform Riccati: if the JD correction $\hat{z}$ is good, then one of the possible corrections in the Riccati method will be very close to $\hat{z}$.

• In case v is *not* a good approximation of the eigenvector in which one is interested, the Jacobi-Davidson correction $\hat{q}$ may not make sense in the same way as a step of the Newton method may not make sense far away from the objective solution. Since in the Riccati method there are other possible expansion vectors (also called corrections) to choose from, this may lead to substantial improvement.

In solving practical problems with the Jacobi-Davidson method, it is not unusual that the expansion vector $\hat{q}$ is computed from (3) by using a few iterations of a Krylov subspace method like Conjugate Gradients or GMRES. We have just argued that the Krylov subspace built in those methods could better be used to project the Riccati equation (2) on, resulting in a small eigenvalue problem (6). We will now support this claim by numerical experiments with matrices from the Matrix Market test collection.

For information on the matrices from Matrix Market, see the webpage

http://math.nist.gov/MatrixMarket/index.html

4 Numerical Experiments

We will now list some results in comparing Jacobi-Davidson with the Riccati method. In order to keep things as simple as possible and to illustrate the main idea, we did not include sophisticated features that could enhance both methods equally well. Therefore, we did not consider restart techniques, harmonic projections, and preconditioning.

For each matrix, we selected a random start vector. This start vector was used for each of the experiments with the same matrix. JD and Riccati were compared in two aspects: cpu-time and number of iteration steps needed to reduce the initial residual by a factor 10^{10}. Since the absolute numbers are not important for the comparison, we list, in the tabular below, the relative numbers only. As an example, the number 0.31 that is listed for the matrix plat1919 for $\ell = 5$ and belonging to cpu-time, indicates that Riccati needed only 0.31 of the cpu-time that JD needed to attain the same residual reduction.

For two of the larger matrices, we give the convergence histories in Figure 1. On the horizontal axis is the iteration number, on the vertical axis the 10-log of the relative residual reduction.

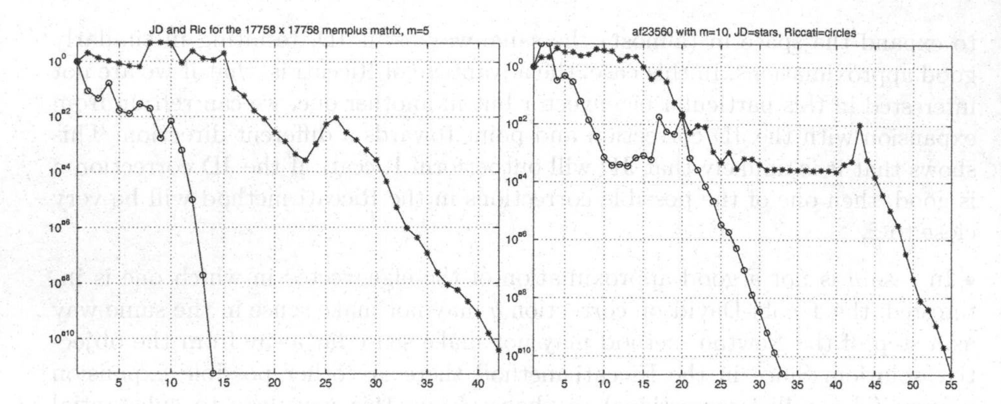

Fig. 1. Comparing JD (asterisks) and Riccati (circles) in convergence behaviour for the real nonsymmetric 17758 × 17758 matrix memplus for $\ell = 5$ (left) and for the real nonsymmetric 23560 × 23560 matrix af23560 with $\ell = 10$ (right)

matrix name:	size	ℓ	5	10	20
sherman4	1104	cpu	0.68	0.10	0.05
real nonsymmetric		its	0.71	0.29	0.11
nnc1374	1374	cpu	0.55	0.28	0.12
real nonsymmetric		its	0.80	0.52	0.22
plat1919	1919	cpu	0.31	0.08	0.02
symmetric indefinite		its	0.53	0.17	0.06
utm3060	3060	cpu	1.07	0.81	0.43
real nonsymmetric		its	0.97	0.68	0.37
lshp3466	3466	cpu	0.79	1.09	1.07
symmetric indefinite		its	0.90	0.92	0.76
bcsstm24	3562	cpu	0.26	0.14	0.06
symmetric positive definite		its	0.44	0.22	0.10
rw5151	5151	cpu	0.99	1.03	0.82
real nonsymmetric		its	1.00	1.00	0.81
cry10000	10000	cpu	0.34	0.10	0.05
real nonsymmetric		its	0.44	0.16	0.07
memplus	17758	cpu	0.17	0.19	0.08
real nonsymmetric		its	0.33	0.23	0.10
af23560	23560	cpu	0.83	0.49	0.91
real nonsymmetric		its	0.89	0.60	0.90
bcsstk32	44609	cpu	1.28	0.90	0.25
symmetric indefinite		its	0.65	0.39	0.13

5 Conclusions

As appears from the numerical experiments, the plain Riccati method almost always outperforms the plain Jacobi-Davidson method, and in many cases by a large factor. When Riccati loses, the difference is almost negligible. This suggests to incorporate the Riccati idea in existing JD codes which use Krylov subspaces to solve the Jacobi-Davidson correction equation.

As suggestions for further research one could think of recursiveness of this idea, since it is in principle a nested subspace method with inner and outer loop, i.e. the space $\mathcal{V}$ is the space built in the outer loop, whereas for each expansion of $\mathcal{V}$ in one direction, as space $\mathcal{U}$ is constructed. It may be of interest to develop general theory for nested subspace methods for eigenvalue problems, which do not take into account that $\mathcal{U}$ is a Krylov subspace. Other choices may be of interest as well.

For readers more interested in the theoretical aspects of the method, and in its adaptation as a block method for the computation of invariant subspaces we refer to [2]. In contrast to [2], the underlying paper presents clear heuristics for its success and is therefore particularly suitable for people from computational science, computer science, and physics.

Acknowledgments

The research leading to this paper and its corresponding presentation at the Fifth International Conference on Numerical Methods and Applications in Borovetz, Bulgaria, August 20-24, 2002, has been made possible through a Fellowship of the Royal Netherlands Academy of Arts and Sciences (KNAW). The author gratefully acknowledges the support.

References

1. Arnoldi, W.E.: The Principle of Minimized Iteration in the Solution of the Matrix Eigenvalue Problem. Quart. Appl. Math., 9 (1951) 17–29.
2. Brandts, J.H.: The Riccati Method for Eigenvalues and Invariant Subspaces of Matrices with Inexpensive Action. Linear Algebra Appl., (2002) accepted.
3. Demmel, J.: Three Methods for Refining Estimates of Invariant Subspaces. Computing, 38 (1987) 43–57.
4. Sleijpen, G.L.G. and van der Vorst, H.A.: Jacobi-Davidson Iteration Method for Linear Eigenvalue Problems. SIAM J. Matrix Anal. Applic., 17 (1996) 401–425.

On a Multigrid Adaptive Refinement Solver for Saturated Non-Newtonian Flow in Porous Media

Willy Dörfler[1], Oleg Iliev[2], Dimitar Stoyanov[2], and Daniela Vassileva[3]

[1] Institut für Angewandte Mathematik II, Universität Karlsruhe (TH), D-76128 Karlsruhe, Germany
doerfler@math.uni-karlsruhe.de

[2] Fraunhofer Institut für Techno- und Wirtschaftsmathematik (ITWM), Europaallee 10, D-67657 Kaiserslautern, Germany
{iliev,stoyanov}@itwm.fhg.de

[3] Institute of Mathematics and Informatics, Bulgarian Academy of Sciences, Acad. G. Bonchev str., bl. 8, BG-1113 Sofia, Bulgaria
vasileva@math.bas.bg

Abstract. A multigrid adaptive refinement algorithm for non-Newtonian flow in porous media is presented. The saturated flow of non-Newtonian fluid is described by continuity equation and generalized Darcy law. The resulting second order nonlinear elliptic equation is discretized by finite volume method on cell-centered grid. A nonlinear full-multigrid, full-approximation-storage algorithm is implemented. Singe grid solver, based on Picard linearization and Gauss-Seidel relaxation, is used as a smoother. Further, a local refinement multigrid algorithm on a composite grid is developed. A residual based error indicator is used in the adaptive refinement criterion. A special implementation approach is used, which allows us to perform unstructured local refinement in conjunction with the finite volume discretization. Several results from numerical experiments are presented in order to examine the performance of the solver.

Key words: nonlinear multigrid, adaptive refinement, non-Newtonian flow in porous media.

1 Introduction

Numerical simulation of flow in porous media is essential for better understanding and proper control in many environmental and technological processes. In the case of saturated flows of Newtonian fluids, a lot of attention is paid to developing efficient multilevel and local refinement solvers. At the same time, numerical algorithms for flow of non-Newtonian fluid in porous media are not so well developed. A reason for this is that the problem in this case is essentially nonlinear one. Main difficulties in solving such problems are related to the (strong) nonlinearity of the coefficients, stronger local effects (compare to the linear case), and possible degeneration of the type of equation. In the current work we consider a numerical algorithm, which allows to efficiently solve non-degenerating problems using a nonlinear multigrid algorithm and adaptive

I. Dimov et al. (Eds.): NMA 2002, LNCS 2542, pp. 174–181, 2003.

local mesh refinement within the multigrid procedure. The degenerating case is a subject of a current research, and we plan to present results in a forthcoming paper.

The reminder of the paper is organized as follows. Section 2 concerns the governing equations and their discretization. The third section describes the multigrid procedure and the adaptive local refinement algorithm. The results from numerical experiments are presented in the fourth section.

2 Governing Equation and Single-Grid Solution Method

The governing equations are the first order system consisting of the generalized Darcy law for single phase (power law non-Newtonian fluid) and the continuity equation [1]. It can be rewritten as a nonlinear second order equation:

$$\sum_{d=1}^{D} \frac{\partial}{\partial x_d}\left(K_d \left| \frac{\partial p}{\partial x_d} \right|^m \frac{\partial p}{\partial x_d} \right) = f, \qquad K_d > 0, \ m = \frac{1}{n} - 1 \geq 0, \ D = 2, 3, \quad (1)$$

where p is the pressure, K_d are coefficients of a diagonal permeability tensor, n is a parameter of the medium. For example, $n < 1$ for pseudoplastic fluids such as polymers and heavy oils.

The finite volume method is used for discretization of Eq. (1) on a cartesian rectangular cell-centered grid (i.e., the unknowns are related to the centers of the rectangles). For brevity we consider here the 2D case, but the 3D case is also implemented in our solver. For discretization in the case of local refinement we refer also to [2]. We let

$$h_2 \left[a_{i+1/2,j} \frac{p_{i+1,j} - p_{ij}}{h_1} - a_{i-1/2,j} \frac{p_{ij} - p_{i-1,j}}{h_1} \right] +$$
$$h_1 \left[b_{i,j+1/2} \frac{p_{i,j+1} - p_{ij}}{h_2} - b_{i,j-1/2} \frac{p_{ij} - p_{i,j-1}}{h_2} \right] = h_1 h_2 f_{ij} \,,$$

where the coefficients $k_d \left(\frac{\partial p}{\partial x_d} \right) := K_d \left| \frac{\partial p}{\partial x_d} \right|^m$ are approximated using arithmetic averaging

$$a_{i+1/2,j} := 0.5(k_{1,ij} + k_{1,i+1,j}), \quad b_{i,j+1/2} := 0.5(k_{2,ij} + k_{2,i,j+1}),$$

where $k_{1,ij} := k_1 \left(\frac{p_{i+1,j} - p_{i-1,j}}{2h_1} \right)$, $k_{2,ij} := k_2 \left(\frac{p_{i,j+1} - p_{i,j-1}}{2h_2} \right)$.

A nonlinear (FAS) multigrid iterative method (to be described below) is used to solve the above system of nonlinear algebraic equations. On each grid, a single grid solver based on the Picard method (i.e., the nonlinear coefficients are computed using the previous iterate p) is used. In this case the linearized system looks as follows:

$$h_2 \left[a^{(s-1)}_{i+1/2,j} \frac{p^{(s)}_{i+1,j} - p^{(s)}_{ij}}{h_1} - a^{(s-1)}_{i-1/2,j} \frac{p^{(s)}_{ij} - p^{(s)}_{i-1,j}}{h_1} \right] +$$
$$h_1 \left[b^{(s-1)}_{i,j+1/2} \frac{p^{(s)}_{i,j+1} - p^{(s)}_{ij}}{h_2} - b^{(s-1)}_{i,j-1/2} \frac{p^{(s)}_{ij} - p^{(s)}_{i,j-1}}{h_2} \right] = h_1 h_2 f^{(s-1)}_{ij} \,.$$

The resulting system of linear algebraic equations is iterated using the symmetric Gauss-Seidel Method (i.e. two Gauss-Seidel sweeps are performed on each linear iteration, in the second sweep the unknowns are updated in the reverse order).

3 Multigrid — Adaptive Local Refinement Algorithm

A nonlinear full-multigrid, full-approximation-storage (FMG-FAS) algorithm [3, 4], which uses the aforementioned single-grid solver as a smoother, has been implemented. The algorithm is briefly described below: we solve on coarser grids in order to provide a good initial guess on the fine grids. In order to account for the nonlinearity of the problem, the FMG is combined with the FAS algorithm The consecutive steps of a two-grid algorithm are listed below:

- Perform ν_1 nonlinear pre-relaxations on $N^h(u^h) = f^h$ on the fine grid;
- Compute the residual $r^h = f^h - N^h(u^h)$ and transfer it to the coarse grid $r^{2h} = I_h^{2h} r^h$;
- Transfer the fine grid solution to the coarse grid $\bar{u}^{2h} = I_h^{2h} u^h$;
- Solve (exactly, or perform ν_{cg} nonlinear iterations for) the coarse grid equation $N^{2h}(u^{2h}) = r^{2h} + N^{2h}(\bar{u}^{2h})$ in order to obtain the correction $c^{2h} = u^{2h} - \bar{u}^{2h}$;
- Interpolate the correction to the fine grid and correct the fine-grid solution $u^h = u^h + I_{2h}^h c^{2h}$.
- Perform ν_2 nonlinear post-relaxations on $N^h(u^h) = f^h$ on the fine grid;

The restriction operator for the residuals is obtained by the volumes' weighted summation of the residuals over the underlying fine-grid control volumes (CVs), while the projection of the solution is obtained as a mean value. Bilinear and trilinear prolongation operators are employed for two- and three-dimensional cases, respectively.

A local refinement multigrid algorithm on a *composite grid* is used. The algorithm is suggested in [5] for vertex based nested grids, it is extended in [2, 6] to the case of cell centered nested grids, and is discussed in [7, 8] for the case of cell centered non-nested grids. Here we do a mass conserving discretization on the composite grid, i.e. the flux over an irregular coarse-grid volume wall is sum of the fluxes over the corresponding walls at the neighbouring fine-grid irregular volumes. In order to approximate the fluxes at the fine-grid irregular volumes, we introduce additional (auxiliary) points, in which the values of the solution are obtained by bilinear interpolation on the coarse grid. The fine grid–smoother works only within the refined subdomain, but on the lower grid levels the additional source term in FAS has to be calculated for the irregular coarse-grid CVs, as well.

Adaptive refinement criterion. A residual based error indicator [9, 10] is used

$$g^2(T) = h \, \| \, \frac{1}{k_{h,E}^{1/2}} \left[(k_h \nabla p_h) \cdot \mathbf{n} \right] \|^2_{\partial T / \partial \Omega} + h^2 \, \| \, \frac{1}{k_h^{1/2}} \left(\nabla \cdot (k_h \nabla p_h) - f_h \right) \|^2_T \, ,$$

where T stands for the current CV, p_h is a bilinear interpolant to the solution defined by its values in the nodes of T (and these values are obtained by bilinear interpolation using the values of the discrete solution in the centers of the CVs), $k_h := \max(k_{1,h}, k_{2,h})$, $k_{d,h} = k_d\left(\frac{\partial p_h}{\partial x_d}\right)$, $d = 1, 2$ and $[\,.\,]_E$ denotes the difference between limits from either side of the edge $E \subset \partial T/\partial\Omega$ (with fixed normal direction $\mathbf{n}$). Note, that in our case the term $\nabla \cdot (k_h \nabla p_h)$ is identically equal to zero, because we use a bilinear interpolant to solution in each element.

Maximum strategy. The adaptive refinement at the end of stage l, for some $1 \leq l \leq l_{\max} - 1$, is done for all CVs T of the composite grid Ω_l, satisfying

$$g(T) > \gamma \max_{\Omega_l} g(T)\,.$$

Fixed energy fraction strategy. The adaptive refinement at the end of stage l, for some $1 \leq l \leq l_{\max} - 1$, is done in a subdomain $\omega_l \subset \Omega_l$ (as small as possible), satisfying

$$\sum_{\omega_l} g^2(T) \geq \gamma^2 \sum_{\Omega_l} g^2(T)\,.$$

4 Numerical Experiments

In the numerical experiments presented below, the boundary conditions (Dirichlet) and the right hand side $f(x, y)$ correspond to an exact solution of the form:

$$p(x, y; a, b) = \exp\left(-x^2/a^2 - y^2/b^2\right),$$

where a, b are parameters, influencing the steepness of the solution and its symmetry.

In all cases, $\nu_1 = 1$ nonlinear pre-smoothing and $\nu_2 = 2$ nonlinear post-smoothings are performed on the finer grid levels, while on the coarsest grid $\nu_{cg} = 30$ nonlinear smoothings are done. Five symmetric Gauss-Seidel linear iterations are performed within each nonlinear smoothing. The multigrid sweeps on a fixed grid are performed until the algebraic residuals fall 10^{-4} times, followed by a refinement and solving on the refined grid. Maximum six levels of refinement are used, the refinement on the first two grids is always global, i.e. within the whole domain.

Some results for the multigrid performance in the case of global refinement are presented in Table 1. The following notations are used. Stage corresponds to the level of refinement, with stage=1 denoting the coarsest grid, 10×10 control volumes (CVs). Sweeps stands for the number of nonlinear multigrid iterations. $\|\text{error}\|_C$ stands for the discrete maximum norm of the error, while $\|\text{error}\|_{L_2}$ denotes the discrete L_2 norm. Results are obtained for $K_1 = K_2 = 1$, and two sets of a and b. As can be seen, the maximum norm of the error and the L_2 one decrease four times when the grid is twice refined, which indicates second order of convergence. The number of multigrid sweeps on each multigrid stage is almost constant in the linear ($m = 0$) case and for the nonlinear case with $m = 0.5$,

Table 1. Multigrid performance.

stage	CVs	sweeps ($a=b=0.15$)	$\|error\|_C$ ($a=b=0.15$)	$\|error\|_{L_2}$ ($a=b=0.15$)	sweeps ($a=0.45,\ b=0.15$)	$\|error\|_C$ ($a=0.45,\ b=0.15$)	$\|error\|_{L_2}$ ($a=0.45,\ b=0.15$)
$m=0$							
2	400	1	2.8e-2	2.4e-3	2	2.4e-2	3.5e-3
3	1 600	2	7.1e-3	5.9e-4	2	6.3e-3	8.3e-4
4	6 400	2	1.8e-3	1.5e-4	2	1.6e-3	2.1e-4
5	25 600	2	4.4e-4	3.7e-5	2	4.1e-4	5.1e-5
6	102 400	2	1.1e-4	9.6e-6	2	1.0e-4	1.3e-5
$m=0.5$							
2	400	3	5.7e-2	4.3e-3	3	4.6e-2	6.3e-3
3	1 600	3	1.4e-2	1.0e-3	3	1.2e-2	1.5e-3
4	6 400	3	3.5e-3	2.5e-4	3	3.0e-3	3.6e-4
5	25 600	3	8.5e-4	6.2e-5	3	7.6e-4	8.8e-5
6	102 400	3	2.1e-4	1.6e-5	3	1.9e-4	2.2e-5
$m=1$							
2	400	5	6.7e-2	5.4e-3	8	5.7e-2	8.3e-3
3	1 600	6	1.6e-2	1.4e-3	9	1.4e-2	2.4e-3
4	6 400	7	3.9e-3	4.4e-4	8	3.5e-3	7.2e-4
5	25 600	6	1.1e-3	1.4e-4	9	9.5e-4	2.2e-4
6	102 400	7	2.9e-4	4.6e-5	12	2.6e-4	6.7e-5

which means that the amount of computations is proportional to the number of CVs. For $m = 1$ (stronger nonlinearity) the number of multigrid sweeps increases with the number of the multigrid stage (i.e. with the number of CVs). Let us note that we do not solve the linear systems obtained on each nonlinear smoothing till convergence, we only do a fixed number of linear iterations. A better MG performance can be achieved in this case using more nonlinear and linear smoothings.

Results using the multigrid adaptive refinement (MG-AR) algorithm are presented in Table 2 and Table 3 for $\gamma = 0.1$ and $\gamma = 0.025$, respectively. Because of the limited length of the paper, we discuss here only the maximum strategy for adaptive refinement. As it can be seen, the MG-AR solutions for $\gamma = 0.025$ have the same accuracy ($\|error\|_C$) as the MG ones, but significantly less resources (memory, i.e. active CVs, and CPU time) are used in this time. The choice of γ is still an open question. If too less CVs are selected for refinement (e.g., $\gamma = 0.1$, $m = 1$), the accuracy on the composite grid can be not so nice as for $\gamma = 0.025$. We still use less CVs (compare to full refinement case) to obtain certain accuracy, but we can not achieve second order convergence in this case.

The last composite grids for $\gamma = 0.025$ are shown on Fig.1 - the left column is for the case $a = b = 0.15$, the right – for $a = 0.45$, $b = 0.15$.

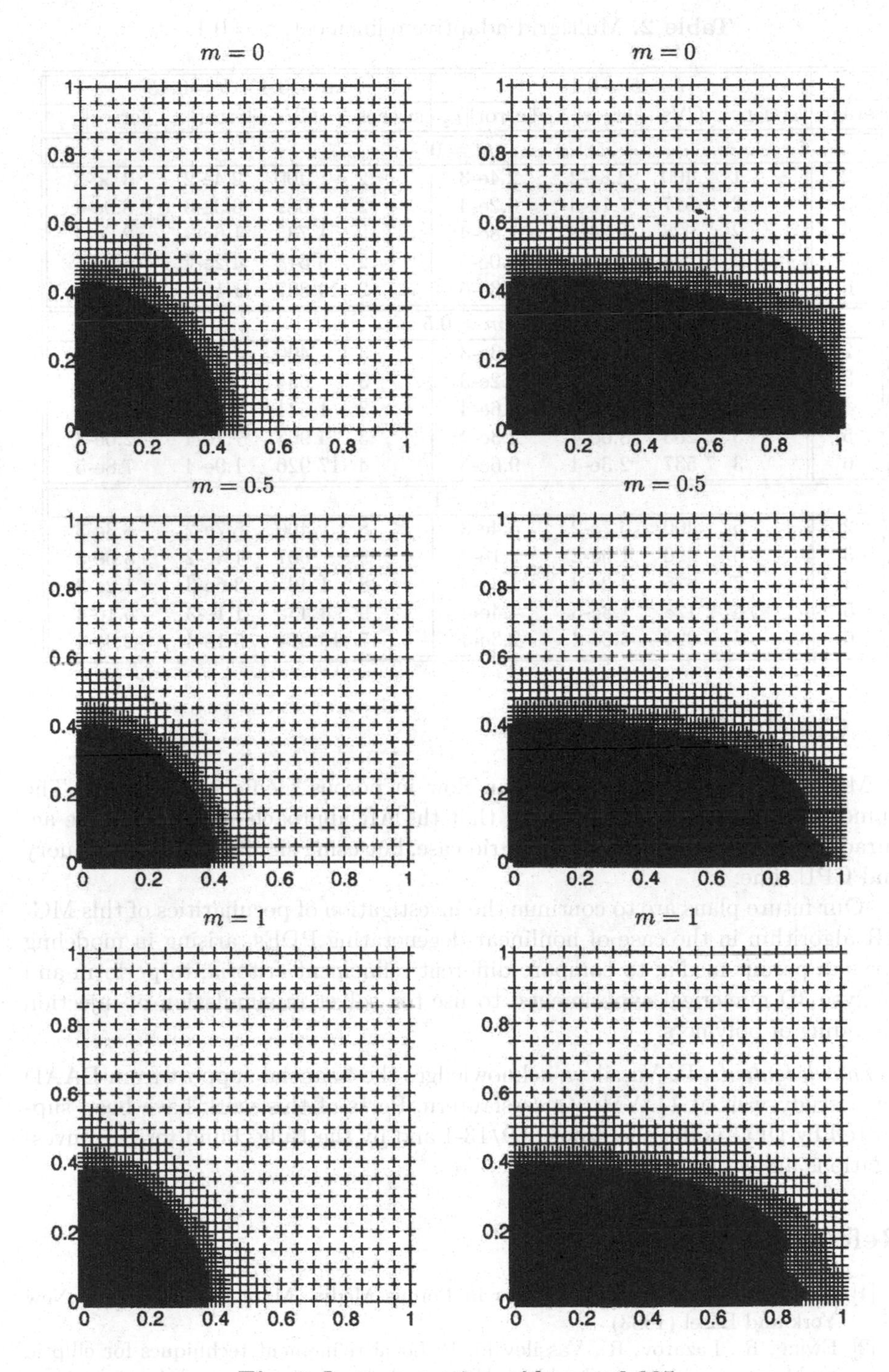

Fig. 1. Last composite grids, $\gamma = 0.025$

Table 2. Multigrid adaptive refinement, $\gamma = 0.1$.

	$a = b = 0.15$				$a = 0.45,\ b = 0.15$			
stage	sweeps	CVs	$\|error\|_C$	$\|error\|_{L_2}$	sweeps	CVs	$\|error\|_C$	$\|error\|_{L_2}$
				$m = 0$				
2	1	400	2.8e-2	2.4e-3	2	400	2.4e-2	3.5e-3
3	2	517	7.1e-3	6.2e-4	2	682	6.3e-3	8.9e-4
4	2	928	1.8e-3	1.8e-4	2	1 735	1.6e-3	2.6e-4
5	2	2 398	4.4e-4	6.0e-5	2	5 518	4.2e-4	8.7e-5
6	2	7 951	1.1e-4	2.3e-5	2	19 867	1.1e-4	3.1e-5
				$m = 0.5$				
2	3	400	5.7e-2	4.3e-3	3	400	4.6e-2	6.3e-3
3	3	502	1.4e-2	1.2e-3	3	634	1.2e-2	1.6e-3
4	3	865	3.5e-3	4.6e-4	3	1 531	3.0e-3	5.4e-4
5	3	2 263	8.6e-4	2.3e-4	3	4 963	7.7e-4	2.0e-4
6	3	7 537	2.3e-4	9.6e-5	4	17 926	1.9e-4	7.8e-5
				$m = 1$				
2	5	400	6.7e-2	5.4e-3	8	400	5.7e-2	8.3e-3
3	6	496	1.6e-2	2.1e-3	9	607	1.4e-2	3.0e-3
4	7	835	4.2e-3	1.3e-3	8	1 405	3.6e-3	1.2e-3
5	6	2 128	1.3e-3	4.4e-4	9	4 435	1.1e-3	5.1e-4
6	7	7 021	5.6e-4	2.3e-4	15	15 874	5.1e-4	2.0e-4

5 Concluding Remarks

A MG-AR solver for non-Newtonian flow in porous media is presented. The numerical experiments demonstrate that the AR approach allows the same accuracy to be obtained, as in the full grid case, but using significantly less memory and CPU time.

Our future plans are to continue the investigation of peculiarities of this MG-AR algorithm in the case of nonlinear degenerating PDEs, arising in modeling flows in porous media; to compare different refinement criteria; to perform and analyse 3D numerical experiments; to use the solver in simulation of injection moulding of polymers.

Acknowledgments. D. Vassileva acknowledges the financial support from DAAD for a study visit at ITWM, Kaiserslautern. Parts of this work have been supported by DFG under the grant 269/13-1 and by the Bulg. Fund for Sci. Investigations under the grant MM-811.

References

[1] Greenkorn, R.: Flow Phenomena in Porous Media. Marcel Dekker, Inc., New York and Basel (1983)

[2] Ewing, R., Lazarov, R., Vassilevski, P: Local refinement techniques for elliptic problems on cell-centered grids. I. Error analysis. Mathematics of Computations, **56** (1991) 437–461

Table 3. Multigrid adaptive refinement, $\gamma = 0.025$.

	$a = b = 0.15$				$a = 0.45,\ b = 0.15$			
stage	sweeps	CVs	$\Vert error\Vert_C$	$\Vert error\Vert_{L_2}$	sweeps	CVs	$\Vert error\Vert_C$	$\Vert error\Vert_{L_2}$
$m = 0$								
2	1	400	2.8e-2	2.4e-3	2	400	2.4e-2	3.5e-3
3	2	568	7.1e-3	5.9e-4	2	817	6.3e-3	8.4e-4
4	2	1 144	1.8e-3	1.5e-4	2	2 338	1.6e-3	2.2e-4
5	2	3 271	4.4e-4	4.3e-5	2	8 002	4.2e-4	5.9e-5
6	2	11 503	1.1e-4	1.3e-5	2	29 947	1.1e-4	1.7e-5
$m = 0.5$								
2	3	400	5.7e-2	4.3e-3	3	400	4.6e-2	6.3e-3
3	3	541	1.4e-2	1.0e-3	3	745	1.2e-2	1.5e-3
4	3	1 060	3.5e-3	2.9e-4	3	2 044	3.0e-3	3.9e-4
5	3	2 968	8.5e-4	9.7e-5	4	6 994	7.6e-4	1.1e-4
6	3	10 345	2.1e-4	3.8e-5	4	26 113	1.9e-4	3.7e-5
$m = 1$								
2	5	400	6.7e-2	5.4e-3	8	400	5.7e-2	8.3e-3
3	6	529	1.6e-2	1.5e-3	9	712	1.4e-2	2.4e-3
4	7	997	3.9e-3	6.3e-4	8	1 840	3.5e-3	8.2e-4
5	6	2 710	1.1e-3	2.2e-4	9	6 160	9.6e-4	3.0e-4
6	7	9 331	3.0e-4	1.0c-4	13	22 795	2.6e-4	1.1e-4

[3] Hackbusch, W.: Multi-Grid Methods and Applications. Springer-Verlag, Berlin Heidelberg New York Tokyo (1985)

[4] Wesseling, P.: An Introduction to Multigrid Methods. New York, Wiley (1991)

[5] McCormick, S.F.: Multilevel Adaptive Methods for Partial Differential Equations. SIAM, Philadelphia (1989)

[6] Ewing, R., Lazarov, R., Vassilevski, P: Local refinement techniques for elliptic problems on cell-centered grids, II: Optimal order two-grid iterative methods. Numer. Linear Algebra Appl., **1** (1994) 337–368

[7] Iliev, O., Stoyanov, D.: On a multigrid, local refinement solver for incompressible Navier-Stokes equations. Mathematical Modelling, **13** (2001) 95–106

[8] Iliev, O., Stoyanov, D.: Multigrid – adaptive local refinement solver for incompressible flows. Large-Scale Scientific Computing, eds. Margenov S. et al., Lect. Notes Comp. Sci., **2179** (2001) 361–368

[9] Dörfler, W.: A convergent adaptive algorithm for Poisson equation. SIAM J. Numer. Anal., **33** (1996) 1106-1124

[10] Dörfler, W., Wilderotter, O: An adaptive algorithm finite element method for a linear Elliptic Equation with Variable Coefficients. Z. Angev. Math. Mech., **80** (2000) 481-491

On a Multigrid Eigensolver for Linear Elasticity Problems

Maxim Larin

Institut für Geometrie und Praktische Mathematik,
Rheinisch-Wesfälische Technische Hochschule Aachen,
Templergraben 55, D-52056 Aachen, Germany,
larin@igpm.rwth-aachen.de

Abstract. In the early eighties the direct application of a multigrid technique for solving the partial eigenvalue problem of computing few of the smallest eigenvalues and their corresponding eigenvectors of a differential operator was proposed by Brandt, McCormick and Ruge [1]. In the present paper an experimental study of the method for model linear elasticity problems is carried out. Based on these results we give some practical advices for a good choice of multigrid-related parameters.

1 Introduction

Eigenvalue problems for differential equations occur in many branches of science and engineering such as mechanics, chemistry, acoustics or optics. The usual methods for solving eigenvalue problems are based on the effect of domination of the smallest eigenvalues in result of repeated multiplication of the inverse matrix A^{-1} on a vector. This applies to such popular techniques as a subspace iteration, the Rayleigh quotient and the Lanczos method [5]. However, with large–scale finite element problems it is often desirable to avoid a costly inversion or, to be more precise, exact factorization of the matrix A . The simplest way is to use some preconditioned iterative procedure instead of the direct method for solving the linear system with A whenever it is required in the algorithm. In particular, multigrid methods allow us to construct *optimal* preconditioners for a sufficiently wide number of industrial applications [6].

Today among of a lot of literature published about efficient eigensolvers one can extract two main directions. The first one is based on the application of multigrid methods to construct an approximated inverse of A and use them as a preconditioner, whereas the second one directly applies the main multigrid ideas of the fine grid relaxation and the coarse grid correction (see [2,3] and references therein).

However, all of these methods treat this eigenvalue problem purely algebraic, and hence, we lose some valuable information about the desired eigenpairs. For example, the smallest eigenvectors of the Laplace operator are very smooth, i.e. they can be well approximated on coarser grids, and hence, one can use this information during the solution process. The nested iteration multigrid process,

I. Dimov et al. (Eds.): NMA 2002, LNCS 2542, pp. 182–191, 2003.

which takes the advantage of this smoothness, was proposed by Brandt, McCormick and Ruge [1]. The idea of this method is that the eigenvalue problems are solved on the sequence of finer grids using an interpolant of the solution on each level as the initial guess for the next one and improving it by an inner nonlinear multigrid method, i.e. suppressing the high frequency oscillations arising as a result of the interpolation process. The present work is an experimental study to understand the main points of the method, the accuracy of computed eigenpairs and regarding its computational complexity.

The paper is organized as follows. In Section 2 the formal definition of the linear elasticity eigenvalue problem is given. The full multigrid method for finding p smallest eigenvalues and their eigenvectors or, shortly, the FMG-EV(p) method will be discussed in Section 3. Based on numerical results we will make some recommendations for chosing effective user-defined parameters.

2 Problem Formulation

The linear elasticity eigenvalue problem in a domain Ω in $\mathbf{R}^3$ with boundary Γ can be formulated in terms of the displacement vector $\overline{\mathbf{u}} = (u_1, u_2, u_3)$, stress tensor $\sigma = (\sigma_{ij})$ and strain tensor $\epsilon = (\epsilon_{ij})$ as follows

$$\begin{aligned} &div\,\sigma = \lambda \overline{\mathbf{u}}, \qquad \epsilon_{ij} = \frac{1}{2}\Big(\frac{\partial u_i}{\partial x_j} + \frac{\partial u_j}{\partial x_i}\Big), \qquad \sigma_{ij}(\mathbf{u}) = \sum_{k,l=1}^{3} c_{ijkl}(x)\,\epsilon_{kl}(\overline{\mathbf{u}}), \\ &\overline{\mathbf{u}} = 0 \text{ on } \Gamma_D, \qquad \sigma \cdot \overline{n} = 0 \text{ on } \Gamma_N = \Gamma \backslash \Gamma_D. \end{aligned} \tag{1}$$

Here $c_{ijkl}(x)$ is elasticity tensor depending on positive material (Lamé) coefficients, $\overline{n}$ is the outward pointing normal on Γ_N and Γ_D, Γ_N denote Dirichlet and Neumman parts of the boundary, respectively. We assume that $\Gamma_D \neq \emptyset$ and the Korn inequality holds. These assumptions are needed to ensure that all eigenvalues λ of the problem (1) are positive. For more details regarding the problem formulation we refer to [4] or other books on elasticity.

The variational formulation of the linear elasticity eigenvalue problem (1) can be derived in a standard way and is stated as follows:

Find $\lambda_k \in \mathbf{R}$ and $\mathbf{u}_k \in [H_0^1(\Omega)]^3 = \{\mathbf{v} \in [H^1(\Omega)]^3 \,:\, \mathbf{v} = 0 \text{ on } \Gamma_D\}$ such that

$$\begin{aligned} a(\mathbf{u}_k, \mathbf{v}) &= \lambda_k b(\mathbf{u}_k, \mathbf{v}),\ k = 1, 2, \ldots, p, \ldots \\ b(\mathbf{u}_k, \mathbf{u}_l) &= \delta_{kl}, \qquad\quad l = 1, 2, \ldots, p, \ldots \end{aligned} \tag{2}$$

for all $\mathbf{v} \in [H_0^1(\Omega)]^3$, where $H^1(\Omega)$ is the usual Sobolev space, δ_{ij} is the Kronecker symbol and two bilinear forms $a(\mathbf{u}, \mathbf{v})$ and $b(\mathbf{u}, \mathbf{v})$ are defined by

$$a(\mathbf{u}, \mathbf{v}) = \int_{\Omega} \left[\sum_{i,j,k,l=1}^{3} c_{ijkl} \frac{\partial u_i}{\partial x_j} \frac{\partial v_k}{\partial x_l} \right] d\Omega, \qquad b(\mathbf{u}, \mathbf{v}) = \int_{\Omega} \left[\sum_{i=1}^{3} u_i v_i \right] d\Omega.$$

Let us assume that the domain Ω is decomposed by a set of finite elements Υ. Introducing $V_h \subset H^1(\Omega)$ the space of vector functions with local support, associated with the vertices of Υ, and applying to (2) the standard Galerkin procedure, we obtain the following eigenvalue problem

$$\begin{array}{ll} A\,\mathbf{u}_k = \lambda_k \mathbf{u}_k, \quad A = A^T \in \mathbf{R}^{n\times n}, \quad 0 < \lambda_1 \leq \lambda_2 \leq \ldots \leq \lambda_p, \\ (\mathbf{u}_k, \mathbf{u}_l) = \delta_{kl}, \quad k,l = 1,\ldots,p, \end{array} \tag{3}$$

where the matrix A corresponds to the bilinear form $a(\cdot,\cdot)$ and $(\mathbf{u},\mathbf{v}) = \mathbf{u}^T\mathbf{v}$.

3 The FMG-EV p Method

Let $\Upsilon_L \subset \Upsilon_{L-1} \subset \cdots \subset \Upsilon_1 \subset \Upsilon_0 = \Upsilon$ be a sequence of nested finite element grids, such that all grids have a similar structure and the (fine) grid Υ_k is defined from the (coarse) grid Υ_{k+1} by an uniform refinement in all spatial directions. The Galerkin procedure applied on each grid leads to the sequence of eigenvalue problems with the corresponding matrices $A^{(k)}$ of the decreasing order. Moreover, the standard (geometrical) linear restriction R_k^{k+1} and interpolation P_{k+1}^k operators are defined.

$\{\mu_i^{(0)}, \mathbf{v}_i^{(0)}\}_{i=1}^p = \mathbf{FMG\text{-}EV}(p, A^{(L)}, \ldots, A^{(0)})$:

$i = 0, \; k = L, \; \mu_{i-1}^{(k)} = 0$

1 $\quad i = i + 1$

Compute the coarse level approximation $\{\mu_i^{(k)}, \mathbf{v}_i^{(k)}\}$,
which is orthogonal to computed ones $\{\mu_j^{(k)}, \mathbf{v}_j^{(k)}\}_{j=1}^{i-1}$

if $((i < cn_k)$ **and** $(i < p))$ go to 1

$\{\mu_j^{(k)}, \mathbf{v}_j^{(k)}\}_{j=1}^i = \mathbf{RITZ}\,(A^{(0)}, \{\mathbf{v}_j^{(k)}\}_{j=1}^i)$

2 $\quad$ **if** $(k = 0)$ **stop**

$k = k - 1$

for $s = 1, \ldots, i$

$\quad \mathbf{v}_{s,0}^{(k)} = P_{k+1}^k \mathbf{v}_s^{(k+1)}$

$\quad \mu_{s,0}^{(k)} = \mu_s^{(k+1)}$

$\quad$ **for** $it = 1, \ldots, q$ $\quad$ *(inner iteration)*

$\quad\quad \{\mu_{s,it}^{(k)}, \mathbf{v}_{s,it}^{(k)}\} = \mathbf{FAS\text{-}EV}\,(A^{(k)}, \mu_{s,it-1}^{(k)}, \mathbf{v}_{s,it-1}^{(k)}, \{\mathbf{v}_j^{(k)}\}_{j=1}^{s-1})$

$\{\mu_j^{(k)}, \mathbf{v}_j^{(k)}\}_{j=1}^i = \mathbf{RITZ}\,(A^{(0)}, \{\mathbf{v}_{j,q}^{(k)}\}_{j=1}^i)$

if $(i < p)$ **then** go to 1
else go to 2

Assume that we have an approximated solution $\mathbf{v}_i^{(k+1)}$ on the (coarse) grid and using their interpolants $\mathbf{v}_{i,0}^{(k)} = P_{k+1}^k \mathbf{v}_i^{(k+1)}$ as the initial guess for some inner iterative process on the next (fine) grid. Then, to eliminate the high frequency oscillations, which arise as a result of the interpolation process, we apply q times the inner multigrid method, which is based on the Full Approximation Scheme (FAS) approach and uses the same sequence of matrices $\{A^{(k)}\}$. Following [1] we proceed vector-by-vector through the inner nonlinear method using an orthogonalization condition to keep them separate. Usually, one or two inner multigrid sweeps are enough to suppress all (or nearly all) undesired frequencies in the approximated solution. It is confirmed by numerical results. Finally, we improve the eigenvector approximations by the Rayleigh-Ritz projection method and repeat this process on the next finer grid until the finest grid is reached.

$\{\mu_i^{(l)}, \mathbf{v}_i^{(l)}\} = \textbf{FAS-EV}\ (A^{(l)}, \mu_{i,0}^{(l)}, \mathbf{v}_{i,0}^{(l)}, \{\mathbf{v}_j^{(k)}\}_{j=1}^{i-1})$:

$\mathbf{b}_i^{(l)} = \mathbf{0}$

for $k = l, \ldots, L-1$

for $it = 1, \ldots, \nu_1$ *(presmoothing)*

$\{\mu_{i,it}^{(k)}, \mathbf{v}_{i,it}^{(k)}\} = \textbf{RELAX}\ (A^{(k)}, \mu_{i,it-1}^{(k)}, \mathbf{v}_{i,it-1}^{(k)}, \{\mathbf{v}_j^{(k)}\}_{j=1}^{i-1}, \mathbf{b}_i^{(k)})$

$\{\mu_i^{(k)}, \mathbf{v}_i^{(k)}\} = \{\mu_{i,\nu_1}^{(k)}, \mathbf{v}_{i,\nu_1}^{(k)}\}$

$\mathbf{b}_i^{(k+1)} = R_k^{k+1}\mathbf{b}_i^{(k)} + \left(A^{(k+1)} R_k^{k+1} - R_k^{k+1} A^{(k)}\right) \mathbf{v}_i^{(k)}$

$\mathbf{v}_{i,0}^{(k+1)} = R_k^{k+1}\mathbf{v}_i^{(k)}$

$\mu_{i,0}^{(k+1)} = \mu_i^{(k)}$

Solve the coarse level problem $A^{(L)}\mathbf{v}_i^{(L)} = \mu_i^{(L)}\mathbf{v}_i^{(L)} + \mathbf{b}_i^{(L)}$

with $(\mathbf{v}_i^{(L)}, R_l^L\mathbf{v}_j^{(l)}) = (R_{L-1}^L\mathbf{v}_i^{(L-1)}, R_l^L\mathbf{v}_j^{(l)}), j = 1, \ldots, i-1$

for $k = L-1, \ldots, l$

$\mathbf{v}_{i,0}^{(k)} = \mathbf{v}_i^{(k)} + P_{k+1}^k(\mathbf{v}_i^{(k+1)} - R_k^{k+1}\mathbf{v}_i^{(k)})$

$\mu_{i,0}^{(k)} = \mu_i^{(k+1)}$

for $it = 1, \ldots, \nu_2$ *(postsmoothing)*

$\{\mu_{i,it}^{(k)}, \mathbf{v}_{i,it}^{(k)}\} = \textbf{RELAX}\ (A^{(k)}, \mu_{i,it-1}^{(k)}, \mathbf{v}_{i,it-1}^{(k)}, \{\mathbf{v}_j^{(k)}\}_{j=1}^{i-1}, \mathbf{b}_i^{(k)})$

$\{\mu_i^{(k)}, \mathbf{v}_i^{(k)}\} = \{\mu_{i,\nu_2}^{(k)}, \mathbf{v}_{i,\nu_2}^{(k)}\}$

There are two main problems with this approach. First of all, there is no exact correspondence between eigenvectors on different levels, i.e. the i-th eigenvalue and its eigenvector on the coarse level can be a *good* approximation to j-th eigenpair on the fine level with $i \neq j$. This is not a problem if j is less than p, and hence, during the following Ritz projection we find both of the desired eigenvectors on the fine level. However, if j is greater than p, then the further efforts to improve this approximation in the direction of i-th eigenvector are vain since FAS works good only in a small neighbourhood of the solution [6]. Thus, on the coarse level we can find a fixed number of approximation vectors, which are a good approximation to desired eigenvectors. This number depends on the order of the coarse level problem, and hence, can be defined as cn_L, where c is a coefficient less than 1. Now if $p > cn_L$, then we find the coarse level approximations only for first cn_L eigenvectors and start our nested iteration process with this smaller number of vectors. After transfering and processing these approximated eigenvectors on the next finer level we compute the deficient coarse level approximations using the relaxation steps followed by orthogonalization with respect to previous ones, and if p is still greater than cn_{L-1}, then we repeat this process on the next finer level and so on. Note that the coarsest level used in the inner multilevel method for ith eigenvector is one on each $\mathbf{v}_i^{(k)}$ first appeared.

Secondly, since we apply the inner nonlinear multigrid method sequentially to each approximated eigenvector, then we have to orthogonalize the current eigenvector approximation to previously computed ones. On the other side, the main FAS idea is in the construction of the coarse grid problem so that its solution is the fine level solution transfered to the coarse grid. Using it we obtain the following sequence of problems on level l

$$\begin{aligned} A^{(k)}\mathbf{w}_i^{(k)} &= \hat{\mu}_i^{(k)}\mathbf{w}_i^{(k)} + \mathbf{b}_i^{(k)}, \\ \mathbf{b}_i^{(k)} &= R_{k-1}^k\mathbf{b}_i^{(k-1)} + \left(A^{(k)}R_{k-1}^k - R_{k-1}^k A^{(k-1)}\right)\mathbf{v}_i^{(k-1)}, \\ \phi_k(\mathbf{w}_i^{(k)}) &= \sigma_k, \qquad \mathbf{b}_i^{(l)} = 0, \qquad k = l, l+1, \ldots, L, \quad i = 1, \ldots, p. \end{aligned} \tag{4}$$

where $\phi_k(\cdot)$ is a normalization condition, which guarantees the uniqueness of (4), and specifies the size of the solutions σ_k. Note that it is not actually necessary that σ_k be equal to 1 since a solution of any reasonable size is acceptable. The specific nature of $\phi_k(\cdot)$ will be discussed below.

Let $\mathbf{u}_j^{(l)}$ is the solution on level l, then $R_l^{l+1}\mathbf{u}_j^{(l)}$ is the solution on next level, and hence, if we want to find $\mathbf{u}_i^{(l)}$, $i \neq j$, we have to orthogonalize the current eigenvector approximation $\mathbf{v}_i^{(l+1)}$ to $R_l^{l+1}\mathbf{u}_j^{(l)}$. Unfortunately, we do not know the exact solution $\mathbf{u}_j^{(l)}$ (it is our original problem!), but we have $\mathbf{v}_j^{(l)}$, $j < i$, which is a good approximation to $\mathbf{u}_j^{(l)}$. Thus, from a practical point of view $R_l^{l+1}\mathbf{v}_j^{(l)}$ is an approximation to $R_l^{l+1}\mathbf{u}_j^{(l)}$, which one can use to keep orthogonalization on the level $l+1$. Now we define

$$\phi_k(\mathbf{v}_i^{(k)}) = (\mathbf{v}_i^{(k)}, R_l^k\mathbf{v}_j^{(l)}), \quad \sigma_k = (R_{k-1}^k\mathbf{v}_i^{(k)}, R_l^k\mathbf{v}_j^{(l)}) \quad j = 1, \ldots, i. \tag{5}$$

Nevertheless, we have to note that the condition (5) does not guarantee a uniqueness of the solution (4), since we work with approximations rather than exact solutions. In our case it may happen that the sequence of test vectors $\{R_l^k\mathbf{v}_j^{(l)}\}_{j=1}^{i-1}$ used for the orthogonalization can be linear dependent, and hence, on the level k we will compute $\mathbf{v}_i^{(k)}$, which is a good approximation to $\mathbf{v}_j^{(l)}, j < i$, and which we already know, instead the desired approximation to $\mathbf{v}_i^{(l)}$. It is confirmed by numerical results.

4 Numerical Results

To test the method we first consider the eigenvalue problem (3) for finding the smallest ($p = 1$) eigenpair $\{\lambda_1, \mathbf{u}_1\}$. We assume the piecewise–linear finite–element discretization of the elasticity problem (1) in the cubical domain $\Omega = [0,1]^3$ on a uniform Cartesian mesh $\mathcal{T}_h = N \times N \times N$ with meshsize $h = N^{-1}$. Consider homogeneous Dirichlet boundary conditions, i.e., $\Gamma_D = \Gamma, \Gamma_N = \emptyset$.

In all tables of this paper $\nu = \nu_1 = \nu_2$ is the number of pre- and post-smoothing iterations on each level, q is the number of inner FAS-EV iterations applied on each level, $Lvls$ is the number of levels used and $Time$ is the CPU

Table 1. Accuracy of the computed eigenpair $\{\mu_1, \mathbf{v}_1\}$ as a function of ν and q, $p = 1$

N=16		$q=1$			$q=2$		
Lvls	ν	$\|\mu_1 - \lambda_1\|$	$\|\mathbf{A}\mathbf{v}_1 - \mu_1\mathbf{v}_1\|$	Time	$\|\mu_1 - \lambda_1\|$	$\|\mathbf{A}\mathbf{v}_1 - \mu_1\mathbf{v}_1\|$	Time
4	1	$0.18766 \cdot 10^{-2}$	$0.37333 \cdot 10^{-1}$	8.60	$0.13093 \cdot 10^{-3}$	$0.89867 \cdot 10^{-2}$	15.17
	2	$0.13861 \cdot 10^{-3}$	$0.90307 \cdot 10^{-2}$	13.56	$0.13880 \cdot 10^{-5}$	$0.67454 \cdot 10^{-3}$	25.12
	3	$0.90189 \cdot 10^{-5}$	$0.19036 \cdot 10^{-2}$	18.46	$0.18045 \cdot 10^{-7}$	$0.75458 \cdot 10^{-4}$	34.99
	4	$0.13531 \cdot 10^{-5}$	$0.67833 \cdot 10^{-3}$	23.45	$0.13159 \cdot 10^{-8}$	$0.15376 \cdot 10^{-4}$	44.89
	5	$0.11025 \cdot 10^{-6}$	$0.19556 \cdot 10^{-3}$	28.42	$0.25888 \cdot 10^{-9}$	$0.55911 \cdot 10^{-5}$	54.69
	10	$0.14618 \cdot 10^{-8}$	$0.12328 \cdot 10^{-4}$	53.03	$0.34772 \cdot 10^{-11}$	$0.65020 \cdot 10^{-6}$	104.07
3	1	$0.19283 \cdot 10^{-2}$	$0.39311 \cdot 10^{-1}$	9.50	$0.10903 \cdot 10^{-3}$	$0.81701 \cdot 10^{-2}$	16.04
	2	$0.12757 \cdot 10^{-3}$	$0.85886 \cdot 10^{-2}$	14.19	$0.18321 \cdot 10^{-5}$	$0.78770 \cdot 10^{-3}$	25.79
	3	$0.13519 \cdot 10^{-4}$	$0.23281 \cdot 10^{-2}$	18.89	$0.72569 \cdot 10^{-7}$	$0.10573 \cdot 10^{-3}$	34.82
	4	$0.18275 \cdot 10^{-5}$	$0.77967 \cdot 10^{-3}$	23.55	$0.21767 \cdot 10^{-7}$	$0.26623 \cdot 10^{-4}$	44.57
	5	$0.28713 \cdot 10^{-6}$	$0.24979 \cdot 10^{-3}$	28.32	$0.10061 \cdot 10^{-7}$	$0.16873 \cdot 10^{-4}$	54.17
	10	$0.36640 \cdot 10^{-7}$	$0.32665 \cdot 10^{-4}$	52.02	$0.54730 \cdot 10^{-9}$	$0.37209 \cdot 10^{-5}$	101.09
2	1	$0.18780 \cdot 10^{-2}$	$0.39511 \cdot 10^{-1}$	13.08	$0.11463 \cdot 10^{-3}$	$0.84105 \cdot 10^{-2}$	22.25
	2	$0.15070 \cdot 10^{-3}$	$0.97192 \cdot 10^{-2}$	17.95	$0.19856 \cdot 10^{-5}$	$0.75905 \cdot 10^{-3}$	31.70
	3	$0.19502 \cdot 10^{-4}$	$0.25037 \cdot 10^{-2}$	23.58	$0.17259 \cdot 10^{-5}$	$0.21648 \cdot 10^{-3}$	41.76
	4	$0.21043 \cdot 10^{-5}$	$0.81073 \cdot 10^{-3}$	25.97	$0.10470 \cdot 10^{-5}$	$0.15833 \cdot 10^{-3}$	49.52
	5	$0.13009 \cdot 10^{-5}$	$0.32103 \cdot 10^{-3}$	31.10	$0.11669 \cdot 10^{-5}$	$0.17195 \cdot 10^{-3}$	57.79
	10	$0.15593 \cdot 10^{-5}$	$0.19889 \cdot 10^{-3}$	51.08	$0.51628 \cdot 10^{-6}$	$0.10705 \cdot 10^{-3}$	96.00

time of the whole iterative process. All calculations were performed on an IBM-SP2 in double precision. By μ_i and $\mathbf{v}_i$ we denote the approximation to the exact eigenvalues and eigenvectors, λ_i and $\mathbf{u}_i$, respectively. Correspondingly, $|\mu_i - \lambda_i|$ measures the difference between the exact and the computed eigenvalues and $\|\mathbf{A}\mathbf{v}_i - \mu_i\mathbf{v}_i\|$ measures the quality of the approximated pair $\{\mu_i, \mathbf{v}_i\}$. The exact eigenvalues λ_i have been computed by the Gaus-Seidel iteration method followed by standard Gramm-Schmidt orthogonalization process until the following stopping criterion was satisfied

$$\|\mathbf{r}_i^{(k)}\|/\|\mathbf{r}_i^{(0)}\| < 10^{-12}, \quad \mathbf{r}_i^{(k)} = A\mathbf{v}_i^{(k)} - \mu_i\mathbf{v}_i^{(k)},$$

where $\mathbf{r}_i^{(0)}$ and $\mathbf{r}_i^{(k)}$ are the initial and the final residuals, respectively.

From the numerical results in Table 1 one can see that by increasing the number of relaxation steps ν per level without changing other parameters, the accuracy of the approximated solution is improved. Comparing results for $q = 1$ with corresponding ones for $q = 2$ in Table 1 one can also see that increasing the number of inner FAS-EV steps leads to an improved accuracy of the approximated eigenpair $\{\mu_i, \mathbf{v}_i\}$. On the other hand, the accuracy of the approximated solution does not depend sensitively on the number of levels used. Due to the discretization error for this problem is $O(h^2) = O(10^{-3})$ one can see that the FMG-EV(p) method can compute the first eigenvalue with a reasonable computational cost, when $q = 1$ and $\nu = 3$.

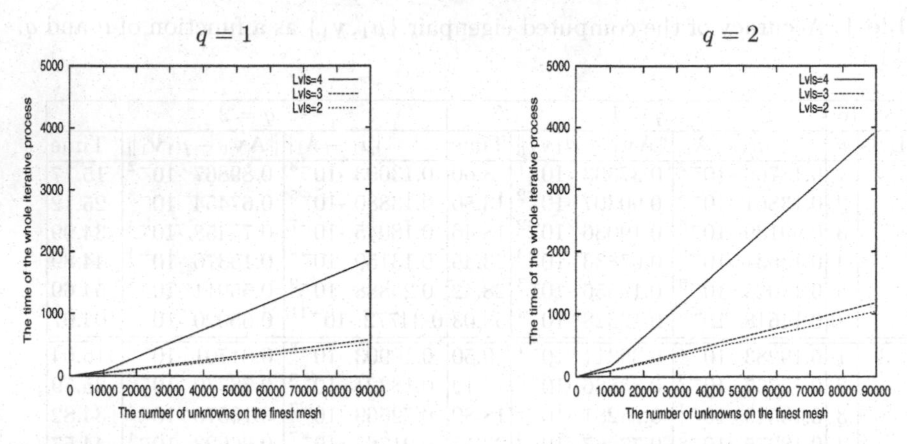

Fig. 1. The time of the whole iterative process **vs.** the number of unknowns.

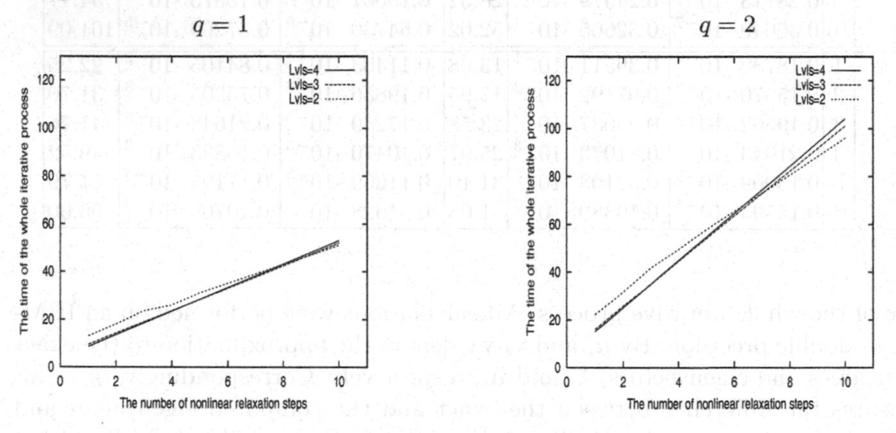

Fig. 2. The time of the whole iterative process **vs.** the number of smoothing steps.

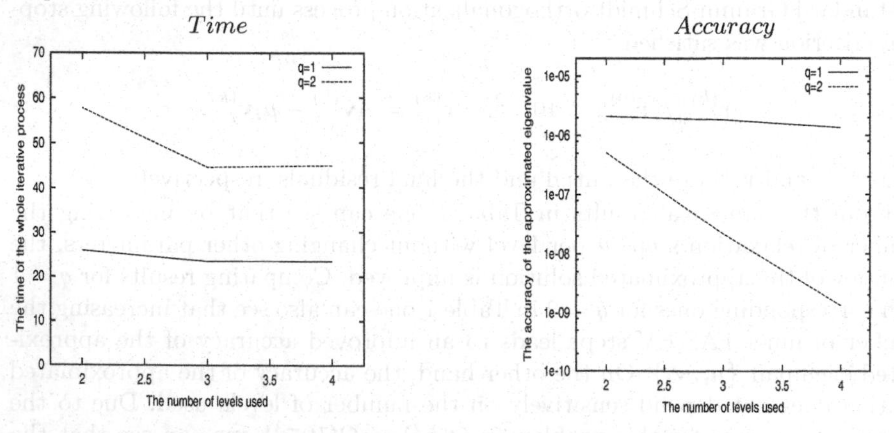

Fig. 3. The time of the iterative process and the accuracy **vs.** the number of levels.

Total computational times for various values of $Lvls$ versus the number of unknowns on the finest mesh and the number of relaxation steps are given in Figures 1 and 2, respectively. Based on the presented numerical results we can see that the FMG-EV(p) method has a nearly optimal computational complexity, i.e. the time of the whole iterative process does not depend on the number of unknowns on the finest level and only slightly depend on the number of levels used.

Moreover, in Figure 3 we present the dependency of the time of eigenvalue solver and the accuracy of the computed eigenvalue as a function of the number of levels. Here we want to note a jumping of the accuracy for the case $q = 2$. It is shown that not all high frequencies can be eliminated during the first step of the FAS-EV method, and hence, we need the second one.

The results of the first five ($p = 5$) eigenvalues and their corresponding eigenvectors on a $(16 \times 16 \times 16)$-grid are given in Tables 2 and 3. Based on the results we conclude that the FMG-EV(p) method behaves similarly as described above for the case $p = 1$. We make only a few remarks.

First, a comparison of the accuracies in Tables 2 and 3 for the first eigenpair $\{\mu_1, \mathbf{v}_1\}$ with these shown in Table 1 shows that when more eigenvectors, whether or not they converge, are included in the process, all approximations are improved. Taking into account the whole computational complexity, which is growing rapidly when ν and q are increasing, a good choice of the parameters seems to be $q = 1$ and $\nu = 3$. This is similar to the results for $p = 1$.

Moreover, the present experiments show that the FAML-EV(p) method is capable of dealing with multiple eigenvalues. Indeed, we have $\lambda_1 = \lambda_2 = \lambda_3$ and one can see from Tables 2 and 3 that we find this eigenvalue and its eigensubspace without problems.

Finally, we want to mention a problem with the eigenvector orthogonalization for the inner nonlinear method. Indeed, when we try to solve the coarse level problem for the second, third or other eigenvectors during the FAS method with some given accuracy, we always perform the maximal number of iterations, i.e. we could not reach the desired accuracy. It seems that it is not a strong restriction on the suggested method since the method is still convergent. However, the proposed technique with corresponding block-type modifications for the inner nonlinear method can be done. The further investigation will be directed in this way.

Summarizing, we want to note that the present method can be used instead of the direct eigensolver when a small number of eigenpairs is required, i.e., $p < \sqrt{n}$. However, if we wanted to find a series of eigenvectors with $p > \sqrt{n}$, then the Lanzos method with LU-factorization would be less time consuming.

References

1. Brandt, A., McCormick, S. and Ruge, J.: Multigrid Methods for Differential Eigenproblems. SIAM J. Sci. Stat. Comput., 4(2) (1983) 244–260.
2. Hackbusch, W.: Multigrid Methods and Applications. Springer-Verlag, Berlin - Heidelberg - New York (1985)

Table 2. Accuracy of computed eigenvalues μ_i as a function of ν and q, $p = 5$

q	ν	$\|\mu_1 - \lambda_1\|$	$\|\mu_2 - \lambda_2\|$	$\|\mu_3 - \lambda_3\|$	$\|\mu_4 - \lambda_4\|$	$\|\mu_5 - \lambda_5\|$
Lvls = 4, NFine = 10125, NCoarse = 3						
1	1	$0.17121 \cdot 10^{-2}$	$0.18504 \cdot 10^{-2}$	$0.23162 \cdot 10^{-2}$	$0.21263 \cdot 10^{-2}$	$0.24166 \cdot 10^{-2}$
	2	$0.87427 \cdot 10^{-4}$	$0.18120 \cdot 10^{-3}$	$0.18459 \cdot 10^{-3}$	$0.14426 \cdot 10^{-3}$	$0.16316 \cdot 10^{-3}$
	3	$0.59710 \cdot 10^{-5}$	$0.15262 \cdot 10^{-4}$	$0.23290 \cdot 10^{-4}$	$0.16597 \cdot 10^{-4}$	$0.20785 \cdot 10^{-4}$
	4	$0.64233 \cdot 10^{-6}$	$0.18220 \cdot 10^{-5}$	$0.27387 \cdot 10^{-5}$	$0.22954 \cdot 10^{-5}$	$0.26639 \cdot 10^{-5}$
	5	$0.62436 \cdot 10^{-7}$	$0.27586 \cdot 10^{-6}$	$0.33646 \cdot 10^{-6}$	$0.29719 \cdot 10^{-6}$	$0.81938 \cdot 10^{-6}$
	10	$0.12715 \cdot 10^{-8}$	$0.16685 \cdot 10^{-8}$	$0.54289 \cdot 10^{-8}$	$0.13347 \cdot 10^{-6}$	$0.34387 \cdot 10^{-6}$
2	1	$0.66509 \cdot 10^{-4}$	$0.13746 \cdot 10^{-3}$	$0.14220 \cdot 10^{-3}$	$0.11944 \cdot 10^{-3}$	$0.13899 \cdot 10^{-3}$
	2	$0.56575 \cdot 10^{-6}$	$0.17661 \cdot 10^{-5}$	$0.25094 \cdot 10^{-5}$	$0.16908 \cdot 10^{-5}$	$0.20703 \cdot 10^{-5}$
	3	$0.52939 \cdot 10^{-8}$	$0.36231 \cdot 10^{-7}$	$0.47925 \cdot 10^{-7}$	$0.13661 \cdot 10^{-6}$	$0.40596 \cdot 10^{-6}$
	4	$0.52667 \cdot 10^{-9}$	$0.87679 \cdot 10^{-9}$	$0.23866 \cdot 10^{-8}$	$0.69039 \cdot 10^{-7}$	$0.18090 \cdot 10^{-6}$
	5	$0.92139 \cdot 10^{-10}$	$0.16693 \cdot 10^{-9}$	$0.27081 \cdot 10^{-8}$	$0.17478 \cdot 10^{-7}$	$0.11907 \cdot 10^{-6}$
	10	$0.11360 \cdot 10^{-9}$	$0.11232 \cdot 10^{-9}$	$0.42050 \cdot 10^{-10}$	$0.65845 \cdot 10^{-8}$	$0.18503 \cdot 10^{-7}$
Lvls = 3, NFine = 10125, NCoarse = 81						
1	1	$0.17126 \cdot 10^{-2}$	$0.18520 \cdot 10^{-2}$	$0.23164 \cdot 10^{-2}$	$0.21153 \cdot 10^{-2}$	$0.22651 \cdot 10^{-2}$
	2	$0.86727 \cdot 10^{-4}$	$0.18121 \cdot 10^{-3}$	$0.18583 \cdot 10^{-3}$	$0.15580 \cdot 10^{-3}$	$0.16274 \cdot 10^{-3}$
	3	$0.59921 \cdot 10^{-5}$	$0.15343 \cdot 10^{-4}$	$0.23375 \cdot 10^{-4}$	$0.12978 \cdot 10^{-4}$	$0.16919 \cdot 10^{-4}$
	4	$0.63646 \cdot 10^{-6}$	$0.18893 \cdot 10^{-5}$	$0.27678 \cdot 10^{-5}$	$0.22854 \cdot 10^{-5}$	$0.28454 \cdot 10^{-5}$
	5	$0.70249 \cdot 10^{-7}$	$0.27376 \cdot 10^{-6}$	$0.34354 \cdot 10^{-6}$	$0.38226 \cdot 10^{-6}$	$0.91077 \cdot 10^{-6}$
	10	$0.13113 \cdot 10^{-8}$	$0.20920 \cdot 10^{-8}$	$0.71929 \cdot 10^{-8}$	$0.16827 \cdot 10^{-6}$	$0.23884 \cdot 10^{-6}$
2	1	$0.66521 \cdot 10^{-4}$	$0.13756 \cdot 10^{-3}$	$0.14223 \cdot 10^{-3}$	$0.11527 \cdot 10^{-3}$	$0.13875 \cdot 10^{-3}$
	2	$0.56377 \cdot 10^{-6}$	$0.17248 \cdot 10^{-5}$	$0.25207 \cdot 10^{-5}$	$0.15277 \cdot 10^{-5}$	$0.22036 \cdot 10^{-5}$
	3	$0.68877 \cdot 10^{-8}$	$0.36520 \cdot 10^{-7}$	$0.81525 \cdot 10^{-7}$	$0.82430 \cdot 10^{-7}$	$0.36574 \cdot 10^{-6}$
	4	$0.78704 \cdot 10^{-9}$	$0.22278 \cdot 10^{-8}$	$0.15379 \cdot 10^{-7}$	$0.86195 \cdot 10^{-7}$	$0.16180 \cdot 10^{-6}$
	5	$0.83009 \cdot 10^{-10}$	$0.23116 \cdot 10^{-9}$	$0.94830 \cdot 10^{-8}$	$0.62724 \cdot 10^{-7}$	$0.87238 \cdot 10^{-7}$
	10	$0.11371 \cdot 10^{-9}$	$0.11276 \cdot 10^{-9}$	$0.36793 \cdot 10^{-9}$	$0.69298 \cdot 10^{-8}$	$0.12069 \cdot 10^{-7}$
Lvls = 2, NFine = 10125, NCoarse = 1029						
1	1	$0.16598 \cdot 10^{-2}$	$0.17743 \cdot 10^{-2}$	$0.22529 \cdot 10^{-2}$	$0.20921 \cdot 10^{-2}$	$0.22736 \cdot 10^{-2}$
	2	$0.89283 \cdot 10^{-4}$	$0.17879 \cdot 10^{-3}$	$0.18547 \cdot 10^{-3}$	$0.14831 \cdot 10^{-3}$	$0.16975 \cdot 10^{-3}$
	3	$0.68256 \cdot 10^{-5}$	$0.15841 \cdot 10^{-4}$	$0.23290 \cdot 10^{-4}$	$0.15277 \cdot 10^{-4}$	$0.19748 \cdot 10^{-4}$
	4	$0.10038 \cdot 10^{-5}$	$0.21974 \cdot 10^{-5}$	$0.32588 \cdot 10^{-5}$	$0.17504 \cdot 10^{-5}$	$0.24626 \cdot 10^{-5}$
	5	$0.16654 \cdot 10^{-6}$	$0.33157 \cdot 10^{-6}$	$0.22759 \cdot 10^{-5}$	$0.19340 \cdot 10^{-6}$	$0.65046 \cdot 10^{-6}$
	10	$0.13450 \cdot 10^{-8}$	$0.56302 \cdot 10^{-8}$	$0.20701 \cdot 10^{-5}$	$0.18521 \cdot 10^{-7}$	$0.10047 \cdot 10^{-6}$
2	1	$0.66248 \cdot 10^{-4}$	$0.13791 \cdot 10^{-3}$	$0.14184 \cdot 10^{-3}$	$0.11067 \cdot 10^{-3}$	$0.15442 \cdot 10^{-3}$
	2	$0.56097 \cdot 10^{-6}$	$0.18273 \cdot 10^{-5}$	$0.56997 \cdot 10^{-5}$	$0.16318 \cdot 10^{-5}$	$0.25670 \cdot 10^{-5}$
	3	$0.25852 \cdot 10^{-7}$	$0.37555 \cdot 10^{-7}$	$0.17063 \cdot 10^{-5}$	$0.26033 \cdot 10^{-7}$	$0.30782 \cdot 10^{-7}$
	4	$0.13821 \cdot 10^{-8}$	$0.25116 \cdot 10^{-8}$	$0.94665 \cdot 10^{-6}$	$0.46884 \cdot 10^{-8}$	$0.39074 \cdot 10^{-7}$
	5	$0.14959 \cdot 10^{-9}$	$0.46594 \cdot 10^{-8}$	$0.15020 \cdot 10^{-5}$	$0.28829 \cdot 10^{-7}$	$0.83986 \cdot 10^{-7}$
	10	$0.11348 \cdot 10^{-9}$	$0.28841 \cdot 10^{-10}$	$0.40120 \cdot 10^{-6}$	$0.18245 \cdot 10^{-8}$	$0.48712 \cdot 10^{-8}$

3. Knyazev, A. V.: Preconditioned Eigensolvers – an Oxymoron? Electronic Transaction on Numerical Analysis, Vol. 7 (1998) 104–123.
4. Lur'e, A. I.: Theory of Elasticity. Moscow, Nauka (1970)
5. Parlett, B.: The Symmetric Eigenvalue Problem. Prentice-Hall, Inc. (1980)
6. Trottenberg, U., Oosterlee, C. and Schüller, A.: Multigrid. Academic Press (2001)

Table 3. Accuracy of computed eigenpairs $\{\mu_i, \mathbf{v}_i\}$ as a function of ν and q, $p = 5$

q	ν	$\lVert\mathbf{A}\mathbf{v}_1 - \mu_1\mathbf{v}_1\rVert$	$\lVert\mathbf{A}\mathbf{v}_2 - \mu_2\mathbf{v}_2\rVert$	$\lVert\mathbf{A}\mathbf{v}_3 - \mu_3\mathbf{v}_3\rVert$	$\lVert\mathbf{A}\mathbf{v}_4 - \mu_4\mathbf{v}_4\rVert$	$\lVert\mathbf{A}\mathbf{v}_5 - \mu_5\mathbf{v}_5\rVert$
		Lvls = 4, NFine = 10125, NCoarse = 3				
1	1	$0.35633 \cdot 10^{-1}$	$0.35849 \cdot 10^{-1}$	$0.48551 \cdot 10^{-1}$	$0.40670 \cdot 10^{-1}$	$0.46340 \cdot 10^{-1}$
	2	$0.69620 \cdot 10^{-2}$	$0.11551 \cdot 10^{-1}$	$0.10216 \cdot 10^{-1}$	$0.96870 \cdot 10^{-2}$	$0.96702 \cdot 10^{-2}$
	3	$0.16195 \cdot 10^{-2}$	$0.23696 \cdot 10^{-2}$	$0.31153 \cdot 10^{-2}$	$0.25545 \cdot 10^{-2}$	$0.28774 \cdot 10^{-2}$
	4	$0.54673 \cdot 10^{-3}$	$0.73684 \cdot 10^{-3}$	$0.10033 \cdot 10^{-2}$	$0.81480 \cdot 10^{-3}$	$0.88640 \cdot 10^{-3}$
	5	$0.15609 \cdot 10^{-3}$	$0.28865 \cdot 10^{-3}$	$0.33740 \cdot 10^{-3}$	$0.27417 \cdot 10^{-3}$	$0.36506 \cdot 10^{-3}$
	10	$0.12719 \cdot 10^{-4}$	$0.14326 \cdot 10^{-4}$	$0.17823 \cdot 10^{-4}$	$0.84203 \cdot 10^{-4}$	$0.13971 \cdot 10^{-3}$
2	1	$0.61016 \cdot 10^{-2}$	$0.89436 \cdot 10^{-2}$	$0.98460 \cdot 10^{-2}$	$0.83752 \cdot 10^{-2}$	$0.91747 \cdot 10^{-2}$
	2	$0.51795 \cdot 10^{-3}$	$0.72641 \cdot 10^{-3}$	$0.96730 \cdot 10^{-3}$	$0.71808 \cdot 10^{-3}$	$0.83226 \cdot 10^{-3}$
	3	$0.42454 \cdot 10^{-4}$	$0.10305 \cdot 10^{-3}$	$0.11552 \cdot 10^{-3}$	$0.11879 \cdot 10^{-3}$	$0.17028 \cdot 10^{-3}$
	4	$0.86842 \cdot 10^{-5}$	$0.15120 \cdot 10^{-4}$	$0.19860 \cdot 10^{-4}$	$0.59069 \cdot 10^{-4}$	$0.10028 \cdot 10^{-3}$
	5	$0.53047 \cdot 10^{-5}$	$0.55883 \cdot 10^{-5}$	$0.10770 \cdot 10^{-4}$	$0.31108 \cdot 10^{-4}$	$0.77286 \cdot 10^{-4}$
	10	$0.87844 \cdot 10^{-6}$	$0.14166 \cdot 10^{-5}$	$0.19605 \cdot 10^{-5}$	$0.16655 \cdot 10^{-4}$	$0.31198 \cdot 10^{-4}$
		Lvls = 3, NFine = 10125, NCoarse = 81				
1	1	$0.35638 \cdot 10^{-1}$	$0.35845 \cdot 10^{-1}$	$0.48545 \cdot 10^{-1}$	$0.39561 \cdot 10^{-1}$	$0.45533 \cdot 10^{-1}$
	2	$0.69323 \cdot 10^{-2}$	$0.11575 \cdot 10^{-1}$	$0.10219 \cdot 10^{-1}$	$0.97138 \cdot 10^{-2}$	$0.98802 \cdot 10^{-2}$
	3	$0.16205 \cdot 10^{-2}$	$0.23698 \cdot 10^{-2}$	$0.31158 \cdot 10^{-2}$	$0.22178 \cdot 10^{-2}$	$0.26073 \cdot 10^{-2}$
	4	$0.54086 \cdot 10^{-3}$	$0.74379 \cdot 10^{-3}$	$0.10048 \cdot 10^{-2}$	$0.88062 \cdot 10^{-3}$	$0.81750 \cdot 10^{-3}$
	5	$0.15830 \cdot 10^{-3}$	$0.28768 \cdot 10^{-3}$	$0.33659 \cdot 10^{-3}$	$0.28437 \cdot 10^{-3}$	$0.34140 \cdot 10^{-3}$
	10	$0.12587 \cdot 10^{-4}$	$0.16523 \cdot 10^{-4}$	$0.17521 \cdot 10^{-4}$	$0.98736 \cdot 10^{-4}$	$0.10736 \cdot 10^{-3}$
2	1	$0.61022 \cdot 10^{-2}$	$0.89445 \cdot 10^{-2}$	$0.98459 \cdot 10^{-2}$	$0.82619 \cdot 10^{-2}$	$0.91492 \cdot 10^{-2}$
	2	$0.52500 \cdot 10^{-3}$	$0.71841 \cdot 10^{-3}$	$0.96738 \cdot 10^{-3}$	$0.66363 \cdot 10^{-3}$	$0.82692 \cdot 10^{-3}$
	3	$0.46287 \cdot 10^{-4}$	$0.10465 \cdot 10^{-3}$	$0.11551 \cdot 10^{-3}$	$0.99123 \cdot 10^{-4}$	$0.16709 \cdot 10^{-3}$
	4	$0.13839 \cdot 10^{-4}$	$0.19808 \cdot 10^{-4}$	$0.21173 \cdot 10^{-4}$	$0.66386 \cdot 10^{-4}$	$0.94793 \cdot 10^{-4}$
	5	$0.52278 \cdot 10^{-5}$	$0.62587 \cdot 10^{-5}$	$0.16354 \cdot 10^{-4}$	$0.57611 \cdot 10^{-4}$	$0.66844 \cdot 10^{-4}$
	10	$0.70692 \cdot 10^{-6}$	$0.79362 \cdot 10^{-6}$	$0.41566 \cdot 10^{-5}$	$0.19018 \cdot 10^{-4}$	$0.23745 \cdot 10^{-4}$
		Lvls = 2, NFine = 10125, NCoarse = 1029				
1	1	$0.35119 \cdot 10^{-1}$	$0.35115 \cdot 10^{-1}$	$0.47902 \cdot 10^{-1}$	$0.39215 \cdot 10^{-1}$	$0.45660 \cdot 10^{-1}$
	2	$0.70299 \cdot 10^{-2}$	$0.11518 \cdot 10^{-1}$	$0.10113 \cdot 10^{-1}$	$0.96555 \cdot 10^{-2}$	$0.10124 \cdot 10^{-1}$
	3	$0.16783 \cdot 10^{-2}$	$0.23134 \cdot 10^{-2}$	$0.31116 \cdot 10^{-2}$	$0.25111 \cdot 10^{-2}$	$0.28154 \cdot 10^{-2}$
	4	$0.58759 \cdot 10^{-3}$	$0.79815 \cdot 10^{-3}$	$0.94454 \cdot 10^{-3}$	$0.69161 \cdot 10^{-3}$	$0.91486 \cdot 10^{-3}$
	5	$0.21834 \cdot 10^{-3}$	$0.33317 \cdot 10^{-3}$	$0.33348 \cdot 10^{-3}$	$0.26339 \cdot 10^{-3}$	$0.35363 \cdot 10^{-3}$
	10	$0.13607 \cdot 10^{-4}$	$0.25534 \cdot 10^{-4}$	$0.22342 \cdot 10^{-3}$	$0.30763 \cdot 10^{-4}$	$0.72578 \cdot 10^{-4}$
2	1	$0.60697 \cdot 10^{-2}$	$0.89510 \cdot 10^{-2}$	$0.98347 \cdot 10^{-2}$	$0.81740 \cdot 10^{-2}$	$0.96936 \cdot 10^{-2}$
	2	$0.52357 \cdot 10^{-3}$	$0.74812 \cdot 10^{-3}$	$0.98926 \cdot 10^{-3}$	$0.79192 \cdot 10^{-3}$	$0.90675 \cdot 10^{-3}$
	3	$0.88952 \cdot 10^{-4}$	$0.10400 \cdot 10^{-3}$	$0.21830 \cdot 10^{-3}$	$0.89080 \cdot 10^{-4}$	$0.86215 \cdot 10^{-4}$
	4	$0.15535 \cdot 10^{-4}$	$0.18265 \cdot 10^{-4}$	$0.14572 \cdot 10^{-3}$	$0.19477 \cdot 10^{-4}$	$0.45988 \cdot 10^{-4}$
	5	$0.58039 \cdot 10^{-5}$	$0.21082 \cdot 10^{-4}$	$0.18240 \cdot 10^{-3}$	$0.40298 \cdot 10^{-4}$	$0.63035 \cdot 10^{-4}$
	10	$0.59914 \cdot 10^{-6}$	$0.40553 \cdot 10^{-5}$	$0.96543 \cdot 10^{-4}$	$0.74399 \cdot 10^{-5}$	$0.14618 \cdot 10^{-4}$

On a Two-Level Parallel MIC(0) Preconditioning of Crouzeix-Raviart Non-conforming FEM Systems*

Raytcho D. Lazarov[1] and Svetozar D. Margenov[2]

[1] Department of Mathematics, Texas A& M University, College Station, TX, USA
Raytcho.Lazarov@radon.math.tamu.edu

[2] Central Laboratory of Parallel Processing, Bulgarian Academy of Sciences, Bulgaria
margenov@parallel.bas.bg

Abstract. In this paper we analyze a two-level preconditioner for finite element systems arising in approximations of second order elliptic boundary value problems by Crouzeix-Raviart non-conforming triangular linear elements. This study is focused on the efficient implementation of the modified incomplete LU factorization MIC(0) as a preconditioner in the PCG iterative method for the linear algebraic system. A special attention is given to the implementation of the method as a scalable parallel algorithm.

Key words: non-conforming FEM, preconditioning, parallel algorithms
AMS subject classifications: 65F10, 65N30

1 Introduction

In this paper we consider the elliptic boundary value problem

$$\begin{aligned} Lu \equiv -\nabla \cdot (a(x)\nabla u(x)) = f(x) &\text{ in } \Omega, \\ u = 0 &\text{ on } \Gamma_D, \\ (a(x)\nabla u(x)) \cdot n = 0 &\text{ on } \Gamma_N, \end{aligned} \tag{1}$$

where Ω is a convex polygonal domain in $\mathbb{R}^2$, $f(x)$ is a given function in $L^2(\Omega)$, $a(x) = [a_{ij}(x)]_{i,j=1}^2$ is a symmetric and uniformly positive definite matrix in Ω, n is the outward unit vector normal to the boundary $\Gamma = \partial\Omega$, and $\Gamma = \bar{\Gamma}_D \cup \bar{\Gamma}_N$. We assume that the entries $a_{ij}(x)$ are piece-wise smooth functions on $\bar{\Omega}$. In the paper we use the terminology of the flow in porous media and we refer to u as a pressure and $-a(x)\nabla u(x)$ as a velocity vector.

The problem (1) can be discretized by the finite volume method, the Galerkin finite element method (conforming or non-conforming) or the mixed finite element method. Each of these methods has its advantages and disadvantages

* This work was supported in part by the National Science Foundation under grant DMS 9973328 and by the gift-grant by Saudi Aramco Oil Co. The second author has been also supported by the Bulgarian NSF Grants MM-801.

I. Dimov et al. (Eds.): NMA 2002, LNCS 2542, pp. 192–201, 2003.

when the problem (1) is used in a particular application. For example, for application related to highly heterogeneous porous media the finite volume and mixed finite element methods have proven to be accurate and locally mass conservative. While applying the mixed FEM to problem (1) the continuity of the velocity normal to the boundary between two adjacent finite element could be enforced by Lagrange multipliers. In [2] Arnold and Brezzi have demonstrated that after the elimination of the unknowns representing the pressure and the velocity from the algebraic system the resulting Schur system for the Lagrange multipliers is equivalent to a discretization of (1) by Galerkin method using linear non-conforming finite elements. Namely, in[2] is shown that the lowest-order Raviart-Thomas mixed finite element approximations are equivalent to the usual Crouzeix-Raviart non-conforming linear finite element approximations when the non-conforming space is augmented with cubic bubbles. Further, such a relationship between the mixed and non-conforming finite element methods has been studied and simplified for various finite element spaces (see, e.g. [1,6]). The work in this direction resulted also in construction of efficient iterative methods for solving mixed FE systems (see, e.g., [7,8,9]).

Galerkin method based on non-conforming Crouzeix-Raviart linear triangular finite elements has been also used in the construction of so called *locking-free* approximations for parameter-dependent problems. Furthermore, the stiffness matrix has a regular sparsity structure such that in each row the number of non-zero entries is at most five.

Therefore, the development of efficient and parallelizable solution methods for non-conforming finite element (FEM) systems is an important problem with a range of applications in scientific computations and engineering. In this paper we construct and study a preconditioner for the algebraic system obtained from discretization of (1) by non-conforming finite element. Our preconditioner is based on MIC(0) factorization of the modified finite element stiffness matrix so that the condition number of the preconditioned system is independent of the possible jumps in the coefficients of the differential equation. Our analysis is done for problems in 2-D domains under the condition that the jumps are aligned with the finite element partition. The study uses the main ideas of the recently proposed highly parallelizable and efficient preconditioners based on MIC(0) for linear conforming and rotated bilinear non-conforming finite elements (see, e.g. [5,10]).

The rest of the paper is organized as follows. In Sections 2 and 3 we introduce the finite element approximation and the two-level algorithm. In Section 4 we propose a locally modified sparse approximation of the Schur complement and prove that the preconditioned Schur system has a condition number that is bounded uniformly with respect to both the problem size and the possible jumps of the coefficients. The algorithm has been analyzed in the case of coefficient and mesh isotropy. Further, in Section 5 we derive estimates for the execution time on a multiprocessor computer system which shows a good parallel scalability for large scale problems. Finally, in Section 6 we present numerical results on a test problem that shows that the proposed parallel preconditioner preserves the robustness and the computational efficiency of the standard MIC(0) factorization algorithm.

2 Finite Element Discreti ation

The domain Ω is partitioned using triangular elements. The partition is denoted by $\mathcal{T}_h$ and is assumed to be quasi-uniform with a characteristic mesh-size h. Most of our analysis is valid for general tensors $a(x)$, but here we restrict our considerations to the $a(x)$ being a scalar function. The partition $\mathcal{T}_h$ is aligned with the discontinuities of the coefficient $a(x)$ so that over each element $e \in \mathcal{T}_h$ the function $a(x)$ is smooth. Further, we assume that $\mathcal{T}_h$ is generated by first partitioning Ω into quadrilaterals Q and then splitting each quadrilateral into two triangles by one of its diagonals, see Figure 1. To simplify our considerations we assume that the splitting into quadrilaterals is topologically equivalent to a square mesh. The weak formulation of the above problem reads as follows: given $f \in L^2(\Omega)$ find $u \in H_D^1(\Omega) = \{v \in H^1(\Omega) : v = 0 \text{ on } \Gamma_D\}$, satisfying

$$\mathcal{A}(u,v) = (f,v) \quad \forall v \in H_D^1(\Omega), \text{ where } \mathcal{A}(u,v) = \int_\Omega a(x)\nabla u(x) \cdot \nabla v(x) dx. \quad (2)$$

We shall discretize this problem by using Crouzeix-Raviart non-conforming linear triangular finite elements. The finite element space V_h consists of piece wise linear functions over $\mathcal{T}_h$ determined by their values in the midpoints of the edges of the triangles. The nodal basis functions of V_h have a support on not more than two neighboring triangles where the corresponding node is the midpoint of their common side. Then the finite element formulation is: find $u_h \in V_h$, satisfying

$$\mathcal{A}_h(u_h, v_h) = (f, v_h) \;\; \forall v_h \in V_h, \text{ where } \mathcal{A}_h(u_h, v_h) = \sum_{e \in \mathcal{T}_h} \int_e a(e)\nabla u_h \cdot \nabla v_h dx. \quad (3)$$

Here $a(e)$ is defined as the integral averaged value of $a(x)$ over each $e \in \mathcal{T}_h$. We note that we allow strong coefficient jumps across the boundaries between the adjacent finite elements. Now, the standard computational procedure leads to the linear system of equations

$$A\mathbf{u} = \mathbf{f}, \quad (4)$$

where A is the corresponding global stiffness matrix and $\mathbf{u} \in \mathbb{R}^N$ is the vector of the unknown nodal values of u_h. The matrix A is sparse, symmetric and positive definite. For large scale problems, the preconditioned conjugate gradient (PCG) method is known to be the best solution method. The goal of this study is to present a **robust and parallelizable** preconditioner for the system (4).

3 The Two-Level Algorithm

Since the triangulation $\mathcal{T}_h$ is obtained by diagonal-wise subdividing each cell Q into two triangles, see Figure 1 (a), we can partition the grid nodes into two groups. The first group contains the centers of the quadrilateral super-elements $Q \subset \Omega$ (the midpoints of the diagonals that split Q into two triangles) and the second group contains the rest of the nodes. With respect to this splitting, A admits the following two-by-two block partitioning that can be written also in a block-factored form

$$A = \begin{bmatrix} A_{11} & A_{12} \\ A_{21} & A_{22} \end{bmatrix} = \begin{bmatrix} A_{11} & 0 \\ A_{21} & S \end{bmatrix} \begin{bmatrix} I & A_{11}^{-1} A_{12} \\ 0 & I \end{bmatrix}, \tag{5}$$

where S stands for the related Schur complement. Obviously, A_{11} is a diagonal matrix so that the Schur complement S can be assembled from the corresponding super-element Schur complements $S_Q = A_{22;Q} - A_{21;Q} A_{11;Q}^{-1} A_{12;Q}$, i.e.

$$S = \sum_{Q \in \mathcal{T}_h} L_Q^T S_Q L_Q,$$

where L_Q stands for the restriction mapping of the global vector of unknowns to the local one corresponding to a macroelement Q containing two triangles. Such a procedure is called *static condensation*. We now introduce S_Q, the local stiffness matrix for Q (in fact, this is the local Schur complement matrix), and its approximation B_Q:

$$S_Q = \begin{bmatrix} s_{11} & s_{12} & s_{13} & s_{14} \\ s_{21} & s_{22} & s_{23} & s_{24} \\ s_{31} & s_{32} & s_{33} & s_{34} \\ s_{41} & s_{42} & s_{43} & s_{44} \end{bmatrix}, \quad B_Q = \begin{bmatrix} b_{11} & s_{12} & 0 & s_{14} \\ s_{21} & b_{22} & s_{23} & 0 \\ 0 & s_{32} & b_{33} & s_{34} \\ s_{41} & 0 & s_{43} & b_{44} \end{bmatrix}. \tag{6}$$

Here $b_{11} = s_{11} + s_{13}$, $b_{22} = s_{22} + s_{24}$, $b_{33} = s_{33} + s_{31}$, $b_{44} = s_{44} + s_{42}$, which ensures that S_Q and B_Q have equal rowsums. The definition of B_Q corresponds to the node numbering as shown in Figure 1. Here the dash lines represent the connectivity pattern of (b) the local Schur complement S_Q and (c) its locally modified sparse approximation B_Q. Assembling the local matrices B_Q we get

$$B = \sum_{Q \in \mathcal{T}_h} L_Q^T B_Q L_Q. \tag{7}$$

The structure of B could be interpreted as a skewed five point stencil. In a very general setting S and B are spectrally equivalent.

After the *static condensation* step, the initial problem (4) is reduced to the solution of a system with the Schur complement matrix S. At this point we apply the PCG method with a preconditioner $\mathcal{C}$ defined as a MIC(0) factorization (see, e.g., [4]) of B, that is, $\mathcal{C} = \mathcal{C}_{MIC(0)}(B)$. This needs of course B to allow for a stable MIC(0) factorization, which will be shown in Section 4.

4 Condition Number Analysis of the Model Problem

The model problem we analyze in this section is set on a uniform square mesh. Then the element stiffness matrix corresponding to the triangle element $e \in \mathcal{T}_h$

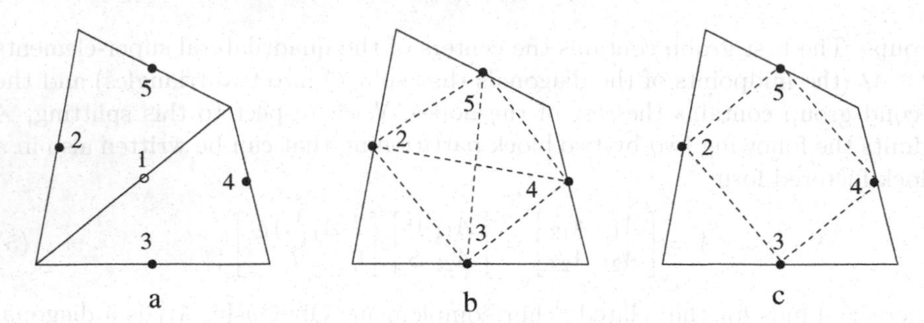

Fig. 1. (a) Node numbering of the super-element Q; (b) Connectivity pattern of S_Q; (c) Connectivity pattern of B_Q.

has the form

$$A_e = 2a_e \begin{bmatrix} 2 & -1 & -1 \\ -1 & 1 & 0 \\ -1 & 0 & 1 \end{bmatrix}. \tag{8}$$

Let us assume now that the square super-element Q consists of the triangles e_1 and e_2 where the element-wise averaged diffusion coefficients are respectively a_1 and a_2. Then the matrices needed for our local analysis are as follows:

$$S_Q = \frac{1}{a_1 + a_2} \begin{bmatrix} a_1^2 + 2a_1a_2 & -a_1^2 & -a_1a_2 & -a_1a_2 \\ -a_1^2 & a_1^2 + 2a_1a_2 & -a_1a_2 & -a_1a_2 \\ -a_1a_2 & -a_1a_2 & a_2^2 + 2a_1a_2 & -a_2^2 \\ -a_1a_2 & -a_1a_2 & -a_2^2 & a_2^2 + 2a_1a_2 \end{bmatrix}, \tag{9}$$

$$B_Q = \frac{1}{a_1 + a_2} \begin{bmatrix} a_1^2 + a_1a_2 & -a_1^2 & 0 & -a_1a_2 \\ -a_1^2 & a_1^2 + a_1a_2 & -a_1a_2 & 0 \\ 0 & -a_1a_2 & a_2^2 + a_1a_2 & -a_2^2 \\ -a_1a_2 & 0 & -a_2^2 & a_2^2 + a_1a_2 \end{bmatrix}. \tag{10}$$

Consider now the local eigenvalue problem: find $\lambda \in R$, $\mathbf{0} \neq \mathbf{w} \in \mathbb{R}^4$ such that

$$S_Q\mathbf{w} = \lambda B_Q\mathbf{w}. \tag{11}$$

Obviously $Ker(S_Q) = Ker(B_Q) = Span\{\mathbf{e}\}$ where $\mathbf{e} = (1,1,1,1)^t$. Thus, (11) reduces to a 3×3 eigenvalue problem. Then using the substitution $\nu = a/b$, $\mu = 1 - \lambda$ we get the following characteristic equation for μ

$$det \begin{bmatrix} \nu + (\nu^2 + \nu)\mu & -\nu^2\mu & -\nu \\ -\nu^2\mu & \nu + (\nu^2 + \nu)\mu & -\nu\mu \\ -\nu & -\nu\mu & \nu + (1 + \nu)\mu \end{bmatrix} = 0. \tag{12}$$

A further simple computation shows that $\mu_1 = 0$ and $\mu_2 = \mu_3 = -1$, and therefore $\lambda_1 = 1$, $\lambda_2 = \lambda_3 = 2$. The global condition number estimate follows directly from the presented local analysis. Namely, we have

$$\mathbf{v}^T S\mathbf{v} = \sum_{Q\in\mathcal{T}_h} \mathbf{v}^T L_Q^T S_Q L_Q \mathbf{v} \leq 2 \sum_{Q\in\mathcal{T}_h} \mathbf{v}^T L_Q^T B_Q L_Q \mathbf{v} = 2\mathbf{v}^T B\mathbf{v}$$

and, similarly, $\mathbf{v}^T S\mathbf{v} \leq \mathbf{v}^T B\mathbf{v}$. The result of our local analysis is summarized in the following theorem:

Theorem 1. *Consider the non-conforming FEM problem (3) on a square mesh. Then:*

(i) the sparse approximation B of the Schur complement S satisfies the conditions for a stable MIC(0) factorization;
(ii) the matrices B and S are spectrally equivalent, namely the following estimate for the relative condition number holds uniformly with respect to any possible jumps of the diffusion coefficients

$$\kappa\left(B^{-1}S\right) \leq 2. \tag{13}$$

5 Parallel Preconditioning Algorithm

In this section we study the possibility to parallelize the proposed method. The first step, i.e. the static condensation, is local and therefore can be performed fully in parallel. This is the reason to focus our attention on the PCG solution of the reduced system with the Schur complement S. Let us recall that the preconditioner was introduced as $\mathcal{C} = \mathcal{C}_{MIC(0)}(B)$. Each PCG iteration consists of one solution of a system with the matrix $\mathcal{C}$, one matrix vector multiplication with the original matrix S, two inner products, and three linked vector triads of the form $\mathbf{v} := \alpha\mathbf{v} + \mathbf{u}$. Therefore the computational complexity of one PCG iteration is given by $\mathcal{N}_{PCG}^{it} \approx \mathcal{N}(\mathcal{C}^{-1}\mathbf{v}) + \mathcal{N}(S\mathbf{v}) + 10N$. Then, for the algorithm introduced in Section 3 we find $\mathcal{N}(\mathcal{C}^{-1}\mathbf{v}) \approx 11N$, $\mathcal{N}(S\mathbf{v}) \approx 13N$, and finally

$$\mathcal{N}_{PCG}^{it} \approx 34N. \tag{14}$$

As we see, the algorithm under consideration is relatively cheap where the solution of the preconditioned system takes less then one third of the total cost.

It is well known that MIC(0) is an inherently sequential algorithm. In the general case, the solution of the arising triangular systems is typically recursive. Below we overcome this disadvantage by a special construction of the matrix B. For simplicity of the presentation we consider the model problem in a square $\Omega = (0,1)^2$ on a square mesh with a mesh size $h = 1/n$ (subsequently each square is split into two triangles to get $\mathcal{T}_h$). The structure of the matrices S and B is illustrated in Figure 2 where each of the diagonal blocks corresponds to one vertical line of the mesh if a column-wise numbering of the unknowns has been used. The big advantage of the introduced matrix B is that all of its diagonal blocks are diagonal. In this case, the implementation of the PCG solution step $\mathcal{C}^{-1}\mathbf{v}$ is fully parallel within each of these blocks.

To establish the theoretical performance characteristics of the preconditioner, a simple general model for the arithmetic and the communication times is applied (see, e.g., [11]). We assume that the computations and communications do not overlap, and therefore, the parallel execution time is the sum of the computation and communication times. We also assume that the execution of M arithmetic

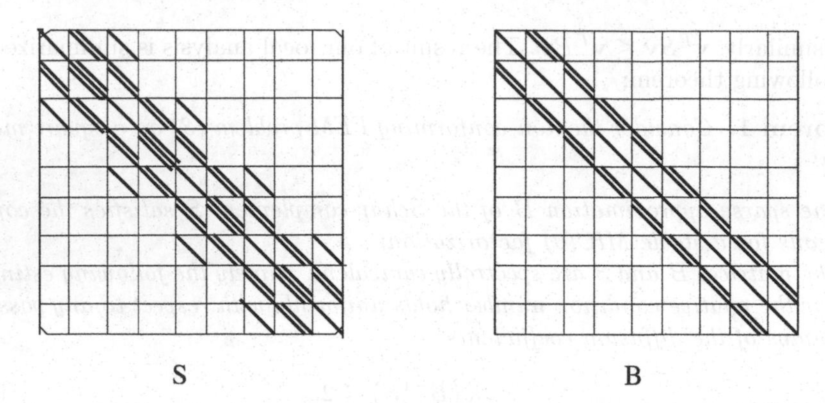

Fig. 2. Sparsity pattern of the matrices S and B, $\Omega = (0,1)^2$.

operations on one processor takes time $T_a = Mt_a$, where t_a is the average unit time to perform one arithmetic operation on one processor (no vectorization). We assume that the communication time to transfer M data elements from one processor to another can be approximated by $T_{com} = \ell(t_s + Mt_c)$, where t_s is the start-up time and t_c is the incremental time necessary for each of M elements to be sent, and ℓ is the graph distance between the processors.

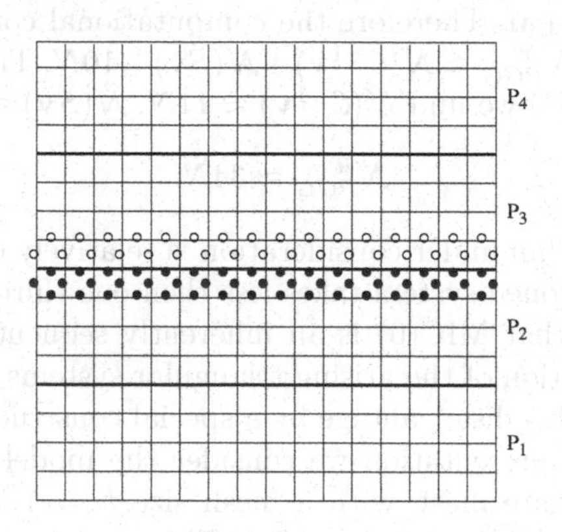

Fig. 3. Strip-wise data distribution between the processors.

Further, we consider a distributed memory parallel algorithm where the number of processors is p, and $n = mp$ for some natural number m. The computational domain is split in p equally sized strips where the processor P_k is responsible for computations related to the k-th strip. Then, we get the following expressions for the communication times related to $\mathcal{C}^{-1}\mathbf{v}$ and $S\mathbf{v}$

$$T_{com}(\mathcal{C}^{-1}\mathbf{v}) = 6n(t_s + t_c), \qquad T_{com}(S\mathbf{v}) = 2t_s + (3n+1)t_c.$$

Note that the above communications are completely local and do not depend on the number of processors p assuming that P_k and P_{k+1} are neighbors. The linked triads are free of communications. The inner product can be performed using one broadcasting and one gathering global communication but they do not contribute to the leading terms of the total parallel time and will not be considered in our analysis. This setting leads to the following expression for the parallel time per one PCG iteration

$$T_p = T_p^{it} \approx \frac{2n(n+1)}{p} t_a + 6nt_s + 9nt_c. \qquad (15)$$

From (15) we conclude that the parallel algorithm is asymptotically optimal and

$$\lim_{n\to\infty} S_p = p, \qquad \lim_{n\to\infty} E_p = 1,$$

where the parallel speed-up and the parallel efficiency are given in the usual form $S_p = T_1/T_p$, and $E_p = S_p/p$.

Remark 1. A more realistic analysis of the parallel performance needs some specific information about the behavior of the introduced average timing parameters t_a, t_s and t_c. The key point here is that a good parallel scalability could be expected only if $n \gg pt_s/t_a$.

6 Numerical Tests

The numerical tests presented below illustrate the PCG convergence rate when the size of the discrete problem and the coefficient jumps are varied. A relative stopping criterion $(\mathcal{C}^{-1} r^{n_{it}}, r^{n_{it}})/(\mathcal{C}^{-1} r^0, r^0) < \varepsilon$ is used in the PCG algorithm, where r^i stands for the residual at the i-th iteration step, $(\cdot,\cdot)$ is the standard Euclidean inner product, and $\varepsilon = 10^{-6}$. The computational domain is the unit square $\Omega = (0,1)^2$ where homogeneous Dirichlet boundary conditions are assumed at the bottom side. A uniform mesh is used, where $h = 1/n$, and the size of the discrete problem is $N = 2n(n+1)$. Let $\Omega = \Omega_1 \bigcup \Omega_2$,

$$\Omega_2 := \{\frac{n-1}{2}h \le x_1 \le \frac{n+1}{2}h,\ x_2 > \frac{n+1}{4}h\},$$

and let a_i be the problem coefficient corresponding to Ω_i, $i = 1,2$. In what follows $a_1 = 1$. This test problem allows us to examine the influence of the coefficient jumps on the number of iterations. Note that the coefficient jumps are highly localized since the width of the domain Ω_2 is just one mesh-size. Two model tests are reported in Table 1 where: (a) $a_2 = 1$ or the differential operator L is the Laplacian $-\Delta$ and (b) $a_2 = 10^3$. We investigate also the influence of approximation of the Schur complement matrix S by the introduced sparse approximation B. We denote by n_{it}^{SS} and n_{it}^{SB} the number of iterations obtained when MIC(0) factorization of S and B are used as preconditioners of S. The qualitative analysis of the results given in Table 1 shows that: (a) the number

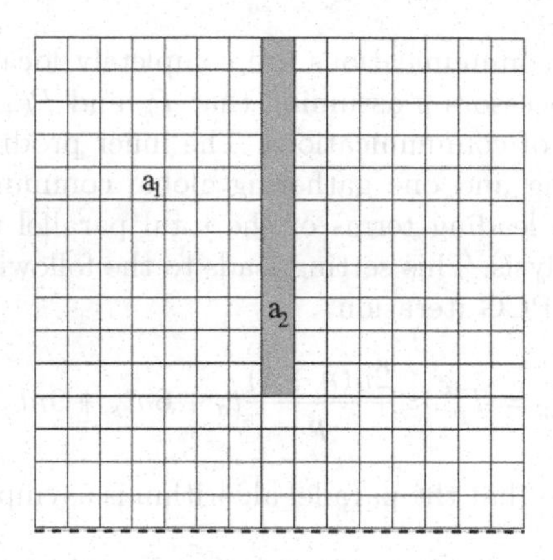

Fig. 4. Test problem: $n = 15$, $\Omega_2 := \{7/15 \leq x_1 \leq 8/15, \quad x_2 > 4/15\}$.

Table 1. *PCG* iterations: *MIC*(0) factorization of S and B.

$h = 1/n$	problemsize	n_{it}^{SS}		n_{it}^{SB}	
n	N	$L \equiv -\Delta$	$a_2 = 10^3$	$L \equiv \Delta$	$a_2 = 10^3$
7	112	10	16	11	17
15	480	16	29	17	30
31	1984	23	47	24	52
63	8064	34	73	35	81
127	32512	50	117	49	129

of iterations in all cases is $O(n^{1/2}) = O(N^{1/4})$, namely, it grows proportionally to $\sqrt{n}$ in agreement with the properties of the MIC(0) factorization of S; (b) the number of iterations n_{it}^{SS} and n_{it}^{SB} are practically the same for both the model problem and for the problem with large jumps in the coefficients. Note that the obtained results are considerably better than what we have as a prediction from the uniform estimate of Theorem 1.

The impact of the coefficient jump on the number of PCG iterations for a fixed 63×63-mesh is presented in Table 2. We see some increase of the iterations with a_2. Nevertheless, the obtained results can be viewed as very promising taking into account that the jump is not only very large, but it is also highly localized within a strip of width h.

Table 2. *PCG* iterations for $n = 65$, varying a_2, and *MIC*(0) factorization of B

a_2	1	10	10^2	10^3	10^4
n_{it}^{SB}	35	45	62	81	93

7 Concluding Remarks

In this paper we have proposed a new two-level preconditioner for Crouzeix-Raviart non-conforming finite element approximation of second order elliptic equations. Our study is motivated by two factors. First, the Crouzeix-Raviart non-conforming finite elements produce algebraic systems that are equivalent to the Schur complement system for the Lagrange multipliers arising from the mixed finite element method for Raviart-Thomas elements (see, e.g. [1,2,6]). Second, a class of highly parallelizable and efficient preconditioners based on MIC(0) have been proposed recently for linear conforming and rotated bilinear non-conforming finite elements (see, e.g. [5,10]). Our further plans include a generalization to 3-D problems on tetrahedral meshes and problems with orthotropy. These are much more complicated problems but we expect to extend our study to such problems and to be able to construct, test, and implement efficient preconditioners with similar theoretical and computational properties.

References

1. Arbogast, T. and Chen, Z.: On the Implementation of Mixed Methods as Nonconforming Methods for Second Order Elliptic Problems. Math. Comp, 64 (1995) 943–972.
2. Arnold, D.N. and Brezzi, F.: Mixed and Nonconforming Finite Element Methods: Implementation, Postprocessing and Error Estimates. RAIRO, Model. Math. Anal. Numer., 19 (1985) 7–32.
3. Axelsson, O., Laayyonni, L., Margenov, S.: On Multilevel Preconditioners Which Are Optimal with Respect to Both Problem and Discretization Parameters. SIAM J. Matr. Anal. Appl. (submitted).
4. Blaheta, R.: Displacement Decomposition — Incomplete Factorization Preconditioning Techniques for Linear Elasticity Problems. Numer. Lin. Alg. Appl., 1 (1994) 107–126.
5. Bencheva, G., Margenov, S.: Parallel Incomplete Factorization Preconditioning of Rotated Linear FEM Systems. Computer & Mathematics with Applications (2002) (submitted).
6. Chen, Z.: Analysis of Mixed Methods Using Conforming and Nonconforming Finite Element Methods. RAIRO, Math. Model. Numer. Anal., 27 (1993) 9–34.
7. Chen, Z.: Equivalence between Multigrid Algorithms for Mixed and Nonconformning Methods for Second Order Elliptic Problems. East-West J. Numer. Math., 4 (1996) 1–33.
8. Chen, Z., Ewing, R.E., Kuznetsov, Yu.A., Lazarov, R.D., and Maliassov, S.: Multilevel Preconditioners for Mixed Methods for Second Order Elliptic Problems. Numer. Lin. Alg. with Appl., 3(5) (1996) 427–453.
9. Chen, Z., Ewing, R.E., and Lazarov, R.D.: Domain Decomposition Algorithms for Mixed Methods for Second Order Elliptic Problems. Math. Comp., 65(214) (1996) 467–490.
10. Gustafsson, I., Lindskog, G.: On Parallel Solution of Linear Elasticity Problems: Part I: Theory. Numer. Lin. Alg. Appl., 5 (1998) 123–139.
11. Saad, Y., Schultz, M.H.: Data Communication in Parallel Architectures. Parallel Comput., 11 (1989) 131–150.

Part IV

Control and Uncertain Systems

Discrete Methods for Optimal Control Problems

Ion Chryssoverghi

Department of Mathematics
National Technical University of Athens
Zografou Campus, 15780 Athens, Greece,
ichriso@math.ntua.gr

Abstract. We consider a constrained optimal control problem, which we formulate in classical and in relaxed form. In order to approximate this problem numerically, we apply various discretization schemes on either of these two forms and study the behavior in the limit of discrete optimality and necessary conditions for optimality. We then propose discrete mixed gradient penalty methods that use classical or relaxed discrete controls and progressively refine the discretization, thus reducing computing time and memory. In addition, when the discrete adjoint state is not defined or difficult to calculate, we propose discrete methods that use approximate adjoints and derivatives. The result is that in relaxed methods accumulation points of generated sequences satisfy continuous strong relaxed optimality conditions, while in classical methods they satisfy weak optimality conditions.

1 The Continuous Optimal Control Problems

Consider the following classical optimal control problem. The state equation is

$$y'(t) = f(t, y(t), w(t)) \text{ in } I := [0, T], \quad y(0) = y^0, \tag{1}$$

where $y := y_w$, $y(t)$ in $\mathbb{R}^p$, the constraints on the control w are $w(t) \in U$ in I, where U is a compact subset of $\mathbb{R}^q$, the state constraints are $G_1(w) := g_1(y(T)) = 0$, $\quad G_2(w) := g_2(y(T)) \leq 0$, where $g_1 : \mathbb{R}^p \to \mathbb{R}^{m_1}$, $g_2 : \mathbb{R}^p \to \mathbb{R}^{m_2}$, and the cost functional to be minimized is $G_0(w) := g_0(y(T))$.

The set of classical controls is defined by

$$W := \{w : I \to U |\ w \text{ measurable}\} \subset L^2(I, \mathbb{R}^q), \tag{2}$$

and the set of relaxed controls (for relevant theory, see [11] and [9]) by

$$R := \{r : I \to M_1(U) |\ r \text{ weakly meas.}\} \subset L^\infty_w(I, M(U)) \equiv L^1(I, C(U))^*, \tag{3}$$

where $M_1(U)$ is the set of probability measures on U. The set W (resp. R) is endowed with the relative strong (resp. weak star) topology, and R is convex, metrizable and compact. If each classical control $w(\cdot)$ is identified with its associated Dirac relaxed control $r(\cdot) := \delta_{w(\cdot)}$, then W may be considered as a subset of R, and W is dense in R. For given $\phi \in L^1(I, C(U))$ and $r \in R$, we

I. Dimov et al. (Eds.): NMA 2002, LNCS 2542, pp. 205–212, 2003.

write $\phi(t,r(t)) := \int_U \phi(t,u)r(t)(du)$. We can now define the relaxed problem. The state equation is

$$y'(t) = f(t,y(t),r(t)) \text{ in } I, \quad y(0) = y^0, \tag{4}$$

where $y := y_r$, the control constraint is $r \in R$, the state constraints are $G_1(r) := g_1(y(T)) = 0$, $G_2(r) := g_2(y(T)) \leq 0$, and the cost functional is $G_0(r) := g_0(y(T))$.

We suppose in the sequel that the functions f, f_y, f_u are continuous w.r.t. (t,y,u) and Lipschitz continuous w.r.t. (y,u), and that the functions g_l, $l = 0,1,2$, are Lipschitz continuous.

Theorem 1. *The functionals G_l are continuous on R and on W. If the relaxed problem is feasible, then it has a solution.*

Note that in the classical problem $y'(t) \in f(t,y(t),U)$ (velocity set), while in the relaxed problem $y'(t) \in \text{co}[f(t,y(t),U)]$. The classical problem may have no solution, and since $W \subset R$, we generally have

$$c' := \min_{r\in R,\ G_1(r)=0,\ G_2(r)\leq 0} G_0(r) \leq \inf_{w\in W,\ G_1(w)=0,\ G_2(w)\leq 0} G_0(w) := c, \tag{5}$$

where the equality holds, in particular, if there are no state constraints. Since usually approximation methods slightly violate the state constraints, approximating the relaxed problem, hence c', is not a drawback in practice (see [11], p. 248). Note also that approximating sequences of classical controls may converge to relaxed ones.

Theorem 2. *We drop here the index l. For $r, \bar{r} \in R$, the relaxed directional derivative of the functional G is given by*

$$DG(r,\bar{r}-r) := \lim_{\alpha\to 0^+} \frac{G(r+\alpha(\bar{r}-r)) - G(r)}{\alpha} = \int_I z(t)f(t,y(t),\bar{r}(t)-r(t))dt, \tag{6}$$

where the adjoint state $z := z_r$, a row vector ($l = 0$) or a matrix ($l = 1,2$), is defined by

$$\begin{aligned} z'(t) &= -z(t)f_y(t,y(t),r(t)) \text{ in } I, \\ z(T) &= g_y(y(T)), \text{ with } y := y_r. \end{aligned} \tag{7}$$

For $w, \bar{w} \in W$, the classical directional derivative of G is

$$\begin{aligned} \mathbf{D}G(w,\bar{w}-w) &:= \lim_{\alpha\to 0^+} \tfrac{G(w+\alpha(\bar{w}-w))-G(w)}{\alpha} \\ &= \int_I z(t)f_u(t,y(t),\bar{w}(t))(\bar{w}(t)-w(t))dt, \end{aligned} \tag{8}$$

where $z := z_r$, with $r := w$, and the controls are purely classical (not Dirac). The functional DG (resp. $\mathbf{D}G$) is continuous on $R \times R$ (resp. $W \times W$).

Theorem 3. *(Necessary conditions for optimality) If the control $r \in R$ (resp. $r \equiv w \in W$) is optimal for the relaxed (resp. classical) problem, then r is strongly* *relaxed extremal, i.e. there exist* $\lambda_0 \in \mathbb{R}$ *and row vectors* $\lambda_1 \in \mathbb{R}^{m_1}$, $\lambda_2 \in \mathbb{R}^{m_2}$, *with* $\lambda_0 \geq 0$, $\lambda_2 \geq 0$, λ_l *not all zero, such that* $DG(r, \bar{r} - r) \geq 0$, *for every* $\bar{r} \in R$, *where* $G := \lambda_0 G_0 + \lambda_1 G_1 + \lambda_2 G_2$, *and that* $\lambda_2 G_2(r) = 0$ *(transversality condition). The above inequality is equivalent to the pointwise strong relaxed minimum principle*

$$z(t)f(t, y(t), r(t)) = \min_{u \in U}[z(t)f(t, y(t), u)], \text{ a.e. in } I, \tag{9}$$

where the adjoint z *is defined with* $g := \lambda_0 g_0 + \lambda_1 g_1 + \lambda_2 g_2$. *If* U *is convex, then this minimum principle implies the pointwise weak relaxed minimum principle*

$$z(t)f_u(t, y(t), r(t))r(t) = \min_{\phi}[z(t)f_u(t, y(t), r(t))\phi(t, r(t))], \text{ a.e. in } I, \tag{10}$$

where the minimum is taken over the set $B(I, U; U)$ *of Caratheodory functions (in the sense of Warga [11])* $\phi : I \times U \to U$, *which, in turn, is equivalent to the global weak relaxed minimum principle (r is weakly relaxed extremal)*

$$\int_I z(t)f_u(t, y(t), r(t))[\phi(t, r(t)) - r(t)]dt \geq 0, \text{ for every } \phi \in B(I, U; U). \tag{11}$$

If the control $w \in W$ *is optimal for the classical problem and* U *is convex, then* w *is weakly classical extremal, i.e. there exist* λ_l *as above such that* $\mathbf{D}G(w, \bar{w} - w) \geq 0$, *for every* $\bar{w} \in W$, *and* $\lambda_2 G_2(r) = 0$. *The above inequality is equivalent to the pointwise weak classical minimum principle*

$$z(t)f_u(t, y(t), w(t))w(t) = \min_{u \in U}[z(t)f_u(t, y(t), w(t))u], \text{ a.e. in } I. \tag{12}$$

2 Discreti ations

Let $(N_n)_{n=0}^{\infty}$ be an increasing sequence of positive integers, with $N_n \to \infty$. Set $N := N_n$, $h_n := T/N$, $t_{ni} = ih_n$, $i = 0, ..., N$, $I_{ni} := [t_{n,i-1}, t_{n,i})$, $i = 1, ..., N-1$, $I_{n,N} := [t_{n,N-1}, t_{nN}]$. We define the set of discrete relaxed controls

$$R_n = \{r_n \in R |\ r_n(t) := r_{ni} \in M_1(U) \text{ on } I_{ni},\ i = 1, ..., N\}, \tag{13}$$

and the set of discrete classical controls

$$W_n = \{w_n \in W |\ w_n(t) := w_{ni} \in U \text{ on } I_{ni},\ i = 1, ..., N\} \subset R_n. \tag{14}$$

The relaxed state equation is discretized by some explicit Runge-Kutta scheme, for example, the first order Euler scheme

$$y_{ni} = y_{n,i-1} + h_n f(t_{n,i-1}, y_{n,i-1}, r_{ni}), \quad y_{n0} = y^0, \tag{15}$$

or the second order trapezoidal Runge-Kutta scheme

$$\begin{aligned} y_{ni} &= y_{n,i-1} + (h_n/2)[f(t_{ni}, \tilde{y}_{ni}, r_{ni}) + f(t_{n,i-1}, y_{n,i-1}, r_{ni})], \\ \tilde{y}_{ni} &= y_{n,i-1} + h_n f(t_{n,i-1}, y_{n,i-1}, r_{ni}), \end{aligned} \tag{16}$$

or by some implicit Runge-Kutta scheme, for example, the second order pure trapezoidal scheme

$$y_{ni} = y_{n,i-1} + (h_n/2)[f(t_{ni}, y_{ni}, r_{ni}) + f(t_{n,i-1}, y_{n,i-1}, r_{ni})]. \tag{17}$$

The discrete control constraint is $r_n \in R_n$ for the discrete relaxed problem and $r_n \equiv w_n \in W_n$ for the discrete classical problem. For both problems, the perturbed discrete state constraints are defined by either (i) $|G_{1n}(r_n)| := |g_1(y_{nN})| \leq \delta_{1n}$ or (ii) $G_{1n}(r_n) := g_1(y_{nN}) = \delta_{1n}$, and $G_{2n}(r_n) := g_2(y_{nN}) \leq \delta_{2n}$, where $\delta_{1n}, \delta_{2n} > 0$ are chosen perturbations, and the discrete cost by $G_{0n}(r_n) := g_0(y_{nN})$.

Theorem 4. *The functionals G_{ln} are continuous on R and W. If any of the above discrete problems is defined and feasible, then it has a solution.*

Theorem 5. *(Euler case) We drop the index l. The relaxed directional derivative of G_n is*

$$DG_n(r_n, \bar{r}_n - r_n) = h_n \sum_{i=1}^{N} z_{ni} f(t_{n,i-1}, y_{n,i-1}, \bar{r}_{ni} - r_{ni}), \; r_n, \bar{r}_n \in R_n, \tag{18}$$

where the discrete adjoint state z_n is defined by

$$z_{n,i-1} = z_{ni} + h_n z_{ni} f_y(t_{n,i-1}, y_{n,i-1}, r_{ni}), \; i = N, ..., 1, \; z_{nN} = g_y(y_{nN}), \tag{19}$$

and the classical directional derivative of G_n

$$\mathbf{D}G_n(w_n, \bar{w}_n - w_n) = h_n \sum_{i=1}^{N} z_{ni} f_u(t_{n,i-1}, y_{n,i-1}, w_{ni})(\bar{w}_{ni} - w_{ni}), \; w_n, \bar{w}_n \in W_n, \tag{20}$$

where z_n corresponds to $r_n \equiv w_n$, and the controls are purely classical. The functional DG_n (resp. $\mathbf{D}G_n$) is continuous on $R_n \times R_n$ (resp. $W_n \times W_n$).

Theorem 6. *(Necessary conditions for discrete optimality, Euler case) If the control $r_n \in R_n$ is optimal for the discrete relaxed problem (constraint case (ii)), then r_n is strongly relaxed extremal, i.e. there exist $\lambda_{0n} \in \mathbb{R}$, $\lambda_{1n} \in \mathbb{R}^{m_1}$, $\lambda_{2n} \in \mathbb{R}^{m_2}$, with $\lambda_{0n} \geq 0$, $\lambda_{2n} \geq 0$, λ_{ln} not all zero, such that*

$$DG_n(r_n, \bar{r}_n - r_n) \geq 0, \; \text{for every } \bar{r}_n \in R_n, \tag{21}$$

where $G_n := \lambda_{0n} G_{0n} + \lambda_{1n} G_{1n} + \lambda_{2n} G_{2n}$, and that $\lambda_{2n}[G_{2n}(r_n) - \delta_{2n}] = 0$. The above inequality is equivalent to the discrete strong relaxed minimum principle

$$z_{ni} f(t_{n,i-1}, y_{n,i-1}, r_{ni}) = \min_{u \in U} [z_{ni} f(t_{n,i-1}, y_{n,i-1}, u)], \; i = 1, ..., N, \tag{22}$$

where z_n is defined with $g := \lambda_{0n} g_0 + \lambda_{1n} g_1 + \lambda_{2n} g_2$. If the control $w_n \in W_n$ is optimal for the discrete classical problem (case (ii)) and U is convex, then w_n is weakly classical extremal, i.e. there exist λ_{ln} as above such that

$$\mathbf{D}G_n(w_n, \bar{w}_n - w_n) \geq 0, \; \text{for every } \bar{w}_n \in W_n, \tag{23}$$

and $\lambda_{2n}[G_{2n}(r_n) - \delta_{2n}] = 0$. *The above inequality is equivalent to the discrete weak classical minimum principle*

$$z_{ni} f_u(t_{n,i-1}, y_{n,i-1}, w_{ni}) w_{ni} = \min_{u \in U} [z_{ni} f_u(t_{n,i-1}, y_{n,i-1}, w_{ni}) u], \; i = 1, ..., N, \tag{24}$$

where z_n *is defined as above, with* $r_n \equiv w_n$.

When the discrete relaxed adjoint is not defined or difficult to calculate, we can define an approximate discrete one. For example, for the explicit trapezoidal Runge-Kutta scheme, we can use the approximate discrete adjoint scheme (with $r_n \in R_n$ or $r_n \equiv w_n \in W_n$)

$$\begin{aligned} z_{i-1}^n &= z_i^n + (h_n/2)[z_{ni} f_y(t_{ni}, y_{ni}, r_{ni}) + \bar{z}_{i-1}^n f_y(t_{n,i-1}, y_{n,i-1}, r_{ni})], \\ \bar{z}_{i-1}^n &= z_i^n + z_i^n f_y(t_{ni}, y_{ni}, r_{ni}), \; i = N, ..., 1, \; z_N^n = g_y(y_{nN}), \end{aligned} \tag{25}$$

which amounts to using backward the same scheme for approximating the continuous adjoint equation. In this case, we define the approximate discrete relaxed functional derivative by

$$\begin{aligned} &D_n G(r_n, \bar{r}_n - r_n) := \\ &(h_n/2) \sum_{i=1}^{N} [z_i^n f(t_{n,i}, y_{n,i}, \bar{r}_{ni} - r_{ni}) + z_{i-1}^n f(t_{n,i-1}, y_{n,i-1}, \bar{r}_{ni} - r_{ni})], \end{aligned} \tag{26}$$

and the approximate discrete classical functional derivative by

$$\begin{aligned} &\mathbf{D}_n G(w_n, \bar{w}_n - w_n) := \\ &(h_n/2) \sum_{i=1}^{N} [z_i^n f_u(t_{n,i}, y_{n,i}, w_{ni}) + z_{i-1}^n f_u(t_{n,i-1}, y_{n,i-1}, w_{ni})](\bar{w}_{ni} - w_{ni}). \end{aligned} \tag{27}$$

These approximations will be used in Algorithms 3 and 4, Section 5.

3 Behavior in the Limit

Theorem 7. *(Control approximation) For every* $r \in R$ *(resp* $w \in W$*), there exists a sequence* $(w_n \in W_n)$ *that converges to* r *in* R *(resp.* w *in* W*).*

Theorem 8. *(Consistency) If* $r_n \to r$ *and* $\bar{r}_n \to \bar{r}$ *in* R*, then* $y_n \to y_r, z_n \to z_r$ *and* $z^n \to z_r$ *in* $C(I)$*,* $G_n(r_n) \to G(r)$*,* $DG_n(r_n, \bar{r}_n - r_n) \to DG(r, \bar{r} - r)$*,* $D_n G(r_n, \bar{r}_n - r_n) \to DG(r, \bar{r} - r)$*.*
Similar properties hold also for classical controls converging in W*.*

Theorem 9. *Let* $\tilde{r} \in R$ *(resp.* $\tilde{w} \in W$*, if it exists) be some optimal (hence admissible) control of the continuous relaxed (resp. classical) problem and* $(\tilde{w}_n \in W_n)$ *a sequence converging to* $\tilde{r}$ *in* R *(resp. to* $\tilde{w}$ *in* W*). We suppose that the perturbations* δ_{1n}*,* δ_{2n} *satisfy* $\delta_{1n}, \delta_{2n} \to 0$*, with* $\delta_{1n} \geq |G_{1n}(\tilde{w}_n)|$*,* $\delta_{2n} \geq G_{2n}(\tilde{w}_n)$*. Let* $(r_n \in R_n)$ *(resp.* $(w_n \in W_n)$*) be a sequence of optimal discrete relaxed* *(resp. classical) controls for the relaxed (resp. classical) problem (constraint case (i)). Then, (a) every accumulation point of the sequence (r_n) (resp. (w_n)) in R (such points always exist) is optimal for the continuous relaxed problem, (b) every accumulation point of the sequence (w_n) in W (if it exists) is optimal for the continuous classical problem.*

Theorem 10. *We suppose that the exact discrete adjoint z_n exists. Let $(r_n \in R_n)$ (resp. $(w_n \in W_n)$) be a sequence of strongly relaxed (resp. weakly classical) extremal and admissible discrete controls (constraint case (ii)), where the perturbations δ_{1n}, δ_{2n} are chosen such as to satisfy minimal feasibility (see [2]) for each n. Then, (a) every accumulation point of (r_n) in R (such points always exist) is strongly relaxed extremal and admissible for the continuous relaxed problem, (b) with U convex, every accumulation point of (w_n) in R (such points always exist) is weakly relaxed extremal and admissible for the continuous relaxed problem, (c) every accumulation point (if it exists) of (w_n) in W is weakly classical extremal and admissible for the continuous classical problem.*

4 Discrete Optimi ation Methods

The usual procedure is to discretize first the optimal control problem, for some large n, and then to apply some optimization method for solving the discrete problem. Alternatively, we propose here discrete optimization methods that progressively refine the discretization during the iterations.

We suppose that for each n, either $N_{n+1} = N_n$, or N_{n+1} is a strict multiple of N_n, which implies that $W_n \subset W_{n'}$ and $R_n \subset R_{n'}$, for every $n' > n$. Let (ε^k) and (β^k) be positive decreasing sequences that converge to zero. Define the penalized discrete functionals on R_n or W_n

$$G_n^k(r) := G_{0n}(r) + (1/\varepsilon^k)[\|G_{1n}(r)\|_2^2 + \|\mathbf{max}(0, G_{2n}(r))\|_2^2]. \tag{28}$$

The first Algorithm is a discrete relaxed conditional gradient penalty method. It can be applied if the discrete adjoint is defined, with U not necessarily convex.
Algorithm 1.
Step 1. Set $n := 0$, $j := 0$ and choose $r_{00} \in R_0$. Step 2. Find $\bar{r}_{nj} \in R_n$ such that

$$d_j := DG_n^n(r_{nj}, \bar{r}_{nj} - r_{nj}) = \min_{r'_n \in R_n} DG_n^n(r_{nj}, r'_n - r_{nj}). \tag{29}$$

Step 3. If $|d_j| \le \beta^n$, set $r_n := r_{nj}$, $n := n + 1$ and go to Step 2.
Step 4. Find $\alpha_j \in [0, 1]$ such that

$$G_n^n(r_{nj} + \alpha_j(\bar{r}_{nj} - r_{nj})) = \min_{\alpha \in [0,1]} G_n^n(r_{nj} + \alpha(\bar{r}_{nj} - r_{nj})). \tag{30}$$

Step 5. Choose any $r'_{nj} \in R_n$ such that

$$G_n^n(r'_{nj}) \le G_n^n(r_{nj} + \alpha_j(\bar{r}_{nj} - r_{nj})). \tag{31}$$

Set $r_{n,j+1} := r'_{nj}$, $j := j + 1$ and go to Step 2.

The next Algorithm is a discrete classical projected gradient penalty method. It can be applied if the discrete adjoint is defined and U is convex.

Algorithm 2.

Step 1. Set $n := 0$, $j := 0$ and choose $w_{00} \in W_0$, $\gamma > 0$. Step 2. Find $\bar{w}_{nj} \in W_n$ such that

$$\begin{aligned} e_j &:= \mathbf{D}G_n^n(w_{nj}, \bar{w}_{nj} - w_{nj}) + (\gamma/2)\|\bar{w}_{nj} - w_{nj}\|^2_{L^2(I)} \\ &= \min_{w'_n \in W_n} [\mathbf{D}G_n^n(w_{nj}, w'_n - w_{nj}) + (\gamma/2)\|w'_n - w_{nj}\|^2_{L^2(I)}]. \end{aligned} \tag{32}$$

Step 3. If $|e_j| \le \beta^n$, set $w_n := w_{nj}$, $n := n + 1$, and go to Step 2.
Step 4. Find $\alpha_j \in [0, 1]$ such that

$$G_n^n(w_{nj} + \alpha_j(\bar{w}_{nj} - w_{nj})) = \min_{\alpha \in [0,1]} G_n^n(w_{nj} + \alpha(\bar{w}_{nj} - w_{nj})). \tag{33}$$

Step 5. Set $w_{n,j+1} := w_{nj} + \alpha_j(\bar{w}_{nj} - w_{nj})$, $j := j + 1$ and go to Step 2.

The third Algorithm is an approximate discrete relaxed conditional gradient penalty method that uses approximate discrete adjoints and functional derivatives (see trapezoidal schemes, Section 3). It can be applied even if the discrete adjoint is not defined, with U not necessarily convex.

Algorithm 3.

Step 1. Set $n := 0$, $k := 0$, and choose $r_0 \in R_0$.
Step 2. Find $\bar{r}_n \in R_n$ such that

$$d_n := D_n G^k(r_n, \bar{r}_n - r_n) = \min_{r'_n \in R_n} D_n G^k(r_n, r'_n - r_n). \tag{34}$$

Step 3. If $|d_n| \le \beta^k$, set $n_k := n$, $r^k_{n_k} := r_n$, $k := k + 1$ and go to Step 2.
Step 4. Find $\alpha_j \in [0, 1]$ such that

$$G_n^k(r_n + \alpha_n(\bar{r}_n - r_n)) = \min_{\alpha \in [0,1]} G_n^k(r_n + \alpha(\bar{r}_n - r_n)). \tag{35}$$

Step 5. Choose any $r'_n \in R_n$ such that $G_n^k(r'_n) \le G_n^k(r_n + \alpha_n(\bar{r}_n - r_n))$. Set $r_{n+1} := r'_n$, $n := n + 1$ and go to Step 2.

Algorithm 4.

Classical projected gradient version of Algorithm 3 (cf. Algorithm 2).

Theorem 11. *Under appropriate assumptions:*

a) Every accumulation point in R (such points always exist) of the sequence constructed in Step 3 of Algorithms 1, 3 (resp. 2) is strongly (resp. weakly) relaxed extremal and admissible for the continuous relaxed problem.

b) Every accumulation point in W (if it exists) of the sequence constructed in Step 3 of Algorithms 2 and 4 is weakly classical extremal and admissible for the continuous classical problem.

In Algorithms 1 and 3, Step 2 reduces to the minimization of a Hamiltonian $H := z\,f$ on U, for each i. In Algorithm 2 and 4, Step 2 reduces to the projection of a vector $v := w - (1/\gamma)H_u^T$ onto U, for each i. Also, it can be shown that in Algorithms 1 and 3, if the initial control is classical (Dirac), then the generated controls can be chosen to be of Gamkrelidze type, i.e. convex combinations of Dirac measures concentrated on $p+1$ points in U, for each i. These controls can then be approximated by piecewise constant classical controls using a standard procedure (see [5], [6]). In Step 3 of Algorithms 1-4, the calculation of the optimal step can be replaced by more efficient and easily implementable procedures such as Armijo's step search. The proof of Theorem 5.1 relies on Theorem 4.1 and the continuity and consistency of functionals and derivatives. For Algorithms 3 and 4, we use also the equicontinuity of functionals w.r.t. α and discretization error estimates (see [6]).

References

1. Chryssoverghi, I., Bacopoulos, A.: Discrete approximation of relaxed optimal control problems. Journal of Optimization Theory and Applications, 65, 3 (1990) 395–407.
2. Chryssoverghi, I., Bacopoulos, A.: Approximation of relaxed nonlinear parabolic optimal control problems. JOTA, 77, 1 (1993) 31–50.
3. Chryssoverghi, I., Bacopoulos, A., Kokkinis, B., Coletsos, J.: Mixed Frank-Wolfe penalty method with applications to nonconvex optimal control problems. JOTA, 94, 2 (1997) 311–334.
4. Chryssoverghi, I., Bacopoulos, A., Coletsos, J., Kokkinis, B.: Discrete approximation of nonconvex hyperbolic optimal control problems with state constraints. Control & Cybernetics, 27, 1 (1998) 29–50.
5. Chryssoverghi, I., Coletsos, J., Kokkinis, B.: Discrete relaxed method for semilinear parabolic optimal control problems. Control & Cybernetics, 28, 2 (1999) 157–176.
6. Chryssoverghi, I., Coletsos, J., Kokkinis, B.: Approximate relaxed descent method for optimal control problems. Control & Cybernetics, 30, 4 (2001) 385–404.
7. Dontchev, A.L., Hager, W. Veliov, V.: Second-order Runge-Kutta approximations in control constrained optimal control. SIAM J. Numer. Anal., 38 (2000) 202–226.
8. Polak, E.: Optimization: Algorithms and Consistent Approximations. Springer, Berlin (1997)
9. Roubíček, T.: A convergent computational method for constrained optimal relaxed control problems. JOTA, 69 (1991) 589–603.
10. Roubíček, T.: Relaxation in Optimization Theory and Variational Calculus. Walter de Gruyter, Berlin (1997)
11. Warga, J.: Optimal Control of Differential and Functional Equations. Academic Press, New York (1972)
12. Warga, J.: Steepest descent with relaxed controls. SIAM J. on Control, 15, 4 (1977) 674–689.

Numerical Schemes of Higher Order for a Class of Nonlinear Control Systems

Lars Grüne and Peter E. Kloeden

Fachbereich Mathematik, J.W. Goethe–Universität
Postfach 111932, 60054 Frankfurt am Main, Germany
gruene,kloeden@math.uni-frankfurt.de
http://www.math.uni-frankfurt.de/~numerik/

Abstract. We extend a systematic method for the derivation of high order schemes for affinely controlled nonlinear systems to a larger class of systems in which the control variables are allowed to appear nonlinearly in multiplicative terms. Using an adaptation of the stochastic Taylor expansion to control systems we construct Taylor schemes of arbitrary high order and indicate how derivative free Runge-Kutta type schemes can be obtained.

1 Introduction

Traditional numerical schemes for ordinary differential equations, such as Runge–Kutta schemes, usually fail to attain their asserted order when applied to ordinary differential control equations due to the measurability of the control functions. A similar situation occurs with stochastic differential equations due to the nondifferentiability of the driving noise processes. To construct higher order numerical schemes for stochastic differential equations, one needs to start with an appropriate stochastic Taylor expansion to ensure consistency with the less robust stochastic calculus as well as a higher order of convergence. This is the opposite procedure to that used for numerical schemes for ordinary differential equations, where heuristic arguments are typically used to derive a scheme and the Taylor expansion is then used to establish its local discretization order.

In [9] it was shown that this approach for stochastic differential equations carries over to control systems with affine control (for these systems the stochastic Taylor expansion is essentially the same as the Fliess expansion [11]). In the present paper we will extend the results from [9] to a larger class of control systems allowing also nonlinearities in the control input. More precisely, we consider d–dimensional controlled nonlinear system with n–dimensional control functions of the form

$$\frac{dx}{dt} = f^0(t,x) + \sum_{j=1}^{m} f^j(t,x)\, g^j(t,u(t)), \tag{1}$$

where $t \in [t_0, T]$, $x = (x^1, \ldots, x^d) \in \mathbb{R}^d$, the vector fields $f^j : \mathbb{R} \times \mathbb{R}^d \to \mathbb{R}^d$ are sufficiently smooth in order to apply our expansion, the functions $g^j : \mathbb{R} \times \mathbb{R}^n \to$

I. Dimov et al. (Eds.): NMA 2002, LNCS 2542, pp. 213–220, 2003.

$\mathbb{R}$ are continuous and the control functions $u(t)$ are measurable and take values in a compact set $\mathbf{U} \subset \mathbb{R}^n$.

Numerical schemes for such systems play an important role in the numerical analysis of nonlinear control systems since in many algorithms the approximation of trajectories appears as a subproblem, see, e.g., the monographs [2] and [8].

The organization of this paper is as follows. We start with the introduction of the necessary notation in Section 2 and the precise statement of the Taylor expansion in Section 3. In Section 4 we explain how numerical Taylor and derivative free (i.e., Runge–Kutta type) schemes can be obtained, and finally in Section 5 we show a numerical example.

2 Setup and Notation

In the following sections we shall refer to the nonautonomous d–dimensional controlled differential equation (1), which we rewrite in the equivalent compact integral form

$$x(t) = x(t_0) + \sum_{j=0}^{m} \int_{t_0}^{t} f^j(s, x(s))\, g^j(s, u(s))\, ds \tag{2}$$

where we set $g^0(t,u) \equiv 1$ so that the first integral term can be included in the summation.

We call a row vector $\alpha = (j_1, j_2, \ldots, j_l)$, where $j_i \in \{0, 1, \ldots, m\}$ for $i = 1, \ldots, l$, a *multi–index* of length $l := l(\alpha) \geq 1$ and for completeness we write $\odot$ for the multi–index of length zero, that is, with $l(\odot) = 0$. We denote the set of all such multi–indices by $\mathcal{M}_m$.

For any $\alpha = (j_1, j_2, \ldots, j_l) \in \mathcal{M}_m$ with $l(\alpha) \geq 1$, denote by $-\alpha$ and $\alpha-$ for the multi–index in $\mathcal{M}_m$ obtained by deleting the first and the last component, respectively, of α, thus $-\alpha = (j_2, \ldots, j_l)$ and $\alpha- = (j_1, \ldots, j_{l-1})$.

For a multi–index $\alpha = (j_1, j_2, \ldots, j_l) \in \mathcal{M}_m$, some integrable control function $u : \mathbb{R} \to \mathbf{U}_m$ and an integrable function $f : [t_0, T] \to \mathbb{R}$ we define the *multiple integral* $I_\alpha[f(\cdot)]_{t_0,t}$ recursively by

$$I_\alpha[f(\cdot)]_{t_0,t} := \begin{cases} f(t) & : \quad l = 0 \\ \int_{t_0}^{t} I_{\alpha-}[f(\cdot)]_{t_0,s}\, g^{j_l}(s, u(s)) ds & : \quad l \geq 1 \end{cases}. \tag{3}$$

We note that $I_\alpha[f(\cdot)]_{t_0,\cdot} : [t_0, T] \to \mathbb{R}$ is continuous, hence integrable, so the integrals are well defined. For example, we obtain

$$I_{(0)}[f(\cdot)]_{t_0,t} = \int_{t_0}^{t} f(s)\, ds, \quad I_{(0,1)}[f(\cdot)]_{0,t} = \int_0^t \int_0^{s_2} f(s_1) g^1(s_2, u(s_2))\, ds_1 ds_2.$$

For simpler notation, we shall often abbreviate $I_\alpha[f(\cdot)]_{t_0,t}$ to $I_{\alpha,t}$ or just I_α when $f(t) \equiv 1$ and shall explicitly write $I_{\alpha,u}[f(\cdot)]_{t_0,t}$, $I_{\alpha,u,t}$ or $I_{\alpha,u}$ when we want to emphasize a specific control function u.

For each $\alpha = (j_1, \ldots, j_l) \in \mathcal{M}_m$ and a function $F : [t_0, T] \times \mathbb{R}^d \to \mathbb{R}$, the *coefficient function* F_α is defined recursively by $F_\odot = F$ and

$$F_\alpha = L^{j_1} F_{-\alpha}, \quad \text{if } l(\alpha) \geq 1,$$

where the partial differential operators are defined by

$$L^0 = \frac{\partial}{\partial t} + \sum_{k=1}^{d} f^{0,k} \frac{\partial}{\partial x^k}, \qquad L^j = \sum_{k=1}^{d} f^{j,k} \frac{\partial}{\partial x^k}, \quad j = 1, \ldots, m.$$

This definition requires the functions F, f^0, f^1, ..., f^m to be sufficiently smooth.

For example, in the autonomous scalar case with $d = m = 1$ for the identity function $F(t, x) \equiv x$ we have

$$F_{(0)} = f^0, \quad F_{(j_1)} = f^{j_1}, \quad F_{(j_1,j_2)} = f^{j_1} f^{j_2\prime},$$

where the dash $'$ denotes differentiation with respect to x. When the function F is not explicitly stated in the text we shall always take it to be the identity function $F(t, x) \equiv x$.

Since different integrals can be expanded in forming a Taylor expansion, the terms with constant integrands cannot be written down completely arbitrarily. Rather, the sets of corresponding multi–indices must form *hierarchical* and *remainder* sets. These sets can be defined in a very general way, see [13]. Here we only need the hierarchical and remainder sets defined by

$$\Gamma_N = \{\alpha \in \mathcal{M}_m : l(\alpha) \leq N\} \quad \text{and} \quad \mathcal{B}(\Gamma_N) = \{\alpha \in \mathcal{M}_m : l(\alpha) = N + 1\}.$$

3 Taylor Expansions and Approximations

We now formulate the Taylor expansion for the d–dimensional controlled system (2) using the terminology from the preceding section.

Theorem 1. *Let $F : \mathbb{R}^+ \times \mathbb{R}^d \to \mathbb{R}$. Then for each $N \geq 0$ the following* Taylor expansion

$$F(t, x(t)) = \sum_{\alpha \in \Gamma_N} I_\alpha \left[F_\alpha (t_0, x(t_0))\right]_{t_0,t} + \sum_{\alpha \in \mathcal{B}(\Gamma_N)} I_\alpha \left[F_\alpha(\cdot, x(\cdot)), \right]_{t_0,t}$$

holds, provided all of the partial derivatives of F, f^0, f^1, ..., f^m and all of the multiple control integrals appearing here exist.

For the proof we refer to [9, Theorem 1], whose proof straightforwardly carries over to our class of systems.

Based on Theorem 1 we can now construct Taylor approximations of arbitrary higher order. In the general multi-dimensional case d, m= 1, 2, ... the *Taylor approximation* for $N = 1, 2, 3, \ldots$ is defined by

$$F_N (t_0, x(t_0), \Delta) := \sum_{\alpha \in \Gamma_N} F_\alpha (t_0, x(t_0)) \, I_{\alpha,t_0,t_0+\Delta} \tag{4}$$

$$= F(t_0, x(t_0)) + \sum_{\alpha \in \Gamma_N \setminus \{\odot\}} F_\alpha (t_0, x(t_0)) \, I_{\alpha,t_0,t_0+\Delta} \tag{5}$$

with the coefficient functions F_α corresponding to the function $F(t, x)$.

When the function $F(t,x)$ is $N+1$ times continuously differentiable and the drift and control coefficients $f^0, f^1, \ldots, f^m$ of the controlled differential equation (2) are N times continuously differentiable, then each of the integrals $I_{\alpha,t_0,t_0+\Delta}\left(F_\alpha(\cdot,x(\cdot))\right)$ for α in the remainder set $\mathcal{B}(\Gamma_N)$ is of order Δ^{N+1}. Since there are only finitely many, specifically $(m+1)!$, remainder integrals, the truncation error here is

$$\left|F_N\left(t_0,x(t_0),\Delta\right)-F\left(t_0+\Delta,x(t_0+\Delta)\right)\right| \le K\,\Delta^{N+1}, \tag{6}$$

where the constant K depends on N as well as on a compact set containing the initial value $(t_0,x(t_0))$ and the solution of the controlled differential equation.

For the function $F(t,x)\equiv x^k$, the kth component of the vector x, and $N=1$, 2 and 3, respectively, the solution $x(t_0+\Delta)$ of the controlled differential equation (2) satisfies the componentwise approximations

$$x^k(t_0+\Delta)=x^k(t_0)+\sum_{j=0}^{m} f^{j,k}(t_0,x(t_0))\,I_{(j)}+O(\Delta^2),$$

$$x^k(t_0+\Delta)=x^k(t_0)+\sum_{j=0}^{m} f^{j,k}(t_0,x(t_0))I_{(j)}+\sum_{j_1,j_2=0}^{m} L^{j_1}f^{j_2,j}\,I_{(j_1,j_2)}+O(\Delta^3)$$

and

$$\begin{aligned}x^k(t_0+\Delta)=x^k(t_0)&+\sum_{j=0}^{m} f^{j,k}(t_0,x(t_0))I_{(j)}+\sum_{j_1,j_2=0}^{m} L^{j_1}f^{j_2,j}\,I_{(j_1,j_2)}\\&+\sum_{j_1,j_2,j_3=0}^{m} L^{j_1}L^{j_2}f^{j_3,k}(t_0,x(t_0))\,I_{(j_1,j_2,j_3)}+O(\Delta^4)\end{aligned}$$

for $k=1,\ldots,d$, where we have written $I_{(j)}$ for $I_{(j),t_0,t_0+\Delta}$ and so on.

4 Numerical Schemes

Using the Taylor approximation from the previous section we now construct numerical schemes by iterating Taylor approximations, or suitable derivative free approximations of those, over a partition of the time interval under interest. Schemes of arbitrary higher order $N=1,2,\ldots$ can be constructed by truncating the Taylor approximation corresponding to the the hierarchical set Γ_N. Here we assume that the multiple control integrals I_α are at our disposal, which is often feasible e.g. by using symbolic manipulators like MAPLE. For a numerical approximation of such integrals see [9, Section 9].

Let $\{t_0,t_1,\ldots,t_n,\ldots,\}$ be a partition of the time interval $[t_0,T]$ with stepsizes $\Delta_n=t_{n+1}-t_n$ and maximal step size $\Delta:=\max_n\Delta_n$. In the general multi-dimensional case d, $m=1,2,\ldots$ for $N=1,2,3,\ldots$ we define the *Taylor scheme of order* N for the controlled differential equation (2) is given componentwise by

$$X_{n+1}^k = X_n^k + \sum_{\alpha \in \Gamma_N \setminus \{\odot\}} F_\alpha^k(t_n, X_n)\, I_{\alpha, t_n, t_{n+1}}$$

with the coefficient functions F_α^k corresponding to $F(t,x) \equiv x^k$ for $k = 1, \ldots, d$ and the multiple control integrals from (3). By standard arguments (see [12] or [10]) it follows from (6) that the global discretization error is of order N when the coefficients f^j of the differential equation (2) are N times continuously differentiable.

Below, we write out the Taylor schemes for $N = 1$ and 2, where we distinguish the purely uncontrolled integrals, that is with multi–indices (0) and (0, 0) from the others, since no special effort is required for their evaluation.

The simplest nontrivial Taylor scheme is the Euler approximation with convergence order $N = 1$. It is given componentwise by

$$X_{n+1}^k = X_n^k + f^{0,k}(t_n, X_n)\, \Delta_n + \sum_{j=1}^m f^{j,k}(t_n, X_n)\, I_{(j), t_n, t_{n+1}} \tag{7}$$

for $k = 1, \ldots, d$, where $\Delta_n = t_{n+1} - t_n = I_{(0), t_n, t_{n+1}}$. The kth component of the *Taylor scheme of order* $N = 2$ is given by

$$X_{n+1}^k = X_n^k + f^{0,k}(t_n, X_n)\, \Delta_n + \sum_{j=1}^m f^{j,k}(t_n, X_n)\, I_{(j), t_n, t_{n+1}} \tag{8}$$

$$+ \frac{1}{2} L^0 f^{0,k}(t_n, X_n)\, \Delta_n^2 + \sum_{\substack{j_1, j_2 = 0 \\ j_1 + j_2 \neq 0}}^{m} L^{j_1} f^{j_2,k}(t_n, X_n)\, I_{(j_1, j_2), t_n, t_{n+1}}$$

for $k = 1, \ldots, d$. For $N = 3$ we refer to [9].

A disadvantage of Taylor schemes is that the derivatives of various orders of the drift and control coefficients must be first derived and then evaluated at each step. Although nowadays symbolic manipulators [3] facilitate the computation of the derivatives in the schemes, it is useful to have approximations and schemes that avoid the use of derivatives of the drift and control coefficients in much the same way that Runge–Kutta schemes do in the more traditional setting since these often have other computational advantages.

Since the Euler or Taylor scheme of order 1 contains no derivatives of the f^j, we illustrate this procedure for the second order Taylor scheme (8). In the autonomous case, from the ordinary Taylor expansion for f^j we obtain

$$L^i f^{j,k}(X_n) = \frac{1}{\Delta_n} \left(f^{j,k}\left(X_n + f^i(X_n)\, \Delta_n\right) - f^{j,k}(X_n) \right) + O(\Delta_n).$$

Since $O(\Delta_n)\, I_{(i,j), t_n, t_{n+1}} = O(\Delta_n^3)$, the remainder is of the same order as the local discretization error if we replace the $L^i f^{j,k}$ by this approximation.

In this way we obtain the *second order derivative–free scheme* in the autonomous case

$$X_{n+1}^k = X_n^k + \frac{1}{2} f^{0,k}(X_n)\,\Delta_n + \sum_{j=1}^{m} f^{j,k}(X_n)\, I_{(j),t_n,t_{n+1}}$$
$$+\frac{1}{2} f^{0,k}\left(X_n + f^0(X_n)\,\Delta_n\right)\Delta_n \tag{9}$$
$$+\frac{1}{\Delta_n} \sum_{\substack{i,j=0 \\ i+j\neq 0}}^{m} \left(f^{j,k}\left(X_n + f^i(X_n)\,\Delta_n\right) - f^{j,k}(X_n)\right) I_{(i,j),t_n,t_{n+1}}$$

for $k = 1, \ldots, d$. In the usual ODE case, that is with $f^j(x) \equiv 0$ for $j = 1, \ldots, m$, this is just the second order Runge–Kutta scheme known as the Heun scheme.

This principle can be extended to obtain higher order derivative–free schemes. See [13] for analogous higher order derivative–free schemes for the stochastic case.

Note that all these schemes simplify considerable when the coefficients f^j of the controlled differential equation (2) satisfy special properties. For example, if the control coefficients $f^1, \ldots, f^m$ are all constants or depend just on t, then all of the spatial derivatives of these control coefficients vanish and, hence, so do the corresponding higher order terms.

Another major simplification occurs under *commutative control,* that is when the f^i satisfy $L^i f^{j,k}(t,x) \equiv L^j f^{i,k}(t,x)$ for all $i, j = 0, 1, \ldots, m$. Then, by the generalized integration–by–parts identities

$$I_{(i,j),t_n,t_{n+1}} + I_{(j,i),t_n,t_{n+1}} = I_{(i),t_n,t_{n+1}}\, I_{(j),t_n,t_{n+1}}, \quad i, j = 0, 1, \ldots, m, \tag{10}$$

we obtain

$$L^i f^{j,k}(t_n, X_n)\, I_{(i,j),t_n,t_{n+1}} + L^j f^{i,k}(t_n, X_n)\, I_{(j,i),t_n,t_{n+1}}$$
$$= L^i f^{j,k}(t_n, X_n)\, I_{(i),t_n,t_{n+1}}\, I_{(j),t_n,t_{n+1}},$$

which involves more easily computed multiple control integrals of lower multiplicity. Note that this condition is similar to the one considered in [14], where the effect of time discretization of the control function is investigated and a second order scheme for the approximation of the reachable set is obtained.

5 A Numerical Example

We have tested the Euler (7) and Heun (9) Schemes from Section 4 with

$$\frac{dx(t)}{dt} = f^0(x(t)) + g(u(t)) f^1(x(t)) := \begin{pmatrix} x_2(t) \\ 0 \end{pmatrix} + (u(t) + u(t)^3) \begin{pmatrix} -x_2(t) \\ 1 \end{pmatrix}$$

with control function $u(t) = \sin(100/t)$ and initial value $x_0 = (0,0)^T$. The resulting schemes have been simplified using the identity (10) such that the only remaining control integrals were $I_{(1),0,t}$ and $I_{(0,1),0,t}$, which have been evaluated using MAPLE. Note that the exact solution for this equation is easily verified to be $x_1(t) = I_{(1,0),0,t} - I_{(1,1),0,t}$, $x_2(t) = I_{(1),0,t}$.

The equation was solved on the interval $[0, 1]$ with timestep $\Delta = 1/N$ and $N = 50, 100, \ldots, 500$. Figure 1 shows the resulting errors $\sup_{n=1,\ldots,N} \|x_n - x(n\Delta)\|$ for the Heun and the Euler scheme. The left figure shows the error over N in a linear scale, the right figure shows the error over Δ in a log-log scale.

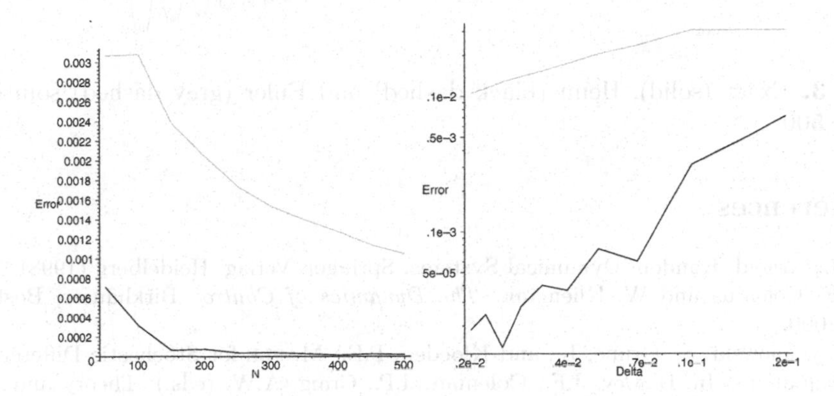

Fig. 1. Global error for Heun (black) and Euler (grey) schemes, linear and log-log

The Figures 2 and 3 show the x_1 component of the exact solution, of the Heun and of the Euler scheme for $N = 100$ and $N = 500$, respectively.

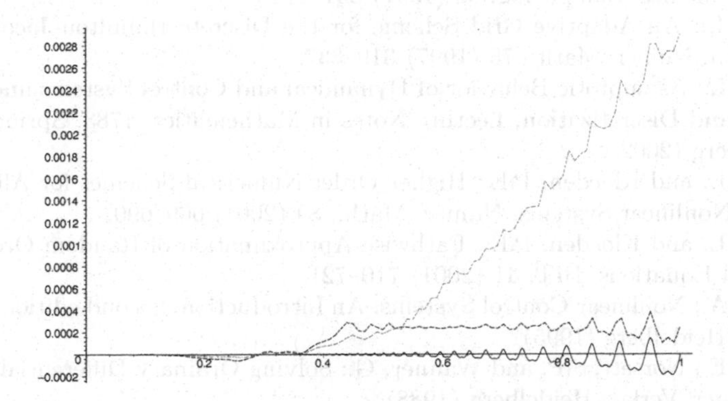

Fig. 2. Exact (solid), Heun (black dashed) and Euler (grey dashed) solution, $N = 100$

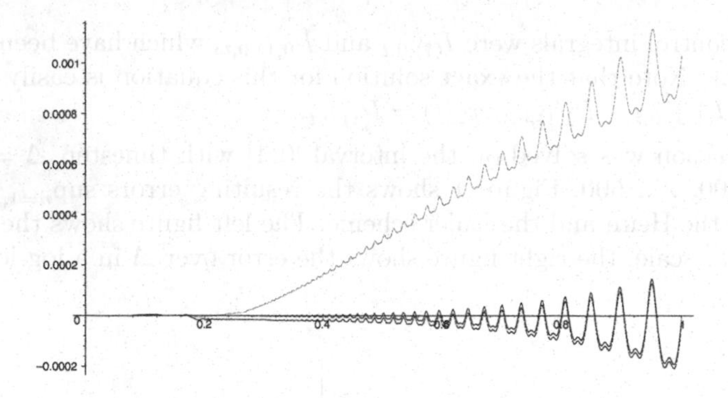

Fig. 3. Exact (solid), Heun (black dashed) and Euler (grey dashed) solution, $N = 500$

References

1. L. Arnold, Random Dynamical Systems. Springer–Verlag, Heidelberg (1998)
2. F. Colonius and W. Kliemann, *The Dynamics of Control*, Birkhäuser, Boston, 2000.
3. Cyganowski, S., Grüne, L., and Kloeden, P.E.: MAPLE for Stochastic Differential Equations. In: Blowey, J.F., Coleman, J.P., Craig, A.W. (eds.): Theory and Numerics of Differential Equations, Springer–Verlag, Heidelberg (2001) 127–178.
4. Deuflhard, P.: Stochastic versus Deterministic Numerical ODE Integration. In: Platen, E. (ed.): Proc. 1st Workshop on Stochastic Numerics, Berlin, WIAS Berlin, Preprint Nr. 21 (1992) 16–20.
5. Falcone, M. and Ferretti, R.: Discrete Time High-Order Schemes for Viscosity Solutions of Hamilton-Jacobi-Bellman Equations. Numer. Math., 67 (1994) 315–344.
6. Ferretti, R.: Higher–Order Approximations of Linear Control Systems via Runge–Kutta Schemes. Computing, 58 (1997) 351–364.
7. Grüne, L.: An Adaptive Grid Scheme for the Discrete Hamilton-Jacobi-Bellman Equation. Numer. Math., 75 (1997) 319–337.
8. Grüne, L.: Asymptotic Behavior of Dynamical and Control Systems under Perturbation and Discretization, Lecture Notes in Mathematics, 1783, Springer–Verlag, Heidelberg (2002)
9. Grüne, L. and Kloeden, P.E.: Higher Order Numerical Schemes for Affinely Controlled Nonlinear Systems. Numer. Math., 89 (2001) 669–690.
10. Grüne, L. and Kloeden, P.E.: Pathwise Approximation of Random Ordinary Differential Equations. BIT, 41 (2001) 710–721.
11. Isidori, A.: Nonlinear Control Systems. An Introduction. Second edition, Springer–Verlag, Heidelberg (1995)
12. Hairer, E., Norsett, S.P. and Wanner, G.: Solving Ordinary Differential Equations I. Springer–Verlag, Heidelberg (1988)
13. Kloeden, P.E. and Platen, E.: Numerical Solution of Stochastic Differential Equations. Springer–Verlag, Heidelberg (1992) (3rd revised and updated printing, 1999)
14. Veliov, V.: On the Time Discretization of Control Systems. SIAM J. Control Optim., 35 (1997) 1470–1486.

A Size Distribution Model Applied to Fish Farming

Juan Hernández[1] and Eucario Gasca-Leyva[2]

[1] Universidad de Las Palmas, c/Saulo Torón s/n, 35017 Las Palmas, Spain
juanh@empresariales.ulpgc.es,
http://www.fcee.ulpgc.es/~juanh/

[2] Center of Research and Advanced Studies of the IPN, Mérida, México

Abstract. This paper presents an age-structured control model applied to the size variability problem in fish farming. This phenomenon is common in commercial culture due to the principal of competition among individuals in ecosystems with high density. The usual consequences are higher differences in fish growth, together with an increase in mortality rate affecting mainly the smaller sizes. Control models are particularly appropriate for aquacultural system, where a great amount of factors can be in fact predetermined by farmers. The formulation of the model involves a series of hypothesis for the system. However, we can induce some practical managerial recommendations from the model analysis. The numerical solution of the related optimal control problem is a challenging issue that we also discuss.

1 Introduction

The industrial activity of a fish farm consists basically on feeding fingerlings until they reach a proper commercial harvesting weight. In this process, fish is affected by a lot of factors that can modify its growth and survival. Some of them, as ration size, diet quality or water temperature have been extensively studied for many commercial species during the last decades, consequently improving managerial practices in fish farming. This paper is centered on other essential element influencing growth and mortality that has not been so extensively studied, as it is the size distribution of fishes.

Fishes are usually fed in ponds inside the coast or in floating cages in the sea, depending on the characteristics of the area or species. One common phenomenon observed in both types of culture is a growth depensation, that is, fishes with identical initial size reach different weights. This variability in fish growth is due to multiple causes, being those related with social behavior of each species, as hierarchy or competition for food, some of the most relevants [1,2,3].

The size variance in a cage carries on negative consequences for a lot of commercial species. The culture is usually conducted with high density levels in order to reach economic efficiency, originating a lower dissolved oxygen in cages and stress in fishes. This carries on usually a decrease in growth together with an increase in mortality rate for some fish species [4,5], affecting principally to

I. Dimov et al. (Eds.): NMA 2002, LNCS 2542, pp. 221–229, 2003.

smaller sizes. Cannibalism is other social behavior that can be observed in situations of size variability, together with food limitations. In particular, cannibalism could be the cause of high mortality observed in culture of salmon [6], gilthead seabream [3] and snakehead [7]. The juvenile mortality due to high density levels is a direct consequence of the logistic-type growth rate in biological systems, and it has been observed in other populations as an endogenous method of growth control [8].

This paper presents an age-structured control model applied to the size variability problem. The age-structured control systems has been extensively applied to population distribution during the last decades [9,10,11]. Control models are particularly appropiated for aquacultural system, where a great amount of factors can be in fact predetermined by farmers. In order to formulate the model, it has been necessary to establish a series of hypothesis over the system. However, although with logical caution, we can induce some practical managerial recommendations from the model analysis.

2 The Dynamical Model

The model we present in this section describes the evolution of a fish population in captivity, focusing in the effect of the size variability over the mortality rate. For the sake of simplicity, we have assumed the following hypothesis:

1. We initially consider the existence of only one cage in the farm. Consequently, the process of hauling and grading is not included. Thus, harvesting will be the only method of reducing the negative effects of size variance.
2. Fish are fed optimally. Differences in fish growth between fishes are not considered, and mortality is the only adversing effect for the culture.[1]

Following the line of the previous age-structured models [10], we call $N(t,a) \equiv$ "number of individuals in time t of age a in the cage". Individuals of age a leave the population due to mortality $m(t,a)$ in time t or due to harvesting $h(t,a)$ in the same time. Thus, the dynamic evolution is determined by the following equation:

$$(\frac{\partial}{\partial t} + \frac{\partial}{\partial a})N(t,a) = -m(t,a)N(t,a) - h(t,a), \tag{1}$$

with the initial and boundary conditions

$$\begin{aligned} N(0,a) &= N_0(a), \\ N(t,0) &= s(t), \end{aligned} \tag{2}$$

where $N_0(a)$ is the initial age distribution in the cage, and $s(t)$ represents the stocking of new fingerlings in the cage.

[1] This hypothesis can be considered not realistic, as density or temperature are usually observed as factors influencing the growth rate. In order to avoid an excesive complexity in the model we omit the inclusion of these factors.

We define the mortality rate $m(t,a)$ as a function of the density level and the fish age.[2] As the cage capacity is fixed, the density is determined by the total biomass existing in the cage, represented by the following expression,

$$B(t) = \int_0^{\bar{a}} w(a)N(t,a)da, \quad (3)$$

where $w(a)$ is the weight corresponding to fish of age a, and $\bar{a}$ is the maximum age achieved by a fish. In common farming, survival is stable for low values of density. However, in case of higher levels of density, a loss of dissolved oxygen is produced, together with an increment of stress, carry on a significant increase in mortality, principally affecting the smaller sizes [4]. Then, we propose the following function,

$$m(a, B(t)) = I(a)\varphi(B(t)), \quad (4)$$

where $\varphi(\cdot)$ indicates the mortality rate in the total population for each level of biomass. It is almost constant around the natural mortality rate m_n for low density values, but heavily increase when the biomass reachs values above a percentage over the total carrying capacity, called $\bar{B}$. The following function incorporates these hypothesis,

$$\varphi(B) = \begin{cases} m_n & \text{if } B \leq \delta\bar{B}, \\ m_n + \mu(B - \delta\bar{B})\tan(\frac{\pi}{2(1-\delta)\bar{B}}(B - \delta\bar{B})) & \text{if } B > \delta\bar{B}, \end{cases} \quad (5)$$

where $\delta \in [0,1]$ represents the percentage of the maximum carrying capacity where the natural mortality start to increase. Clearly, $\varphi \in C^1(0,\bar{B})$, non decreasing, and has an asymptote in $B = \bar{B}$. Figure 1 shows a representation of this function for different values of the parameter μ.

The function $I(a)$ represents the distribution of the mortality rate over the different ages in the cage. Necessarily this distribution function has to verify

$$\int_0^{\bar{a}} I(a)da = \bar{a}. \quad (6)$$

The constant function $I(a) = 1$ represents the case of a homogeneuos survival across the different ages. In order to incorporate another distribution in a culture we have to define $I(a)$ as a decreasing function, that is, $I > 0$, $I' < 0$. With respect to the second derivative, several options can be choosen. We assume that mortality is concentrated in a range of small ages, decreasing dramatically for the range of larger ages. In this case there exists an inflexion point that separates the two groups. A function preserving this hypothesis is

[2] In order to describe the cannibalism impact over mortality, we have to consider also the difference between the ages of the predator and the prey, $(a' - a)$. The mortality rate is null for differences below a determined value, that corresponds with the capacity of the predator to eat the prey [7]. Nevertheless, there is a straight positive dependence between cannibalism and density conditions, so we can assume that certain cannibalism effect is included in mortality due to high density.

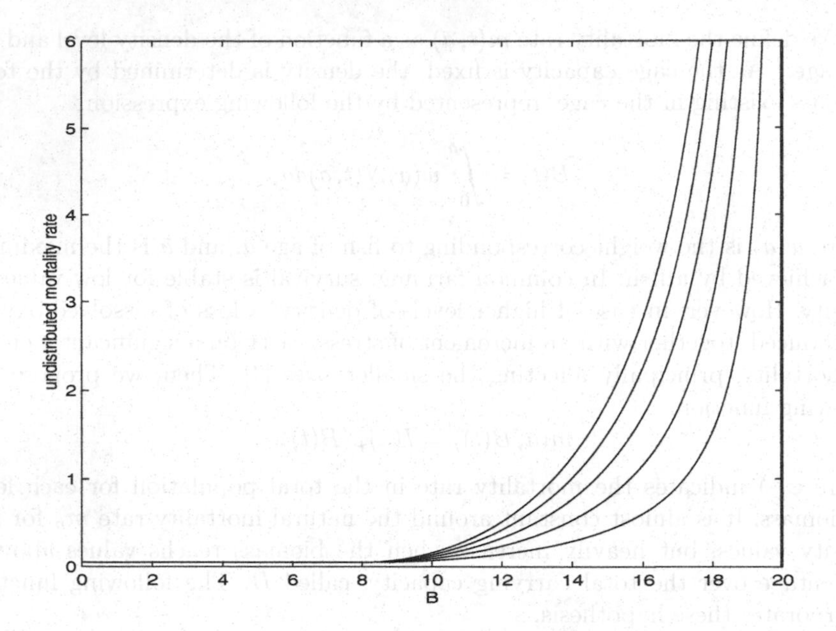

Fig. 1. Graphical representation of the undistributed mortality rate function $\varphi(B)$. The parameter values are $\delta = 0.4, m_n = 0.05$, and $\bar{B} = 20$. Parameter μ ranges in the interval [0.04, 0.2].

$$I(a) = \beta \frac{1 + e^{-\alpha a^*}}{1 + e^{-\alpha(a^* - a)}}, \tag{7}$$

where a^* indicates the age delimiting high and low mortality groups and α determines the gap in the curve. The parameter β is obtained from the general condition over $I(a)$ and the rest of the parameters. Figure 2 shows different possible representations for this function.

The parameter values defined above are determined starting from field data.

The harvesting function $h(t, a)$ can also be described as a rate $e(t, a)$ over the total individuals of that age in time t. Thus,

$$h(t, a) = e(t, a)N(t, a), \text{ with } 0 \le e(t, a). \tag{8}$$

Therefore, the model above can be rewritten as

$$(\frac{\partial}{\partial t} + \frac{\partial}{\partial a})N(t, a) = -(m(B(t), a) + e(t, a))N(t, a). \tag{9}$$

For the weight function there are different alternatives. The von Bertalanffy growth curve [12] has been traditionally accepted as a good representation of the fish growth in the natural environments, and it is represented by the expression

$$w(a) = w_0 + (\bar{w} - w_0)(1 - e^{-ka})^3, \tag{10}$$

where $\bar{w}$ is the maximum weight that fish can reach. The parameter k determines the conversion between age and weight of fish, and it depends on each species.

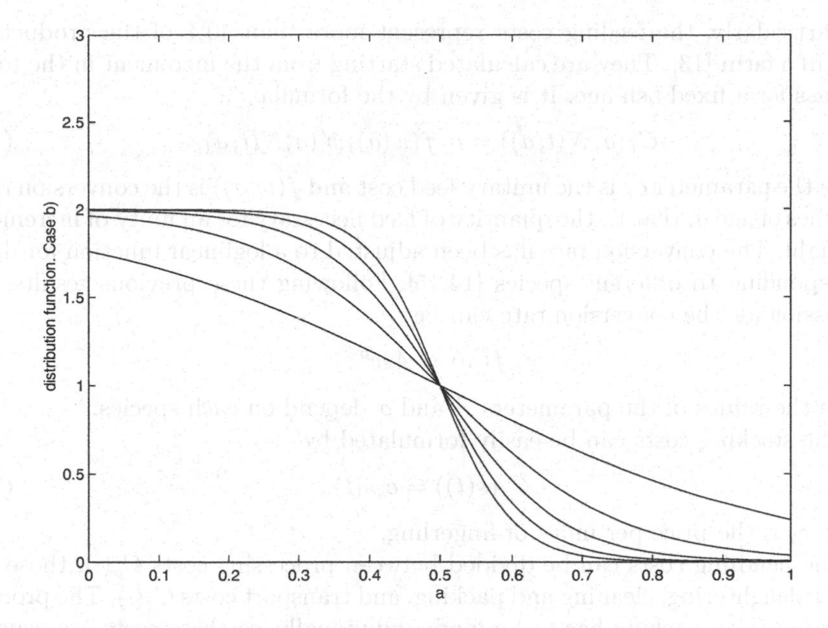

Fig. 2. Graphical representation of the distribution function; case b). The parameter values are $\bar{a} = 1$ and α ranges in the interval [4,20].

3 The Optimi ation Problem

The main objective in a fish farm is to optimize the firm rentability during the period of the industrial activity T, that we assume finite. In particular, for the production process described above, the farmer problem is to determine the stocking intensity $s(t)$ and harvesting rate $e(t,a)$ in order to maximize the firm profit, that is

$$\begin{aligned} \max & \int_0^{\bar{a}} e^{-rT} w(a) R(a) N(T,a) da \\ & + \int_0^T \Big(\int_0^{\bar{a}} e^{-rt} \{ w(a) R(a) e(t,a) N(t,a) - C_f(a, N(t,a)) \} da \\ & - e^{-rt} \{ C_h(t) + C_s(s(t)) \} \Big) \, dt. \end{aligned} \tag{11}$$

The first term is the salvage value obtained with all the biomass remained in the cage after the culture period T. It is measured in current values, where r is the discount rate. The function $R(\cdot)$ is the revenue obtained per unity of weight. For the sake of simplicity, we consider initially this revenue as constant ($R(a) = \theta$, where θ is the market price of the fish per kilo).

We have not included fixed cost in the model, and the variable costs are represented by three different components: the stocking costs $C_s(\cdot)$, the feeding costs $C_f(\cdot)$ and the handling costs $C_h(\cdot)$, that includes the cost around harvesting, cleaning, packing and transporting of fishes.

Particularly, the feeding costs represent more than 40% of the production costs in a farm [13]. They are calculated starting from the increment in the total biomass for a fixed fish age. It is given by the formula

$$C_f(a, N(t,a)) = c_f f(w(a))w'(a)N(t,a), \tag{12}$$

where the parameter c_f is the unitary feed cost and $f(w(a))$ is the conversion rate for fishes of age a, that is, the quantity of feed necessary for an unity of increment in weight. The conversion rate has been adjusted to a loglinear function for data corresponding to different species [14,15]. Following these previous results, an expression for the conversion rate can be

$$f(w) = A w^{\sigma}, \tag{13}$$

where the values of the parameters A and σ depend on each species.

The stocking costs can be easily formulated by

$$C_s(s(t)) = c_s s(t), \tag{14}$$

where c_s is the price per unity of fingerling.

The handling costs can be divided between processing costs $C_p(\cdot)$, those involved slaughtering, cleaning and packing, and transport costs $C_q(\cdot)$. The process of cleaning and packing has to be made individually, so these costs are usually determined by the total number of fishes processes [14]. The second ones are dependent on the total biomass transported. Thus, we can describe the handling costs as

$$C_h(t) = C_p(P(t)) + C_q(Q(t)) = c_p P(t) + c_q Q(t), \tag{15}$$

where c_p, c_q are the unitary cost of processing and transport, respectively. The functions $P(t)$ and $Q(t)$ are the total amount of harvested fish and harvested biomass in time t, respectively. So,

$$P(t) = \int_0^{\bar{a}} h(t,a)da, \; Q(t) = \int_0^{\bar{a}} w(a)h(t,a)da. \tag{16}$$

Summaryzing, the complete control optimum problem can be enounced as

$$\begin{aligned}
\max_{s(t),e(t,a)} & \int_0^{\bar{a}} e^{-rT}\theta\, w(a)N(T,a)da \\
& + \int_0^T \Big(\int_0^{\bar{a}} e^{-rt}\{\theta\, w(a)e(t,a)N(t,a) - C_f(a, N(t,a))\}da \\
& - e^{-rt}\{C_p(P(t)) + C_q(Q(t)) + C_s(s(t))\}\Big)\, dt
\end{aligned}$$

$$\begin{aligned}
s.t. \quad & \Big(\frac{\partial}{\partial t} + \frac{\partial}{\partial a}\Big)N(t,a) = -(m(a, B(t)) + e(t,a))N(t,a), \qquad (17) \\
& B(t) = \int_0^{\bar{a}} w(a)N(t,a)da, \\
& P(t) = \int_0^{\bar{a}} e(t,a)N(t,a)da, \\
& Q(t) = \int_0^{\bar{a}} w(a)e(t,a)N(t,a)da,
\end{aligned}$$

with initial and boundary conditions

$$N(0,a) = N_0(a),$$
$$N(t,0) = s(t),$$

and restrictions on the control variables

$$0 \le e(t,a) \le \bar{e},$$
$$0 \le s(t) \le \bar{s},$$

where $\bar{e}$ and $\bar{s}$ are upper bounds for the two controls.

4 The Ad oint System and Necessary Conditions

The optimality conditions for problems as presented above have been described in [16]. In case of solution, there exist functions $\xi(t,a)$, $\zeta_1(t)$, $\zeta_2(t)$, $\zeta_3(t)$, verifying

$$\begin{aligned} -\left(\frac{\partial}{\partial t} + \frac{\partial}{\partial a}\right)\xi(t,a) &= -(m(a,B) + e(t,a))\xi(t,a) \\ &\quad + e^{-rt}\left[\theta\, w(a)e(t,a) - c_f f(w(a))w'(a)\right] \\ &\quad + \zeta_1(t)w(a) + \zeta_2(t)e(t,a) + \zeta_3(t)w(a)e(t,a), \end{aligned} \tag{18}$$

$$\begin{aligned} \zeta_1(t) &= -\varphi'(B(t))\int_0^{\bar{a}} \xi(t,a)I(a)N(t,a)da, \\ \zeta_2(t) &= -c_p e^{-rt}, \\ \zeta_3(t) &= -c_q e^{-rt}, \end{aligned}$$

where

$$\xi(T,a) = e^{-rT}\theta\, w(a),$$
$$\xi(t,\bar{a}) = 0.$$

The part of the Hamiltonian depending on the control $e(t,a)$ is

$$H_e = \left\{e^{-rt}[\theta\, w(a)N(t,a) - c_p N(t,a) - c_q w(a)N(t,a)] - \xi(t,a)N(t,a)\right\} e(t,a),$$

which is linear in $e(t,a)$. The part of the Hamiltonian depending on the control $s(t)$ is also linear:

$$H_s = (\xi(t,0) - e^{-rt}c_s)s(t).$$

We call $\lambda(t,a) = e^{rt}\xi(t,a)$. Substituting in the equation above, the adjoint system becomes

$$\begin{aligned} \left(\frac{\partial}{\partial t} + \frac{\partial}{\partial a}\right)\lambda(t,a) &= \{r + m(a,B(t)) + e(t,a)\}\,\lambda(t,a) + c_f f(w(a))w'(a) \\ &\quad + c_p e(t,a) + c_q w(a)e(t,a) - \theta\, w(a)e(t,a) \\ &\quad + \left(\int_0^{\bar{a}} \lambda(t,a)I(a)N(t,a)da\right)\varphi'(B(t))w(a), \end{aligned}$$

$$\lambda(T,a) = \theta\, w(a),$$
$$\lambda(t,\bar{a}) = 0.$$

To solve the boundary value problem consisting of the original and the adjoint system, completed with the condition for maximization of the Hamiltonian, is a challenging numerical problem, for which appropriate methods are under development.

In Table 1 we present the values for the parameters in the model and the initial condition.

Table 1. List of parameters, description and values.

Parameter	Description	Value
$\bar{a}$	maximum age	2
$\bar{B}$	maximum biomass (kg)	480.000
$\bar{w}$	maximum weight (kg)	1
m_n	natural mortality	0.05
μ	mortality rate function parameter	0.2
δ	percentage in maximum biomass	0.7
α	gap in mortality rate parameter	10
a^*	low and high mortality border	0.5
w_0	initial weight (kg)	0.01
k	age-weight conversion parameter	10
θ	fish price per kg (euros)	5
r	interest rate (annual)	0.06
A	feed conversion rate parameter	1.7
σ	feed conversion rate parameter	0.11
c_f	feed cost (euros/kg)	0.6
c_s	fingerling cost (euros/unity)	0.3
c_p	processing cost (euros/unity)	0.3
c_q	transport cost (euros/kg)	1.2

The initial condition is

$$N_0(a) = \begin{cases} 590.000 & \text{if} \quad a = 0, \\ 0 & \text{if} \ \ 0 < a \leq 1. \end{cases}$$

References

1. Brett, J.R.: Environmental Factors and Growth. In: Hoar, W.S., Randall, D.J., and Brett, J.R. (eds.): Fish Physiology, Vol. 8. Academic Press, London. (1979) 599–667.
2. Olla, B.L., Davis, M.W., Ryer C.H.: Behavioural Deficits in Hatchery-Reared Fish: Potential Effects on Survival Following Release. Aquaculture and Fishery Management, 25 (Supp. 1) (1995) 19–34.
3. Goldan, O., Popper, D., Karplus, I.: Management of Size Variation in Juvenile Gilthead Sea Bream (*Sparus Aurata*). I: Particle Size and Frequency of Feeding Dry and Live Food. Aquaculture, 152 (1997) 181–190.
4. Yi, Y., Kwein Lin, C.: Effects of Biomass of Caged Nile Tilapia (*Oreochromis Niloticus*) and Aeration on the Growth and Yields in an Integrated Cage-Cum-Pond System. Aquaculture, 195 (2001) 253–267.

5. Bjørnsson, B.: Effects of Stocking Density on Growth Rate of Halibut (*Hippoglossus Hippoglossus L.*) Reared in Large Circular Tanks for Three Years. Aquaculture, 123 (1994) 259–270.
6. Gunnes, K.: Effect of Size Grading Young Atlantic Salmon (*Salmo Salar*) on Subsequent Growth. Aquaculture, 9 (1976) 381–386.
7. Qin, J., Fast, A.W.: Size and Feed Dependent Cannibalism with Juvenile Snakehead *Channa Striatus*. Aquaculture, 144 (1996) 313–320.
8. Calhoun, B.F.: Population Density and Social Pathology. Scientific American, 206 (1962) 139.
9. Hoppensteadt, F.: Mathematical Theories of Populations: Demographics, Genetics and Epidemics. Society for Industrial and Applied Mathematics. Philadelphya, Pennsylvania (1975)
10. Gurtin, M.E., Murphy, L.F.: On the Optimal Harvesting of Persistent Age-Structured Populations. Journal of Mathematical Biology, 13 (1981) 131–148.
11. Murphy, L.F., Smith, S.J.: Optimal Harvesting of an Age-Structured Population. Journal of Mathematical Biology, 29 (1990) 77–90.
12. Clark, C.W.: Bioeconomic Modelling and Fisheries Management. John Wiley & Sons (1985)
13. Cho, C.Y., Bureau, D.P.: Development of Bioenergetic Models and the Fish-PrFEQ Software to Estimate Production, Feeding Ration and Waste Output in Aquaculture. Aquatic Living Resources, 11 (1998) 199–210.
14. Forsberg, O.I.: Optimal Harvesting of Farmed Antlantic Salmon at Two Cohort Management Strategies and Different Harvest Operation Restrictions. Aquaculture Economics and Management, 3 (1999) 143–158.
15. León, C.J., Hernández, J.M., Gasca-Leyva, E.: Cost Minimization and Input Substitution in the Production of Gilthead Seabream. Aquaculture Economics and Management, 5 (2001) 147–170.
16. Veliov, V.M., Feichtinger, G., Tragler, G.: Optimality Conditions for Age-Structured Control Systems. Personal communication (2001)

A Stabilizing Feedback of an Uncertain Control System*

Mikhail Krastanov and Neli Dimitrova

Institute of Mathematics and Informatics, Bulgarian Academy of Sciences
Acad. G. Bonchev str. bl.8, 1113 Sofia
krast@math.bas.bg, nelid@bio.bas.bg

Abstract. A nonlinear control system, describing the continuous methane fermentation is considered. By means of a nonlinear coordinate change the control system is transformed into an equivalent form, involving only measurable quantities of the real process. Assuming that the parameters of the model are not exactly known but bounded within intervals, the set of optimal static points according to a practical criterion is computed. A continuous feedback control is proposed, which stabilizes asymptotically the dynamic system towards a reasonable subregion of this set. Outputs from computer simulation in Maple are also reported.

1 Introduction

The "single substrate/single biomass" model of methane fermentation (cf. [1,7] etc.) is based on two nonlinear ordinary differential equations and one nonlinear algebraic equation:

$$\frac{ds}{dt} = -k_1 \frac{\mu_{\max} s}{k_s + s} x + u(s_{\text{in}} - s) \tag{1}$$

$$\frac{dx}{dt} = \frac{\mu_{\max} s}{k_s + s} x - ux \tag{2}$$

$$Q = k_2 \frac{\mu_{\max} s}{k_s + s} x, \tag{3}$$

where s [mg/dm^3] is the substrate concentration, x [mg/dm^3] is the biomass concentration, u [1/day] is the dilution rate, s_{in} [mg/dm^3] is the influent substrate concentration, $\mu_{\max}$ [1/day] is the maximum specific growth rate of microorganisms, k_s [mg/dm^3] is the saturation constant, k_1 is the yield coefficient, k_2 [l^2/mg] is coefficient and Q [dm^3/day] is the methane gas flow rate.

It is known that only the substrate concentrations s and s_{in}, and the effluent gas flow Q are measurable [2,5]. Some recent studies report that the variable x can be estimated (cf. for example [4]). Practical experiments and parameter estimation results (cf. [7]) show that most of the coefficients of the model

* This work has been partially supported by the Bulgarian National Science Fund under grant No. MM807/98 and MM1104/01.

I. Dimov et al. (Eds.): NMA 2002, LNCS 2542, pp. 230–237, 2003.

(1)–(3) are not exactly known but bounded within intervals. Our goal is to compute the set of all optimal static points when the model parameters vary in the corresponding intervals and to construct a continuous feedback (according to a suitable criterion) stabilizing asymptotically the uncertain control system to this set. It should be mentioned that the proposed control low depends only on the measurable quantities s, s_{in} and Q.

The paper is organized as follows. In Section 2, we describe the steady states of the methane fermentation model with respect to the measurable variables s and Q when the coefficients vary in given intervals. In Section 3, we present a nonlinear continuous stabilizing feedback control of the uncertain system. Some numerical experiments are reported in the last Section 4 using the computer algebra system Maple.

2 Steady State Analysis nder ncertainties

As mentioned above, the substrate concentration $s(t)$ and the output flow rate $Q(t)$ are measurable. Denoting

$$y(t) = \frac{Q(t)}{s(t)}$$

(notc that $y(t)$ is a measurable quantity), the dynamics (1)–(3) takes the form

$$\frac{ds}{dt} = -\frac{k_1}{k_2} sy + u(s_{\text{in}} - s) \tag{4}$$

$$\frac{dy}{dt} = \frac{y}{k_{\text{s}} + s}\left(\frac{k_1}{k_2} sy + \mu_{\max} s - u(k_{\text{s}} + s_{\text{in}})\right) \tag{5}$$

$$Q = sy. \tag{6}$$

For biological reasons, the state variables s, y, the control u, as well as the coefficients $\mu_{\max}$, k_{s}, k_1 and k_2 and s_{in} are positive; additionally, the inequalities $s < s_{\text{in}}$ and $u < \mu_{\max}$ hold true as well.

The steady states of the process satisfy the nonlinear system

$$\frac{k_1}{k_2} sy - u(s_{\text{in}} - s) \;=\; 0 \tag{7}$$

$$\frac{y}{k_{\text{s}} + s}\left(\frac{k_1}{k_2} sy + \mu_{\max} s - u(k_{\text{s}} + s_{\text{in}})\right) \;=\; 0. \tag{8}$$

It is easy to see that the system (7)–(8) possesses an unique positive solution $(s(u), y(u))$, where

$$s(u) = \frac{k_{\text{s}} u}{\mu_{\max} - u}, \quad y(u) = \frac{k_2}{k_1} u \frac{s_{\text{in}} - s(u)}{s(u)} \quad \text{for all } u \in U \subset U_0 := \left(0, \frac{\mu_{\max} s_{\text{in}}}{k_{\text{s}} + s_{\text{in}}}\right).$$

After substituting $s = s(u)$, $y = y(u)$ in the expression for Q in (6) we obtain $Q(u) = s(u)y(u)$, which is defined on U. There is an unique point $\hat{u} \in U$ where $Q(u)$ achieves its maximum, that is

$$\max_{u \in U} Q(u) = Q(\hat{u}) \text{ and } \hat{u} = \mu_{\max}\left(1 - \sqrt{\frac{k_s}{k_s + s_{in}}}\right).$$

Further we compute

$$s^* = s(\hat{u}) = \sqrt{k_s(k_s + s_{in})} - k_s, \quad y^* = y(\hat{u}) = \frac{k_2}{k_1}\hat{u}\frac{s_{in} - s^*}{s^*} = \frac{k_2}{k_1}\mu_{\max}\frac{s_{in} - s^*}{k_s + s^*}.$$

The point (s^*, y^*) is called optimal static point of (4)–(6). Assume now that the four model coefficients k_1, k_2, k_s and $\mu_{\max}$ vary in given compact intervals $[k_1] = [k_1^-, k_1^+]$, $[k_2] = [k_2^-, k_2^+]$, $[k_s] = [k_s^-, k_s^+]$ and $[\mu_{\max}] = [\mu_{\max}^-, \mu_{\max}^+]$ respectively. The admissible interval U for the steady states becomes $\hat{U}$, where

$$\hat{U} \subset \hat{U}_0 := \left(0, \frac{\mu_{\max}^- s_{in}}{k_s^+ + s_{in}}\right).$$

Consider $s^* = s^*(k_s)$ as function of k_s, defined on the interval $[k_s]$ and compute its range $\{s^*(k_s) | \; k_s \in [k_s]\} = [s_1, s_2]$, where

$$s_1 = \sqrt{k_s^-(k_s^- + s_{in})} - k_s^-, \qquad s_2 = \sqrt{k_s^+(k_s^+ + s_{in})} - k_s^+.$$

For any arbitrary but fixed $s \in [s_1, s_2]$ consider

$$y^* = y^*(s) = \frac{k_2}{k_1}\mu_{\max}\frac{s_{in} - s}{k_s + s}$$

as function of $\mu_{\max}$, k_s, k_1 and k_2, defined on $[\mu_{\max}]$, $[k_s]$, $[k_1]$ and $[k_2]$, respectively. The range of $y^*(s) = y^*(s; k_1, k_2, k_s, \mu_{\max})$ on $[k_1]$, $[k_2]$, $[k_s]$, $[\mu_{\max}]$ is presented by $[y^-(s), y^+(s)]$ with

$$y^-(s) = \frac{k_2^-}{k_1^+}\mu_{\max}^-\frac{s_{in} - s}{k_s^+ + s}, \qquad y^+(s) = \frac{k_2^+}{k_1^-}\mu_{\max}^+\frac{s_{in} - s}{k_s^- + s}.$$

The set $\mathcal{Y} = \{(s, y) | \; s_1 \le s \le s_2, \; y^-(s) \le y \le y^+(s)\}$ is called optimal static set of the process (4)–(6) involving intervals in the coefficients. It is illustrated on Figure 1.

3 Feedback Control in the s, y -plane

Our basic assumption is that the four model coefficients k_1, k_2, k_s and $\mu_{\max}$ vary in given compact intervals $[k_1]$, $[k_2]$, $[k_s]$ and $[\mu_{\max}]$, respectively. For simplicity, we shall consider the above intervals to be given in the following form

$$\begin{aligned} &[\mu_{\max}] = [\tilde{\mu}_{\max}(1 - r), \tilde{\mu}_{\max}(1 + r)], \; [k_s] = [\tilde{k}_s(1 - r), \tilde{k}_s(1 + r)], \\ &[k_1] = [\tilde{k}_1(1 - r), \tilde{k}_1(1 + r)], \qquad\qquad [k_2] = [\tilde{k}_2(1 - r), \tilde{k}_2(1 + r)], \end{aligned} \tag{9}$$

where $\tilde{\mu}_{\max}$, $\tilde{k}_s$, $\tilde{k}_1$ and $\tilde{k}_2$ are the centers and $r \cdot \tilde{\mu}_{\max}$, $r \cdot \tilde{k}_s$, $r \cdot \tilde{k}_1$ and $r \cdot \tilde{k}_2$ with $0 \le r < 1$ are the radii of the corresponding intervals. The general case

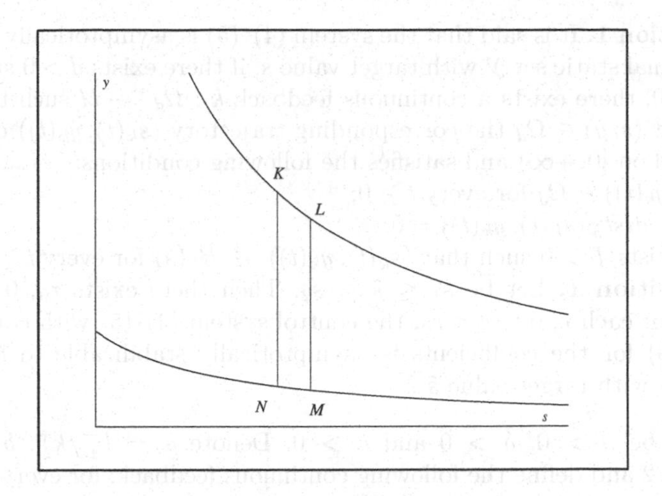

Fig. 1. The optimal static set $\mathcal{Y} = KLMN$ in the (s, y)-plane. The boundary curves are $y^-(s) = (MN)$ and $y^+(s) = (KL)$.

with different values of r for different parameters of the model could be studied in the same way.

We study the asymptotic stabilizability of the system (4)–(6) in a suitable neighbourhood of the optimal static set $\mathcal{Y} = KLMN$. To state the problem, we introduce some assumptions, notions and notations. By $dist_{\mathcal{Y}}(P)$ we denote the distance between the point P and the set $\mathcal{Y}$. Let us fix a positive real number $d > 0$. By Ω_d we denote a compact neighbourhood of the optimal static set $\mathcal{Y}$ consisting of all points $P = (s, y)$ such that $dist_{\mathcal{Y}}(P) \leq d$. Let the set $\mathcal{U}$ of all admissible values of the control is a compact interval such that $\hat{U}_0 \supset \mathcal{U} \supset \hat{U}$. Any continuous function $k : \Omega_d \to \mathcal{U}$ is called a continuous feedback.

Since $\mathcal{Y}$ is a two-dimensional set with nonempty interior, it is reasonable to choose (using a suitable criterion) a value $\tilde{s}$, $s_1 \leq \tilde{s} \leq s_2$, of the substrate concentration and to try to stabilize the control system to the intersection of the sets $\mathcal{Y}$ and $\mathcal{B}(\tilde{s}) := \{(s, y) |\ s = \tilde{s}\}$ (which is a small subset of the set of optimal static points $\mathcal{Y}$). For example, the target point $\tilde{s} \in [s_1, s_2]$ can be chosen so that the minimal value of the methane gas flow rate $Q = Q(s, y)$, $(s, y) \in \mathcal{Y}$, is maximal, i. e.

$$\tilde{s} = \underset{s_1 \leq s \leq s_2}{\operatorname{argmax}} \quad \min_{y^-(s) \leq y \leq y^+(s)} Q(s, y). \tag{10}$$

Since the model coefficients k_1, k_2, k_s and $\mu_{\max}$ vary in given compact intervals and we do not know their exact values (which means that the considered control system is uncertain), it is impossible (cf. [6]) to stabilize it to the set $\mathcal{Y} \cap \mathcal{B}(\tilde{s})$. For that reason, for each $h > 0$ we introduce a neighbourhood $\mathcal{B}_h(\tilde{s}) := \{(s, y) | \ |s - \tilde{s}| \leq h\}$ of the set $\mathcal{B}(\tilde{s})$ and give the following definition:

Definition 1. It is said that the system (4)–(5) is asymptotically stabilizable to the optimal static set $\mathcal{Y}$ with target value $\tilde{s}$, if there exists $d > 0$ such that for every $h > 0$, there exists a continuous feedback $k : \Omega_d \rightarrow \mathcal{U}$ such that for any initial point $(s, y) \in \Omega_d$ the corresponding trajectory $(s_k(t), y_k(t))$ of (4)–(5) is well defined on $[0, +\infty)$ and satisfies the following conditions:
(a) $(s_k(t), y_k(t)) \in \Omega_d$ for every $t \geq 0$;
(b) $\lim_{t\to\infty} dist_{\mathcal{Y}}(s_k(t), y_k(t)) = 0$;
(c) there exists $T > 0$ such that $(s_k(t), y_k(t)) \in \mathcal{B}_h(\tilde{s})$ for every $t \geq T$.

Proposition 1. Let be $s_1 \leq \tilde{s} \leq s_2$. Then there exists r_0, $0 < r_0 < 1$, such that for each r, $0 \leq r < r_0$, the control system (4)–(5) with corresponding intervals (9) for the coefficients is asymptotically stabilizable to the optimal static set $\mathcal{Y}$ with target value $\tilde{s}$.

Proof. Let be $h > 0$, $\delta > 0$ and $d > 0$. Denote $a = k_1^-/k_2^+$, $b = k_1^+/k_2^-$, $c = (a + b)/2$ and define the following continuous feedback: for every $(s, y) \in \Omega_d$ we set

$$k(s,y) = \begin{cases} b\dfrac{sy}{s_{\text{in}} - s} + \delta\dfrac{h}{s_{\text{in}} - s}, & \text{if } s \leq \tilde{s} - h, \\[2ex] \left(b\dfrac{\tilde{s} - s}{h} + c\left(1 - \dfrac{\tilde{s} - s}{h}\right)\right)\dfrac{sy}{s_{\text{in}} - s} + \delta\dfrac{\tilde{s} - s}{s_{\text{in}} - s}, & \text{if } \tilde{s} - h < s \leq \tilde{s}, \\[2ex] a\dfrac{sy}{s_{\text{in}} - s} - \delta\dfrac{h}{s_{\text{in}} - s}, & \text{if } s \geq \tilde{s} + h, \\[2ex] \left(a\dfrac{s - \tilde{s}}{h} + c\left(1 - \dfrac{s - \tilde{s}}{h}\right)\right)\dfrac{sy}{s_{\text{in}} - s} - \delta\dfrac{s - \tilde{s}}{s_{\text{in}} - s}, & \text{if } \tilde{s} \leq s < \tilde{s} + h. \end{cases}$$

It is easy to see that for $\delta = 0$ the values of the feedback $k(s, y)$ belong to the interval $\hat{U}$ when (s, y) lies on the boundary of the set $\mathcal{Y}$. The continuity of $k(\cdot, \cdot)$ with respect to δ, s and y implies that there exist $\delta > 0$ and $d > 0$ such that the values of $k(s, y)$ are still admissible control values for $(s, y) \in \Omega_d$.

Denote for simplicity $F(s, y; u) = (\varphi(s, y; u), \psi(s, y; u))^T$, where

$$\varphi(s, y; u) = -\frac{k_1}{k_2}sy + u(s_{\text{in}} - s)$$

$$\psi(s, y; u) = \frac{y}{k_{\text{s}} + s}\left(\frac{k_1}{k_2}sy + \mu_{\max}s - u(k_{\text{s}} + s_{\text{in}})\right).$$

It can be directly verified that

$$\frac{d}{dt}s(t) \geq h\delta > 0, \quad \text{for } s \leq \tilde{s} - h \quad \text{and} \quad \frac{d}{dt}s(t) \leq -h\delta < 0, \quad \text{for } s \geq \tilde{s} + h. \tag{11}$$

These inequalities imply the existence of $T > 0$ such that

$$\tilde{s} - h \leq s_k(t) \leq \tilde{s} + h \ \text{ for } t \geq T, \tag{12}$$

i. e. $(s_k(t), y_k(t)) \in \mathcal{B}_h(\tilde{s})$ for every $t \geq T$. Hence (c) from Definition 1 holds true.

Consider the sets $Y_m = \{(s,y)|\ s > 0, y^-(s) \leq y \leq y^+(s)\}$, $Y_l = \{(s,y)|\ s > 0, y < y^-(s)\}$ and $Y_u = \{(s,y)|\ s > 0, y > y^+(s)\}$.

Case 1. First, we prove that for every point $(s,y) \in \Omega_d \cap Y_m$ and for every $p \in N^P(s,y)$ the following inequality holds true

$$\langle p, F(s,y;k)\rangle \leq 0, \tag{13}$$

where $N^P(s,y)$ is the proximal normal cone to the set Ω_d at the point (s,y) [3],

$$N^P(s,y) := \left\{p \in R^2 |\ \text{there exists } t > 0 \text{ such that } dist_{\Omega_d}((s,y)+tp) = t\|p\|\right\}.$$

It can be directly checked that if $(s,y) \in \Omega_d \cap Y_m$ then every element p from $N^P(s,y)$ is a convex combination of the vectors $(0, \pm 1)^T$ and $\nu^{\pm}(s,y)$, where

$$\nu^{\pm}(s,y) = \left(-k_1^{\mp} y - k_2^{\pm} \mu_{\max}^{\pm}\,, -k_1^{\mp}(k_s^{\mp} + s)\right).$$

The relations (11) imply that

$$\langle(-1,0), F(s,y;k)\rangle \leq 0 \quad \text{for } s < s_1 \tag{14}$$

$$\langle(1,0), F(s,y;k)\rangle \leq 0 \quad \text{for } s > s_2. \tag{15}$$

Next, it can be verified that there exist $\delta_0 > 0$ and r_0, $0 < r_0 < 1$, such that for $0 < \delta < \delta_0$ and $0 \leq r < r_0$ the inequalities

$$\langle \nu^+(s,y), F(s,y;k(s,y))\rangle \leq 0 \quad \text{for } y = y^+(s),\ s_1 - d \leq s \leq s_2 + d;$$

$$\langle \nu^-(s,y), F(s,y;k(s,y))\rangle \leq 0 \quad \text{for } y = y^-(s),\ s_1 - d \leq s \leq s_2 + d$$

hold true. These inequalities, as well as (14), (15) imply that for every point $(s,y) \in \partial\, \Omega_d \cap Y_m$ and for every $p \in N^P(s,y)$ (p is a convex combination of the vectors $(0, \pm 1)^T$ and $\nu^{\pm}(s,y)$)

$$\langle p, F(s,y;k)\rangle \leq 0, \tag{16}$$

Applying Theorem 3.8 from [3] (see also [8]), we can conclude that for every initial point $(s,y) \in \Omega_d \cap Y_m$ the corresponding trajectory $(s_k(t), y_k(t))$ of (4)–(6) is well defined on $[0, +\infty)$ and $(s_k(t), y_k(t)) \in \Omega_d \cap Y_m$ for every $t \geq 0$. Moreover, the inequality (12) implies that $(s_k(t), y_k(t)) \in \Omega_d \cap Y_m \cap \mathcal{B}_h(\tilde{s})$ for all $t \geq T$.

Case 2. Let be $(s,y) \in Y_l \bigcap \Omega_d$. We define the function

$$\zeta_1(t) = k_2^- \mu_{\max}^- (s_{\text{in}} - s_k(t)) - k_1^+ (k_s^+ + s_k(t)) y_k(t).$$

Since the feedback $k(s,y)$ is a positive continuous function on the compact interval $\mathcal{U}$, there is a constant $\kappa > 0$ such that $|k(s,y)| \geq \kappa$. It can be easily seen that $\dot{\zeta}_1(t) \leq -k(s_k(t), y_k(t))\zeta_1(t) \leq -\kappa\zeta_1(t)$ which implies $\zeta_1(t) \leq \zeta_1(0)e^{-\kappa t}$. The last inequality means that $\lim_{t\to\infty} \zeta_1(t) = 0$. The latter and the inequalities (12) imply that $\lim_{t\to\infty} dist_Y(s_k(t), y_k(t)) = 0$.

Case 3. If $(s, y) \in Y_u \bigcap \Omega_d$ we consider the function

$$\zeta_2(t) = k_1^-(k_s^- + s_k(t))y_k(t) - k_2^+ \mu_{\max}^+(s_{\text{in}} - s_k(t))$$

and obtain similarly $\zeta_2(t) \leq \zeta_2(0)e^{-\kappa t}$ and $\lim_{t\to\infty} dist_Y(s_k(t), y_k(t)) = 0$. Hence, (b) from Definition 1 holds true. The proof of Proposition 1 is completed.

4 Computer Simulation

For our computer simulation we take from [1] the following average values for the parameters in the model (1)–(3):

$$\mu_{\max} = 0.4, \ \ k_s = 0.4, \ \ k_1 = 27.4, \ \ k_2 = 75, \ \ s_{\text{in}} = 3. \tag{17}$$

We consider these values for $\mu_{\max}$, k_s, k_1 and k_2 as centers of the corresponding intervals according to (9).

All computations and graphic outputs are performed using the computer algebra system Maple. The point $\tilde{s}$, which satisfies (10) is chosen to be $\tilde{s} = s_2 \approx 0.7785$. Let $(s(0), y(0))$ be an arbitrary point from Ω_d which is taken as a starting point. We use the stabilizing feedback $k(s, y)$. With randomly chosen points for $\mu_{\max}$, k_s, k_1, k_2 from the corresponding intervals we solve numerically the system (4)–(5) on a mesh $t_i = ih$, $i = 1, 2, \ldots, n$; thereby at any point t_i we pick out the appropriate feedback. After n steps we choose new random values for the coefficients and repeat the process N times. For the numerical solution we use the procedure `dsolve` from the Maple library. Figures 2 and 3 visualize the numerical outputs for $\delta = 0.001$, $r = 0.05$ with different initial conditions.

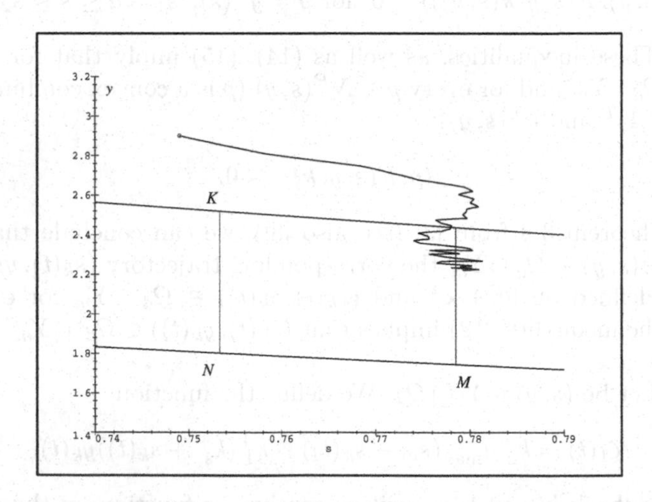

Fig. 2. Feedback control for the uncertain system in the (s, y)-plane for initial conditions $s(0) = 0.749$, $y(0) = 2.9$, with $h = 0.01$, $n = 10$ and $N = 35$.

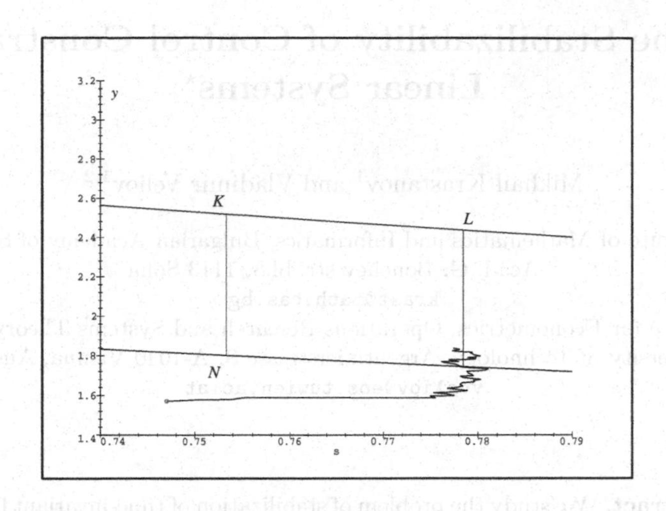

Fig. 3. Feedback control for the uncertain system in the (s, y)-plane for initial conditions $s(0) = 0.747$, $y(0) = 1.57$, with $h = 0.008$, $n = 10$ and $N = 55$.

References

1. Bastin, G., Dochain, D.: On-line Estimation and Adaptive Control of Bioreactors. Els. Sci. Publ., N. Y. (1991)
2. Ciccarella, C., Dalla, M., Germani, A.: A Luenberger-like Observer for Nonlinear Systems. Int. J. Control, 57 (1993) 536–556.
3. Clarke, F., Ledyaev, Yu., Stern, R., Wolenski, P.: Nonsmooth Analysis and Control Theory. Graduate Text in Mathematics, 178, Springer (1998)
4. Hadj-Sadok, M. Z., Gousé, J.-L.: Bounds Estimations for Uncertain Models of Wastewater Treatment. to appear in IEEE
5. Petersen, I., Savkin, A.: Robust Kalman Filtering for Signals and Systems with Large Uncertainties. Birkhäuser, Eds. Boston (1999)
6. Quincampoix, M., Seube, N.: Stabilization of Uncertain Control Systems through Piecewise Constant Feedback. J. of Math. Analysis and Applications, 218 (1998) 240–255.
7. Simeonov, I.: Modelling and Control of Anaerobic Digestion of Organic Waste. Chem. Biochem. Eng. Q., 8 (1994) 45–52.
8. Veliov, V.: Sufficient Conditions for Viability under Imperfect Measurement. Set-valued analysis, Vol. 1 (1993) 305–317.

On the Stabilizability of Control Constrained Linear Systems*

Mikhail Krastanov[1] and Vladimir Veliov[1,2]

[1] Institute of Mathematics and Informatics, Bulgarian Academy of Sciences
Acad. G. Bonchev str. bl.8, 1113 Sofia
krast@math.bas.bg
[2] Institute for Econometrics, Operations Research and Systems Theory, Vienna University of Technology, Argentinierstrasse 8, A-1040 Vienna, Austria,
vveliov@eos.tuwien.ac.at

Abstract. We study the problem of stabilization of time-invariant linear systems with controls which are constrained in a cone. Under small-time local controllability conditions we propose a simple construction of a Lipschitz piecewise linear stabilizing feedback. Moreover, we show that the stabilizing feedback can be chosen in such a way that the number of switchings from one linear form to another, that may occur along a trajectory of the closed-loop system, is uniformly bounded.

1 Introduction

In this note we consider the control system:

$$\dot{x} \in Ax + U, \tag{1}$$

where $x \in \mathrm{R}^n$, A is a matrix of dimension $n \times n$ and U is a closed convex cone. Given a feedback function $k : \mathrm{R}^n \to U$, the closed-loop system has the form

$$\dot{x}(t) = Ax(t) + k(x(t)), \tag{2}$$

In the sequel, the feedback $k(\cdot)$ is a Lipschitz function on R^n, therefore the notion of asymptotic stability for (2) has the standard meaning.

The main result is the following theorem. Some of its corollaries are given discussed in Section 4.

Theorem 1. *Let the control system (1) is small-time locally controllable at the origin. Then there exists a continuous positively homogeneous piecewise linear feedback* $k : \mathrm{R}^n \to U$ *such that the closed-loop system (2) is asymptotically stable at the origin.*

---

* This research was partially supported by the Austrian Science Foundation under contract N0. 14060-OEK and by the Ministry of Science and Higher Education - National Fund for Science Research under contracts MM-807/98 and MM-1104/01.

I. Dimov et al. (Eds.): NMA 2002, LNCS 2542, pp. 238–245, 2003.

The proof of the theorem is constructive. It is based on a reduction of the original system to an auxiliary system with unconstrained controls, which however, contains sign-type nonlinearities with respect to a part of the controls. Thanks to its simple structure, the auxiliary system allows decomposition to a cascade of single-input systems for which stabilizing controls are easy to obtain, applying the Ackermann formula ([1,2]), for example. The resulting stabilizing control for the original system is piecewise linear. More precisely, each component of the stabilizer is a sum of a linear function and absolute values of linear functions. We also show that the stabilizing feedback can be constructed in such a way that the number of switchings from one linear form to another, that may occur along a trajectory of the closed-loop system, depends on the dimension of the state only.

The stabilization problem with cone-constrained controls is studied (under weaker controllability conditions) also in [4], but the constructions proposed there are much more complicated and the resulting feedback lows are not piecewise linear.

For comparison we also mention that for the multiple integrators on R^n a bounded stabilizing feedback that can be represented as a composition of a "saturation functions" and a linear function does not exist ([8]). This was the motivation in [5] for developing a complicated construction of a locally Lipschitz stabilizing feedback in R^n for linear systems for which U is a box containing the origin in its interior, the matrix A has not eigenvalues with positive real parts and its all uncontrollable modes have strictly negative real parts. Our result is weaker than that in [5] in the respect that the feedback that we construct is only locally bounded (piecewise linear), but on the other hand we consider a set U that need not contain the origin in its interior, and this makes the problem different.

In the next section we introduce some notations and auxiliary results. In Section 3 we prove Theorem 1, which is discussed in more details in the last section.

2 Preliminaries

We recall that admissible trajectories of (1) defined on $[0, T]$ are all absolutely continuous functions on $[0, T]$ that satisfy (1) for almost every t.

We assume that the system is small-time local controllable (STLC) at the origin. In the case of a cone U this means that for every $T > 0$ and for every point $x \in \mathrm{R}^n$ there exists an admissible trajectory $x(\cdot)$ such that $x(0) = x$ and $x(T) = 0$. Necessary and sufficient algebraic conditions for STLC of (1) are proved in [3,7] and will be used further.

Below co B denotes the convex hull of the set $B \subset \mathrm{R}^n$, $\mathrm{rec}(V)$ stays for the maximal subspace contained in the convex cone $V \subset \mathrm{R}^n$, $\mathrm{Lin}(Z)$ is the minimal linear subspace of R^n that contains the set Z and $\mathrm{Inv}(Z)$ is the minimal linear subspace of R^n that contains the set Z and is invariant with respect to the matrix A.

In the proof of Theorem 1 we use the following assertions:

Lemma 1. *Suppose that L is a proper subspace of R^n, U is a convex cone and* rec(*co* $(L \cup U)) \neq L$. *Then there exists elements $u_i \in U \setminus L$, $i = 1, 2$, such that $u_1 + u_2 \in L$.*

Lemma 2. *Let $\lambda > 0$ and let D be an $n \times n$-matrix whose eigenvalues have real parts less than $-\lambda$. Then there exists a constant C such that for every $\eta(\cdot) \in \mathcal{L}_\infty(0, +\infty)$ the solution of*

$$\dot{x}(t) = Dx(t) + \eta(t), \quad x(0) = x_0$$

satisfies for all $t \geq 0$ and $T \geq t$ the inequality

$$|x(T)| \leq C \left[e^{-\lambda T}|x_0| + \frac{1}{\lambda} e^{-\lambda(T-t)} \operatorname*{essup}_{s \in [0,t]} |\eta(s)| + \frac{1}{\lambda} \operatorname*{essup}_{s \in [t, +\infty)} |\eta(s)| \right].$$

The proofs of these lemmas are standard.

3 Proof of the Main Result

The following is an equivalent formulation of the STLC condition from [3,6,7]: Introduce successively the subspaces

$$\begin{aligned} L_1 &= \mathrm{Inv}(\mathrm{rec}(U)), \\ L_2 &= \mathrm{Inv}(\mathrm{rec}(\mathrm{co}(L_1 \cup U))), \\ \ldots & \ldots\ldots\ldots \\ L_{i+1} &= \mathrm{Inv}(\mathrm{rec}(\mathrm{co}(L_i \cup U))), \\ \ldots & \ldots\ldots\ldots \end{aligned}$$

Then the system (1) is STLC if and only if $L_k = \mathrm{R}^n$ for some (minimal) k. In fact, if the system is STLC, then $L_{i+1} = L_i$ for $i \geq n$, therefore $k \leq n$. If the set U is a subspace, then $k = 1$. In this case the claim of the theorem is well known. To avoid notational complications we assume below that $k = 2$. Since the construction is a proof are inductive, they go in the same way for $k > 2$.

Let us fix $b_1, \ldots, b_p \in \mathrm{rec}(U)$ and the numbers $n_1, \ldots, n_p$ in such a way that the vectors

$$b_1, Ab_1, \ldots, A^{n_1-1}b_1, \ldots, b_p, Ab_p, \ldots A^{n_p-1}b_p \tag{3}$$

are linearly independent, each of the subspaces

$$\begin{aligned} L_{1,1} &= \mathrm{Lin}(L_{2,1}, d_2, Ad_2, \ldots, A^{m_2-1}d_2), \ldots \\ L_{1,p} &= \mathrm{Lin}(L_{1,p-1}, b_p, Ab_p, \ldots, A^{n_p-1}b_p) \end{aligned}$$

is invariant with respect to A, and

$$\mathrm{Inv}(b_1, \ldots b_p) = \mathrm{Inv}(\mathrm{rec}(U)) = L_1.$$

Since we have assumed that $L_1 \neq \mathrm{R}^n$, using Lemma 1 we obtain pairs

$$d_1, d_1^-, \ldots, d_q, d_q^- \in U \quad \text{for which} \quad d_i + d_i^- \in L_1,$$

and appropriate numbers $m_1, \ldots, m_q$ such the set of vectors (3) together with

$$d_1, Ad_1, \ldots, A^{m_1-1}d_1, \ldots, d_q, Ad_q, \ldots A^{m_q-1}d_q \tag{4}$$

constitute a basis in L_2 (therefore in R^n, according to the assumption that $k = 2$), and moreover the subspaces

$$\begin{aligned} L_{2,1} &= \mathrm{Lin}(L_1, d_1, Ad_1, \ldots, A^{m_1-1}d_1), \\ L_{2,2} &= \mathrm{Lin}(L_{2,1}, d_2, Ad_2, \ldots, A^{m_2-1}d_2), \\ &\ldots \\ L_{2,q} &= \mathrm{Lin}(L_{2,q-1}, d_q, Ad_q, \ldots, A^{m_q-1}d_q) \end{aligned}$$

are invariant with respect to A. (If $k > 2$ we continue in the same way applying again Lemma 1.)

Let us denote $l_i = -(d_i + d_i^-) \in L_1$, and

$$s(w) = \begin{cases} 0, & \text{if } w \geq 0, \\ w, & \text{if } w < 0. \end{cases}$$

Then for every measurable functions $v_i(t)$ and $w_j(t)$ with real values, any trajectory of the system

$$\dot{x} = Ax + \sum_{i=1}^{p} b_i v_i(t) + \sum_{j=1}^{q} d_j w_j(t) + \sum_{j=1}^{q} l_j s(w_j(t)) \tag{5}$$

is also a trajectory of the original system (1). Indeed, $\pm b_i \in U$. Moreover, if $w_j \geq 0$ we have also $d_j w_j + l_j s(w_j) = d_j w_j \in U$, otherwise $d_j w_j + l_j s(w_j) = (d_j + l_j) w_j = -d_j^- w_j = d_j^- |w_j| \in U$. Then for any control v_i, w_j of (5) we may define the control u which has values in U by the expression

$$u = \sum_{i=1}^{p} b_i v_i + \sum_{j \in J_+} d_j w_j + \sum_{j \in J_-} d_j^- |w_j|, \tag{6}$$

where $J_+ = \{j : \; w_j \geq 0\}$, $J_- = \{j : \; w_j < 0\}$. When applied to (1) the control u gives the same trajectory as v_i, w_i applied to (5). Thus we reduce the problem of stabilization of (1) to the problem of stabilization of (5), with unconstrained controls v_i, w_j. The latter system, however contains the simple nonlinearity $s(\cdot)$ with respect to a part of the controls.

Now we proceed with the stabilization of (5). Denote by M the matrix whose columns are the vectors (3), (4). The construction of the vectors (3), (4) makes this matrix invertible. Passing to the new variable $y = M^{-1}x$ the system (5) takes the form

$$\dot{y} = M^{-1}AMy + \sum_{i=1}^{p} M^{-1}b_i v_i(t) + \sum_{j=1}^{q} M^{-1}d_j w_j(t) + \sum_{j=1}^{q} M^{-1}l_j s(w_j(t)). \quad (7)$$

Let us denote by $e_1, e_2, \ldots, e_n$ the canonical basis in R^n. Using the equalities

$$M^{-1}b_1 = e_1, \quad M^{-1}Ab_1 = e_2, \ldots, M^{-1}A^{n_p-1}b_p = e_{k_1}, \quad k_1 = \dim(L_1),$$
$$M^{-1}d_1 = e_{k_1+1}, \quad M^{-1}Ad_1 = e_{k_1+2}, \ldots, M^{-1}A^{m_q-1}d_q = e_n,$$

the invariance of the spaces $L_{1,i}, L_{2,j}$ with respect to A, and the inclusions $l_i \in L_1$, one can check that system (7) has the following 'cascade' form:

$$\begin{aligned} \dot{y}_1 &= Ey_1 + Fy_2 + Bv \qquad + Cs(w) \\ \dot{y}_2 &= \qquad Gy_2 \qquad + Dw \end{aligned} \quad (8)$$

where y_1 is k_1-dimensional, y_2 is $(n-k_1)$-dimensional, the rest of the matrices have the corresponding dimensions, $s(w)$ should be understood componentwise. Let $\hat{w}(y_2)$ be any linear feedback that stabilizes the second subsystem, and let $\hat{v}(y_1)$ be any linear feedback that stabilizes the first subsystem for fixed $y_2 = 0$, $w = 0$. Then $(\hat{v}(y_1), \hat{w}(y_2))$ stabilizes the overall system (8), according to Lemma 2. Hence, $v(x) = \hat{v}([M^{-1}x]_1)$, $w(x) = \hat{w}([M^{-1}x]_2)$ is a linear stabilizing feedback for (5). The proof is complete.

4 Consequences and Comments

1. According to the proof of Theorem 1 the state space R^n can be split into a finite number of polyhedral convex cones $S_1, \ldots S_M$ (with faces determined by the equations $w_i(x) = 0$), such that the stabilizing feedback $u(x)$ is linear on each S_k and continuous on R^n. Thanks to the continuity of $u(\cdot)$ we may assume all S_k closed without any ambiguity.

Theorem 2. *The stabilizing feedback in Theorem 1 can be chosen in such a way that there is a number N such that for every trajectory $x(\cdot)$ of the closed-loop system (2) there are $0 = t_0 < t_2 < \ldots < t_N = +\infty$ such that on each subinterval (t_i, t_{i+1}) the trajectory $x(t)$ belongs to some of the sets S_k.*

Proof. As in the proof of the main result, we set $y = M^{-1}x$ and consider the control system (8). Let us choose the linear feedback laws $\hat{v}(y_1) = Py_1$ and $\hat{w}(y_2) = Qy_2$ (here P and Q are matrices of appropriate dimensions) such that the eigenvalues λ_i, $i = 1, \ldots, k$ of the matrix R are negative real numbers, where

$$R = \begin{pmatrix} E + BP & 0 \\ 0 & G + DQ \end{pmatrix}.$$

It is well known that each component $q_{ij}(t)$ of the exponent matrix $\exp((G + DQ)t)$ is a quasipolynomial which has at most $n-1$ zeros. Let y_2^0 be an arbitrary point. Then $y_2(t) = \exp((G+DQ)t)y_2^0$ is the solution of the second subsystem of (8) starting from y_2^0 at the moment of time 0. Clearly, each component $y_2^j(t)$ of $y_2(t)$ is also a quasipolinomial with at most $n-1$ zeros (as is a linear combination of quasipolinomials of this type). Let $t_0 = 0$, $t_N = +\infty$ and $0 < t_1 < t_2 < \cdots < t_{N-1}$ be the zeros of all components $y_2^j(t)$, $j = 1, 2, \ldots, n$, i.e.

$$\{t_0, t_1, \ldots, t_{N-1}\} = \{t_0\} \cup_{j=1}^{n} \{t_1^j, t_2^j, \ldots t_{n_j}^j\}.$$

Thus,

$$N \leq n(n-1)$$

and each component of $y_2(t)$ does not change its sign on every open interval (t_{i-1}, t_i), $i = 1, \ldots, N$, i.e. the trajectory $x(t) = My(t)$ belongs to some of the sets S_k on each of these open intervals. This completes the proof in the case $k = 2$.

Since each component of $y_2(t)$ does not change its sign on every open interval (t_{i-1}, t_i), $i = 1, \ldots, N$, we obtain (according to the Cauchy formula) that each component of $y_1(t)$ is also a quasipolinomial on (t_{i-1}, t_i), $i = 1, \ldots, N$, and hence it has also a finite number of zeros (it can be shown that the number of zeros of each component of $y_2(t)$ can be estimated by a constant which does not depend on the starting point y_1^0). Thus, the same argument can be also applied in the case $k > 2$.

2. The stabilizing feedback in Theorem 1 is defined trough the one defined for the cascade system (8). The matrices E, F, G, B, D in this system, however, have its own specific structures. This structure is implied by the invariance of the spaces $L_{1,i}$ and $L_{2,j}$ with respect to A. One can easily check that the second subsystem in (8) has a cascade form, where each subsystem depends on a single control w_j, and does not depend on the state variables from the upper levels. This makes it possible, utilizing again Lemma 2, to construct the feedback controls $\hat{w}_j$ independently of each other, each of them stabilizing a corresponding part of y_2. The explicit Ackermann formula ([1,2]) can be utilized for this purpose, since each subsystem has a single control.

3. As shown in [7], in the case of a cone U that is not a subspace the STLC property need not be robust: arbitrarily small perturbations in the data may destroy the controllability. Nevertheless, below we shall show that the stabilizability property of the control system (1) is robust with respect to small perturbations.

We denote by $\mathcal{B}$ the unit ball of R^n centered at the origin, by $\|A\|$ - the standard norm of the matrix A and by $H(V_1, V_2)$ - the Hausdorff distance between the sets V_1 and V_2.

Proposition 1. *Let the assumptions of Theorem 1 hold true. Then there exists $\tilde{\delta} > 0$ such that for every $n \times n$ matrix $\tilde{A}$ and for every closed convex cone $\tilde{U} \subset \mathrm{R}^n$ for which*

$$\|\tilde{A} - A\| \leq \tilde{\delta} \quad \textit{and} \quad H(\tilde{U} \cap \mathcal{B}, U \cap \mathcal{B}) \leq \tilde{\delta}, \tag{9}$$

the perturbed system

$$\dot{x}(t) \;=\; \tilde{A}x(t) \;+\; \tilde{U}, \tag{10}$$

is stabilizable by a Lipschitz continuous positively homogeneous feedback control $\tilde{u}(x) \in \tilde{U}$.

Proof. We shall use the notations from the proof of Theorem 1. The stabilizability property does not depend on the choice of the coordinates. Thus, for simplicity we shall consider the original system (2) and the system (10) with respect to the above introduced y-variables, i.e. we consider the following representations of the both systems (10) in y-coordinates:

$$\dot{y}(t) \;=\; Py(t) \;+\; W \;\text{ and }\; \dot{y}(t) \;=\; \tilde{P}y(t) \;+\; \tilde{W}, \tag{11}$$

where $y = (y_1, y_2)^T$, $\delta \le \min\left(1, \|M^{-1}\|, \|M^{-1}\|^2\right)\tilde{\delta}$,

$$W = M^{-1}U,\; P = M^{-1}AMy \;=\; \begin{pmatrix} E & F \\ 0 & G \end{pmatrix},\quad \|\tilde{P} - P\| \le \delta \;\text{ and }\; \|\tilde{W} - W\| \le \delta.$$

Let $\tilde{w}(y_2) = Ry_2$ be a linear feedback that stabilizes the second subsystem in (8), and let $\tilde{v}(y_1) = Qy_1$ be a linear feedback that stabilizes the first subsystem in (8) for fixed $y_2 = 0$ and $w = 0$. Then there exist positive definite symmetric matrices L_1 and L_2 and a positive real number ε such that for every point $y = (y_1, y_2)^T$

$$\langle L_1 y_1, Ey_1 + BQy_1\rangle \le -2\varepsilon\|y_1\|^2 \;\text{ and }\; \langle L_2 y_2, Gy_2 + DRy_2\rangle \le -2\varepsilon\|y_2\|^2. \tag{12}$$

Let M be an arbitrary positive number. For every point $y = (y_1, y_2)^T$ of R^n we set

$$L(y_1, y_2) := \frac{1}{2}\langle L_1 y_1,\; y_1\rangle + \frac{M}{2}\langle L_2 y_2,\; y_2\rangle,$$

and

$$\mathcal{F}(y) = \begin{pmatrix} BQy_1 \;+\; Cs(Ry_2) \\ DRy_2 \end{pmatrix}.$$

Clearly, $\mathcal{F}(\cdot)$ is a positively homogeneous piecewise linear function such that $\mathcal{F}(y) \in W$ for each $y \in \mathrm{R}^n$ for which there exists a conctant $C > 0$ such that

$$\|\mathcal{F}(y)\| \;\le\; C\|y\|. \tag{13}$$

Then for every $y = (y_1, y_2)^T \in \mathrm{R}^n$, we define the feedback $\tilde{\mathcal{F}}(y)$ as the metric projection of $\mathcal{F}(y)$ on the set $\tilde{W} \cap \mathcal{B}$. Clearly, $\tilde{\mathcal{F}}(y)$ is a positively homogeneous Lipschitz continuous functions with values from $\tilde{W}$ ($F(\cdot)$ is a function of this kind and $\tilde{W}$ is a closed convex cone). According to the definition of $\tilde{\mathcal{F}}(y)$ we obtain that

$$\|\tilde{\mathcal{F}}(y) - \mathcal{F}(y)\| \le \|\mathcal{F}(y)\| \left\| \frac{\tilde{\mathcal{F}}(y)}{\|\mathcal{F}(y)\|} - \frac{\mathcal{F}(y)}{\|\mathcal{F}(y)\|} \right\| \le \|\mathcal{F}(y)\|\,\delta. \tag{14}$$

Let $y = (y_1, y_2)^T$ be an arbitrary nonzero vector. Then

$$\left\langle \text{grad } L(y), \tilde{P}y + \tilde{\mathcal{F}}(y) \right\rangle = \langle \text{ grad } L(y), Py + \mathcal{F}(y)\rangle +$$

$$\left\langle \text{ grad } L(y), (\tilde{P} - P)y + \tilde{\mathcal{F}}(y) - \mathcal{F}(y) \right\rangle.$$

Taking into account (12) we obtain that

$$\langle \text{ grad } L(y), Py + \mathcal{F}(y) \rangle =$$
$$= \langle L_1 y_1, \ Ey_1 + Fy_2 + BQy_1 + Cs(y_2) \rangle + M \langle L_2 y_2, \ Gy_2 + DRy_2 \rangle \leq$$
$$\leq -2\varepsilon \left(\|y_1\|^2 + M\|y_2\|^2 \right) + \langle L_1 y_1, Fy_2 + Cs\,(y_2) \rangle \leq$$
$$\leq -\varepsilon(\|y_1\|^2 + M\|y_2\|^2) \tag{15}$$

for all sufficiently large $M > 0$. According to (14) and (13) we obtain that

$$\left\langle \text{ grad } L(y), (\tilde{P} - P)y + \tilde{\mathcal{F}}(y) - \mathcal{F}(y) \right\rangle \leq$$
$$\leq \|L\| \ \|y\| \delta(\|y\| + \|\mathcal{F}(y)\|) \leq \delta \ \|L\| \ \ \|y\|^2 \ (1 + C).$$

The last inequality and (15) imply that for suffisiently large $M > 0$ and sufficiently small $\delta > 0$

$$\left\langle \text{ grad } L(y), \tilde{P}y + \tilde{\mathcal{F}}(y) \right\rangle < 0 \tag{16}$$

for all nonzero vectors y, i.e. the function L is a Lyapunov function for the closed loop control system (16). Thus, the stabilizability of the system (1) is robust with respect to small perturbations.

4. The stabilizing feedback $u(x)$ that we propose is Lipschitz in x and $u(0) = 0$. This makes it possible to apply Theorem 1 also for stabilization of a nonlinear system

$$\dot{x} = f(x, u), \quad u \in W \tag{17}$$

with f having Lipschitz derivatives in x and u and W – a convex cone. We can prove that the stabilizing feedback for (1) with $A = f'_x(0)$, $U = f'_u(0)W$ (provided that the linear system is STLC) stabilizes *locally* also (17).

References

1. Ackermann, J.: Sampled-Data Control Systems. Springer-Verlag, Berlin (1985)
2. Ackermann, J., Utkin, V.: Sliding Mode Control Design Based on Ackermann's Formula. IEEE Transactions on Automatic Control, 43 (1998) No. 2, 234–237.
3. Bianchini, R.M.: Instant Controllability of Linear Autonomous Systems. J. Optimiz. Theory Appl., 39 (1983) 237–250.
4. Smirnov, G.V.: Stabilization by Constrained Controls. SIAM J. Control and Optimization, 34 (1996) No 5 1616–1649.
5. Sontag, E., Yang, Y., Sussmann, H.: A General Result on the Stabilization of Linear Systems Using Bounded Controls. IEEE Transactions on Automatic Control, 39 (1994) No. 12 2411–2425.
6. Sussmann, H.: Small-time Local Controllability and Continuity of the Optimal Time Function for Linear Systems. J. Optimization Theory Appl., 53 (1987) 281–296.
7. Veliov, V.: On the Controllability of Control Constrained Systems. Mathematica Balkanica, New series, 2 (1988) No. 2–3 147–155.
8. Yang, Y., Sussmann, H.: On the Stabilizability of Multiple Integrators by Means of Bounded Feedback Controls. Proc. of the 30-th IEEE Conference on Decision and Control, Brighton, UK, Dec. 1991, IEEE Publications, New York, 70–72.

An Interval Analysis Based Algorithm for Computing the Stability Margin of Uncertain Systems

Aurelio Piazzi[1] and Antonio Visioli[2]

[1] Dipartimento di Ingegneria dell'Informazione, University of Parma,
aurelio@ce.unipr.it
[2] Dipartimento di Elettronica per l'Automazione, University of Brescia,
visioli@ing.unibs.it

Abstract. A new algorithm for robust stability analysis of linear time-invariant systems which depend on continuously and nonlinearly by uncertain parameters is proposed. The algorithm computes the stability margin as the maximal l_∞ domain in the parameter space compatible with stability. An interval procedure is used to check positivity on an annular domain centered on the nominal parameter vector.

1 Introduction

Consider linear time-invariant systems whose stability is investigated by examining its characteristic polynomial $Q(s;p)$:

$$Q(s;p) = s^n + a_1(p)s^{n-1} + a_2(p)s^{n-2} + \cdots + a_{n-1}(p)s + a_n(p).$$

Here $p := [p_1, p_2, \ldots, p_q]^T \in \mathbb{R}^q$ is the uncertain parameter vector and $a_i(p)$, $i = 1, 2, \ldots, n$ are known nonlinear continuous coefficients depending on the uncertain parameters. Given the nominal parameter vector p^o and a weighted l_∞ norm in the parameter space ($||p||_\infty^w = \max_{i=1,\ldots,q}\{|p_i|/w_i\}$, $w_i > 0$), define the l_∞ domain $\mathcal{B}(\rho)$, centered on p^o, as $\mathcal{B}(\rho) := \{p \in \mathbb{R}^q : ||p - p^o||_\infty^w \leq \rho\}$. Then, the stability margin ρ^* of the uncertain system is defined as $\rho^* := \sup \mathcal{M}$ with $\mathcal{M} := \{\rho \in \mathbb{R} : Q(s;p) \text{is stable} \forall p \in \mathcal{B}(\rho)\}$. Practically, the knowledge of ρ^* gives the meaningful information on how far the uncertain parameters can deviate from the nominal ones without suffering a loss of system stability. We consider general nonlinear dependencies on the uncertain parameters and, by using a new algorithm based on an ad hoc interval procedure whose global convergence is assured under mild technical restrictions, we can compute arbitrarily good lower and upper bounds of ρ^*. Proofs will be omitted for brevity.

2 Preliminaries

Denote with $\mathcal{D}$ any compact and convex set of $\mathbb{R}^q$ containing p^o.

I. Dimov et al. (Eds.): NMA 2002, LNCS 2542, pp. 246–254, 2003.

Theorem 1. *$Q(s;p)$ is (Hurwitz) stable for all $p \in \mathcal{D}$ if and only if: a) $Q(s;p^o)$ is (Hurwitz) stable; b) $a_n(p) > 0\, \forall p \in \mathcal{D}$; c) $H_{n-1}(p) > 0\, \forall p \in \mathcal{D}$.*

This theorem was first presented, in slightly different terms, by Frazer and Duncan (1929) who derived it by means of a resultant-based variable elimination starting from boundary crossing conditions (see [1]).

Stability of the nominal system — $Q(s;p^o)$ is stable — is assumed throughout. As a consequence $\mathcal{M}$ is not empty and it can be rewritten, by virtue of Theorem 1, as $\mathcal{M} = \{\rho : a_n(p) > 0\,,\, H_{n-1}(p) > 0 \,\text{for all}\, p \in \mathcal{B}(\rho)\}$. This suggests to compute the stability margin by solving the following optimization problem:

$$\rho^* = \sup \rho \tag{1}$$

subject to

$$a_n(p) > 0 \text{and} H_{n-1}(p) > 0 \quad \forall p \in \mathcal{B}(\rho) \tag{2}$$

This nonstandard semi-infinite optimization problem will be solved exactly by the positivity-based algorithm presented in the next section. Define $\mathcal{M}_a := \{\rho \in \mathbb{R} : a_n(p) > 0\, \forall p \in \mathcal{B}(\rho)\}$ and $\mathcal{M}_h := \{\rho \in \mathbb{R} : H_{n-1}(p) > 0\, \forall p \in \mathcal{B}(\rho)\}$. Taking into account the continuity property of $a_n(p)$ and $H_{n-1}(p)$ it follows that $\mathcal{M}_a = [0, \rho_a^*)$ and $\mathcal{M}_h = [0, \rho_h^*)$ with $\rho_a^* = \sup \mathcal{M}_a$ and $\rho_h^* = \sup \mathcal{M}_h$.

Property 1. (Vicino, Tesi and Milanese, 1990) [2]

$$\rho^* = \min\{\rho_a^*, \rho_h^*\}. \tag{3}$$

Property 2. The following two statements hold:

1. If for a given real ρ_{al} it is verified that $a_n(p) > 0\, \forall p \in \mathcal{B}(\rho_{al})$ then $\rho_{al} < \rho_a^*$, i.e. ρ_{al} is a strictly lower bound of ρ_a^*.
2. Let a point $p_c \in \mathbb{R}^q$ be such that $a_n(p_c) \leq 0$. Then, having defined $\rho_{au} := ||p_c - p^o||_\infty^w$, it follows that $\rho_{au} \geq \rho_a^*$, i.e. ρ_{au} is an upper bound of ρ_a^*.

Remark 1. A property, which is perfectly analogous to Property 2, holds in relation to set $\mathcal{M}_h$. Indeed Property 2 still holds if we substitute respectively $a_n(p)$, ρ_{al} and ρ_a^* with $H_{n-1}(p)$, ρ_{hl} and ρ_h^*.

Property 3. Let be given real values ρ_{al}, ρ_{hl}, ρ_{au} and ρ_{hu} satisfying

$$\rho_{al} < \rho_a^*\,,\, \rho_{hl} < \rho_h^*\,,\, \rho_{au} \geq \rho_a^*\,,\, \rho_{hu} \geq \rho_h^* \tag{4}$$

Then it follows that

$$\min\{\rho_{al}, \rho_{hl}\} < \rho^* \text{ and } \rho^* \leq \min\{\rho_{au}, \rho_{hu}\} \tag{5}$$

The above Property 3 specifies how to construct lower and upper bounds of ρ^* from the knowledge of lower and upper bounds of ρ_a^* and ρ_h^*.

3 The Positivity-Based Algorithm

Nomenclature and notation 1:

ρ_l strictly lower bound of ρ^*: $\rho_l < \rho^*$
ρ_u upper bound of ρ^*: $\rho_u \geq \rho^*$
ρ_a^* $:= \sup\{\rho \in \mathbb{R} : a_n(p) > 0 \, \forall p \in \mathcal{B}(\rho)\}$
ρ_{al} strictly lower bound of ρ_a^*: $\rho_{al} < \rho_a^*$
ρ_{au} upper bound of ρ_a^*: $\rho_{au} \geq \rho_a^*$
ρ_h^* $:= \sup\{\rho \in \mathbb{R} : H_{n-1}(p) > 0 \, \forall p \in \mathcal{B}(\rho)\}$
ρ_{hl} strictly lower bound of ρ_h^*: $\rho_{hl} < \rho_h^*$
ρ_{hu} upper bound of ρ_h^*: $\rho_{hu} \geq \rho_h^*$
$\mathcal{A}(\rho_1, \rho_2)$ $:= \{p \in \mathbb{R}^q : ||p - p^o||_\infty^w \geq \rho_1 \text{ and } ||p - p^o||_\infty^w \leq \rho_2\}$ with $\rho_2 > \rho_1 \geq 0$ (l_∞ annular domain centered on p^o)
ϵ given required precision to compute ρ^* ($\epsilon > 0$)
ϵ_p numerical threshold to be used with the IPTEST procedure ($\epsilon_p > 0$)
ϵ_z numerical threshold to be used with the ZSEARCH procedure ($\epsilon_z > 0$)
$f(p)$ real continuous function defined over the uncertain parameter space for which $f(p^o) > 0$ (it can be $a_n(p)$ or $H_{n-1}(p)$)
ρ_f^* $:= \sup\{\rho \in \mathbb{R} : f(p) > 0 \, \forall p \in \mathcal{B}(\rho)\}$
ρ_{fl} strictly lower bound of ρ_f^*: $\rho_{fl} < \rho_f^*$
ρ_{fu} upper bound of ρ_f^*: $\rho_{fu} \geq \rho_f^*$
$f^*(\mathcal{D})$ global minimum of function $f(p)$ over any compact domain $\mathcal{D} \subseteq \mathbb{R}^q$.
$\mathcal{P}_a$ $:= \{p \in \mathbb{R}^q : a_n(p) > 0\}$, positive region of the parameter space relative to $a_n(p)$.
$\mathcal{P}_h$ $:= \{p \in \mathbb{R}^q : H_{n-1}(p) > 0\}$, positive region of the parameter space relative to $H_{n-1}(p)$.
$\mathcal{P}_f$ $:= \{p \in \mathbb{R}^q : f(p) > 0\}$, positive region of the parameter space relative to $f(p)$.
$\partial\mathcal{P}$ $:=$ boundary of (any) set $\mathcal{P} \subseteq \mathbb{R}^q$.
$\mathcal{S}(p; r)$ $:= \{p' \in \mathbb{R}^q : ||p' - p|| \leq r\}$ ($r > 0$), sphere of radius r and center p ($||\cdot||$ denotes any norm in $\mathbb{R}^q$).

The positivity-based stability margin computation algorithm herein presented, denoted shortly as algorithm PBSMC, is composed of three parts, denoted as *Phase* I, *Phase* II and *Phase* III. The input-output description of the overall algorithm is the following.

Input of algorithm PBSMC: $a_n(p)$, $H_{n-1}(p)$, thresholds ϵ_p, ϵ_z, precision ϵ.

Output of algorithm PBSMC: ρ_l and ρ_u satisfying $\rho_u - \rho_l \leq \epsilon$.

The variables ϵ_p, ϵ_z and ϵ has to be considered as global variables so that they are available to any phase or algorithm fragment. The aim of the following Phase I is to determine a positive lower bound of ρ^*, i.e. $\rho_l > 0$.

Phase I

1. $\rho_{al} := 0$, $\rho_{au} := +\infty$.
2. $\rho_{hl} := 0$, $\rho_{hu} := +\infty$.

3. $\rho_a := 1$, $\rho_h := 1$.
4. Apply procedure LBIMPROVEMENT with arguments $a_n(p)$, ρ_{al}, ρ_a to obtain $\rho_{al} > 0$ and possibly a finite ρ_{au}.
5. Apply procedure LBIMPROVEMENT with arguments $H_{n-1}(p)$, ρ_{hl}, ρ_h to obtain $\rho_{hl} > 0$ and possibly a finite ρ_{hu}.
6. $\rho_l := \min\{\rho_{al}, \rho_{hl}\}$, $\rho_u := \min\{\rho_{au}, \rho_{hu}\}$.
7. End.

The role of procedure LBIMPROVEMENT, applied twice at steps 4 and 5, is the improvement of the current lower bounds of ρ_a^* and ρ_h^*. It follows a formal description of this procedure.

Input of LBIMPROVEMENT: $f(p)$, ρ_1 and ρ_2 s. t. $\rho_1 < \rho_f^*$ and $\rho_2 > \rho_1 \geq 0$.

Output of LBIMPROVEMENT: ρ_{fl} s. t. $\rho_1 < \rho_{fl} < \rho_f^*$, and if possible a ρ_{fu} s. t. $\rho_{fu} \geq \rho_f^*$.

Procedure LBIMPROVEMENT

1. `flagub` := "false".
2. Apply procedure IPTEST to $f(p)$ over $\mathcal{A}(\rho_1, \rho_2)$ and obtain ξ_f.
3. In case $\xi_f = -1$ then $\rho_2 := (\rho_1 + \rho_2)/2$, `flagub` := "true" and go to 2.
4. In case $\xi_f = 0$ then $\epsilon_p := \epsilon_p/2$, $\rho_2 := (\rho_1 + \rho_2)/2$ and go to 2.
5. In case $\xi_f = +1$ then $\rho_{fl} := \rho_2$.
6. If `flagub` = "true" then apply procedure ZSEARCH to obtain ρ_{fu}.
7. End.

Procedure LBIMPROVEMENT uses, at step 2, the IPTEST procedure to check the positivity of $f(p)$ over the annular domain $\mathcal{A}(\rho_1, \rho_2)$. This positivity test, for which a suitable interval application is proposed in Section 4, has to satisfy this input-output definition.

Input of IPTEST: $f(p)$ and $\mathcal{A}(\rho_1, \rho_2)$.

Output of IPTEST: an integer $\xi_f \in \{-1, 0, +1\}$ satisfying these statements:

a) if $\xi_f = +1$ then it has been proved that $f(p) > 0 \ \forall p \in \mathcal{A}(\rho_1, \rho_2)$;
b) if $\xi_f = -1$ then it has been found a point $p_c \in \mathcal{A}(\rho_1, \rho_2)$ such that $f(p_c) \leq 0$;
c) if $\xi_f = 0$ then it has been proved that $|f^*(\mathcal{A}(\rho_1, \rho_2))| < \epsilon_p$. In case $\xi_f = -1$ the procedure output includes the point $p_c \in \mathcal{A}(\rho_1, \rho_2)$ and $f_c := f(p_c)$.

At step 6 of procedure LBIMPROVEMENT if the logical variable `flagub` is set to "true" we could define directly ρ_{fu} as $||p_c - p^o||_\infty^w$ since $||p_c - p^o||_\infty^w \geq \rho_f^*$. In order to obtain a better upper bound of ρ_f^*, and in such a way to speed up the convergence of the overall algorithm, it is applied the procedure ZSEARCH which performs a zero search on the segment line connecting p^o with p_c: $p(\alpha) := (1-\alpha)p^o + \alpha p_c$, $\alpha \in [0,1]$. Indeed considering that $f(p^o) > 0$ and $f(p_c) \leq 0$ and taking into account the continuity of $f(p)$ it is possible to determine, given a small positive threshold ϵ_z, a point of the segment $p(\alpha')$ for which $f(p(\alpha')) \leq 0$ and $f(p(\alpha')) + \epsilon_z > 0$. Therefore, in general, $\rho_{fu} := ||p(\alpha') - p^o||_\infty^w \leq ||p_c - p^o||_\infty^w$. This can be accomplished by means of a simple bisection method, or Brent's method or any other globally convergent zero search method. The input-output definition of procedure ZSEARCH is the following.

Input of ZSEARCH: $f(p)$, p_c, f_c with $f_c = f(p_c) \leq 0$.

Output of ZSEARCH: a ρ_{fu} such that $\rho_{fu} \leq ||p_c - p^o||_\infty^w$ and for which there exists $p' \ni \rho_{fu} = ||p' - p^o||_\infty^w$, $f(p') \leq 0$ and $f(p') + \epsilon_z > 0$.

The aim of the following Phase II is to determine a finite upper bound of ρ^*.

Phase II

1. If $\rho_u < +\infty$ then terminate Phase II.
2. $\rho := \max\{\rho_{al}, \rho_{hl}\}$.
3. $\rho := 2\rho$.
4. Apply procedure IPTEST to $a_n(p)$ over $\mathcal{A}(\rho_{al}, \rho)$ and obtain ξ_a.
5. In case $\xi_a = -1$ apply procedure ZSEARCH to determine ρ_{au} and terminate Phase II.
6. In case $\xi_a = 0$ set $\epsilon_p := \epsilon_p/2$ and apply procedure IPTEST to $H_{n-1}(p)$ over $\mathcal{A}(\rho_{hl}, \rho)$ and obtain ξ_h.
   6.1. In case $\xi_h = -1$ apply procedure ZSEARCH to determine ρ_{hu} and terminate Phase II.
   6.2. In case $\xi_h = 0$ set $\epsilon_p := \epsilon_p/2$ and go to 3.
   6.3. In case $\xi_h = +1$ set $\rho_{hl} := \rho$ and go to 3.
7. In case $\xi_a = +1$ set $\rho_{al} := \rho$ and apply procedure IPTEST to $H_{n-1}(p)$ over $\mathcal{A}(\rho_{hl}, \rho)$ and obtain ξ_h.
   7.1. In case $\xi_h = -1$ apply procedure ZSEARCH to determine ρ_{hu} and set $\rho_l := \rho_{hl}$, $\rho_u := \rho_{hu}$; then apply procedure FINALP, with arguments $H_{n-1}(p)$, ρ_l, ρ_u, and terminate the algorithm (Phase III has not to be activated).
   7.2. In case $\xi_h = 0$ set $\epsilon_p := \epsilon_p/2$ and go to 3.
   7.3. In case $\xi_h = +1$ set $\rho_{hl} := \rho$ and go to 3.
8. End.

Remark 2: If Phase I has also determined a finite ρ_{au} or ρ_{hu} then Phase II is not necessary (see step 1 of Phase II).

At step 7 of Phase II if $\xi_h = -1$ this proves that the critical constraint of problem (1) is that relative to $H_{n-1}(p)$. Hence $\rho^* = \rho_h^*$ and it is not necessary to proceed with Phase III. Indeed it suffices to activate the procedure FINALP to improve ρ_l and ρ_u until the required precision is reached.

Input of FINALP: $f(p)$, ρ_{fl} and ρ_{fu}.

Output of FINALP: new values of ρ_{fl} and ρ_{fu} satisfying $(\rho_{fu} - \rho_{fl}) \leq \epsilon$.

Procedure FINALP

1. If $(\rho_{fu} - \rho_{fl}) \leq \epsilon$ then terminate.
2. $\rho := (\rho_{fl} + \rho_{fu})/2$.
3. Apply procedure IPTEST to $f(p)$ over $\mathcal{A}(\rho_{fl}, \rho)$ and obtain ξ_f.
4. In case $\xi_f = -1$ apply procedure ZSEARCH to determine ρ_{fu} and go to 1.

5. In case $\xi_f = 0$ set $\epsilon_p := \epsilon_p/2$ and $\rho := (\rho_{fl} + \rho)/2$; apply procedure LBIMPROVEMENT with arguments $f(p)$, ρ_{fl} and ρ to improve ρ_{fl} and possibly ρ_{fu}. Go to 1.
6. In case $\xi_f = +1$ set $\rho_{fl} := \rho$ and go to 1.
7. End.

It follows the description of the last Phase III which computes ρ_l and ρ_u with required precision.

Phase III

1. $\rho_u := \min\{\rho_{au}, \rho_{hu}\}$.
2. $\rho_l := \min\{\rho_{al}, \rho_{hl}\}$.
3. If $(\rho_u - \rho_l) \leq \epsilon$ then terminate.
4. $\rho := (\rho_l + \rho_u)/2$.
5. If $(\rho_{al} \geq \rho_{hl})$ then go to 11.
6. If $(\rho_{au} \leq \rho_{hl})$ then apply procedure FINALP with arguments $a_n(p)$, ρ_l, ρ_u and terminate.
7. Apply procedure IPTEST to $a_n(p)$ over $\mathcal{A}(\rho_{al}, \rho)$ and obtain ξ_a.
8. In case $\xi_a = -1$ apply procedure ZSEARCH to update ρ_{au} and set $\rho_u := \rho_{au}$.
   8.1. If $(\rho_{au} \leq \rho_{hl})$ then apply procedure FINALP with arguments $a_n(p)$, ρ_l, ρ_u and terminate, else go to 3.
9. In case $\xi_a = 0$ set $\epsilon_p := \epsilon_p/2$ and $\rho := (\rho_l + \rho)/2$; apply also procedure LBIMPROVEMENT with arguments $a_n(p)$, ρ_{al}, ρ to improve ρ_{al} and possibly ρ_{au}.
   9.1. If the application of LBIMPROVEMENT has also improved ρ_{au} then $\rho_u := \rho_{au}$. Go to 2.
10. In case $\xi_a = +1$ set $\rho_{al} := \rho$ and go to 2.
11. If $(\rho_{hu} \leq \rho_{al})$ then apply procedure FINALP with arguments $H_{n-1}(p)$, ρ_l, ρ_u and terminate.
12. Apply procedure IPTEST to $H_{n-1}(p)$ over $\mathcal{A}(\rho_{hl}, \rho)$ and obtain ξ_h.
13. In case $\xi_h = -1$ apply procedure ZSEARCH to update ρ_{hu} and set $\rho_u := \rho_{hu}$.
    13.1. If $(\rho_{hu} \leq \rho_{al})$ then apply procedure FINALP with arguments $H_{n-1}(p)$, ρ_l, ρ_u and terminate, else go to 3.
14. In case $\xi_h = 0$ set $\epsilon_p := \epsilon_p/2$ and $\rho := (\rho_l + \rho)/2$; apply also procedure LBIMPROVEMENT with arguments $H_{n-1}(p)$, ρ_{hl}, ρ to improve ρ_{hl} and possibly ρ_{hu}.
    14.1. If the application of LBIMPROVEMENT has also improved ρ_{hu} then $\rho_u := \rho_{hu}$. Go to 2.
15. In case $\xi_h = +1$ set $\rho_{hl} := \rho$ and go to 2.
16. End.

4 The Interval Positivity Procedure IPTEST

Nomenclature and notation 2:

- "box" of $\mathbb{R}^q$: a finite multidimensional interval which can be defined as $[p_1^-, p_1^+] \times [p_2^-, p_2^+] \times \cdots \times [p_q^-, p_q^+] \subseteq \mathbb{R}^q$ with $-\infty < p_i^- \leq p_i^+ < +\infty$, $i = 1, 2, \ldots, q$.
- $\text{mid}(\mathcal{B}) := ((p_1^- + p_1^+)/2, (p_2^- + p_2^+)/2, \ldots, (p_q^- + p_q^+)/2) \in \mathbb{R}^q$, "midpoint" of box $\mathcal{B}$.
- $\text{lb}(f, \mathcal{B})$: lower bound of the global minimum $f^*(\mathcal{B})$ computed as lower endpoint of an "inclusion function" (associated to $f(p)$) evaluated at $\mathcal{B}$ [4].

The input-output definition of procedure IPTEST poses the problem of devising a suitable algorithmic method which has to deal with the special l_∞ annular domain $\mathcal{A}(\rho_1, \rho_2)$. We propose therefore to adapt the interval positivity test over a box of $\mathbb{R}^q$ presented in [3]. Indeed it is always possible to decompose $\mathcal{A}(\rho_1, \rho_2)$ in $2q$ boxes of $\mathcal{R}^q$. Consider for example the bidimensional case $q = 2$. Then $\mathcal{A}(\rho_1, \rho_2) = \mathcal{B}_1 \cup \mathcal{B}_2 \cup \mathcal{B}_3 \cup \mathcal{B}_4$ with

$$\begin{aligned}
\mathcal{B}_1 &= [p_1^o - w_1\rho_2, p_1^o - w_1\rho_1] \times [p_2^o - w_2\rho_1, p_2^o + w_2\rho_1], \\
\mathcal{B}_2 &= [p_1^o + w_1\rho_1, p_1^o + w_1\rho_2] \times [p_2^o - w_2\rho_1, p_2^o + w_2\rho_1], \\
\mathcal{B}_3 &= [p_1^o - w_1\rho_2, p_1^o + w_1\rho_2] \times [p_2^o - w_2\rho_2, p_2^o - w_2\rho_1], \\
\mathcal{B}_4 &= [p_1^o - w_1\rho_2, p_1^o + w_1\rho_2] \times [p_2^o + w_2\rho_1, p_2^o + w_2\rho_2].
\end{aligned} \tag{6}$$

Formulas (6) can be easily generalized for the q-dimensional case.

Procedure IPTEST

1. $u := +\infty$.
2. Decompose $\mathcal{A}(\rho_1, \rho_2)$ into $2q$ boxes $\mathcal{B}_1, \mathcal{B}_2, \ldots, \mathcal{B}_{2q}$.
3. For $i = 1, 2, \ldots, 2q$.
   3.1. If $f(\text{mid}(\mathcal{B}_i)) \leq 0$ then $p_c := \text{mid}(\mathcal{B}_i)$, $f_c := f(p_c)$, $\xi_f := -1$ and terminate.
   3.2. If $\text{lb}(f, \mathcal{B}_i) \leq 0$ then put pair $(\mathcal{B}_i, \text{lb}(f, \mathcal{B}_i))$ into *List* in such a way to preserve the nondecreasing ordering of the lower bounds and set $u := \min\{u, f(\text{mid}(\mathcal{B}_i))\}$.
   3.3. End of i-loop.
4. If *List* is empty set $\xi_f := +1$ and terminate.
5. $l :=$ the second member of the first element of *List*.
6. If $(u - \epsilon_p) < 0$ and $(l + \epsilon_p) > 0$ then $\xi_f := 0$ and terminate.
7. Bisect, thus getting boxes $\mathcal{D}_1$ and $\mathcal{D}_2$, the box of the first element of *List* on its maximal dimension.
8. If $f(\text{mid}(\mathcal{D}_1)) \leq 0$ then $p_c := \text{mid}(\mathcal{D}_1)$, $f_c := f(p_c)$, $\xi_f := -1$ and terminate.
9. Repeat for the box $\mathcal{D}_2$ the same action performed (for box $\mathcal{D}_1$) at step 8.
10. Discard the first element from *List*.
11. If $\text{lb}(f, \mathcal{D}_1) \leq 0$ then put $(\mathcal{D}_1, \text{lb}(f, \mathcal{D}_1))$ into *List* in such a way to preserve the nondecreasing ordering of the lower bounds; set $u := \min\{u, f(\text{mid}(\mathcal{D}_1))\}$.
12. Repeat for the box $\mathcal{D}_2$ the same action performed (for box $\mathcal{D}_1$) at step 11.

13. Go to 4.
14. End.

Definition 1. *The positive region $\mathcal{P}_f$ is said to be not degenerate if one of the following conditions holds: i) $\partial\mathcal{P}_f$ is empty; ii) for every $p \in \partial\mathcal{P}_f$ and for every sphere $\mathcal{S}(p;r)$ there exists a point $\tilde{p} \in \mathcal{S}(p;r)$ such that $f(\tilde{p}) < 0$.*

Theorem 2. *Assume that $\mathcal{P}_a$ and $\mathcal{P}_h$ are not degenerate and $\rho^* < +\infty$. Then algorithm PBSMC converges with certainty, satisfying the exposed input-output definition, for every positive values of ϵ_p and ϵ_z.*

5 An Illustrative Example

The overall algorithm has been implemented both in C language by exploiting the Profil/Bias interval routine for C language made by Knuppel [5]. It has been chosen to use, in procedure ZSEARCH, the simple bisection method and, in procedure IPTEST for computing $\text{lb}(f, \mathcal{B})$, the Baumann meanvalue form which is the optimal inclusion function in the class of meanvalue forms (see [4]).

Consider the following characteristic polynomial

$$Q(s;p) = s^4 + p_1^3 p_2 s^3 + p_1^2 p_2^2 p_3 s^2 + p_1 p_2^3 p_3^2 s + p_3^3 \tag{7}$$

where p_1, p_2, and p_3 are the uncertain parameters. The nominal system takes the values $p_1^o = 1.4$, $p_2^o = 1.5$, and $p_3^o = 0.8$; $Q(s;p^o)$ is stable. To perform the robust stability analysis choose $w_1 = 0.25$, $w_2 = 0.20$, and $w_3 = 0.20$ as the l_∞ norm weights. From (7) derive

$$\begin{aligned} a_4(p) &= p_3^3 \\ H_3(p) &= p_1^6 p_2^6 p_3^3 - p_1^6 p_2^2 p_3^3 - p_1^2 p_2^6 p_3^4. \end{aligned}$$

With $\epsilon_p = 10^{-3}$, $\epsilon_z = 10^{-3}$ and choosing the precision $\epsilon = 0.01$ algorithm PBSMC gives the results $\rho_l = 1.083378$, $\rho_u = 1.091306$. With better precision $\epsilon = 0.001$, maintaining $\epsilon_p = 10^{-3}$ and $\epsilon_z = 10^{-3}$, the obtained results are: $\rho_l = 1.089324$, $\rho_u = 1.089963$.

6 Conclusions

A new algorithm for computing without conservativeness the stability margin ρ^* of linear time-invariant systems depending on uncertain parameters has been presented. The algorithm uses the positivity of functions $a_n(p)$ and $H_{n-1}(p)$ over annular domains centered on p^o to develop a branch and bound strategy which permits efficiently to discard from computations, as soon as possible, one of the two functions. The task of testing positivity over the annular domains of the parameter space is committed to an interval procedure. In such a way it is possible to perform a robust stability analysis for systems whose characteristic polynomials depend on uncertain parameters through general nonlinear functions. In practice, the coefficients $a_i(p)$ can be any continuous functions.

References

1. Ackermann, J.: Uncertainty structures and robust stability analysis. Proceedings of the First European Control Conference (1991) 2318–2327.
2. Vicino, A., Tesi, A., Milanese, M.: Computation of nonconservative stability perturbation bounds for systems with nonlinearly correlated uncertainties. IEEE Transactions on Automatic Control, 35 (1990) 835–841.
3. Piazzi, A., Marro, G.: On computing the robust decay rate of uncertain systems. Proceedings of the IFAC Symposium on Robust Control Design (1994) 46–51.
4. Ratschek, H., Rokne, J.: New computer methods for global optimization. Ellis Horwood Limited, Chichester, GB (1988)
5. Knüppel, O.: BIAS - Basic Interval Arithmetic Subroutines. Technische Universität Hamburg-Harburg, Germany (1993)

Evaluation of Crisis, Reversibility, Alert Management for Constrained Dynamical Systems Using Impulse Dynamical Systems

Patrick Saint-Pierre

Centre de Recherche Viabilité, Jeux, Contrôle
Université Paris IX - Dauphine
saint-pierre@viab.dauphine.fr

Abstract. Considering constrained dynamical systems characterized by a differential inclusion $x' \in F(x)$, we are interested in studying the situation where, for various reasons, the state leaves the constrain domain K either because the initial position does not belong to the Viability Kernel of K for F or it belongs to a "sustainable or tolerable" but not "comfortable" domain. This question appears in numerous models in Social Sciences or in Genetics as well as for controlling security in Automatics and Robotics, like Aircraft landing, rolling and taking off. After recalling basic concepts in Viability Theory and using hybrid calculus, we show how to evaluate and manage crisis in general cases.

1 Introduction

We consider a dynamical impulse system describing the evolution of a state variable $x \in X$ which should remain in a constrained domain. Research efforts are devoted to the study of such systems mainly in extending the mathematical tools of the Control Theory ([5, Bensoussan & Menaldi] in order to take into account as well constraints (qualitative properties) as optimality performance (quantitative properties), and in developing methods for non linear hybrid systems (see for instance [19, Sastry], [2, Aubin]), [4, Tomlin, Crück, Bayen, and ref.]).

On the other hand the study of controlled system with state constraints has been widely developed this last decade using the main concepts of Set-Valued Analysis [3, Aubin & Frankowska] and Numerical Set-Valued Analysis [16, Saint-Pierre] which can be regarded as a "Tool Box" for viability theory and numerical viability (viability, equilibria, stability, reachability, controllability, optimal control (minimal time-to-reach problem, infinite horizon optimal control [6, Cardaliaguet, Quincampoix & Saint-Pierre], differential games (discriminating and leadership kernels, conditional and guaranteed strategies, minimal hitting time function, robustness, qualitative analysis [13, Leitmann], [18, Seube], [9, Dordan] and their applications to mathematical economics, finance, demography or biology as well as to enginery and automatics when taking into account constraints is unescapable.

I. Dimov et al. (Eds.): NMA 2002, LNCS 2542, pp. 255–263, 2003.

Crisis time function has been introduced in [10, Doyen & Saint-Pierre]), allowing the characterization of reversibility and irreversibility of the lapse for constraints. In particular these concepts can be found for instance in Environmental Protection models (preservation of future generation, precaution principle) or in Biodiversity (maintaining polymorphism).

The Viability Kernel Algorithm originally designed for computing viability kernels in the frame of set-valued analysis has been widen to approximate the smallest lower semicontinuous subsolution of the Hamilton-Jacobi-Belmann equation [11, Frankowska], [6, Cardaliaguet, Quincampoix & Saint-Pierre]. Moreover viability techniques have been extended to hybrid systems first for computing the Hybrid Kernel which broadens the concept of Viability Kernel, second for computing the Minimal Cost function in non smooth contexts and in the presence of uncertainty [8, Crück]. The scope of this paper is to show how Viability Theory applied in the context of hybrid systems allows to manage crisis that is to say situations where precisely viability fails.

2 Continuous Discrete and Hybrid Systems

2.1 Definitions

Let x denote the state vector belonging to a finite dimensional vector space X.

• A continuous evolution is given by a set-valued map $F : X \rightsquigarrow X$ with convex compact values, closed graph defined on a closed set $D \subset X$.

$$x'(t) \in F(x(t)), \text{ for almost all } t \in \mathbb{R}^+. \tag{1}$$

For instance, $F(x) = \{f(x,u), \text{ where } u \in U(x)\}$.

• A discrete (or impulsive) evolution is given by a set-valued map Φ with compact values, closed graph defined on a closed set $C \subset X$.

$$x^{n+1} \in \Phi(x^n). \tag{2}$$

A dynamical system is said to be continuous if $C = \emptyset$, discrete if $D = \emptyset$ and hybrid if $C \cap D \neq \emptyset$.

We denote $\mathcal{S}^c_F(x)$ the set of all absolutely continuous solutions to the differential inclusion (1) with initial condition $x(0) = x$ and by $\mathcal{S}^d_\Phi(x)$ the set of all sequences solutions to the recursive inclusion (2) with initial condition $x^0 = x$.

A *run* of an impulse system (F, Φ) is a sequence $\overrightarrow{x}(\cdot) := \{(\tau_i, x_i(\cdot))\}_{i \in I} \in (\mathbb{R}^+ \times \mathcal{C}(0, \infty; X))^{\mathbb{N}}$, where $x_i(\cdot) \in \mathcal{S}^c_F(x_i(0))$ is the ith motive which is an absolutely continuous solution to (1) starting from $x_i = x_i(0)$ at time 0 until time τ_i and $x_{i+1} \in \Phi(x_i(\tau_i))$.

We denote $\mathcal{S}^h_{F,\Phi}(x)$ the set of all runs associated with both differential and recursive inclusions with initial condition $x_0 = x$.

3 Viability ernels

Let $D \subset X$. We denote $\mathcal{D}$ the set of all functions defined on R with values in D and $\overrightarrow{\mathcal{D}}$ the set of all sequences contained in D. Let us recall the following definitions and properties.

A closed set D is a viability domain for F if and only if $\forall x \in D$, $S_F^c(x) \cap \mathcal{D} \neq \emptyset$ (Qualitative Property). This is equivalent to the following properties:

i) - $\forall x \in D,\ F(x) \cap T_D(x) \neq \emptyset$ (Geometrical Property)

ii) - $\forall x \in D,\ \forall p \in NP_D(x),\ H(x,p) \leq 0$ (Dual Property)

where $T_D(x)$ is the contingent cone to D at x, $NP_D(x)$ is the set of proximal normal to D at x and $H(x,p) := \min_{y \in F(x)} < y, p >$ is the associated Hamiltonian. The largest closed viability domain contained in a closed set $K \subset X$ exists ([1]). It is called the Viability Kernel of K for F and is denoted $Viab_F(K)$.

A set D is a discrete viability domain for Φ if and only if $\forall x \in D$, $S_\Phi^d(x) \cap \overrightarrow{D} \neq \emptyset$ (Qualitative Property). This is equivalent to the following property:

$\forall x \in D,\ \Phi(x) \cap D \neq \emptyset$ (Geometrical Property)

The largest closed discrete viability domain contained in K exists ([16]). It is called the Discrete Viability Kernel of K for Φ and is denoted $\overrightarrow{Viab_\Phi}(K)$.

3.1 Impulse Systems with State Constraints and Hybrid Viability Kernels

Let $\overrightarrow{x}(\cdot) := \{(\tau_i, x_i(\cdot))\}_{i \in I} \in (\mathbb{R}^+ \times X \times \mathcal{C}(0,\infty; X))^{\mathbb{N}}$ be a run associated with the hybrid dynamical system (1,2). We set $T = \sum_{i=0}^{n} \tau_i$.

Let be $\delta > 0$ and consider sequences $(t_i)_{i \in I}$ and $(\vartheta_i)_{i \in I}$ given by $t_0 = \vartheta_0 = 0$, $t_{i+1} = t_i + \tau_i$ and $\vartheta_{i+1} = \vartheta_i + \tau_i + \delta$. We have $\vartheta_{i+1} = t_i + i\delta$. With any $t \in [0,T]$ we associate i_t satisfying $t_{i_t} \leq t < t_{i_t+1}$, with any $\vartheta \in \mathbb{R}$ we associate i_ϑ satisfying $\vartheta_{i_\vartheta} \leq \vartheta < \vartheta_{i_\vartheta+1}$. and with a run $\overrightarrow{x}(\cdot)$ we associate

$$x^e(\vartheta) = x_0 + \sum_{i=0}^{i_\vartheta} \left(\int_0^{\tau_i^\star} x_i'(\tau) d\tau + \alpha_i^\star (x_{i+1} - x_i(\tau_i)) \right)$$

where the motive $x_i(\tau)$ starts from $x_i = x_i(0)$ at time 0 until time τ_i, $\tau_i^\star = \min(\tau_i, \vartheta - \vartheta_i)$ and $\alpha_i^\star = \max(0, \min(1, \vartheta - \vartheta_i - \tau_i))$.

We call <u>*hybrid*</u> solution associated with a run $\overrightarrow{x}(\cdot)$ the map $t \to x(t)$ given by $x(t) = x^e(t + i_t\delta)$.

<u>Definition</u> - The *hybrid viability kernel* of K for the impulse system (F, Φ, K) is the largest closed subset of initial states belonging to K from which starts at least one hybrid viable solution. We denote this set $Hyb_{(F,\Phi)}(K)$.

The Hybrid Viability Kernel can be approximated by a sequence of discrete viability kernels associated with suitable discrete systems [17, Saint-Pierre].

3.2 Assumptions and Approximated Impulse Systems

- C is compact and $\Phi : C \rightsquigarrow X$ is upper semi-continuous compact valued

$0 < \delta_{inf} \leq \inf_{x \in C} \inf_{y \in \Phi(x)} \|x - y\| \leq \sup_{x \in C} \sup_{y \in \Phi(x)} \|x - y\| \leq \delta_{sup}$

so that Φ has no fix point and its graph is closed. We set $\forall x \notin C, \Phi(x) = \emptyset$.

• K is a compact set and F is a Marchaud map satisfying $\sup_{x \in K} \sup_{y \in F(x)} \|y\| \le M$.

This assumption implies that F is closed. As usual in the context of set-valued numerical analysis we assume that a "good" approximation F_ρ of F exists in the sense that $Graph(F_\rho)$ remains in a not too large neighborhood of F [16, Saint-Pierre].

4 Management of Crisis

4.1 Reversibility and Irreversibility

The notion of Crisis corresponds to the situation where the evolution of the state described by a differential inclusion violates the constraints.

Viability Theory discusses conditions allowing the state maintained in the constraint set. However important questions arise:

1 - What to say and to do when the state leaves the constraint set?

2 - Can we predict for a given initial position a future crisis?

3 - If a crisis occurs or if it is unavoidable in the future, what to say about its reversibility, its duration and how to recover Viability?

4 - Is it possible to define a scale of viability measuring seriousness of crisis to come and to numerically evaluate crisis?

Consider a dynamical system $x' \in F(x)$ and two closed subsets $K^s \subset K^h \subset X$ representing "soft" constraints *(viable, tolerable, comfortable,...)* and "hard" constraints *(unsustainable, unbearable, fatal,...)*.

4.2 The Reversibility Domain

We look for the set of initial position in K_h from which there exists at least one solution which, if it leaves for a while the constraint set K_s, remains in $K_h \backslash K_s$ during a finite laps of time before coming back and forever remaining in K_s. We denote this set $Rever_F(K_h; K_s)$.

The set $Rever_F(K_h; K_s)$ is the capture domain of target $Viab_F(K_s)$) for F under constraint K_h ([15]).

Indeed any solution which comes back and definitively remains in K_s necessarily comes back in the Viability Kernel of K_s.

Moreover, if $Viab_F(K_h) = Viab_F(K_s)$, $Rever_F(K_h; K_s)$ is the Viability Kernel of K_h with target $Viab_F(K_s)$ [15, Quincampoix & Veliov].

4.3 The Minimal Crisis Time Function

Consider the map which evaluates the minimal time the state will spend out of K^s while remaining in K^h when starting from an initial position x:

$$\begin{aligned} C_{K^s,F}^{K^h}(x) &= \inf_{x(\cdot)\in\mathcal{S}_F(x))\cap\mathcal{K}^h} \mu(t \mid x(t) \notin K^s) \\ &= \inf_{x(\cdot)\in\mathcal{S}_F^c(x)\cap\mathcal{K}^h} \int_0^{+\infty} \mathcal{X}_{X\backslash K^s}(x(s))(1+\mathcal{I}_{\mathcal{K}^h}(x(s)))ds, \end{aligned}$$

where μ denotes the Lebesgue measure in $\mathbb{R}$, $\mathcal{X}$ and $\mathcal{I}$ stand for the characteristic and the indicatrix functions. We impose that $C^{K^h}_{K^s,F}(x) = +\infty$ whenever every trajectory $x(\cdot)$ starting from x violates the constraints K^s during an infinite time.

Theorem 1. *[Doyen & Saint-Pierre] If $F : X \rightsquigarrow X$ is a Marchaud set-valued map, then*
a) $\forall x \in \mathrm{Dom}(C^{K^h}_{K^s,F})$, the infimum is reached: $\exists\, x^*(\cdot) \in \mathcal{S}^{K^h}_F(x)$ such that

$$C^{K^h}_{K^s,F}(x) = \int_0^{+\infty} \mathcal{X}_{X\setminus K^s}(x^*(s))(1 + \mathcal{I}_{K^h}(x^*(s)))ds,$$

b) We have $\mathrm{Epi}(C^{K^h}_{K^s,F}) = \mathrm{Viab}_{\widetilde{F}}(K^h \times \mathbb{R}^+)$, where $\widetilde{F}$ corresponds to the extended dynamic with closed convex righthand side

$$\begin{cases} x' \in F(x) \\ y' \in -\overline{co}(\mathcal{X}_{X\setminus K^s})(x)), \end{cases}$$

c) $C^{K^h}_{K^s,F}(\cdot) = \vartheta^{K^h}_{Viab_F(K^s)}(\cdot)$: the crisis map is lower semi-continuous and coincides with the Minimal Time-to-reach function where the constraint is K^h and the target is $Viab_F(K^s)$.
d) • $Viab_F(K^s)$ is the 0-level set of $C^{K^h}_{K^s,F}$: $\{x \in K^h \mid C^{K^h}_{K^s,F}(x) = 0\}$.

Proof. a) Existence of the infimum derives from an Aubin's theorem which states that the graph of the restriction of $\mathcal{S}_F(\cdot)$ to any compact is compact and from Fatou's Lemma.

b) Marchaud property of the set-valued map $\mathcal{X}^{\natural}_{K^s}(\cdot)$ implies existence of a Viability Kernel. Then we prove that $\mathrm{Epi}(C^{K^h}_{K^s,F}) \subset Viab_{\widetilde{F}}(K^h \times \mathbb{R}^+)$.

Conversely we prove that for any $(x, y) \in Viab_{\widetilde{F}}(K^h \times \mathbb{R}^+)$ there exists a solution $x(\cdot)$ such that $\int_0^\infty \mathcal{X}_{X\setminus K^s}(x(s))ds \leq y$ and deduce that $C^{K^h}_{K^s,F}(x) = \inf_{x(\cdot)\in\mathcal{S}^c_F(x)\cap\mathcal{K}^h} \int_0^{+\infty} \mathcal{X}_{X\setminus K^s}(x(t))dt \leq y$. Therefore $Viab_{\widetilde{F}}(K^h \times \mathbb{R}^+) \subset \mathrm{Epi}(C^{K^h}_{K^s,F})$.

4.4 Approximation Algorithm

From Theorem 1, the Viability Algorithm provides an algorithm to approximate $C^{K^h}_{K^s,F}(\cdot)$. Let $\rho > \tau > 0$ be the time and the space steps such that $\frac{\tau}{\rho} \to 0^+$.

Let X_τ be a grid of X with mesh of size τ and $K^h_\tau := (K^h + \tau\mathcal{B}) \cap X_\tau$.

We set $\mathcal{K}^{h,0}_\tau = \mathcal{K}^h_\tau := K^h_\tau \times \tau\mathbb{Z}^+$ and $C^0_\tau : K^h_\tau \to \tau\mathbb{Z}^+$, the lower semicontinuous function defined by $Epi(C^0_\tau) = \mathcal{K}^{h,0}_\tau$.

With the sequence $\mathcal{H}^{h,n}_\tau$ defined by the Viability Kernel Algorithm ([16]):

$$\mathcal{H}^{h,n+1}_\tau := \mathcal{H}^{h,n}_\tau \cap (\mathcal{I} + \tau\widetilde{F}_\tau)^{-1}(\mathcal{H}^{h,n}_\tau)$$

we associate the functions $C^n_\tau : K^h_\tau \to \tau\mathbb{Z}$ which epigraph coincides with $\mathcal{H}^{h,n}_\tau$.

Proposition 2 *Functions C^n_τ are lower semicontinuous and satisfy:*

$$C^{n+1}_\tau(x) := \begin{cases} \rho - \tau + \min_{v\in F(x)\, b\in\mathcal{B}} C^n_\tau(x+\rho(v+\phi(\rho))b)), & \text{if } x \in K^h,\ d_C(x) > M\rho + \tau \\ C^n_\tau(x) & \text{if not,} \end{cases}.$$

The limit exists $\forall x \in K^h_\tau$: $C^\infty_\tau(x) := \lim_{n\to+\infty} C^n_\tau(x)$ and $Epi(C^\infty_\tau) = \overrightarrow{Viab}_{\Gamma_{\rho,\tau}}(\mathcal{H}^h_\tau)$ where $\Gamma_{\rho,\tau}(x,y) := \{(x,y) + \rho\Phi_\rho(x,y)\}\bigcap(X_\tau \times \tau\mathbb{Z})$.

Let $\tau \to 0$. Thanks to the Refinement Principle [16] we have:

Proposition 3 *The sequence $C^\infty_\tau(\cdot)$ converges to $C^{K^h}_{K^s,F}(\cdot)$ in the epigraphic sense: $\mathcal{E}pi(C^{K^h}_{K^s,F}) = \lim_{\tau\to 0}\mathcal{E}pi(C^\infty_\tau)$. Moreover $\forall x \in X_\tau$, $C^\infty_\tau(x) \le C^{K^h}_{K^s,F}(x)$ and C^∞_τ converges pointwisely to $C^{K^h}_{K^s,F}$*

$$\forall x \in K,\ C^{K^h}_{K^s,F}(x) = \lim_{p\to+\infty} \min_{x\in(x+\tau\mathcal{B})\cap X_\tau} C^\infty_\tau(x)$$

Epigraphical convergence follows from the Convergence Theorem of discrete viability kernels to the viability kernel and pointwise convergence follows from that viability kernels are epigraphs of l.s.c. functions $C^{K^h}_{K^s,F}$ and C^∞_τ.

4.5 Example: The Goodwin Employment-Wages Model

We illustrate this approach by considering a production function $Y = f(K,L,t) = f(K,L,t) = min\left[\frac{K}{k}, Le^{qt}\right]$ depending on the capital K, the firm's ask for labor L and the time t where $\frac{K}{k}$ is the return of the capital and Le^{qt} is the amount of working time necessary for producing Y. The growth rate of mean productivity for each worker q is assumed to be constant (see [12, Goodwin]).

Production is shared between wages w and benefits $\pi : Y = wL + \pi$. Benefits $\pi = I$ are fully reinvested and wages $wL = C$ are fully consummated so that at each time $Y = C + I$. Saving $S = rK$ is reinvested and increases capital K: $\frac{dK}{dt} = rK$. Ask for labor N increases with constant rate, following a demographic evolution of malthusian type: $N = N_o e^{nt}$.

Let be $x = \frac{L}{N}$ the employment rate. Phillips curve expresses the dependance of wages on the employment rate: $\frac{w'}{w} = \beta x + (\alpha - \beta)$, $\beta > 0$ and $y = \frac{wL}{Y} = \frac{C}{Y}$ denotes the part of the production intended for wages.

Assuming $\pi = I = S$, we have $\frac{rK}{Y} = 1 - y$ and $r = \frac{Y}{K}(1-y) \ge \frac{1}{k}(1-y)$.

Let us introduce an anticipative inflation/deflation control $\rho_a \in [\rho_{min}; \rho_{max}]$ allowing Public Authority to intervene. The Goodwin model reads:

$$x' = x[\tfrac{1}{k}(1-y) - q - n]$$
$$y' \in y[\beta(x-1) - q] + \alpha[\rho_{min}; \rho_{max}]$$

Constraint $x \in [0,8; 1,2]$ expresses reasonable unemployment or call to foreign manpower and $y \in [0; 0,5]$ limits the part of the production activity devoted to wages.

An evolution may leave the constraint set and remain out for some lapse of time. One want to define an alert process to help in choosing a good policy in order to recover "as well as possible" acceptable economical situation, to "prevent" or to "minimize" future crisis. An answer to this problem is given thanks to the computation of the Minimal Crisis Time function.

But this is not necessarily the best measure crisis as shown on figure 4.5. We can observe that the minimal time is reached for a trajectory which crosses a very high wages level area.

Also, introducing Alert levels in the frame of hybrid dynamical system, we can extend the Minimal Crisis Time approach and introduce qualitative criteria.

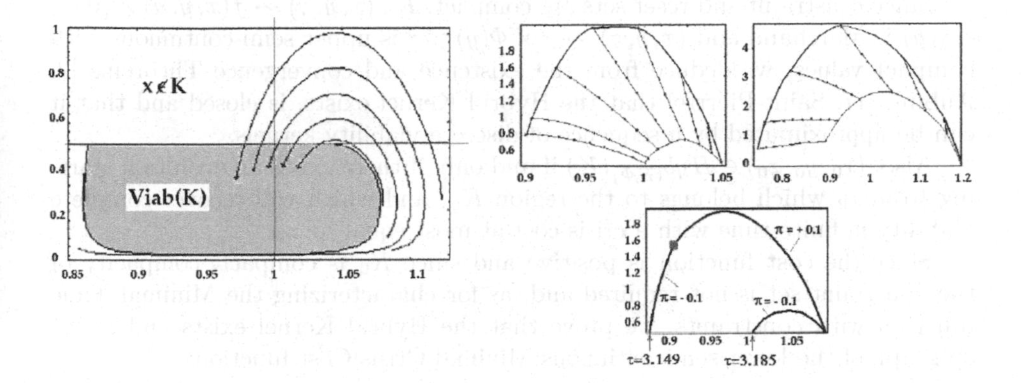

Fig. 1. Crisis Trajectories and isochronal curves. The Minimal time trajectory is obtained by taking $\pi = 0.1$ in a first period and $\pi = -0.1$ in a second period.

It consists in minimizing a crisis cost function in order to take into account as well crisis time as seriousness degree of crisis. It also allows to apply specific procedures depending on the position of the state x in or out of K_s.

4.6 Hybrid System Approach and Alert Management

Let us consider a finite partition (alert levels) $\{K_i\}_{i=0,\ldots,p}$ of X with $K_0 = K^s$ and $K_p = \overline{X \backslash K^h}$ and an index variable $y \in I = \{0, \ldots, p\}$, a cost variable $z \in R^+$ and a transition set-valued map $\Phi : I \rightsquigarrow I$ defined on $\cup_{i \in I} \partial K_i$ by $\Phi(i) = \{j \in I \mid \partial K_j \cap \partial K_i \neq \emptyset\}$.

Consider the hybrid system:

$$\begin{array}{c}\text{Continuous Dynamic:} \\ X' \in F(X)\end{array} \begin{cases} x' = f(x, y, u) \\ y' = 0 \\ z' = -\gamma(y) \end{cases} ; \quad \begin{array}{c}\text{Discrete Dynamic:} \\ X^{n+1} \in \Phi(X^n)\end{array} \begin{cases} x^{n+1} = x^n \\ y^{n+1} \in \phi(y^n) \\ z^{n+1} = z^n \end{cases}$$

where the constraint set $K := \{(x, y, z) \in X \times I \times R,\ x \in K_y,\ z \geq 0\}$ and the reset set $C := \{(x, y, z) \in X \times I \times R,\ x \in \cup_{i \in I} \partial K_i\}$.

Equation $y' = 0$ means that index y does not change whenever the trajectory does not reach ∂K_y.

The function $\gamma(y)$ represents a "cost" (or penalties, weight,...) generated by the membership of x to the alert level K_y with $\gamma(0) = 0$ and $\gamma(n) = +\infty$.

The variable $z(t)$ measures the total cost of the lapse from constraint K_0 from time 0 to t associated with the partition and the cost function $\gamma(\cdot)$.

Conclusion - Thanks to the extension of viability techniques to hybrid calculus we can evaluate for any initial position how serious will be a present or future crisis taking into account the predetermined scale of seriousness and approximate the optimal control which corresponds to the Minimal Cost trajectory.

Since constraint and reset sets are compact, $F : (x, y, z) \rightsquigarrow f(x, y, u) \times \{0\} \times -\gamma(y)$ is Marchaud and $(x, y, z) \rightsquigarrow x \times \Phi(y) \times z$ is upper semi-continuous with compact values, we deduce from the existence and convergence Theorems [2, Aubin], [17, Saint-Pierrre] that the Hybrid Kernel exists, is closed and that it can be approximated by a sequence of discrete viability kernels.

Also, $(x_0, y_0, z_0) \in Hyb_{(F,\Phi)}(K)$ if and only if there exists an evolution starting from x_0 which belongs to the region K_{y_0} and which will recover complete viability in finite time with a crisis cost at most equal to z_0.

Since the cost function is positive and since K_0 is compact, compacity of the constraint set is not required and, as for characterizing the Minimal Time function with constraints, we prove that the Hybrid Kernel exists and is the epigraph of the lower semicontinuous Minimal Crisis Cost function:

$$\Gamma(x) := \inf\{z \text{ such that } (x, y, z) \in Hyb_{F,\Phi}(K),\ y \in i(x)\}.$$

Remark - The minimal Crisis Time function is a particular case where $p = 2$, $K_0 = K^s$, $K_1 = \overline{K^h \backslash K^s}$, $K_2 = \overline{X \backslash K^h}$, $\gamma(0) = 0$, $\gamma(1) = 1$ and $\gamma(2) = +\infty$.

References

1. Aubin, J.-P.: Viability Theory. Birkhäuser, Boston, Basel, Berlin (1991)
2. Aubin J.-P.: Impulse Differential Inclusions and Hybrid Systems: A Viability Approach. Lecture Notes, University of California at Berkeley (1999)
3. Aubin J.-P., Frankowska, H.: Set-Valued Analysis. (1990)
4. Bayen, A., Crück, E., Tomlin, C.: Guaranteed Overapproximations of Unsafe Sets for Continuous and Hybrid Systems: Solving the Hamilton-Jacobi Equation Using Viability Techniques. LNCS, Vol. 2289, Springer (2002)
5. Bensoussan, A., Menaldi: Hybrid Control and Dynamic Programming, Dynamics of Continuous. Discrete and Impulse Syst., 3 (1997) 395–442.
6. Cardaliaguet P., Quincampoix, M., Saint-Pierre, P.: Set-valued Numerical Methods for Optimal Control and Differential Games. In: Stochastic and Differential Games. Theory and Numerical Methods, Annals of the Intern. Soc. of Dynamical Games, Birkhäuser (1999) 177–247.
7. Crück, E.: Problèmes de cible sous contrainte d'état pour des systèmes non linéaires avec sauts d'état. C.R.A.S. Paris, t.333 Série I (2001) 403–408.
8. Crück, E.: Thèse en préparation. (2002)
9. Dordan, O.: Analyse qualitative, Masson (1995)

10. Doyen, L., Saint-Pierre, P.: Scale of Viability and Minimal Time of Crisis, Set-Valued Analysis, 5 (1997) 227–246.
11. Frankowska, H.: Lower Semicontinuous Solutions to Hamilton-Jacobi-Bellman Equations. Proc. of 30th CDC Conference, IEEE, Brighton (1991)
12. Goodwin, R.M.: A Growth Cycle. In: Feinstein (ed.) Socialism, Capitalism and Economic Growth. (1967)
13. Leitmann, G.: Guaranteed Asymptotic Stability for a Class of Uncertain Linear Dynamical Systems. J. of Optim. Theory & Appl., 27(1) (1979)
14. Quincampoix, M.: Differential Inclusions and Target Problems. SIAM J. Control and Optimization, 30 (1992) 324–335.
15. Quincampoix, M., Veliov V.: Viability with a Target: Theory and Applications. Applic. of math. in engineering, Heron Press (1998) 47–54.
16. Saint-Pierre, P.: Approximation of the Viability Kernel. Applied Mathematics & Optimization, 29 (1994) 187–209.
17. Saint-Pierre, P.: Hybrid Kernels and Capture Basins for Impulse Constrained Systems. Proceedings of Hybrid Systems: Computation and Control, Lecture Notes in Computer Science, 2289, Springer (2002)
18. Seube, N.: Robust Stabilization of Uncertain Systems. Journal of Math. Analysis and Applications, (1995) 452–466.
19. Sastry, S.: Non Linear Systems. Analysis, Stability and Control. Springer-Verlag (1999)

Part V

Numerical Methods for Sensor Data Processing

Digital Quadrature Demodulation of LFM Signals Obtained by Lowpass Filtering*

Vera Behar[1] and Christo Kabakchiev[2]

[1] Dep. of Biomedical Engineering, Institute of Technology-Technion,
32000 Haifa, Israel
vera@biomed.technion.ac.il

[2] Institute of Information Technologies, Bulgarian Academy of Sciences,
Acad. G. Bonchev Str., Bl. 2, 1113 Sofia, Bulgaria
ckabakchiev@iit.bas.bg

Abstract. An effective digital demodulation system based on the technique of Ho, Chan and Inko is designed for quadrature demodulation of signals with a linear frequency modulation (LFM). The influence of LFM-signal parameters on the quality of demodulation is studied. Two variants of a sampling frequency are chosen, and a set of lowpass filters for Q-component generation is designed for each variant of a sampling frequency. The quality of demodulation is estimated in terms of absolute amplitude and phase errors evaluated as a function of system parameters. The results obtained in simulations show that the optimal choice of a sampling frequency (48 MHz) enables to reduce the imbalance phase errors to minimum (0.0012°) if the optimal equiripple lowpass filter with only 6 elements is chosen for Q-component generation.

1 Introduction

There are a number of ways to obtain quadrature components of LFM signals. The conventional method is to multiply the original signal with reference signals having the same frequency but shifted by 90° in phase relative to one another. Each of two baseband signals that remain after low-pass filtering and referred as the in-phase (I) and quadrature (Q) signals is sampled at rate Fs in an analog-to-digital (A/D) converter. Phase errors due to imperfect matching of the I- and Q-channels can be of 2°-3° that limits the performance achievable from signal processors. The general approach to solve the imbalance problem is to directly sample the input signal before mixing with reference signals and producing the I- and Q-samples through digital signal processing. This approach requires only one A/D converter instead of two, but in that case the sampling frequency is determined by the highest frequency component of the input signal. Presently there are available a number of techniques that allow directly sampling at rate determined by the signal bandwidth, not by the highest frequency component in

* Research supported under IIT-BAS Grant No 010044, MPS Ltd. Grant, BNF"SR" Grant No I-902/99.

I. Dimov et al. (Eds.): NMA 2002, LNCS 2542, pp. 267–273, 2003.

the input signal. According to these methods, the sampling is properly chosen in relation to the center frequency and bandwidth [1,2].

The paper presents a set of variants of a demodulation system based on the technique of Ho, Chan and Inko described in [3]. The signal processing is intended for quadrature demodulation of signals with a linearly modulated frequency (LFM-signals). A set of lowpass filters is designed to generate the Q-component samples of the LFM signal with concrete parameters – carrier frequency, bandwidth, and duration. Three approaches are exploited in the low-pass filter design – "ideal" lowpass filtering with truncated length of the filter impulse response, weighted lowpass filtering by using well-known window functions and, finally, optimal equiripple filtering with an frequency response approximated with Chebyshev polynoms according to the Parks-McClellan algorithm and Chebyshev approximation theory [4]. The influence of basic demodulation system parameters (sampling frequency, filter type, filter and decimation rate) on the quality of quadrature demodulation is studied. The quality of quadrature demodulation is estimated in terms of absolute amplitude and phase errors of signals computed at the output of the system as a function of system parameters. The numerical results obtained in simulations show that properly chosen sampling frequency greater than $2\Delta f$ and properly designed lowpass filter allow achieving high quality of demodulation of LFM signals even if the length of a lowpass filter is small (4 or 6). Examples of tasks, which require processing of radar signals in such a form, are presented in [5,6].

2 Demodulation System

Consider a continuous radio signal with a linearly modulated frequency:

$$s(t) = S_0(t)\cos[2\pi(f_0 - \Delta f/2)t + bt^2] \tag{1}$$

where f_0 is the carrier frequency, Δf is the bandwidth, b is the frequency slope and $S_0(t)$ is the envelope function non-zero only over the duration of the signal ($S_0(t) = 1$ for $0 \le t \le T$, 0 otherwise). The frequency slope, b, is chosen such that $b \cdot T = \Delta f$. The principle scheme of a demodulation system based on the technique of Ho, Chan and Inko is shown in Fig.1. The demodulation system involves only one A/D converter to digitize the input LFM signal $s(t)$ at rate Fs. According to [3] the sampling frequency is determined as:

$$F_S = 4f_0/(4M+1) \tag{2}$$

where M is any positive integer.

The samples of the LFM signal $s(t)$ at time instant $t_k = kT_s$, are:

$$\begin{aligned} s[kT_S] &= \cos(2\pi f_0 kT_S + bk^2T_S^2 - \pi\Delta f kT_S) = \\ &= I[kT_S]\cos(2\pi f_0 kT_S) - Q[kT_S]\sin(2\pi f_0 kT_S) \end{aligned} \tag{3}$$

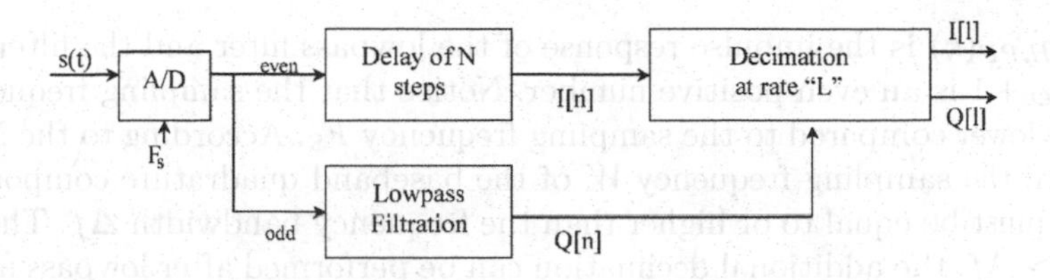

Fig. 1. Quadrature demodulation system

After substituting f_0 with (2) the expression (3) takes the form:

$$s[mT_S] = \begin{cases} (-1)^{\frac{i}{2}} I[mT_S], \ m = 2k \\ (-1)^{\frac{i-1}{2}} Q[mT_S], \ m = 2k+1 \end{cases}, \quad (4)$$
$$i = (2M-1)m \text{ and } k = 0, 1, 2, \ldots, N_i$$

It can be seen that the I-component is derived from samples of the input LFM signal taken at even time instants. The corresponding Q-component is derived from samples of the input LFM signal taken at odd time instants. Let us assume that the I-component is generated through decimating the input LFM signal at rate of 2:

$$I[nT_W] = (-1)^{\frac{-(4M+1)i}{2}} s[iT_S], \ i = 2(n+1) - 2 \text{ and } n = 0, 1, 2, \ldots N_W \quad (5)$$

The sampling period of the I-component is $T_W = 2T_s$ that corresponds to the sampling frequency $W = F_s/2$. The number of samples of the I-component is $N_W = T_i W$. Further processing should be made to produce the Q-component at the same time instants. Multiplying both sides of the equation (3) by $-2\sin(k\pi/2)$ gives the signal $s_1[kT_s]$:

$$\begin{aligned} s_1[kT_S] &= \sin(k\pi + bk^2T_S^2 + \pi\Delta f kT_S) + \sin(bk^2T_S^2 + \pi\Delta f kT_S) = \\ &= s_2[kT_S] + s_3[kT_s] \end{aligned} \quad (6)$$

where $k = 0, 1, 2, \ldots, N_i$. It can be seen in (4), that the spectrum of $s_2[kT_s]$ is centered at a frequency $F_s/2$, and the spectrum of $s_3[kT_s] = Q[kT_s]$ is centered at a zero frequency. Consequently, the desired samples of the Q-component $Q[kT_s]$ can be produced by low-pass filtering of $s_1[kT_s]$ with a cutoff frequency of $F_s/4$. In other words, the samples of both I- and Q-components of the LFM signal can be obtained as follows:

$$I[nT_W] = (-1)^n s[2nT_S] \text{ and } Q[nT_W] = s_3[nT_W], \text{ where } k = 0, 1, 2, \ldots, N_i \quad (7)$$

From equations (6) and (7) it follows that the estimates of $Q[kT_s]$ can be generated as:

$$\hat{Q}[nT_W] = -\sum_{k=0}^{N_Q} (-1)^{(n-1-k+\frac{N_Q+1}{2}} g_{LPF}(k) s[(2n + N_Q - 2k)T_W], \quad (8)$$
$$n = 0, 1, 2, \ldots, N_W$$

where $g_{LPF}(k)$ is the impulse response of the lowpass filter and the filter length $N = N_Q + 1$ is an even positive number. Notice that the sampling frequency W is twice lower compared to the sampling frequency F_S. According to the Nyquist theorem, the sampling frequency W of the baseband quadrature components, I and Q, must be equal to or higher than the frequency bandwidth Δf. Therefore, if $W >> \Delta f$, the additional decimation can be performed after lowpass filtering. In that case, the decimation rate Lcan be found as $L = F_s/2\Delta f$. Decimation of a signal at rate L gives:

$$I[lT_{\Delta f}] = I[LlT_W] \text{ and } Q[lT_{\Delta f}] = Q[LlT_W], \text{ where } l = 0, 1, 2, \ldots, N_{\Delta f} \quad (9)$$

3 Lowpass Filter Description

We consider three approaches for lowpass filter design, the ideal lowpass filter with truncated length, the windowed filter, and the optimal equiripple filter, to generate the Q-samples of the LFM signal.

The Unwindowed FIR Filter is an "ideal" lowpass filter with limited length. The "ideal" filter impulse response is determined by an inverse Fourier transform of its transfer function. The filter coefficients of the unwindowed lowpass FIR filter,$g_{as}(k)$, are obtained by truncation of the impulse response of the ideal lowpass filter. In real application, however, the filter length is limited by computational constrains and it is very important to design a lowpass filter with small number of coefficients as possible.

The Windowed FIR Filter is a weighted "ideal" lowpass filter. The filter coefficients $g_W(k)$ are obtained as a product of the unwindowed filter coefficients, $g_{as}(k)$, and any window function $W(k)$. The nature of the window function $W(k)$determines the quality of the filter (main lobe width, side lobe level). In the lowpass filter design, we have used a set of well-know window functions.

The Optimal Equiripple FIR Filter is a lowpass filter whose frequency impulse is approximated with Chebyshev polynoms and has an equiripple behavior in the frequency domain. The most popular Parks-McClellan algorithm is used to design the optimal equiripple lowpass filters [4]. The lowpass filters designed are optimal in the sense that they minimize the maximal error between the desired frequency response and the actual frequency response. In the filter design are used the parameters: cutoff frequency – F_1; stopband frequency – F_2; passband ripple level – δ_1; stopband ripple level – δ_2. The ripple ratio $k = \delta_1/\delta_2$ is assumed to be 1. Both quadrature components of the LFM-signal are sampled at rate $W = F_s/2$, because the filter performs only odd samples of the input signal.

4 Simulation Results

The influence of such demodulation system parameters – sampling frequency of the input LFM signal, filter type and filter length – on the quality of quadrature demodulation can be estimated by calculating the phase and amplitude errors of the demodulated signal. The calculated errors are: absolute Q-component error; absolute envelope error; absolute phase error. In order to design the concrete demodulation system we should know the parameters of the LFM signal. With this aim, we consider the LFM signal whose carrier frequency f_0) is 60 MHz, bandwidth Δf) is 4.8 MHz, and pulse duration (T_i) is 10μs. According to the technique of Ho, Chan and Inko the sampling frequency F_S can be found from (2). The possible values of the sampling frequency F_S calculated for given parameters of the LFM signal are presented in Table 1.

Table 1. Possible values of sampling frequency

M	1	2	3	4	5	6
F_S [MHz]	48	240/9	240/13	240/17	240/21	9.6

Further, for the sake of clearness we consider only both borderline variants (**M=1** and **M=6**) corresponding to the maximal and minimal sampling rate. They are:

Variant 1: The sampling rate is $F_S = 48$ MHz and the rate of decimation is $L = 5$.

Variant 2: The sampling rate is $F_S = 9.6$ MHz and no decimation is required.

In Table 2 are presented the absolute errors of the demodulated LFM signal, which are calculated for the case when the Q-component is generated by filtering in an unwindowed or windowed filter.

The windowed lowpass filters are designed for both variants of a demodulation system. Several well-known window functions are used in the filter design - Hamming, Hanning, Chebyshev, Kaiser, Blackman and Bartlett. For comparison, the unwindowed lowpass filters are also designed. The errors are calculated for both variants of sampling, 48 MHz and 9.6 MHz and for three variants of a filter length (N =6,10,20). In order to decrease the errors due to transient processes in the lowpass filter, the absolute errors are estimated over a time interval shorter than LFM signal duration, i.e. without several initial and several final samples of the demodulated signal. The computation of the signal errors is realized by software in MATLAB.

The numerical results presented in Table 2 show that the sampling frequency of 9.6 MHz is not acceptable for implementation of the demodulation system because the absolute phase errors are very high even if the filter length is of 20. In that case, the phase errors are comparable to these of the analog implementation of the demodulation system (2°). On the contrary, the sample frequency of 48

Table 2. Signal errors at the output of a demodulation system

Window	Errors	Sampling frequency 48 MHz			Sampling frequency 9.6 MHz		
		Lowpass filter length			Lowpass filter length		
		6	10	20	6	10	20
Unwindowed	Δe	0.1	0.06	0.03	0.18	0.08	0.048
	$\Delta\theta°$	2.5	1.75	1	9	5.5	2.75
	ΔQ	0.1	0.06	0.03	0.2	0.13	0.06
Hamming	Δe	0.013	0.0055	0.0025	0.45	0.2	0.009
	$\Delta\theta°$	0.35	0.125	0.08	20	6.5	0.8
	ΔQ	0.014	0.0055	0.0025	0.5	0.2	0.01
Hanning	Δe	0.0021	0.0003	$7{\cdot}10^{-6}$	0.33	0.14	0.015
	$\Delta\theta°$	0.06	0.01	0.002	15	5.5	0.9
	ΔQ	0.0018	0.0003	$8{\cdot}10^{-6}$	0.35	0.15	0.02
Chebyshev	Δe	0.062	0.032	0.01	0.4	0.225	0.026
	$\Delta\theta°$	1.8	0.9	0.28	27	9.5	4
	ΔQ	0.06	0.03	0.01	0.4	0.23	0.07
Kaiser	Δe	0.027	$2.5{\cdot}10^{-6}$	$6{\cdot}10^{-6}$	0.08	0.065	0.035
	$\Delta\theta°$	0.8	0.001	0.0003	7.5	4	2
	ΔQ	0.03	$3{\cdot}10^{-6}$	$8{\cdot}10^{-6}$	0.15	0.075	0.042
Blackman	Δe	0.014	$1.1{\cdot}10^{-4}$	$3.5{\cdot}10^{-6}$	0.55	0.3	0.006
	$\Delta\theta°$	0.42	0.003	0.0015	27.5	9	0.6
	ΔQ	0.015	$1.2{\cdot}10^{-4}$	$4{\cdot}10^{-6}$	0.6	0.3	0.012
Bartlett	Δe	0.15	0.08	0.035	0.5	0.27	0.055
	$\Delta\theta°$	4.5	2.2	1	25	9	1.7
	ΔQ	0.15	0.08	0.038	0.5	0.3	0.06

MHz is desirable because the absolute phase errors can be reduced to minimal if the properly lowpass filter is implemented in the demodulation system.

Table 3 gives the results for the case when the Parks-McClellan algorithm is used for optimal equiripple lowpass filter design. The optimal filters designed have the parameters: cutoff frequency – $F_1 = 2.4$ MHz; stopband frequency – $F_1 = 11.88$ MHz; filter length – $N = 4$ and 6.

Table 3. Optimal filter coefficients and signal errors (Fs=48 MHz)

N	$g_{opt}(1)$	$g_{opt}(2)$	$g_{opt}(3)$	$g_{opt}(4)$	$g_{opt}(5)$	$g_{opt}(4)$	Δe	$\Delta\theta°$	ΔQ
4	–0.067	0.567	0.567	–0.067	–	–	$3{\times}10^{-4}$	0.013	$4.5{\times}10^{-4}$
6	0.013	–0.102	0.589	0.589	–0.102	0.013	$8{\times}10^{-7}$	0.001	$2{\times}10^{-6}$

The filter coefficients and signal errors presented in Table 3 are calculated only for the case when the sampling frequency of the LFM signal equals to 48 MHz (**Variant 1**). If the sampling frequency is chosen to be of 9.6 MHz (**Variant 2**), it is not possible to design the optimal lowpass filter with the Parks-McClellan algorithm because the transition frequency band of the lowpass filter is zero. The numerical results given in Table 3 show that the use of the optimal lowpass filter designed with the Park-McClellan algorithm is the most effective even if the filter length is very small (only 4 elements).

5 Conclusions

A direct A/D demodulation system based on the technique of Ho, Chan and Inko is considered for the case of quadrature demodulation of LFM signals with given parameters. Two variants of a sampling frequency, 48 MHz and 9.6 MHz, are chosen, and a set of lowpass filters is designed for each variant of a sampling frequency. The quality of quadrature demodulation is estimated by calculating of absolute amplitude and phase errors as a function of system parameters. The analysis of results shows that the optimal choice of a sampling frequency (48 MHz) enables to reduce the imbalance phase errors to minimum (0.0012°) if the optimal equiripple filter with only 4 or 6 elements is chosen for Q-component generation.

References

1. Rice, D. W., Wu, K. H.: Quadrature Sampling with High Dynamic Range. IEEE Trans., AES-18 (4) (1982) 736–739.
2. Watters, W. M., Jarrett, B. R.: Bandpass Signal Sampling and Coherent Detection. IEEE Trans., AES-18, 4 (1982) 740–749.
3. No, K. C., Chan, Y. I., Inkol, R.: A Digital Quadrature Demodulation System. IEEE Trans., AES-32, 4 (1996) 1218–1226.
4. Rabiner, L. R., Gold, B.: Theory and Application of Digital Signal Processing. Prentice-Hall, New Jersey (1975)
5. Kabakchiev, C., Behar, V.: CFAR Radar Image Detection in Pulse Jamming. IEEE Fourth Int. Symp. ISSSTA'96, Mainz (1996) 182–185.
6. Behar, V., Kabakchiev, C., Doukovska, L.: Adaptive CA CFAR Processor for Radar Target Detection in Pulse Jamming. Journal of VLSI Signal Processing Systems for Signal, Image, and Video Technology, Vol. 26, 3 (2000) 386–396.

An Accelerated IMM JPDA Algorithm for Tracking Multiple Manoeuvring Targets in Clutter*

Ljudmil V. Bojilov, Kiril M. Alexiev, and Pavlina D. Konstantinova

Central Laboratory for Parallel Processing,
Bulgarian Academy of Sciences
Acad. G. Bonchev Str., Bl. 25-A, 1113 Sofia, Bulgaria
bojilov@bas.bg, alexiev@bas.bg, pavlina@bas.bg

Abstract. Theoretically the most powerful approach for tracking multiple targets is known to be Multiple Hypothesis Tracking (MHT) method. The MHT method, however, leads to combinatorial explosion and computational overload. By using an algorithm for finding the K-best assignments, MHT approach can be considerably optimized in terms of computational load. A much simpler alternative of MHT approach can be the Joint Probabilistic Data Association (JPDA) algorithm combined with Interacting Multiple Models (IMM) approach. Even though it is much simpler, this approach can overwhelm computations as well. To overcome this drawback an algorithm due to Murty and optimized by Miller, Stone and Cox is embedded in IMM-JPDA algorithm for determining a ranked set of K-best hypotheses instead of all feasible hypotheses. The presented algorithm assures continuous maneuver detection and adequate estimation of manoeuvring targets in heavy clutter. This affects in a good target tracking performance with limited computational and memory requirements. The corresponding numerical results are presented.

Keywords. tracking, manoeuvring, cluttered environment, assignment

1 Introduction

The most complicated case in target tracking is undoubtedly to track multiple manoeuvring targets in heavy clutter. Numerous methods and algorithms have been devoted to this problem and for any one of them *pro* and *cons* can be pointed out. For example, the theoretically most powerful approach for tracking multiple manoeuvring targets in clutter is known to be MHT method. This method often leads to combinatorial explosion and computational overload that restricts its implementation. In recent years, numerous papers have been devoted to algorithms, which are capable to compute a ranked set of assignments of measurements to targets. Such algorithms make MHT approach practically implementable for the first time.

* This work was supported by the Center of Exellence BIS21 under Grant ICA1-2000-70016.

I. Dimov et al. (Eds.): NMA 2002, LNCS 2542, pp. 274–282, 2003.

Another less complicated approach, especially for tracking manoeuvring targets is known to be Multiple Models (MM) approach. The most promising algorithm based on this approach is Interacting Multiple Models algorithm. At the price of some suboptimality this algorithm reaches the best implementation in terms of speed and stability. In the presence of clutter, however, the IMM algorithm most often fails. In this case the PDA (and JPDA) approach can be implemented. When tracking multiple closely spaced targets, the JPDA algorithm can be implemented successfully even in the presence of heavy clutter. In the previous paper [1], we have proposed an algorithm unifying features of IMM and JPDA algorithms at the same time. This algorithm proved to be good alternative of MHT approach for clusters containing up to 4 targets. When the number of targets in the cluster exceeds this limit, however, the total number of all feasible hypotheses increases exponentially. In this paper we propose an extension of the algorithm in [1] where instead of enumeration of all feasible hypotheses we use ranked assignment approach to find the first K-best hypotheses only. The value of K has to be such that the weight of scores sum of this K-best hypotheses prevail over the total sum.

This paper is organized as follows. In the next section we expose our motivation for this paper as well as the problem formulation. Here a brief outline of IMM-JPDA algorithm is given and the need of its extension is discussed. In the 3rd section the extended algorithm is described. Here the stress is made over the extension part of the algorithm. In the last, 4th section simulation results are presented. These results reveal that the extended algorithm shows better performance in terms of speed than cited in [1] IMM-JPDA algorithm.

2 Motivation and Problem Formulation

When several closely spaced targets form a cluster, the JPDA algorithm generates all feasible hypotheses and computes their scores. The set of all feasible hypotheses includes such hypotheses as 'null' hypothesis and all its 'derivatives'. Consideration of all feasible assignments is important for optimal calculation of assignment probabilities [6]. If, for example, the score of every one of these hypotheses differs from any of the others by no more than an order, it should not be possible to truncate some significant part of all hypotheses. If, however, prevailing share of the total score is concentrated in a small percent of the total number of all hypotheses, then temptation to consider only this small percent of all hypotheses becomes very high.

In order to investigate this idea a typical example with five closely spaced targets is used with overlapping validation regions and shared measurements. In the first run (1st scenario) 17 measurements are disposed in the targets' gates, and in the second run (2nd scenario) 9 measurements are disposed. At every run, all feasible hypotheses are generated and their scores are computed and summarized. The results are given in Figure 1 and Figure 2. These two examples are chosen out of numerous experiments as typical ones.

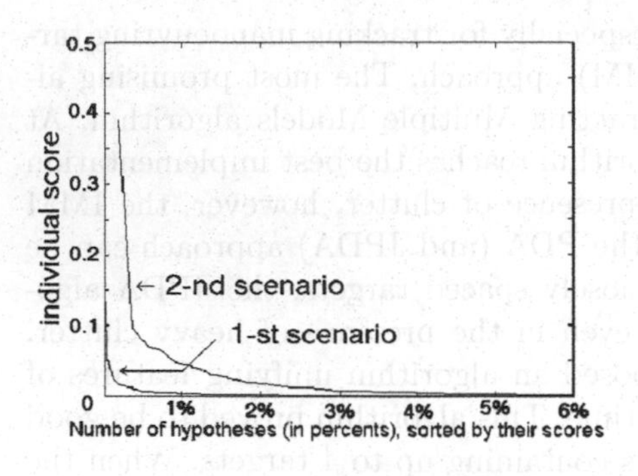

Fig. 1. Hypotheses' score distribution

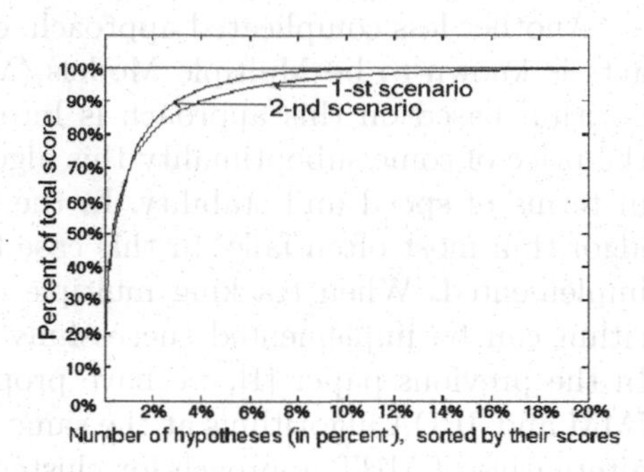

Fig. 2. Cumulative score distribution

Two plots of Figure 1 show sorted individual scores of the feasible hypotheses. Only first six percents of all hypotheses are depicted on this figure both scenarios. It can be seen that the scores of the hypotheses dramatically reduce their values. A more informative presentation of this result is given in Figure 2, where the cumulative score's distributions of the two scenarios are given. This figure confirms our expectations that in most cases of practical importance only small number of all hypotheses concentrates the prevailing part of their total score sum.

Next, a brief description of proposed in [1] algorithm follows. For simplicity and without losing generality two models are assumed.

2.1 IMM-JPDA Algorithm Description

The IMM JPDA algorithm starts with the same step as IMM PDA algorithm [5], but in cycle for every particular target in the cluster.

Step 1. Computation of the mixed initial conditions $\hat{x}_i^{0t}$ for every target i and for the filter, matched to model t:

a) mixed state estimate

$$\hat{x}_i^{0t}(k-1|k-1) = \sum_{s=1}^{2} \hat{x}_i^s(k-1|k-1)\mu_{s|t}^i(k-1|k-1)t = 1,2 \tag{1}$$

Here, it is supposed that mixing probabilities $\mu_{s|t}^i$ are already computed.

b) mixed covariance estimate

$$P_i^{0t}(k-1|k-1) = \sum_{s=1}^{2} \mu_{s|t}^i(k-1|k-1)\{P_i^s(k-1|k-1) + [\hat{x}_i^s - \hat{x}_i^{0t}][\hat{x}_i^s - \hat{x}_i^{0t}]'\} \tag{2}$$

Here P_i^s is covariance update of model s for target i.

Next, some JPDA steps follow.

Step 2. State predictions $\hat{x}_i^{0t}(k|k-1)$ and covariance predictions $P_i^{0t}(k|k-1)$ for the next scan k for every target and for every model are calculated.

Step 3. In this step, after receiving the set of measurements at scan k, a clustering is performed. Further on, it is assumed that the algorithm will proceed with every particular cluster.

At this point, in the traditional JPDA algorithm, hypotheses generation have to be performed. However, to avoid combinatorial explosion we include here our innovation.

Step 4. Calculating 'predicted model probabilities':

$$\mu_i^t(k|k-1) = \sum_{s=1}^{2} p_{st}\mu_i^t(k-1) \tag{3}$$

where $\mu_i^t(k-1)$ is model probability and p_{st} are Markovian switching probabilities. Now, the individual model state predictions are merged for every particular target:

$$\hat{x}_i^0(k|k-1) = \sum_{i=1}^{2} \hat{x}_i^{0t}(k|k-1)\mu_i^t(k|k-1) \tag{4}$$

The algorithm continues with hypotheses generation and next computation of hypotheses scores and association probabilities. Our description terminates with the last two JPDA steps.

Step 5. After association probabilities computation, the JPDA algorithm continues as a PDA algorithm for every individual target. For every target the 'merged' combined innovation is computed

$$\nu_i(k) = \sum_{j=1}^{m_i(k)} p_{ij}\nu_{ij}(k) \tag{5}$$

Step 6. At this step, our algorithm returns to the multiple model case by splitting 'merged' combined innovation from the previous equation. For every individual target and for every particular model the combined innovations are computed:

$$\nu_i^t(k) = \nu_i(k) + H_i(k)\hat{x}_i^0(k|k-1) - H_i^t(k)\hat{x}_i^{0t}(k|k-1) \tag{6}$$

The next few steps of this algorithm fully coincide with the well-known IMM PDA algorithm [5] and will be omitted.

3 Accelerating Extension to IMM PDA Algorithm

Our extension to IMM JPDA algorithm is directed to the most time consuming part of the algorithm - hypotheses generation and their scores computation. If we take as a simple example a cluster with 4 targets and 10 measurements distributed in their validation regions (Table 1), the total number of all feasible hypotheses for this example approaches 400. When, however, the number of targets in the cluster exceeds 5 or 6 and there are more than 15 measurements in their gates, the number of all hypotheses to be generated reaches thousands. To avoid this overwhelming of computations we propose the next trade-off: to take into consideration only little part of all feasible hypotheses with the highest scores and concentrating the prevailing share of the total score sum.

In order to find out the first K-best hypotheses we use an algorithm due to Murty [2] and optimized by Miller et al. [3]. This algorithm gives a set of assignments to the assignment problem, ranked in order of increasing cost. As a first step in solving this problem we have to define the cost matrix of the assignment problem. It can be seen that the score of any particular hypothesis (equation (3) from [1])

$$P'(H_i) = \beta^{[N_M-(N_T-N_{nD})]}(1-P_D)^{N_{nD}} P_D^{(N_r-N_{nD})} g_{ij} \dots g_{mn} \tag{7}$$

is an expression of multipliers. The score of every feasible hypothesis (i.e. the probability of being true) can be calculated using a table similar to Table 1. In this table, at the place of indices in the boxes of the Table 1 we put multipliers equal to probability to assign the given measurement to the corresponding target (Table 2).

Table 1. Indices of the measurements falling in the gates of corresponding targets

T1	T2	T3	T4
0	0	0	0
4	6	3	1
8	7	4	2
9	8	5	3
		6	4
		9	

Table 2. Multipliers of the corresponding measurements

T1	T2	T3	T4
$\beta(1-P_D)$	$\beta(1-P_D)$	$\beta(1-P_D)$	$\beta(1-P_D)$
$g_{14}P_D$	$g_{26}P_D$	$g_{33}P_D$	$g_{41}P_D$
$g_{18}P_D$	$g_{27}P_D$	$g_{34}P_D$	$g_{42}P_D$
$g_{19}P_D$	$g_{28}P_D$	$g_{35}P_D$	$g_{43}P_D$
		$g_{36}P_D$	$g_{44}P_D$
		$g_{39}P_D$	

Now, combining indices from Table 1 in an admissible manner, and so, generating every one of feasible hypotheses we can at the same time multiply corresponding elements from Table 2, obtaining the score of the so generated hypothesis (equation (7)).

On the other side, every solution of the assignment problem represents a sum of elements of the cost matrix. We have to define this cost matrix in such way, that the value of every possible solution of the assignment can be potentially a score of some feasible hypothesis. Let us take logarithm from both sides of (7). From the left-hand side we obtain logarithm of hypothesis probability, and, from the right-hand side, a sum of logarithms of elements from Table 2. The correspondence between multipliers in equation (7) and sum of their logarithms gives a hint of how to construct the cost matrix and to solve the problem mentioned above.

The considered cost matrix contains negative logarithms of the elements of Table 2. If we find the optimal solution (in this particular case - the minimal) of assignment problem with this cost matrix it will coincide with the hypothesis with highest probability, i.e., both the optimal solution and the highest probability hypothesis will connect the targets with the same measurements. The cost matrix of a cluster from Table 1 appears in Table 3 where $ln^0 = -ln[(1-P_D)\beta]$, $ln_{ij} = -ln[g_{ij}P_D]$.

Table 3. The cost matrix of the example

	f1	**f2**	**f3**	**f4**	**z1**	**z2**	**z3**	**z4**	**z5**	**z6**	**z7**	**z8**	**z9**
T1	ln^0	×	×	×	×	×	×	ln_{14}	×	×	×	ln_{18}	ln_{19}
T2	×	ln^0	×	×	×	×	×	×	×	ln_{26}	ln_{27}	ln_{28}	×
T3	×	×	ln^0	×	×	×	ln_{33}	ln_{34}	ln_{35}	ln_{36}	×	×	ln_{39}
T4	×	×	×	ln^0	ln_{41}	ln_{42}	ln_{43}	ln_{44}	×	×	×	×	×

The symbol × represents arbitrary chosen quantity greater than the greatest element out of the set of elements denoted with ln . In order to use the algorithm in [3] for finding the K-best hypotheses, the matrix from Table 3 have to be added up to square matrix filling in the remaining rows with the same value ×. After finding the first K-best solution (i.e. generation of the first K-best hypotheses) the algorithm continues its work as in [1].

4 Simulation Results

At the beginning we have to choose the number of first K-best hypotheses to be generated. The value of K has to be, in some sense, optimal so that: a) to be small enough to ensure accelerated algorithm, and, in the same time, b) do not be so small that to give rise to distortion in computing assignment probabilities.

As it can be seen from Figure 1 the scores of feasible hypotheses decrease very rapidly and some 5-10 percents of them (Figure 2) cover more than 95 percents of the total score sum. However, as we know neither the total number, nor the total sum, we try to derive indirect criterion for determining the value of K. One possible expression can be:

$$H(k) < \alpha \cdot H(l) . \tag{8}$$

In order to tune experimentally the value of α a range of experimental runs have been carried out. Every one run is performed with scenario with the same number of 6 targets and 12 measurements but with different reciprocal (relative) location. Averaging over 1000 runs we find the values 0.01 and 0.005 for α equally appropriate.

We compare the algorithm presented in this paper with the same algorithm but without acceleration discussed in previous section (algorithm from [1]) These presented algorithms were tested extensively on variety of scenarios involving different number of manoeuvring and closely spaced targets and in presence of heavy clutter. We construct a set of scenarios with 2,3,4 and 5 targets in a cluster and in presence of moderate and heavy clutter. The used scenarios are similar to scenarios from [1] where we look for the limit of IMM JPDA algorithm in terms of number of target in the cluster.

As in [1] we start experiments with 3-targets scenario. For every scenario we include two level of clutter: moderate and heavy. The clutter is modeled with a Poison distribution and for the two levels we choose the values of Poison parameter to be $\beta V = 1$ and $\beta V = 2$.

A. Scenario with 3 closely spaced targets.

Table 4. Time per cluster in seconds for 3-targets scenario

	All hypotheses computation		First k-best hypotheses only	
Targets in a cluster	2 targets	3 targets	2 targets	3 targets
$\beta V = 1$	0.016	0.062	0.02	0.26
$\beta V = 2$	0.011	0.136	0.09	0.68

Comparison with the algorithm where all feasible hypotheses are computed gives unexpected results. Obviously the program frame for finding out first K-best hypotheses is heavy and unsuitable for simple cases. Even so, the both approaches give results far below the real time implementation threshold

B. Scenario with 4 closely spaced targets.

Table 5. Time per cluster in seconds for 4-targets scenario

	All hypotheses computation		First k-best hypotheses only	
Targets in a cluster	3 targets	4 targets	3 targets	4 targets
$\beta V = 1$	0.03	3.94	0.79	3.39
$\beta V = 2$	0.22	124.7	3.42	9.86

It can be seen in this case (Figure 3) that when scenario becomes denser the results become comparable (especially for clusters with 4-targets) and for the most heavy case ($\beta V = 2$) the processing time for the first algorithm increases almost exponentially (Table 5). In the same time, the processing time for the new algorithm increases polynomially.

C. Scenario with 5 closely spaced targets.

For this scenario (Figure 4) only the proposed algorithm has been tested. For the most dense case, when five closely spaced targets have to be tracked in

Table 6. Time per cluster in seconds for 5-targets scenario

	First K-best hypotheses computation		
Targets in a cluster	3 targets	4 targets	5 targets
$\beta V = 1$	0.35	1.16	8.2
$\beta V = 2$	1.58	6.36	15.4

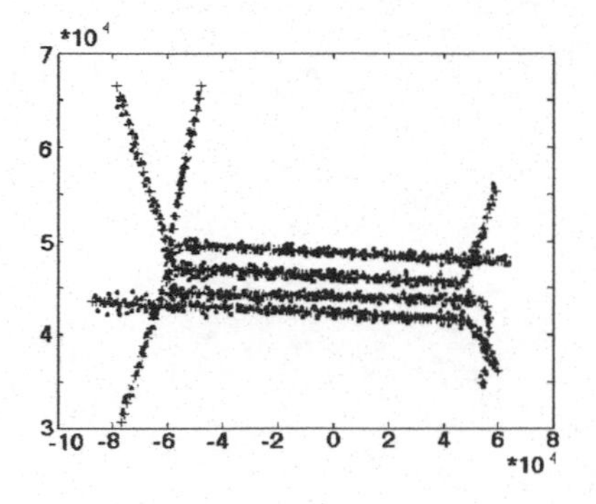

Fig. 3. Four-target scenario with $\beta\, V = 2$

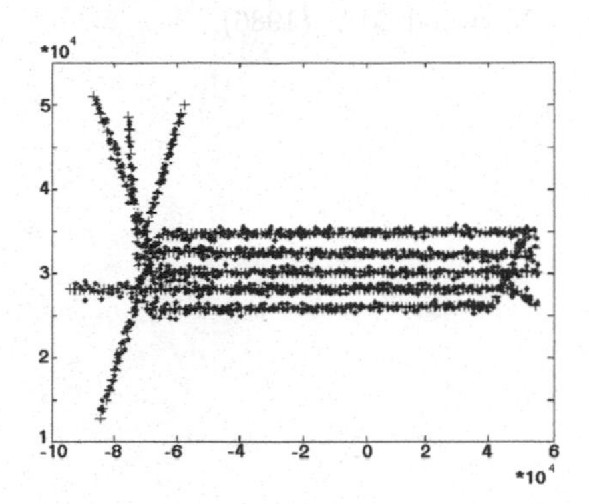

Fig. 4. Five-target scenario with $\beta\, V = 2$

heavy clutter we compute average time per scan t=8.7 sec. But as it can be seen from the Table 6, when in a given scan all five targets fall into the cluster the processing time becomes twice the average time. It can be stated that this case is the limit of algorithm implementation.

5 Conclusions

In this paper a new algorithm is presented for tracking closely spaced targets in moderate and heavy clutter. The algorithm is improved version of an algorithm previously presented by the authors. In the new algorithm instead of all feasible hypotheses only part of them are generated. By means of an algorithm for finding the first K-best solutions of the assignment problem we generate the first K-best feasible hypotheses in terms of their probability of being true. This trade-off do not lead to observable assignment probability degradation and in the same time definitely speedup the algorithm processing.

References

1. Bojilov, L., Alexiev, K., Konstantinova, P.,: An Algorithm Unifying IMM and JPDA Approaches. Comptes Randue de l'Academie Bulgare des Sciences (to be published).
2. Murty, K.,: An Algorithm for Ranking All the Assignment in Order of Increasing Cost. Operations Research. **16** (1968) 682–687

3. Miller, M., Stone, H., Cox, I.: Optimizing Murty's Ranked Assignment Method. IEEE Transactions on AES. **33**(1997) 851–862
4. Jonker, R., Volgenant, A.: A shortest Augmenting Path Algorithm for Dens and Sparse Assignment Problems. Computing. **38** (1987) 325–340
5. Yaakov Bar-Shalom: Multitarget-Multisensor Tracking: Advanced Applications. Artech House. Norwood. MA. (1990)
6. Blackman, S.: Multiple-Target Tracking with Radar Applications. Artech House. Norwood. MA. (1986)

Tracking Algorithms Represented as Classes*

Emanuil Djerassi and Pavlina Konstantinova

Central Laboratory for Parallel Processing,
Bulgarian Academy of Sciences
Acad. G. Bonchev Str., Bl. 25-A, 1113 Sofia, Bulgaria
djerassi@bol.bg, pavlina@bas.bg

Abstract. In this paper we consider some possibilities for applying object oriented approach to developing Target Tracking (TT) programs. We examine one of the main parts of TT algorithms — track evaluation — and propose a structure of classes that may simplify and alleviate creating and testing TT programs. These classes implement tracks, and consist of data and methods describing track kinematics - vectors, matrices and filtering algorithms. The track hierarchy contains classes for Linear Kalman Filter, Extended Kalman Filter and Probabilistic Data Association Filter. An example shows how Interacting Multiple Model (IMM) filter can be implemented using objects of these classes.

1 Introduction

There are many algorithms for target tracking (TT) which differ in complexity, but they all use two main types of data processing: data association and track evaluation. The typical tracking process consists of repetitive sets of measurements (e. g. scans) from some sensors (e. g. radars), which produce measurement data (usually called observations), originating from unknown quantity of real objects (targets) or clutter (false alarms). Then the observations are assigned (associated) to some assumptive set of trajectories (tracks) and some probability measures of these tracks are being evaluated [6,1]. The aim is to maintain a set of tracks, matching as close as possible the real set of targets. It is evident that tracks are principal elements in all TT algorithms. During the tracking process they may appear and disappear, changing their state and properties. That's why it seems natural to define them in terms of object oriented programming as classes, and use them as objects.

2 Problem Formulation

In general, a track is a set of measurements from same target at different times. However, in most tracking algorithms, each the track is approximated at each time step by a difference equation in the form [3]:

$$x(k+1) = F(k)x(k) + G(k)u(k) \tag{1}$$

* The research reported in this paper is supported by Center of Excellence BIS21 grant ICA1-2000-70016.

I. Dimov et al. (Eds.): NMA 2002, LNCS 2542, pp. 283–290, 2003.

where $x(k)$ is the n-dimensional target state vector at time k, consisting of the quantities to be estimated, F is a transition matrix, G is a control matrix, and u is a control vector. $x(k+1)$ is the state vector prediction for time $k+1$. The measurement vector received from the sensor is:

$$z(k) = Hx(k) \tag{2}$$

where H is the measurement matrix.

Because of the measurement errors and false alarms, the real state vector x is never known. Instead, we have to work with its estimation $\hat{x}$. The process of estimating is usually called filtering, and corresponding algorithms are called filters. Nowadays, common filters used for this purpose are based on Kalman filter.

2.1 Linear Kalman Filter

When eq. (1) and (2) are linear, the linear Kalman filter is used. Its basic form is:

$$\hat{x}(k+1|k) = F(k)\hat{x}(k|k) + G(k)u(k) \tag{3}$$
$$\hat{z}(k+1|k) = H(k+1)\hat{x}(k+1|k) \tag{4}$$
$$\nu(k+1) = z(k+1) - \hat{z}(k+1|k) \tag{5}$$
$$P(k+1|k) = F(k)P(k|k)F(k)' + Q(k) \tag{6}$$
$$S(k+1) = H(k+1)P(k+1|k)H(k+1)' + R(k) \tag{7}$$
$$W(k+1) = P(k+1|k)H(k+1)'S(k+1)^{-1} \tag{8}$$
$$\hat{x}(k+1|k+1) = \hat{x}(k+1|k) + W(k+1)\nu(k+1) \tag{9}$$
$$P(k+1|k+1) = P(k+1|k) - W(k+1)S(k+1)W(k+1)' \tag{10}$$

where $\hat{x}$ is the target's state vector estimation, W is the gain matrix, S is the innovation covariance matrix, Q is the noise covariance matrix, R is the measurement covariance matrix, ν is the innovation vector, and P is the covariance matrix.

2.2 Nonlinear (Extended) Kalman Filter

When equations (1) and/or (2) are nonlinear, extended Kalman filter is used. Its equations are the same as those of the linear Kalman filter (3)-(10), but the matrices $F(k)$ and/or $H(k)$ are Jacobians, based on the first-order Taylor expansion of the nonlinear functions (1) and/or (2) respectively. Hence, after the Jacobians are calculated, the nonlinear filter estimation can be reduced to a linear filter estimation.

2.3 Probabilistic Data Association (PDA) Filter

If the observations from a single target are mixed with clutter, Probabilistic Data Association filter is usually applied instead of the classic Kalman filter [4]. It is also called "all neighbors method" because the updated track estimate contains contributions from all N observations within the track's gate. The probability of the hypothesis $H_j (j = 1, 2, ...N)$ that the observation is a valid return for the track i, is proportional to the likelihood function g_{ij}:

$$g_{ij} = \frac{exp^{-\frac{d_{ij}^2}{2}}}{(2\pi)^{M/2}\sqrt{|S_i|}} \tag{11}$$

where $d_{ij}^2 = \nu_{ij}^T S_i^{-1} \nu_{ij}$, ($\nu_{ij}$ is the measurement residual for track i and measurement j, according to (5)), and M is the measurement vector size.

Then

$$p'_{ij} = \beta^{N-1} P_D g_{ij}, \quad j = 1, 2, ..., N$$

where β is the extraneous return density, and P_D is the detection probability. The probabilities (p_{ij}), associated with the N+1 hypotheses, are computed through the normalization equation:

$$p_{ij} = \frac{p'_{ij}}{\sum_{l=0}^{N} p'_{il}}$$

The residual for use in the Kalman Filter update equation, is a weighted sum of the residuals associated with the N observations:

$$\tilde{y}_i = \sum_{j=1}^{N} p_{ij} \tilde{y}_{ij}(k) \tag{12}$$

where $\tilde{y}_{ij}(k) = y_j(k) - H\hat{x}_i(k|k-1)$, and $y_j(k)$ is the observation j received at scan k. The covariance P is updated according to the equations:

$$P(k|k) = P^0(k|k) + dP(k) \tag{13}$$

where

$P^0(k|k) = p_{i0} P(k|k-1) + (1 - p_{i0}) P^*(k|k)$

$dP(k) = W(k)[\sum_{j=1}^{N} p_{ij} \tilde{y}_{ij} \tilde{y}_{ij}^T - \tilde{y}_i \tilde{y}_i^T] W^T(k)$

and

$P^*(k|k) = [I - W(k)H] P(k|k-1)$.

3 Description of the Classes

We take Equations (2) and the data in them as a basis of structure of classes, describing tracks. At the root of the hierarchy is an abstract class containing all the vectors and matrices from (2), the method KFiltering, implementing the equations, and some virtual methods for track initiation and nonlinear filter calculations. Over it a chain of descendent classes is created, implementing linear Kalman filter, extended Kalman filter and probabilistic data association filter.

3.1 Abstract Class for Kalman Filter

```
class ClsAKFTrack
{ protected:
   int Label1;
  static int Nsize;      // state vector size
  static int Msize;      // measurement vector size
  static Matrix Q; // noise covariance matrix
  static Matrix R; // measurement covariance matrix
  static Matrix G; // control matrix
  static Vector U; // control vector
  Matrix F;     // transition matrix
  Matrix H;     // measurement matrix
  Vector X;     // object state vector
  Matrix P;     // covariance matrix
  Matrix S;     // innovation covariance matrix
  Vector ZPrediction; // measurement prediction vector
  Vector Zmeasurement; // measurement vector
     public:
  virtual void CreateModel(float * Sigma, float Tscan)=0;
  virtual int  CheckGating(Vector Zmeasurement);
  virtual void InitTrack();
  virtual void DefineH(){};      // specific for nonlinear H
  virtual void DefineF(){};      // specific for nonlinear F
  virtual void MeasurementPrediction();
  virtual void Innovation();
  virtual void Covariance Update();
  void KFiltering();
};
```

It should be noted that the data for Q, R, G and U are declared static because they are the same for all objects of that class [5,7]. The method KFiltering consists of the following steps (some of them are also implemented as methods):

- DefineF - calculates the Jacobian of F in the case of extended Kalman filter; for linear Kalman filter it does nothing.
- State Prediction - implements Equation (3).
- Covariance Prediction - implements Equation (6).
- DefineH - calculates the Jacobian of H in the case of extended Kalman filter; for linear Kalman filter it does nothing.
- MeasurementPrediction - For the linear case implements Equation (4). This method is declared virtual. For the nonlinear case it is defined according to the correspondent measurement and state vectors.
- Innovation - Implements Equation (5). In some specific cases like PDAF, this method is defined to calculate combined innovation according to eq.(12).
- Filter Gain - Implements Equations (7), (8).
- State Update - Implements Equation (9).

- Covariance Update - Implements Equations (10). In the case of PDAF, this method is defined to implement equation (13).

3.2 Linear Kalman Filter

The declaration of the class for linear Kalman filter is:

```
class ClsLKFTrack : public ClsAKFTrack
{   public:
   void CreateModel(float * Sigma, float Tscan);
   virtual void InitTrack();
   virtual void DefineH(){};
   virtual void DefineF(){};
   virtual void MeasurementPrediction();
   virtual int  CheckGating(Vector Zmeasurement);
   virtual void Innovation();
   virtual void CovarianceUpdate();
};
```

This class inherits the data and the methods of the abstract class and implements the virtual functions. The method MeasurementPrediction calculates (4), Innovation calculates (5), and CovarianceUpdate calculates (10). The function CreateModel should be executed only once to set the matrices Q, R, G and the vector U. Its parameters are Sigma - process noise, and Tscan — the scan period. This class is not abstract and can be used for creating objects.

3.3 Nonlinear Kalman Filter

The declaration of the class for nonlinear Kalman filter is:

```
class ClsEKFTrack : public ClsLKFTrack
{   public:
   virtual void DefineH();  // specific for nonlinear H
   virtual void DefineF();  // specific for nonlinear F
   virtual void MeasurementPrediction();
   virtual int  CheckGating(Vector Zmeasurement);
};
```

This class inherits the data of the ClsLKFTrack class, and its virtual methods DefineF and DefineH are defined to implement specific functions that calculate the Jacobians, as stated earlier.

3.4 Probabilistic Data Association (PDA) Filter

A new class, derived from ClsEKFTrack, can be used for tracking targets in clutter. For this purpose, the number of measurements in the gate NumOfObsInTrackGate, and an array with the observation scores ObsInTrackGate, are added. The following virtual functions are defined for this class: CheckGating fills the array of measurements in the gate and their scores; Innovation is defined to compute combined innovation, according to (12); CovarianceUpdate updates covariance matrix P, according to (13). The class declaration is:

```
class ClsPDAFTrack : public ClsEKFTrack
 {  int NumOfObsInTrackGate;
    NumAndScore ObsInTrackGate[MaxNumberOfObs];
  public:
   virtual void DefineH();  // specific for nonlinear H
   virtual void DefineF();  // specific for nonlinear F
   virtual void MeasurementPrediction();
   virtual void Innovation();
   virtual int  CheckGating(Vector Zmeasurement);
   virtual void CovarianceUpdate();
};
```

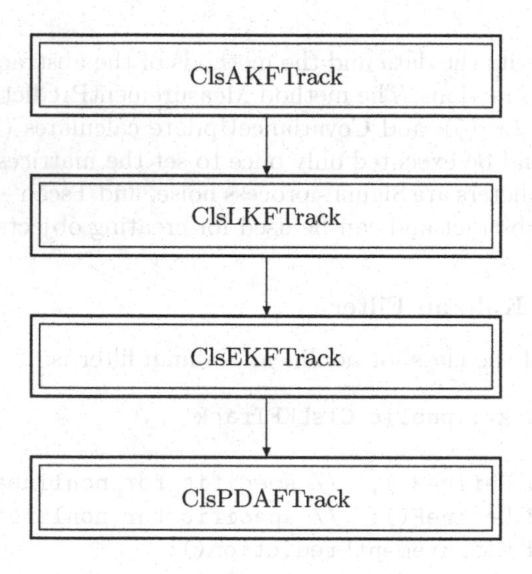

Fig. 1. Class hierarchy

4 sing the Classes An Example

The classes described above can be used for creating new classes for more complex filters and tracking algorithms. For example, they can be used for implementing the Interacting Multiple Models (IMM) target tracking algorithm. The IMM estimator is often considered to be the most significant advancement in target tracking since the Kalman filter [1].

The stages of this approach are [2]:

1. Calculating the mixing probabilities p_{ij} using Markov transition probabilities for r models

$$\mu_{i|j}(k-1|k-1)=\frac{1}{\bar{c}_j}p_{ij}\mu_i(k-1) \quad j=1,...,r \tag{14}$$

where the normalizing constants are $\bar{c}_j=\sum_{i=1}^{r}p_{ij}\mu_i(k-1) \quad j=1,...,r$.

2. Calculating the mixed initial conditions for the filter matched to the model j

$$\hat{x}^{0j}(k-1|k-1)=\sum_{i=1}^{r}\mu_{i|j}(k-1|k-1)\hat{x}^i(k-1|k-1) \qquad j=1,...,r \tag{15}$$

$$\begin{aligned}P^{0j}(k-1|k-1)=&\sum_{i=1}^{r}\mu_{i|j}(k-1|k-1)\left\{P^i(k-1|k-1)\right.\\&+\left[\hat{x}^i(k-1|k-1)-\hat{x}^{0j}(k-1|k-1)\right]\\&\left.\left[\hat{x}^i(k-1|k-1)-\hat{x}^{0j}(k-1|k-1)\right]^T\right\} \qquad j=1,\ldots,r\end{aligned} \tag{16}$$

3. $\hat{x}^{0j}$ and P^{0j} are used as input to the filter matched to $M_j(k)$, which uses $z(k)$ to yield state vector $\hat{x}^j(k|k)$, matched to model j, and covariance matrix for this model $P^j(k|k)$.

4. Mode probability update

The likelihood functions Λ_j, corresponding to r filters are computed using (11). Then

$$\mu_j(k)=\frac{1}{c}\Lambda_j(k)\bar{c}_j \qquad j=1,\ldots,r \tag{17}$$

where r is the number of models, μ_j is the probability that the model j is correct given the measurements received until scan k.

As an example, we outline here one possible implementation of the IMM algorithm, using some of our classes and taking advantage of polymorphism. We choose an algorithm with two embedded models having different process noise and define two new classes ClsModel1 and ClsModel2 for each model derived from nonlinear Kalman filtering class ClsEKFTrack. By the virtual method CreateModel we define their specific process noise matrices Q.

The main steps of the algorithm are:

- Calculate mixed state and covariance values for the KFiltering method by a new method IMMPreparation, which implements:
  a) Calculating the model probabilities according to (8) by the method WeightPrediction;
  b) Computing combined innovation and state covariance matrix according to (9) and (10) by two friend functions MeansMixing and CovariancesMixing, using as parameters the objects of the classes ClsModel1 and ClsModel2.
- Apply KFiltering to each object with mixed values calculated above.

– Compute the updated model probabilities according to (11) using the method WeightUpdate.

In the program we create one object from each of the classes and an array ObPtr with two pointers to the base class, and then initialize the first and the second pointer with the addresses of the objects from the two classes. Then we call `ObPtr[j]->IMMPreparation()` and `ObPtr[j]->KFiltering()` in a loop for each model, applying Kalman filtering with the specific data for the two models. Finaly, the function WeightUpdate computes new values for the next processing cycle.

5 Conclusions

In this paper we propose a set of classes for implementing tracking algorithms. The track hierarchy contains classes for Linear Kalman Filter, Extended Kalman Filter and Probabilistic Data Association Filter. These classes can be used for creating more complex filters and tracking algorithms.

References

1. Bar-Shalom Y., Blair W.: Multitarget-Multisensor Tracking: Applications and Advances, vol 3. Artech House (2000)
2. Bar-Shalom Y., Xiao-Rong Li: Estimation and Tracking: Principles, Techniques and Software. Artech House. (1993).
3. Bar-Shalom Y., Xiao-Rong Li: Multitraget-Multisensor Tracking: Principles and Techniques. YBS (1995)
4. Blackman S.S.: Multiple-Target Tracking with Radar Applications. Artech House (1986)
5. Georgiev Vl., Georgieva J. et all: Manual for using of computers and programming in C++. Sofia (2000) (in Bulgarian)
6. Karaivanova A., Djerassi E.: Track Formation Using MHT Approach. Proc. of the International Conference on Mathematical Modeling and Scientific Computations. DATECS Publishing. Sofia (1993) 31–33
7. Simov G.: Programming in C++. SIM. Sofia (1993) (in Bulgarian)

CFAR Processors in Pulse Jamming*

Ivan Garvanov[1], Vera Behar[2], and Christo Kabakchiev[1]

[1] Institute of Information Technologies, Bulgarian Academy of Sciences
Acad. G. Bonchev Str., Bl. 2, 1113 Sofia, Bulgaria
igarvanov@iit.bas.bg, ckabakchiev@iit.bas.bg

[2] Dep. of Biomedical Engineering, Institute of Technology, Haifa 3200
vera@biomed.technion.ac.il

Abstract. In this paper we study the efficiency of CA CFAR BI, EXC CFAR BI and Adaptive CFAR PI detectors in strong pulse jamming. We achieve new results for the average decision threshold (ADT) using the minimum detectable signal (Pd=0.5). For comparison we use also the approach with Monte-Karlo simulation for estimation of the ADT of the studied CFAR detectors. Differently from other authors, we consider the entire range (0 to 1) of the probability for the appearance of pulse jamming in range cells.

1 Introduction

Cell-Averaging Constant False Alarm Rate (CA CFAR) signal processing proposed by Finn and Johnson in [1] is often used for radar signal detection. The detection threshold is determined as a product of the noise level estimate in the reference window and a scale factor to achieve the desired probability of false alarm. The presence of strong pulse jamming (PJ) in both, the test resolution cell and the reference cells, can cause drastic degradation in the performance of a CA CFAR processor as shown in [12].

In such situations it would be desirable to know the CFAR losses, depending on the parameters of PJ, for rating the radar behavior. There are two approaches for the calculation of CFAR losses offered by Rolling and Kassam in [2,3]. The conventional method, used in [5,7-12], is to compute the additional SNR needed for the CFAR processing scheme beyond that for the optimum processor, to achieve a fixed detection probability (e.g. 0.5). For a particular CFAR scheme losses obviously vary with detection probability. Alternatively, the authors in [2,3] use another criterion based on the average decision threshold (ADT), since the threshold and the detection probability are closely related to each other. Then the difference between the two CFAR systems is expressed by the ratio between the two ADTs measured in dB, as shown in [2,3].

The false alarm rate of the postdetection integrator (PI) is extremely sensitive to pulse jamming, and the binary integrator (BI) which uses a K-out-of-M

* This work is supported by IIT – 010044, MPS Ltd. Grant "RDR" and Bulgarian NF "SR" Grant No I – 902/99.

I. Dimov et al. (Eds.): NMA 2002, LNCS 2542, pp. 291–298, 2003.

decision rule is insensitive to at most (K-1) interfering pulses [7]. For keeping of constant false alarm rate in PJ the CA CFAR processor presented in [9,12] is used. But this method is not as effective as the conventional method for the calculation of CFAR losses. For the minimization of CFAR losses in case of pulse jamming postdetection integration (PI) or binary integration (BI) is implemented in CFAR processors as shown in [5,8,10]. The use of excision CFAR detectors, supplemented by a postdetection integrator or a binary integrator as shown in [6,7,10], increases the CFAR losses. Minimum CFAR losses in PJ are obtained in [5,11] with a CFAR adaptive postdetection integration (API) processor with adaptive selection on PJ in reference windows and appriory selection in test windows as shown in [5], and adaptive censoring in reference and test windows as presented in [11].

We assume in this paper that the noise in the test cell is Rayleigh envelope distributed and target returns are fluctuating according to Swerling II model as in [3,5]. As a difference from the authors in [5], we assume that the samples of PJ are distributed according to the compound exponential law, where weighting coefficients are the probabilities of corrupting and non-corrupting of the samples. Differently from [7-11], we consider the entire range (0 to 1) of the probability for the appearance of pulse jamming in range cells. For values of the weighting coefficients higher than 0.3, the Poisson process model is used, but it is rough [15]. The binomial distribution is correct in this case.

In this paper we study different CFAR techniques in the presence of strong pulse jamming, similarly to [5]. We use the average decision threshold (ADT) approach for comparison of the processors. The analytical expressions for the probability functions of CA CFAR BI, EXC CFAR BI and API CFAR detectors are achieved in [7,8,10,11]. We achieve in these paper new results for the ADT using the SNR approach. The SNR of the minimum detectable signal (Pd=0.5) is approximately the same as the ADT of each CFAR system. For comparison we use also the approach with Monte-Karlo simulation for estimation of the ADT of the studied CFAR detectors. The experimental results show that the API CFAR processors are most effective for values of the probability of appearance of pulse jamming in the interval (0 to 0.5). For $\varepsilon_0 > 0.5$ we recommend binary integration after the CFAR processor.

2 Performance of CA CFAR BI and Excision CFAR BI Processors in the Presence of Pulse amming

2.1 Probability of Detection and False Alarm of CA CFAR BI Detectors

After filtration the signal is applied to a square-law detector and sampled in range by the $(N+1)$ resolution cells resulting in a vector of $(N+1)$ observations. The sampling rate is such that the samples are statistically independent. It is assumed that L broadband pulses hit the target and one sample of the cell under test suffices to detect a pulse of target returns. In conditions of pulse jamming the background environment includes the random interfering pulses and the receiver noise. Therefore the samples surrounding the cell under test (a reference window) may be drawn from two classes. One class represents the interference-plus-noise situation, which may appear at the output of the receiver with the probability ε_0. This probability can be expressed as $\varepsilon_0 = t_c F_j$, where F_j is the average repetition frequency of PJ and t_c is the length of pulse transmission. The other class represents the noise only situation, which may appear at the outputs of the receiver with the probability $(1 + \varepsilon_0)$. The set of samples from the test resolution cell at the input of the binary integrator (x_l, where $l = 1, \ldots, L$) is assumed to be distributed according to Swerling II case [9,15] with the probability density function (pdf) given by:

$$f(x) = \frac{1-\varepsilon_0}{\lambda_0 (1+s)} \exp\left(\frac{-x}{\lambda_0 (1+s)}\right) + \frac{\varepsilon_0}{\lambda_0 (1+r_j+s)} \exp\left(\frac{-x}{\lambda_0 (1+r_j+s)}\right) \quad (1)$$

where λ_0 is the average power of the receiver noise, r_j is the average interference-to-noise ratio (INR) of pulse jamming, s is the per pulse average signal-to-noise ratio (SNR) and N is the number of observations in a reference window. In this case the Poisson process model is used, and it is valid only for function (pdf) of the reference window outputs can be defined as (1), setting $s = 0$.

All samples (x_l) are compared with the threshold $H_D$according to the rule:

$$\Phi_l(x_l) = \begin{cases} 1, \text{ if } x_l \geq H_D \\ 0, \quad \text{otherwise} \end{cases} \quad (2)$$

The binary integrator performs summing of L decisions Φ_l. The target radar image detection is declared if this sum exceeds the second threshold M.

$$\begin{cases} H_1 : (t\arg et \text{ is present}), \text{ if } \sum_{l=1}^{L} \Phi_l \geq M \\ H_0 : (\text{no } t \arg et), \qquad \text{otherwise} \end{cases} \quad (3)$$

In a conventional CA CFAR detector the noise level estimate is formed as a sum of all the outputs of the reference window: $V = \sum_{i=1}^{N} x_i$. In this case the mgf of the estimate V is defined to be $M_V(U) = M_x^N(U)$, where $M_x(U)$ is the mgf of the random variable x_i. The mgf of the estimate V is obtained in [9]:

$$M_V(U) = \sum_{i=1}^{N} \frac{C_N^i \varepsilon_0^i (1-\varepsilon_0)^{N-i}}{(1+\lambda_0 U)^{N-i} (1+\lambda_0 (1+r_j) U)^i} \quad (4)$$

According to (3) the probability of target detection for CA CFAR BI processor as in [10] is computed by the expression:

$$P_D = \sum_{l=M}^{L} C_L^l Pd_1^l (1-Pd_1)^{L-l} \quad (5)$$

where

$$Pd_1 = (1-\varepsilon_0) M_V\left(\frac{T}{\lambda_0 (1+s)}\right) + \varepsilon_0 M_V\left(\frac{T}{\lambda_0 (1+r_j+s)}\right) \quad (6)$$

The probability of false alarm is evaluated by (5, 6, 4), setting $s = 0$.

2.2 Probability of Detection and False Alarm of Excision CFAR BI Detectors

In an excision CFAR BI processor the noise level estimate V is formed as an average mean of nonzero samples at the output of the excisor $\{y_i\}_N$, that is: $V = \frac{1}{k}\sum_{i=1}^{K} y_i$. According to [6] the operation of the excisor is defined as follows:

$$y_i = \begin{cases} x_i : x_i \le B_E \\ 0 : \ \text{othewise} \end{cases} \tag{7}$$

where B_E is the excision threshold.

The probability that a sample x_i survives at the output of the excisor, is given as:

$$P_E = 1 - (1-\varepsilon_0)\exp\left(-B_E/\lambda_0\right) - \varepsilon_0 \exp\left(-B_E/\lambda_0\left(1+r_j\right)\right) \tag{8}$$

The probability that k out of N samples of the reference window survive at the output of the excisor is given as: $v(k) = C_N^k P_E^k (1-P_E)^{N-k}$.

In this article we use the moment generating function on the excision CFAR processor from [7]

$$M_V(U) = \sum_{k=1}^{N} C_N^k P_E^k (1-P_E)^{N-k} M_V(U,k) \tag{9}$$

where

$$\begin{aligned} M_V(U,k) = \sum_{i=0}^{k} C_k^i \left\{ \frac{\varepsilon_0(1-\exp(R_1 - B_E U/k))}{(1-\exp(R_1))(1+U\lambda_0(1+r_j)/k)} \right\}^i \cdot \\ \cdot \left\{ \frac{(1-\varepsilon_0)(1-\exp(R_2 - B_E U/k))}{(1-\exp(R_2))(1+U\lambda_0(1+r_j)/k)} \right\}^{k-i} \end{aligned} \tag{10}$$

and $R_1 = -B_E/\lambda_0(1+r_j)$; $R_2 = -B_E/\lambda_0$.

The probability of target detection for excision CFAR BI processor in [7,10] is computed by the expression:

$$P_D = \sum_{l=M}^{L} C_L^l Pd_2^l \left(1 - Pd_2\right)^{L-l} \tag{11}$$

where

$$\begin{aligned} Pd_2 = \sum_{k=1}^{N} C_N^k P_E^k (1-P_E)^{N-k} \left\{ (1-\varepsilon_0) M_V\left(\frac{T}{\lambda_0(1+s)}, k\right) + \right. \\ \left. + \varepsilon_0 M_V\left(\frac{T}{\lambda_0(1+r_j+s)}, k\right) \right\} \end{aligned} \tag{12}$$

The probability of false alarm is evaluated by (11, 12), setting $s = 0$.

3 Performance of API CFAR Processors in the Presence of Pulse amming

Let us assume that L pulses hit the target, which is modeled according to Swerling II case. The received signal is sampled in range by using $(M+1)$ resolution cells resulting in a matrix with $(M+1)$ rows and L columns. Each column of the data matrix consists of the values of the signal obtained for L pulse intervals in one range resolution cell. Let us also assume that the first $M/2$ and the last $M/2$ rows of the data matrix are used as a reference window in order to estimate the "noise-plus-interference" level in the test resolution cell of the radar. In this case the samples of the reference cells result in a matrix X of the size $M \times L$. The test cell or the target includes the elements of the $(M/2+1)$ row of the data matrix and is a vector Z of length L. The elements of the reference window are independent random variables with the compound exponential distribution law (1), setting $s=0$. In the presence of a desired signal from a target the elements of the test resolution cell are independent random variables with distribution law (1).

3.1 Probability of Detection and False Alarm of API CFAR Detectors

We use the adaptive censoring algorithm, proposed by Himonas and Barkat in [4], before pulse-to-pulse integration for censoring of the elements of pulse jamming in the reference window and the test resolution cells, in order to form the detection algorithm.

The expression for the probability of target detection for an API CFAR processor is achieved in [11,14]. The authors in [11,14] study the probability of target detection only for $\varepsilon_0 \leq 0.5$ and calculate only the first member of the expression (13). We calculate the probability of target detection for $\varepsilon_0 \in (0;1)$ and calculate the value of P_D by using the following expression:

$$
\begin{aligned}
P_D = &\sum_{k=1}^{N} \binom{N}{k} (1-\varepsilon_0)^k \varepsilon_0^{N-k} \sum_{l=1}^{L} \binom{L}{l} (1-\varepsilon_0)^l \varepsilon_0^{L-l} \sum_{i=0}^{l-1} \binom{k+i-1}{i} \frac{T^s(1+s)^k}{(T+1+s)^{k+i}} + \\
&\sum_{k=1}^{N} \binom{N}{k} (1-\varepsilon_0)^k \varepsilon_0^{N-k} \varepsilon_0^{L} \sum_{i=0}^{L-1} \binom{k+i-1}{i} T^i (1+r_j+s)^i (T+1+r_j+s)^{-(k+i)} + \\
&\sum_{l=1}^{L} \binom{L}{l} (1-\varepsilon_0)^l \varepsilon_0^{L-l} \varepsilon_0^{N} \sum_{i=0}^{l-1} \binom{N+i-1}{i} T^i \left(\frac{1+s}{1+r_j}\right)^N \left(T + \frac{1+s}{1+r_j}\right)^{-(N+i)} + \\
&\varepsilon_0^N \varepsilon_0^L \sum_{i=0}^{L-1} \binom{N+i-1}{i} T^i \left(\frac{1+r_j+s}{1+r_j}\right)^N \left(T + \frac{1+r_j+s}{1+r_j}\right)^{-(N+i)}
\end{aligned}
\tag{13}
$$

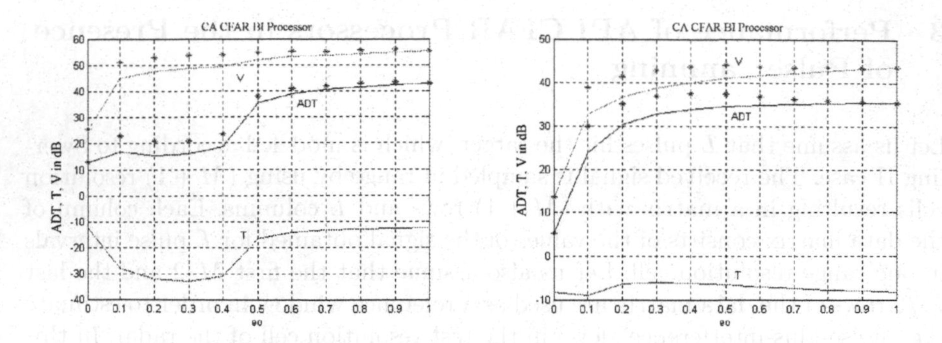

Fig. 1. CA CFAR BI $M = 16$, $L = 16$ **Fig. 2.** CA CFAR BI $M = 10$, $L = 16$

4 Average Decision Threshold of CFAR Detectors

Deviating from the methods usually described in radar literature, we use the average decision threshold (ADT) for comparison of various CFAR processors. We compute the SNR , needed for the CFAR processing scheme, to achieve a fixed detection probability (e.g. 0.5). The losses or the difference between the two CFAR systems can be expressed by the ratio of the two ADT's measured in dB [2].

$$\Delta[dB] = \frac{ADT_1}{ADT_2} = 10\log\frac{E(T_1V_1)}{E(T_2V_2)} \text{ for } P_{fa1} = P_{fa2},\ P_{D1} = P_{D2} = 0.5 \quad (14)$$

5 Numerical Results

The experimental results are obtained for the following parameters: average power of the receiver noiseλ_0=1, average interference-to-noise ratio (INR) r_j=30 [dB], probability for the appearance of pulse jamming with average length in the range cells ε_0 from 0 to 1. Probability of false alarm $P_{fa} = 10^{-6}$ and excision threshold $B_E = 2$. The size of the testing sample is 16 and the reference window is of the size 16×16. The results for the ADT are received by using Monte-Karlo method and the probability functions (SNR). They are marked as follows: Monte-Karlo (*) and SNR (continuous line).

The ADT, T and V of CA CFAR BI processors with a binary rule M-out-of-L (16/16 and 10/16) are showed on Fig.1 and Fig.2. It can be seen that CFAR BI processors with the binary rule M-out-of-L=16/16 are better in cases of lower values $\varepsilon_0 \leq 0.5$ of the probability for the appearance of pulse jamming. For higher values of the probability for the appearance of pulse jamming $\varepsilon_0 > 0.5$, the using of the binary rule M-out-of-L=10/16 results in lower losses.

The ADT, T and V of excision CFAR BI processors with a binary rule M-out-of-L (16/16 and 10/16) are showed on Fig.3 and Fig.4. It can be seen that excision CFAR BI detectors have the same behavior as CFAR BI detectors.

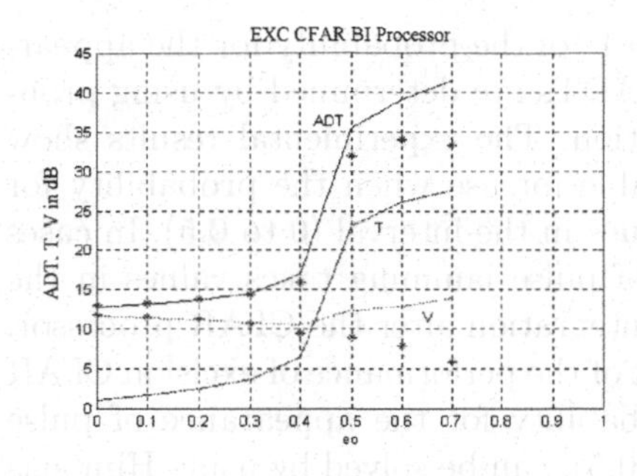

Fig. 3. EXC CFAR BI $M = 16$, $L = 16$

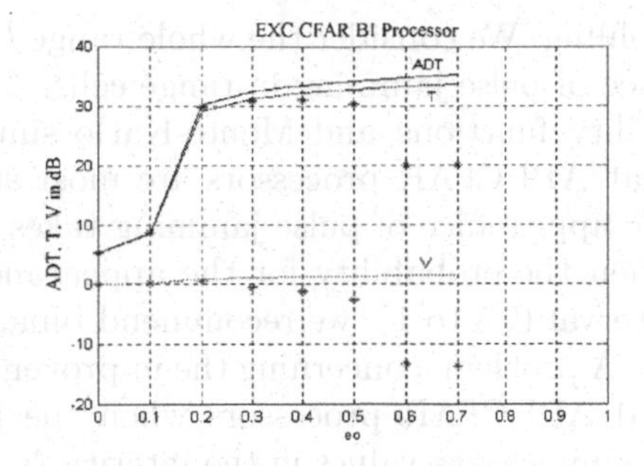

Fig. 4. EXC CFAR BI $M = 10$, $L = 16$

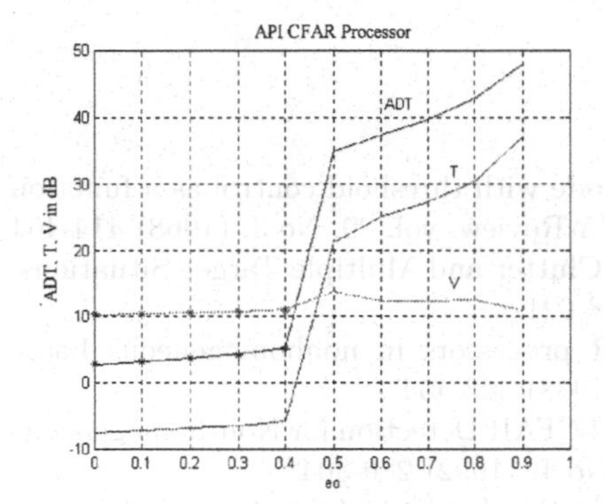

Fig. 5. API CFAR $M = 16$ $L = 16$

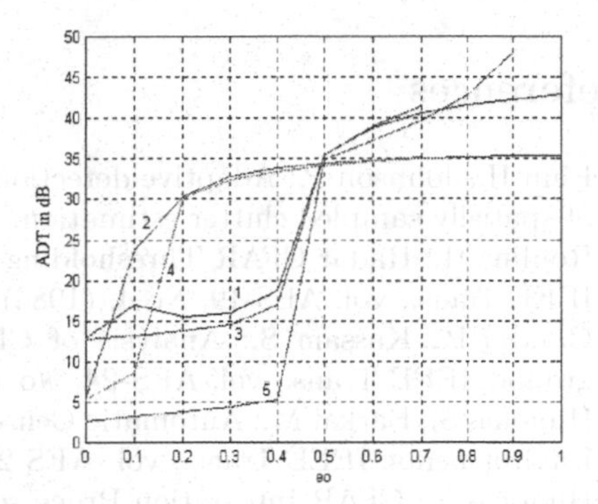

Fig. 6. ADT from fig.1 to fig.5

The ADT, T and V of an API CFAR processor are shown on Fig.5. In this case the results for the ADT achieved by using the probability functions (SNR) are identical with the results achieved by using Monte-Karlo simulation for values of ε_0 up to 0.4. The suggested algorithm is not working for higher values of ε_0 due to the fact, that the hypothesis for censoring in the test cell is disturbed. In such cases the big difference in power between the background and the pulse jamming is disturbed and the automatic censoring of pulse jamming is impossible.

The ADTs of all studied processors are shown on Fig.6. The numbers from 1 to 5 correspond to the detectors from Fig.1 to Fig.5. The API CFAR processor is the most suitable one to use for values of the probability for the appearance of pulse jamming $\varepsilon_0 \leq 0.5$. When the probability for the appearance of pulse jamming ε_0 takes value between 0.5 and 1 both, CA CFAR BI and EXC CFAR BI processors with M-out-of-L=10/16 rule can be successfully used.

6 Conclusions

We investigate in this paper the efficiency of different CFAR techniques in the presence of strong pulse jamming by using the ADT approach suggested by

Rohling. We consider the whole range (0 to 1) of the probability for the appearance of pulse jamming in range cells. The ADTs are determined by using probability functions and Monte-Karlo simulation. The experimental results show that API CFAR processors are most suitable for use when the probability for the appearance of pulse jamming takes values in the interval (0 to 0.5). In cases when the probability for the appearance of pulse jamming takes values in the interval (0.5 to 1), we recommend binary integration after the CFAR processor.

A problem, concerning the improvement of the performance of excision CFAR and API CFAR processors when the probability for the appearance of pulse jamming takes values in the interval ($\varepsilon_0 > 0.5$), can be solved by using Himonas approach [4]. In such cases the threshold estimation is achieved by using the cells with pulse jamming.

References

1. Finn H., Johnson P.: Adaptive detection mode with threshold control as a function of spatially sampled clutter estimation, RCA Review, vol. 29, No 3, (1968) 414-464
2. Rohling H.: Radar CFAR Thresholding in Clutter and Multiple Target Situations, IEEE Trans., vol. AES-19, No 4, (1983) 608-621
3. Gandhi P., Kassam S.: Analysis of CFAR processors in nonhomogeneous background, IEEE Trans., vol. AES-24, No 4, (1988) 443-454
4. Himonas S., Barkat M.: Automatic Censored CFAR Detection for Non-homogeneous Environments, IEEE Trans., vol. AES-28, No 1, (1992) 286-304
5. Himonas S.: CFAR Integration Processors in Randomly Arriving Impulse Interference, IEEE Trans., vol. AES-30, No 3, (1994) 809-816
6. Goldman H.: Analysis and application of the excision CFAR detector, IEE Proceedings, vol.135, No 6, (1988) 563-575
7. Behar V., Kabakchiev C.: Excision CFAR Binary Integration Processors, Compt. Rend. Acad. Bulg. Sci., vol. 49, No 11/12, (1996) 45-48
8. Kabakchiev C., Behar V.: CFAR Radar Image Detection in Pulse Jamming, IEEE Fourth Int. Symp. ISSSTA'96, Mainz, Germany, (1996) 182-185
9. Behar V.: CA CFAR radar signal detection in pulse jamming, Compt. Rend. Acad. Bulg. Sci., vol. 49, No 7/8, (1996) 57-60
10. Kabakchiev C., Behar V.: Techniques for CFAR Radar Image Detection in Pulse Jamming, IEEE Fourth Int. Symp. EUM'96, Praga,Cheh Republic, (1996) 347-352
11. Behar V., Kabakchiev C., Dukovska L.: Adaptive CFAR Processor for Radar Target Detection in Pulse Jamming, Journal of VLSI Signal Processong, vol. 26, No 11/12, (2000) 383-396
12. Kabakchiev C., Dukovska L., Garvanov I.: Comparative Analysis of Losses of CA CFAR Processors in Pulse Jamming, CIT, No 1, (2001) 21-35
13. Garvanov I., Kabakchiev C.: Average decision threshold of CA CFAR and excision CFAR detectors in the presence of strong pulse jamming, German Radar Symposium 2002, (GRS 2002), Bonn, Germany, September (2002), (submitted for presentation)
14. Chakarov V.: Adaptive CFAR processor for radar target detection design of SHARC signal processor, PhD thesis, Technical University - Sofia, (1999)
15. Akimov P., Evstratov F., Zaharov S.: Radio Signal Detection, Moscow, Radio and Communication, 1989, pp. 195-203, (in Russian)

Statistical Classifier of Radar Returns*

Miroslava Ivanova and Boriana Vassileva

Central Laboratory for Parallel Processing
Bulgarian Academy of Sciences
"Acad G. Bonchev" St., Bl. 25-A, 1113 Sofia, Bulgaria
bvassil@bas.bg

Abstract. For the last years many algorithms for radar return identification based on wideband or imaging radar have been developed. However, because of its all-weather performance, microwave radar is still the most reliable sensor for surveillance.
This paper considers the problem of radar return identification using microwave radar. Radar return is classified into several different classes, namely ground, weather, birds and aircraft, using Stehwien-Haykin classifier based on features derived from maximum entropy spectral analysis. To evaluate the practicality and effectiveness of this classifier its performance is tested by Monte Carlo simulation. Clutter samples, are generated by passing a white noise sequence through linear digital filter. The power spectral density is reasonably represented by second or third order Batterworth. The results show that the classifier correctly identifies the radar return with mean error rate less than 4 %.

1 Introduction

One of the most important applications of radar is the monitoring of the air traffic environment to aid in save aircraft operation. The objective of these surveillance systems is to ensure orderly air traffic flow, including the avoidance of bird hits, flying into meteorological disturbance and collisions with other aircraft. Therefore, capability of detecting aircraft and distinctively classifying the various situations likely to arise in practice is necessary.

In the past years many algorithms for moving target detection have been developed. Radar return identification is usually based on wideband or imaging radar. Not much attention has been paid to the development of radar return identification for surveillance using microwave radar. Because of its all-weather performance such radar is still the most reliable sensor for surveillance [5]. Two classes of microwave radar returns classification are widely used. The first class, represented by the nearest neighbor algorithm, includes nonparametric techniques that may require only partial measurement information [2,3]. The resulting classifiers usually achieve good performance over a large variety of distribution functions. The second class includes parametric techniques that

* Partially supported by Center of Excellence BIS21 grant ICA1-2000-70016 and Bulgarian National Foundation for Scientific Investigations grant No I-902/99.

I. Dimov et al. (Eds.): NMA 2002, LNCS 2542, pp. 299–306, 2003.

require a complete statistical model of the radar measurement [3,5]. Bayesian and Dempster-Shafer methods are applied for such statistical identification. The Dempster-Shafer method is developed to overcome a perceived inability of the Bayesian method to represent uncertain evidence [1]. However, with the correct formulation, the Bayesian method is computationally simpler and has higher probabilities for the correct decision [5]. In this work one Bayesian radar returns classifier suggested by Stehwien and Haykin [6] is realized and investigated. To evaluate the practicality and effectiveness of this classifier its performance is tested by Monte Carlo simulation.

Five sections follow this introduction. Spectral characteristics of different classes of radar return are briefly discussed in Section 2. Burg's algorithm for extracting all spectral information into a set of parameters is described in Section 3. Section 4 represents the classifier algorithm. In Section 5 Monte Carlo simulation results are presented. Concluding remarks are given in Section 6.

2 Spectral Characteristics of Radar Returns

Classification of radar returns requires that a set of independent and ideally completely separable features be defined to achieve a low probability of misclassification. There are various feature sets which may be defined, such as spatial amplitude distributions, parameters relating movement from scan to scan, and spectral parameters. The spectral information is of particular interest in this work.

Radar return is produced by energy returning from any object within the range of the system. Four object types are of primary interest in air traffic control centers: aircraft, ground, weather, and birds. The most important characteristic (excluding positioning information) of the coherent pulse radar returns is the Power Spectral Density (PSD). Figure 1 shows typical shapes of the normalized PSD of ground, storm, birds flock and aircraft radar returns.

In general ground return can be easily identified by its near-zero doppler frequencies, and its relatively narrow spectral peak. Widening of the peak, as well as any non-zero doppler component, usually arises from the antenna scanning motion. Wind induced motion of foliage on trees and vegetation on hillsides will contribute to non-zero doppler frequencies. Weather radar return is expected to display some internal motion depending of the type of system. For instance the violent updrafts and downdrafts within thunderstorm cells are expected to result in a wider than ground spectral peak. Birds flocks typically have the widest spectral peak. It is due to the relative motion of the individual birds with respect to each other, as well as their wingbeats and other body motion. The overall doppler frequency is not likely to be very meaningful for weather or bird since its relative velocity with respect to the radar site is not unique. Aircraft returns will likely have a rather narrow spectral peak, especially jets. The overall characteristics of aircraft are so different from these of ground, weather and bird, that little difficulty is anticipated in their identification.

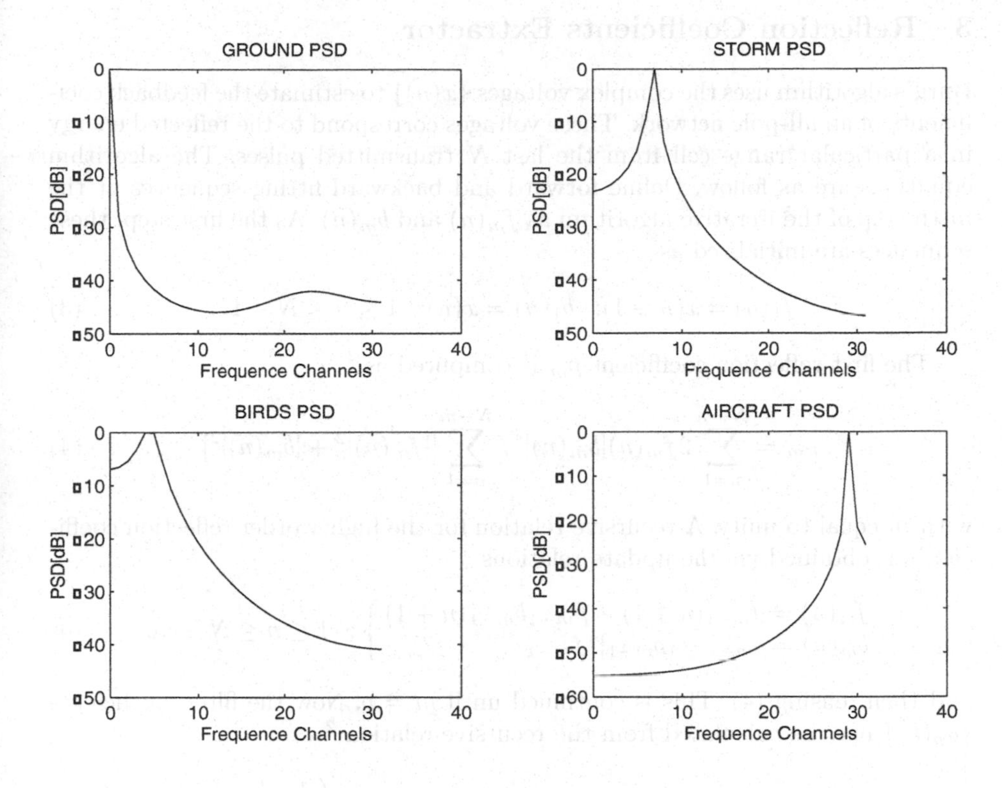

Fig. 1. Radar returns PSD

It is obviously that PSD of radar return is usefull for object classification because it provides a measure of the speed of the radar object as well as its relative internal motion. Results of analysis on real-live radar returns show that their PSD $\Phi(\omega)$ may be adequately modeled as an all-pole function [2]

$$\Phi(\omega) = \frac{\sigma_p^2}{2\pi|\sum_{k=0}^{p} a(k)\exp(-j\omega k)|^2}, \quad -\pi \leq \omega \leq \pi \tag{1}$$

where ω denotes spectral frequency and σ_p^2 is the process variance. Formula (1) represents the estimated PSD of Autoregressive (AR) process $\{x(n)\}$ of order p, i. e.

$$x(n) = \Sigma_{k=1}^{p} a(k)x(n-k) + \varepsilon(n), \tag{2}$$

where $\{\varepsilon(n)\}$ is zero mean white Gaussian noise. Burg's harmonic mean algorithm [2] is one of the AR coefficient $\{a(k)\}$ estimators with strongest theoretical background. It has a good performance even for relatively short data records, which are typical for microwave radar. This algorithm extracts all spectral information into a set of ordered parameters (from which the PSD may then be computed) and is thus ideally suited as feature extractor for radar return classifier.

3 Reflection Coefficients Extractor

Burg's algorithm uses the complex voltages $\{x(n)\}$ to estimate the feedback coefficients of an all-pole network. These voltages correspond to the reflected energy in a particular range cell from the last N transmitted pulses. The algorithm equations are as follow. Define forward and backward fitting sequences at the m-th step of the iterative algorithm as $f_m(n)$ and $b_m(n)$. As the first step, these sequences are initialized as

$$f_1(n) = x(n+1);\;\; b_1(n) = x(n);\;\; 1 \leq n \leq N-1. \tag{3}$$

The first reflection coefficient ρ_m is computed as [1]

$$\rho_m = \sum_{n=1}^{N-m} 2f_m(n)[b_m(n)]^* / \sum_{n=1}^{N-m} [|f_m(n)|^2 + |b_m(n)|^2] \tag{4}$$

with m equal to unity. A recursive relation for the higher order reflection coefficients is obtained via the update relations

$$\left.\begin{array}{l} f_m(n) = f_{m-1}(n+1) - \rho_{m-1} b_{m-1}(n+1) \\ b_m(n) = b_{m-1} - [\rho_{m-1}]^* f_{m-1} \end{array}\right\};\;\; 1 \leq n \leq N-m. \tag{5}$$

and then reusing (4). This is continued until $m = p$. Now the filter coefficients $\{a_p(k)\}$ may be calculated from the recursive relation [2]

$$a_p(k) = a_{p-1}(k) + \rho_k a_{p-1}^*(p-k);\;\; a_p(k) = \begin{cases} 1, & k = 0 \\ \rho_k, & k = p \\ 0, & k > p \end{cases}. \tag{6}$$

The reflection coefficients represent the incremental predictable information extracted from the input data at each stage. If the order p of the AR process is correct, any further application of the (4) and (5) is expected to result in zero valued coefficients since the prediction error sequence is white.

Unfortunately, a consequence of the short data length available is an excessively high variance in the estimate of the higher order coefficients. To reduce this variance, and to find a better estimate of the true value for the reflection coefficients, the multi-segment version of the Burg's formula may be used

$$\rho_m = \sum_{i=1}^{I} \sum_{n=1}^{N-m} 2f_{m,i}(n)[b_{m,i}(n)]^* / \sum_{i=1}^{I} \sum_{n=1}^{N-m} [|f_{m,i}(n)|^2 + |b_{m,i}(n)|^2]. \tag{7}$$

This equation yields the coefficients which minimize the spatial average of the forward and backward prediction error powers over all I segments [2]. In this

[1] (*) denotes the complex conjugate

[2] The subscript p is used because the algorithm employs recursive relations that successively approximates the data by a one-pole network, two-pole network, et cetera, up to a p-pole network

way time series from the same return source, but offset in space a small amount, are combined to form a low variance estimate. Clearly, the lower the variance within the features, the greater their separability.

As the overall doppler shift is a part of spectral information, it is clearly necessary to reference all spectra to the same frequency before a classification can be made. In the case of a single spectral peak, the phase angle of the first reflection coefficient represents the frequency of the peak, and as such may be used as a basis for shifting

$$\acute{\rho}_m = \rho_m \exp(-i \arg(\rho_1)) \tag{8}$$

4 Statistical Classifier

For the purpose of the classifier the real and imaginary parts of the shifted reflection coefficients are treated as independent features. For the p coefficient case this results in a $2p-1$ dimensional feature vector [3] $X = \{X_1, X_2, \ldots, X_{2p-1}\}$. The problem is to assign this feature vector to one of classes H_i $i = 1, 2, \ldots, L$ in such a way that the total number of misclassifications is minimized. The Bayesian method selects the most likely class H_d according to its likelihood. That is

$$H_d = argmax_i P(H_i|X), \;\; i = 1, 2, \ldots, L. \tag{9}$$

The posterior probability $P(H_i|X)$, can be written as

$$P(H_i|X) = \frac{p(X|H_i)P(H_i)}{\sum_{i=1}^{L} p(X|H_i)P(H_i)}, \tag{10}$$

where the class conditional density functions $p(X|H_i)$ are assumed to be multivariate Gaussian [4], the joint density function $\sum_{i=1}^{L} p(X|H_i)P(H_i)$ serves as a normalizing factor, and $P(H_i)$ is the prior probability of class H_i being correct. As the prior information is not available, propositions are assumed to have the same probability $P(H_i) = 1/L$. After taking the log of the (10) and removing the constants, the likelihood ratio test (9) is reduced to choosing the class with the largest discriminant function

$$G_i = -(X - M_i)^T R_i^{-1} (X - M_i) - \ln(det R_i), \;\; i = 1, 2, \ldots, L, \tag{11}$$

$$H_d = argmax_i \, G_i \tag{12}$$

where R_i are the respective class covariance matrices and M_i are the class mean vectors. Since no a priori knowledge about M_i and R_i is available, their respective maximum likelihood estimates are used

$$\hat{M}_i = \sum_{k=1}^{K} X(k)/K; \;\; \hat{R}_i = \sum_{k=1}^{K} (X(k) - \hat{M}_i)(X(k) - \hat{M}_i)^T/K; \tag{13}$$

where K is the number of the time series which are included in the class prototypes (M_i and R_i) computation, and the sample features $X(k)$ are known to belong to class H_i.

[3] the first coefficient is real valued due to spectrum shifting

5 Monte Carlo Simulation Results

The performance of the described classifier is evaluated by Monte Carlo computer simulation.

The example radar has beamwith of 0.85 degrees and selectable pulse repetition frequency 500, 1000 or 2000 Hz. The antenna rotates at 4.5 rev/min and thus yields 16, 32 or 64 hits per point target per beamwidth.

Radar returns are generated by passing a white noise sequence with zero mean and specified variance, through a linear digital filter. The PSD of the resulting output sequence is given by (1), where $1/(2\pi|\sum_{k=0}^{p} a(k)\exp(-j\omega k)|)$ is the amplitude response of the filter. One reasonable PSD model is Butterworth filter [2]. Design parameters for four types of radar returns are given in Table 1, where W_s and W_p are normalized band-edge frequencies,R_p is the maximum ripple allowed in the passband, and R_s is the minimum stopband attenuation. Ground returns are simulated using lowpass filter whereas a bandpass filter is utilized for storm, birds flock and aircraft returns. The modeled received signals

Table 1.

Return source	W_p	W_s	R_p [dB]	R_s [dB]
Ground	0.0005	0.001	1	20
Storm	0.18; 0.22	0.10; 0.32	4	30
Birds flock	0.09; 0.16	0.01; 0.24	4	30
Aircraft	0.9; 0.91	0.86; 0.95	4	30

are additive sums of radar returns and thermal noise. The amplitudes of some generated signals are shown in Figure 2 [4]. Class prototypes (mean vectors and covariance matrices) are estimated using $K = 100$ time series for each type of radar returns. These time series represent an ergodic stationary process, i.e. the ensemble average is equal with probability one to the time average.

Table 2. $N = 64$

Filter order	Ground	Storm	Birds flock	Aircraft
$p = 2$	8.02	17.1	10.1	19.6
$p = 3$	8.2	7.1	2.9	9.2
$p = 4$	5.3	4.9	2.5	5.8
$p = 5$	3.9	2.7	2.3	3.8
$p = 6$	3.5	2.07	2.2	2.5
$p = 7$	4.3	1.6	1.4	1.2

[4] The corresponding normalized PSD are shown in Figure 1

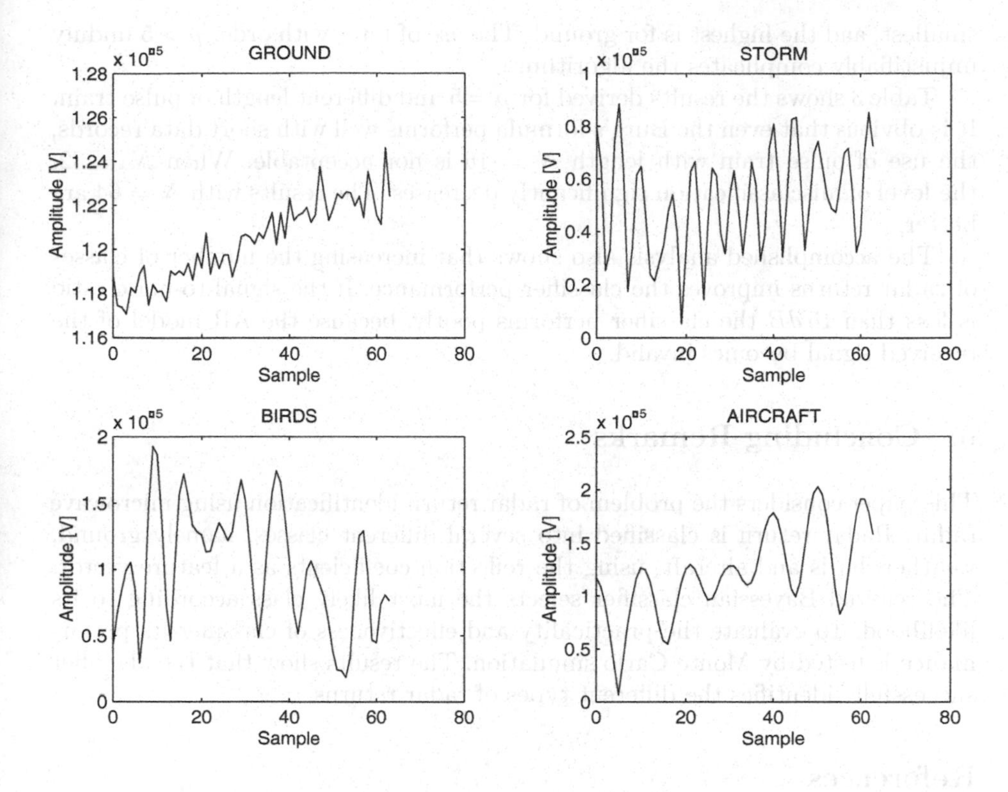

Fig. 2. Radar return amplitudes

The multi-segment Burg's formula (7) is used as the feature extractor for spatial extended ground, storm and birds flock returns. We assume $N = 16$, 32 or 64 pulse sequence in azimuth along $I = 4$ consecutive range rings. The aircraft returns (point target returns) are received from one ($I = 1$) range ring. The classifier performance is described by the percentage of misclassification for each class of signals. The test results for $N = 64$ and signal-to-noise ratio 25 dB are shown in Table 2. It can be seen that the higher the filter order, the better the classifier works. Acceptable level of misclassification, namely 3.2%, is achieved when $p = 5$. In this case the percentage of misclassification for birds flock is the

Table 3. $p = 5$

Pulse train	Ground	Storm	Birds flock	Aircraft
$N = 16$	26.8	38	29	9
$N = 32$	8.9	14.1	11.7	8.6
$N = 64$	3.9	2.7	2.3	3.8

smallest, and the highest is for ground. The use of filter with order $p > 5$ unduly unjustifiably complicates the algorithm.

Table 3 shows the results derived for $p = 5$ and different length of pulse train. It is obvious that even the Burg's formula performs well with short data records, the use of pulse train with length $N = 16$ is not acceptable. When $N = 32$, the level of misclassification significantly decreases. The results with $N = 64$ are better.

The accomplished analysis also shows that increasing the number of classes of radar returns improves the classifier performance. If the signal-to-noise ratio is less than $15\,dB$ the classifier performs poorly, because the AR model of the received signal becomes invalid.

6 Concluding Remarks

This paper considers the problem of radar return identification using microwave radar. Radar return is classified into several different classes, namely ground, weather, birds and aircraft, using the reflection coefficients as a feature vector. The realized Bayessian classifier selects the most likely class according to its likelihood. To evaluate the practicality and effectiveness of classifier its performance is tested by Monte Carlo simulation. The results show that the classifier successfully identifies the different types of radar returns.

References

1. Blackman S.: Multiple-Target Tracking with Radar Application. Artech House (1986)
2. Haykin S., Currie B., Kesler S.: Maximum-Entropy Spectral Analysis of Radar Clutter. Proc. IEEE, 70 (1982) 953–962.
3. Jouny I., Garber F., Ahalt S.: Classification of Radar Targets Using Synthetic Neural Networks. IEEE Trans. on AES, 29 (1993) 336–343.
4. Kay S., Makhool J.: On the Statistics of the Estimated Reflection Coefficients of an Autoregressive Process. IEEE Trans. on ASSP, 31 (1983) 1447–1455.
5. Leung H., Wu J.: Bayesian and Dempster-Shafer Target Identification for Radar Surveilance. IEEE Trans. on AES, 36 (2000) 432–447.
6. Stehwien W., Haykin S.: Statistical Classification of Radar Clutter. Proc. IEEE, CH2270-7/86/0000-0101 (1986) 101–106.

Estimation of Markovian Jump Systems with Unknown Transition Probabilities through Bayesian Sampling*

Vesselin P. Jilkov[1], X. Rong Li[1], and Donka S. Angelova[2]

[1] Department of Electrical Engineering, University of New Orleans
New Orleans, LA 70148, USA
{vjilkov, xli}@uno.edu

[2] Central Laboratory for Parallel Processing, Bulgarian Academy Of Sciences
25A Acad. G. Bonchev St, 1113 Sofia, Bulgaria
donka@bas.bg

Abstract. Addressed is the problem of state estimation for dynamic Markovian jump systems (MJS) with *unknown* transitional probability matrix (TPM) of the embedded Markov chain governing the system jumps. Based on recent authors' results, proposed is a new TPM-estimation algorithm that utilizes stochastic simulation methods (viz. Bayesian sampling) for finite mixtures' estimation. Monte Carlo simulation results of TMP-adaptive interacting multiple model algorithms for a system with failures and maneuvering target tracking are presented.

1 Introduction

Markovian jump systems (MJS) evolve in a jump-wise manner by switching among predefined dynamic models, according to a finite Markov chain. Theoretically, in the estimation for MJS (or more generally, *hybrid system* [1]) the TPM is almost always assumed known. In practice it is a "design parameter" whose choice is done *a priori*. For most applications, however, a priori information about the TPM may be inadequate or even lacking. Using an inaccurate TPM may lead to performance degradation of the state estimation. The uncertainty regarding the TPM is a major issue in the application of MM to real-life problems, e.g., maneuvering target tracking, systems' fault detection and isolation, and many others. Thus, it is highly desirable to have algorithms which can identify the TPM or estimate it recursively in the course of processing measurement data. There exist only a few publications in the engineering literature devoted to solving this important and difficult problem [2,3,4,5] (See [6] for a review of previous contributions).

Our work, initiated in [7,6] and furthered in this paper, is conceptually based on the assumption that the unknown TPM is a *random constant* matrix with a

* Research Supported in part by ONR grant N00014-00-1-0677, NSF grant ECS-9734285, NASA/LEQSF grant (2001-4)-01, Center of Excellence BIS21 grant ICA1-2000-70016, and Bulgarian NSF grant I-1202/02

I. Dimov et al. (Eds.): NMA 2002, LNCS 2542, pp. 307–315, 2003.

given prior distribution defined over the *continuous* set of valid TPMs. Within this Bayesian framework we have obtained a recursion for the TPM's posterior PDFs in terms of an MM estimator's model probabilities and likelihoods and developed an optimal estimator for the case of a two-state Markov chain. Since the optimal solution has (linearly) increasing computation, we also proposed (for the general case of a multiple-state Markov chain) three computationally feasible, approximate algorithms which utilize alternative approximation techniques — viz. second moment approximation, quasi-Bayesian (QB) approximation, and numerical integration. In this paper we continue with development of a new approximate algorithm for TPM estimation. It has been stimulated by the recent advances in the stochastic sampling (*Markov chain Monte Carlo* (MCMC)) methods for highly nonlinear estimation/filtering problems, e.g., [8]. Specifically, we utilize the *stochastic expectation maximization* (SEM) (also referred to as *data augmentation* (DA)) for finite mixture estimation [9] to implement our approach [7,6] for joint TPM and MM state estimation. To illustrate and compare the performance of the proposed sampling-based TPM-adaptive MM estimation algorithm, we also provide simulation results from two most typical applications of MM filtering.

2 Problem Formulation

Consider the following model of a Markovian jump system in discrete time $k = 1, 2, \ldots$

$$x(k) = f\left[k, m(k), x(k-1)\right] + g\left[k, m(k), w\left[k, m(k)\right]\right] \quad (1)$$

$$z(k) = h\left[k, m(k), x(k)\right] + v\left[k, m(k)\right] \quad (2)$$

where x is the *base (continuous) state* vector with transitional function f, w is the random input vector, z is the measurement vector with measurement function h, and v is the random measurement error vector. All vectors are assumed of appropriate dimensions. The *modal (discrete) state* $m(k) \in \mathbb{M} \triangleq \{1, 2, \ldots, m\}$ is a Markov chain with initial and transition probabilities respectively

$$P\{m_j(0)\} = \mu_j(0) \quad (3)$$

$$P\{m_j(k)\,|m_i(k-1)\} = P\{m_j(k)\,|m_i(k-1), z^k\} = \pi_{ij}, \; i, j = 1, \ldots m \quad (4)$$

where $m_i(k)$ stands for the event $\{m(k) = i\}$ and $z^k = \{z(1), \ldots, z(k)\}$.

Provided all parameters of the model (1)–(4) are *known* the MMSE-optimal estimate $\hat{x}(k) = E\left[x(k)\,|z^k\right]$ of the base state x can be obtained by the Bayesian *full-hypothesis-tree* (FHT) estimator [10]. The FHT is however infeasible because of its exponentially growing computation and thus suboptimal approximations with limited complexity are of interest in practice. There exist a number of such approximations, commonly referred to as *multiple model* (MM) algorithms. For the class of Bayesian MM estimation a good trade-off between performance and computational cost is provided by the popular *Interacting MM* (IMM) estimator

[11,12]. To be more specific we will apply our further development to the case of IMM algorithm as an example.

Let us consider now the state estimation problem for the above hybrid system model (1)–(4) without the presumed knowledge of the transition probability matrix $\Pi = [\pi_1, \pi_2, \ldots, \pi_m]'$ with $\pi_i = [\pi_{i1}, \pi_{i2}, \ldots, \pi_{im}]'$, $i = 1, \ldots, m$, defined by (4). We assume that Π is a *random* time-invariant matrix with given prior distribution defined over the simplex of valid TPMs. In such a Bayesian framework we consider the problem of finding recursively the posterior MMSE estimate $\overline{\Pi}(k) = E\left[\Pi | z^k\right]$ of Π for $k = 1, 2, \ldots$, and formulate the following.

Bayesian TPM-adaptive MM State Estimation. On receipt of a new measurement z_k:

– Run the MM algorithm with the previous state/covariance estimates, model probabilities $\mu(k-1)$ and $\overline{\Pi}(k-1)$ to update the current state/ covariance estimates, model probabilities

$$\mu(k) = [\mu_1(k), \ldots, \mu_m(k)]' \text{ with } \mu_i(k) \triangleq P\left\{ m_i(k) \mid \Pi, z^{k-1}\right\} \tag{5}$$

and model likelihoods

$$\mathbf{\Lambda}(k) = [\Lambda_1(k), \ldots, \Lambda_m(k)]' \text{ with } \Lambda_j(k) \triangleq p\left[z(k) \mid m_j(k), \Pi, z^{k-1}\right] \tag{6}$$

– Update the TPM estimate $\overline{\Pi}(k)$ based on $\overline{\Pi}(k-1)$ and the new measurement information contained in the TPM likelihood $p\left[z(k) | \Pi, z^{k-1}\right]$.

Note that this scheme is not restricted to the IMM algorithm only. Provided a feasible recursive TPM estimator, as specified above, is available the adaptation scheme is applicable to various MM state estimation algorithms in a straightforward manner.

3 Posterior PDF of TPM

Bayesian estimation of the TPM is based on the recursive relationships for updating the TPM's posterior PDF in terms of the model probabilities and likelihoods, defined by (5)–(6). As obtained in [6]

$$p\left(\Pi | z^k\right) = \frac{\mu'(k-1)\,\Pi\,\mathbf{\Lambda}(k)}{\mu'(k-1)\,\overline{\Pi}(k-1)\,\mathbf{\Lambda}(k)} p\left(\Pi | z^{k-1}\right) \tag{7}$$

where $\mu(k-1)$ and $\mathbf{\Lambda}(k)$ are computed by replacing the unknown Π with its best estimate $\overline{\Pi}(k-1)$ (conditioned on z^{k-1}) available at that time. Analogically, the marginal PDF $p\left(\pi_i | z^k\right)$ of each row π_i', $i = 1, \ldots, m$ of Π is given by the following.

Theorem. *If*

$$\int \pi_{l} p\left(\pi_1, \ldots, \pi_m | z^{k-1}\right) d\pi_1 \ldots d\pi_{i-1} d\pi_{i+1} \ldots d\pi_m = \overline{\pi}_l(k-1)\, p\left(\pi_i | z^{k-1}\right) \tag{8}$$

for $l, i = 1, 2, \ldots m$; $l \neq i$ then

$$p\left(\pi_i|z^k\right) = \left\{1 + \eta_i\left(k\right)\left[\pi_i - \overline{\pi}_i\left(k-1\right)\right]' \mathbf{\Lambda}\left(k\right)\right\} p\left(\pi_i|z^{k-1}\right) \tag{9}$$

where

$$\eta_i\left(k\right) = \frac{\mu_i\left(k-1\right)}{\mu'\left(k-1\right)\overline{\Pi}\left(k-1\right)\mathbf{\Lambda}\left(k\right)} \tag{10}$$

Proof. See [6].

Remark 1. The condition (8) is satisfied if π_i, $i = 1, \ldots, m$ are independent. However, it is in general looser than the independence assumption.

The recursion (9) has been used in [6] for derivation of four MMSE TPM estimators. In this work it is the basis of the sampling-based TPM estimator, developed next.

4 TPM Estimation by Stochastic Simulation

An immediate use of the Theorem is that it allows (under the stated assumptions) to decompose the estimation of the TPM into estimation of its rows. Indeed, from (9) the likelihood of each π_i, $i = 1, \ldots, m$, is represented in the following *mixture* form [6]

$$f\left[z\left(k\right)|\pi_i\right] = p\left[z\left(k\right)|\pi_i, z^{k-1}\right] = \sum_{j=1}^{m} \pi_{ij} g_{ij}\left(k\right) \quad \text{with} \tag{11}$$

$$g_{ij}\left(k\right) = g_{ij}\left[z\left(k\right)\right] = 1 + \eta_i\left(k\right)\left[\Lambda_j\left(k\right) - \overline{\pi}_i'\left(k-1\right)\mathbf{\Lambda}\left(k\right)\right] \tag{12}$$

Thus for each π_i, $i = 1, \ldots, m$, our TMP-estimation problem can be boiled down to a well known classification problem referred to as *prior probability estimation* of finite mixtures [13,14], or as *unsupervised learning* [15].

4.1 Mixture Estimation by Data Augmentation

The prior probability estimation problem can be stated as follows: *For the finite mixture model (11)–(12) with known mixture component PDFs $g_j\left[z(k)\right]$, estimate the unknown probabilistic weights π_j, $j = 1, \ldots, m$ given a sequence of independent observations $z(1)$, $z(2)$, ..., $z(K)$*

There exist a number of publications and results in solving this problem, see e.g. [13,14,16,15]. One of the most powerful and promising recent approaches in this direction is based on the idea of stochastic simulation — represent (approximately) the distributions (prior and posterior) by random samples rather than by some analytical, parametric (e.g. by moments), or numerical (by deterministic samples) descriptions. Bayesian sampling is loosely speaking a numerical mechanization (simulation) of the Bayesian update equation. In the context of the prior probability estimation problem its particular interpretation is known as *data augmentation* method [9]. It can be interpreted as a Bayesian extension of the well known *expectation-maximization* (EM) method [17] and is therefore also referred to as *stochastic* EM (SEM).

Specifically, represent first the mixture model (11) (the index i is dropped) in terms of missing (incomplete) data. That is, define vectors $\delta(k) = (\delta_1(k), \delta_2(k), \ldots, \delta_m(k))'$, $k = 1, 2, \ldots, K$ with components $\delta_j(k) \in \{0,1\}$, $j = 1, 2, \ldots, m$, which indicate that the measurement z_k has density $g_j[z(k)]$. Then the SEM algorithm, similarly to the regular EM, implements a two-step iterative scheme. In the first step, instead of estimating the "missing data" $\delta(k)$ by their expectation (as in the EM), the missing data are simulated by sampling from their conditional distribution, i.e., $\delta^{(s)} \sim p(\delta|\pi^{(s)})$. Then, in the second step of the iteration, the unknown mixture proportions π are sampled from their posterior, conditioned on the simulated missing parameters, i.e. $\pi^{(s+1)} \sim p(\pi|z, \delta^{(s)})$. Thus, by iterating with $s = 1, 2, \ldots$ this procedure, as shown in [9], converges to simulate from $p(\pi|z)$. In a more general setting this method is well known as Gibbs (or Bayesian) sampling and belongs to the class of the Markov chain Monte Carlo (MCMC) methods. Generally speaking, the Bayesian sampling produces an ergodic Markov chain $\pi^{(s)}$ with stationary distribution $p(\pi|z)$. For a sufficiently large s_0 a set of samples $\pi^{(s_0+1)}, \ldots, \pi^{(s_0+S)}$ is approximately distributed as $p(\pi|z)$ and, due to ergodicity, averaging can be made with respect to time.

4.2 TPM-adaptive MM Algorithm

By applying the outlined above SEM algorithm for mixture densities [9] to the TPM estimation formulated as a set of mixture estimation problems for each π_i, $i = 1, \ldots, m$ via (11)–(12), we obtain the following SEM-TPM estimator. The algorithm operates within a generic Bayesian TPM-adaptive MM estimation scheme (e.g. IMM) as specified in Section 2.

Algorithm *(SEM-TPM Estimator)*

For $k = 1, 2, \ldots$

For $i = 1, 2, \ldots m$ *(possibly in parallel)*

- *Conditional PDFs Computation*

$$g_{ij}(k) = 1 + \frac{1}{c_k}\mu_i(k-1)\left[\Lambda_j(k) - \overline{\pi}_i'(k-1)\,\mathbf{\Lambda}(k)\right], \quad j = 1, 2, \ldots m$$

$$\text{with } c_k = \mu'(k-1)\,\overline{\Pi}(k-1)\,\mathbf{\Lambda}(k)$$

- *Data Augmentation (Stochastic EM)*
  - *Initialization*

$$\pi_i^{(0)} = \overline{\pi}_i(k-1)$$

- *Iterations* $(s = 0, 1, \ldots, s_0 + S - 1)$
  * *Missing Data Conditional PMFs*

$$q_{ij}^{(s)}(l) = \frac{\pi_{ij}^{(s)} g_{ij}(l)}{\sum_{j=1}^{m} \pi_{ij}^{(s)} g_{ij}(l)}, \quad l = 1, 2, \ldots k, \quad j = 1, 2, \ldots m$$

  * *Missing Data Generation (Multinomial Sampling)*

$$\delta_i^{(s)}(l) = (0, \ldots, 0, 1, 0, \ldots, 0) \sim \left\{ q_{ij}^{(s)}(l) \right\}_{j=1}^{m}, \quad l = 1, 2, \ldots k$$

  * *Parameter Evaluation (Dirichlet Distribution Sampling)*

$$\pi_i^{(s+1)} \sim \mathcal{D}\left(\pi; \alpha_{i1} + \sum_{l=1}^{k} \delta_{i1}^{(s)}(l), \ldots, \alpha_{im} + \sum_{l=1}^{k} \delta_{im}^{(s)}(l) \right)$$

- *Output Estimates*

$$\overline{\pi}_i(k) = \frac{1}{S} \sum_{\sigma=1}^{S} \pi_i^{(s_0+\sigma)}$$

Remark 2. The use of the *Dirichlet distribution* (DD) [18] $\mathcal{D}(\pi; \alpha_1, \ldots, \alpha_m)$, $\alpha_j \geq 0$, $j = 1, \ldots, m$ for a mixture's prior probabilities $\pi = (\pi_1, \ldots, \pi_m)$ in the DA algorithm is justified by the fact that the conjugate priors on π are DDs [9]. The SEM-TPM estimator requires knowledge of the prior parameters of α_{ij}. This however is not a shortcoming of the algorithm, because if $\alpha_{i1} = \alpha_{i2} = \ldots = \alpha_{im} = 1$ the DD reduces to an uniform distribution and the algorithm can be initialized with noniformative (uniform) prior.

5 Simulation

In this section we illustrate the performances of the proposed TPM-adaptive IMM algorithm. Two typical applications of Markov switching models are simulated – *a system subject to measurement failures* and *maneuvering target tracking*.

5.1 Example 1: A System with Failures

Under consideration is a scalar dynamic system described for $k = 1, 2, \ldots$ by

$$x(k) = x(k-1) + w(k)$$
$$z(k) = m(k)x(k) + [100 - 90m(k)]\, v(k)$$

with $x(0) \sim \mathcal{N}(0, 20^2)$, $w(k) \sim \mathcal{N}(0, 2^2)$, $v(k) \sim \mathcal{N}(0, 1)$. The Markov model of switching is: $m(k) \in \left\{ m^{(1)}, m^{(2)} \right\} = \{0, 1\}$, $\mu_1(0) = \mu_2(0) = 0.5$, $\pi_{ij} =$

$P\{m(k+1) = m^{(j)}|m(k) = m^{(i)}\}$, $i, j = 1, 2$. Note that $m(k) = 0$ corresponds to a measurement failure at k.

Three IMM algorithms were implemented: an IMM knowing the true TPM (*TPM-exact* IMM), an IMM with a typical design of fixed TPM with diagonally-dominant elements ($\pi_{11} = \pi_{22} = 0.9$) (*non-adaptive* IMM), and an IMM using the SEM-TPM estimator (*TPM-adaptive* IMM).

In the scenario simulated the true TPM was with $\pi_{11} = 0.6$, $\pi_{22} = 0.15$. Results from 100 Monte Carlo runs are shown in Figure 1: (a) Accuracy/convergence of the SEM-TPM estimator and (b) comparative accuracy of the three IMM algorithms. The estimation accuracies of TPM and state estimation are evaluated in terms of *mean absolute error* – the absolute value of the actual error, averaged over runs. Figure 1 (a) illustrates the convergence of the TPM estimator. Figure 1 (b) shows that in case of a mismatch regarding the TPM the adaptive algorithms perform better than the typical IMM design (with π_{22} fairly largely mismatched in this scenario).

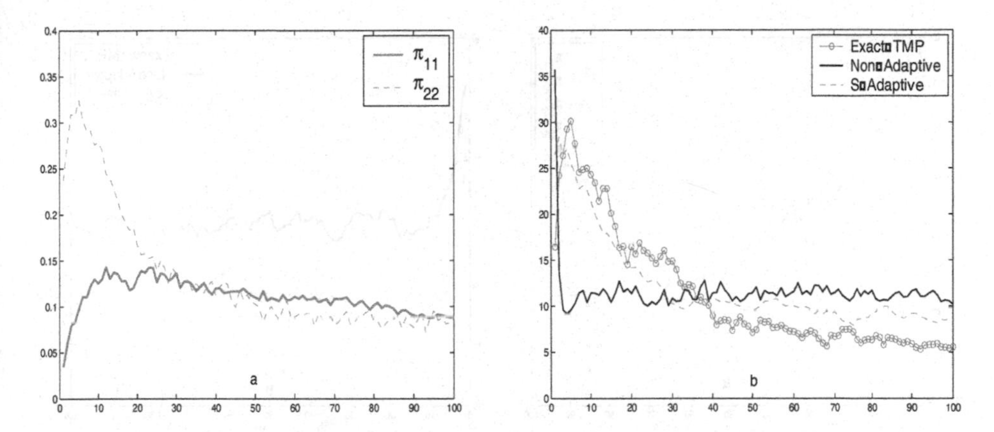

Fig. 1. Deterministic TPM Scenario: Error of (a) TPM and (b) IMM estimation

5.2 Example 2: Maneuvering Target Tracking

We considered a simplified illustrative example of maneuvering target with Markov jumping acceleration. The one-dimensional target dynamics is given by

$$\begin{bmatrix} p \\ v \end{bmatrix}(k+1) = \begin{bmatrix} 1 & T \\ 0 & 1 \end{bmatrix}\begin{bmatrix} p \\ v \end{bmatrix}(k) + \begin{bmatrix} T^2/2 \\ T \end{bmatrix}[a(k+1) + w(k+1)]$$

where p, v and a stand for the target position, velocity and acceleration respectively and w models "small" perturbations in the acceleration. $p(0) \sim \mathcal{N}[80000, 100^2]$, $v(0) \sim \mathcal{N}[400, 100^2]$ and $w(k) \sim \mathcal{N}[0, 2^2]$ is white noise sequence independent of $p(0)$ and $v(0)$. The acceleration process $a(k)$ is a

Markov chain with three states $a_1 = 0$, $a_2 = 20$, $a_2 = -20$ and initial probabilities $\mu_1(0) = 0.8$, $\mu_2(0) = \mu_3(0) = 0.1$. The sampling period is $T = 10$. The measurement equation is $z(k) = x(k) + \nu(k)$ where the measurement error $\nu(k) \sim \mathcal{N}\left[0, 100^2\right]$ is white noise.

Again three IMM algorithms were implemented: TPM-exact, non-adaptive (with $\pi_{ii} = 0.9, \pi_{ij} = 0.05$ $j \neq i$, $i, j = 1, 2, 3$), and TPM-adaptive.

To obtain an overall, scenario independent evaluation we ran the algorithms over an ensemble of TPMs, *randomly* (uniformly) chosen from the set of *all* valid TPMs. Results from 100 MC runs of such *randomly generated TPM* scenarios are presented in Figure 2. The adaptive IMMs show better overall performance than the non-adaptive one (Figure 2 (b)). Clearly, in quite a lot of the runs the fixed TPM of the latter is mismatched with the true one, while the adaptive algorithms provide better estimated TPM. This indicates that for situations where the target behavior can not be described by one TPM it is preferable to use the adaptive TPM.

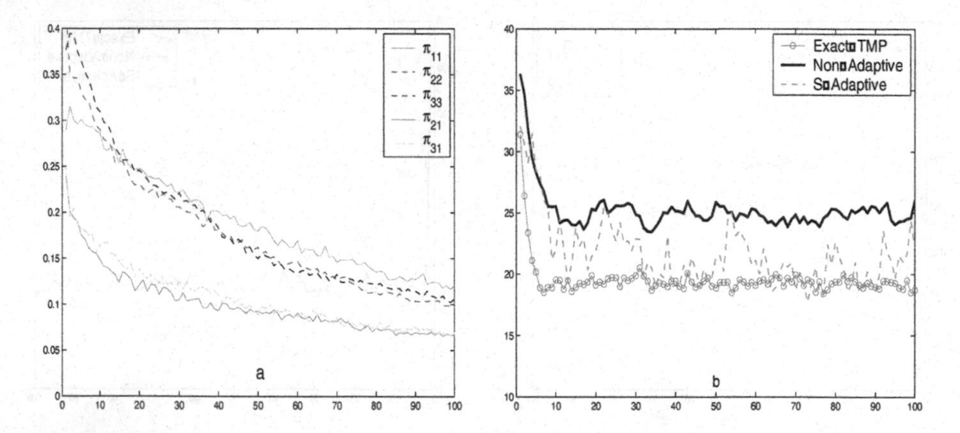

Fig. 2. Random-TPM Scenario: Accuracy of (a) TPM and (b) IMM velocity estimation

6 Conclusion

A new, stochastic sampling-based algorithm for estimating the *unknown* transition probabilities of a Markovian jumping dynamic system within the multiple model state estimation paradigm has been proposed in this paper. As illustrated by simulation examples, the proposed algorithm can effectively serve for off-line identification of hybrid (Markovian jump) dynamic systems, or for on-line adaptation of the transition probability matrix in multiple model state estimation algorithms.

References

1. Li, X.R.: Hybrid Estimation Techniques. In: Leondes, C.T. (ed.): Control and Dynamic Systems: Advances in Theory and Applications. Vol. 76. Academic Press, New York (1996) 213–287.
2. Sawaragi, Y., Katayama, T., Fujichige, S.: Adaptive Estimation for a Linear System with Interrupted Observations. IEEE Trans. Automatic Control(1973) 152–154.
3. Tugnait, J.K., Haddad, A.H.: Adaptive Estimation in Linear Systems with Unknown Markovian Noise Statistics. IEEE. Trans. Information Theory26 (1980) 66–78.
4. Tugnait, J.K.: Adaptive Estimation and Identification for Discrete Systems with Markov Jump Parameters. IEEE Trans. Automatic ControlAC-27 (1982) 1054–1065.
5. Goutsias, J., Mendel, J.M.: Optimal Simultaneous Detection and Estimation of Filtered Discrete semi-Markov Chains. IEEE. Trans. Information TheoryIT-34 (1988) 551–568.
6. Jilkov, V.P., Li, X.R.: Bayesian Estimation of Transition Probabilities for Markovian Jump Systems. IEEE Trans. Signal Processing(2002) Submitted.
7. Jilkov, V.P., Li, X.R.: Adaptation of Transition Probability Matrix for Multiple Model Estimators. In: Proc. Fourth Intern. Conf. on Information Fusion, Montreal, QC, Canada (2001)
8. Doucet, A., de Frietas, N., Gordon, N. (eds.): Sequential Monte Carlo Methods in Practice. Statistics for Engineering and Information Science. Springer-Verlag, New York (2001)
9. Diebolt, J., Robert, C.P.: Estimation of Finite Mixture Distributions through Bayesian Sampling. J. R. Statist. Soc., 56 (1994) 363–375.
10. Bar-Shalom, Y., Li, X.R., Kirubarajan, T.: Estimation with Applications to Tracking and Navigation: Theory, Algorithms, and Software. Wiley, New York (2001)
11. Blom, II.A.P., Bar-Shalom, Y.: The Interacting Multiple Model Algorithm for Systems with Markovian Switching Coefficients. IEEE Trans. Automatic ControlAC-33 (1988) 780–783.
12. Mazor, E., Averbuch, A., Bar-Shalom, Y., Dayan, J.: Interacting Multiple Model Methods in Target Tracking: A Survey. IEEE Trans. Aerospace and Electronic SystemsAES-34 (1996) 103–123.
13. Kazakos, D.: Recursive Estimation of Prior Probabilities Using a Mixture. IEEE. Trans. Information TheoryIT-23 (1977) 203–211.
14. Dattatreya, G.R., Kanal, L.N.: Estimation of Mixing Probabilities in Multiclass Finite Mixtures. IEEE Trans. Systems, Man, and Cybernetics, SMC-20 (1990) 149–158
15. Titterington, D., Smith, A., Makov, U.: Statistical Analysis of Finite Mixture Distributions. John Wiley & Sons (1985)
16. Smith, A.F.M., Makov, U.E.: A Quasi-Bayesian Sequential Procedures for Mixtures. J. R. Statist. Soc., 40 (1978) 106–112.
17. Dempster, A.P., Liard, N.M., Rubin, D.B.: Maximum Likelihood from Incomplete Data Via the EM Algorithm. J. R. Statist. Soc., B 39 (1977) 1–38.
18. Port, S.C.: Theoretical Probability for Applications. John Wiley & Sons (1994)

A Comparison of Decision Making Criteria and Optimization Methods for Active Robotic Sensing

Lyudmila Mihaylova, Tine Lefebvre, Herman Bruyninckx, Klaas Gadeyne, and Joris De Schutter

Katholieke Universiteit Leuven, Celestijnenlaan 300B, B-3001 Heverlee Belgium,
Lyudmila.Mihaylova@mech.kuleuven.ac.be

Abstract. This work presents a comparison of decision making criteria and optimization methods for active sensing in robotics. *Active sensing* incorporates the following aspects: (*i*) where to position sensors, and (*ii*) how to make decisions for next actions, in order to maximize information gain and minimize costs. We concentrate on the second aspect: "Where should the robot move at the next time step?". Pros and cons of the most often used statistical decision making strategies are discussed. Simulation results from a new multisine approach for active sensing of a nonholonomic mobile robot are given.

1 Introduction

One of the features of robot intelligence is to deal robustly with uncertainties. This is only possible when the robot is equipped with sensors, e.g., contact sensors, force sensors, distance sensors, cameras, encoders, gyroscopes. To perform a task, the robot first needs to know: "Where am I now?". After that the robot needs to decide "What to do next?", weighting future information gain and costs. The latter decision making process is called *active sensing*. Distinction is made sometimes between active sensing and active localization. "Active localization" refers to robot motion decisions (e.g. velocity inputs), "active sensing" to sensing decisions (e.g. when a robot is allowed to use only one sensor at a time). In this paper we refer to both strategies as "active sensing". Choosing actions requires to trade off the immediate with the long-term effects: the robot should take both actions to bring itself closer to its *task completion* (e.g. reaching a goal position within a certain tolerance) and actions for the purpose of *gathering information*, such as searching for a landmark, surrounding obstacles, reading signs in a room, in order to keep the uncertainty small enough at each time instant and assure a good task execution. Typical tasks where active sensing is useful are performed in less structured environments. The uncertainties are so important that they influence the task execution: industrial robot tasks in which the robot is uncertain about the configuration (positions and orientation) of its tool and work pieces [1]; mobile robot navigation in a known map (indoor and outdoor) [2,3] where starting from an uncertain initial configuration the robot has to move

I. Dimov et al. (Eds.): NMA 2002, LNCS 2542, pp. 316–324, 2003.

to a desired goal configuration within a preset time; vision applications with active selection of camera parameters such as focal length and viewing angle to improve the object recognition procedures [4,5]; reinforcement learning [6]: the robot needs to choose a balance between its localization (*exploiting*) and the new information it can gather about the environment (*exploring*).

Estimation, control and active sensing. Next to an active sensing module, intelligent robots should also include an estimator and a controller:

- *Estimation.* To overcome the uncertainty in the robot and environment models, as well as the sensor data, estimation techniques [7,8] compute the system state after fusing the data in an optimal way.
- *Control.* Knowing the desired task, the controller is charged with following the task execution as closely as possible. Motion execution can be achieved either by feedforward, feedback control or a combination of both [9].
- *Active sensing* is the process of determining the inputs by optimizing a function of costs and utilities. These inputs are then sent to the controller.

Active sensing is challenging for various reasons: (*i*) The robot and sensor models are *nonlinear*. Some methods linearize these models, but many nonlinear problems cannot be treated this way and impose the necessity to develop special techniques for action generation. (*ii*) The task solution depends on an *optimality criterion* which is a *multi-objective* function weighting the information gain and some costs. It is related to the *computational load* especially important for *on-line* task execution. (*iii*) *Uncertainties* in the robot and environment models, the sensor data need to be dealt with. (*iv*) Often measurements do not supply information about all variables, i.e. the system is *partially observable.*

The remainder of the paper is organized as follows. In Section 2, the active sensing problem is described. The most often used decision making criteria are compared and results for active sensing of a nonholonomic mobile robot are presented. Section 3 gives the main groups of optimization algorithms for active sensing. Section 4 terminates with the conclusions.

2 Active Sensing Problem Formulation

Active sensing can be considered as a trajectory generation for a *stochastic dynamic* system described by the model

$$\boldsymbol{x}_{k+1} = \boldsymbol{f}(\boldsymbol{x}_k, \boldsymbol{u}_k, \boldsymbol{\eta}_k) \tag{1}$$

$$\boldsymbol{z}_{k+1} = \boldsymbol{h}(\boldsymbol{x}_{k+1}, \boldsymbol{s}_{k+1}, \boldsymbol{\xi}_{k+1}) \tag{2}$$

where $\boldsymbol{x}$ is the system state vector, $\boldsymbol{f}$ and $\boldsymbol{h}$ nonlinear system and measurement functions, $\boldsymbol{z}$ is the measurement vector, $\boldsymbol{\eta}$ and $\boldsymbol{\xi}$ are respectively system and measurement noises. $\boldsymbol{u}$ stands for the input vector of the state function, $\boldsymbol{s}$ stands for a sensor parameter vector as input of the measurement function (an example is the focal length of a camera). The subscript k denotes discrete time. The system's states and measurements are influenced by the inputs $\boldsymbol{u}$ and $\boldsymbol{s}$. Further, we make no distinction and denote both inputs to the system with $\boldsymbol{a}$ (actions). Conventional systems consisting only of control and estimation components assume that these inputs are given and known. Intelligent systems should be able to adapt the inputs in a way to get the "best" estimates and in the meanwhile to perform the *active sensing* task "as good as possible".

So, an appropriate *multi-objective performance criterion* (often called *value function*) is needed to quantify for each sequence of actions $\boldsymbol{a}_1, \ldots, \boldsymbol{a}_N$ (also called *policy*) both the information gain and the gain in task execution:

$$J = \min_{\boldsymbol{a}_1, \ldots, \boldsymbol{a}_N} \{ \sum_j \alpha_j \mathcal{U}_j + \sum_l \beta_l \mathcal{C}_l \} \tag{3}$$

This criterion is composed by a weighted sum of *rewards*: (*i*) j terms $\mathcal{U}_j$ characterizing the minimization of *expected uncertainties* (maximization of *expected information extraction*) and (*ii*) l terms $\mathcal{C}_l$ specifying other *expected costs and utilities*, e.g. travel distance, time, energy, distances to obstacles. Both $\mathcal{U}_j$ and $\mathcal{C}_k$ are function of the policy $\boldsymbol{a}_1, \ldots, \boldsymbol{a}_N$. The weighting coefficients α_j and β_l give different impact to the two parts, and are arbitrarily chosen by the designer. When the state at the goal configuration fully determines the rewards, the terms $\mathcal{U}_j$ and $\mathcal{C}_l$ are computed based on this state only. When attention is paid to both the goal configuration and the intermediate time evolution, the terms $\mathcal{U}_j$ and $\mathcal{C}_l$ are a function of the $\mathcal{U}_{j,k}$ and $\mathcal{C}_{l,k}$ at different time steps k. Criterion (3) is to be minimized with respect to the sequence of actions under *constraints*

$$\boldsymbol{c}(\boldsymbol{x}_1, \ldots, \boldsymbol{x}_N, \boldsymbol{a}_1, \ldots, \boldsymbol{a}_N) \leq \boldsymbol{c}_{thr}. \tag{4}$$

$\boldsymbol{c}$ is a vector of physical variables that can not exceed some threshold values $\boldsymbol{c}_{thr}$. The thresholds express for instance maximal allowed velocities and acceleration, maximal steering angle, minimum distance to obstacles, etc.

2.1 Action Sequence

The description of the sequence of actions $\boldsymbol{a}_1, \ldots, \boldsymbol{a}_N$ can be done in different ways and has a major impact on the optimization problem that needs to be solved afterwards (Section 3).

- The actions can be described as lying on a reference trajectory plus a *parameterized* deviation of it (e.g. by a finite sine/cosine series, or by an elastic band or elastic strip formulation, [9,10]). In this way, the optimization problem is reduced to a finite-dimensional optimization problem on the parameters.
- The most general way to present the policy is a sequence of freely chosen actions, not restricted to a certain form of trajectory. Constraints, such as maximal acceleration and maximal velocity, can be added to produce executable trajectories. This active sensing problem is called a *Markov Decision Process* (MDP) for systems with fully observable states and *Partially Observable Markov Decision Process* (POMDP) for systems where measurements do not fully observe the states or for systems with measurement noise.

2.2 Performance Criteria Related to Uncertainty

The terms $\mathcal{U}_j$ represent (*i*) the expected uncertainty of the system about its state; or (*ii*) this uncertainty compared to the accuracy needed for the task completion. In a Bayesian framework, the characterization of the uncertainty of the estimate is based on a scalar loss function of its probability density function. Since no scalar function can capture all aspects of a matrix, no function suits the needs of every experiment. Common used functions are:

- ***based on the covariance matrix***: The covariance matrix $\boldsymbol{P}$ of the probability distribution of state $\boldsymbol{x}$ is a measure for the uncertainty on the estimate. Minimizing $\boldsymbol{P}$ corresponds to minimizing the uncertainty. Active sensing is looking for actions which minimize the *posterior* covariance matrix $\boldsymbol{P} = \boldsymbol{P}_{post}$ or the inverse of the *Fisher information matrix* $\boldsymbol{I}$ [11] which describes the posterior covariance matrix of an efficient estimator $\boldsymbol{P} = \boldsymbol{I}^{-1}$. Several scalar functions of $\boldsymbol{P}$ can be applied [12]:
  - *D-optimal design*: minimizes the matrix determinant, $det(\boldsymbol{P})$, or the logarithm of it, $log(det(\boldsymbol{P}))$. The *minimum* is *invariant* to any transformation of the state vector $\boldsymbol{x}$ with a non-singular Jacobian such as scaling. Unfortunately, this measure does not allow to verify task completion: the determinant of the matrix being smaller than a certain value does not impose any of the covariances of the state variables to be smaller than their toleranced value.
  - *A-optimal design*: minimizes the trace $tr(\boldsymbol{P})$. Unlike D-optimal design, A-optimal design does not have the invariance property. The measure does not even make sense physically if the target states have inconsistent units. On the other hand, this measure allows to verify task completion.
  - *L-optimal design*: minimizes the weighted trace $tr(\boldsymbol{W}\boldsymbol{P})$. A proper choice of the weighting matrix $\boldsymbol{W}$ can render the L-optimal design criterion *invariant* to transformations of the variables $\boldsymbol{x}$ with a non-singular Jacobian: $\boldsymbol{W}$ has units and is also transformed. A special case of L-optimal design is the tolerance-weighted L-optimal design [1], which proposes a natural choice of $\boldsymbol{W}$ depending on the desired standard deviations (tolerances) at task completion. The value of this scalar function has a *direct relation to the task completion.*
  - *E-optimal design*: minimizes the maximum eigenvalue $\lambda_{max}(\boldsymbol{P})$. Like A-optimal design, this is not invariant to transformations of $\boldsymbol{x}$, nor does the measure makes sense physically if the target states have inconsistent units, but the measure allows to verify task completion.
- ***based on the probability density function***: Entropy [13] is a measure of the uncertainty of a state estimate containing more information about the probability distribution than the covariance matrix, at the expense of more computational costs. The entropy based performance criteria are:
  - the *entropy* of the posterior distribution: $E[-\log p_{post}(\boldsymbol{x})]$. $E[.]$ indicates the expected value.
  - the *change in entropy* between two distributions $p_1(\boldsymbol{x})$ and $p_2(\boldsymbol{x})$: $E[-\log p_2(\boldsymbol{x})] - E[-\log p_1(\boldsymbol{x})]$. For active sensing, $p_1(\boldsymbol{x})$ and $p_2(\boldsymbol{x})$ can be the prior and posterior or the posterior and the goal distribution.

- the *Kullback-Leibler distance* or *relative entropy* [14] is a measure for the goodness of fit or closeness of two distributions: $E[\log \frac{p_2(\boldsymbol{x})}{p_1(\boldsymbol{x})}]$. The expected value is calculated with respect to $p_2(\boldsymbol{x})$. The relative entropy and the change in the entropy are *different* measures. The *change in entropy* only quantifies how much the form of the probability distributions changes whereas the *relative entropy* also represents a measure of how much the distribution has moved. If $p_1(\boldsymbol{x})$ and $p_2(\boldsymbol{x})$ are the same distributions, translated by different mean values, the change in entropy is zero, while the relative entropy is not.

Example. Distance and orientation sensing of a mobile robot to known beacons is considered. The sequence of motions of a nonholonomic wheeled mobile robot (WMR) [15], moving from a starting to a goal configuration, is restricted to a *parameterized trajectory.* The optimal trajectory is searched in the class $\mathcal{Q} = \mathcal{Q}(\boldsymbol{p}), \boldsymbol{p} \in \mathcal{P}$, of harmonic functions, where $\boldsymbol{p}$ is a vector of parameters obeying to preset physical constraints. With N the number of functions, the new (modified) robot trajectory is generated on the basis of a reference one by the lateral deviation l_k (*lateral* is called the orthogonal robot motion deviation from a straight line reference trajectory in y direction) as a linear superposition

$$l_k = \sum_{i=1}^{N} A_i sin(i\pi \frac{s_{r,k}}{s_{r,total}}), \tag{5}$$

of sinusoids, with constant amplitudes A_i, $s_{r,k}$ is the path length up to instant k, $s_{r,total}$ is the total path length, and r refers to the reference trajectory. In this formulation active sensing is a *global optimization problem* (on the whole robot trajectory) with a criterion to be minimized

$$J = \min_{A_{i,k}} \{\alpha_1 \mathcal{U} + \alpha_2 \mathcal{C}\} \tag{6}$$

under *constraints* (for the robot velocity, steering and orientation angles). α_1 and α_2 are dimensionless positive weighting coefficients. Here $\mathcal{U}$ is in the form

$$\mathcal{U} = tr(\boldsymbol{W}\boldsymbol{P}), \tag{7}$$

where $\boldsymbol{P}$ is the covariance matrix of the estimated states (at the goal configuration), computed by an Unscented Kalman filter [16] and $\boldsymbol{W}$ is a weighting matrix). The cost term $\mathcal{C}$ is assumed to be the relative time $\mathcal{C} = t_{total}/t_{r,total}$, where t_{total} is the total time for reaching the goal configuration on the modified trajectory, $t_{r,total}$ the respective time over the reference trajectory. The weighting matrix $\boldsymbol{W}$ represents a product of a normalizing matrix $\boldsymbol{N}$, and a scaling matrix $\boldsymbol{M}$, $\boldsymbol{W} = \boldsymbol{M}\ \boldsymbol{N}$. The matrix $\boldsymbol{N} = diag\{1/\sigma_1^2,\ 1/\sigma_2^2, \ldots, \sigma_n^2\}$. σ_i, $i = 1, \ldots, n$, are assumed here to be the standard deviations at the goal configuration on the reference trajectory. Depending on the task, they could be chosen otherwise. The scaling matrix $\boldsymbol{M}$ here is the identity matrix. Simulation results obtained both with (7), and with the averaged criterion $\mathcal{U}_a = \frac{1}{k_b - k_a} \sum_{k=k_a}^{k_b} tr(\boldsymbol{W_k}\boldsymbol{P_k})$

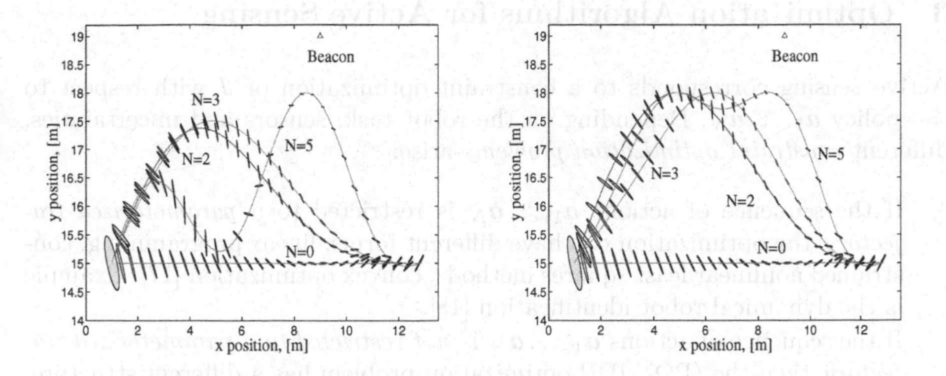

Fig. 1. Trajectories, generated with : (a) $\mathcal{U} = tr(\boldsymbol{W}P)$ (b) the averaged criterion $\mathcal{U}_a = \frac{1}{k_b - k_a} \sum_{k=k_a}^{k_b} tr(\boldsymbol{W_k P_k})$

with optimization over the interval $[k_a, k_b] = [30sec, 100sec]$ are given on Figs. 1 and 2. The modified trajectory, generated with different number of sinusoids N (in accordance with (5)), and the reference trajectory are plotted together with the uncertainty ellipses Figs. 1,2. As it is seen from Figs. 1,2 the most accurate results at the goal configuration for $\mathcal{U}$ and J are obtained with $N = 5$ sinusoids. Better accuracy is provided with bigger N, at the cost of increased computational load. Through active sensing the robot is approaching to the beacons (Fig. 1), that is a distinction from a movement over a reference trajectory. Faster increase of the information at the beginning of the modified trajectories and higher accuracy, is obtained than those on the straight-line. From other side, trajectories generated by the averaged criterion $\mathcal{U}_a$ are characterized with better general performance then those generated with (7) (Fig. 2).

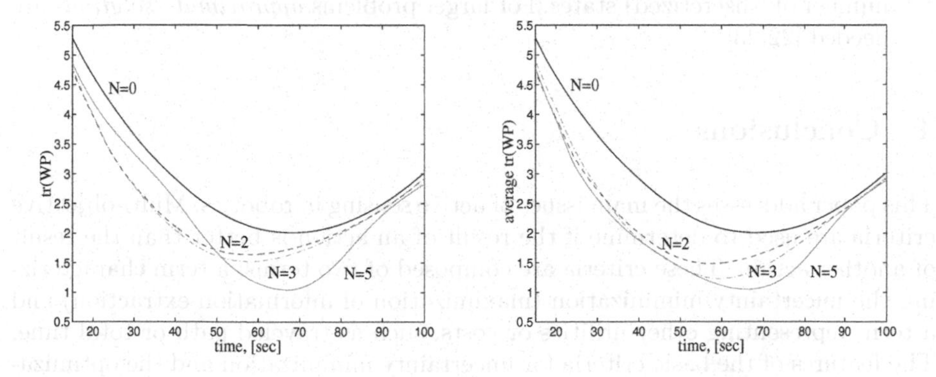

Fig. 2. Performance criteria: (a) $\mathcal{U} = tr(\boldsymbol{W}P)$, (b) $\mathcal{U}_a = \frac{1}{k_b - k_a} \sum_{k=k_a}^{k_b} tr(\boldsymbol{W_k P_k})$

3 Optimi ation Algorithms for Active Sensing

Active sensing corresponds to a constraint optimization of J with respect to the policy $\boldsymbol{a}_1, \ldots \boldsymbol{a}_N$. Depending on the robot task, sensors and uncertainties, different *constraint optimization problems* arise:

- If the sequence of actions $\boldsymbol{a}_1, \ldots \boldsymbol{a}_N$ is restricted to a *parameterized trajectory*, the optimization can have different forms: linear programming, constrained nonlinear least squares methods, convex optimization [17]. Example is the dynamical robot identification [18].
- If the sequence of actions $\boldsymbol{a}_1, \ldots \boldsymbol{a}_N$ is *not restricted to a parameterized trajectory*, then the (PO)MDP optimization problem has a different structure. This could be a finite-horizon, i.e. over a fixed finite number of time steps (N is finite), or an infinite-horizon problem ($N = \infty$). For every state it is rather straightforward to know the immediate reward being associated to every action (1 step policy). The goal however is to find the policy that maximizes the reward over a long term (N steps). Different optimization procedures exist for this kind of problems, examples are:
  - *Value iteration:* due to the sequential structure of the problem, the optimization can be performed as subsequent solution of problems with only 1 (of the N) variables a_i. The *value iteration* algorithm, a *dynamic programming* algorithm, calculate recursively the optimal value function and policy [19] for finite and infinite horizon problems.
  - *Policy iteration:* an iterative technique similar to dynamic programming, is introduced by Howard [20] for infinite horizon systems.
  - *Linear programming:* an infinite horizon problem can be represented and solved as a linear program [21].
  - *State based search methods* represent the system as a graph whose nodes correspond to states and can handle finite and infinite horizon problems [22]. *Tree search* algorithms search for the optimal path in the graph.

  Unfortunately, exact solutions can only be found for (PO)MDPs with a small number of (discretized) states. For larger problems *approximate solutions* are needed [22,23].

4 Conclusions

This paper addresses the main issues of active sensing in robotics. Multi-objective criteria are used to determine if the result of an action is better than the result of another action. These criteria are composed of two terms: a term characterizing the uncertainty minimization (maximization of information extraction) and a term representing other utilities or costs, such as traveled path or total time. The features of the basic criteria for uncertainty minimization and the optimization procedures are outlined. Simulation results for active sensing of a wheeled mobile robot with a parameterized sequence of actions are presented.

Acknowledgments. Herman Bruyninckx and Tine Lefebvre are, respectively, Postdoctoral and Doctoral Fellows of the Fund for Scientific Research-Flanders (F.W.O–Vlaanderen) in Belgium. Lyudmila Mihaylova is a Postdoctoral Fellow at Katholieke Universiteit Leuven, on leave from the Bulgarian Academy of Sciences. Financial support by the Center of Excellence BIS21 grant ICA1-2000-70016 and the K. U. Leuven's Concerted Research Action GOA/99/04 are gratefully acknowledged.

References

1. Geeter, J. D., De Schutter, J., Bruyninckx, H., Brussel, H. V., and Decrton, M.: Tolerance-Weighted L-optimal Experiment Design: a New Approach to Task-Directed Sensing. Advanced Robotics, Vol. 13, no.4 (1999) 401–416.
2. Roy, N., Burgard, W., Fox, D., and Thrun, S.: Coastal Navigation — Mobile Robot Navigation with Uncertainty in Dynamic Environments. Proc. of ICRA (1999)
3. Cassandra, A., Kaelbling, L., and Kurien, J.: Acting under Uncertainty: Discrete Bayesian Models for Mobile Robot Navigation. Proc. of the IEEE/RSJ Int. Conf. on Intelligent Robots and Systems (1996)
4. DeSouza, G. N. and Kak, A.: Vision for Mobile Robot Navigation: A Survey. IEEE Trans. on Pattern Analysis and Machine Intel., Vol. 24, no. 2 (2002) 237–267.
5. Denzler, J. and Brown, C.: Information Theoretic Sensor Data Selection for Active Object Recognition and State Estimation. IEEE Transactions on Pattern Analysis and Machine Intelligence, Vol. 24, (2002) 145–157.
6. Sutton, R. and Barto, A.: Reinforcement Learning, An introduction. MIT (1998)
7. Arulampalam, M., Maskell, S., Gordon, N., and Clapp, T.: A Tutorial on Particle Filters for Online Nonlinear/non-Gaussian Bayesian Tracking. IEEE Trans. on Signal Proc., Vol. 50, no. 2 (2002) 174–188.
8. Bar-Shalom, Y. and Li, X.: Estimation and Tracking: Principles, Techniques and Software. Artech House (1993)
9. Laumond, J.-P.: Robot Motion Planning and Control. Guidelines in Nonholonomic Motion Planning for Mobile Robots, by Laumond, J.-P, Sekhavat, S., and Lamiraux, F., available at http://www.laas.fr/~jpl/book.html: Springer-Verlag (1998)
10. Brock, O. and Khatib, O.: Elastic Strips: A Framework for Integrated Planning and Execution. Proc. of 1999 Int. Symp. of Experim. Robotics, (1999) 245–254.
11. Fisher, R.: On the Mathematical Foundations of Theoretical Statistics. Phylosophical Trans. of the Royal Society of London, Series A, Vol. 222 (1922) 309–368.
12. Fedorov, V.: Theory of Optimal Experiments. Academic press, New York ed. (1972)
13. Shannon, C.: A Mathematical Theory of Communication, I and II. The Bell System Technical Journal, Vol. 27 (1948) 379–423 and 623–656.
14. Kullback, S.: On Information and Sufficiency. Annals of mathematical Statistics, Vol. 22 (1951) 79–86.
15. Mihaylova, L., De Schutter, J., and Bruyninckx, H.: A Multisine Approach for Trajectory Optimization Based on Information Gain. Proc. of IROS Conf. (2002)
16. Julier, S., Uhlman, J., and Durrant-Whyte, H.: A New Method for the Transformation of Means and Covariances in Filters and Estimators. IEEE Trans. on AC, Vol. 45, no. 3 (2000) 477–482.
17. NEOS: Argonne national laboratory and northwestern university, optimization technology center. http://www-fp.mcs.anl.gov/otc/Guide/.
18. Swevers, J., Ganseman, C., Tukel, D., De Schutter, J., and Brussel, H. V.: Optimal Robot Excitation and Identification. IEEE Trans. on AC, Vol. 13, no. 5 (1997) 730–740.

19. Bellman, R.: Dynamic Programming. New Jersey: Princeton Univ. Press (1957)
20. Howard, R. A.: Dynamic Programming and Markov Processes. MIT Press (1960)
21. Schweitzer, P. and Seidmann, A.: Generalized Polynomial Approximations in Markovian Decision Processes. J. of MA and Appl., Vol. 110 (1985) 568–582.
22. Boutilier, C., Dean, T., and Hanks, S.: Decision-Theoretic Planning: Structural Assumptions and Computational Leverage. J. of AI Research, Vol. 11 (1999) 1–94.
23. Lovenjoy, W. S.: A Survey of Algorithmic Methods for Partially Observed Markov Decision Processes. Annals of Operations Research, Vol. 18 (1991) 47–66.

Gaze-Contingent Multi-modality Displays of Multi-layered Geographical Maps

S.G. Nikolov[1], D.R. Bull[1], and I.D. Gilchrist[2]

[1] Image Communications Group, Centre for Communications Research
University of Bristol, Merchant Venturers Building
Woodland Road, Bristol BS8 1UB, UK
{Stavri.Nikolov,Dave.Bull}@bristol.ac.uk

[2] Department of Experimental Psychology, University of Bristol
8 Woodland Road, Bristol BS8 1TN, UK
I.D.Gilchrist@bristol.ac.uk

Abstract. In this work a system for construction of gaze-contingent multi-modality displays of multi-layered geographical maps is proposed. Such displays can be very useful for designing various map viewing and reading experiments and hopefully for computation of the visual span in map processing for different tasks, e.g. target detection or navigation. In gaze-contingent displays a window centred around the observer's fixation point is modified while the observer moves their eyes around the display. In the proposed technique this window, in the central part of vision, is taken from one of the input images, which in this case contains information from a number of map layers, while the rest of the display, in peripheral vision, contains information from another set of possibly different map layers. Thus different map information can be channelled to the central and the peripheral vision. The human visual system fuses the information from these two sets of layers into a single percept.

1 Introduction

In the last thirty years, various display techniques have been developed and used to present an observer or interpreter with multiple 2-D images coming from different instruments, sensors or modalities. Choosing an optimal display technique is critical for successful image fusion. To a large extent it determines how the human observer will be able to fuse the information coming from the input images. Some of the approaches still most commonly used in software systems include an adjacent display with or without a linked cursor, a 'chessboard' display and a transparency weighted display. In previous work we demonstrated how gaze-contingent multi-modality displays (GCMMD) can be successfully used for fusion of 2-D images [9] and for fusion of 3-D (volumetric) images [8]. In the proposed technique a gaze-contingent window, in the central part of vision, is taken from one of the input modalities, while the rest of the display, in peripheral vision, comes from the other one. If the display parameters are carefully set, the human visual system can fuse these two images into a single percept. Various examples

I. Dimov et al. (Eds.): NMA 2002, LNCS 2542, pp. 325–332, 2003.

were considered in [9] including registered medical images (CT and MR), remote sensing images, partially-focused images, and multi-layered geographical maps. In this work we will study in more detail how gaze-contingent multi-modality displays of multi-layered geographical maps can be constructed. Another issue which is discussed in the paper is how to select and channel different information from the multi-layered map to the central and to the peripheral human vision.

2 Multi-layered Geographical Maps

With more and more information being gathered and displayed on geographical maps, multi-layered information representations are increasingly favoured in digital map production. Having such representations, different map layers can be 'switched on' or 'off' depending on the task at hand, and thus determining the visual complexity of the map display. An excellent example of an online atlas containing multi-layered geographical maps is provided by Environment Australia. Their web site [2] allows users to construct, view and print multi-layered maps of Australia's coastal zones. After choosing a particular area, it is possible to select data layers from several pre-classified schemes and from a long list of more specific data to create customised maps. The display characteristics of these data layers on the map can also be modified. An example showing the interface of the atlas and Tasmania'a marine farming map is given in Figure 1.

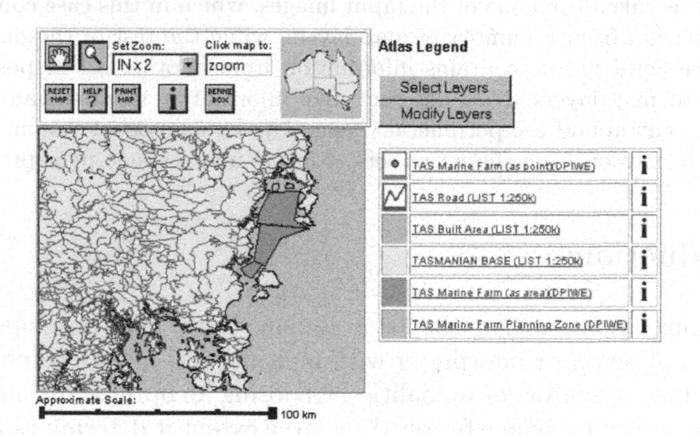

Fig. 1. Tasmania's multi-layered marine farming map from Environment Australia.

3 Studies of Eye Movements in Map Reading

A few map reading studies using gaze-tracking have been published in the literature. Some early results on eye movements in map reading can be found in

[11]. In several of their studies Phillips and Noyle [11,10] point out that the main difficulty in using eye movement recording for studying map reading behaviour is the disparity between fixation and attention. The same authors suggest that this disparity may be quite large for symbols such as colour codes, which can be processed several degrees of visual angle away from the fovea, but typographical symbols, e.g. place and road names, must fall close to the fovea if they are to be read. For this reason eye movement recording is especially informative for studying map typography [11,10]. In [6] Doyle and Moran investigated whether different map-types elicit different eye-fixation patterns. In their experiment participants actively processed routes of differing lengths and complexity on both 2-D and depicted 3-D 'you-are-here' maps. The findings reported in [6] are that eye fixation duration was not affected by 3-D depiction but by route complexity.

4 Ga e-Contingent Displays

The gaze-contingent window technique, offering a high degree of experimental control, has been widely used in reading, scene perception and visual search studies [7,14,13,12]. In gaze-contingent displays (GCD) a window centred around the observer's fixation point is modified while the observer moves their eyes around the display (see Figure 2, left). In its classical form, this technique obscures all objects from view except those within a the window. In reading research the moving mask and moving window paradigms have proven to be invaluable in determining the chronometric and spatial characteristics of processing written text [15]. Due to technical limitations, gaze-contingent window paradigms have frequently been applied to reading studies, but more rarely to scene perception [15]. Gaze-contingent displays require high sampling rates and very low delay.

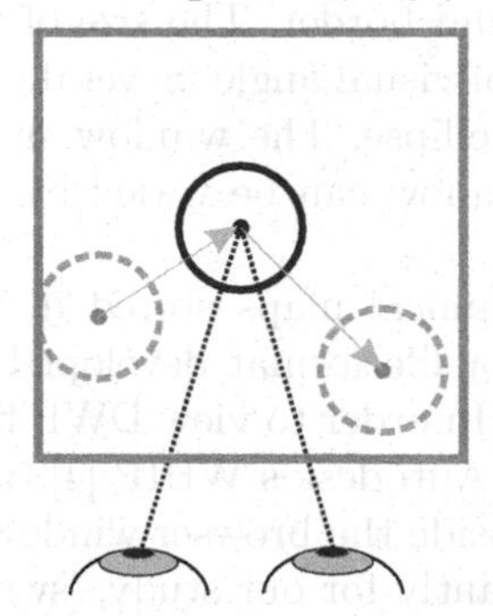

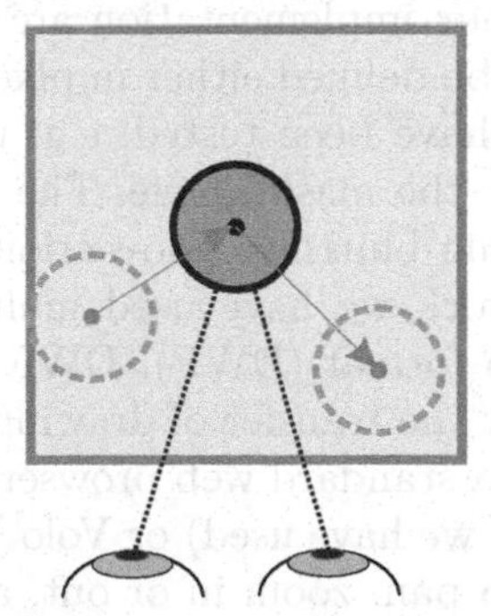

Fig. 2. Gaze-contingent display (left). Gaze-contingent multi-modality display (right): the window information (dark grey) is taken from the foreground image (FI), while the background information (light grey) comes from the background image (BI).

5 Ga e-Contingent Multi-modality Displays

Several hardware and software gaze-contingent display solutions have been proposed in the literature. For information about some gaze-contingent display implementations see for example [15,12]. Our implementation is described in detail in [9]. Here we will briefly repeat the main points.

An EyeLink I system from SR Research Ltd. was used to record eye movements. This system operates at a sampling rate of 250 Hz (4 ms temporal resolution) and measures an observer's gaze position with an average error of less than 0.5 degrees of visual angle. The headband has three cameras, allowing simultaneous tracking of both eyes and head position for head-motion compensation. The EyeLink system uses an Ethernet link between the eye-tracker and the display computer for real-time gaze-position and saccade data transfer. In the present study all displays were presented on a 22-inch Silicon Graphics monitor with 85 Hz refresh rate and 1024×768 pixels screen resolution. The displays were generated by an Silicon Graphics Visual Workstation having dual 933MHz Intel processors and running Microsoft Windows NT 4.0.

A real-time gaze-contingent window is implemented, having different shapes, sizes, and borders. Our system uses two input images, one as the foreground image (FI) and the other as the background image (BI). This is illustrated in Figure 2 (right). Differential window updates are performed in order to speed up window redraw and to avoid flicker. Since the Windows graphics device interface (GDI) provides only routines for fast rectangular bitmap display, the following implementation is used. A virtual screen (memory device context) is created and used to prepare the content of a rectangular window. A mask image (MI) is generated which defines the size and shape of the window. The 'MaskBlt' function from the Win32 application programming interface (API) then combines the colour data from the FI and the BI using the MI and sends its output to the virtual screen. Then, differential window updates can be used as if having a rectangular window. The gaze-contingent window parameters which can be changed in our implementation are its size, shape, and border. The size of the window can be defined either in pixels or in degrees of visual angle. Several window shapes have been tested, e.g. rectangle, circle, ellipse. The window shape is defined by the mask image. The border of the window can be varied by, for example, using blurring or foveation pyramids.

In our work we have used multi-layered geographical maps stored in the drawing web format (DWF). DWF is an open vector file format developed by Autodesk for the transfer of drawings over networks. In order to view DWF files one can use a standard web browser plug-in, such as Autodesk's WHIP [4] (this is the viewer we have used) or Volo View [3]. Then inside the browser window it is possible to pan, zoom in or out, and most importantly for our study, 'switch on' or 'off' different map layers.

6 Ga e-Contingent Multi-modality Displays of Maps

In order to use the gaze-contingent multi-modality display system described above, foreground and background images are generated and stored as bitmap files, each one containing information from a number of map layers (see Figure 3). By careful selection of the information to be presented in the gaze-contingent window and in the background, one can design various map viewing and reading experiments. There is no need for spatial registration of the input images, since one and the same WHIP viewing parameters, i.e. position and magnification, are used.

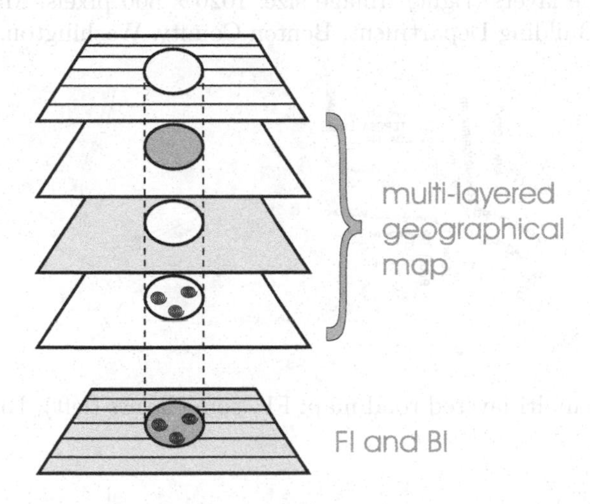

Fig. 3. Construction of the FI (to provide information inside the window) and BI (to provide information outside the window) from the multiple layers of the map.

An example multi-layered road map of Benton County (Washington) is shown in Figure 4 (left). Several layers of the map have been 'switched off' and the result is displayed in Figure 4 (right). Figure 6 (left) presents an GCMMD of the images in Figure 4. A different FI and BI constructed by selecting different layers from the multi-layered map are shown in Figure 5. An GCMMD of the two images is given in Figure 6 (right). Another example of a multi-layered geological map of an area in Southeastern British Columbia is presented in Figure 7. Figure 8 shows two GCMMD of these images.

7 Discussion

In this paper we have demonstrated how the gaze-contingent multi-modality displays proposed in [9] can be used for perceptual studies of multi-layered

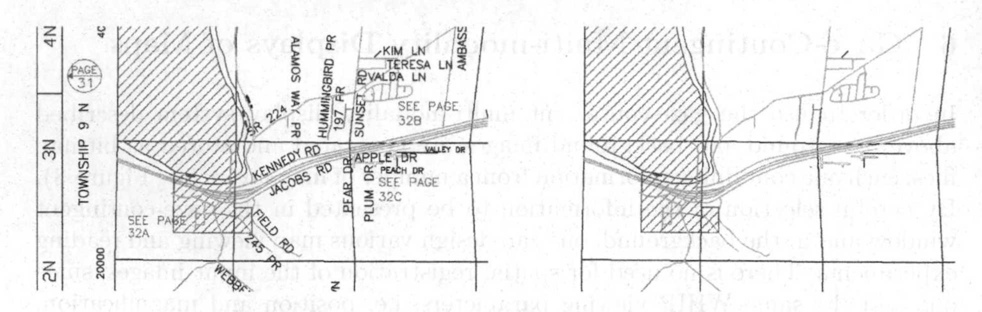

Fig. 4. Benton (N. Carywood Road) multi-layered road map: FI - all layers (left); BI - some layers (right). Image size: 1020 × 600 pixels. Map courtesy of the Planning/Building Department, Benton County, Washington, USA.

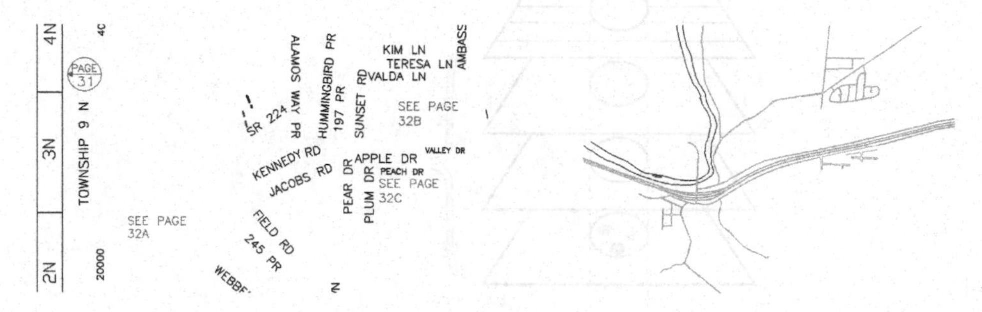

Fig. 5. Benton multi-layered road map: FI - some layers (left); BI - other layers (right).

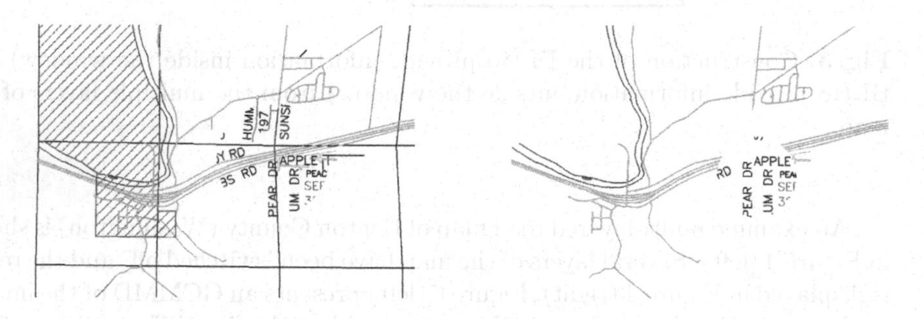

Fig. 6. Benton multi-layered road map: left - GCMMD of the FI and BI in Figure 4; right - GCMMD of the FI and BI in Figure 5. Circular window diameter: 250 pixels (left); 180 pixels (right).

geographical maps. Most of the initial observations made when using gaze-contingent multi-modality displays reported in [9] are very much valid for map displays, i.e. : (a) in most cases we are able to fuse the information from the two input images, i.e. the two sets of map layers, but only if the display parameters are carefully set; (b) as expected a circular gaze-contingent window was felt

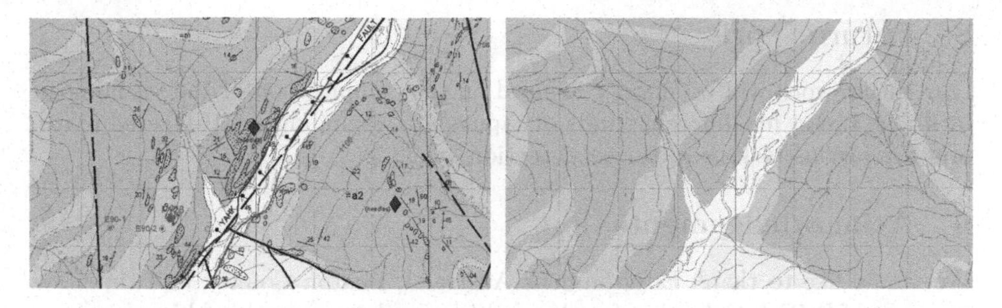

Fig. 7. British Columbia multi-layered geological map: FI - all layers (left); BI - some layers (right). Image size: 1020 × 600 pixels. Geoscience Map 1998-2: Geological Compilation of the Yahk (East Half) and Yahk River (West Half) Map Areas, Southeastern British Columbia. Map courtesy of the Government of British Columbia.

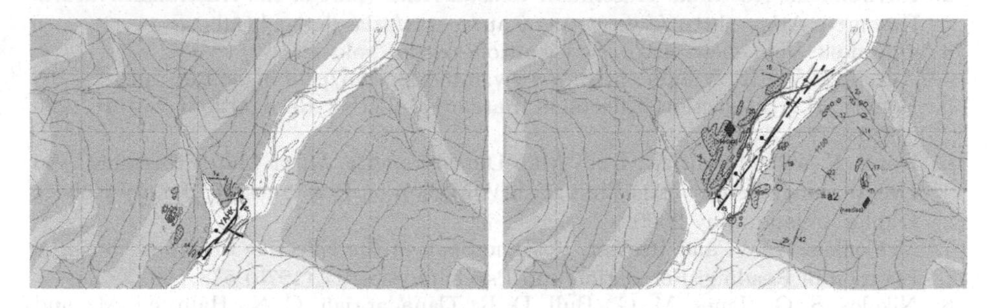

Fig. 8. British Columbia multi-layered geological map: two GCMMD. Circular window diameter: 200 pixels (left); 420 pixels (right).

to provide best results; (c) any window update delays, flicker or gaze-position estimation inaccuracies, e.g. due to poor eye-tracker calibration, even if minimal, were found to be very distracting and preventing the fusion process. An additional interesting property of GCMMD of multi-layered maps when the BI is generated from a set of map layers which is a subset of the layers used to generate the FI (as in Figure 4 for example), is that the gaze-contingent window looks 'transparent' (see Figure 6, left), i.e. the 'background' pixels in the window are 'transparent'. Human visual resolution ability drops off dramatically away from the axis of fixation [1,5]. Thus, if the visual content that is in the BI is fine scale then this information will not be easily processed by peripheral vision and may give no additional benefit when compared to a blank BI. Importantly, however, visual acuity is dramatically affected in peripheral vision by the presence of additional information [5]. If the amount of peripheral information is reduced (as shown in Figures 6 and 8) and only that information that is (task) relevant is presented this should give a real performance advantage.

8 Future ork

Our future work will be mainly focused on the computation of the visual span in image fusion for different kinds of input data, including multi-layered maps, and for various tasks, e.g. target detection or navigation.

9 Acknowledgements

We would like to thank Environment Australia, Benton County, and the Government of British Columbia for the multi-layered maps used in this paper.

References

1. Anstis, S. M.: A Chart Demonstrating Variation in Acuity with Retinal Position. Vision Research, 14 (1974) 589–592.
2. Environment Australia: Australian Coastal Atlas (part of the Australian Natural Resources Atlas). http://www.ea.gov.au/coasts/atlas/index.html.
3. Autodesk: Volo View. http://www.autodesk.com/voloview.
4. Autodesk: WHIP! Viewer. http://www.autodesk.com/cgi-bin/whipreg.pl.
5. Bouma, H.: Interaction Effects in Parafoveal Letter Recognition. Nature, 226 (1970) 177–178.
6. Doyle, F. and Moran, A.: Does Map-Dimensionality Affect Map Processing? An Eye-Movement Study. The First Irish Workshop on Eye-Tracking, Trinity College Dublin, Ireland, May 2000.
7. McConkie, G. W. and Rayner, K.: The Span of the Effective Stimulus During a Fixation in Reading. Perception and Psychophysics, 17 (1975) 578–586.
8. Nikolov, S. G., Jones, M. G., Bull, D. R., Canagarajah, C. N., Halliwell, M., and Wells, P. N. T.: Focus+Context Visualisation for Image Fusion. 4th International Conference on Information Fusion (Fusion 2001), Montreal, Canada, 7-10 August, Vol. I, pages WeC3–3 – WeC3–9. International Society of Information Fusion (ISIF) (2001) (invited lecture).
9. Nikolov, S. G., Jones, M. G., Gilchrist, I. D., Bull, D. R., and Canagarajah, C. N.: Multi-Modality Gaze-Contingent Displays for Image Fusion. Proceedings of the 5th International Conference on Information Fusion (Fusion 2002), Annapolis, MD, USA, 2002, (invited lecture, to appear).
10. Noyle, L.: The Positioning of Type on Maps: the Effect of Surrounding Material on Word Recognition Time. Human Factors, 22 (1980) 353–360.
11. Phillips, R. J.: Searching for a Target in a Random Arrangement of Names: an Eye Fixation Analysis. Canadian Journal of Psychology, 35(4) (1981) 330–346.
12. Pomplun, M., Reingold, E. M., and Shen, J.: Investigating the Visual Span in Comparative Search: the Effects of Task Difficulty and Divided Attention. Cognition, 81 (2001) B57–B67.
13. Rayner, K.: Eye Movements in Reading and Information Processing: 20 Years of Research. Psychological Bulletin, 124 (1998) 372–422.
14. Saida, S. and Ikeda, M.: Useful Visual Field Size for Pattern Perception. Perception and Psychophysics, 25 (1979) 119–125.
15. van Diepen, P. M. J., De Graef, P., and Van Rensbergen, J.: On-line Control of Moving Masks and Windows on a Complex Background Using the ATVista Videographics Adapter. Behavior Research Methods, Instruments and Computers, 26 (1994) 454–460.

About the Specifics of the IMM Algorithm Design*

Iliyana Simeonova and Tzvetan Semerdjiev

Central Laboratory for Parallel Processing,
Bulgarian Academy of Sciences
Acad G. Bonchev, Bl. 25-A, 1113 Sofia, Bulgaria
iliyana@bas.bg

Abstract. It is well known, that interacting multiple model (IMM) state estimation algorithm is one of the most cost-effective filters for tracking maneuvering targets. The present paper is related to the specifics of the IMM algorithm design. It combines the results, conclusions and experience of different authors considered in their papers. The results discussed and depicted here are root mean square errors and the filters ability to distinct the various flight phases. This paper helps the air traffic control experts fast and easy to make a decision which IMM configuration is suitable for a given problem.

1 Introduction

The Interacting Multiple Model estimator is a suboptimal hybrid filter that has been shown to be one of the most cost-effective hybrid state estimation schemes [9]. The model of hybrid system and the IMM algorithm, initially proposed in (Blom 1984) are a base for a synthesis of more efficient filters for tracking maneuvering aircraft. In this paper we present the special features of the IMM design procedure. In section 2 we give a brief description of the hybrid system (aircraft dynamics) and measurement models and of the IMM algorithm principle. The models used in the IMM configuration to describe different flight phases and the measurement model used for our application are presented in section 3 . Section 4 gives a discussion on the selection of filter design parameters. Finally, an example of IMM tracking filter design and performance evaluation are presented in section 5.

2 A Hybrid Model Based Algorithm - IMM

2.1. Hybrid System and Measurement Models. An aircraft trajectory can be subdivided into distinct segments, corresponding to modes of flight [9](for instance uniform motion and maneuvers). The multiple model or hybrid system approach assumes the system to be in one of a finite number of modes

* Partially supported by Center of Excellence BIS21 grant ICA1-2000-70016

I. Dimov et al. (Eds.): NMA 2002, LNCS 2542, pp. 333–341, 2003.

[2]. So the aircraft motion can be modeled by a hybrid system. Hybrid system is characterized by two state variables: continuous state variable $x(k) \in R^{n_x}$ (aircraft position, speed, acceleration, at cetera) and a discrete regime variable $j = m_j(k) \in M_r = 1, 2....r$, which describes the distinct segment of the aircraft trajectory. The transitions (jumps) of the mode variable are modeled with a Markov chain. The nonlinear hybrid system is usually described by the equations:

$$x(k) = f_j[(k-1), x(k-1)] + g_j[k-1, x(k-1), v_j(k-1)] \quad \forall\, j \in M_r \ . \quad (1)$$

$$z(k-1) = h_j[(k-1), x(k-1)] + w_j(k-1) \quad \forall\, j \in M_r \ . \quad (2)$$

where $x(k) \in R^{n_x}$ is the system state vector, $z(k-1) \in R^{n_z}$ is measurement vector, $v_j(.)$ and $w_j(.)$ are mode dependent process noise and measurement noise sequences, assumed to be white, zero-mean and mutually independent with covariances $Q(m_j(k))$ and $R(m_j(k-1))$ respectively. The available measurements for estimation process can be: [range and azimuth]; [range, azimuth and range rate] at cetera. The $f_j[.], g_j[.], h_j[.]$ are known functions. The system at time (k) is assumed to be among r possible modes [7], i.e. $j = m_j(k) \in M_r = 1, 2....r$, so $j = m_j(k)$ denotes that the j-th submodel is in effect during the sampling period T ending at time (k). $m_j(k)_{k=1,2....}$ is a Markov chain with completely known initial $P_j = P[m_j(0)] = j$ and transitional probabilities $p_{ij} = P[m_j(k)|m_i(k-1)]$.

2.2. IMM Algorithm. The problem of hybrid estimation is to assess the system state and behavior mode based on the sequence of noisy measurements. The IMM algorithm is one of the most effective recent suboptimal Bayesian filters for hybrid system estimation . In each cycle it consists of three basic steps: interaction (mixing), filtering, and combination. The IMM algorithm provides the combined system state estimate: $\hat{x}(k|k) = \sum_{j=1}^{r} \hat{x}^j(k|k)\mu_j(k)$ and estimates its associated covariance matrix $P(k|k)$ as a weighted sum of the estimates : $\hat{x}^j(k|k)$ and there covariances: $P^j(k|k)$, formed by r-mode conditional parallel Kalman filters. The posterior mode probabilities $\mu_j(k)$ are calculated on the base of the likelihood of the measurement received at the current time step. As we mentioned above mode switching is govern by a Markov chain.

The IMM algorithm has three desirable properties: it is recursive, modular and has fixed computational requirements per cycle. A detailed explanation of the IMM can be found in [1,2,5,9].

3 The Aircraft Motion and Sensor Modeling

3.1. Aircraft Motion Modeling. Civilian aircraft in air traffic control (ATC) systems have two basic modes of flight:

- Uniform motion (UM) - the straight and level flight with a constant speed and heading and
- Maneuver - turning or climbing/descending.[1]

Let us consider the linear version of (1) to aircraft trajectory modeling in two Cartesian coordinates (x, y):

$$X(k) = F_j.X(k-1) + G_j.v_j(k-1) \ . \qquad (3)$$

where $v_j(k-1) = \left[v_{jx}(k-1)\ v_{jy}(k-1)\right]^T$ are white noise sequences used to model uncertain accelerations. The model extension in (x, y, z)- coordinates is straightforward. In this section we will present some of the most used aircraft motion models.

Notations: $X \equiv X(k)$; F_j/G_j - transition/noise gain matrix for both coordinates (x, y); f_j/g_j - transition/noise gain matrix for each coordinate x/y.

3.1.1. Piecewise Constant White Acceleration Model is a second order (nearly constant velocity) model with the following parameters:

• State space vector: $X = \left[x\ \dot{x}\ y\ \dot{y}\ \right]^T$

• Transition matrix: $F_1 = diag[f_1, f_1]$, where $f_1 = \begin{bmatrix} 1 & T \\ 0 & 1 \end{bmatrix}$

• Noise gain: $G_1 = diag[g_1, g_1]$ where $g_1 = \left[T * T/2 \quad T\right]^T$

This model assumes the variations in velocity components for each coordinate are piesewise constant zero-mean white noise accelerations [1]. The process noise variances in each coordinate are assumed to be equal: $\sigma^2_{\nu x} = \sigma^2_{\nu y} = q$ and $\sigma_{\nu x \nu y} = 0$.

• A "nearly constant velocity motion model" (M_1) for the UM modeling is obtained by the choice of "small" noise values: $q = q_1$ [1,6].

• The same model, but with higher levels of process noise $q = q_2$ can model "rough" maneuvers. This model is denoted as M_2.

3.1.2. Piecewise Constant Wiener Process Acceleration Model is a third order (nearly constant acceleration) model used to describe the maneuvering phase of flight. It has the following parameters:

• State space vector: $X = \left[x\ \dot{x}\ \ddot{x}\ y\ \dot{y}\ \ddot{y}\right]^T$

• Transition matrix: $F_3 = diag[f_3, f_3]$ where $f_3 = \begin{bmatrix} 1 & T & T*T/2 \\ 0 & 1 & T \\ 0 & 0 & 1 \end{bmatrix}$

• Noise gain: $G_3 = diag[g_3, g_3]$ where $g_3 = \left[g_1 \quad 1\right]^T$

This model assumes the acceleration increments for each component during k-th sampling period are zero-mean white sequence [2].

• A "nearly constant acceleration motion model" M_3 is obtained by the choice of "small" $q = q_3$ [1,6].

• The model M_4 is the same but with higher levels of process noise $q = q_4$[6].

3.1.3. Coordinated Turn Models (CTM) are another way to describe the maneuvering mode of flight. The turning of civilian aircraft usually follows a pattern known as a "coordinated turn": that means the target is moving with constant speed and turning with constant turn rate. There are two basic coordinated turns models:

Constant turn rate models: here turn rate ω is a completely known design parameter. The models M_5 and M_6 are used for a left-hand turn ($\omega > 0$) and for a right-hand turn ($\omega < 0$), respectively. This assumption is suitable for a civilian aircraft [5]. The state space vector and noise gain coincide with those of models M_1 and M_2, but the transition matrix is different:

• Transition matrix : $F_{5,6} = \begin{bmatrix} 1 & \omega^{-1}\sin(\omega.T) & 0 & \omega^{-1}(\omega - \cos(\omega.T)) \\ 0 & \cos(\omega.T) & 0 & -\sin(\omega.T) \\ 0 & \omega^{-1}(\omega - \cos(\omega.T)) & 1 & \omega^{-1}\sin(\omega.T) \\ 0 & \sin(\omega.T) & 0 & \cos(\omega.T) \end{bmatrix}$

• The process noise variances in each coordinate are q_5 and q_6 respectively.

For military aircraft the above assumption is less natural, so the model M_7 presents the case where ω is not known. What we need here is to augment the state space vector with unknown turn rate ω.

• The discrete time state space equation is nonlinear, because the transition matrix is a function of the state component (ω):

$$X(k) = F_7(\omega).X(k-1) + G_7.v_7(k-1) \ . \qquad (4)$$

• State space vector: $X = \begin{bmatrix} x \ \dot{x} \ y \ \dot{y} \ \omega \end{bmatrix}^T$

• Transition matrix: $F_7 = \left[\begin{array}{c|c} F_{5,6} & 0 \\ \hline 0 & 1 \end{array}\right]$

• Noise gain: $G_7 = diag[g_7, g_7]$ where $g_7 = \begin{bmatrix} g_1 & 0 \end{bmatrix}^T$

Here $v_7(k-1) = \begin{bmatrix} v_{7x}(k-1) \ v_{7y}(k-1) \end{bmatrix}^T$ are white noise sequences used to model uncertain accelerations in x and y coordinates due to uncertainty in ω.

• The process noise variance is q_7.

The reader is referred to [1,2,5,6], for a comprehensive presentation of aircraft motion modeling.

3.2. Measurement Model. Let us consider the case when (2) is not mode dependent:

$$z(k-1) = h[(k-1), X(k-1)] + w(k-1) \ . \qquad (5)$$

For our application the nonlinear measurement function h[.] of (5) converts Cartesian target state space vector coordinates (x,y) to z=[range and azimuth]^T, h[.]=[h_1 h_2]^T where : $h_1 = \sqrt{x^2 + y^2}$ and $h^2 = arctan\frac{x}{y}$.

Due to the nonlinearity of the measurement(sensor) equation (5), the Extended Kalman filter is used in the IMM configuration. It requires linearization of h[.] about the position prediction, that yields Yacobian.

4 IMM Estimator Design

In this section we will give the practitioners some practical rules how to chose the best IMM filter design parameters.

To obtain the best possible results, the IMM algorithm has to be properly designed. According to [1] the designer should take into account:

• the estimation accuracy in both position and velocity. A trade-off between the peak errors and the errors during uniform motion is desirable;

• the timeliness of the maneuver detection and termination;

• the complexity of the implementation.

According to [1] the design of an IMM estimator consist of the following steps:

- selection of the set of models describing aircraft dynamics and their structure;
- selection of the process noise intensivities for the various models;
- selection of the Markov chain transition probabilities.

4.1. The Model Set Selection. It should consider the complexity as well as the quality of the model. Typically the models used in the IMM configuration include one nearly constant velocity motion model for non-maneuvering regime of flight and a set of exact or approximate maneuver models for maneuvering phases. The models for uniform and maneuvering motion are presented in the previous section. It should be mentioned that: increasing the number of conditional submodels to cover the uncertain behavior of highly maneuvering objects increases considerably the computations but does not guarantee better performance due to the competition among models [3]. The precise modeling of every aircraft trajectory segment will lead to more accurate results especially in speed estimation [5].

4.2. Process Noise Selection. Selection of process noise standard deviations for each model is an important part of the estimator design. The process noise levels are selected based on the expected disturbances and target maneuvering magnitudes [1]. Let us consider some of the presented in the previous section models.

- The **small** process noise of model M_1 accounts for air turbulence, winds aloft changes [2], slow turns as well as small linear accelerations.
- The **high level** process noise of model M_2 allows for target acceleration and is applicable (with limited degree of success) to tracking maneuvering target [2]. The process noise range is usually selected as follows: $0.5.a_{max} \leq q_2 \leq a_{max}$ [2], where a_{max} is the maximum acceleration magnitude. According to [2] the *IMM configuration* with M_1 and M_2 models does obtain acceptable results for maneuvers with turn rates up to 3 deg/s, it does not however, yield good estimates for faster turns.
- M_3 model with **low process** noise provides more accurate estimation during a maneuver.
- M_4 model with **high level** process noise can model more precisely maneuver on/off set. The noise range should be of the order of the magnitude of the maximum acceleration increment over a sampling period: $0.5\triangle a_{max} \leq q_4 \leq \triangle a_{max}$ [2]. The *IMM configuration* with one second order model and two third order models with different noise levels [3,5,7] is best suitable for estimating more intensive ($a_n = 7$g) and short durable maneuvers and longitudinal acceleration as well. But it provides a considerable errors for moderate turns (1-5g).
- The right choice of the noise levels - $q_{5,6}$ of the models M_5 and M_6 depends on what turn rate is expected and how many models are to be used for the maneuvers. The standard deviation of the process noise can be selected as: $q_{5,6} = 0.5.(\omega_{i+1} - \omega_i).V = 0.5.\triangle U$ [4], where ω_{i+1}, ω_i are turn rates of two adjacent models and V - expected linear speed . The *IMM configuration* could include one uniform motion model and a set of different constant turn

rate models. According to [4] the use of coordinated turn models with known-ω is better than IMM with estimated ω when the models in the former fully "covered" the turn rate of target motion and vice-versa.

• Because of the delay in estimating ω at the onset of the maneuver the use of model M_7 produces rather large peaks, however once the ω estimate has converges, a very good tracking performance is obtained during turns [5]. According to [3] it is the best suitable for tracking aircraft performing maneuvers with moderate, apriori unknown normal acceleration ($a_n = 1 \div 5$g), and for more complex maneuvers with longitudinal and transversal accelerations.

4.3. Transition Probabilities. The Markov chain transition probabilities are related to the expected sojourn time in the various modes [1]. These probabilities are chosen according to the designer's beliefs about the frequency of the change from one mode to the rest. They can be subsequently adjusted by means of Monte Carlo simulations. The guideline for a proper choice is to match roughly the transition probabilities with the actual mean sojourn time (τ_i) of each mode [2]. The matrix diagonal coefficients can be determined by using the following equality: $p_{ii} = 1 - \frac{1}{E[\tau_i]}$. The choice of the transition probabilities provides a certain degree of trade-off between the peak estimation errors at the onset of the maneuver and the maximum reduction of the estimation errors during the uniform motion [2].

5 Performance Analysis

5.1. Target Motion Scenario. In order to test the capabilities of different IMM configurations we consider a class of maneuvering aircraft performing sweep maneuvers with normal acceleration up to 7g (g $\approx$ 9.8). Its motion scenario is given in Fig.1

5.2. Sensor Parameters. The simulation involve a single track while scan (TWS) radar with scanning period of 1s. The sensor parameters considered here are : range and azimuth accuracy: $\sigma_{\mathrm{D}} = 120$ m and $\sigma_\beta = 0.2$ deg, respectively.

5.3. IMM Tracking Filter Design. Here we introduce three different sets of models (resp. IMM2, IMM3, IMM–CT) to describe target motion scenario. There design parameters are:

• IMM2 - **mode set**:$[M_1, M_{3I}]$;**-process noises:**$\sigma_1 = 2.5\,\mathrm{m/s^2}, \sigma_{3I} = 20.5\,\mathrm{m/s^2}$-I - intermediate noise;**-transition probabilities** $p_{11} = 0.9, p_{21} = 0.20$

• IMM3 - **mode set**:$[M_1, M_3, M_4]$;**-process noises:**$\sigma_1 = 2.5\,\mathrm{m/s^2}, \sigma_3 = 7\,\mathrm{m/s^2}$, $\sigma_4 = 40\,\mathrm{m/s^2}$;**-transition probabilities** $p_{11} = 0.9, p_{12} = 0.05, p_{21} = 0.15$, $p_{22} = 0.75, p_{31} = 0.20, p_{32} = 0.05$

• IMM-CT - **mode set**:$[M_1, M_5, M_6]$;**-process noises:**$\sigma_1 = 2.5\,\mathrm{m/s^2}, \sigma_{5,6} = 3.5\,\mathrm{m/s^2}, \omega_{5,6} = \pm 0.233\,\mathrm{rad/s}$;**-transition probabilities** $p_{11} = 0.9, p_{12} = 0.05$, $p_{21} = 0.20, p_{22} = 0.80, p_{31} = 0.20, p_{32} = 0.00$

5.4. Performance Evaluation and Analysis. Here the IMM filters' efficiency was evaluated according to the root mean-square(RMS) error both in position and in velocity. The results presented here are based on Nr = 100 Monte Carlo runs. RMS is defined as $\sigma_{\mathrm{pos}} = \sqrt{\frac{1}{N} \cdot \sum_{i=1}^{N} [x^i(k) - \hat{x}^i(k|k)]^2 + [y^i(k) - \hat{y}^i(k|k)]^2}$

where superscript i denote the results from run i, and x(k),y(k) are true target positions(x,y).The equation for velocity RMS error is straightforward.

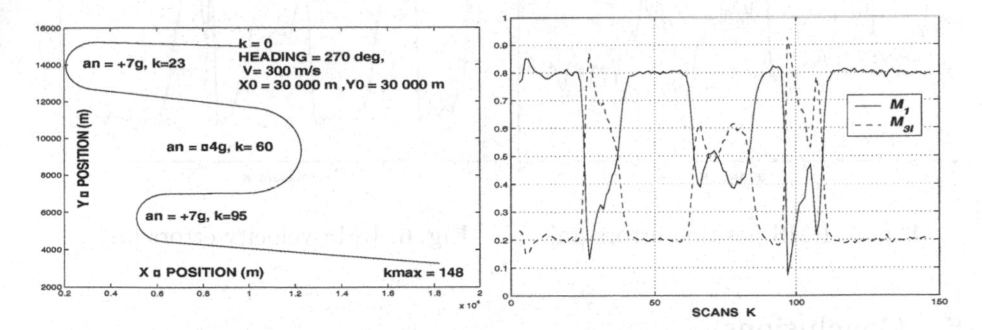

Fig. 1. Target Trajectory

Fig. 2. IMM2. Mode probabilities

• The average mode probabilities for each case over scans are depicted in Figs. (2-4). The correct regime has the largest probability during each segment [5].
• Comparison of RMS position and velocity errors are shown in Figs. 5 and 6, respectively. The general behavior of these error curves is quite typical of IMM algorithms. Natural transients are observed at the onset and termination of maneuvers. The peaks at the start and end of a maneuver are caused by the delay of mode probabilities switching from one mode to another. After this switching the slower decrease on the errors corresponds to the convergence of maneuver filter [5]. The estimation accuracy is gained by using coordinated turn models.

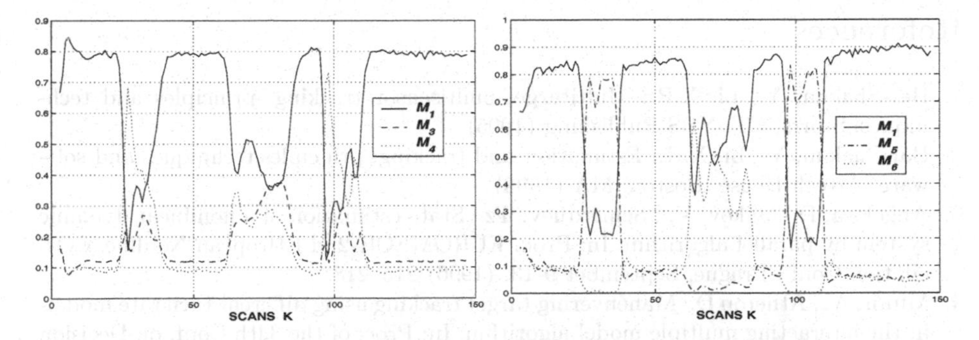

Fig. 3. IMM3. Mode probabilities

Fig. 4. IMM–CT. Mode probabilities

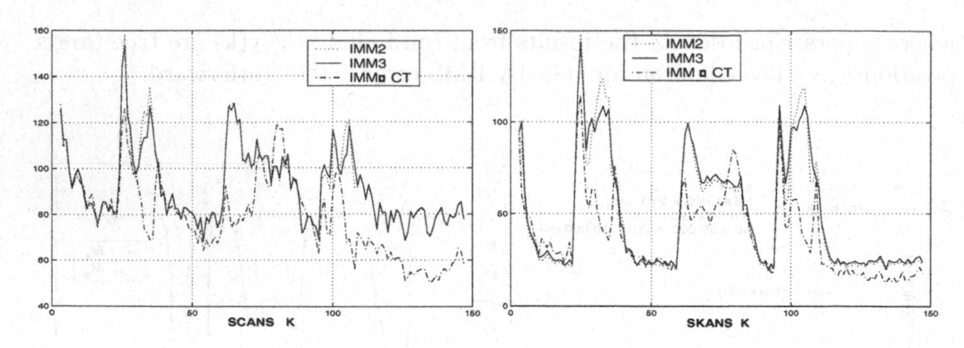

Fig. 5. RMS position errors [m]

Fig. 6. RMS velocity errors [m/s]

6 Conclusions

There are a lot of articles devoted to the IMM tracking filter design procedure. In this contribution we have gerneralized the results conclusions and experience of different authors considered in their papers. So that the practitioner is able fast and easy to determine the advantages and capacity of different IMM structures, given the target motion scenario. This is an objective not addressed in the available bibliography on IMM tracking filters. Besides the simulation results are obtained through a comprehensive MATLAB tool. It could be use as a base for a synthesis of many other motion scenarios and IMM tracking filters structures. For the sake of brevity, the hole version of the paper is not presented here. Details may be found through the `iliyana@bas.bg`

Acknowledgements. We would like to thank Donka Angelova for the final paper review and for providing the comprehensive MATLAB algorithms.

References

1. Bar-Shalom, Y., Li, X.R.: Multitarget-multisensor tracking: principles and techniques. Storrs, CT: YBS Publishing (1995)
2. Bar-Shalom, Y., Li, X.R.: Estimation and tracking: principles,techniques and software. Artech House, Boston, MA (1993)
3. Angelova, D. , Jilkov, V. , Semerdjiev, Tz.: State estimation of a nonlinear dynamic system by parallel algorithm. In: Proc. EUROMECH-2nd EUropean Nonlinear Oscilation Conf., Prague, September 9-13, (1996) 215–218
4. Munir, A. , Atheron D.: Maneuvering target tracking using different turn rate models in the interacting multiple model algorithm. In: Proc. of the 34th Conf. on Decision and Control, New Orleans, LA-Desember (1995) 2747–2751
5. Bar-Shalom, Y.:Multitarget-Multisensor Tracking: Aplications and Advances, Volume II. Artech House, Boston, MA (1993)
6. Lero, D. , Bar-Shalom Y.: Interactive Multiple Model Tracking with Target Amplitude Features. IEEE Trans. Aerospace and electronic systems. **AES-29** (1993) 495–508

7. Angelova, D. , Jilkov, V. , Semerdjiev, Tz.: Tracking Maneuvering Target by Interacting Multiple model. Comtes rendus de l'Academie bulgare des Sciences **Tome 49** (1996) 37–40
8. Herrero, J. , Portas, J. , López, J. , Vela, Garcia, J.: Interactive Multiple Model Filter Optimisation Tool for Air Traffic Control Applications. In Proc.Fourth Annual Conf. on Information Fusion, Montréal, Québec, Canada (2001) TuB2-19–TuB2-26
9. Mazor, E. , Averbuch, A. , Bar-Shalom Y., Dayan, J.: Interacting Multiple Model Methods in Target Tracking: A Survey. IEEE Trans. Aerospace and electronic systems **AES-34** (1998) 103–123

Target Behaviour Tendency Estimation Using Fuzzy Logic*

Albena Tchamova and Tzvetan Semerdjiev

Central Laboratory for Parallel Processing,
Bulgarian Academy of Sciences
Acad G. Bonchev Str., Bl. 25-A, 1113 Sofia, Bulgaria
tchamova@bas.bg, signal@bas.bg

Abstract. The objective of this paper is to present an approach for targets' behaviour tendency estimation. An algorithm for target behaviour tracking is developed and evaluated. It is based on Fuzzy Logic principles applied to conventional passive radar amplitude measurements. A set of fuzzy models is used to describe the tendencies of target behaviour. A noise reduction procedure is applied. The performance of the developed algorithm in the presence of noise is estimated on the base of computer simulations.

1 Introduction

The objective of this work is to present an approach for targets' behavior tendency estimation based on Fuzzy Logic principles applied to conventional passive radars. It utilizes the measured emitter's amplitude value in consecutive time moments and uses a set of developed particular filters with respective set of possible target behavior models.

2 Statement of the Problem

In order to track targets using passive sensors it is necessary to compensate the unknown ranges by using additional information received from the emitter. In our case we suppose that the observed target emits constant signal. It is perceived by the sensor with a non-constant, but a varying strength (referred as amplitude). The augmented measurement vector at the end of each time interval $k = 1, 2, \dots$ is $Z = \{Z_\Theta, Z_A\}$, where: Z_Θ denotes the measured local angle with zero-mean Gaussian noise ν_Θ,and $Z_A = A + \nu_A$ denotes corresponding amplitude value with zero-mean Gaussian noise $\nu_A = N(0, \sigma_{\nu_A})$ and covariance σ_{ν_A}. The variation of amplitude value is because of the cluttered environment and the varying unknown distance to the object. It is conditioned by modes of target behaviour (approaching or descending). Our goal is to utilize received

* The research reported in this paper is partially supported by Center of Excellence BIS21 grant ICA1-2000-70016

I. Dimov et al. (Eds.): NMA 2002, LNCS 2542, pp. 342–349, 2003.

amplitude feature measurement, for predicting and estimating the tendency of target behaviour. The block diagram of target's behavior tracking system is shown on Fig. 1. Two single-model-based filters running in parallel and using two models for target behaviour (*Approaching* and *Receding*) are maintained. At initial time moment k the target is characterized by the fuzzified amplitude state estimates according to the two models $A^{App}(k/k)$ and $A^{Rec}(k/k)$. The new observation at time is assumed to be the true value, corrupted by additive measurement noise. It is fuzzified according to the chosen fuzzification interface. A weighting procedure for noise reduction is developed and applied. Particular

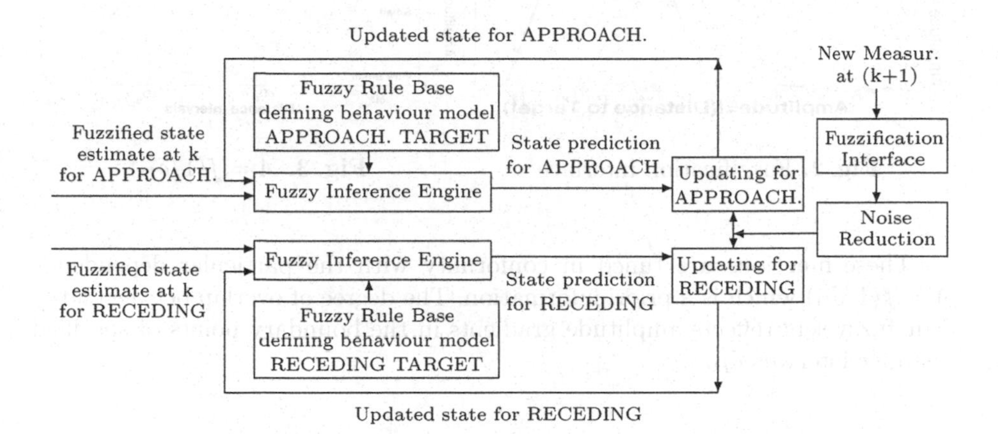

Fig. 1. Block diagram of target's behavior tracking system

tendency prediction and updating methods are used to estimate present and future target behaviour states. In general this diagram resembles the commonly used approaches in standard tracking systems [1]. But the peculiarity consists in the implemented fuzzy logic approach [2] to realize the main steps.

3 Fu y Approach for Target Behaviour Tracking

3.1 Fuzzification Interface

Fuzzification transforms each received from a sensor numerical measurement into fuzzy set in accordance to the a priori defined fuzzy partition of input space (the input frame Θ). It comprises all considered linguistic values related to particular important input variables and their associated membership functions. An important variable in our case is the transmitted from the emitter amplitude value $A(k)$, received at consecutive time moments $k = 1, 2, \ldots$. The fuzzification interface (Fig. 2) maps $A(k)$ into four fuzzy sets, which define four linguistic values $\Theta = \{VerySmall(VS), Small(S), Big(B), VeryBig(VB)\}$.Their membership functions are not arbitrarily chosen, but rely on the well-known inverse proportion dependency between the measured amplitude value and corresponding distance to target (Fig. 3).

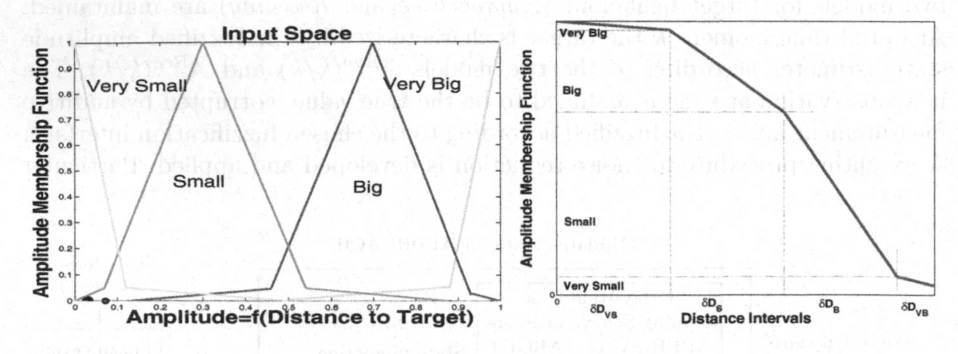

Fig. 2. Fuzzification interface **Fig. 3.** $A = f(1/\delta_D)$

These functions are tuned in conformity with the particular dependency $A = f(1/\delta_D)$ which is a priori information. The degree of overlap between adjacent fuzzy sets reflects amplitude gradients in the boundary points of specified distance intervals δ_D.

3.2 Fuzzy Model Identification

A set of fuzzy mapping IF-THEN rules forms our fuzzy models. The foundation of fuzzy mapping rules is the fuzzy graph g, which is an union of Cartesian products involving linguistic input-output associations. It is described by a set of i number fuzzy rules in the form of: 'IF x is A_i THEN y is B_i '. It is expressed mathematically as:

$$g = \bigcup_i A_i \times B_i \tag{1}$$

where A and B are the linguistic values, describing input and output variables. The Cartesian product of A and B is defined as:

$$\mu_{A\times B}(u,v) = \mu_A(u) \otimes \mu_B(v) \tag{2}$$

where $\otimes$ denotes a fuzzy conjunction (t-norm) operator; $\mu_{A\times B}(u,v)$ is a membership function, which measures the degree of truth of the implication relation between corresponding antecedents and consequents.

Relying on the cornerstone principle of modelling [2], our model is derived as a fuzzy graph, in which Larsen product operator is used for fuzzy conjunction and "maximum" for fuzzy union operator:

$$g = \max_i(\mu_{A_i\times B_i}(u,v)) = \max_i(\mu_{A_i}(u) \cdot \mu_{B_i}(v)) \tag{3}$$

The inference of such a collection is based on compositional rule of inference:

$$B' = A' \circ g = A' \circ \bigcup_i A_i \times B_i \tag{4}$$

We apply the most commonly used Zadeh max-min compositional rule [2,3]. If input "x is A' " is given, the inferred output is:

$$\mu_{B'}(y) = \max_{x_i}(\min(\mu_{A'}(x_i), \mu_{A\times B}(x_i, y_i))) \tag{5}$$

Bearing in mind that our considered fuzzy rules have one and the same antecedents and consequents, we consider two essential models of possible target behaviour: (a) *Approaching Target* - which behaviour in time is characterized by an increase of the amplitude, described by a set of transitions: $VS \to VS \to S \to S \to B \to B \to VB \to VB$; (b) *Receding Target* - which behaviour in time is characterized by a decrease of the amplitude value, described by a set of transitions: $VB \to VB \to B \to B \to S \to S \to VS \to VS$. To comprise appropriately these models the following fuzzy rule bases have to be carried out:

Behaviour 1: APPROACHING TARGET	**Behaviour 2: RECEDING TARGET**
Rule1:IF $A(k)$ **is** VS **THEN** $A(k+1)$ **is** VS	**Rule1:IF** $A(k)$ **is** VB **THEN** $A(k+1)$ **is** VB
Rule2:IF $A(k)$ **is** VS **THEN** $A(k+1)$ **is** S	**Rule2:IF** $A(k)$ **is** VB **THEN** $A(k+1)$ **is** B
Rule3:IF $A(k)$ **is** S **THEN** $A(k+1)$ **is** S	**Rule3:IF** $A(k)$ **is** B **THEN** $A(k+1)$ **is** B
Rule4:IF $A(k)$ **is** S **THEN** $A(k+1)$ **is** B	**Rule4:IF** $A(k)$ **is** B **THEN** $A(k+1)$ **is** S
Rule5:IF $A(k)$ **is** B **THEN** $A(k+1)$ **is** B	**Rule5:IF** $A(k)$ **is** S **THEN** $A(k+1)$ **is** S
Rule6:IF $A(k)$ **is** B **THEN** $A(k+1)$ **is** VB	**Rule6:IF** $A(k)$ **is** S **THEN** $A(k+1)$ **is** VS
Rule7:IF $A(k)$ **is** VB **THEN** $A(k+1)$ **is** VB	**Rule7:IF** $A(k)$ **is** VS **THEN** $A(k+1)$ **is** VS

In conformity with equations (1-3) and by using the specified membership functions, we obtain the resulting fuzzy graphs as fuzzy relations:

Relation1:Approaching Target

$k \to k+1$	VS	S	B	VB
VS	1	1	0.15	0.02
S	0.15	1	1	0.15
B	0.02	0.15	1	1
VB	0	0.02	0.15	1

Relation1:Receding Target

$k \to k+1$	VS	S	B	VB
VS	1	0.15	0.02	0.0
S	1	1	0.15	0.02
B	0.15	1	1	0.15
VB	0.02	0.15	1	1

Then we are able to realize our models' based filters running in parallel.

3.3 Models' Conditioned Amplitude State Tendency Prediction

At initial time moment k the target is characterized by the fuzzified amplitude values according to the models $\mu_{A^{A}pp}(k/k)$ and $\mu_{A^{R}ec}(k/k)$. Using these fuzzified amplitudes and applying the described above Zadeh max-min compositional rule equation(5) to relation 1-*App*($k \to k+1$) and relation 2 - *Rec*($k \to k+1$), we obtain modes conditioned amplitude state tendency prediction for time moment $k+1$, i.e.:

$$\mu_{A^{A}pp}(k+1/k) = \max(\min(\mu_{A^{A}pp}(k/k), \mu_{A}pp(k \to k+1))) \tag{6}$$

$$\mu_{A^{R}ec}(k+1/k) = \max(\min(\mu_{A^{R}ec}(k/k), \mu_{R}ec(k \to k+1))) \tag{7}$$

3.4 Weighting Procedure for Noise Reduction

In order to reduce the influence of measurement noise over the amplitude tendency prediction, a weighting procedure is applied to make the measurement more informative. This procedure can be considered as an adaptive linear combiner as follows:

• we compute the degree to which the new fuzzified measurement intersects each of the linguistic terms in the frame $\Theta = \{VS, S, B, VB\}$. Actually that way we make considerations about the likelihoods of receiving particular observation on condition that it originates from each of these terms, i.e.:

$$L_i(A(k+1)/\Theta(i)) = hgt[A(k+1) \cap \Theta(i)] = \sup\{\min(\mu_{A(k+1)}, \mu_{\Theta(i)})\}, i = 1 \div 4 \quad (8)$$

where the operator hgt denotes the height of a resulting fuzzy sets, obtained after intersection between fuzzified new amplitude value and membership function of each of the linguistic terms in the frame Θ;

• using these likelihoods as respective weighting coefficients we form the convex combination of the linguistic terms. Thus we take into account the degree of their influence over the received measurement. A normalization procedure is applied. The new fuzzy set represents the weighted measurement with a following membership function:

$\mu_{A^W}(x) = \sum_i L_i^N \cdot \mu_{\Theta(i)}$, where $L_i^N = L_i/\Sigma L_i$; $L_i^N \geq 0$; $\sum_i L_i^N = 1$.

Example At scan 4 the new crisp amplitude measurement is $A = 0.7487$.

• *after applying fuzzification procedure one obtains:*

$\mu_{VS}(A) = 0.0$; $\mu_S(A) = 0.0189$; $\mu_B(A) = 0.7854$; $\mu_{VB}(A) = 0.0373$.

• *bearing in mind the a priori defined input feature frame θ, it is possible to define:* $L_1(A/VS) = hgt[A \cap VS] = \max\{\min(\mu_A, \mu_{VS})\} =$

$= \max\{\min(0,1), \min(0.0189, 0.15), \min(0.7854, 0), \min(0.0373, 0)\} = 0.0189$

the application of the above procedure according to the other linguistic values yields: $L_2(A/S) = 0.15$; $L_3(A/B) = 0.7854$, $L_4(A/VB) = 0.15$.

• *a normalization procedure is applied to L_i:* $L_i^N = L_i/\Sigma L_i$, $i = 1 \div 4$. *It yields:* $L_1^N = 0.0172$; $L_2^N = 0.1358$; $L_3^N = 0.7112$; $L_4^N = 0.1358$.

• *the weighted measurement is formed as a convex combination:* $\mu_A^W = L_1^N * \mu_{VS} + L_2^N * \mu_S + L_3^N * \mu_B + L_4^N * \mu_{VB}$. *As a result it is obtained:* $\mu_{VS}(A^W) = 0.0499$; $\mu_S(A^W) = 0.3259$; $\mu_B(A^W) = 1.0$; $\mu_{VB}(A^W) = 0.3225$.

3.5 State Updating

The updated states are obtained through a fuzzy set intersection between the weighted new measurement and corresponding modes conditioned amplitude state predictions:

$$\mu_{A^A{}_{pp}}(k+1/k+1) = \min(\mu_{A^W}, \mu_{A^A{}_{pp}}(k+1/k)) \quad (9)$$

$$\mu_{A^R{}_{ec}}(k+1/k+1) = \min(\mu_{A^W}, \mu_{A^R{}_{ec}}(k+1/k)) \quad (10)$$

4 Simulation Study

A simulation scenario is developed for a simple target trajectory (Fig. 4) in plane coordinates (X, Y) and for constant velocity movement. The target's starting point and the velocities are: $(X_0 = 5km, Y_0 = 10km)$, $\dot{X} = 100m/s, \dot{Y} = 100m/s$ and $\dot{X} = -100m/s, \dot{Y} = -100m/s$. The time sampling rate is $T = 5s$. Target dynamic process is modeled by a simple equations:

$$x(k) = x(k-1) + \dot{x} \cdot T; \quad y(k) = y(k-1) + \dot{y} \cdot T \tag{11}$$

The measured by passive radar amplitude value $Z_A(k) = A(k) + \nu_A(k)$, is random Gaussian distributed process (Fig. 5), with $A(k) = 1/D(k)$ mean, and covariance $\sigma_A(k) = 0.3 \cdot rand(1,1)/D(k)$. $D(k) = \sqrt[2]{x(k)^2 + y(k)^2}$ is the distance to the target, $\{x(k), y(k)\}$ is the corresponding vector of coordinates, and $\nu_A(k)$is the measurement noisy. Each amplitude value (true one and corresponding noisy one) received at time (scan) $k = 1, 2, \ldots$ is processed according to the block diagram of our target's behavior tracking system (Fig. 1).

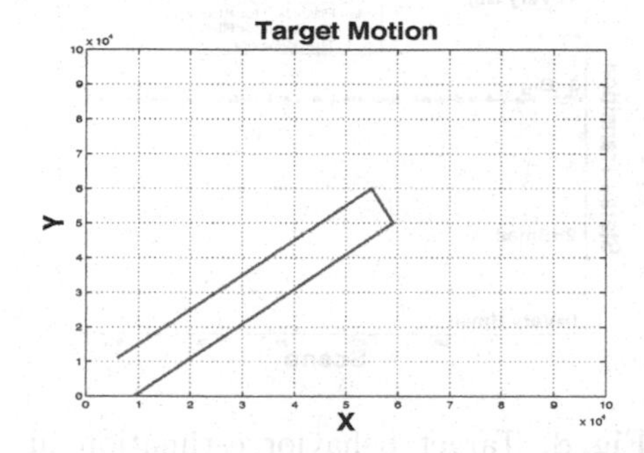

Fig. 4. Target trajectory

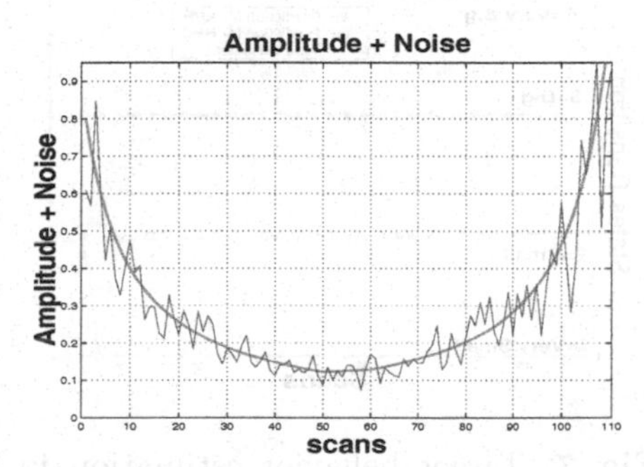

Fig. 5. Measurements dynamics

Figures 6–10 show the results obtained during the whole motion of the observed target (descending and approaching directions). They represent the tendency in target behavior, which is described via the time (scan) consecutive transitions of amplitude value $VB \rightarrow VB \rightarrow B \rightarrow B \rightarrow S \rightarrow S \rightarrow VS \rightarrow VS$ and respectively $VS \rightarrow VS \rightarrow S \rightarrow S \rightarrow B \rightarrow B \rightarrow VB \rightarrow VB$.

Fig. 6 represents the case when the measured amplitude values are without measurement's noise, i.e. $Z_A(k) = A(k)$.

Two models - *Approaching* and *Receding* are maintained in parallel. As a result of the developed algorithm (Fig. 1) implementation, it becomes possible to make a correct decision about the considered models' plausibility. It could be seen that between scans 1 and 90 target motion is supported by the correct for that case *Descending* model. In the same time the *Approaching* model has no reaction to the measurements dynamics, because it does not match the real

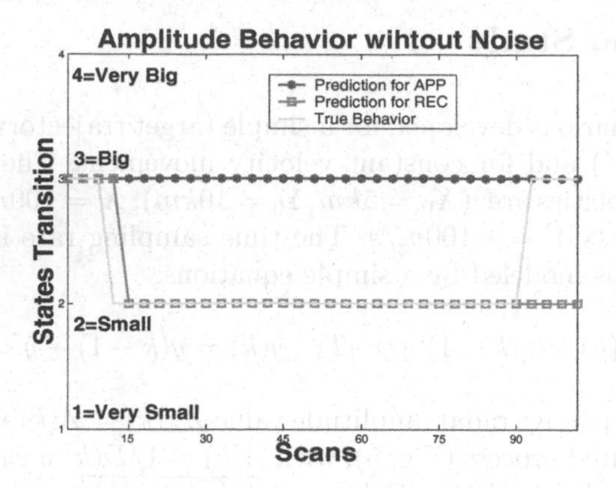

Fig. 6. Target behavior estimation (without measurement noise)

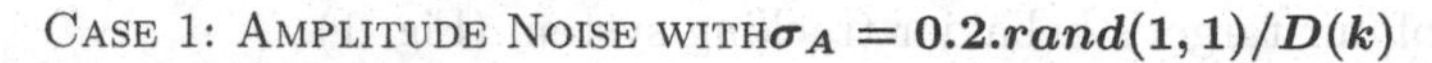

CASE 1: AMPLITUDE NOISE WITH$\boldsymbol{\sigma_A = 0.2.rand(1,1)/D(k)}$

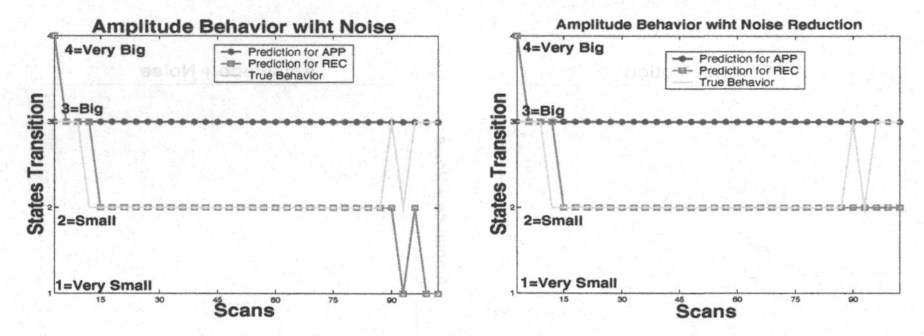

Fig. 7. Target behavior estimation in case of noise

Fig. 8. Target behavior estimation in case of noise reduction

target behaviour *Receding*. Taking into account Fig. 5, the amplitude measurements dynamics between scans 10 and 90 could be analized as relatively weak from the point of view of fuzzification interface (Fig. 2). Such a transition area is contingent on the assumed possibility for sojourning time, when the measured amplitude values during consecutive scans stable reside in one and the same regions of that interface. It is characterized with a latency delay before switching to the opposite behaviour mode. After scan 90 and until scan 115 it is obvious that the *Descending* model misses the amplitude changes, while the *Approaching* model becomes the plausible one. Fig. 7 represents the case when the measured amplitude values are corrupted by noise with $\sigma_A = 0.2.rand(1,1)/D(k)$.Some disorder and discrepancy between predicted behaviour tendency and true amplitude behaviour take place, and it is difficult to make a firm decision about the target behaviour tendency. As it is shown in Fig. 8 the application of noise re-

CASE 2: AMPLITUDE NOISE WITH $\sigma_A = 0.4.rand(1,1)/D(k)$

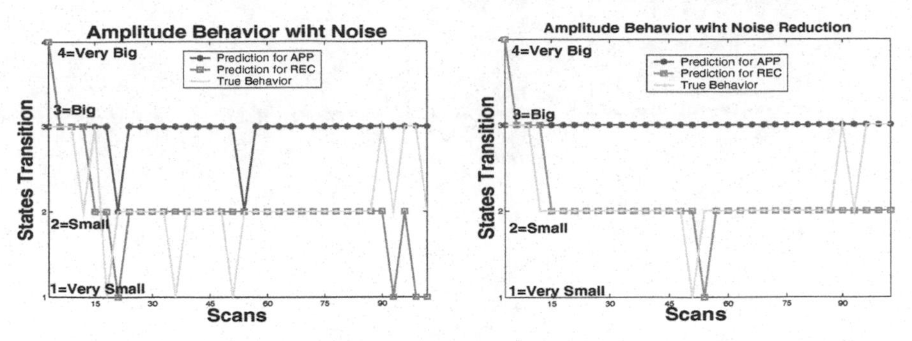

Fig. 9. Target behavior estimation in case of noise

Fig. 10. Target behavior estimation in case of noise reduction

duction procedure produces a "smoothed" predicted behaviour tendency, and it becomes possible to make a robust decision about the target behaviour tendency.

The effect of that procedure is more essential when input measurements are corrupted by a bigger noise - for example with $\sigma_A = 0.4.rand(1,1)/D(k)$ (Fig. 9). In that case some chaotic behaviour is detected. In such a critical situation the noise reduction procedure assures a more consistent process of amplitude tendency prediction (Fig. 10).

5 Conclusions

An approach for targets' behavior tendency estimation was proposed and evaluated. It is based on Fuzzy Logic principles applied to conventional passive radar measurements. A particular real-time running algorithm was developed. It was evaluated using computer simulation. Dealing simultaneously with numerical and linguistic data an opportunity for robust reasoning becomes possible. The additionally applied weighting procedure for noise reduction improves the overall process of target behavior tendency estimation. The developed algorithm is suitable and adapted for noisy amplitude measurements processing. It works with modest computational load and provides simple and robust decisions about target behavior tendency.

References

1. Blackman S., Populi R.: Design and Analysis of Modern Tracking Systems (1999)
2. Mendel J.M.: Fuzzy Logic Systems for Engineering: A Tutorial, Proc. of the IEEE, Vol. 83, No.3, March, 1995, 345–377.
3. Zadeh L.: From Computing with Numbers to Computing with Words – From Manipulation of Measurements to Manipulation of Perceptions, IEEE Transactions on Circuits and Systems-Vol.45, No.1, January 1999.

Part VI

Contributed Papers

Two Approaches to the Finite Element Analysis of the Stiffened Plates

Andrey B. Andreev, Jordan T. Maximov, and Milena R. Racheva

Technical University
5300 Gabrovo, Bulgaria
{Andreev, Maximov}@tugab.bg, MRacheva@mail.com

Abstract. The goal of this study is to investigate and to compare two finite element approaches for presenting the stiffeners of the rectangular bending plates. The first approach is when the stiffness and mass matrices are obtained by superpositioning the plate and the beam elements. The model of the second one is realized only by the plate finite elements, but we give an account of the different stiffness of the stiffeners by means of elements with different thickness. The plates are subjected to a transversal dynamic load. We consider the corresponding variational forms.

1 Introduction

Stiffened plates represent a significant structural component of Civil Engineering structures, as well as ship and aircraft structures. This type of plate is used extensively because of its high strength to weight ratio [7,8]. For detailed studies on bending, vibration and stability analysis of stiffened plates see for example [4,5,10].

The objective of this research is to model the stiffeners of rectangular plates by two ways and to compare these approaches. We use the finite element method (FEM) in combination with the normal shapes method [7,8]. Dynamic loaded plates are considered and some mechanical and dynamical characteristics of the stiffened plates are analyzed.

A thin plate implies that its thickness is small compared to its other dimensions. We suppose that the displacements of the points at the middle plane of the plate in the normal direction to this plane are small compared to the thickness of the plate and also that the transverse normal stresses are negligible. It is to be noted here that for stiffened plate structures, the spacing of the stiffening ribs has to be small compared to the dimensions of the plate so that beam action will not be predominant.

This paper is a follow-up of the research of the authors, which were initiated in [1].

I. Dimov et al. (Eds.): NMA 2002, LNCS 2542, pp. 353–362, 2003.

2 Formulation of the Problem

Consider a rectangular plate with thickness $\widetilde{h}$ and let the linear dimensions of the plate parallel to the x and $y-$direction be a and b, respectively. The flexural rigidity of this isotropic plate is $D=\frac{E.\widetilde{h}^3}{12(1-\nu)}$, where E is the modulus of elasticity, ν is the Poisson ratio. We denote by Ω the correspondent rectangular domain with boundary Γ.

Let us consider the bending stiffened plate subjected to a transversal dynamic load $\mathcal{P}(x,y;t)$. The stiffeners are placed parallel to its sides. The general orthotropic plate equation will be reduced to the fourth-order hyperbolic differential equation with some terms containing Dirac $\delta-$function [1,5]:

$$D.\Delta^2 W+\sum_{k=1}^{n_x}\delta(y-b_k)\left[EI_{xk}\frac{\partial^4 W}{\partial x^4}\right]+\sum_{k=1}^{n_y}\delta(x-a_k)\left[EI_{yk}\frac{\partial^4 W}{\partial y^4}\right] \tag{1}$$

$$=-m.\frac{\partial^2 W}{\partial t^2}+\mathcal{P}-\sum_{k=1}^{n_x}\delta(y-b_k).m_{xk}\frac{\partial^2 W}{\partial t^2}-\sum_{k=1}^{n_y}\delta(x-a_k).m_{yk}\frac{\partial^2 W}{\partial t^2}.$$

The unknown $W=W(x,y;t)$ of (1) represents the vertical displacement of the plate. The mass of the plate per unit area is m and m_{xk} (m_{yk}) is the mass per unit length of the $k-$th $x-$directional ($y-$directional) stiffener and I_{xk} (I_{yk}) is the second moment of inertia of the $k-$th $x-$directional ($y-$directional) stiffener. The numbers n_x and n_y represent the total number of $x-$directional and $y-$directional stiffeners respectively and Δ^2 is the biharmonic operator. We suppose that $a_k\in[0,a]$, $k=1,\dots,n_y$ and $b_k\in[0,b]$, $k=1,\dots,n_x$.

The equation (1) with Dirichlet or mixed homogeneous boundary conditions will be solved at initial conditions (W_0 and W_1 are given functions):

$$W(x,y,;0)=W_0(x,y),\quad \frac{\partial W}{\partial t}=W_1(x,y)\quad \text{in}\ \Omega. \tag{2}$$

Now we modify the presented approach to model the stiffened plate. Let from any stiffener, $\widetilde{H}$, ($\widetilde{H}\gg\widetilde{h}$) be the thickness of the stiffener and let it be made from the same material as the plate.

Reporting on the bending of the stiffened sections, we replace EI_{xk} and EI_{yk} by the flexural rigidity $\widetilde{D}=\frac{E.\widetilde{H}^3}{12(1-\nu)}$. The masses of the stiffeners in the model equation (1) augment, i.e. $\overline{m}_{xk}>m_{xk}$, $\overline{m}_{yk}>m_{yk}$. Let ε be a fixed positive number. With 2ε we associate the width of any stiffener. We denote

$$\xi_{a_k}(x)=\eta(x-a_k+\varepsilon)-\eta(x-a_k-\varepsilon),\quad k=1,\dots,n_y,$$

$$\xi_{b_k}(y)=\eta(y-b_k+\varepsilon)-\eta(y-b_k-\varepsilon),\quad k=1,\dots,n_x,$$

where η is the Heviside function. The functions $\xi_{a_k}(x)$, $\xi_{b_k}(y)$ are zero on Ω except for the corresponding rectangle with width 2ε. Thus we obtain:

$$D.\Delta^2 W + \sum_{k=1}^{n_x} \xi_{b_k}(y)\widetilde{D}.\Delta^2 W + \sum_{k=1}^{n_y} \xi_{a_k}(x)\widetilde{D}.\Delta^2 W \tag{3}$$

$$= -m.\frac{\partial^2 W}{\partial t^2} + \mathcal{P} - \sum_{k=1}^{n_x} \xi_{b_k}(y).\overline{m}_{xk}\frac{\partial^2 W}{\partial t^2} - \sum_{k=1}^{n_y} \xi_{a_k}(x).\overline{m}_{yk}\frac{\partial^2 W}{\partial t^2}.$$

We can derive the free harmonic vibrations to the problems (1) and (3). We put $\mathcal{P} = 0$ and separate the variables $W(x, y; t) = \overline{W}(x, y)e^{i\lambda t}$, where $\overline{W}(x, y)$ is the shape function and λ is the natural frequency of the vibration.

Thus we get the following spectral problems corresponding to the equations (1) and (3) respectively:

$$D.\Delta^2 \overline{W} + \sum_{k=1}^{n_x} \delta(y - b_k).EI_{xk}\frac{\partial^4 \overline{W}}{\partial x^4} + \sum_{k=1}^{n_y} \delta(x - a_k).EI_{yk}\frac{\partial^4 \overline{W}}{\partial y^4} \tag{4}$$

$$= \lambda^2 \left(m\overline{W} + \sum_{k=1}^{n_x} \delta(y - b_k).m_{xk}\overline{W}(x, b_k) + \sum_{k=1}^{n_y} \delta(x - a_k).m_{yk}\overline{W}(a_k, y) \right),$$

$$D.\Delta^2 \overline{W} + \sum_{k=1}^{n_x} \xi_{b_k}(y).\widetilde{D}.\Delta^2 \overline{W} + \sum_{k=1}^{n_y} \xi_{a_k}(x).\widetilde{D}.\Delta^2 \overline{W} \tag{5}$$

$$= \lambda^2 \left(m\overline{W} + \sum_{k=1}^{n_x} \xi_{b_k}(y).\overline{m}_{xk}\overline{W}(x, b_k) + \sum_{k=1}^{n_y} \xi_{a_k}(x).\overline{m}_{yk}\overline{W}(a_k, y) \right).$$

The eigenfunctions $\overline{W}(x, y)$ satisfies homogeneous boundary conditions.

3 Variational Formulation

The considered problem can be recasted in weak form, in order to apply numerical variational methods. Let $H^s(\Omega)$ be the usual Sobolev space for positive integer s provided with the norm $\|\cdot\|_{s,\Omega}$ [3]. Consider the problems (1) and (3), $(x, y; t) \in \Omega \times (0, T)$ with initial conditions (2). Supposing homogeneous boundary conditions we multiply the equation (1) or (3) by $z(x, y) \in H^2(\Omega)$ and integrate by parts. We denote the $L_2(\Omega)-$ inner product by $(\cdot,\cdot)$ and we shall drop the argument (x, y) for notational convenience. Using the multi-index notation we define the corresponding a and $b-$forms for the problems (1) and (3) respectively. We have for any $u, v \in H^2(\Omega); i = 1; 2$:

$$a_i(u, v) = a(u, v) + a_{ix}(u, v) + a_{iy}(u, v),\ b_i(u, v) = b_{ix}(u, v) + b_{iy}(u, v),$$

where

$$a(u, v) = \int_0^a \int_0^b D \sum_{|\alpha|=2} \partial^\alpha u.\partial^\alpha v\, dx\, dy,$$

$$a_{1x}(u, v) = \sum_{k=1}^{n_x} EI_{xk} \int_0^a \left(\frac{\partial^2 u}{\partial x^2}.\frac{\partial^2 v}{\partial x^2} \right)_{y=b_k} dx,$$

$$a_{1y}(u,v) = \sum_{k=1}^{n_y} EI_{yk} \int_0^b \left(\frac{\partial^2 u}{\partial y^2} . \frac{\partial^2 v}{\partial y^2} \right)_{x=a_k} dy,$$

$$a_{2x}(u,v) = \sum_{k=1}^{n_x} \widetilde{D} \int_0^a \int_{b_k-\varepsilon}^{b_k+\varepsilon} \sum_{|\alpha|=2} \partial^\alpha u . \partial^\alpha v \, dx \, dy,$$

$$a_{2y}(u,v) = \sum_{k=1}^{n_y} \widetilde{D} \int_{a_k-\varepsilon}^{a_k+\varepsilon} \int_0^b \sum_{|\alpha|=2} \partial^\alpha u . \partial^\alpha v \, dx \, dy,$$

$$b_{1x}(u,v) = \sum_{k=1}^{n_x} m_{xk} \int_0^a (u.v)_{y=b_k} dx, \quad b_{1y}(u,v) = \sum_{k=1}^{n_y} m_{yk} \int_0^b (u.v)_{x=a_k} dy,$$

$$b_{2x}(u,v) = \sum_{k=1}^{n_x} \overline{m}_{xk} \int_0^a \int_{b_k-\varepsilon}^{b_k+\varepsilon} u.v \, dx \, dy, \quad b_{2y}(u,v) = \sum_{k=1}^{n_y} \overline{m}_{yk} \int_{a_k-\varepsilon}^{a_k+\varepsilon} \int_0^b u.v \, dx \, dy.$$

The weak formulation of the problem (1), (2) or (3), (2) is: *Find a function* $W : t \in [0,T] \to W(t) \in H^2(\Omega)$ *such that*

$$a_i(W(t), z) = -m \left(\frac{d^2 W}{dt^2}, z \right) + (\mathcal{P}(t), z) - b_i(W(t), z) \quad \forall z \in H^2(\Omega), \quad i = 1; 2, \tag{6}$$

$$W(0) = W_0, \quad \frac{dW}{dt}(0) = W_1.$$

The weak formulation corresponded to the eigenvalue problem (4) or (5) is: *Find* $\lambda \in \mathbf{R}$ *and* $\overline{W} \in H_0^2(\Omega)$ *such that:*

$$a_i(\overline{W}, z) = \lambda^2 \left[m(\overline{W}, z) + b_i(\overline{W}, z) \right], \quad \forall z \in H_0^2(\Omega), \quad i = 1; 2. \tag{7}$$

4 Finite Element Analysis

The finite element formulation for the two approaches are much the same as the case of the stiffened plate considered in [1]. The solution of the hyperbolic equation (1) or (3) and of the problem (6) is of a time varying quantity defined over the spatial domain Ω. In such case, finite element discretization is commonly applied only in the spatial variables, leaving a coupled system of ordinary differential equations to be solved numerically.

We approximate the considered problems by the rectangular finite elements. The approximating polynomials are of degree less than or equal to $p \geq 3$ with respect to each variables x, y. The finite element space V_h is a subspace of $H^2(\Omega)$ and it has to satisfy the C^1-condition [3,6]. The parameter h represents the maximal diameter of all finite elements.

The semidiscrete approximate problems of (6) are: *Find* $W_h \in V_h$ *such that for any* $z_h \in V_h$

$$a_i(W_h(t), z_h) = -m \left(\frac{d^2 W_h}{dt^2}, z_h \right) + (\mathcal{P}(t), z_h) - b_i(W_h(t), z_h) \quad , i = 1; 2, \tag{8}$$

with initial conditions $W_h(0) = W_{0,h}$, $\frac{dW_h}{dt}(0) = W_{1,h}$.

Expanding the approximate displacement function $W_h(t)$ as well as $W_{0,h}$ and $W_{1,h}$ in terms of a chosen basis of V_h, we have

$$[M]_i.\left\{\ddot{X}\right\} + [K]_i.\{X\} = \{\mathcal{P}(t)\}, \quad i = 1;2, \tag{9}$$

where $[M]_i$ and $[K]_i$ are global mass and stiffness matrices. $\{\mathcal{P}(t)\}$ is the loading vector and superposed dot of the unknown vector $\{X\}$ denotes the time derivative. Similarly as in [1] we can consider systems with damping [4,10], then the term $[C].\left\{\dot{X}\right\}$ is added in the equation (9), where the damping matrix $[C]$ is supposed proportional to the matrix $[K]$.

We present two approaches to the FE analysis:

- ($i = 1$) The stiffness and the mass matrices are generated from the contributions of both the plate and the stiffeners. Then the element stiffness and mass matrices obtained in the local co-ordinate system are transformed into global co-ordinate system for plate and beam members and assembled in order to obtain global stiffness and mass matrices.
- ($i = 2$) In this case only the plate finite elements are used. Thus the double bending of the stiffened plate is reported on the $x-$ and $y-$axes (see the next section).

Now we sketch the proof of stability of the approximations (8) subject to the homogeneous boundary conditions.

Proposition 1. *The approximate displacement solution of the problem (8) satisfies a stability estimate.*

Proof. The assertion is valid for both cases ($i = 1;2$). Taking $z_h = \dot{W}_h = \frac{dW_h}{dt}$ for each time–level t, we integrate with respect to time from $t = 0$ to $t = T$:

$$\frac{m}{2}\int_0^T \frac{d}{dt}\,\|\dot{W}_h(t)\|^2_{0,\Omega}\,dt + \frac{1}{2}\int_0^T \frac{d}{dt}\,a_i\left(W_h(t), W_h(t)\right) dt$$

$$= \int_0^T \left(\mathcal{P}(t), \dot{W}_h(t)\right) dt - \frac{1}{2}\int_0^T \frac{d}{dt} b_i\left(W_h(t), W_h(t)\right) dt,$$

$$\frac{m}{2}\left(\|\dot{W}_h(T)\|^2_{0,\Omega} - \|\dot{W}_h(0)\|^2_{0,\Omega}\right) + \frac{1}{2}\left(b_i\left(W_h(T), W_h(T)\right) - b_i\left(W_h(0), W_h(0)\right)\right)$$

$$+\frac{1}{2}\int_0^T \frac{d}{dt} a_i\left(W_h(t), W_h(t)\right) dt = \int_0^T \left(\mathcal{P}(t), \dot{W}_h(t)\right) dt.$$

Obviously $a_i(\cdot,\cdot)$ is V_h-elliptic [3]. Using the ellipticity of $a_i(\cdot,\cdot)$ with constant $\rho > 0$ and the Cauchy-Schwartz inequality, we obtain:

$$m\|\dot{W}_h(T)\|^2_{0,\Omega} + \rho\|W_h(T)\|^2_{2,\Omega} + b_i\left(W_h(T), W_h(T)\right) \leq m\|\dot{W}_h(0))\|^2_{0,\Omega}$$

$$+\rho\|W_h(0)\|^2_{2,\Omega} + b_i\left(W_h(0), W_h(0)\right) + \int_0^T \left(\|\mathcal{P}(t)\|^2_{0,\Omega} + \|\dot{W}_h(t)\|^2_{0,\Omega}\right) dt.$$

Setting $m = 1$, stability follows from Gronwall's lemma [9]. Then, we have

$$\|W_h(T)\|_{0,\Omega}^2 + \rho\|W_h(T)\|_{2,\Omega}^2 + b_i(W_h(T), W_h(T)) \leq \rho\|W_{0,h}\|_{2,\Omega}^2$$

$$+e^T\|W_{1,h}\|_{0,\Omega}^2 + b_i(W_{0,h}, W_{0,h}) + \int_0^T \|\mathcal{P}(t)\|_{0,\Omega}^2 . e^{(T-t)}\, dt,$$

where $W_{0,h}$ and $W_{1,h}$ are the FE approximations of the given functions in (2).

Consequently, we can prove the convergence of W_h to W as $h \to 0$ and an estimate of the order of this convergence for the cases considered by standard argument (see [6]).

Next we apply the method of normal shapes [8]. First we consider the matrix equations corresponding to (7). It has to solve the lowest eigenvalues and corresponding eigenvectors satisfying the equation $[K]_i.\{\overline{X}\} = \lambda^2[M]_i.\{\overline{X}\}$, where λ is the natural frequency of the plate ($\lambda > 0$).

Next we define the matrix $[\overline{X}]_i$ which columns are the first eigenvectors. This matrix determines the vector of general co-ordinates $\{X_G\}_i$ by means of the equation $\left[\overline{X}\right]_i^{-1}\{X\} = \{X_G\}_i$.

Thus, the matrices $[K]_i$ and $[M]_i$ can be diagonalized using the orthogonality of the eigenvectors: $\left[\overline{X}\right]_i^T [M]_i \left[\overline{X}\right]_i = [M_G]_i$; $\left[\overline{X}\right]_i^T [K]_i \left[\overline{X}\right]_i = [K_G]_i$. We obtain lumped mass schemes for both approaches (see for ex. [1,4,5]).

Determining the vector $\{X_G\}_i$, we easily transform the solution in the original (physical) co-ordinates [1,8]: $\{X\}_i = \left[\overline{X}\right]_i \{X_G\}_i$.

5 Numerical Results

Consider a cantilever supported rectangular plate with thickness 10 mm, presented in Figure 1. The external load, distributed along the surface, acts on the square with side length 100 mm. Its intensity changes with respect to the time (Figure 2). This function corresponds approximately to the instantly placed load. A concentrated mass, equal to 3 kg, is located in point B. The plate is stiffened. The transverse section of any stiffener is a rectangle with the measures 30×20 mm.

Material properties of the plate are: $E = 2.10^{11}$Pa; $\nu = 0.3$; density $\rho = 7850\text{kg/m}^3$.

The stiffeners can be placed in two manners:

(A) Symmetrically (Figure 3a), in which case the middle plane of the plate is the plane of the material symmetry;

(B) Asymmetrically (Figure 3b), in which case the construction is no longer a plane.

In both possibilities, the mass of the stiffened plate is the same.

The following problems are presented:

- Analysis of the possibilities for FE modeling of every case.
- Comparison of the strained and deformed state (SDS) of every case.

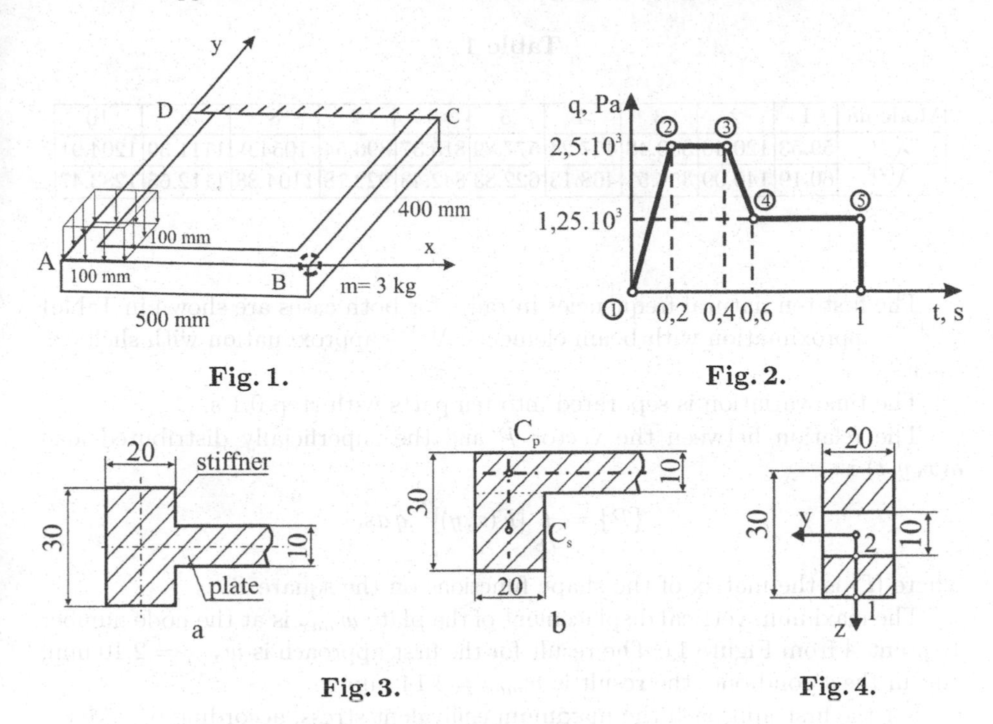

Fig. 1.

Fig. 2.

Fig. 3.

Fig. 4.

Analysis of Case *(A)* (Figure 3a)

The presented construction of the stiffened plate is plane with respect to the SDS. Two ways are expedient here:
- the stiffeners are modeled as beam elements;
- the stiffeners with thickness 30 mm are modeled as parts of the plate.

The finite element procedure is implemented. The plate domain is discretized using an uniform triangulation with rectangular elements for both cases.

The stiffeners are approximated by finite elements type BEAM3D (see for ex. [2]). Their transverse section is shown on Figure 4.

The global stiffness matrix $[K]$ of the construction is:

$$[K] = [K_b] + [K_p],$$

where $[K_b]$ is a stiffness matrix of all beam elements and $[K_p]$ of all plate elements.

If we adopt, that the section of the beam elements is dense rectangle, in any common node between beam and plate FE, there would be a duplication between the stiffness of the plate and the stiffness of the beam FE with transverse section 10×20 mm. The physical equivalent is a sum of nonexistent in reality material of the stiffener with that of the plate. As a result, there would be an error in the augmentation of the stiffness of the construction.

In the second case only plate FE are used. The stiffeners have a thickness of 30 mm while the plate part has a thickness of 10 mm. In both cases the lumped mass is accounted through the application of the one-nod FE type MASS [2].

Table 1.

Mode no.	1	2	3	4	5	6	7	8	9	10
$\lambda^{(1)}$	59.53	120.40	330.48	457.24	577.89	815.37	896.54	1054.9	1111.39	1204.91
$\lambda^{(2)}$	60.19	140.09	357.51	468.13	622.83	842.43	922.18	1104.88	1112.65	1283.47

The first ten natural frequencies in rad/s for both cases are shown in Table1 ($\lambda^{(1)}$ – approximation with beam elements; $\lambda^{(2)}$ – approximation with shell elements).

The time variation is separated into ten parts with step 0.1 s.

The relation between the vector $\mathcal{P}$ and the superficially distributed load $q(x, y, t)$ is:

$$\{\mathcal{P}\} = \int_S [F(x, y)]^T .q\, ds,$$

where $[F]$ is the matrix of the shape functions on the square S.

The maximum vertical displacement of the plate w_{max} is at the node number 1 (point A from Figure 1). The result for the first approach is $w_{max} = 2.10$ mm, and in the second one, the result is $w_{max} = 2.14$ mm.

For the first approach the maximum equivalent stress, according to v.Mises, is at point 1 (see Figure 4) from the clamped section (point D from Figure 1) of the beam FE: $\sigma_{v.M.} = |\sigma_x| = 143.3$ MPa, i.e. the strained state (SS) at this point is one-dimensional with a non-zero principal stress $\sigma_3 = -149.3$ MPa. The maximum equivalent stress for the plate at the same nod is $\sigma_{v.M.} = 48.4$ MPa (see point 2, Figure 4). In this case SS is two-dimensional with non-zero principal stresses: $\sigma_1 = 54.52$ MPa, $\sigma_3 = 16.78$ MPa.

In the second approach $\max \sigma_{v.M.} = 146.65$ MPa is at point 1 from Figure 4, for which we receive two-dimensional SS with non-zero stresses: $\sigma_1 = 164.4$ MPa and $\sigma_3 = 47.24$ MPa. This approach gives more accurate results, because the discretization of the stiffeners with plate FE allows for the estimation of their bending, both around X and Y axis.

Analysis of Case *(B)* (Figure 3b)

In this case the construction is spatial. The stiffeners discretization with beam FE requires additional usage of one-dimensional FE (type rigid-bar for steady state problems and type BEAM3D for dynamic researches) for relation between the nods C_s and C_p (Figure 3b). In this case we witness the result of the duplicated stiffness. This approach is justified for the stiffeners with high and narrow transverse section.

The approach of discretization of the stiffeners by shell finite elements (type shell 4) leads to the construction of type panel shell. Only one type of FE is used, but the effect of the duplicated stiffness stays.

Plates which are stiffened with massive (homogeneous) stiffeners, is reasonable to be moderated as massive bodies. The discretization is made by FE type SOLID [2]. The SS in all points is three-dimensional. The contribution to the maximum equivalent stress has the normal stress σ_z. In this aspect the model of the massive body gives the best results.

The comparison between possibilities (A) and (B) from Figure 3 shows that the greater stiffness of the construction is obtained in the second case. In the case of equal masses for both cases, the eigenvalues in the second case are higher, according to Rayleigh' Theorem [8].

6 Conclusions

- Plates, for which its middle planes coincide with the principal plane of inertia of the stiffeners, form a plane construction, which has the same feature as the one of a plate in aspect of SDS;
- The stiffeners can be approximated with FE of type BEAM3D for the defined above plates. This approximation is achieved through an empty transverse section, in the cases of large height and small width. The error of using such finite elements decreases with the augmentation the ratio between H (height) and $\widetilde{b}$ (width), because SS in any rib point of the beam elements is one-dimensional. It is better to use shell finite elements in the case of larger $\widetilde{b}$. In this manner both σ_x and σ_y will exercise influence on the point with maximal equivalent stresses. This is due to the double bending of the plate, i.e. for a more accurate solution.
- Plates with asymmetrically situated stiffeners and with relatively large measures H and $\widetilde{b}$ are better defined as massive constructions. In the cases of narrow, but high transverse sections the stiffeners can be approximated with beam or shell finite elements. In this manner an error is introduced through the effect of the duplicated stiffness. This effect is as weakly noticeable as the ratio $H : \widetilde{h}$ is greater, where $\widetilde{h}$ is the thickness of the plate.

References

1. Andreev, A.B., Maximov, J.T., Racheva, M.R.: Finite element method for plates with dynamic loads. In: Margenov, S., Wasniewski, J., and Yalamov, P. (eds.): Large-Scale Scientific Computing. Lecture Notes in Computer Science, Vol. 2179, Springer-Verlag (2001) 445–453.
2. Argyris, J., Mlejnek, H.P.: Finite Element Methods. Friedr. Vieweg & Sohn Braunschweig/Wiesbaden (1986)
3. Ciarlet, P.G.: The Finite Element Method for Elliptic Problems. North-Holland Amsterdam (1978)
4. Mukherjee, A., Mikhopadhyay, M.: Finite element free vibration analysis of stiffened plates. The Aeronautical Journal (1986) 267–273.
5. Mikhopadhyay, M.: Vibration and stability analysis of stiffened plates by semi-analytic finite difference method, Part I: Consideration of bending displacements only. J. of Sounds and Vibration, Vol. 130, 1 (1989) 27–39.
6. Raviart, P.A., Thomas, J.-M.: Introduction a l'Analyse Numerique des Equations aux Derivées Partielles. Masson Paris (1988)
7. Szilard, R.: Theory of Plates. Prentice Hall New York (1975)

8. Timoshenko, S.P., Young, D.H., Weaver, W.: Vibration Problems in Engineering. John Wiley & Sons, New York (1977)
9. Thomee, V.: Galerkin Finite Element Methods for Parabolic Problems. Springer (1997)
10. Wah, T.: Vibration of Stiffened Plates. Aero Quarterly XV (1964) 285–298.

On the Postprocessing Technique for Eigenvalue Problems

Andrey B. Andreev and Milena R. Racheva

Technical University
5300 Gabrovo, Bulgaria
Andreev@tugab.bg, MRacheva@mail.com

Abstract. We present a new strategy of accelerating the convergence rate for the finite element solutions of the large class of linear eigenvalue problems of order $2m$. The proposed algorithms have the superconvergence properties of the eigenvalues, as well as of the eigenfunctions. This improvement is obtained at a small computational cost. Solving a more simple additional problem, we get good finite element approximations on the coarse mesh. Different ways for calculating the postprocessed eigenfunctions are considered. The case where the spectral parameter appears linearly in the boundary conditions is discussed. The numerical examples, presented here, confirm the theoretical results and show the efficiency of the postprocessing method.

1 Introduction and Setting of the Problem

Let Ω be a bounded domain in $\mathbf{R}^d, d = 1, 2, 3$ and the boundary $\partial\Omega \equiv \Gamma$ will be assumed sufficiently smooth in order to avoid technical difficulties. For positive integer k we denote by $H^k(\Omega)$ and $H_0^k(\Omega)$ the usual Sobolev spaces provided with the norm $\| \cdot \|_{k,\Omega}$ (see [1,2]).

Let $\mathcal{A}$ be a formally self-adjoint and positive elliptic differential operator of order $2m$ and let $\mathcal{B}$ be a bounded and positive elliptic operator of order $2r$, $m > r \geq 0$. In the case $r = 0$, $\mathcal{B} \equiv b(x)$ and $b(x)$ is a smooth on $\overline{\Omega}$ function bounded below by a positive constant.

Consider the following linear eigenvalue problem:

$$\mathcal{A}u(x) = \lambda \mathcal{B}u(x), \quad x \in \Omega, \tag{1}$$

subject to the homogeneous bondary conditions

$$\mathcal{C}u(x) = 0, \quad x \in \Gamma, \tag{2}$$

where $\mathcal{C} = \{C_j\}_{j=1}^m$, $C_j u(x) = \sum_{|\alpha| \leq 2m-1} s_{j,\alpha}(x).D^\alpha u(x)$, $j = 1, \ldots, m$, $x \in \Gamma$ are linearly independent conditions. The one-dimensional case ($d = 1$ with $\Omega = (a, b)$) allows $2m$ more general homogeneous boundary conditions (see for ex. [3]). We shall consider later in §3 the case when the eigenvalue parameter appears in the boundary conditions. Of particular note is that discussion here is restricted to self-adjoint problems.

I. Dimov et al. (Eds.): NMA 2002, LNCS 2542, pp. 363–371, 2003.

We denote the $L_2(\Omega)$–inner product by $(\cdot,\cdot)$ and we shall throughout use the letter C for a generic positive constant which may vary with the context.

Let D be the space of all functions $u(x) \in C^{2m}(\overline{\Omega})$ satisfying the boundary conditions (2). We assume also that the coefficients of operator $\mathcal{C}$ are smooth enough and such that, integrating by parts we define the symmetric bilinear forms:

$$(\mathcal{A}u, v) = (u, \mathcal{A}v) \equiv a(u,v); \quad (\mathcal{B}u, v) = (u, \mathcal{B}v) \equiv b(u,v) \quad \forall\, u, v \in D.$$

We may define the Hilbert spaces $H_{\mathcal{A}}$ and $H_{\mathcal{B}}$ with $a(u,u) = \|u\|^2_{\mathcal{A}}$ and $b(u,u) = \|u\|^2_{\mathcal{B}}$ respectively. Thus $H_{\mathcal{A}}$ and $H_{\mathcal{B}}$ are completions of D with respect to $\|\cdot\|_{\mathcal{A}}$ and $\|\cdot\|_{\mathcal{B}}$ respectively, and $H_{\mathcal{A}} \subset H_{\mathcal{B}}$.

We denote by V a closed subspace of $H^m(\Omega) \cap H_{\mathcal{A}}$. We assume that the bilinear form $a(u,v)$ is elliptic, i.e. there exists a constant $\rho > 0$ such that $\forall\, v \in V,\ \rho\|v\|^2_{m,\Omega} \le a(v,v)$. Moreover, the a and b–forms are continuous [1,2].

The variational form of the problem (1), (2) is: *find* $\lambda \in R,\ u(x) \in V$ *such that*

$$a(u,v) = \lambda b(u,v) \quad \forall v \in V. \tag{3}$$

For the problem (3) there is a countable infinite set of eigenvalues, all being strictly positive (see [3,4,5]).

For any function $w \in H^r(\Omega)$ consider the following elliptic problem:

$$a(u,v) = b(w,v) \quad \forall v \in V.$$

Then the operator $\mathcal{T} : H^r(\Omega) \to V$ defined by $u = \mathcal{T}w,\ u \in V$ is the solution operator for boundary value (source) problem.

It follows immediately from their definitions that $a(u,v)$ and $b(u,v)$ are symmetric (or Hermitian) forms. Thus $a(u,v)$ is an inner product on $H_{\mathcal{A}}$ that is equivalent to the inner product on $H^m(\Omega)$. In a similar way we see that $b(u,v)$ is an inner product on $H_{\mathcal{B}}$ that is equivalent to the inner product on $H^r(\Omega)$. Consequently $a(\mathcal{T}u, v) = a(u, \mathcal{T}v)\ \forall u, v \in H^m(\Omega)$; $b(\mathcal{T}u, v) = b(u, \mathcal{T}v)\ \forall u, v \in H^r(\Omega)$.

Thus the operator $\mathcal{T}$ is symmetric and positive. It follows by the Riesz representation theorem ($a(\cdot,\cdot)$ is an inner product on V), that $\mathcal{T}$ is bounded. As it is shown in [6], λ is an eigenvalue and u is the corresponding eigenfunction if and only if $u - \lambda\mathcal{T}u = 0,\ u \ne 0$.

We are interested in the approximation of the eigenpairs of (3) by the finite element method. Consider a family of regular finite element partitions τ_h of $\overline{\Omega}$ which fulfill standard assumptions (see [2], Chapter 3) and assume that this family satisfies an inverse assumption, i.e. there exists a constant $\nu > 0$ such that $\forall\, h, \forall\, e \in \tau_h\ \frac{h}{h_e} \le \nu$, where $h_e =$ diameter of $e, h = \max_{e \in \tau_h} h_e$.

Using the partition τ_h we associate the finite-dimensional subspaces V_h and $\widetilde{V}_h$ of $V \cap C^{m-1}(\overline{\Omega})$ such that the restriction of every function of these spaces over every finite element $e \in \tau_h$ is a polynomial of degree n and n_1 at most, respectively, and $n_1 > n$. As a matter of fact, the value of n_1 and how it depends on n, m and r will be determined later.

The approximate eigenpair (λ_h, u_h) obtained by the finite element method based on V_h are determined by:

$$a(u_h, v) = \lambda_h b(u_h, v) \quad \forall v \in V_h. \tag{4}$$

It is well-known that the rate of convergence of finite element approximation to the eigenvalues and eigenfunctions is given by the following estimates [3,7]:

$$\|u - u_h\|_{m,\Omega} \leq C.h^{n+1-m} \|u\|_{n+1,\Omega}, \tag{5}$$

$$\|u - u_h\|_{r,\Omega} \leq C.h^{n+1-r} \|u\|_{n+1,\Omega}, \tag{6}$$

$$|\lambda - \lambda_h| \leq C.h^{2(n+1-m)} \|u\|^2_{n+1,\Omega}, \tag{7}$$

where C is independent of the mesh size and it depends on the continuous problem (see [7]).

2 Superconvergent Postprocessing Technique

The postprocessing method we propose gives better accuracy for eigenvalues as well as for eigenfunctions when a $2m-$order finite element eigenvalue problem (4) is solved. This method globally approximates the eigenfunctions in the energy norm and the postprocessing deals with solving a corresponding linear elliptic problem.

Let $u_h \in V_h$ be any approximate eigenfunction of (4) with $b(u_h, u_h) = 1$.

Since the finite element solution u_h is already known, we consider the following elliptic problems:

$$a(\widetilde{u}, v) = b(u_h, v) \quad \forall v \in V, \tag{8}$$

$$a(\widetilde{u}_h, v_h) = b(u_h, v_h) \quad \forall v_h \in \widetilde{V}_h. \tag{9}$$

Then we define the numbers $\widetilde{\lambda} = \frac{1}{b(\widetilde{u}, u_h)}$, $\widetilde{\lambda}_h = \frac{1}{b(\widetilde{u}_h, u_h)}$, where $\widetilde{u}$ and $\widetilde{u}_h$ are the solutions of (8) and (9), respectively. We suppose also that the bilinear form $a(\cdot, \cdot)$ on $V \times V$ is regular (see [2], p.138).

Theorem 1. *Let the finite element subspaces V_h and $\widetilde{V}_h$ contain polynomials of degree n and n_1 respectively, such as $n \geq 2m - 1$ and $n_1 \geq n + m - r$. If (u, λ) is an eigenpair of problem (3), $u \in H^{n+1}(\Omega)$ and (u_h, λ_h) is the corresponding solution of (4) and eigenfunctions being normalized $b(u, u) = b(u_h, u_h) = 1$, then we have the following superconvergence estimate:*

$$|\lambda - \widetilde{\lambda}_h| \leq C.h^{2(n+1-r)} \|u\|^2_{n+1,\Omega}. \tag{10}$$

Proof. Using the properties of $\mathcal{T}$ we get $a(\mathcal{T}u, v) = b(u, v) \ \forall v \in V$.

On the other hand u is a solution of the eigenvalue problem, i.e. $a(u, \mathcal{T}u) = \lambda b(u, \mathcal{T}u)$. From the symmetry of the operator $\mathcal{T}$ we have

$$\frac{1}{\lambda} - \frac{1}{\widetilde{\lambda}} = b(\mathcal{T}u, u) - b(\mathcal{T}u_h, u_h)$$

$$= b(\mathcal{T}u, u) - b(\mathcal{T}u_h, u_h) + b(\mathcal{T}(u - u_h), u - u_h) - b(\mathcal{T}(u - u_h), u - u_h),$$

consequently

$$\frac{1}{\lambda} - \frac{1}{\widetilde{\lambda}} = 2b(\mathcal{T}u, u - u_h) - b(\mathcal{T}(u - u_h), u - u_h). \tag{11}$$

Now we will estimate the two terms in the right-hand side of (11). Using the continuity of the b–form, we get:

$$2b(\mathcal{T}u, u - u_h) = \frac{2}{\lambda}(1 - b(u, u_h)) = \frac{1}{\lambda}(b(u, u) - 2b(u, u_h) + b(u_h, u_h))$$

$$= \frac{1}{\lambda} b(u - u_h, u - u_h) \leq C\, \|u - u_h\|^2_{r,\Omega}. \tag{12}$$

Having in mind that the operator $\mathcal{T}$ is bounded, we have:

$$|b(\mathcal{T}(u - u_h), u - u_h)| \leq C\, \|u - u_h\|^2_{r,\Omega}. \tag{13}$$

Finally, from (11),(12) and (13) we obtain $|\lambda - \widetilde{\lambda}| \leq C\, \|u - u_h\|^2_{r,\Omega}$, and in view of (6) we get:

$$|\lambda - \widetilde{\lambda}| \leq C.h^{2(n+1-r)}\, \|u\|^2_{n+1,\Omega}. \tag{14}$$

Let us estimate $|\widetilde{\lambda} - \widetilde{\lambda}_h|$:

$$\frac{1}{\widetilde{\lambda}} - \frac{1}{\widetilde{\lambda}_h} = b(\widetilde{u}, u_h) - b(\widetilde{u}_h, u_h) = a(\widetilde{u}, \widetilde{u}) - a(\widetilde{u}_h, \widetilde{u}_h)$$

$$= a(\widetilde{u} - \widetilde{u}_h, \widetilde{u}) + a(\widetilde{u}_h, \widetilde{u}) - a(\widetilde{u}_h, \widetilde{u}_h) = a(\widetilde{u} - \widetilde{u}_h, \widetilde{u}) - a(\widetilde{u} - \widetilde{u}_h, \widetilde{u}_h),$$

consequently $\frac{1}{\widetilde{\lambda}} - \frac{1}{\widetilde{\lambda}_h} = a(\widetilde{u} - \widetilde{u}_h, \widetilde{u} - \widetilde{u}_h) \leq C\, \|\widetilde{u} - \widetilde{u}_h\|^2_{m,\Omega}$.

The standard assumptions of the smoothness of $\widetilde{u}$ give

$$|\widetilde{\lambda} - \widetilde{\lambda}_h| \leq C\, h^{2(n_1+1-m)}.$$

Combining this inequality with (14) we prove the estimate (10). In the last step, we take $n_1 = n + m - r$ in order to retain the superconvergence property.

Now we shall improve the estimate (5) by means of the presented postprocessing argument. Let $\widetilde{\mathcal{P}}_h : V \to \widetilde{V}_h$ be the elliptic projection operator defined by

$$\forall u \in V,\ \forall v_h \in \widetilde{V}_h,\ \ a(u - \widetilde{\mathcal{P}}_h u, v_h) = 0.$$

Using this definition one can easily derive the well-known approximate properties of $\widetilde{\mathcal{P}}_h$ (see, e.g. [2] and [7]):

$$\forall u \in H^{n_1+1}(\Omega) \cap V, \quad \|u - \widetilde{\mathcal{P}}_h u\|_{m,\Omega} = \inf_{v_h \in \widetilde{V}_h} \|u - v_h\|_{m,\Omega}.$$

We define for any approximate solution u_h of (4) the functions:

$$\widetilde{w} = \widetilde{\lambda}_h.\widetilde{u} = \widetilde{\lambda}_h.\mathcal{T}u_h \ \text{ and } \ \widetilde{w}_h = \widetilde{\mathcal{P}}_h \widetilde{w} = \widetilde{\lambda}_h.\widetilde{\mathcal{P}}_h \circ \mathcal{T}u_h.$$

Theorem 2. *Let the conditions of Theorem1 be fulfilled. Then we have the following superconvergence estimate of finite element eigenfunctions:*

$$\|u - \widetilde{w}_h\|_{m,\Omega} \leq C.h^{n+1-r} \; \|u\|_{n+1,\Omega}. \tag{15}$$

Proof. First we estimate $\|u - \widetilde{w}\|_{m,\Omega}$. Consider the equality

$$a(u - \widetilde{w}, u - \widetilde{w}) = a(u, u) - 2a(u, \widetilde{w}) + a(\widetilde{w}, \widetilde{w}).$$

We transform all the members in the right-hand side using (3), the definition of $\widetilde{w}$ and the properties of the operator $\mathcal{T}$:

$$a(u, u) = \lambda b(u, u) = \lambda a(\mathcal{T}u, u) = \lambda^2 a(\mathcal{T}u, \mathcal{T}u),$$

$$a(u, \widetilde{w}) = \lambda b(u, \widetilde{w}) = \lambda \widetilde{\lambda}_h a(\mathcal{T}u, \mathcal{T}u_h), \quad a(\widetilde{w}, \widetilde{w}) = \widetilde{\lambda}_h^2 a(\mathcal{T}u_h, \mathcal{T}u_h).$$

Taking into account that the eigenfunctions u and u_h are normalized with respect to the b−norm, we obtain:

$$a(u - \widetilde{w}, u - \widetilde{w}) = \lambda^2 \; a(\mathcal{T}u, \mathcal{T}u) - 2\lambda\widetilde{\lambda}_h \; a(\mathcal{T}u, \mathcal{T}u_h) + \widetilde{\lambda}_h^2 \; a(\mathcal{T}u_h, \mathcal{T}u_h)$$

$$= 2\widetilde{\lambda}_h - 2\widetilde{\lambda}_h \, b(u, u_h) + \lambda - \widetilde{\lambda}_h + \frac{\widetilde{\lambda}_h^2}{\widetilde{\lambda}} - \widetilde{\lambda}_h = \widetilde{\lambda}_h \, b(u - u_h, u - u_h) + (\lambda - \widetilde{\lambda}_h) + \frac{\widetilde{\lambda}_h}{\widetilde{\lambda}}(\widetilde{\lambda}_h - \widetilde{\lambda}).$$

We estimate the last terms using the results ot Theorem1 as well as

$$|b(u - u_h, u - u_h)| \leq C.\|u - u_h\|_{r,\Omega}^2.$$

Finally, in accordance with V−ellipticity of $a(\cdot,\cdot)$ with $n_1 \geq n + m - r$, we now arrive at

$$\|u - \widetilde{w}\|_{m,\Omega}^2 \leq C(\lambda).h^{2(n+1-r)}. \tag{16}$$

Next, the approximation properties of the operator $\widetilde{\mathcal{P}}_h$ and the standard assumptions of the smoothness of $\widetilde{w}$ give the inequality

$$\|\widetilde{w} - \widetilde{w}_h\|_{m,\Omega}^2 \leq C(\lambda).h^{2(n_1+1-m)}.$$

Combining this result and (16) we get thc estimate (15).

3 Superconvergent Postprocessing for the Problems with Eigenvalue on the Boundary Conditions

Let us consider the case when the eigenvalue occurs linearly in the boundary conditions. Thus for the equation (1) we associate the boundary conditions:

$$\mathcal{C}u(x) = \lambda \mathcal{D}u(x) \; , \quad x \in \Gamma, \tag{17}$$

where the operator $\mathcal{C}$ is determined in (2) and

$$\mathcal{D} = \{D_j\}_{j=1}^m, \quad D_j u(x) = \sum_{|\alpha| \leq m-1} t_{j,\alpha}(x) D^\alpha u(x).$$

The coefficients $t_{j,\alpha}(x)$ are nonnegative and belong to $C^r(\Gamma)$. For the problem (1), (17) we make the following two assumptions:

– This problem in weak formulation is symmetric and it has countably many real eigenvalues;
– The corresponding elliptic (source) problem admits a solution operator which is self-adjoint, bounded and compact.

Let us determine the bilinear forms in accordance with the more general elliptic and self-adjoint operators $\mathcal{A}$ and $\mathcal{B}$ of order $2m$ and $2r$ $(m > r \geq 0)$ respectively and with the considered problem (1), (17):

$$(\mathcal{A}u, v) = (u, \mathcal{A}v) = a_1(u,v) + a_2\langle u,v\rangle + \lambda a_3\langle u,v\rangle, \quad \forall u, v \in \widetilde{\mathcal{D}},$$

$$(\mathcal{B}u, v) = (u, \mathcal{B}v) = b_1(u,v) + b_3\langle u,v\rangle, \quad \forall u, v \in \widetilde{\mathcal{D}},$$

where $\widetilde{\mathcal{D}}$ is the set of all functions $u(x) \in C^{2m}(\overline{\Omega})$, satisfying the boundary conditions (17). We denote $b_2\langle u,v\rangle = b_3\langle u,v\rangle - a_3\langle u,v\rangle$. Then

$$a_1(u,v) = \sum_{|\alpha|\leq m} \int_\Omega p_\alpha^1 D^\alpha u . D^\alpha v \, dx, \quad a_2\langle u,v\rangle = \sum_{|\alpha|\leq s} \int_\Gamma p_\alpha^2 D^\alpha u . D^\alpha v \, d\gamma,$$

$$b_1(u,v) = \sum_{|\alpha|\leq r} \int_\Omega q_\alpha^1 D^\alpha u . D^\alpha v \, dx, \quad b_2\langle u,v\rangle = \sum_{|\alpha|\leq s} \int_\Gamma q_\alpha^2 D^\alpha u . D^\alpha v \, d\gamma,$$

where m, r, s are integer with $0 \leq s \leq r < m$.

The coefficients p_α^k and q_α^k, $k = 1, 2$ are supposed such that we can define the Hilbert spaces $H_\mathcal{A}$ and $H_\mathcal{B}$ with $a(u,u) = \|u\|_\mathcal{A}^2$ and $b(u,u) = \|u\|_\mathcal{B}^2$. Hence $H_\mathcal{A}$ and $H_\mathcal{B}$ are completions of $\widetilde{\mathcal{D}}$ with respect to the norms $\|\cdot\|_\mathcal{A}$ and $\|\cdot\|_\mathcal{B}$ respectively. Let V be the closed subspace of $H^m \cap H_\mathcal{A}$. We put

$$a\langle u,v\rangle = a_1(u,v) + a_2\langle u,v\rangle, \quad b\langle u,v\rangle = b_1(u,v) + b_2\langle u,v\rangle.$$

Thus the considered problem is: *find* $\lambda \in R$, $u(x) \in V$, $u \not\equiv 0$ *such that*

$$a\langle u,v\rangle = \lambda b\langle u,v\rangle \quad \forall \ v \in V. \tag{18}$$

As in the previous homogeneous case we define the finite element spaces V_h and $\widehat{V}_h$ which are subspaces of $H^m(\Omega) \cap C^{m-1}(\overline{\Omega})$. The approximating polynomials on every finite element are of degree n and n_1 respectively, where $n \geq 2m - 1$ and $n_1 \geq n + m - r$.

The approximate eigenpair (λ_h, u_h) obtained by the finite element method based on V_h is determined by: *find* $\lambda_h \in R$, $u_h(x) \in V_h$ *such that*

$$a\langle u_h, v_h\rangle = \lambda_h b\langle u_h, v_h\rangle \quad \forall v_h \in V_h. \tag{19}$$

Let (λ_h, u_h) be any solution of (19) with $b\langle u_h, u_h\rangle = 1$. Consider the elliptic problems:

$$a\langle \widehat{u}, v\rangle = b\langle u_h, v\rangle, \quad \forall v \in V,$$
$$a\langle \widehat{u}_h, v_h\rangle = b\langle u_h, v_h\rangle \quad \forall v_h \in \widehat{V}_h.$$

Then we define the numbers: $\widehat{\lambda} = \frac{1}{b\langle \widehat{u}, u_h\rangle}$, $\widehat{\lambda}_h = \frac{1}{b\langle \widehat{u}_h, u_h\rangle}$, where $\widehat{u}$ and $\widehat{u}_h$ are the solutions of the corresponding elliptic problems.

Theorem 3. *Let for the eigenvalue problem of order* $2m$ (λ, u) *and the corresponding* (λ_h, u_h) *be any solutions of (18)and (19) respectively. We assume:*
(i) $u \in H^{n+1}(\Omega)$, $\widehat{u} \in H^{n_1+1}(\Omega)$, $n_1 \geq n+m-r$;
(ii) the eigenfunctions are normalized $b(u,u) = b(u_h, u_h) = 1$;
(iii) the restrictions of u *and* u_h *on the boundary* Γ *belong to the space* $H^s(\Gamma)$, $m > r \geq s \geq 0$. *Then*

$$|\lambda - \widehat{\lambda}| \leq C.h^{\sigma} \|u\|^2_{n+1,\Omega},$$

$$|\widehat{\lambda} - \widehat{\lambda}_h| \leq C.h^{2(n+1-r)} \|u\|^2_{n_1+1,\Omega},$$

where $\sigma = \min\{2(n+1-r); 2(n-s)+1\}$.

The proof of this theorem is similar to the proof of Theorem1. Here we use the regularity of the domain Γ (see [8]) as well as the Trace Theorem [9]. As a consequence from the results of Theorem3 we have:

$$|\lambda - \widehat{\lambda}_h| = \mathcal{O}(h^{\sigma}). \tag{20}$$

The final estimate (20) is of type superconvergence (compare with (7)). The order of convergence increases with $2(m-r)$ or $2(m-s)-1$ according to the considered problem.

The acceleration of the convergence is due to the solving of an additional elliptic problem by the finite element method. The degree of the approximating polynomials increases with $(m-r)$.

Let us define the functions $\widehat{w} = \widehat{\lambda}_h.\widehat{u}$ and $\widehat{w}_h = \widehat{\lambda}_h.\widehat{u}_h$.

Obviously $\widehat{w}_h = \widehat{\mathcal{P}}_h \widehat{w}$, where $\widehat{\mathcal{P}}_h$ is the orthogonal projection operator on the finite element space $\widehat{V}_h$ with respect to the inner product $a\langle\cdot,\cdot\rangle$.

Theorem 4. *Let the conditions of Theorem3 be fulfilled. Then we have the following a posteriori error estimates for eigenfunctions:*

$$\|u - \widehat{w}\|_{m,\Omega} = \mathcal{O}(h^{\frac{\sigma}{2}}) \quad \text{and} \quad \|u - \widehat{w}_h\|_{m,\Omega} = \mathcal{O}(h^{\frac{\sigma}{2}}),$$

where $\sigma = \min\{2(n+1-r); 2(n-s)+1\}$.

The results of the last two theorems show the applicability of the postprocessing algorithm to a large class of self-adjoint eigenvalue problems.

4 Algorithm

We present our postprocessing method for accelerating the convergence of the eigenvalues and the eigenfunctions when the selfadjoint elliptic problem of order $2m$ is considered:

- Find any eigenfunction u_h and eigenvalue λ_h from (4) with $V_h = V_h(n)$, $n \geq 2m-1$ and $b(u_h, u_h) = 1$.
- Construct the finite element space $\widetilde{V}_h = \widetilde{V}_h(n_1)$ where $n_1 = n+m$ using the same partition τ_h.
- Find the solution $\widetilde{u}_h$ of the associate elliptic problem (9).
- Determine the approximate eigenvalue $\widetilde{\lambda}_h = \frac{1}{b(u_h, \widetilde{u}_h)}$.
- Determine the approximate eigenfunction by $\widetilde{w}_h = \widetilde{\lambda}_h.\widetilde{u}_h$.

5 Numerical Results

This example illustrates the case when the eigenvalue parameter appears on a boundary condition. Consider a long thin bar of length l which is supported at the one endpoint and the other endpoint is elastically restricted. The problem of determining the limit axial load P which acts on the elastical endpoint leads to the following problem: find the smallest natural frequencies and the corresponded shape functions of the eigenvalue problem

$$(\alpha y'')'' = -\lambda y'', \quad x \in (0,l),$$

$$y(0) = y'(0) = 0,$$

$$y''(l) = 0, \quad \Big[(\alpha y'')' - cy + \lambda y\Big]_{x=l} = 0,$$

where $\alpha(x)$ is the flexural rigidity, the density of the bar is unit and c is the coefficient of stiffness for the elastical contact.

The bilinear forms are ($m = 2$; $r = 1$):

$$a\langle y, z\rangle = \int_0^l \alpha y'' z''\,dx + cy(l)z(l); \quad b(y,z) = \int_0^l y'z'\,dx.$$

We emphasize the appearance of a boundary condition in the $a-$form.

When $\alpha \equiv 1$, the eigenvalues can be obtained by finding the positive roots k of the equation ($\lambda = k^2$): $c \sin kl = (ckl - k^3)\cos kl$. The corresponded eigenfunctions are:

$$y_j(x) = C_j\left[\sin\sqrt{\lambda_j}(l-x) + x\,\sqrt{\lambda_j}\cos\sqrt{\lambda_j}l - \sin\sqrt{\lambda_j}l\right], \quad j = 1,2,\ldots.$$

We determine C_j having in mind that the eigenfunctions are normalized $b(y_j, y_j) = 1,\ j = 1,2,\ldots$.

When $l = 1$ and $c = 0.7$, the first three eigenvalues are: $\lambda_1 = 3.0326255054$, $\lambda_2 = 22.2714621611$, $\lambda_3 = 61.7079731773$.

The numerical results we present in Table1 and Table2 below are calculated by one-dimensional cubic (Hermite polynomials) finite elements ($n = 3$). The solution of the corresponding elliptic problem has been solved on the same grid by the interpolation polynomials of degree five ($n_1 = 5$).

It is readily seen that a considerable acceleration of convergence due to the postprocessing arises to the coarse grid, i.e. on the grid with 16 or 32 finite elements.

The results in Table2 show that the postprocessing calculation with 16 elements gives more accurate results in $H^2(\Omega)$ norm than the calculation with 128 elements without postprocessing procedure.

Table 1.

elem.		$j=1$	$j=2$	$j=3$
16	$\widetilde{\lambda}_{h,j}$	3.0326255054	22.2714626400	61.7080009909
32	$\widetilde{\lambda}_{h,j}$	3.0326255070	22.2714621692	61.7079736213
64	$\widetilde{\lambda}_{h,j}$	3.0326254788	22.2714621692	61.7079731885
64	$\lambda_{h,j}$	3.0326255073	22.2714621648	61.7079926084
128	$\lambda_{h,j}$	3.0326255055	22.2714622172	61.7079743951

Table 2.

j	$\Vert y_{h,j}-\widetilde{w}_{h,j}\Vert^2_{2,\Omega}$ 16 elem.	$\Vert y_j-\widetilde{w}_{h,j}\Vert^2_{2,\Omega}$ 32 elem.	$\Vert y_j-y_{h,j}\Vert^2_{2,\Omega}$ 128 elem.	$\Vert y_j-y_{h,j}\Vert^2_{2,\Omega}$ 64 elem.
1	$3,96.10^{-14}$	$2,27.10^{-15}$	$1,18.10^{-10}$	$1,88.10^{-9}$
2	$1,05.10^{-9}$	$4,02.10^{-12}$	$5,7.10^{-8}$	$9,12.10^{-7}$
3	$1,73.10^{-7}$	$7,06.10^{-10}$	$1,22.10^{-6}$	$1,9.10^{-5}$

References

1. Adams, R.A.: Sobolev Spaces. New York Academic Press (1975)
2. Ciarlet, P.G.: The Finite Element Method for Elliptic Problems. North-Holland Amsterdam New York Oxford (1978)
3. Pierce, J.G., Varga, R.S.: Higher order convergence results for the Rayleigh-Ritz method applied to eigenvalue problems: Improved error bounds for eigenfunctions. Numer Math. 19 (1972) 155–169.
4. Babuška, I., Osborn, J.: Eigenvalue Problems. Handbook of Numer. Anal., Vol.II, North-Holland, Amsterdam (1991)
5. Chatelin, F.: The spectral approximation of linear operators with applications to the computation of eigenelements of differential and integral operators. SIAM Rev. 23 (1981) 495–522.
6. Rektorys, K.: Variational Methods in Mathematics. Science and Engineering SNTL–Publishers in Technical Literature, Prague (1980)
7. Strang, G., Fix, G.J.: An Analysis of the Finite Element Method. Prentice-Hall, Englewood, Cliffs, NJ (1973)
8. Grisvard, P.: Singularities in Boundary Value Problems. Masson Springer-Verlag (1992)
9. Brener, S.C., Scott, L.R.: The Mathematical Theory of Finite Elelment Methods. Texts in Appl. Math., Springer-Verlag (1994)

The Kachanov Method for a Rigid-Plastic Rolling Problem

Todor Angelov Angelov

Institute of Mechanics, Bulgarian Academy of Sciences,
"Acad.G.Bonchev" str., bl.4, Sofia 1113, Bulgaria
taa@imbm.bas.bg

Abstract. In this work, the method of successive linearization, proposed by L. M. Kachanov for solving nonlinear variational problems, arizing in the deformation theory of plasticity, is applied to a steady state, hot strip rolling problem. The material behaviour is described by a rigid-plastic, incompressible, strain rate dependent material model and for the roll-workpiece interface a constant friction law is used. The problem is stated in the form of a variational inequality with strongly nonlinear and nondifferentiable terms. The equivalent minimization problem is also given. Under certain restrictions on the material characteristics, existence and uniqueness results are obtained and the convergence of the method is proved.

1 Introduction

It has been found that the flow theory of plasticity, with a rigid-plastic (viscoplastic) material model, gives a good approximation of the material behaviour for most of the metal-forming processes [1–5]. For example the linear rigid-viscoplastic or Bingham material model [6], has been successfully applied in [7] and references therein, to steady state wire-drawing and extrusion problems. Depending on the process conditions, at high temperatures and loading for example, the strain rate dependent material models, are more appropriate. For solving these nonlinear problems, the iterative methods developed in [8–12] could be used. A method, in which the nonlinear problem is replaced by a sequence of "physically close" linear problems, has been proposed by Kachanov, for solving the nonlinear problems in the deformation theory of plasticity [13]. Abstract convergence resuls of this method have been presented in [14] and references therein. They have been extended to a variational inequality of second kind in [15]. In this work we apply the method of Kachanov to a steady state rolling of a hot strip, supposed isotropic, rigid-viscoplastic, incompressible metallic body, frictionally interacting with a roll, through a constant friction law. The problem is formulated as a variational inequality with strongly nonlinear and nondifferentiable terms. The equivalent constrained minimization problem is also given. Under certain restrictions on the secant modules, existence and uniqueness results are obtained and the convergence of the Kachanov's iterations is proved.

I. Dimov et al. (Eds.): NMA 2002, LNCS 2542, pp. 372–378, 2003.

2 Statement of the Problem

We suppose that the workpiece occupies the domain $\Omega \subset \mathrm{R}^k (k = 2,3)$, with sufficiently regular boundary Γ, constituting of six open, disjoint subsets. Γ_1 and Γ_5 are the vertical back and front end of the workpiece. The contact boundary is denoted by Γ_3. The boundaries Γ_1, Γ_2, Γ_3 and Γ_4 are assumed free of tractions. Due to the symmetry, only one half of the workpiece is considered, as Γ_6 is the boundary of symmetry. The points of $\bar{\Omega} = \Omega \cup \Gamma$ are identified by their cartesian coordinates $\mathbf{x} = \{x_i\}$, and the standard indicial notation and summation convention is used. Let $\mathbf{u}(\mathbf{x}) = \{u_i(\mathbf{x})\}$, $\sigma(\mathbf{x}) = \{\sigma_{ij}(\mathbf{x})\}$, $\dot{\varepsilon}(\mathbf{x}) = \{\dot{\varepsilon}_{ij}(\mathbf{x})\}$, $(1 \leq i, j \leq k)$, denote the velocity vector, stress and strain rate tensors respectively. Let

$$\bar{\sigma} = \sqrt{\frac{3}{2} s_{ij} s_{ij}}, \qquad \dot{\bar{\varepsilon}} = \sqrt{\frac{2}{3} \dot{e}_{ij} \dot{e}_{ij}}, \tag{1}$$

be the equivalent stress and strain rate, where $s_{ij} = \sigma_{ij} - \sigma_H \delta_{ij}$, $\dot{e}_{ij} = \dot{\varepsilon}_{ij} - \dot{\varepsilon}_v \delta_{ij}/3$, are the components of the deviatoric stress and the strain rate tensors and $\sigma_H = \sigma_{ii}/3$, $\dot{\varepsilon}_v = \dot{\varepsilon}_{ii}$ are the hydrostatic pressure and the volume dilatation strain rate. Consider the following problem:

- Find the velocity $\mathbf{u}$ and stress σ fields, satisfying the following equations and relations:

- equation of equilibrium

$$\sigma_{ij,j} = 0 \quad \text{in} \quad \Omega, \tag{2}$$

- incompressibility condition

$$\dot{\varepsilon}_v = 0 \quad \text{in} \quad \Omega, \tag{3}$$

- strain rate - velocity relations

$$\dot{\varepsilon}_{ij} = \frac{1}{2}(u_{i,j} + u_{j,i}), \tag{4}$$

- yield criterion and flow rule

$$F(\sigma_{ij}, \dot{\bar{\varepsilon}}) \equiv \bar{\sigma}^2 - \sigma_p^2(\dot{\bar{\varepsilon}}) = 0, \qquad \dot{e}_{ij} = \frac{3}{2} \frac{\dot{\bar{\varepsilon}}}{\bar{\sigma}} s_{ij}, \tag{5}$$

- boundary conditions

$$\sigma_{ij} n_j = 0 \quad \text{on} \quad \Gamma_1 \cup \Gamma_2 \cup \Gamma_4 \cup \Gamma_5, \tag{6}$$

$$\sigma_T = \mathbf{0}, \quad u_N = 0 \quad \text{on} \quad \Gamma_6, \tag{7}$$

$$\begin{gathered} u_N = 0 \quad \text{and} \\ \text{if} \quad |\sigma_T(\mathbf{u})| < \tau_f(\mathbf{x}), \quad \text{then} \quad \mathbf{u}_T - \mathbf{u}_{TR} = \mathbf{0}, \\ \text{if} \quad |\sigma_T(\mathbf{u})| = \tau_f(\mathbf{x}), \quad \text{then} \quad \exists \quad \text{const.} \quad \lambda \geq 0, \\ \text{such that} \quad \mathbf{u}_T - \mathbf{u}_{TR} = -\lambda \sigma_T(\mathbf{u}) \quad \text{on} \quad \Gamma_3. \end{gathered} \tag{8}$$

Here δ_{ij} is Kronecker symbol; $\mathbf{n} = \{n_i\}$ is the unit normal vector outward to Γ; u_N, $\mathbf{u}_T$ and σ_N, σ_T are the normal and tangential components of the velocity and the stress vector; $\mathbf{u}_{TR}$ is the roll velocity; $\tau_f(\mathbf{x})$ is the shear strength limit for Γ_3,

$$\tau_f(\mathbf{x}) = min\Big(\mu_f(\mathbf{x})p_N(\mathbf{x}),\ \ m_f(\mathbf{x})\tau_{p0}(\mathbf{x})\Big). \tag{9}$$

where $\mu_f(\mathbf{x})$ is the coefficient of friction, $p_N(\mathbf{x})$ is a priori known normal pressure distribution on Γ_3, $m_f(\mathbf{x})$ is the friction factor; $\sigma_p(\mathbf{x}, \dot{\bar{\varepsilon}})$ is the strain rate dependent, uniaxial yield limit, assumed in the form

$$\sigma_p(\mathbf{x}, \dot{\bar{\varepsilon}}) = \sigma_{p0}(\mathbf{x}) + \sigma_{p1}(\dot{\bar{\varepsilon}}), \tag{10}$$

where $\sigma_{p0}(\mathbf{x}) \geq 0$ is the initial yield limit, $\tau_{p0}(\mathbf{x}) = \sigma_{p0}(\mathbf{x})/\sqrt{3}$ is the initial shear yield limit, $\sigma_{p1}(\dot{\bar{\varepsilon}})$, with $\sigma_{p1}(0) = 0$, is assumed increasing, continuous differentiable function of $\dot{\bar{\varepsilon}}$. Let us denote

$$G(\dot{\bar{\varepsilon}}^2) = \frac{\sigma_{p1}(\dot{\bar{\varepsilon}})}{\dot{\bar{\varepsilon}}} \tag{11}$$

and suppose that

$$c_1 \leq G(\dot{\bar{\varepsilon}}^2) \leq c_2, \quad G(\dot{\bar{\varepsilon}}^2) + 2\dot{\bar{\varepsilon}}^2 G'(\dot{\bar{\varepsilon}}^2) \geq c_3, \quad G'(\dot{\bar{\varepsilon}}^2) \leq 0, \quad \forall\, \dot{\bar{\varepsilon}} \in [0, \infty), \tag{12}$$

where prime denotes a derivative with respect to the argument and c_1, c_2, c_3 are positive constants.

3 Variational Formulation

Let us introduce the space

$$\mathbf{V} = \left\{\mathbf{v}:\ \ \mathbf{v} \in \mathbf{H}^1(\Omega),\ \ v_{i,i} = 0 \ \text{ in } \ \Omega,\ \ v_N = 0 \ \text{ on } \ \Gamma_3 \cup \Gamma_6\right\},$$

of admissible velocities with the usual $\mathbf{H}^1(\Omega)$ inner product and norm, where

$$\mathbf{H}^1(\Omega) = (H^1(\Omega))^k, \quad \mathbf{H}(\Omega) = (L_2(\Omega))^k, \quad \mathbf{V} \subset \mathbf{H} \equiv \mathbf{H}^* \subset \mathbf{V}^*,$$

and the dual spaces are denoted by a star. Let us suppose that

$$\tau_f(\mathbf{x}) \in L_\infty(\Gamma_3), \quad \sigma_{p0}(\mathbf{x}) \in L_\infty(\Omega).$$

Then the following variational problem is associated with the problem (2)-(9):
- Find $\mathbf{u} \in \mathbf{V}$, satisfying:

$$a(\mathbf{u}; \mathbf{u}, \mathbf{v} - \mathbf{u}) + j(\mathbf{v}) - j(\mathbf{u}) \geq 0, \quad \forall \mathbf{v} \in \mathbf{V}. \tag{13}$$

Here

$$a(\mathbf{w}; \mathbf{u}, \mathbf{v}) = \int_\Omega \frac{2}{3} G(\dot{\bar{\varepsilon}}^2(\mathbf{w}))\dot{\varepsilon}_{ij}(\mathbf{u})\dot{\varepsilon}_{ij}(\mathbf{v})\mathrm{d}x, \quad \mathbf{w}, \mathbf{u}, \mathbf{v} \in \mathbf{V} \tag{14}$$

$$j(\mathbf{v}) = \int_{\Omega} \sigma_{p0}(\mathbf{x})\dot{\bar{\varepsilon}}(\mathbf{v})\mathrm{d}\mathbf{x} + \int_{\Gamma_3} \tau_f(\mathbf{x})|\mathbf{v}_T - \mathbf{u}_{TR}|\mathrm{d}\Gamma, \quad \mathbf{v} \in \mathbf{V}. \tag{15}$$

Let us introduce the functional

$$J_0(\mathbf{v}) = \int_{\Omega} \frac{1}{2} \int_0^{\dot{\bar{\varepsilon}}^2(\mathbf{v})} G(s)\mathrm{d}s\mathrm{d}x, \tag{16}$$

and state the problem:
- Find $\mathbf{u} \in \mathbf{V}$, such that

$$J(\mathbf{u}) = \inf_{\mathbf{v}\in\mathbf{V}} J(\mathbf{v}), \quad J(\mathbf{v}) = J_0(\mathbf{v}) + j(\mathbf{v}), \quad \mathbf{v} \in \mathbf{V}. \tag{17}$$

Proposition 1. For any fixed $\mathbf{w} \in \mathbf{V}$, $a(\mathbf{w};\mathbf{u},\mathbf{v}) : \mathbf{V} \times \mathbf{V} \to \mathrm{R}$ is a symmetric, bilinear form and there exist positive constants α_0 and α_1 such that

$$a(\mathbf{w};\mathbf{u},\mathbf{u}) \geq \alpha_0 \|\mathbf{u}\|_1^2, \tag{18}$$

$$|a(\mathbf{w};\mathbf{u},\mathbf{v})| \leq \alpha_1 \|\mathbf{u}\|_1 \|\mathbf{v}\|_1. \tag{19}$$

Proof: This result follows from the Korn's inequality and the first condition in (12), see e.g. [6, 9, 11].

Proposition 2. The functional $J_0(\mathbf{u}) : \mathbf{V} \to \mathrm{R}$ is proper, continuous, and Gâteaux differentiable

$$\langle J_0'(\mathbf{u}), \mathbf{v}\rangle = a(\mathbf{u};\mathbf{u},\mathbf{v}), \ \forall \mathbf{u}, \mathbf{v} \in \mathbf{V}, \tag{20}$$

with strongly monotone

$$\langle J_0'(\mathbf{v}) - J_0'(\mathbf{u}), \mathbf{v} - \mathbf{u}\rangle \geq m\|\mathbf{v} - \mathbf{u}\|_1^2, \quad \forall \mathbf{u}, \mathbf{v} \in \mathbf{V}, \tag{21}$$

and Lipschitz continuous Gâteaux derivative

$$\|J_0'(\mathbf{v}) - J_0'(\mathbf{u})\|_* \leq M\|\mathbf{v} - \mathbf{u}\|_1, \quad \forall \mathbf{u}, \mathbf{v} \in \mathbf{V}, \tag{22}$$

where m and M are positive constants, and satisfies the inequality

$$J_0(\mathbf{v}) - J_0(\mathbf{u}) \leq \frac{1}{2}\Big(a(\mathbf{u};\mathbf{v},\mathbf{v}) - a(\mathbf{u};\mathbf{u},\mathbf{u})\Big), \quad \forall \mathbf{u}, \mathbf{v} \in \mathbf{V}. \tag{23}$$

Proof: The properties (20) - (22) can be checked directly using (16) and the second condition in (12). The inequality (23) follows from the third condition in (12). Let us define the function

$$S(\dot{\bar{\varepsilon}}^2(\mathbf{v})) = \int_0^{\dot{\bar{\varepsilon}}^2(\mathbf{v})} G(s)\mathrm{d}s. \tag{24}$$

Then from (23) we have

$$S(\dot{\bar{\varepsilon}}^2(\mathbf{v})) - S(\dot{\bar{\varepsilon}}^2(\mathbf{u})) \leq \Big(G(\dot{\bar{\varepsilon}}^2(\mathbf{u}))\dot{\bar{\varepsilon}}^2(\mathbf{v}) - G(\dot{\bar{\varepsilon}}^2(\mathbf{u}))\dot{\bar{\varepsilon}}^2(\mathbf{u})\Big), \tag{25}$$

which is true if $S(\dot{\varepsilon}^2(\mathbf{v}))$ is a concave function, i.e. when $S''(\dot{\varepsilon}^2(\mathbf{v})) = G'(\dot{\varepsilon}^2(\mathbf{v})) \leq 0$.

Proposition 3. $j(\mathbf{v}) : \mathbf{V} \to \mathrm{R}$ is proper, nondifferentiable, convex and continuous functional on $\mathbf{V}$.

Proof: This result is easily checked on (15).

Summarizing the above results we have:

Proposition 4. The functional $J(\mathbf{v}) : \mathbf{V} \to \mathrm{R}$ is proper, strictly convex, weakly lower semicontinuous and coercive

$$\lim_{\|\mathbf{v}\|_1 \to +\infty} J(\mathbf{v}) = +\infty, \tag{26}$$

and therefore there exists a unique solution $\mathbf{u} \in \mathbf{V}$ of the problem (17), which is also the unique solution of the problem (13).

Proof: The strict convexity follows from the fact that $J_0(\mathbf{v})$ is Gâteaux differentiable and strongly monotone. The weakly lower semicontinuity follows from the continuity and convexity of $J(\mathbf{v})$ and Gâteaux differentiability of $J_0(\mathbf{v})$. Coercivity can be checked directly. The other follows from the classical results in the convex analysis [12].

4 The achanov Method

Let $\mathbf{u}_0 \in \mathbf{V}$ be an arbitrary element and state the problem:

- Find $\mathbf{u}_{n+1} \in \mathbf{V}, n = 0, 1, ...,$ satisfying

$$a(\mathbf{u}_n; \mathbf{u}_{n+1}, \mathbf{v} - \mathbf{u}_{n+1}) + j(\mathbf{v}) - j(\mathbf{u}_{n+1}) \geq 0, \quad \forall \mathbf{v} \in \mathbf{V}. \tag{27}$$

We also state the problem:

- Find $\mathbf{u}_{n+1} \in \mathbf{V}, n = 0, 1, ...,$ such that

$$J_n(\mathbf{u}_{n+1}) = \inf_{\mathbf{v} \in \mathbf{V}} J_n(\mathbf{v}), \quad J_n(\mathbf{v}) = \left(\frac{1}{2} a(\mathbf{u}_n; \mathbf{v}, \mathbf{v}) + j(\mathbf{v}) \right). \tag{28}$$

Then the following result holds:

Proposition 5. The problem (27) has a unique solution $\mathbf{u}_{n+1} \in \mathbf{V}$, which is also the unique solution of the problem (28).

Proof: This follows from the classical results for the elliptic variational inequalities of second kind, see e.g. [10, 11].

Then analogously to [14, 15], the following theorem can be proved:

Theorem 1. Assume that the properties established by Proposition 1 - Proposition 5 hold. Then the sequence $\{\mathbf{u}_n\}$, defined by (27) or (28) is such that

$$\lim_{n \to \infty} \|\mathbf{u} - \mathbf{u}_n\|_1 = 0, \tag{29}$$

where $\mathbf{u}$ is the unique solution of problem (13) or (17).

Proof: From the inequality (23) with $\mathbf{v} = \mathbf{u}_{n+1}$ and $\mathbf{u} = \mathbf{u}_n$ and using (27) we get

$$J(\mathbf{u}_{n+1}) \leq J_n(\mathbf{u}_{n+1}) \leq J(\mathbf{u}_n), \tag{30}$$

which implies that the sequences $\{J(\mathbf{u}_n)\}$ and $\{J_n(\mathbf{u}_{n+1})\}$ are decreasing and bounded below by $J(\mathbf{u})$ and therefore they are convergent to one and the same limit. Then from (27) with $\mathbf{v} = \mathbf{u}_n$ we have

$$\alpha_0 \|\mathbf{u}_{n+1} - \mathbf{u}_n\|_1^2 \leq a(\mathbf{u}_n; \mathbf{u}_{n+1} - \mathbf{u}_n, \mathbf{u}_{n+1} - \mathbf{u}_n)$$

$$\leq -a(\mathbf{u}_n; \mathbf{u}_n, \mathbf{u}_{n+1} - \mathbf{u}_n) + j(\mathbf{u}_n) - j(\mathbf{u}_{n+1})$$

$$= a(\mathbf{u}_n; \mathbf{u}_n, \mathbf{u}_n) + 2j(\mathbf{u}_n) - a(\mathbf{u}_n; \mathbf{u}_{n+1}, \mathbf{u}_{n+1}) - 2j(\mathbf{u}_{n+1})$$

$$-\Big(a(\mathbf{u}_n; \mathbf{u}_{n+1}, \mathbf{u}_n - \mathbf{u}_{n+1}) + j(\mathbf{u}_n) - j(\mathbf{u}_{n+1})\Big)$$

$$= 2J(\mathbf{u}_n) - 2J_n(\mathbf{u}_{n+1}) - \Big(a(\mathbf{u}_n; \mathbf{u}_{n+1}, \mathbf{u}_n - \mathbf{u}_{n+1}) + j(\mathbf{u}_n) - j(\mathbf{u}_{n+1})\Big)$$

$$\leq 2J(\mathbf{u}_n) - 2J_n(\mathbf{u}_{n+1}), \tag{31}$$

and therefore

$$\lim_{n\to\infty} \|\mathbf{u}_n - \mathbf{u}_{n+1}\|_1 = 0. \tag{32}$$

Further we prove the strong convergence (29). From (13) with $\mathbf{v} = \mathbf{u}_n$ and from (27) with $\mathbf{v} = \mathbf{u}$ and $\mathbf{v} = \mathbf{u}_n$ we consequently have:

$$m\|\mathbf{u} - \mathbf{u}_n\|_1^2 \leq a(\mathbf{u}; \mathbf{u}, \mathbf{u} - \mathbf{u}_n) - a(\mathbf{u}_n; \mathbf{u}_n, \mathbf{u} - \mathbf{u}_n)$$

$$\leq -a(\mathbf{u}_n; \mathbf{u}_n, \mathbf{u} - \mathbf{u}_n) + j(\mathbf{u}_n) - j(\mathbf{u}) + a(\mathbf{u}_n; \mathbf{u}_{n+1}, \mathbf{u} - \mathbf{u}_n) - a(\mathbf{u}_n; \mathbf{u}_{n+1}, \mathbf{u} - \mathbf{u}_n)$$

$$+a(\mathbf{u}_n; \mathbf{u}_{n+1}, \mathbf{u} - \mathbf{u}_{n+1}) - a(\mathbf{u}_n; \mathbf{u}_{n+1}, \mathbf{u} - \mathbf{u}_{n+1}) + j(\mathbf{u}_{n+1}) - j(\mathbf{u}_{n+1})$$

$$= -\Big(a(\mathbf{u}_n; \mathbf{u}_{n+1}, \mathbf{u} - \mathbf{u}_{n+1}) + j(\mathbf{u}) - j(\mathbf{u}_{n+1})\Big)$$

$$+a(\mathbf{u}_n; \mathbf{u}_{n+1}, \mathbf{u}_n - \mathbf{u}_{n+1}) + j(\mathbf{u}_n) - j(\mathbf{u}_{n+1}) + a(\mathbf{u}_n; \mathbf{u}_{n+1} - \mathbf{u}_n, \mathbf{u} - \mathbf{u}_n)$$

$$\leq a(\mathbf{u}_n; \mathbf{u}_{n+1}, \mathbf{u}_n - \mathbf{u}_{n+1}) + j(\mathbf{u}_n) - j(\mathbf{u}_{n+1}) + a(\mathbf{u}_n; \mathbf{u}_{n+1} - \mathbf{u}_n, \mathbf{u} - \mathbf{u}_n)$$

$$\leq C_0 \|\mathbf{u}_n - \mathbf{u}_{n+1}\|_1 \Big(1 + C_1 \|\mathbf{u} - \mathbf{u}_n\|_1\Big), \tag{33}$$

where C_0,C_1 are positive constants, and taking the limit at $n \to \infty$ we obtain (29).

References

1. Washizu, K.: Variational Methods in Elasticity and Plasticity. Pergamon Press (1982)
2. Zienkiewicz, O.C.: Flow formulation for numerical solution of forming processes. In: Pittman, J.F.T., Zienkiewicz, O.C., Wood, R.D., Alexander, J.M. (eds.): Numerical Analysis of Forming Processes. John Wiley & Sons (1984) 1–44.
3. Perzyna, P.: Fundamental problems in viscoplasticity. Adv. Appl. Mech. 9 (1966) 243–377.
4. Cristescu, N., Suliciu, I.: Viscoplasticity. Martinus Nijhoff Publ., Bucharest (1982)
5. Mosolov, P.P., Myasnikov, V.P.: Mechanics of Rigid-Plastic Media. Nauka, Moscow (1981) (in Russian).
6. Duvaut, G., Lions, J.-L.: Inequalities in Mechanics and Physics. Springer-Verlag, Berlin (1976)
7. Awbi, B., Shillor, M., Sofonea, M.: A contact problem for Bingham fluid with friction. Appl. Analysis, 72 (1999) 469–484.
8. Le Tallec, P.: Numerical solution of viscoplastic flow problems by augmented lagrangians. IMA, J. Num. Analysis, 6 (1986) 185–219.
9. Angelov, T.A.: A secant-modulus method for a rigid-plastic rolling problem. Int. J. Nonlinear Mech. 30 (1995) 169–178.
10. Glowinski, R.: Numerical Methods for Nonlinear Variational Problems. Springer–Verlag, Berlin (1984)
11. Kikuchi, N., Oden, J.T.: Contact Problems in Elasticity: A Study of Variational Inequalities and Finite Element Methods. SIAM, Philadephia PA (1988)
12. Fučik, S., Kufner, A.: Nonlinear Differential Equations. Elsevier, Amsterdam (1980)
13. Mikhlin, S.G.: The Numerical Performance of Variational Methods. Walters–Noordhoff, The Netherlands (1971)
14. Nečas, J., Hlavaček, I.: Mathematical Theory of Elastic and Elasto-Plastic Bodies: An Introduction. Elsevier, Amsterdam (1981)
15. Chen, J., Han, W., Huang, H.: On the Kačanov method for a quasi–Newtonian flow problem. Numer. Funct. Anal. and Optimization, 19 (9 & 10) (1998) 961–970.

Implementation of Bilinear Nonconforming Finite Elements in an Eulerian Air Pollution Model: Results Obtained by Using the Rotational Test

Anton Antonov[1], Krassimir Georgiev[2], Emilia Komsalova[2], and Zahari Zlatev[3]

[1] Wolfram Research Inc.,
100 Trade Center Drive, Champaign, IL 61820-7237, USA
antonov@wolfram.com

[2] Central Laboratory for Parallel Processing,
Acad. G. Bonchev, Bl. 25-A, 1113 Sofia, Bulgaria
georgiev@cantor.bas.bg, emilia@cantor.bas.bg

[3] National Environmental Research Institute,
Frederiksborgvej 399, DK-4000 Roskilde, Denmark
zz@dmu.dk

Abstract. The implementation of bilinear nonconforming finite elements in the advection-diffusion part of an Eulerian air pollution model for long-range transport of air pollutants is discussed. The final aim will be to implement such elements in the operational version of a particular air pollution model, the Danish Eulerian Model (DEM). One-dimensional first-order finite element method is currently used during the space discretization of the advection-diffusion part in the operational version of DEM. The application of more accurate methods in the advection part of DEM is desirable. Two different bilinear nonconforming finite elements have been implemented and compared. The rotational test is very popular among researchers in the fields of meteorology and environmental modelling. Numerical results that are obtained in the treatment of the rotational test with the new finite element schemes show that these elements have good qualities and, therefore, it is worthwhile to replace the one-dimensional first-order finite elements with one of the bilinear nonconforming finite elements considered in the paper.

Key words: air pollution modelling, rotational test, nonconforming finite elements

Subject classifications: 65N30, 76R05

1 Introduction

The task for finding reliable and robust control strategies for keeping the air pollution under certain safe levels is of great importance for the modern society. The air pollution is not restricted to the regions where there are big emissions sources, but also in the surrounding areas and even further. In order to estimate and predict the pollution levels and to take efforts for reducing the air

I. Dimov et al. (Eds.): NMA 2002, LNCS 2542, pp. 379–386, 2003.

pollution to some acceptable levels the output results of the air pollution models used have to be as reliable as possible. Large, complex and advanced air pollution models are used to meet this requirement. Such a model is the *Danish Eulerian Model* (DEM) for long transport of air pollutants, which is developed at the National Environmental Research Institute (Roskilde, Denmark) ([16]). All physical and chemical processes included in the phenomenon, which is well known under the name "*Long-Range Transport of Air Pollution*" (LRTAP), are adequately described in it.

LRTAP consists of: emission, transport of the air pollutants due to the wind and transformations during the transport (diffusion, deposition and chemical reactions). The DEM is represented by a system of partial differential equations which is splitted into submodels (the transport of the pollutants due to the wind) is of great importance. An uncertain choice of the spatial discretization may lead to a non-physical solution, i.e. negative concentrations. One-dimensional first-order finite elements are currently used during the space discretization of the advection (and the diffusion) part in the operational version of DEM [2]. Bilinear rectangular finite elements are used in the *Object Oriented version* of DEM (OODEM) ([1]). The application of more accurate elements is desirable. The implementation of two different bilinear nonconforming finite elements is discussed in this paper.

The remainder of the paper is organised as follows. Short description of the DEM can be found in Section 2. Section 3 focuses on the proposed bilinear finite elements and some of their properties. The next section is devoted to the *rotational test* (a popular tool among the researchers in the fields of meteorology and environmental modelling for testing the numerical algorithms) with the new finite element schemes. Some numerical results obtained for the rotational test together with comparisons and discussions are presented in Section 4, too. Finally, Section 5 summarises our conclusions and plans for the future work.

2 Danish Eulerian Model

The Eulerian approach in which the behaviour of the species is described relative to a fixed coordinate system is a common way of treating heat and mass transfer. The Eulerian statistics are readily measurable and the mathematical expressions are directly applicable to the situations in which the chemical reactions take place. Let us denote with q the number of the pollutants included, and let Ω be the model domain. The DEM is described by the following system of partial differential equations (PDE's)

$$\begin{aligned}\frac{\partial c_s}{\partial t} = &-\frac{\partial (uc_s)}{\partial x} - \frac{\partial (vc_s)}{\partial y} - \frac{\partial (wc_s)}{\partial z} + \\ &+\frac{\partial}{\partial x}\left(K_x\frac{\partial c_s}{\partial x}\right) + \frac{\partial}{\partial y}\left(K_y\frac{\partial c_s}{\partial y}\right) + \frac{\partial}{\partial z}\left(K_z\frac{\partial c_s}{\partial z}\right) + \\ &+E_s + Q_s\left(c_1, c_2, \ldots, c_q\right) - \left(k_{1s} + k_{2s}\right) c_s, \qquad s = 1, 2, \ldots, q,\end{aligned} \tag{1}$$

where $c_s = c_s(x, y, z, t)$ are the concentrations of the pollutants; $u = u(x, y, z, t)$, $v = v(x, y, z, t)$ and $w = w(x, y, z, t)$ are the wind velocities; K_x, K_y and K_z are the diffusion coefficients; the emission sources are described by E_s; k_{1s} and k_{2s} are the deposition coefficients, and finally, the chemical reactions are described by $Q_s = Q_s(c_1, c_2, \ldots, c_q)$ (this term couples the system of PDE's and moreover it introduces nonlinearity in the model). In DEM, as usual in the atmospheric models, an operator splitting is used. The main advantage of this approach is that when each process in (1) is treated separately the most efficient numerical integration technique can be chosen. A splitting procedure proposed in [9,8] is used in DEM. It leads to five submodels representing the horizontal advection (2), the horizontal diffusion (3), the chemistry and the emissions (4), the depositions (dry and wet) (5) and the vertical exchange (6).

$$\frac{\partial c_s^{(1)}}{\partial t} = -\frac{\partial (u c_s^{(1)})}{\partial x} - \frac{\partial (v c_s^{(1)})}{\partial y} \tag{2}$$

$$\frac{\partial c_s^{(2)}}{\partial t} = \frac{\partial}{\partial x}\left(K_x \frac{\partial c_s^{(2)}}{\partial x}\right) + \frac{\partial}{\partial y}\left(K_y \frac{\partial c_s^{(2)}}{\partial y}\right) \tag{3}$$

$$\frac{d c_s^{(3)}}{dt} = E_s + Q_s(c_1^{(3)}, c_2^{(3)}, \ldots, c_q^{(3)}) \tag{4}$$

$$\frac{d c_s^{(4)}}{dt} = -(\kappa_{1s} + \kappa_{2s}) c_s^{(4)} \tag{5}$$

$$\frac{d c_s^{(5)}}{dt} = -\frac{\partial (w c_s^{(5)})}{\partial z} + \frac{\partial}{\partial z}\left(K_z \frac{\partial c_s^{(5)}}{\partial z}\right). \tag{6}$$

These five systems of PDE's are solved successively at each time step. The concentrations obtained for each submodel at given time step are used as initial conditions for the next submodel in row. So, solving (6) at the time step t the initial conditions for (2) at the time step $(t + \Delta t)$ are obtained.
The two-dimensional version of DEM is considered in this study, i.e. the vertical echange submodel (6) is skipped and $c_s = c_s(x, y, t)$, $u = u(x, y, t)$ and $v = v(x, y, t)$.

3 Bilinear Rectangular Nonconforming Finite Elements in Advection-Diffusion Submodel

The advection submodel (2) and diffusion submodel (3) are combined when the finite element method is used for a spatial discretization. Let $\tau = \{R\}$ be a regular decomposition of the model domain Ω into quadrilateral elements R with diameters h_R and h is the characteristic diameter, i.e. $h = \max_{R \in \tau} h_R$. One dimensional first order finite elements are currently used during the space discretization of the advection-diffusion part of the operational two-dimensional version of DEM. Bilinear rectangular finite elements are used in the object oriented version of DEM. In order to get more accurate results higher order finite elements are needed. The use of bilinear rectangular non-conforming finite elements seems to be an improvement. The non-conforming methods allow the freedom of being discontinuous, in some sense, without sacrificing the accuracy of the numerical solution. The space function consists of four terms ([12]). They are obtained simply rotating the usual bilinear element to employ $Span\{1, x, y, x^2 - y^2\}$ as the local basis. If the square $\widehat{R} = [-1, 1]^2$ is taken as a reference element then the usual bilinear basis for conforming Galerkin procedure over rectangular elements is based on $Span\{1, x, y, xy\}$ on the reference element. If this basis is rotated through 45 *degrees* a basis built on $\mathcal{R} = Span\{1, x, y, x^2 - y^2\}$ will be obtained. An important feature of this approach is that an unique interpolation at the midpoints of the edges (the nodes of the nonconforming discretization) is valid. The nodal parameters can be chosen by two different ways. The function values at the mid-point of each edge of the rectangle are used in the first version while four mean-values of the shape function along the edges are used in the second version. Both versions are convergent for rectangular meshes which are of interest in DEM.

In fact, the advection and diffusion together are modelled by convection dominated convection-diffusion boundary value problem (or singular perturbation problem). In the two dimensional case (which is of interest in this study)

$$- K\Delta c + \mathbf{v} \cdot \nabla c = 0 \quad \text{in } \Omega \tag{7}$$

$$c = g \quad \text{on } \Gamma = \partial\Omega \tag{8}$$

is considered, where $\Gamma = \partial\Omega$ is the boundary of the model domain, g is given sufficiently smooth function on Ω representing the Dirichlet boundary conditions, $c \in \{c_1, c_2, \dots, c_q\}$ and $\mathbf{v} = (v_1, v_2)$ is the velocity vector.

Let us consider the standard weak formulation for (7)-(8):
find: $c \in V = H_0^1(\Omega)$, such that for all $v \in V$

$$K(\nabla c, \nabla v) + (\mathbf{v} \cdot \nabla c, v) = 0, \tag{9}$$

where $(.,.)$ denotes the inner product in $L^2(\Omega)$. We denote by P_R the bilinear transformation between the reference element $\widehat{R}$ and the element R of our triangulization. Then, our finite element space V_h can be defined by the expression

$$v_h := \left\{ v_h \in L^2(\Omega) : v_{h|R} \in Q_1^{rot}(R), \quad \forall R \right\}, \tag{10}$$

where the space of the rotated bilinear shape functions $Q_1^{rot}(R)$ is defined ([12,7]) by

$$Q_1^{rot}(R) := \left[\widehat{c} \circ P_R^{-1} : \widehat{c} \in Span\{1, \widehat{x}, \widehat{y}, \widehat{x}^2 - \widehat{y}^2\} \right].$$

If the numbering of the nodes (midpoints of the edges) is "*west*→1", "*south*→2", "*north*→3" and "*east*→4", then the basic functions in the first case are:

$$\begin{aligned}\varphi_1 &= \frac{1}{4}[1 - 2x + (x^2 - y^2)],\ \varphi_3 = \frac{1}{4}[1 + 2y - (x^2 - y^2)] \\ \varphi_2 &= \frac{1}{4}[1 - 2y - (x^2 - y^2)],\ \varphi_4 = \frac{1}{4}[1 + 2x + (x^2 - y^2)]\end{aligned} \quad . \tag{11}$$

In the second case (integral midvalue interpolation) the basic functions are :

$$\begin{aligned}\varphi_1 &= \frac{1}{16}[2 - 4x + 3(x^2 - y^2)],\ \varphi_3 = \frac{1}{16}[2 + 4y - 3(x^2 - y^2)] \\ \varphi_2 &= \frac{1}{16}[2 - 4y - 3(x^2 - y^2)],\ \varphi_4 = \frac{1}{16}[2 + 4x + 3(x^2 - y^2)]\end{aligned} \quad . \tag{12}$$

For smooth solutions the convergence in the mesh-dependent *streamline-diffusion norm* is of order $h^{3/2}$ ([11]).

4 Rotational Test

The *rotational test* ([3,10]) is the most popular tool among the researchers in the fields of meteorology and environmental modelling for testing the numerical algorithms used in the large air pollution models. From the mathematical point of view it is pure advection problem which creates difficulties in the numerical treatment. Its governing PDE's system is:

$$\frac{\partial c_s}{\partial t} = -(1 - y)\frac{\partial c_s}{\partial x} - (x - 1)\frac{\partial c_s}{\partial y}, \tag{13}$$

where $s = 1, 2, \ldots, q,\quad 0 \le x \le 2,\quad 0 \le y \le 2$. In the existing operational version of DEM the advection part is additional split to two one-dimensional subproblems ([8]):

$$\frac{\partial c_s^*}{\partial t} = -(1 - y)\frac{\partial c_s^*}{\partial x}, \qquad \frac{\partial c_s^{**}}{\partial t} = -(x - 1)\frac{\partial c_s^{**}}{\partial y}. \tag{14}$$

The first of these submodels is treated on each grid line parallel to the O_x axis, while the second one - on each grid line parallel to the O_y axis. Results and conclusions from the numerical experiments with an advection-chemistry module which is a generalization of the rotational test can be found in [4,5,6].
The initial cone in the rotational test has a center in the point $(0.24, 0.48)$ and radius $1/8$. Its height is 100. One rotation of the cone takes 1600 time steps. All the test experiments presented bellow are inside the OODEM. *Crank-Nicolson method* and *Restart GMRES(30)* are used to solve the discretized problem. In fact, the codes are taken from the *Portable, Extensible Toolkit for Scientific computation* (PETSc), which is a suite of uni- and parallel-processor codes for solving large-scale problems modeled by partial differential equations (see [14,15]). The pre- and post processing procedures, including visualization, are made using *Mathematica* ([13]).

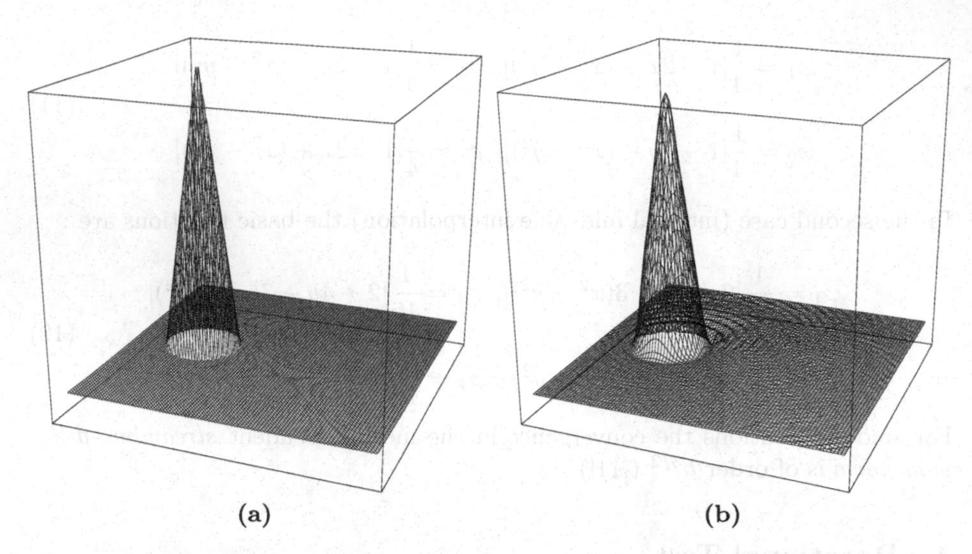

Fig. 1. (a) *Cone in the beginning; hight* = 100; **(b)** *Cone obtained after one rotation with conforming bilinear quadrileteral finite elements; hight = 91.606.*

The results obtained after one full rotation (1600 time steps) are presented on Fig. 1 and Fig. 2. It can be seen that using the nonconforming bilinear quadrilateral finite elements with basic functions defind as in (11) leads to big oscilations almost in the whole model domain (see, Fig. 2 (a)). The situation is complitely different with the second choise of basis functions (see, Fig. 2 (b)). There are still some oscilations but only in the small area just in front of the cone, and almost the same as when the conforming finite elements are used (see, Fig. 1 (b)). It should be mentioned that the implementation of the nonconforming bilinear finite elements (in both cases) are on uniform quadrilateral grid while the grid of the conforming finite elemnts is hybrid, and consists of two lines with triangles neighbour to the boundary of the domain being inside ractangulars. It seems to be essential because the numerical experiments with conforming bilinear finite elements and uniform rectangular grid showd bigger oscilations. It should be expected that using combined grid in both case of the nonconforming elements will improve the quality of the solution. Let us mention that in all numerical solutions discussed here there are some negative values for the concentration. In the cases of the conforming and the second variant of the nonconforming finite elements they are reasonable while in the case of the first variant of the nonconforming finite elements they are considerable. The comparisons between the heights of the cone after one full rotation in the three cases shows that the nonconforming finite elements keep this height better (≈95%) than the conforming ones (<92%).

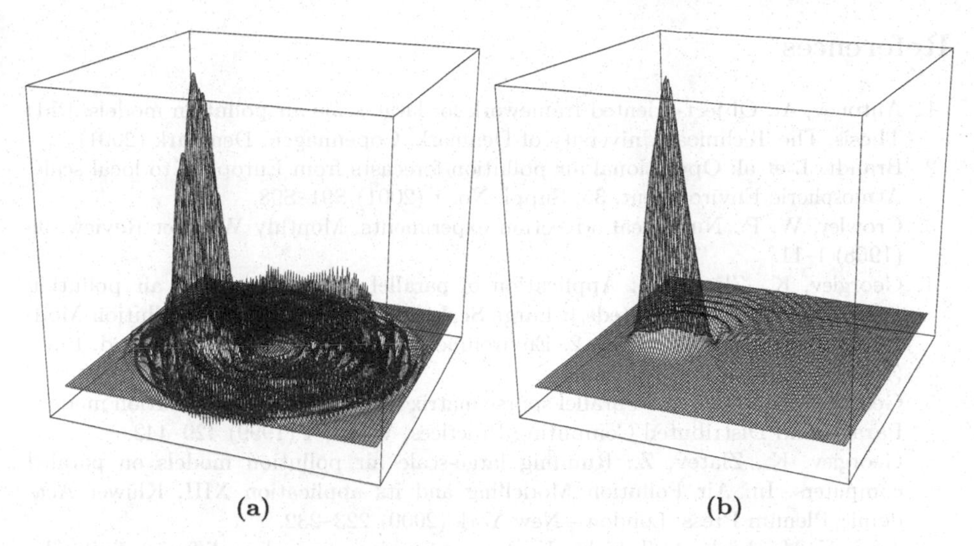

Fig. 2. (a) *Cone obtained after one rotation with nonconforming bilinear quadrileteral finite elements (first choise of basic functions); hight* = 95.453; **(b)** *Cone obtained after one rotation with nonconforming bilinear quadrileteral finite elements (second choise of basic functions); hight* = 94.751.

5 Conclusions

The numerical results that are obtained in the treatment of the rotational test with nonconforming bilinear quadrlateral finite elements (some of them are reported in the paper) show that these elements can replace the one-dimensional first order finite elements using now in the operational two-dimensional version of the Danish Eulerian Model for long range transport of air pollutants. Some additional efforts should be done in order to improve the now existing algorithm (boundary conditions, etc.) as well as some more preliminary tests with adding diffusion and chemical reactions are needed. It is expected that diffusion and chemistry will improve the quality of the numerical solution. The application of the presented in the paper nonconforming bilinear finite elements in the operational version of DEM will improve the accuracy into one of the most important its part - advection-diffusion submodel, and, therefore, the output results from the model should be expected to be more reliable.

Acknowledgements

The research reported in this paper was partly supported by the project of European Commission — BIS 21 under contract ICA1-CT-2000-70016 and by the Bulgarian Ministry of Education and Science under Grants I-811/98 and I-901/99.

References

1. Antonov, A.: Object-oriented framework for large scale air pollution models. PhD Thesis, The Technical University of Denmark, Copenhagen, Denmark (2001)
2. Brandt, J. et al: Operational air pollution forecasts from European to local scale. Atmospheric Environment, 35, Suppl. No. 1 (2001) S91–S98.
3. Crowley, W. P.: Numerical advection experiments. Monthly Weather Review, 96 (1968) 1–11.
4. Georgiev, K., Zlatev, Z.: Application of parallel algorithms in an air pollution model. In: Z.Zlatev et al (eds.): Large Scale Computations in Air pollution Modelling. NATO Science, Series 2. Environmental Security, 57, Kluwer Acad. Publ. (1999) 173–184.
5. Georgiev, K., Zlatev, Z.: Parallel sparse matrix algorithms for air pollution models. Parallel and Distributed Computing Practices, Vol. 2, 4 (1999) 429–442.
6. Georgiev, K., Zlatev, Z.: Running large-scale air pollution models on parallel computers. In: Air Pollution Modelling and its application XIII, Kluwer Academic/Plenum Press, London - New York (2000) 223–232.
7. John, V, Maubach, J., Tobiska, L.: Nonconforming streamline-diffusion finite element methods for convection-diffusion problems. Numer. Math. 78 (1997) 165–188.
8. McRae, G. J., Goodin, W. R., and Seinfeld, J. H.: Numerical solution of the atmospheric diffusion equations for chemically reacting flows. Journal of Computational Physics, 45 (1984) 1–42.
9. Marchuk, G. I.: Mathematical modeling for the problem of the environment. North-Holland, Amsterdam (1985)
10. Molenkampf, C. R.: Accuracy of finite-difference methods applied to the advection equation. Journal of Applied Meteorology, 7 (1968) 160–167.
11. Petrova, S., Tobiska, L., Vassilevski, P.: Multigrid methods based on matrix-dependent coarse spaces for nonconforming streamline-diffusion finite element discretization of convection-diffusion problems. East-West J. Numer. Math. 8 (2000) 223-242.
12. Rannacher, R., Turek, S.: Simple nonconforming quadrilateral Stokes element. Numer. Meth. for PDE's, 8 (1992) 97–111.
13. Wolfram, S.: Mathematica: a system for doing mathematics by computer. Wolfram Media, Cambridge Univ. Press, Fourth Edition (1999)
14. *http://www-unix.Mcs.anl.gov/petsc/petsc-current/docs/manualpages/SLES/SLESSolve.html*
15. *http://www-unix.Mcs.anl.gov/petsc/petsc-current/docs/manualpages/KSP/KSPGMRESSETRestart.html*
16. Zlatev, Z.: Computer treatment of large air pollution models. Kluwer Academic Publishers, Dordrecht-Boston-London (1995)

Contour-Integral Representation of Single and Double Layer Potentials for Axisymmetric Problems

Emilia G. Bazhlekova and Ivan B. Bazhlekov*

Institute of Mathematics, BAS, acad. G. Bonchev str., bl. 8, 1113 Sofia, Bulgaria

Abstract. Based on recently proposed non-singular contour-integral representations of single and double layer potentials for 3D surfaces, formulas in the axisymmetric case are derived. They express explicitly the singular layer potentials in terms of elliptic integrals. The presented expressions are non-singular, satisfy exactly very important conservation principles and directly take into account the multivaluedness of the double layer potential. The results are compared with another method for calculating the single and double layer potentials. The comparison demonstrates higher accuracy and better performance of the presented formulas.

1 Introduction

Significant advance has been made recently in the numerical simulation of complex multiphase Stokes flows using boundary integral methods (BIM), see e.g. [1,2,3]. An essential part of BIM is the calculation of the single and double layer potentials in the case of constant density. The main difficulties are due to the singularity of the kernels:

$$\mathbf{G}(\mathbf{x},\mathbf{x}_0) = \mathbb{I}/|\widehat{\mathbf{x}}| + \widehat{\mathbf{x}}\widehat{\mathbf{x}}/|\widehat{\mathbf{x}}|^3; \quad \mathbf{T}(\mathbf{x},\mathbf{x}_0) = -6\widehat{\mathbf{x}}\widehat{\mathbf{x}}\widehat{\mathbf{x}}/|\widehat{\mathbf{x}}|^5; \; \mathbf{P}(\mathbf{x},\mathbf{x}_0) = 2\widehat{\mathbf{x}}/|\widehat{\mathbf{x}}|^3,$$

where $\widehat{\mathbf{x}} = \mathbf{x} - \mathbf{x}_0$, $(\widehat{\mathbf{x}}\widehat{\mathbf{x}})_{ij} = \widehat{x}_i\widehat{x}_j$ and $\mathbb{I}$ is the unit tensor. Consider also the following auxiliary kernel:

$$\mathbf{Q}(\mathbf{x},\mathbf{x}_0) = \mathbf{T}(\mathbf{x},\mathbf{x}_0) + \mathbb{I}\mathbf{P}(\mathbf{x},\mathbf{x}_0) = 2\mathbb{I}\widehat{\mathbf{x}}/|\widehat{\mathbf{x}}|^3 - 6\widehat{\mathbf{x}}\widehat{\mathbf{x}}\widehat{\mathbf{x}}/|\widehat{\mathbf{x}}|^5. \tag{1}$$

Note that $\mathbf{G}$ has first order singularity at $\mathbf{x} = \mathbf{x}_0$ and $\mathbf{T}$, $\mathbf{P}$ and $\mathbf{Q}$ - second order. The following identities satisfied for arbitrary closed surface S_c with outward unit normal $\mathbf{n}$ (see e.g. [4]) express important conservation properties:

$$\int_{S_c} \mathbf{G}(\mathbf{x},\mathbf{x}_0) \cdot \mathbf{n}(\mathbf{x})\, ds(\mathbf{x}) = \int_{S_c} \mathbf{Q}(\mathbf{x},\mathbf{x}_0) \cdot \mathbf{n}(\mathbf{x})\, ds(\mathbf{x}) = 0; \tag{2}$$

---

* Corresponding author. Present address: Section Materials Technology, Faculty of Mechanical Engineering, Eindhoven University of Technology, P.O. Box 513, 5600 MB Eindhoven, The Netherlands; I.Bazhlekov@tue.nl (This work was supported by the Dutch Polymer Institute)

I. Dimov et al. (Eds.): NMA 2002, LNCS 2542, pp. 387–394, 2003.

$$\int_{S_c} \mathbf{T}(\mathbf{x},\mathbf{x}_0)\cdot\mathbf{n}(\mathbf{x})\,ds(\mathbf{x}) = -c\mathbb{I}; \quad \int_{S_c} \mathbf{P}(\mathbf{x},\mathbf{x}_0)\cdot\mathbf{n}(\mathbf{x})\,ds(\mathbf{x}) = c. \tag{3}$$

The constant c in (3) represents the discontinuity of the potentials at $\mathbf{x} = \mathbf{x}_0$:

$$c = \begin{cases} 8\pi, & \text{if } \mathbf{x}_0 \text{ inside } S_c; \\ 4\pi, & \text{if } \mathbf{x}_0 \text{ on } S_c; \\ 0, & \text{if } \mathbf{x}_0 \text{ outside } S_c. \end{cases}$$

Different approaches to perform the integration of the singular kernels exist in the literature, for example: higher order integration rules combined with a mesh refinement around the singular point (see e.g. [1] and [3]), in this case the identities (2-3) are not satisfied exactly; "near-singularity" subtraction (see e.g. [2]), where the identities (2-3) are satisfied exactly, however, the accuracy could worsen locally around the singular point. This problem is tackled also in our recent study [6], where layer potentials are expressed by contour integrals, (4-6). These representations serve as a starting point for the present work.

Consider a surface, for example drop interface, and let D be a part of it. It is assumed that D and its contour Γ are bounded and piecewise smooth. Let $\mathbf{n}_D$ be the unit normal vector to D, outward to the surface. Let $\mathbf{t}$ be the unit tangential to Γ vector, defined by $\mathbf{t} = \mathbf{b} \times \mathbf{n}_D$, where $\mathbf{b}$ is the unit normal to Γ vector, lying in the tangential plane to D, see figure 1.

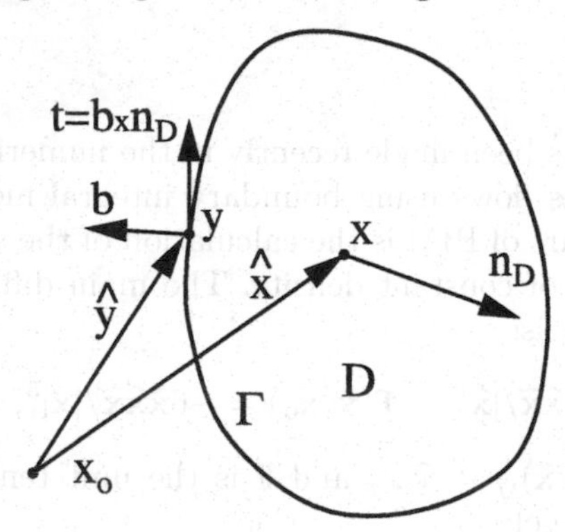

Fig. 1. Schematic sketch of the surface D and its contour Γ in 3D.

The following formulas, proposed in [6], express the single and double layer potentials in the case of density $f \equiv 1$ by means of contour integrals:

$$\int_D \mathbf{G}(\mathbf{x},\mathbf{x}_0)\cdot\mathbf{n}_D(\mathbf{x})\,ds(\mathbf{x}) = \int_\Gamma \frac{\mathbf{t}(\mathbf{y})\times\widehat{\mathbf{y}}}{|\widehat{\mathbf{y}}|}\,dl(\mathbf{y}); \tag{4}$$

$$\int_D \mathbf{Q}(\mathbf{x},\mathbf{x}_0)\cdot\mathbf{n}_D(\mathbf{x})\,ds(\mathbf{x}) = 2\int_\Gamma \frac{\widehat{\mathbf{y}}(\widehat{\mathbf{y}}\times\mathbf{t}(\mathbf{y}))}{|\widehat{\mathbf{y}}|^3}\,dl(\mathbf{y}) \tag{5}$$

$$\int_D \mathbf{P}(\mathbf{x},\mathbf{x}_0)\cdot\mathbf{n}_D(\mathbf{x})\,ds(\mathbf{x}) = 2\int_\Gamma \frac{\mathbf{a}\cdot(\widehat{\mathbf{y}}\times\mathbf{t}(\mathbf{y}))}{|\widehat{\mathbf{y}}|(|\widehat{\mathbf{y}}|+\mathbf{a}\cdot\widehat{\mathbf{y}})}\,dl(\mathbf{y}) + c, \tag{6}$$

where $\widehat{\mathbf{y}} = \mathbf{y} - \mathbf{x}_0$ and $\mathbf{a}$ is an arbitrary unit vector. The constant c in (6) is $0, -4\pi$ or -8π depending of the orientation of $\mathbf{a}$ and Γ, see [6], and guarantees the satisfaction of (3).

Compared to the existing methods for calculation of the single and double layer potentials, the above contour integral representations have some important advantages: they are non-singular, satisfy exactly the identities (2-3) and offer higher accuracy for the numerical integration. In the following section formulas analogous to (4-6) are derived in the case when D is an axisymmertic surface.

2 Contour Integration

In this work we consider an axisymmetric surface D and apply formulas (4-6) to obtain explicit representations of the single and double layer potentials with constant density in terms of complete elliptic integrals of the first, second and third kind.

Introduce a cylindrical coordinate system (z, r, φ). It is known that in the case of axisymmetric surface D none of the integrals (4-6) is a function of the azimuthal angle φ. Thus, a considerable simplification results from analytical integration along rings centered about the axis of symmetry.

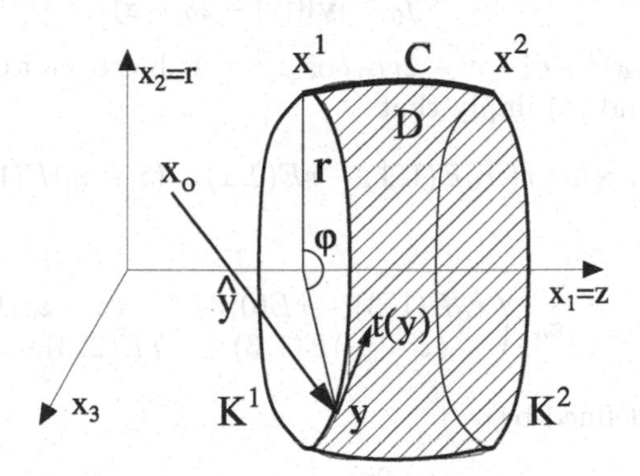

Fig. 2. Schematic sketch of an axisymmetric surface D and its contour $K^1 \cup K^2$.

In the axisymmetric case the contour Γ of D consists of two circles: $\Gamma = K^1 \cup K^2$, see figure 2. Denote the integrals in the r.h.s. of (4), (5) and (6), where Γ is a circle K, by $\mathbf{I}^G(\mathbf{x}_0, K)$, $\mathbf{I}^Q(\mathbf{x}_0, K)$ and $I^P(\mathbf{x}_0, K)$, respectively:

$$\mathbf{I}^G(\mathbf{x}_0, K) = \int_K \frac{\mathbf{t}(\mathbf{y}) \times \widehat{\mathbf{y}}}{|\widehat{\mathbf{y}}|} \, dl(\mathbf{y});$$

$$\mathbf{I}^Q(\mathbf{x}_0, K) = 2 \int_K \frac{\widehat{\mathbf{y}}(\widehat{\mathbf{y}} \times \mathbf{t}(\mathbf{y}))}{|\widehat{\mathbf{y}}|^3} \, dl(\mathbf{y});$$

$$I^P(\mathbf{x}_0, K) = 2 \int_K \frac{\mathbf{a} \cdot (\widehat{\mathbf{y}} \times \mathbf{t}(\mathbf{y}))}{|\widehat{\mathbf{y}}|(|\widehat{\mathbf{y}}| + \mathbf{a} \cdot \widehat{\mathbf{y}})} \, dl(\mathbf{y}) + c.$$

Our goal is to find explicit representations of $\mathbf{I}^G$, $\mathbf{I}^Q$ and I^P as functions of $\mathbf{x}_0$ and $\mathbf{x}$, where $\mathbf{x}$ is the intersection point of K and the half-plane $\varphi = 0$. Let $\mathbf{x} = (z, r, 0)$ and $\mathbf{x}_0 = (z_0, r_0, 0)$ in cylindrical coordinates. Then $\mathbf{y} = (z, r\cos\varphi, r\sin\varphi)$, $\widehat{\mathbf{y}} = (z - z_0, r\cos\varphi - r_0, r\sin\varphi)$, $\mathbf{t}(\mathbf{y}) = (0, -\sin\varphi, \cos\varphi)$, and $dl(\mathbf{y}) = r\,d\varphi$, see figure 2. Therefore, the above integrals depend only on $\mathbf{x}_0$ and $\mathbf{x}$ and read as follows (note that in the axisymmetric case $\mathbf{I}^G$ reduces to a vector with two non-zero components and $\mathbf{I}^Q$ reduces to a 2×2 tensor):

$$\mathbf{I}^G(\mathbf{x}_0, \mathbf{x}) = r\int_0^{2\pi} \frac{(\, r_0\cos\varphi - r \;,\; (z - z_0)\cos\varphi \,)}{|\widehat{\mathbf{y}}|}\, d\varphi, \tag{7}$$

$$\mathbf{I}^Q(\mathbf{x}_0, \mathbf{x}) = -2r\int_0^{2\pi} \frac{\mathbf{A}}{|\widehat{\mathbf{y}}|^3}\, d\varphi, \tag{8}$$

with

$$\mathbf{A} = \begin{pmatrix} (z - z_0)(r_0\cos\varphi - r) & (z - z_0)^2\cos\varphi \\ r_0 r - (r_0^2 + r^2)\cos\varphi + r_0 r\cos^2\varphi & (z - z_0)(r\cos\varphi - r_0)\cos\varphi \end{pmatrix},$$

and

$$I^P(\mathbf{x}_0, \mathbf{x}) = 2r\int_0^{2\pi} \frac{r - r_0\cos\varphi}{|\widehat{\mathbf{y}}|(|\widehat{\mathbf{y}}| + z_0 - z)}\, d\varphi + c, \tag{9}$$

where $|\widehat{\mathbf{y}}| = ((z - z_0)^2 + r_0^2 + r^2 - 2rr_0\cos\varphi)^{1/2}$. It has been taken $\mathbf{a} = (-1, 0, 0)$.

Formulas (7) and (8) imply that

$$\mathbf{I}^G(\mathbf{x}_0, \mathbf{x}) = r(\, r_0 E(1,1) - rE(0,1) \;,\; (z - z_0)E(1,1)\,) \tag{10}$$

and

$$\mathbf{I}^Q(\mathbf{x}_0, \mathbf{x}) = -2r(z - z_0)\begin{pmatrix} r_0 E(1,3) - rE(0,3) & (z - z_0)E(1,3) \\ (z - z_0)E(1,3) & rE(2,3) - r_0 E(1,3) \end{pmatrix}. \tag{11}$$

Here $E(i, m)$ are defined by

$$E(i, m) = \int_0^{2\pi} \frac{\cos^i\phi\, d\phi}{(A^2 - B^2\cos\phi)^{m/2}} \tag{12}$$

with $A = ((z - z_0)^2 + r^2 + r_0^2)^{1/2}$ and $B = (2rr_0)^{1/2}$. Functions $E(i, m)$ can be expressed in terms of elliptic integrals as follows

$$E(0, m) = \frac{4}{C^m} E_{m/2}(k) \tag{13}$$

$$E(1, m) = \frac{4}{C^m B^2}(A^2 E_{m/2}(k) - C^2 E_{(m-2)/2}(k))$$

$$E(2, m) = \frac{4}{C^m B^4}(A^4 E_{m/2}(k) - 2A^2C^2 E_{(m-2)/2}(k) - C^4 E_{(m-4)/2}(k)),$$

$$E(3, m) = \frac{4}{C^m B^6}(A^6 E_{m/2}(k) - 3A^4C^2 E_{(m-2)/2}(k) + 3A^2C^4 E_{(m-4)/2}(k) - C^6 E_{(m-6)/2}(k))$$

where

$$E_{m/2}(k) = \int_0^{\pi/2} \frac{d\phi}{(1 - k^2\cos^2\phi)^{m/2}}, \tag{14}$$

$C = (A^2+B^2)^{1/2} = ((z-z_0)^2+(r+r_0)^2)^{1/2}$ and $k = \sqrt{2B}/C = 2(rr_0)^{1/2}((z-z_0)^2+(r+r_0)^2)^{-1/2}$. Note that $E_{1/2}(k)$ and $E_{-1/2}(k)$ are the complete elliptic integrals of the first and second kind, respectively, and (see [5]):

$$E_{3/2}(k) = \frac{E_{-1/2}(k)}{1-k^2},\ E_{5/2}(k) = \frac{2(2-k^2)}{3(1-k^2)^2}E_{-1/2}(k) - \frac{1}{3(1-k^2)}E_{1/2}(k). \tag{15}$$

In this way, substituting (13) and (15) in (10) and (11), we obtain explicit representations of $\mathbf{I}^G(\mathbf{x}_0,\mathbf{x})$ and $\mathbf{I}^Q(\mathbf{x}_0,\mathbf{x})$ in terms of complete elliptic integrals of the first and second kind.

Concerning $I^P(\mathbf{x}_0,\mathbf{x})$, (9) implies after somewhat longer calculations:

$$I^P(\mathbf{x}_0,\mathbf{x}) = \frac{4(z-z_0)}{((z-z_0)^2+(r+r_0)^2)^{1/2}}\left[E_{1/2}(k) + \frac{r-r_0}{r+r_0}\Pi(a,k)\right], \tag{16}$$

where k and $E_{1/2}(k)$ are defined as above, $a = 2(rr_0)^{1/2}/(r+r_0)$ and $\Pi(a,k)$ is the complete elliptic integral of third kind, defined by:

$$\Pi(a,k) = \int_0^{\pi/2} \frac{d\phi}{(1-a^2\cos^2\phi)(1-k^2\cos^2\phi)^{1/2}}. \tag{17}$$

It remains to recall that $\Gamma = K^1 \cup K^2$ in the axisymmetric case and we obtain from (4-6) the final results:

$$\int_D \mathbf{G}(\mathbf{x},\mathbf{x}_0)\cdot\mathbf{n}_D(\mathbf{x})\,ds(\mathbf{x}) = \mathbf{I}^G(\mathbf{x}_0,\mathbf{x}^2) - \mathbf{I}^G(\mathbf{x}_0,\mathbf{x}^1); \tag{18}$$

$$\int_D \mathbf{Q}(\mathbf{x},\mathbf{x}_0)\cdot\mathbf{n}_D(\mathbf{x})\,ds(\mathbf{x}) = \mathbf{I}^Q(\mathbf{x}_0,\mathbf{x}^2) - \mathbf{I}^Q(\mathbf{x}_0,\mathbf{x}^1); \tag{19}$$

$$\int_D \mathbf{P}(\mathbf{x},\mathbf{x}_0)\cdot\mathbf{n}_D(\mathbf{x})\,ds(\mathbf{x}) = I^P(\mathbf{x}_0,\mathbf{x}^2) - I^P(\mathbf{x}_0,\mathbf{x}^1); \tag{20}$$

where $\mathbf{x}^1$ and $\mathbf{x}^2$ are the boundary points of the trace C of D in $\varphi = 0$, see figure 2, and $\mathbf{I}^G(\mathbf{x}_0,\mathbf{x})$, $\mathbf{I}^Q(\mathbf{x}_0,\mathbf{x})$ and $I^P(\mathbf{x}_0,\mathbf{x})$ are given by (10), (11) and (16), respectively. The representation for the double layer potential follows applying (1) and the above results:

$$\int_D \mathbf{T}(\mathbf{x},\mathbf{x}_0)\cdot\mathbf{n}_D(\mathbf{x})\,ds(\mathbf{x}) = \mathbf{I}^Q(\mathbf{x}_0,\mathbf{x}^2) - \mathbf{I}^Q(\mathbf{x}_0,\mathbf{x}^1) - \mathbb{I}(I^P(\mathbf{x}_0,\mathbf{x}^2) - I^P(\mathbf{x}_0,\mathbf{x}^1)). \tag{21}$$

A numerical verification of the above-derived formulas (18-21) is presented in the following section.

3 Comparison with Other Methods

Different methods exist for calculation of the layer potentials in a vicinity of the singular point, $\mathbf{x}_0 \in D$. In general, they are based on the following representations of the single and double layer potentials for axisymmetric interfaces, see e.g. [4]:

$$\int_D \mathbf{G}(\mathbf{x}, \mathbf{x}_0) \cdot \mathbf{n}_D(\mathbf{x})\, ds(\mathbf{x}) = \int_C M_{ij} n_j(\mathbf{x})\, dl(\mathbf{x}), \tag{22}$$

$$\int_D \mathbf{T}(\mathbf{x}, \mathbf{x}_0) \cdot \mathbf{n}_D(\mathbf{x})\, ds(\mathbf{x}) = \int_C q_{ijk} n_k(\mathbf{x})\, dl(\mathbf{x}), \tag{23}$$

where C is the trace of the surface in the $\varphi = 0$ half-plane. The free index i refers to the r or z component, and the repeated indices j and k are summed over the r and z components. The components of the coefficient matrices $\mathbf{M}$ and $\mathbf{q}$ presented in [4], are given in the Appendix for completeness.

For our comparison we take the following particular case. The surface D is considered here to be a segment of the unit sphere centered at $(0,0,0)$ bounded between the planes $z = \pm \sin \frac{\pi}{72}$. Thus, the trace C of D in $\varphi = 0$ half-plane is the arc of the unit circumference with boundary points $\mathbf{x}^1 = (-\sin \frac{\pi}{72}, \cos \frac{\pi}{72}, 0)$ and $\mathbf{x}^2 = (\sin \frac{\pi}{72}, \cos \frac{\pi}{72}, 0)$. The pole $\mathbf{x}_0$ is chosen to move along the axis r: $\mathbf{x}_0 = (0, r_0, 0)$. This choice of C and $\mathbf{x}_0$ is made to demonstrate the advantages of the proposed formulas (18-21) around the singular point $r_0 = 1$. On figure 3 the present result, (18), for the second component I_2^G of the single layer potential $\mathbf{I}^G$ (the first component is zero) is compared with the numerical ones based on

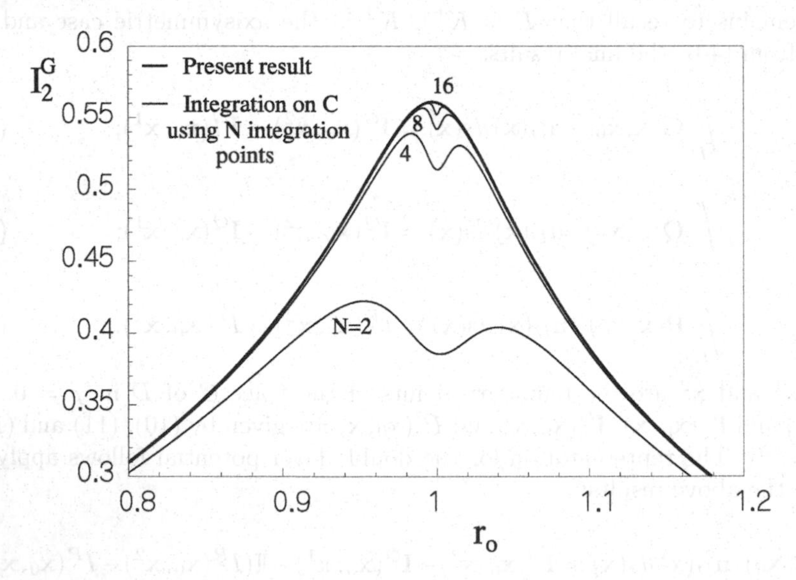

Fig. 3. Comparison for I_2^G between the two representations: (18) obtained using the contour integration (thicker line) and (22) - by integration on the line C (thinner lines).

(22). For the normal vector $\mathbf{n}_D(\mathbf{x})$ in (22-23) the exact value for the sphere D is taken. The integration on C is performed dividing it into $N-1$ even segments and applying the trapezoidal rule in each of them.

The same procedure is used for the first diagonal element I^T_{11} of the double layer potential $\mathbf{I}^T$ (the non-diagonal elements are zero). On figure 4 the r.h.s.

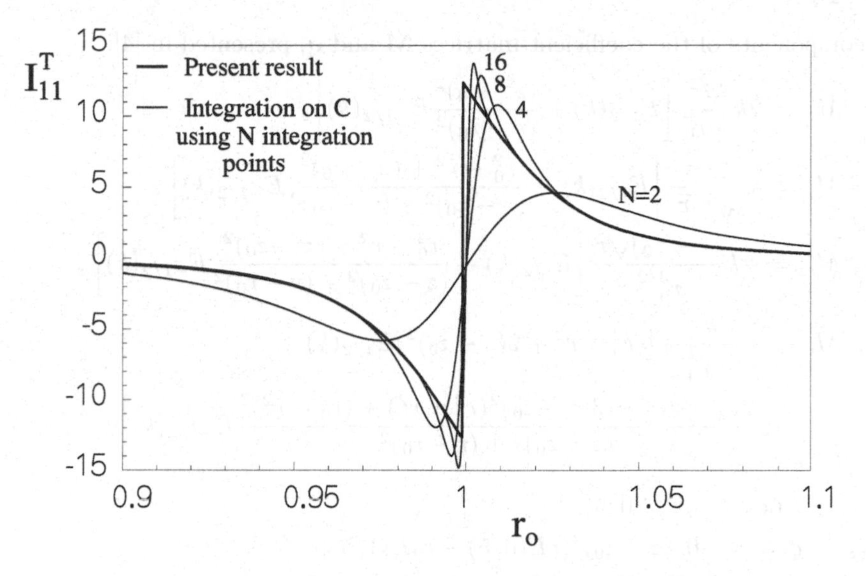

Fig. 4. Comparison for I^T_{11} between the results obtained using the contour integration (21) (thicker line) and the line integration (23) (thinner lines).

of our formula (21) is compared with the r.h.s. of (23). Note that the discontinuity of the diagonal elements of the double layer potential in the singular point $\mathbf{x} = \mathbf{x}_0$ is automatically taken into account in our formula. Regarding the performance of the considered methods: the contour integration is equivalent to the line integration on C for $N = 2$ and the CPU time for the line integrals is proportional to N. The calculation in the present section for both methods: contour and line integration, are performed using the *Mathematica* software.

4 Conclusions

In the present study we offer contour-integral representations of layer potentials with constant density for axisymmetric problems. The proposed formulas (18-21) express explicitly the layer potentials over an axisymmertric surface D as functions of the boundary points, $\mathbf{x}^1$ and $\mathbf{x}^2$, of its trace C in the $\varphi = 0$ half-plane. Thus, they offer an efficient and accurate method for numerical calculation of the layer potentials. They are non-singular when $\mathbf{x}_0 \in D$ in contrast to the surface integrals. Another very important feature of the proposed contour-integral representations is that they satisfy exactly the identities (2-3) for a closed interface.

In addition, the normal vector is automatically taken into account in them. The presented comparisons in the previous section demonstrate the above-mentioned advantages.

5 Appendix

The components of the coefficient matrices **M** and **q**, presented in [4]:

$$M_{zz} = 2k\frac{\sqrt{r}}{\sqrt{r_0}}\left[E_{1/2}(k) + \frac{(z-z_0)^2}{(r-r_0)^2}E_{-1/2}(k)\right],$$
$$M_{zr} = k\frac{z-z_0}{\sqrt{r_0 r}}\left[E_{1/2}(k) - \frac{r_0^2 - r^2 + (z-z_0)^2}{(z-z_0)^2+(r-r_0)^2}E_{-1/2}(k)\right],$$
$$M_{rz} = -k\frac{(z-z_0)\sqrt{r}}{r_0^{3/2}}\left[E_{1/2}(k) - \frac{r_0^2 - r^2 - (z-z_0)^2}{(z-z_0)^2+(r-r_0)^2}E_{-1/2}(k)\right],$$
$$M_{rr} = \frac{k}{r^{1/2}r_0^{3/2}}\left[(r_0^2 + r^2 + 2(z-z_0)^2)E_{1/2}(k)\right.$$
$$\left. - \frac{2(z-z_0)^4 + 3(z-z_0)^2(r_0^2+r^2) + (r_0^2-r^2)^2}{(z-z_0)^2+(r-r_0)^2}E_{-1/2}(k)\right];$$

$$q_{zzz} = -6(z-z_0)^3E(0,5),$$
$$q_{zzr} = q_{zrz} = -6r(z-z_0)^2(rE(0,5) - r_0E(1,5)),$$
$$q_{zrr} = -6r(z-z_0)(r_0^2E(2,5) + r^2E(0,5) - 2rr_0E(1,5)),$$
$$q_{rzz} = -6r(z-z_0)^2(rE(1,5) - r_0E(0,5)),$$
$$q_{rzr} = q_{rrz} = -6r(z-z_0)((r^2+r_0^2)E(1,5) - rr_0(E(0,5)+E(2,5))),$$
$$q_{rrr} = -6r(r^3E(1,5) - r^2r_0(E(0,5)+2E(2,5)) + rr_0^2(E(3,5)+2E(1,5)) - r_0^3E(2,5)).$$

Applying (13) and (15), all components can be written in terms of complete elliptic integrals of the first and second kind.

References

1. Stone H.A., Leal L.G.: Relaxation and breakup of an initially extended drop in an otherwise quiescent fluid. J. Fluid Mech. **198** (1989) 399
2. Davis R.H.: Buoyancy-driven viscous interaction of a rising drop with a smaller trailing drop. Phys. of Fluids **11** no.5 (1999) 1016
3. Kwak S., Fyrillas M.M. and Pozrikidis C.: Effect of surfactants on the instability of a liquid thread. Part II: Extensional flow. Int. J. Multiphase Flow **27** (2001) 39
4. Pozrikidis C.: *Boundary-Integral and Singularity Methods for Linearized Viscous Flow.* Cambridge U.P., Cambridge (1992)
5. Byrd P.F., Friedman M.D.: *Handbook of Elliptic Integrals for Engineers and Scientists.* Springer-Verlag, New York (1971)
6. Bazhlekov I., Bazhlekova E.: Non-singular contour-integral representation of single and double layer potentials. (in preparation)

Uniformly Convergent High-Order Schemes for a 2D Elliptic Reaction-Diffusion Problem with Anisotropic Coefficients

Iliya Brayanov and Ivanka Dimitrova

Department of Mathematics, University of Rousse
7017 Rousse, Bulgaria,
braianov@ami.ru.acad.bg, ita@ami.ru.acad.bg

Abstract. Two dimensional elliptic reaction - diffusion problem with highly anisotropic coefficients is considered. The second order derivative with respect to one of the independent variables is multiplied by a small parameter ε. In this work, we construct and study finite difference schemes, defined on *a priori* Shishkin meshes, uniformly convergent with respect to the small parameter ε, which have order three except for a logarithmic factor. Numerical experiments confirming the theoretical results are given.

1 Introduction

In this paper we will derive HOC difference schemes for the following singularly perturbed elliptic boundary value problem

$$L_\varepsilon u = -\varepsilon^2 \frac{\partial^2 u}{\partial x^2} - \frac{\partial^2 u}{\partial y^2} + a(x,y)u = f(x,y), \quad (x,y) \in \Omega = (0,1) \times (0,1), \tag{1}$$

$$u(x,0) = g_s(x),\ u(x,1) = g_n(x),\ u(0,y) = g_w(y),\ u(1,y) = g_e(y), \tag{2}$$

where $\varepsilon \in (0,1]$ is a small positive parameter. The functions f, g and a are assumed to be sufficiently smooth and $a(x,y) \geq \alpha^2 = constant > 0$.

This problem is the anisotropic model problem [1]. It is also a basic model of singularly perturbation problem [5,12]. For small values of ε, the solution vary rapidly in the elliptic boundary layers $\partial\Omega_1 \equiv \{x = 0, 0 \leq y \leq 1\}$ and $\partial\Omega_2 \equiv \{x = 1, 0 \leq y \leq 1\}$, cf. [5,12].

For small values of ε, it is well known that the classical numerical methods for such problems will produce wild oscillations throughout the whole domain. The solutions are not uniform convergent to u in ε. Special schemes are needed for such problems. Recently a great progress have been made for one-dimensional problems, cf. [2,8,10].

Uniformly convergent schemes of high-order are the most interesting among the ε-uniform schemes. In [2,3] such schemes were constructed on Shishkin meshes for 1D problems, while up to now there are not results for 2D case. Using ideas from [2,11] we shall construct HOC (High Order Compact) schemes for our problem.

I. Dimov et al. (Eds.): NMA 2002, LNCS 2542, pp. 395–402, 2003.

2 Properties of the Continuous Problem

The following theorem gives sufficient conditions for smoothness of the classical solution when $a(x,y) \equiv a = \text{constant}$.

Theorem 1. *Assume that* $f \in C^{4+\alpha}(\overline{\Omega})$ *and* $g_s, g_w, g_n, g_e \in C^{6+\alpha}(0,1)$. *Also, let the compatibility conditions hold:*

$$
\begin{aligned}
&g_w(0) = g_s(0), \; -\varepsilon^2 g_s^{(2)}(0) - g_w^{(2)}(0) + a g_w(0) = f(0,0), \\
&-\varepsilon^4 g_s^{(4)}(0) + g_w^{(4)}(0) + a\left(\varepsilon^2 g_s^{(2)}(0) - g_w^{(2)}(0)\right) = \varepsilon^2 \frac{\partial^2 f(0,0)}{\partial x^2} - \frac{\partial^2 f(0,0)}{\partial y^2}, \\
&-\varepsilon^6 g_s^{(6)}(0) - g_w^{(6)}(0) + a\left(\varepsilon^4 g_s^{(4)}(0) + 2 g_w^{(4)}(0) - a g_w^{(2)}(0)\right) = \\
&\varepsilon^4 \frac{\partial^4 f(0,0)}{\partial x^4} - \varepsilon^2 \frac{\partial^4 f(0,0)}{\partial x^2 \partial y^2} - a \frac{\partial^2 f(0,0)}{\partial y^2} + \frac{\partial^4 f(0,0)}{\partial y^4},
\end{aligned}
\tag{3}
$$

at the vertex $(0,0)$ *and similar ones at the other three vertices of* Ω. *Then the problem (1),(2) has unique classical solution* $u \in C^{6+\alpha}(\overline{\Omega})$ *that satisfies the estimates*

$$
\left| \frac{\partial^{k+l} u(x,y)}{\partial x^k, \partial y^l} \right| \leq C\varepsilon^{-k}, \quad 0 \leq k, l, \; k+l \leq 6, \tag{4}
$$

for some positive constant C *independent of the small parameter* ε.

Sketch of the proof: The compatibility conditions (3) and the smoothness of the solution are obtained similarly as in [5]. The bounds (4) could be proved by means of the maximum principle and appropriate barrier functions, cf. [7,8].

The estimates (4) are not precise enough for the numerical analysis. Below, we construct a Shishkin type decomposition for the solution u of the problem (1), (2) and give estimates of the derivatives up to sixth order. For simplicity we again consider the case $a(x,y)$ =constant.

We start with the smooth part $W(x,y)$ that is a solution of the reduced problem, obtained after setting $\varepsilon = 0$ in (1). We present this part as a sum $W = \sum_{i=0}^{i=5} \varepsilon^i W_i$. Replacing this sum in (1) and equating the coefficients in powers of ε we obtain the following problems for W_i

$$
\begin{aligned}
&-\frac{\partial^2 W_i(x,y)}{\partial y^2} + a W_i(x,y) = F_i(x,y), \quad (x,y) \in \Omega, \\
&F_0(x,y) = f(x,y), \; F_{2i-1}(x,y) = 0, \; F_{2i}(x,y) = \frac{\partial^2 W_{2i-2}(x,y)}{\partial x^2}, \; i \geq 1, \\
&W_0(x,0) = g_s(x), \; W_0(x,1) = g_n(x), \; W_i(x,0) = 0, \; W_i(x,1) = 0, \; i \geq 1.
\end{aligned}
\tag{5}
$$

Therefore $W_i = 0$ for i-odd.

The difference $u - W$ is not small for $x = 0$ and $x = 1$. The boundary layer functions V^k, $k = 1, 2$ defined below are designed to remove these discrepancies.

We introduce new stretch variables: $\eta^1 = x/\varepsilon$ and $\eta^2 = (1-x)/\varepsilon$. Let's make a formal expansion of the equation $LV^k(\eta^k, y) = 0$ in powers of ε, and equate to zero the coefficients of ε^i, to obtain a sequence of functions $V_0^k, V_1^k, \ldots,\ V^k = \sum_{i=0}^{5} \varepsilon^i V_i^k(\eta^k, y)$, that are solutions of the following elliptic problems

$$\begin{aligned}
&-\frac{\partial^2 V_i^k(\eta^k,y)}{\partial(\eta^k)^2} - \frac{\partial^2 V_i^k(\eta^k,y)}{\partial y^2} + aV_i^k(\eta^k, y) = 0,\ \eta^k \in (0,\infty),\ y \in (0,1),\\
&V_i^k(\infty, y) = 0,\ V_i^k(0,y) = \Psi_i^k(y),\ V_i^k(\eta^k, 0) = 0,\ V_i^k(\eta^k, 1) = 0.\\
&\Psi_0^1(y) = g_w(y) - W_0(0,y),\ \Psi_0^2(y) = g_e(y) - W_0(1,y),\\
&\Psi_i^1(y) = -W_i(0,y),\ \Psi_i^1(y) = -W_i(1,y),\ i \geq 1.
\end{aligned} \tag{6}$$

Therefore $V_i^k = 0$ for i-odd. In order the solutions of (6) to be sufficiently smooth additional computability conditions should be posed. To ensure sixth order smoothness of V_0^1 we should have at the corner $(0,0)$ the conditions

$$\begin{aligned}
&f(0,0) + g_w^{(2)}(0) - ag_w(0) = 0, \quad g_w^{(4)}(0) - ag_w^{(2)}(0) + \frac{\partial^2 f(0,0)}{\partial^2 y} = 0,\\
&-g_w^{(6)}(0) + 2ag_w^{(4)}(0) - a^2 g_w^{(2)}(0) = \frac{\partial^4 f(0,0)}{\partial y^4} - a\frac{\partial^2 f(0,0)}{\partial y^2}.
\end{aligned} \tag{7}$$

As we can see they are obtained from (4) after setting $\varepsilon = 0$. Similarly, the conditions giving the smoothness of V_2^1 are obtained from (4) after equating to zero the coefficients of ε^2

$$g_s^{(2)}(0) = 0, \qquad \frac{\partial^2 f(0,0)}{\partial x^2} = 0, \qquad \frac{\partial^4 f(0,0)}{\partial x^2 \partial y^2} = 0, \tag{8}$$

and of V_4^1 after equating to zero the coefficients of ε^4

$$g_s^{(4)}(0) = 0, \qquad \frac{\partial^4 f(0,0)}{\partial x^4} = 0. \tag{9}$$

If the conditions (7)-(9) hold, then by means of the maximum principle and special barrier functions we can prove, that the functions V_i^k, $k = 1,2$, $i = 0,\ldots,5$ satisfy the estimates

$$\left|\frac{\partial^{k+l} V_i^k(\eta^k, y)}{\partial x^k \partial y^l}\right| \leq C\exp(-\alpha\eta), \tag{10}$$

for some positive constant C.

Setting now $R = \varepsilon^6 R_6$, where R is the remainder $R = u - W - V^1 - V^2$ we obtain the following problem for R_6

$$\begin{aligned}
L_\varepsilon R_6(x,y) &= \frac{\partial^2 W_4(x,y)}{\partial x^2},\ (x,y) \in \Omega,\\
R_6(x,y) &= 0, \quad (x,y) \in \partial\Omega.
\end{aligned}$$

The additional condition ensuring smoothness of the derivatives of R_6 up to sixth order are obtained from (4) after equating to zero the coefficients of ε^6

$$g_s^{(6)}(0) = 0. \tag{11}$$

Now we set $u^r = W + R, u^s = V^1 + V^2$. Using (10) we obtain the following theorem

Theorem 2. *Let the coefficients of (1), (2) satisfy the conditions of Theorem 1 and the conditions (7)-(9) and (11). Then the solution u can be decomposed into regular part u^r, that satisfies for all integer k, l, $0 \le k, l, k+l \le 6$ the estimates*

$$\left| \frac{\partial^{k+l} u^r(x,y)}{\partial x^k \partial y^l} \right| \le C, \ (x,y) \in \overline{\Omega}, \tag{12}$$

and singular part u^s that satisfies

$$\left| \frac{\partial^{k+l} u^s(x,y)}{\partial x^k \partial y^l} \right| \le C\varepsilon^{-k} \left(\exp(-\alpha x/\varepsilon) + \exp(-\alpha(1-x)/\varepsilon) \right), \ (x,y) \in \overline{\Omega}, \tag{13}$$

for some positive constant C independent of the small parameter ε.

Scetch of the proof: From Theorem 1 follows that the remainder R has ε-bounded derivatives up to sixth order. The derivatives of the smooth part W are also ε-bounded, since it is a solution of independent of ε problem. The bounds (13) of the singular part follow from the estimates (10).

3 Derivation of High-Order Schemes

To derive high-order compact scheme for problem (1), (2) we shall use ideas from [4,11].

We introduce an uniform mesh in y direction, $w_y = \{y_j : y_j = jk, j = 0, \ldots, J, Jk = 1\}$ and special condensed Shiskin mesh in x direction, that is piecewise uniform in each subinterval $[0, \sigma]$, $[\sigma, 1-\sigma]$, $[1-\sigma, 1]$

$$w_x = \{x_i : x_i = x(i-1) + h_i, \, i = 0, \ldots, N, \, x_0 = 0, x_N = 1\},$$
$$h_i = \begin{cases} 2(1-2\sigma)/N, \, i = N/4+1, \ldots, 3N/4+1, \\ 4\sigma/N, \qquad \text{otherwise} \end{cases}$$

where $\sigma = \min(0.25, \sigma_0\sqrt{\varepsilon}\ln N)$.

The central difference scheme is given by

$$L^1_{\varepsilon,h} U_{ij} \equiv -\varepsilon^2 U_{\bar{x}\hat{x},ij} - U_{\bar{y}y,ij} + a_{ij}U_{ij} = f_{ij}, \ 1 \le i \le N-1, \, 1 \le j \le M-1,$$

$$U_{0j} = g_w(y_j), \, U_{Nj} = g_e(y_j), \, U_{i0} = g_s(x_i), \, U_{iJ} = g_n(x_i), \tag{14}$$

where

$$U_{x,i-1j} = U_{\bar{x},ij} = \frac{U_{ij} - U_{i-1j}}{h_i},\; U_{\hat{x},ij} = \frac{2(U_{i+1j} - U_{ij})}{(h_{i+1} + h_i)},\; U_{y,ij-1} = U_{\bar{y},ij} =$$
$$= \frac{U_{ij} - U_{ij-1}}{k}, U_{\bar{x}\hat{x},ij} = \frac{2(U_{x,ij} - U_{\bar{x},ij})}{h_i + h_{i+1}},\; U_{\bar{y}y,ij} = \frac{(U_{y,ij} - U_{\bar{y},ij})}{k}.$$

Its local truncation error satisfies

$$\tau^1_{ij,u} \equiv L^1_{\varepsilon,h}\left[u(x_i,y_j) - U_{ij}\right] = -\frac{\varepsilon^2(h_{i+1} - h_i)}{3}\frac{\partial^3 u_{ij}}{\partial x^3} - \frac{2\varepsilon^2(h^3_{i+1} + h^3_i)}{4!(h_i + h_{i+1})}\frac{\partial^4 u_{ij}}{\partial x^4}$$
$$- \frac{k^2}{12}\frac{\partial^4 u_{ij}}{\partial y^4} + \mathcal{O}(J^{-4}) + \mathcal{O}(N^{-3}). \tag{15}$$

Therefore to construct a compact scheme of order $\mathcal{O}(J^{-4} + N^{-3})$ (possibly non uniform), we must find compact approximations of the first three terms in (15), that have the same order. For the first term in (15) we only need to analyze the transition points $x = \sigma$ and $x = 1 - \sigma$. Differentiating (1) with respect to x and y, we get

$$-\varepsilon^2\frac{\partial^3 u_{ij}}{\partial x^3} = \frac{\partial f_{ij}}{\partial x} - \frac{\partial a_{ij}u_{ij}}{\partial x} + \frac{\partial^3 u_{ij}}{\partial x\partial y^2} \approx \delta_{i,N/4}\left(f_{\bar{x},ij} + \frac{h_i}{2}f_{\bar{x}\hat{x},ij} - (au)_{\bar{x},ij}\right.$$
$$\left.-\frac{h_i}{2}(au)_{\bar{x}\hat{x},ij} + u_{\bar{x}\bar{y}y,ij} + \frac{h_i}{2}u_{\bar{x}\hat{x}\bar{y}y,ij}\right) + \delta_{i,3N/4}\left(f_{x,ij} - \frac{h_{i+1}}{2}f_{\bar{x}\hat{x},ij} - (au)_{x,ij}\right.$$
$$\left.+\frac{h_{i+1}}{2}(au)_{\bar{x}\hat{x},ij} + u_{x\bar{y}y,ij} - \frac{h_{i+1}}{2}u_{\bar{x}\hat{x}\bar{y}y,ij}\right),\quad \delta_{i,k} = \begin{cases} 0 \text{ if } i \neq k, \\ 1 \text{ if } i = k, \end{cases} \tag{16}$$

$$-\varepsilon^2\frac{\partial^4 u_{ij}}{\partial x^4} = \frac{\partial^2 f_{ij}}{\partial x^2} - \frac{\partial^2 a_{ij}u_{ij}}{\partial x^2} + \frac{\partial^4 u_{ij}}{\partial x^2\partial y^2} \approx f_{\bar{x}\hat{x},ij} - (au)_{\bar{x}\hat{x},ij} + u_{\bar{x}\hat{x}\bar{y}y,ij}, \tag{17}$$

$$-\frac{\partial^4 u_{ij}}{\partial y^4} = \frac{\partial^2 f_{ij}}{\partial y^2} - \frac{\partial^2 a_{ij}u_{ij}}{\partial y^2} + \varepsilon^2\frac{\partial^4 u_{ij}}{\partial x^2\partial y^2} \approx f_{\bar{y}y,ij} - (au)_{\bar{y}y,ij} + \varepsilon^2 u_{\bar{x}\hat{x}\bar{y}y,ij}. \tag{18}$$

Replacing (16)-(18) in (15), we obtain the scheme

$$L^2_{\varepsilon,h} \equiv -\varepsilon^2 U_{\bar{x}\hat{x},ij} - U_{\bar{y}y,ij} + \left(\beta_{2,i} - \beta_{3,i} - \varepsilon^2\beta_4\right)U_{\bar{x}\hat{x}\bar{y}y,ij} + \delta_{i,N/4}\beta_{1,i}V_{\bar{x}\bar{y}y,ij} +$$
$$\delta_{i,3N/4}\beta_{1,i}V_{x\bar{y}y,ij} + Q^h((aU)_{ij}) = Q^h(f_{ij}),$$

$$U_{0j} = g_w(y_j),\; U_{Nj} = g_e(y_j),\; U_{i0} = g_s(x_i),\; U_{iJ} = g_n(x_i), \tag{19}$$

where

$$Q^h(V_{ij}) = V_{ij} + \beta_{1,i}\left(\delta_{i,N/4}V_{\bar{x},ij} + \delta_{i,3N/4}V_{x,ij}\right) + \left(\beta_{3,i} - \beta_{2,i}\right)V_{\bar{x}\hat{x},ij} + \beta_4 V_{\bar{y}y,ij},$$
$$\beta_{1,i} = \frac{(h_{i+1} - h_i)}{3},\; \beta_{2,i} = \beta_{1,i}\left(\delta_{i,3N/4}\frac{h_{i+1}}{2} - \delta_{i,N/4}\frac{h_i}{2}\right),$$
$$\beta_{3,i} = \frac{2(h^3_{i+1} + h^3_i)}{4!(h_i + h_{i+1})},\; \beta_4 = \frac{k^2}{12}.$$

The system (19) could be written in matrix form, with matrix A that is badly conditioned for small values of ε, but the simple diagonal preconditioning $D = diag\{A(i,i)^{-1}\}$, cf. [9], improves the situation and makes the condition number independent of ε.

The truncation error of (19) satisfies the estimates

$$|\tau_{ij,u}^2| \leq \begin{cases} C\left(J^{-4} + J^{-2}N^{-2}\sigma_0^2 \ln^2 N + N^{-4}\sigma_0^4 \ln^4 N\right), \ 1 \leq i < N/4, \\ C\left(J^{-4} + J^{-2}N^{-2}\sigma_0^2 \ln^2 N + N^{-4}\sigma_0^4 \ln^4 N\right), \ 3N/4 < i \leq N, \\ C\left(J^{-4} + N^{-4} + J^{-2}N^{-2} + N^{-\alpha\sigma_0}\right), \ N/4 < i < 3N/4, \\ C\left(J^{-4} + J^{-2}(N^{-2} + \varepsilon^2 N^{-1}) + N^{-3} + d_{N/4}N^{-\alpha\sigma_0}\right), \ i = N/4, \\ C\left(J^{-4} + J^{-2}(N^{-2} + \varepsilon^2 N^{-1}) + N^{-3} + d_{3N/4}N^{-\alpha\sigma_0}\right), \ i = 3N/4, \end{cases} \tag{20}$$

where $0 \leq j \leq J$ and $d_i = \max\{1, \beta_{3,i}/\varepsilon^2, \beta_{1,i}/\varepsilon\}$.

From estimates (20) follows that if we set $\sigma_0 = 4/\alpha$ we obtain a scheme that is almost of fourth order at the uniform points.

4 Uniform Convergence

As a result of the construction and the asymptotic analysis of the previous sections the following theorem holds.

Theorem 3. *Let the coefficients, the right hand side and the boundary data of the problem (1), (2) satisfy the conditions from Theorem 2. Then for the solution $u(x,y)$ of the continuous problem (1), (2) and the solution U of the discrete problem (15)-(17) are valid the estimates*

$$|u_{ij} - U_{ij}| \leq C\left(J^{-4} + J^{-2}(N^{-2} + \varepsilon^2 N^{-1}) + N^{-4}\sigma_0^4 \ln^4 N + N^{-3} + N^{-\alpha\sigma_0}\right), \tag{21}$$

for all i, j, $0 \leq i \leq N$, $0 \leq j \leq J$ and some positive constant C independent of the small parameter ε.

Sketch of the proof: Using the discrete Green function, the embedding theorem for anisotropic meshes [6] and the estimates (20), we can prove the estimate (21).

5 Numerical Results

In this section we will present an example for our method applied to the problem (1), (2), with zero boundary conditions and coefficient $a(x,y) = 2$. The right hand side $f(x,y)$ is chosen so that the solution of (1), (2) is

$$u(x,y) = \left(1 - \frac{\exp(-x/\varepsilon) + \exp(-(1-x)/\varepsilon)}{1 + \exp(-1/\varepsilon)} + x(1-x)\right) y(1-y).$$

Table 1. Error on Shishkin meshes

N	$\varepsilon = 1$	$\varepsilon = 10^{-1}$	$\varepsilon = 10^{-2}$	$\varepsilon = 10^{-3}$	$\varepsilon = 10^{-4}$	$\varepsilon = 10^{-5}$	$\varepsilon = 10^{-6}$
8	$8.26e-7$	$1.20e-4$	$7.18e-4$	$7.33e-4$	$7.34e-4$	$7.35e-4$	$7.35e-4$
$\rho(N)$	16.01	11.46	2.66	2.71	2.72	2.72	2.72
16	$5.16e-8$	$1.04e-5$	$2.70e-4$	$2.70e-4$	$2.70e-4$	$2.70e-4$	$2.70e-4$
$\rho(N)$	16.00	16.21	3.87	3.87	3.87	3.87	3.87
32	$3.22e-9$	$6.42e-7$	$6.98e-5$	$6.98e-5$	$6.98e-5$	$6.98e-5$	6.98e-5
$\rho(N)$	16.00	15.75	5.80	5.80	5.80	5.80	5.80
64	$2.02e-10$	$4.08e-8$	$1.20e-5$	$1.20e-5$	$1.20e-5$	$1.20e-5$	$1.20e-5$
$\rho(N)$	15.99	15.97	8.37	8.37	8.37	8.37	8.37
128	$1.26e-11$	$2.55e-9$	$1.44e-6$	$1.44e-6$	$1.44e-6$	$1.44e-6$	$1.44e-6$
$\rho(N)$	15.24	15.96	9.33	9.33	9.33	9.33	9.33
256	$8.27e-13$	$1.60e-10$	$1.54e-7$	$1.54e-7$	$1.54e-7$	$1.54e-7$	$1.54e-7$

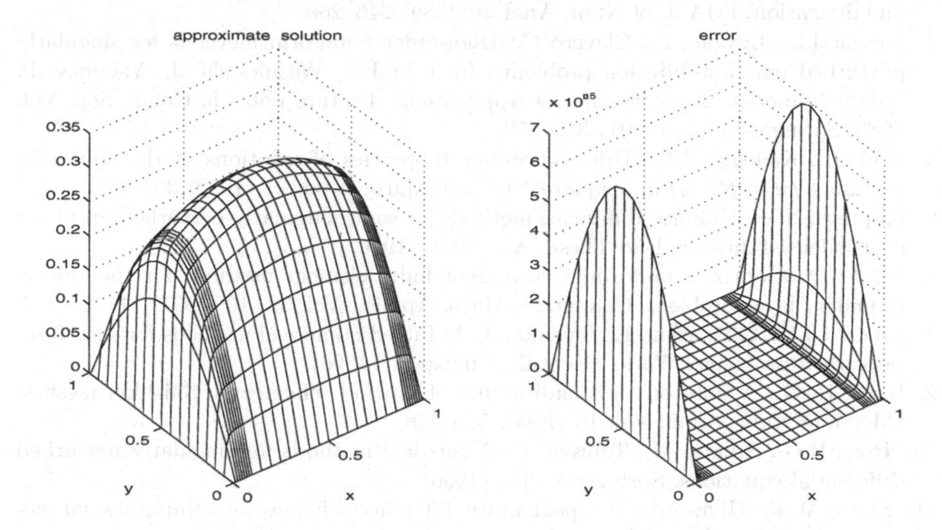

Fig. 1. Approximate solution and error

This is the solution taken from J. Li in [7] to illustrate the finite element method constructed for this problem. This u has the typical boundary layer behavior Here the exact solution is known, so we can measure accurately the solution errors. In Table 1 is given the error in maximum norm on the scheme (17) for different values of N and ε. In the numerical experiments the parameter $\sigma_0 = 4$ and $J = N/2$. The convergence ratio $\rho(N)$ is

$$\rho(N) = \frac{\|u - U\|_\infty^N}{\|u - U\|_\infty^{2N}}.$$

The numerical results agree with the order given by Theorem 3. The error is uniform in ε and for larger number of mesh points the order of convergence becomes higher than 3. Figure 1 shows the approximate solution and the error on the scheme (17), for $N = 32, J = 16$. It illustrates very well the boundary layers and confirms the fact that the largest error appears in the transition points.

References

1. Bastian, P., Wittum, G.: On robust and adaptive multi-grid methods. In: Hemker, P.W. et. al (eds.): Multigrid Methods. Proceed. of the 4-th European Multigrid Conference, Basel (1994) 1–19.
2. Clavero, C., Gracia, J.L., Lisbona, F.: High order methods on Shishkin meshes for singular perturbation problems of convection-diffusion type. Numerical Algorithms 22, 2 (1999), 73–97.
3. Gartland, E.C. Jr.: Compact high-order finite differences for interface problems in one dimension. IMA J. of Num. Anal. 9 (1989) 243–260.
4. Gracia J.L., Lisbona F., Clavero C.: High-order ε-uniform methods for singularly perturbed reaction-diffusion problems. In: L.Vulkov, Waśniewski, J., Yalamov, P. (eds.): Numerical Analysis and its Applications. Lecture notes in Comp. Sci., Vol. 1988. Springer-Verlag (2001) 350–359.
5. Han, H., Kellogg, R.B.: Differentiability properties of solutions of the equations $-\varepsilon^2 \Delta u + ru = f(x, y)$ in a square. SIAM J. Math. Anal. 21 (1990) 394–408.
6. Kopteva, N.V.: Uniform difference methods for some singularly perturbed problems on condensed meshes. Phd Thesis, M.: MGU, 1996.
7. Li, J.: Quasioptimal uniformly convergent finite element methods for the elliptic boundary layer problem. Computers Math. Applic., Vol. 33, 10 (1997) 11–22.
8. Miler, J.J.H., O'Riordan, E., Shishkin, G.I.: Fitted numerical methods for singularly perturbed problems. World Scientific, Singapore (1996)
9. Roos, H.-G.: A note on the conditioning of upwind schemes on Shishkin meshes. IMA J. of Num. Anal., Vol. 16 (1996) 529–538.
10. Ross, H.G., Stynes, M., Tobiska, L.: Numerical methods for singularly perturbed differential equations. Springer Verlag (1996)
11. Spotz, W.F.: High-order compact finite difference schemes for computational mechanics. Ph. D. Thesis, University of Texas at Austin (1995)
12. Vasileva, A., Butusov, V.: Asymptotic methods in singular perturbation theory. Vyshaya Shkola, Moscow (1996) (in Russian).

Kantorovich Method for Solving the Multi-dimensional Eigenvalue and Scattering Problems of Schrödinger Equation

Ochbadrakh Chuluunbaatar[1], Michael S. Kaschiev[2], Vera A. Kaschieva[3], and Sergey I. Vinitsky[1]

[1] Joint Institute for Nuclear Research, 141980 Dubna, Moscow region, Russia,
[2] Present address: South - West University Neofit Rilski, Blagoevgrad, Bulgaria, Permanent address: Institute of Mathematics and Informatics, Bulgarian Academy of Science, Sofia, Bulgaria,
[3] Department of Mathematics, Technical University - Sofia, Bulgaria

Abstract. A Kantorovich method for solving the multi-dimensional eigenvalue and scattering problems of Schrödinger equation is developed in the framework of a conventional finite element representation of smooth solutions over a hyperspherical coordinate space. Convergence and efficiency of the proposed schemes are demonstrated on an exactly solvable model of three identical particles on a line with pair attractive zero-range potentials below three-body threshold. It is shown that the Galerkin method has a rather low rate of convergence to exact result of the eigenvalue problem under consideration.

1 Introduction

Elaboration of stable numerical methods for the elliptic partial differential equation is one of the main problems of computational mathematics. From this point of view, the creation of numerical schemes for solving the Schrödinger equation in a multi-dimensional space is a very important task of computational physics. This is based on the fact that the numerical solution of such equation has a wide application in different quantum-mechanical problems such as the modern calculations of the weakly bound states and elastic scattering in a system of three helium atoms considered as point particles with some short range pair potentials, i.e. a trimer of helium atoms [11], and modern laser physics experiments [9]. So, the above experiments require computer modelling of dynamics of exotic few-body Coulomb systems in external laser pulse fields [7]. There are two conditions for elaborating numerical methods: to be stable and to have a high accuracy.

The main idea of this paper is to formulate Kantorovich method for solving the multi-dimensional eigenvalue and scattering problems for Schrödinger equation MSE. In this method multi-dimensional boundary problem is reduced to a system of ordinary differential equations of second order with variable coefficients on a semi-axis with the help of expansion of the solution by a set of

I. Dimov et al. (Eds.): NMA 2002, LNCS 2542, pp. 403–411, 2003.

orthogonal solutions of an auxiliary parametric eigenvalue problem. Then the finite-element method is applied, to construct numerical schemes for solving corresponding boundary problem for the system of ordinary differential equations with an arbitrary accuracy in the space step. Note that variable coefficients of ordinary differential equations and the corresponding solutions can have a long-range asymptotic behavior. That is why one has to be very careful in the formulation of the boundary problems under consideration. We consider as an example the known exactly solvable model of three identical particles on a line with pair attractive zero-range potentials below three-body threshold, build up adequate formulations and corresponding schemes. We verify an accuracy of these schemes in comparison with theoretical estimations, using sequences of the enclosed meshes and examine rate of convergence to exact results with respect to number of basis functions.

2 Statement of the Problem

We consider three identical particles in the local Jacobi coordinates $\{\xi,\eta\} \in R^2$ in the center-of-mass system, $\eta = 2^{-1/2}(x_1 - x_2)$, $\xi = 6^{-1/2}(x_1 + x_2 - 2x_3)$, where $x_i \in R^1$, $i = 1,2,3$ are Cartesian coordinates of particles on a line. In polar coordinates ρ and θ, $\eta = \rho\cos\theta$, $\xi = \rho\sin\theta$, $-\pi < \theta \le \pi$, the Schrödinger equation for a partial wave function $\Psi(\rho,\theta)$ has the form:

$$\left[-\frac{1}{\rho}\frac{\partial}{\partial\rho}\rho\frac{\partial}{\partial\rho} + h_\rho\right]\Psi(\rho,\theta) = \frac{2m}{\hbar^2}E\Psi(\rho,\theta). \tag{1}$$

Here E is the relative energy in the center-of-mass, $m = (m_1m_2 + m_1m_3 + m_2m_3)/(m_1+m_2+m_3)$ is the effective mass. We choose pair potentials $V_i(\sqrt{2}\eta) = g\delta(|\eta|)/\sqrt{2}$, $i = 1,2,3$ as delta-functions of a finite strength $g = c\bar{\kappa}\sqrt{2}(\hbar^2/m)$, and consider an attractive case $c = -1$ and $\bar{\kappa} = \pi/6$, supports bound state $\phi_0(\eta) = \sqrt{\bar{\kappa}}\exp(-\bar{\kappa}|\eta|)$ with double energy $\epsilon_0^{(0)} = -\bar{\kappa}^2$, so that $2E = q^2 + \epsilon_0^{(0)}$ in units $\hbar = m = 1$, where q is a relative momentum of the pair [4]. The parametric Hamiltonian h_ρ at each fixed value $\rho \in R^1_+$ has the form (in a.e. $\hbar = m = 1$)

$$h_\rho = -\frac{1}{\rho^2}\frac{\partial^2}{\partial\theta^2} + \frac{2c\bar{\kappa}}{\rho}\sum_{n=0}^{5}\delta(\theta - \theta_n),\ \theta_n = n\pi/3 + \pi/6. \tag{2}$$

Using six-fold symmetric representation (2) we formulate the following boundary problem corresponding to equation (1) in the case $E < 0$ [6]

$$-\left[\frac{1}{\rho}\frac{\partial}{\partial\rho}\rho\frac{\partial}{\partial\rho} + \frac{1}{\rho^2}\frac{\partial^2}{\partial\theta^2}\right]\Psi(\rho,\theta) = 2E\Psi(\rho,\theta), \tag{3}$$

with boundary conditions by the angle variable $\theta_n \le \theta < \theta_{n+1}$

$$\begin{aligned}&\frac{1}{\rho}\frac{\partial\Psi(\rho,\theta_i)}{\partial\theta} = (-1)^{i-n}c\bar{\kappa}\Psi(\rho,\theta_i),\quad i = n, n+1,\quad n = 0,1,...5,\\ &\Psi(\rho,\theta_{n+1} - 0) = \Psi(\rho,\theta_{n+1} + 0),\end{aligned} \tag{4}$$

and asymptotic conditions by radial variable

$$\lim_{\rho\to 0} \rho \frac{\partial \Psi(\rho,\theta)}{\partial \rho} = 0,$$
$$\Psi(\rho,\theta)|_{\rho\to\infty} \to \chi_0^{as}(\rho) B_0^{as}(\rho,\theta) + F(k,\theta)\sqrt{\frac{\pi}{2k\rho}} e^{-k\rho}, k = \sqrt{-2E}. \tag{5}$$

For a scattering problem in the open channel $E(q) > \epsilon_0^{(0)}$, i.e. for $0 < q < \bar{\kappa}$

$$\chi_0^{as}(\rho) \approx \sin(q\rho + \delta)/\sqrt{q\rho}, \tag{6}$$

where $\delta = \delta(q)$ is unknown phase shift, and for bound states $(E(q) \leq \epsilon_0^{(0)})$

$$\chi_0^{as}(\rho) \approx e^{-\bar{q}\rho}/\sqrt{\rho}, \tag{7}$$

where $\varepsilon = \bar{q}^2 = -q^2 \geq 0$ is unknown binding energy of the three body system.

3 The antorovich Method

Consider a formal expansion of the solution of Eqs. (1)-(2) using the infinite set of one-dimensional basis functions $B_j(\rho;\theta) \in W_2^1(-\pi,\pi)$, $j = 0,1,2,\ldots$:

$$\Psi(\rho,\theta) = \sum_{j=0}^{\infty} \chi_j(\rho) B_j(\rho;\theta). \tag{8}$$

In Eq. (8), functions $\chi(\rho)^T = (\chi_0(\rho,),\chi_1(\rho),\ldots,)$ are unknown, and surface functions $B(\rho;\theta)^T = (B_0(\rho;\theta), B_1(\rho;\theta),\ldots,)$ form an orthonormal basis for each value of ρ which is treated here as a given parameter. In the Kantorovich approach [10], functions $B_j(\rho;\theta)$ are determined as solutions of the following one-dimensional parametric eigenvalue problem $\theta_n \leq \theta < \theta_{n+1}$:

$$\begin{cases} \frac{1}{\rho^2}\frac{\partial^2 B_j(\rho;\theta)}{\partial\theta^2} = -\epsilon_j(\rho) B_j(\rho;\theta), \\ \frac{1}{\rho}\frac{\partial B_j(\rho;\theta_i)}{\partial\theta} = (-1)^{n-i} c\bar{\kappa} B_j(\rho;\theta_i), \quad i = n, n+1, \quad n = 0,1,...5, \\ B_j(\rho;\theta_{n+1} - 0) = B_j(\rho;\theta_{n+1} + 0). \end{cases} \tag{9}$$

The eigenfunctions of this problem are normalized as follows

$$< B_i(\rho;\theta)|B_j(\rho;\theta) >= \int_{-\pi}^{\pi} B_i(\rho;\theta) B_j(\rho;\theta) d\theta = \delta_{ij}. \tag{10}$$

After substitution of expansion (8) into the Rayleigh-Ritz variational functional (see [3]) and subsequent minimization of the functional, the solution of Eqs. (1)-(2) is reduced to a solution of an eigenvalue problem for the finite set of n_{max} ordinary second-order differential equations for determining energy E and coefficients (radial wave functions) $\chi(\rho)$ of expansion (8)

$$-\mathbf{I}\frac{1}{\rho}\frac{d}{d\rho}\rho\frac{d\chi}{d\rho} + \mathbf{V}\chi - \mathbf{A}\frac{d\chi}{d\rho} - \frac{1}{\rho}\frac{d\rho\mathbf{A}\chi}{d\rho} = 2E\mathbf{I}\chi, \quad \lim_{\rho\to 0}\rho\frac{d\chi(\rho)}{d\rho} = 0. \tag{11}$$

The boundary conditions on $\rho \to \infty$ are (6) or (7) corresponding to the problem. In these expressions matrix $\mathbf{V}$ is symmetric one and $\mathbf{A}$ is anti-symmetric. They are given by the formulas

$$\begin{aligned} V_{ij} = H_{ij} + 0.5(\epsilon_i + \epsilon_j)\delta_{ij}, \quad H_{ij}(\rho) = \left\langle \tfrac{\partial}{\partial\rho} B_i(\rho;\theta) \,\middle|\, \tfrac{\partial}{\partial\rho} B_j(\rho,\theta) \right\rangle, \\ A_{ij}(\rho) = \left\langle B_i(\rho;\theta) \,\middle|\, \tfrac{\partial}{\partial\rho} B_j(\rho;\theta) \right\rangle. \end{aligned} \tag{12}$$

As is shown in paper [8] the problem (9) - (10) has analytical solutions

$$\begin{aligned} B_0(\rho;\theta) &= \sqrt{\tfrac{y_0^2 - x^2}{\pi(y_0^2 - x^2) + |x|}} \cosh\left[6y_0(\theta - n\pi/3)\right], \\ B_j(\rho;\theta) &= \sqrt{\tfrac{y_j^2 + x^2}{\pi(y_j^2 + x^2) - |x|}} \cos\left[6y_j(\theta - n\pi/3)\right], \\ \epsilon_0(\rho) &= -\left(\tfrac{6y_0(\rho)}{\rho}\right)^2, \quad \epsilon_j(\rho) = \left(\tfrac{6y_j(\rho)}{\rho}\right)^2. \end{aligned} \tag{13}$$

The functions $\epsilon_0(\rho), \epsilon_j(\rho), j = 1, 2, \ldots$ can be calculated as the roots of transcendental equations, which follow from the boundary problems (9)-(10) at c=-1,

$$\begin{aligned} y_0(\rho)\tanh(\pi y_0(\rho)) = -x, \ 0 \le y_0(\rho) < \infty, \ x = c\tfrac{\pi}{36}\rho, \\ y_j(\rho)\tan(\pi y_j(\rho)) = x, \ j - \tfrac{1}{2} < y_j(\rho) < j, \ j = 1, 2, 3, \ldots \end{aligned} \tag{14}$$

So using analytical expressions for functions $B_0, \epsilon_0, \ B_j, \epsilon_j$ we have a possibility to calculate the matrix elements $V_{ij}, \ A_{ij}$ analytically.

4 The Galerkin Method

Let us consider the following expansion for the wave function $\Psi(\rho,\theta)$

$$\Psi(\rho,\theta) = \frac{1}{\sqrt{2\pi}}\bar{\chi}_0(\rho) + \frac{1}{\sqrt{\pi}}\sum_{j=1}^{\infty} \bar{\chi}_j(\rho)\cos(6j\theta). \tag{15}$$

In this expansion the basic function $\eta_j(\theta)$ are solutions of the eigenvalue problem $-\dfrac{d^2b}{d\theta^2} = \bar{\epsilon}\rho^2 b$ with six-fold symmetry conditions and functions $\bar{\chi}_j(\rho)$ are unknown coefficients. In this case we have that $b_j(\theta) = \cos(6j\theta)$. Using these functions the corresponding matrix elements in the system of radial equations (12) have the form

$$H_{ij}(\rho) = \begin{cases} c/\rho, \text{ if } i = j = 0, \\ (-1)^j\sqrt{2}c/\rho, \text{ if } i = 0, \ j \ne 0, \\ (-1)^i\sqrt{2}c/\rho, \text{ if } i \ne 0, \ j = 0, \\ (-1)^{i+j}2c/\rho, \text{ if } i \ne 0, \ j \ne 0, \end{cases} \tag{16}$$

$$A_{ij}(\rho) = 0, \ V_{ij}(\rho) = H_{ij}(\rho) + 0.5(\bar{\epsilon}_i(\rho) + \bar{\epsilon}_j(\rho))\delta_{ij}, \ \bar{\epsilon}_i(\rho) = (6i/\rho)^2. \tag{17}$$

Below we will compare the convergence of the Kantorovich method and the Galerkin method on example of a calculation for the ground state of the discrete spectrum problem under consideration.

5 Reducing the Problem to a Finite Interval

We consider a reduction the boundary problem from semi-axis to finite interval using known asymptotic behavior of variable coefficients $H_{ij}(\rho)$, $A_{ij}(\rho)$ and $\varepsilon_j(\rho)$ and solutions $\chi_j(\rho)$ at small and large values of radial variable ρ.

5.1 Discrete Spectrum Problem

In the paper [6] the following asymptotics for components $\chi_0(\rho), \chi_1(\rho), \ldots$ are hold

$$\chi_0(\rho) \approx \frac{\exp(-\bar{q}\rho)}{\sqrt{\rho}}, \quad \chi_j(\rho) \approx \sqrt{\frac{\pi}{2k\rho}} \exp(-k\rho), \; j = 1, 2, \ldots. \tag{18}$$

From these relations we can obtain the homogeneous third type boundary condition for large $\rho_m, \rho_m >> 1$, $j = 1, 2, \ldots$

$$\rho_m \frac{d\chi_0(\rho_m)}{d\rho} = -(\frac{1}{2} + \bar{q}\rho_m)\chi_0(\rho_m), \; \rho_m \frac{d\chi_j(\rho_m)}{d\rho} = -(\frac{1}{2} + k\rho_m)\chi_j(\rho_m). \tag{19}$$

Here $\bar{q}^2 = -2E - \pi^2/36 \geq 0$, where $E < 0$ is the unknown eigenvalue. So the problem has to be nonlinear. But in the considered case we can use for simplicity its known analytical value ($-2E^b_{exact} = \pi^2/9$ for a ground statc, $-2E^{hb}_{exact} = \pi^2/36$ for a half-bound state) and the problem becomes a linear one.

5.2 Continuous Spectrum Problem

In this case the asymptotics of functions χ_j, $j = 0, 1, \ldots$ for large ρ are

$$\begin{aligned} \chi_0(\rho)\Big|_{\rho\to\infty} &\to J_0(q\rho) - \tan\delta' Y_0(q\rho) \\ &= \sqrt{\tfrac{2}{q\pi\rho}} \left(\sin\left(q\rho + \tfrac{\pi}{4}\right) + \tan\delta' \cos\left(q\rho + \tfrac{\pi}{4}\right)\right) + O(\rho^{-3/2}), \\ \chi_j(\rho)\Big|_{\rho\to\infty} &\to (\epsilon_0(\rho) - \epsilon_j(\rho))^{-1} \left[-A_{j0}(\rho)\frac{d}{d\rho} - \frac{1}{\rho}\frac{d}{d\rho}\rho A_{j0}(\rho) + H_{j0}(\rho)\right] \chi_0(\rho). \end{aligned} \tag{20}$$

Now we have the following nonhomogeneous third type boundary conditions for given value of $0 < q < \bar{\kappa}$, $2E = q^2 - \dfrac{\pi^2}{36} < 0$

$$\rho \frac{d\chi_0(\rho)}{d\rho}\Big|_{\rho=\rho_m} = \rho \left(\frac{dJ_0(q\rho)}{d\rho} - \tan\delta' \frac{dY_0(q\rho)}{d\rho} \right)\Big|_{\rho=\rho_m}, \tag{21}$$

$$\tan(\delta') = \frac{\pi}{2} \int_0^{\rho_m} J_0(q\rho) V_0(\rho) \chi(\rho) \rho d\rho, \; \delta = \frac{\pi}{4} - \delta' + \pi. \tag{22}$$

Here $\delta = \delta(q)$ is the value of the required phase shift at a fixed value of momentum $0 < q < \pi/6$ and $J_0(\rho)$ and $Y_0(\rho)$ are the cylindrical Bessel functions of the first and second king, and

$$V_0(\rho)\chi(\rho) = \sum_{j=1}^{\infty}\left(-A_{0j}(\rho)\frac{d}{d\rho} - \frac{1}{\rho}\frac{d}{d\rho}\rho A_{0j}(\rho) + H_{0j}(\rho)\right)\chi_j(\rho) + \left(V_{00}(\rho) + \frac{\pi^2}{36}\right)\chi_0(\rho). \tag{23}$$

For high accuracy calculation of $\tan(\delta')$ we take into account asymptotic correction terms in an exact definition of the phase shift

$$\tan(\delta') = \frac{\pi}{2}\int_0^{\rho_m} J_0(q\rho)V_0(\rho)\chi(\rho)\rho d\rho + v^{as} - \tan(\delta')u^{as}. \tag{24}$$

As a result we have the following formula

$$\tan(\delta') = \left(\frac{\pi}{2}\int_0^{\rho_m} J_0(q\rho)V_0(\rho)\chi(\rho)\rho d\rho + v^{as}\right)/(1 + u^{as}), \tag{25}$$

where the asymptotic correction terms are of the form

$$\begin{aligned} v^{as} &= \tfrac{\pi}{2}\textstyle\int_{\rho_m}^{\infty} J_0(q\rho)\left(V_{00}(\rho) + \frac{\pi^2}{36}\right) J_0(q\rho)\rho d\rho \sim O\left(\frac{1}{\rho_m}\right), \\ u^{as} &= \tfrac{\pi}{2}\textstyle\int_{\rho_m}^{\infty} J_0(q\rho)\left(V_{00}(\rho) + \frac{\pi^2}{36}\right) Y_0(q\rho)\rho d\rho \sim O\left(\frac{1}{\rho_m^2}\right). \end{aligned} \tag{26}$$

Here the exact value of the phase shift δ for each $q \in (0, \pi/6)$ equals

$$\delta_{exact} = \frac{3\pi}{2} - \arctan\frac{8\sqrt{3}q\pi}{\pi^2 - 36q^2}. \tag{27}$$

6 Numerical Method

For numerical solution of one-dimensional eigenvalue problems and boundary value problem (11) subject to the corresponding boundary conditions (6) and (7), the high-order approximations of the finite element method [12,5] elaborated in our previous papers [3,1,2] have been used. One-dimensional finite elements of order $p = 1, 2, \ldots, 10$ have been implemented. Using the standard finite element procedures [5], these problems are approximated by the generalized algebraic eigenvalue problem

$$\mathbf{K}\mathbf{F}^h = E^h\mathbf{M}\mathbf{F}^h. \tag{28}$$

and the system of linear algebraic equations

$$\hat{\mathbf{K}}\mathbf{u}^h = \mathbf{U}. \tag{29}$$

Here $\mathbf{K}$ and $\mathbf{M}$ are the standard stiffness and mass matrices, corresponding to discrete spectrum problem, matrix $\hat{\mathbf{K}}$ and right-hand side vector $\mathbf{U}$ correspond to continuous spectrum problem, E^h and $\mathbf{F}^h$ are the numerical approximation of the corresponding eigenvalue problem and $\mathbf{u}^h$ is the finite element approximation of the continuous wave function on the finite-element grid.

7 Numerical Results

Here we study the convergence rate of the Kantorovich and Galerkin methods (KM and GM) in depending on a number of equations of system (11). The problem under consideration is a good test for numerical methods because it has analytical solutions of discrete and continuous spectrum. First we consider the eigenvalue problem with $\rho_m = 50$. We use the 1000 finite elements of forth order. The finite element grid consists 4001 nodes. We consider the calculations with double and quadruple precision for the KM and double precision of the GM. In Table 1 the differences $\Delta E = -2E^h + 2E^{exact}$ for each case are shown. One can see that if we use the quadruple precision the KM monotonically converges to the exact values while for double precision calculations it holds true only till 35 equations. Note that for the solution of algebraic eigenvalue problem we apply the Subspace iteration method. The main step there is to find the solutions of system of linear algebraic equations with matrix K using the Cholesky decomposition. For $N = 50$ this system consists of 200050 equations. It can be solved stable only if we use the quadruple precision. The last column shows that there is a rather low rate of convergence for the GM, because it is compatible with boundary conditions only in vicinity of small values ρ. So, the GM it can not apply in the scattering problem. For scattering problem we calculate phase shift δ^h at $q\rho_m = 300$ and use 1500 finite elements of forth order. The finite element grid consists 6001 nodes. In Table 2 differences $\Delta\delta = \delta^{exact} - \delta^h$ are shown. One can see that the KM converges monotonically to the exact values δ^{exact} with a rate of order 1/N.

Table 1. The convergence of the Kantorovich method (KM) for the differences $\Delta E = -2E^h + 2E^{exact}$ of energy value of the ground state versus the number of equations. First column shows the number of equations N, second and fourth ones display the accuracy of calculations for quadruple and double precision. Sixth column shows a rather low rate of convergence of the Galerkin method (GM). Thirds, fifth and seventh columns shows correspondingly CPU times for calculations on PC III-750MHz

N	ΔE_{KM}^{quad}	CPU^{quad}	ΔE_{KM}^{double}	CPU^{double}	ΔE_{GM}^{double}	CPU^{double}
1	1.801(-04)	0.225	1.801(-04)	0.023	9.662(-2)	0.020
2	2.762(-06)	1.169	2.762(-06)	0.082	4.116(-2)	0.046
3	2.697(-07)	2.791	2.697(-07)	0.183	2.573(-2)	0.083
4	5.413(-08)	5.142	5.413(-08)	0.325	1.866(-2)	0.154
5	1.594(-08)	8.414	1.594(-08)	0.515	1.462(-2)	0.220
6	5.949(-09)	11.978	5.950(-09)	0.770	1.201(-2)	0.336
10	3.967(-10)	33.765	3.979(-10)	2.194	7.010(-3)	0.880
20	1.099(-11)	137.805	1.245(-11)	8.273	3.431(-3)	3.448
30	1.390(-12)	303.616	3.276(-12)	18.222	2.271(-3)	8.096
35	6.357(-13)	430.993	3.194(-12)	24.608	1.943(-3)	11.556
40	3.232(-13)	571.624	3.361(-12)	31.970	1.697(-3)	15.457
50	1.046(-13)	916.614	4.427(-12)	52.430	1.355(-3)	26.433

Table 2. The differences $\Delta\delta = \delta^{exact} - \delta^{h}$ of exact and numerical results (with the double precision) of phase shift versus the number of equations N and momentum q.

N	q					
	0.002	0.100	0.200	0.300	0.400	0.500
1	6.180(-1)	2.972(-2)	3.946(-2)	3.311(-2)	6.857(-2)	8.513(-2)
2	2.991(-2)	5.716(-3)	1.038(-2)	1.548(-2)	2.064(-2)	2.583(-2)
3	5.279(-3)	3.011(-3)	5.920(-3)	8.869(-3)	1.182(-2)	1.478(-2)
4	1.706(-3)	2.074(-3)	4.128(-3)	6.188(-3)	8.250(-3)	1.031(-2)
5	7.554(-4)	1.587(-3)	3.165(-3)	4.746(-3)	6.329(-3)	7.914(-3)
6	4.034(-4)	1.285(-3)	2.566(-3)	3.848(-3)	5.131(-3)	6.417(-3)
10	8.353(-5)	7.303(-4)	1.459(-3)	2.188(-3)	2.918(-3)	3.662(-3)
16	2.567(-5)	4.435(-4)	8.863(-4)	1.329(-3)	1.774(-3)	2.273(-3)
20	1.627(-5)	3.518(-4)	7.031(-4)	1.054(-3)	1.410(-3)	1.858(-3)

8 Conclusions

The stably numerical schemes for solving MSE with high accuracy with respect to variables ρ are developed. New results are obtained for the long-range potential MSE by using PC without essential computer resources (see Table 1). It is shown that the obtained numerical results strongly correspond to the theoretical ones. This paper opens the way to apply elaborated methods for solving the MSE for the system of second-order ordinary differential equations and realizing the Kantorovich method for multi-dimensional problems [3].

The investigation was carried out under the financial support by RFBR (Grants No-00-01-00617, No-00-02-16337) and a grant of the President of Bulgarian State Agency for Atomic Energy (2000-2002).

References

1. Abrashkevich, A.G., Abrashkevich, D.G., Kaschiev, M.S., Puzynin, I.V.: Finite-element solution of the coupled channel Schrödinger equation using high-order accuracy approximation. Comput. Phys. Commun. 85 (1995) 40–64.
2. Abrashkevich, A.G., Abrashkevich, D.G., Kaschiev, M.S., Puzynin, I.V.: FESSDE, a program for finite-element solution of the coupled channel Schrödinger equation using high-order accuracy approximation. Comput. Phys. Commun. 85 (1995) 65–81.
3. Abrashkevich, A.G., Kaschiev, M.S., Vinitsky, S.I.: A new method for solving an eigenvalue problem for a system of three Coolomb particles within the hyperspherical adiabatic representation. J. Comp. Phys. 163 (2000) 328–348.
4. Amaya-Tapia, A., Larsen, S.Y., Popiel, J.J.: Few-Body Systems, 23 (1997) 87.
5. Bathe, K.J.: Finite Element Procedures in Engineering Analysis. Prentice-Hall, Englewood Cliffs, New York (1982)

6. Chuluunbaatar, O., Puzynin, I.V., Pavlov, D.V., Gusev, A.A., Larsen, S.Y., and Vinitsky, S.I. Preprint JINR, P11-2001-255, Dubna, 2001 (in Russian)
7. Derbov, V.L., Melnikov, L.A., Umansky, I.M., and Vinitsky, S.I.: Multipulse laser spectroscopy of pHe+: Measurement and control of the metastable state population. Phys. Rev. 55 (1997) 3394–3400.
8. Gibson, W., Larsen, S.Y., Popiel, J.J.: Hyperspherical harmonics in one dimension. I. adiabatic effective potentials for three particles with delta-function interactions. Phys. Rev. A, 35 (1987) 4919.
9. Holzscheiter, M.H., Charlton, M.: Ultra-low energy antihydrogen. Rep. Prog. Phys. 62 (1999) 1–60.
10. Kantorovich, L.V., Krylov, V.I.: The approximation methods of higher analysis. Nauka, Moscow (1952) (in Russian)
11. Rudnev, V., Yakovlev, S. Chem. Phys. Lett. 328 (2000) 97.
12. Strang, G., Fix, G.: An Analisys of the Finite Element Method. Printice-Hall, Englewood Cliffs, N.J. (1973)

A Unified Algorithm to Predict Both Compressible and Incompressible Flows

Alexander Churbanov
Institute for Mathematical Modeling, RAS
4-A Miusskaya Square, 125047 Moscow, Russia
chur@imamod.ru

Abstract. An efficient pressure-based algorithm to solve the compressible Navier-Stokes equations at low Mach numbers including the limiting case of fully incompressible computations is developed in the present work. The algorithm is based on two-scale splitting of the pressure into the volume-averaged and dynamic parts and uses the full Navier-Stokes equations without any reductions or preconditioning in order to study flows at all speeds.

1 Introduction

It is well-known that there are essential difficulties in viscous compressible flow predictions at low Mach number due to large distinction between acoustic and convection-diffusion time scales as well as decreasing the pressure-density coupling. Explicit schemes are very inefficient at $\mathrm{M} \to 0$ due to the Courant-Friedrichs-Levi time-step restriction. Moreover, even implicit methods face serious problems in numerical errors for such flows if the density is considered as a main dependent variable in uncoupled numerical procedures.

To avoid these problems, so-called low-Mach-number (LMN) approximation has been developed by various authors [1,2,3] as a suitable remedy. The dimensional pressure is decomposed in this approach into two normalized variables: volume-averaged part $\bar{p}(t)$ (spatially uniform mean pressure) which has thermodynamic sense and depends on time only, and dynamic part $p_d(t,\mathbf{x})$ (pressure fluctuation) varying both in space and in time:

$$P(t,\mathbf{x}) = P_0\bar{p}(t) + \rho_0 V_0^2 p_d(t,\mathbf{x}) , \tag{1}$$

where

$$P_0\bar{p}(t) = \frac{1}{V_\Omega}\int\limits_\Omega P(t,\mathbf{x})\,\mathrm{dv} , \quad V_\Omega = \int\limits_\Omega \mathrm{dv} , \tag{2}$$

or equivalent

$$\int\limits_\Omega p_d(t,\mathbf{x})\,\mathrm{dv} = 0 . \tag{3}$$

Here Ω is a problem domain and two different reference values P_0 and $\rho_0 V_0^2$ are used for two pressure components. Equation (2) (or (3)) specifies a unique

I. Dimov et al. (Eds.): NMA 2002, LNCS 2542, pp. 412–419, 2003.

pressure decomposition of this type. This two-scale splitting of the pressure allows to avoid all singularities at $\mathrm{M} \to 0$.

The dimensionless form of equations (1),(3) after normalization via P_0 seems like this:

$$p(t, \mathbf{x}) = \bar{p}(t) + \gamma \mathrm{M}^2 p_d(t, \mathbf{x}), \tag{4}$$

$$\int_{\Omega} p_d(t, \mathbf{x})\, \mathrm{dv} = 0. \tag{5}$$

It was shown in many applications [4,5,6,7], that this model provides correct numerical results for many slightly compressible gas flows. However, some simplifications are employed in the LMN-approximation: namely, the dynamic pressure is omitted in the energy equation as well as in the equation of state. It is clear, that this is valid only at low enough values of M.

An efficient pressure-based algorithm to handle slightly compressible flows based on two-scale pressure splitting (4),(5) without any reductions or preconditioning of the governing equations is developing in the works of the present authors [8,9]. In contrast to the above LMN-approximation, the full Navier-Stokes equations are used in the algorithm and therefore, it is capable to study viscous flows at all speeds. Evaluation of pressure components is based purely on their definition via (4),(5).

2 Complete System of Governing Equations

The full time-dependent compressible Navier-Stokes equations and the equation of the energy conservation are considered as governing equations. Using reference values of the velocity V_0, length L_0, pressure P_0 as well as temperature drop ΔT and taking into account the above pressure decomposition (1),(3) the following normalization procedure is employed (primed quantities are dimensionless):

$$\begin{gathered} t' = t/(L_0/V_0), \quad x_i' = x_i/L_0, \quad v_i' = v_i/V_0, \\ p_d' = (P(t,\mathbf{x}) - P_0\bar{p}'(t))/(\rho_0 V_0^2), \\ \bar{p}'(t) = \frac{1}{V_\Omega} \int_\Omega \frac{P(t,\mathbf{x})}{P_0} \mathrm{v}, \quad V_\Omega = \int_\Omega \mathrm{dv}, \\ \rho' = \rho/\rho_0, \; \rho_0 = \rho(P_0, T_0), \; \Theta' = (T - T_0)/\Delta T, \\ \mu' = \mu(T)/\mu_0, \; \lambda' = \lambda(T)/\lambda_0, \; C_p' = C_p(T)/C_{p0}, \\ \mu_0 = \mu(T_0), \quad \lambda_0 = \lambda(T_0), \quad C_{p0} = C_p(T_0). \end{gathered} \tag{6}$$

The governing equations are written in the dimensionless form as follows:

$$\frac{\partial \rho}{\partial t} + \mathrm{div}(\rho \mathbf{v}) = 0, \tag{7}$$

$$\frac{\partial(\rho v_i)}{\partial t} + \mathrm{div}(\rho v_i \mathbf{v}) = -\mathrm{grad}_i p_d + \frac{2}{\mathrm{Re}} \left(\mathrm{DIV}_i(\mu \dot{S}) - \frac{1}{3} \mathrm{grad}_i(\mu \mathrm{div}\mathbf{v}) \right) + F_i, \tag{8}$$

$$\frac{\partial(\rho\Theta)}{\partial t} + \mathrm{div}(\rho\mathbf{v}\Theta) = \frac{1}{\mathrm{Re}\cdot\mathrm{Pr}}\mathrm{div}\,(\lambda \mathrm{grad}\Theta) +$$

$$+\frac{\gamma-1}{r_T}\left(\frac{1}{\gamma}\frac{\partial\bar{p}}{\partial t} + \mathrm{M}^2\frac{\partial p_d}{\partial t}\right) + \frac{(\gamma-1)\mathrm{M}^2}{r_T}\left((\mathbf{v}\cdot \mathrm{grad}p_d) + \frac{J_{dis}}{\mathrm{Re}}\right), \tag{9}$$

$$\rho = \frac{\bar{p}(t) + \gamma\mathrm{M}^2 p_d(t,\mathbf{x})}{r_T\Theta + 1}, \tag{10}$$

$$\int_{\Omega} p_d(t,\mathbf{x})\,\mathrm{dv} = 0\,. \tag{11}$$

The dimensionless parameters that define a problem are ratio of the specific heats γ, parameter of temperature variation $r_T = \Delta T/T_0$ and Mach, Reynolds, Prandtl and Grashof numbers:

$$\mathrm{M} = \frac{V_0}{\sqrt{\gamma P_0/\rho_0}}, \quad \mathrm{Re} = \frac{V_0 L_0 \rho_0}{\mu_0}, \quad \mathrm{Pr} = \frac{C_{p0}\mu_0}{\lambda_0}, \quad \mathrm{Gr} = \frac{g r_T L_0^3 \rho_0^2}{\mu_0^2}.$$

Gravitational force $\mathbf{F}$ is defined as

$$\mathbf{F} = \frac{\mathrm{Gr}}{\mathrm{Re}^2 r_T}\rho\mathbf{e}\,.$$

Due to the fact that instead of the pressure we introduced two its parts, additional integral relation (11) is necessary to derive a complete system of equations.

It should be noted that the volume-averaged pressure $\bar{p}$ disappears in all terms containing spatial derivatives — in equation of momentum (8) and equation of energy (9). So, these equations have not now any singularities and it becomes possible to consider M tends to and even equals 0.

3 Finite-Difference Algorithm

In the numerical algorithm discrete approximations are constructed using finite-difference approach and the MAC-type staggered grid. To construct implicit scheme for the time-dependent equations of hydrodynamics, the Douglas-Rachford operator-splitting technique (similar to the well-known prediction-correction SIMPLEC method) is employed. The basic peculiarity is connected with a correct implementation in such an approach integral relation (11) which is nothing but an linear equation added to the standard system of gas dynamic equations.

To construct pressure-correction algorithms, it is suitable to introduce pressure increments $\delta\bar{p}$ and $\delta p_d(\mathbf{x})$ as follows:

$$\delta\bar{p} = \bar{p}^{n+1} - \bar{p}^n\,, \quad \delta p_d = p_d^{n+1} - p_d^n\,, \tag{12}$$

so, that:

$$p^{n+1} = p^n + \delta\bar{p} + \gamma\mathrm{M}^2\delta p_d\,. \tag{13}$$

It is easy to see that these pressure increments satisfy two following properties:
(a) To satisfy integral relation (11) for the dynamic pressure at every time-level, the dynamic increment δp_d must satisfy just the same relation:

$$\int_{\Omega} \delta p_d \, \mathrm{dv} = 0 \,. \tag{14}$$

(b) If we already have some intermediate values $(\delta\bar{p})^*$ and $(\delta p_d)^*$ such that this relation is not satisfied

$$p^{n+1} - p^n = (\delta\bar{p})^* + \gamma \mathrm{M}^2 (\delta p_d)^* \,, \quad \text{but} \quad \int_{\Omega} (\delta p_d)^* \, \mathrm{dv} \neq 0 \,, \tag{15}$$

then the correct values of the pressure increments can be evaluated via the following expressions:

$$\delta\bar{p} = (\delta\bar{p})^* + \frac{\gamma \mathrm{M}^2}{V_\Omega} \int_{\Omega} (\delta p_d)^* \, \mathrm{dv} \,, \quad \delta p_d = (\delta p_d)^* - \frac{1}{V_\Omega} \int_{\Omega} (\delta p_d)^* \, \mathrm{dv} \,, \tag{16}$$

that is nothing but a linear shift of the nonzero value of the integral from the dynamic pressure into the volume-averaged part.

It is evident from equations (7)–(11) that one has the common Navier-Stokes problem if $\bar{p}(t)$ is defined. In the discrete case one obtains a closed problem at every time-level if constant $\delta\bar{p}$ (or $\bar{p}^{n+1}$) is defined in some or another way. The increments $\delta\bar{p}$ and $\delta p_d(\mathbf{x})$ are to be corrected according to expressions (16) at the end of calculations at every time-level in order to satisfy integral relation (11).

It should be noted that constant $\delta\bar{p}$ (or $\bar{p}^{n+1}$) cannot be defined in an arbitrary way. At M= 0 the pressure problem turns into the Neumann problem for the Poisson equation arises if a enclosed flow domain is considered. So, the well-known compatibility constraint must be satisfied. The equation for evaluating $\delta\bar{p}$ is derived from this compatibility constraint (see equation (20) below). The equation may be obtained from the discrete analog of the next equation:

$$\frac{\partial}{\partial t}\left(\bar{p}(t) \int_{\Omega} \frac{1}{T} \, \mathrm{dv} \right) + \int_{\delta\Omega} (\rho \mathbf{v}, \mathbf{n}) \, \mathrm{dl} = 0 \,. \tag{17}$$

This equation is derived from the volume-averaged continuity equation and the state equation at M= 0.

Following the way of standard pressure-based methods, we first approximate the continuity and momentum equations and then derive the pressure equation by algebraic combinations of these discrete equations. The density is evaluated from the pressure and temperature via the equation of state. Substituting this state equation in the continuity one, we can approximate the temporal derivative of the density in it as follows:

$$\frac{\partial \rho}{\partial t} \approx \frac{\delta \bar{p} + \gamma \mathrm{M}^2 \delta p_d}{\tau(\Theta^n r_T + 1)} + \frac{\bar{p}^n + \gamma \mathrm{M}^2 p_d^n}{\tau}\left(\frac{1}{\Theta^{n+1} r_T + 1} - \frac{1}{\Theta^n r_T + 1}\right).$$

The following computations are performed for a transition to a new time-level.

First, prediction of the intermediate mass velocity is performed:

$$\left(\frac{1}{\tau} + L_h(\mathbf{v}^n)\right)(\rho \mathbf{v})^* = \frac{(\rho \mathbf{v})^n}{\tau} - \mathrm{grad}_h(p_d^n) + \mathbf{S}^n, \tag{18}$$

$$l = 0,\ \rho^l = \rho^n, \quad (\delta p_d)^* = 0, \quad (\delta \bar{p})^* = 0, \quad \Theta^l = \Theta^{n+1}.$$

Next, the inner iterative procedure to predict consistently both pressure increments is carried out:

$$\left(\frac{\rho^n}{\tau} + G_h(\rho \mathbf{v}^n)\right)\Theta^{l+1} = \frac{(\rho\Theta)^n}{\tau} + \frac{\gamma - 1}{r_T \gamma}\frac{(\delta \bar{p})^* + \gamma \mathrm{M}^2 (\delta p_d)^*}{\tau} +$$

$$+\frac{(\gamma-1)\mathrm{M}^2}{r_T}\left((\mathbf{v}^n \cdot \mathrm{grad}_h(p_d^n)) + \frac{J_{dis}^n}{\mathrm{Re}}\right), \tag{19}$$

$$(\delta \bar{p})^* = -\left(\sum_\omega \frac{1}{\Theta^n r_T + 1} h_1 h_2 h_3\right)^{-1} \times \left(\tau \sum_{\partial\omega} ((\rho \mathbf{v})^*, \mathbf{n}) h_{\partial\omega}\right.$$

$$\left. + \sum_\omega \frac{(\bar{p}^n + \gamma M^2 p_d^n)}{\tau^2} \times \left(\frac{1}{\Theta^{l+1} r_T + 1} - \frac{1}{\Theta^n r_T + 1}\right) h_1 h_2 h_3\right), \tag{20}$$

$$\left(\mathrm{div_h grad_h} - \frac{\gamma \mathrm{M}^2}{\tau^2(\Theta^n r_T + 1)}\right)(\delta p_d)^* = \frac{\mathrm{div_h}(\rho \mathbf{v})^*}{\tau} + + \frac{(\delta \bar{p})^*}{\tau^2(\Theta^n r_T + 1)} +$$

$$+\frac{(\bar{p}^n + \gamma M^2 p_d^n)}{\tau^2}\left(\frac{1}{r_T \Theta^{l+1} + 1} - \frac{1}{\Theta^n r_T + 1}\right), \tag{21}$$

$$l = l + 1.$$

Here operators $L_h(\mathbf{v}^n)$ and $G_h(\mathbf{v}^n)$ stand for the convective and diffusive terms in the momentum and energy equations. The superscript * denotes the intermediate values of unknowns.

The increment of the volume-averaged pressure is evaluated from equation (20). The increment of the dynamic pressure is evaluated from equation (21). This equation is quite similar to the Poisson equation usually used in pressure correction algorithms.

As the rule two or three inner iterations are performed to achieve the pressure-temperature coupling with a prescribed accuracy at every time-level.

After that, the following correction calculations are carried out:

$$\Theta^{n+1} = \Theta^l, \; \rho^{n+1} = \frac{p^n + (\delta\bar{p})^* + \gamma \mathrm{M}^2 (\delta p_d)^*}{\Theta^{n+1} r_T + 1}, \tag{22}$$

$$\mathbf{v}^{n+1} = \left((\rho\mathbf{v})^* - \tau \mathrm{grad}_h (\delta p_d)^*\right) / \rho^{n+1}, \tag{23}$$

$$\delta\bar{p} = (\delta\bar{p})^* + \gamma \mathrm{M}^2 V_\omega^{-1} \sum_\omega (\delta p_d)^* \, h_1 h_2 h_3, \tag{24}$$

$$\delta p_d = (\delta p_d)^* - V_\omega^{-1} \sum_\omega (\delta p_d)^* \, h_1 h_2 h_3, \tag{25}$$

$$p_d^{n+1} = p_d^n + \delta p_d, \quad \bar{p}^{n+1} = \bar{p}^n + \delta\bar{p}, \tag{26}$$

$$p^{n+1} = \bar{p}^{n+1} + \gamma \mathrm{M}^2 p_d^{n+1}. \tag{27}$$

It includes correction of the intermediate velocity (23) which is standard for the pressure-correction algorithms and correction of both parts of the pressure (volume-averaged and dynamic) involving linear shift in order to satisfy the necessary integral relation.

4 Numerical Results

To examine accuracy and evaluate robustness of the new algorithm, some different test problems have been investigated numerically.

The algorithm has been validated on the standard problem of free convection of air in a 2D square cavity with the thermally-insulated top and bottom and side walls of different temperature. This problem with various aspect ratio has been investigated and compared with numerical results from [4,5] obtained in the framework of LMN-approximation. As it was found in [5], oscillatory heat and fluid flow regimes occurred at high enough Rayleigh numbers and aspect ratios. Our calculations for A=2, Ra=3×10^7 and M=1.7×10^{-7} also indicated oscillatory regime shown in Fig. 1 in the form of instantaneous flow pattern (left figure) and temperature field (right). It is easy to see that complicated vortices occur in this case in the vicinity of the left upper and right lower corners of the cavity. Due to vortex fluctuations, the quasi-oscillatory regime is observed here. These time-dependent results have been proved using different values of the time-step. It was obtained that our results are close enough to the predictions of Paolucci *et al.* [5].

To validate possibilities of the numerical method for studying the limiting case – flows with M=0, an incompressible flow over a square cylinder placed in a free stream has been investigated as a test (see, e.g. [10,11,12]).

Predicted vortex shedding structures for the Reynolds number in the range from 50 up to 1000 have indicated a good agreement with available numerical and experimental data both in the Strouhal number and average drag/lift

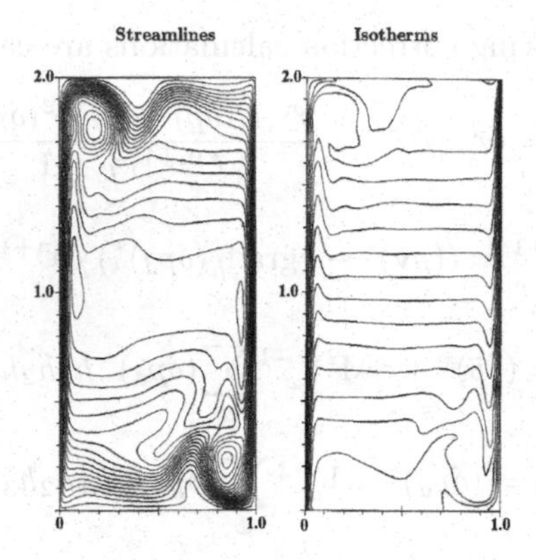

Fig. 1. Instantaneous flow pattern and temperature field for Ra=3 $\times 10^7$

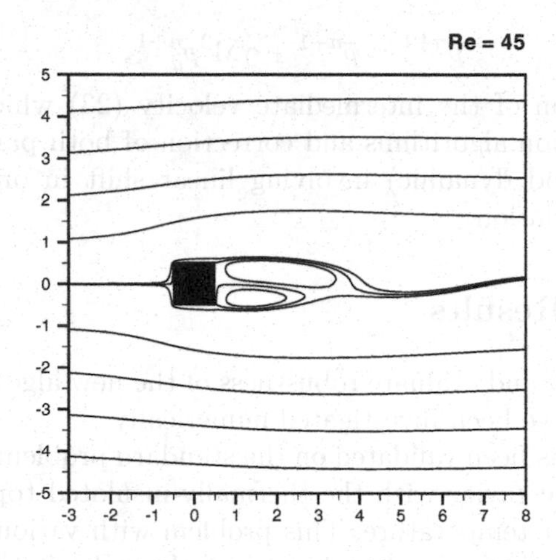

Fig. 2. Vortex shedding from a square cylinder, Re=45

coefficients [10,11,12]). Figure 2 demonstrates instanteneous flow pattern of vortex shedding process for Re=45. It was found from calculations that the critical Reynolds number Re_{cr} determining the onset of vortex shedding from the square cylinder lies about $\mathrm{Re}_{cr} \approx 45$ that is close to results of other researchers (see [11] and cited there data).

The applied problem of practical interest investigated numerically with the present method was the heat and flow transfer in a horizontal chemical reactor with high temperature drops between the incoming flow and the heated parts of the reactor walls. Using the developed algorithm there were conducted predictions of the considered reactor with various values of $\mathrm{Gr}/\mathrm{Re}^2$. It was found that in a small enough range of this governing parameter the heat and fluid flow in the channel can change greatly. At low values of this governing parameter we obtain steady-state solutions very similar to predictions and measurements from

[13]. But further increasing of Gr/Re^2 leads to loss of stability and forming an oscillatory flow regime.

5 Conclusions

The new implicit method for predicting both compressible and incompressible flows is developed and verified. Computations have indicated that the proposed method allows to predict without loss of accuracy and efficiency compressible flows in a wide range of Mach number including the limit case of fully incompressible fluids.

This work was supported in part by Russian Foundation for Basic Research under Grants # 00–01–00290 and # 00–01–00291.

References

1. Rehm, R.G., Baum, H.R.: The equations of motion for thermally driven, buoyant flows. J. Res. Nat. Bur. Stan **83** (1978) 297–308
2. O'Rourke, P.J., Bracco, F.V.: Two scaling transformations for the numerical computation of multidimensional unsteady laminar flames. J. Comput. Phys. **33** (1979) 185–203
3. Paolucci, S.: On the filtering of sound from the Navier-Stokes equations. Sandia National Laboratories Rep. SAND82-8257. Livermore, California (1982)
4. Chenoweth, D.R., Paolucci, S.: Natural convection in an enclosed vertical air layer with large horizontal temperature differences. J. Fluid Mech. **169** (1986) 173–210
5. Paolucci, S., Chenoweth, D.R.: Transition to chaos in a differentially heated vertical cavity. J. Fluid Mech. **201** (1989) 379–410
6. Buffat, M.: Simulation of two- and three-dimensional internal subsonic flows using a finite element method. Int. J. Numer. Methods Fluids **12** (1991) 683–704
7. Fröhlich, J., Peyret, R.: Direct spectral methods for the low Mach number equations. Int. J. Num. Meth. Heat Fluid Flow **2** (1992) 195–213
8. Churbanov, A.G., Pavlov, A.N., Voronkov, A.V., Ionkin, A.A.: Prediction of low Mach number flows: a comparison of pressure-based algorithms. In: Taylor, C., Cross, J.T. (eds.): Numerical Methods in Laminar and Turbulent Flow, Vol. 10. Pineridge Press, Swansea, U.K. (1997) 1099–1110
9. Churbanov, A.G., Pavlov, A.N.: A pressure-based algorithm to solve the full Navier-Stokes equations at low Mach number. In: Papailiou, K.D. et al. (eds.): Computational Fluid Dynamics'98, Vol. 1, Pt. 2. Wiley, Chichester (1998) 894–899
10. Okajima, A., Nagahisa, T., Rokugon, A.: A numerical analysis of flow around rectangular cylinders. JSME Int. J. Ser. II **33** (1990) 702–711
11. Sohankar, A., Norberg, C., Davidson, L.: Low-Reynolds-number flow around a square cylinder at incidence: study of blockage, onset of vortex shedding and outlet boundary conditions. Int. J. Numer. Methods Fluids **26** (1998) 39–56
12. Pavlov, A.N., Sazhin, S.S., Fedorenko, R.P., Heikal, M.R.: A conservative finite difference method and its application for the analysis of a transient flow around a square prism. Int. J. Numer. Methods Heat Fluid Flow **10** (2000) 6–46
13. Visser, E.P., Kleijn, C.R., Govers, C.A.M., Hoogendoorn, C.J., Giling, L.J.: Return flows in horizontal MOCVD reactors studied with use of TiO_2 particle injection and numerical calculations. J. Crystal Growth **94** (1989) 929–946

Parameter Estimation in Size/Age Structured Population Models Using the Moving Finite Element Method

Gabriel Dimitriu
University of Medicine and Pharmacy, Faculty of Pharmacy,
Department of Mathematics and Informatics,
6600 Iasi, Romania,
dimitriu@umfiasi.ro

Abstract. We consider the problem of estimating variable parameters in models describing the evolution in time of populations in which individual size or age are taken into account. Our approach uses a Galerkin technique on a non uniform and time dependent grid for the numerical solution of the equation. A moving finite element method which combines the method of characteristics with finite element techniques is applied. Numerical results to an example problem are presented.

1 Introduction

In this paper we consider a computational technique for the problem of estimating variable (temporally and spatially dependent) coefficients in a class of models describing the evolution in time of size/age structured populations. These models are based on a Markov transition assumption for the growth process and have the form ([1], [3])

$$\frac{\partial}{\partial t} u(t,x) + \frac{\partial}{\partial x}\left(q_1(t,x)u(t,x)\right) = \frac{1}{2}\frac{\partial^2}{\partial x^2}\left(q_2(t,x)u(t,x)\right) - q_3(t,x)u(t,x) + f(t,x), \tag{1}$$

where $x_{min} \leq x \leq x_{max}$, $t > 0$, with boundary conditions

$$\begin{aligned} \left[q_1 u - \tfrac{1}{2}\tfrac{\partial}{\partial x}(q_2 u)\right]_{x=x_{min}} &= \int_{x_{min}}^{x_{max}} q_4(t,\xi)u(t,\xi)\,d\xi, \\ \left[q_1 u - \tfrac{1}{2}\tfrac{\partial}{\partial x}(q_2 u)\right]^{x=x_{max}} &= 0, \end{aligned} \tag{2}$$

and initial condition

$$u(0,x) = u_0(x). \tag{3}$$

Here $u(t,x)$ is the population density at time t and in size class x, q_1 is the mean rate of increase in size (or first moment), q_2 is the second moment of the rate of increase in size and q_3 is the mortality rate. In (2) q_4 represents the fecundity function and $u_0(x)$ in (3) is the initial size distribution. The population flux is given by $q_1 u - \frac{1}{2}\frac{\partial}{\partial x}(q_2 u)$ and hence the left boundary condition (2) defines a

I. Dimov et al. (Eds.): NMA 2002, LNCS 2542, pp. 420–429, 2003.

recruitment rate at the minimum size x_{min}, while the right boundary condition guarantees that x_{max} is the maximum size attainable by any member of the population.

In section 2, we formulate the parameter estimation problem. Section 3 deals with the approximate estimation problem. Here we establish the convergence of the state variable corresponding to any convergent sequence of parameters. In section 4, we give implementation details for our computational technique and present numerical results.

2 Parameter Identification Problem

In this section we formulate the problem of estimating the unknown variable coefficients $q = (q_1, q_2, q_3, q_4) \equiv (q_1(t,x), q_2(t,x), q_3(t,x), q_4(t,x))$ appearing in the equation (1)-(3). We assume that the parameter q belongs to some admissible parameter set Q_{AD}, where Q_{AD} is a given compact subset of

$$Q = \left\{ q \; : \; q \in L^\infty(0,T;W) \times L^\infty(0,T;W) \times L^\infty(0,T;\tilde{W}) \times L^\infty(0,T;\tilde{W}) \right\},$$

where $W \equiv W^{1,\infty}(x_{min}, x_{max})$, $\tilde{W} \equiv L^\infty(x_{min}, x_{max})$ and $0 < \nu_2 \leq q_2$ with ν_2 is a fixed constant.

We first analyze the direct problem (1)-(3). We formulate the system (1)-(3) in a variational or weak form; that is, we seek a solution $t \to u(t)$ on $0 \leq t \leq T$ with $u(t) \in V \equiv H^1(x_{min}, x_{max})$ satisfying

$$\begin{aligned} &\langle u_t, \psi\rangle + \langle D(q_1 u), \psi\rangle + \tfrac{1}{2}\langle D(q_2 u), D\psi\rangle + \langle q_3 u, \psi\rangle + q_1(t, x_{min}) u(t, x_{min}) \psi(x_{min}) \\ &-q_1(t, x_{max}) u(t, x_{max}) \psi(x_{max}) - \psi(x_{min}) \int_{x_{min}}^{x_{max}} q_4(t,\xi) u(t,\xi)\, d\xi - \langle f, \psi\rangle = 0, \end{aligned} \tag{4}$$

for all $\psi \in V$, along with initial condition $u(0) = u_0$. Here $D \equiv \frac{\partial}{\partial x}$ and $\langle \cdot\,, \cdot\rangle$ is the usual inner product in $L^2(x_{min}, x_{max})$. We define a parameter dependent bilinear form $\sigma(t, \cdot\,, \cdot\,; q) : V \times V \to I\!R^1$ by

$$\begin{aligned} \sigma(t, \varphi, \psi; q) &= \langle D(q_1 u), \psi\rangle + \tfrac{1}{2}\langle D(q_2 u), D\psi\rangle + q_1(t, x_{min}) \varphi(x_{min}) \psi(x_{min}) \\ &-q_1(t, x_{max}) \varphi(x_{max}) \psi(x_{max}) - \psi(x_{min}) \int_{x_{min}}^{x_{max}} q_4(t,\xi) u(t,\xi)\, d\xi + \langle q_3 \varphi, \psi\rangle. \end{aligned} \tag{5}$$

Then the variational form (4) can be equivalently rewritten as

$$\langle u_t, \psi\rangle + \sigma(t, u, \psi; q) = \langle f, \psi\rangle, \qquad \psi \in V, \qquad u(0) = u_0. \tag{6}$$

In what follows we will use $|\varphi| = \|\varphi\|_{L^2(x_{min}, x_{max})}$, $\|\varphi\| = \|\varphi\|_V$ and $|\varphi|_\infty = \|\varphi\|_{L^\infty(x_{min}, x_{max})}$. The bilinear form (5) is bounded and satisfies a uniform (in $q \in Q_{AD}$) coercive inequality. To be more precise, we have

Lemma 1. *There exist constants* C_1, C_2, $C_3 > 0$ *such that:* (i) $\sigma(t, \varphi, \varphi; q) + C_1|\varphi|^2 \geq C_2\|\varphi\|^2$; (ii) $|\sigma(t, \varphi, \psi; q)| \leq C_3\|\varphi\|\|\psi\|$, *for* $\varphi, \psi \in V$, $q \in Q_{AD}$.

Using Lemma 1 and the following assumptions on the system (1)-(3): *i)* The function $f \in L^2(0,T;V')$; *ii)* The initial function $u_0 \in L^2(x_{min}, x_{max})$, we can appeal to the weak variational theory of Lions (Chapter III in [6]), to guarantee that (6) has a unique weak solution $u(\cdot)$ satisfying $u \in L^2(0,T;V)$ and $u_t \in L^2(0,T;V')$.

Now, we introduce a class of problems for the estimation of variable coefficients $q = (q_1, q_2, q_3, q_4)$ of the system (1)-(3). That is,

(ID) Given observations $z_i \in L^2(x_{min}, x_{max})$ for the population densities $u(t_i, \cdot)$, $i = \overline{1,n}$, we seek to determine a best estimate $\tilde{q} \in Q_{AD}$ that minimizes a least squares criterion

$$J(q) = \sum_{i=1}^{n} |u(t_i, q) - z_i|^2, \tag{7}$$

where $u = u(q)$ is the solution to (6) corresponding to q.

3 Approximate Parameter Identification Problem

In this section we define the problem of approximating the infinite dimensional inverse problem for (6), (7) by a sequence of finite dimensional parameter estimation problems.

In what follows we describe a class of moving element methods (the specific method considered below was suggested by Harrison in [5]) which appears to be suitable for the numerical solution of the equation (1). In the usual Galerkin or finite element approach we require that our approximate solution u^N lies in a finite dimensional linear subspace with basis functions defined on fixed nodes. Here, we consider a general finite element procedure defined on a space grid, which is not only non-uniform, but also consists of nodes which are time dependent. Let

$$\Delta^N(t) = \{x_i^N(t)\}_{i=1}^{N+1}, \qquad \text{with} \qquad x_{min} = x_1^N(t) < \ldots < x_{N+1}^N(t) = x_{max},$$

be a partition of $[x_{min}, x_{max}]$. Define

$$\underline{\Delta}^N(t) \equiv \min_{1 \le i \le N} (x_{i+1}^N(t) - x_i^N(t)) \qquad \text{and} \qquad \overline{\Delta}^N(t) \equiv \max_{1 \le i \le N} (x_{i+1}^N(t) - x_i^N(t)).$$

Furthermore, we define the sequence of time dependent finite dimensional subspaces $H^N(t) \subset V$, $N = 1, 2, \ldots$, generated by piecewise linear splines corresponding to the nodes $\Delta^N(t)$. Then $H^N(t)$ is the span of $w_i^N(t,x)$, where the w_i^N are the usual linear spline basis functions on $[x_{min}, x_{max}]$ which are implicit functions of time (depending as usual on x and on t through the time dependent nodes $x_i^N(t)$). Let $I^N(t)$ be the usual spline interpolation operator ([7]) relative to the basis w_i^N for $H^N(t)$.

We make the following assumptions on the partitions $\Delta^N(t)$: (H_1) For each $t \in [0,T]$, $\overline{\Delta}^N(t) \to 0$ as $N \to \infty$; (H_2) For each $\varphi \in H^2(x_{min}, x_{max})$ and

$\psi \in L^2(x_{min}, x_{max})$, there exists K (depending possibly on φ but not on ψ or N) such that

$$|\langle \frac{d}{dt} I^N(t)\varphi, \psi\rangle| \le K \left\{ \overline{\Delta}^N(t) \right\}^{\frac{1}{2}} |\psi|.$$

The moving finite element method satisfies these requirements (see [3]).

For any fixed $q \in Q_{AD}$, we now define approximating systems for (6) in $H^N(t)$ as follows: We seek $u^N(t) \in H^N(t)$ such that for all $\psi \in H^N(t)$,

$$\langle u_t^N, \psi\rangle + \sigma(t, u^N, \psi; q) = \langle f, \psi\rangle, \qquad u^N(0) = I^N(0)u_0. \tag{8}$$

This equation can be viewed as a "projection" of (6) into the finite dimensional subspaces $H^N(t)$ and hence, as a consequence, it is also well-posed. For each N, there is an approximating estimation problem associated with (8):

$(ID)^N$ Find $\tilde{q}^N \in Q_{AD}$ that minimizes $J^N(q) = \sum_{i=1}^n |u^N(t_i, q) - z_i|^2$, where u^N is the solution to (8), corresponding to q.

The solution $\tilde{q}^N$ of $(ID)^N$ is meaningful only if $\tilde{q}^N$ approximates the solution $\tilde{q}$ of the original inverse problem (7). We proceed to establish such a result (see also the discussions in [2]).

Theorem 1. *Let $f \in H^1(x_{min}, x_{max})$ and $u \in H^2(x_{min}, x_{max})$. Suppose q^N is any arbitrary sequence in Q_{AD} with $q^N \to q$ and $q \in Q_{AD}$. Then $u^N(t, q^N) \to u(t, q)$ in $L^2(x_{min}, x_{max})$, for each $t \in [0, T]$, where u^N and u are solutions of* (8), *and* (6) *respectively.*

Proof. The proof is based on variational arguments, Gronwall's inequality and the basic spline estimates obtained from Theorems 2.4, 2.5 of [7].

Let $u^N(q^N)$, $u(q)$ denote the corresponding solutions to (8), (6) respectively. Then for $t \in [0, T]$,

$$|u^N(q^N) - u(q)| \le |u^N(q^N) - I^N(t)(u(q))| + |I^N(t)(u(q)) - u(q)|.$$

Using basic spline estimates in [7] and (H_1), the second term converges to zero as $N \to \infty$. Hence it suffices to argue the convergence of $z^N(t) \to 0$ in $L^2(x_{min}, x_{max})$ for each $t \in [0, T]$, where $z^N(t) = u^N(t, q^N) - I^N(t)(u(t, q))$. Since $\langle z_t^N, \psi\rangle = \langle (u^N - u + u - I^N(t)u)_t, \psi\rangle$, we have, after substituting (6) and (8),

$$\langle z_t^N, \psi\rangle = -\sigma(t, u^N, \psi; q^N) + \sigma(t, u, \psi; q) + \langle (I - I^N(t))u_t, \psi\rangle - \langle \dot{I}^N(t)u, \psi\rangle,$$

or $\langle z_t^N, \psi\rangle + \sigma(t, z^N, \psi; q^N) = \sigma(t, u, \psi; q) - \sigma(t, I^N(t)u, \psi; q^N) + \langle (I - I^N(t))u_t, \psi\rangle - \langle \dot{I}^N(t)u, \psi\rangle$, with initial condition $z^N(0) = 0$. Choosing $\psi = z^N(t)$, we obtain

$$\begin{aligned}\frac{1}{2}\frac{d}{dt}|z^N|^2 + \sigma(t, z^N, z^N; q^N) &= \sigma(t, u, z^N; q) - \sigma(t, I^N(t)u, z^N; q^N) \\ &\quad + \langle (I - I^N(t))u_t, z^N\rangle - \langle \dot{I}^N(t)u, z^N\rangle.\end{aligned} \tag{9}$$

We note that by integration by parts $\langle z^N Dq_1^N, z^N\rangle = q_1^N(x_{max})z^N(x_{max})^2 - q_1^N(x_{min})z^N(x_{min})^2 - 2\langle q_1^N Dz^N, z^N\rangle$, so that $\langle D(q_1^N z^N), z^N\rangle = \langle z^N Dq_1^N, z^N\rangle + \langle q_1^N Dz^N, z^N\rangle = q_1^N(x_{max})z^N(x_{max})^2 - q_1^N(x_{min})z^N(x_{min})^2 - \langle q_1^N z^N, Dz^N\rangle$.

Hence, if we introduce the following auxiliary notations

$$\Delta_1^N(t) = q_1 Du - q_1^N DI^N(t)u, \qquad \Delta_{1'}^N(t) = uDq_1 - I^N(t)uDq_1^N,$$

$$\Delta_2^N(t) = q_2 Du - q_2^N DI^N(t)u, \qquad \Delta_{2'}^N(t) = uDq_2 - I^N(t)uDq_2^N,$$

$$\delta_1^N(t) = q_1(t, x_{min})u(t, x_{min}) - q_1^N(t, x_{min})(I^N(t)u)(t, x_{min}),$$

$$\delta_2^N(t) = q_1^N(t, x_{max})(I^N(t)u)(t, x_{max}) - q_1(t, x_{max})u(t, x_{max}),$$

then using (5) we can rewrite (9) as

$$\begin{aligned}
&\tfrac{1}{2}\tfrac{d}{dt}|z^N|^2 - \langle q_1^N z^N, Dz^N\rangle + \tfrac{1}{2}\langle z^N Dq_2^N, Dz^N\rangle + \tfrac{1}{2}\langle q_2^N Dz^N, Dz^N\rangle \\
&= \langle \Delta_1^N(t), z^N\rangle + \langle \Delta_{1'}^N(t), z^N\rangle + \tfrac{1}{2}\langle \Delta_2^N(t), Dz^N\rangle + \tfrac{1}{2}\langle \Delta_{2'}^N(t), Dz^N\rangle \\
&\quad + \delta_1^N(t) z^N(x_{min}) + \delta_2^N(t) z^N(x_{max}) \\
&\quad + z^N(x_{min}) \int_{x_{min}}^{x_{max}} q_4(t,\xi)(u^N - u)(t,\xi)\, d\xi + \langle q_3 u - q_3^N u^N, z^N\rangle.
\end{aligned} \tag{10}$$

Using standard inequalities and the bounds $\nu_i > 0$, $i = 1,2,3,4$ on Q_{AD} of the parameters q_1, q_2, q_3 and q_4 respectively, we obtain the following estimates (where ε, η are positive constants to be chosen later):

i) $\langle(-q_1^N + \frac{1}{2}Dq_2^N)z^N, Dz^N\rangle \geq -\nu_1 \frac{1}{4\varepsilon}|z^N|^2 - \varepsilon\nu_1|Dz^N|^2$,
ii) $\langle q_2^N Dz^N, Dz^N\rangle \geq \nu_2 |Dz^N|^2$, $\quad \langle -\dot{I}^N(t)u, z^N\rangle \leq \frac{K^2}{2}\overline{\Delta}^N(t) + \frac{1}{2}|z^N|^2$,
iii) $\langle \Delta_1^N(t), z^N\rangle + \langle \Delta_{1'}^N(t), z^N\rangle \leq \frac{1}{2}|\Delta_1^N(t)|^2 + \frac{1}{2}|\Delta_{1'}^N(t)|^2 + |z^N|^2$,
iv) $\langle \Delta_2^N(t), Dz^N\rangle + \langle \Delta_{2'}^N(t), Dz^N\rangle \leq \frac{1}{2\varepsilon}|\Delta_2^N(t)|^2 + \frac{1}{2\varepsilon}|\Delta_{2'}^N(t)|^2 + \varepsilon|Dz^N|^2$,
v) $\delta_1^N(t)z^N(x_{min}) \leq \frac{1}{4\eta}|\delta_1^N(t)|^2 + \eta|z^N(x_{min})|^2$,
vi) $\delta_2^N(t)z^N(x_{max}) \leq \frac{1}{4\eta}|\delta_2^N(t)|^2 + \eta|z^N(x_{max})|^2$,
vii) $\langle (I - I^N(t))u_t, z^N\rangle \leq \frac{1}{2}|(I - I^N(t))u_t|^2 + \frac{1}{2}|z^N|^2$.

In the second inequality of ii) we have used (H_2). We note that K depends on $u(t)$, but not on z^N or N. Using the above estimates in (10) we obtain

$$\begin{aligned}
&\tfrac{1}{2}\tfrac{d}{dt}|z^N|^2 + \tfrac{1}{2}\left\{-\tfrac{\nu_1}{2\varepsilon} - 4\right\}|z^N|^2 \\
&\leq \left\{\varepsilon\nu_1 - \tfrac{\nu_2}{2} + \tfrac{\varepsilon}{2}\right\}|Dz^N|^2 + \eta\left\{|z^N(x_{min})|^2 + |z^N(x_{max})|^2\right\} + \tfrac{1}{2}h^N(t),
\end{aligned} \tag{11}$$

where $h^N(t) = |\Delta_1^N(t)|^2 + |\Delta_{1'}^N(t)|^2 + \frac{1}{2\varepsilon}|\Delta_2^N(t)|^2 + \frac{1}{2\varepsilon}|\Delta_{2'}^N(t)|^2 + \frac{1}{2\eta}|\delta_1^N(t)|^2 + \frac{1}{2\eta}|\delta_2^N(t)|^2 + |(I - I^N(t))u_t|^2 + K^2\overline{\Delta}^N(t)$. Now, since $|z^N(x_{min})|^2 \leq |z^N|_\infty^2 \leq |z^N|^2$ (a similar estimate holds for $|z^N(x_{max})|^2$), we can equivalently rewrite (11) as

$$\frac{d}{dt}|z^N|^2 \leq h^N(t) + \left\{\frac{\nu_1}{2\varepsilon} + 2\nu_3 + 5\right\}|z^N|^2 + \{2\varepsilon\nu_1 - \nu_2 + 2\varepsilon + 4\eta\}|Dz^N|^2. \tag{12}$$

Let $C = \frac{\nu_1}{2\varepsilon} + 2\nu_3 + 5$ and choose $\varepsilon = \frac{\nu_2 - 4\eta}{2\nu_1 + 2}$ which is strictly positive if $\eta > 0$ sufficiently small (i.e., we choose $0 < \eta < \frac{\nu_2}{4}$). Then after integrating both sides of (12), we obtain

$$|z^N(t)|^2 \le |z^N(0)|^2 + \int_0^t h^N(\xi)\, d\xi + \int_0^t C|z^N(\xi)|^2\, d\xi.$$

Thus from the above inequality, the fact that $|z^N(0)| = 0$, and Gronwall's inequality, to show that $z^N(t) \to 0$ in L^2 it suffices to argue that $h^N(t) \to 0$ for each $t \in [0, T]$. By the definition of $h^N(t)$, this convergence is easily argued under basic spline estimates ([7]), hypothesis (H_1) and the smoothness assumptions on the state variable u.

Now, we consider the problem of parameter approximation. The parameter identification problems are, in general, infinite dimensional in both the state u and the parameter q and thus one must consider a sequence of computationally solvable approximating problems (which, in most cases, involves two separate approximation levels, one for the state and one for the parameter). We define a family Q_{AD}^M, $M = 1, 2, \ldots$, of finite dimensional sets that approximate Q_{AD} in some appropriate sense. Furthermore, we assume that there are mappings $i^M : Q_{AD} \subset L^2 \to L^2$, such that $Q_{AD}^M = i^M(Q_{AD})$ with the following approximate properties:

(H_3) Q_{AD}^M is a compact subset of Q; for each $q \in Q_{AD}$, $i^M(q) \to q$, as $M \to \infty$, uniformly in q.

Using these two levels of approximation, the next result shows that any sequence of solutions q^N of the approximating inverse problems possesses a convergent subsequence and that the limiting function $\bar{q}$ is a solution to the original parameter estimation problem. The proof can be found in numerous references (e.g., see [2]).

Theorem 2. *Let $Q_{AD}^M = i^M(Q_{AD})$ with properties (H_3). Let q_M^N be a solution of minimizing J^N over Q_{AD}^M. Then, for any convergent subsequence $\{q_{M_j}^{N_i}\}$ with $q_{M_j}^{N_i} \to q^*$, as $N_i \to \infty$ and $M_j \to \infty$, the limit q^* is a solution to the original problem of minimizing J, given by (7), over Q_{AD}.*

4 Numerical Implementation

We discuss implementation details for a specific example of the approximation framework presented in Section 3. Recalling that $H^N(t) = span\{w_j^N(t,x)\}$, we could equivalently consider $H^N(t)$ as the span of $\{B_j^N\}$ where the piecewise linear elements B_j^N are just the w_j^N normalized so that each element B_j^N has area under its graph equal to unity. We shall, in our computational method, use the B_j^N as basis elements (and the w_j^N as test elements in the weak formulation below). The Galerkin approximation u^N of u is then given by

$$u^N(t,x) = \sum_{j=1}^{N+1} B_j^N(t,x) y_j^N(t) \tag{13}$$

Substituting (13) into the weak form (4) with test functions $\psi = w_i^N(t,x)$, $i = \overline{1, N+1}$, we obtain the following system of $N+1$ ordinary differential equations

$$M(t)\frac{d}{dt}y^N(t) = P(t)y^N(t) - R(t)y^N(t) + F(t), \tag{14}$$

where $y^N(t) = \text{col}(y_1^N(t), \ldots, y_{N+1}^N(t))$, $F(t) = \text{col}(\langle f, w_1^N\rangle, \ldots, \langle f, w_{N+1}^N\rangle)$, $M(t)$, $P(t)$ and $R(t)$ are $(N+1)\times(N+1)$ matrices with elements given by

$$M_{ij} = \langle w_i^N, B_j^N\rangle, \quad R_{ij}(t) = \langle w_i^N, \frac{\partial}{\partial t}B_j^N + \frac{\partial}{\partial x}(q_1 B_j^N)\rangle,$$

$$\begin{aligned} P_{ij}(t) = & -\langle\frac{\partial}{\partial x}(q_1 B_j^N), w_i^N\rangle - \frac{1}{2}\langle\frac{\partial}{\partial x}w_i^N, \frac{\partial}{\partial x}(q_2 B_j^N)\rangle \\ & -q_1(x_{min})w_i^N(x_{min})B_j^N(x_{min}) + q_1(x_{max})w_i^N(x_{max})B_j^N(x_{max}) \\ & +w_i^N(x_{min})\int_{x_{min}}^{x_{max}} q_4(t,\xi)B_j^N(\xi)\,d\xi - \langle q_3 B_j^N, w_i^N\rangle. \end{aligned}$$

The basic idea of Harrison's moving finite element scheme ([5]) is to choose the nodes $\{x_i^N\}_{i=1}^{N+1}$ so that the basis functions B_j^N satisfy (approximately) the pure drift equation

$$\frac{\partial}{\partial t}B_j^N = -\frac{\partial}{\partial x}(q_1 B_j^N), \qquad j = 1, \ldots, N+1. \tag{15}$$

Then $R(t) \equiv 0$ and (14) reduces to the following pure diffusion equation

$$M(t)\frac{d}{dt}y^N = P(t)y^N + F(t) \tag{16}$$

which is computationally easier than (14). The construction of the nodes x_j^N and the basis functions B_j^N which approximately satisfy (15) is as follows ([5]): The interval $[x_{min}, x_{max}]$ is initially subdivided into N subintervals with mesh points $x_{min} = v_1 < v_2 < \cdots < v_{N+1} = x_{max}$ and uniform meshes equal to $(x_{max} - x_{min})/N$. The time varying nodes $x_i^N(t)$ are then chosen to satisfy

$$\frac{d}{dt}x_i^N(t) = q_1(t, x_i^N(t)), \qquad x_i^N(0) = v_i, \qquad i = 1, 2, \ldots, N+1. \tag{17}$$

The existence and uniqueness theory of ordinary differential equations guarantees that the moving grids $x_i^N(t)$ do not cross. Therefore, the time dependent partition satisfies $x_i^N(t) < x_{i+1}^N(t)$ for $i = \overline{1, N}$. The basis and test functions $B_j^N(t,x)$ and $w_j^N(t,x)$ respectively, are piecewise linear functions defined by

$$w_j^N(t,x) = \begin{cases} \frac{x - x_{j-1}^N(t)}{x_j^N(t) - x_{j-1}^N(t)} & x_{j-1}^N(t) < x \le x_j^N(t), \\ \frac{x_{j+1}^N(t) - x}{x_{j+1}^N(t) - x_j^N(t)} & x_j^N(t) \le x < x_{j+1}^N(t), \qquad \text{for } j = 2, \ldots, N, \end{cases}$$

$$w_1^N(t,x) = \frac{x_2^N(t) - x}{x_2^N(t) - x_1^N(t)} \qquad w_{N+1}^N(t,x) = \frac{x - x_N^N(t)}{x_{N+1}^N(t) - x_N^N(t)}$$

and

$$B_j^N(t,x) = \frac{2w_j^N(t,x)}{x_{j+1}^N(t) - x_{j-1}^N(t)} \qquad \text{for } j = 2, \ldots, N,$$

$$B_1^N(t,x) = \frac{2w_1^N(t,x)}{x_2^N(t) - x_1^N(t)} \qquad B_{N+1}^N(t,x) = \frac{2w_{N+1}^N(t,x)}{x_{N+1}^N(t) - x_N^N(t)}.$$

The construction of Q_{AD}^M is simplified by assuming that the parameters are separable, i.e. $q_m(t,x) = \alpha_m(t)\beta_m(x)$ for $m = 1,2,3,4$. Then for the approximation of q_1, q_2, q_3 and q_4 in Q_{AD}^M, we define $q_m^M(t,x) = \alpha_m^M(t)\beta_m^M(x)$, where

$$\alpha_m^M(t) = \sum_{i=1}^{M+1} a_{m,i}^M L_i^M(t), \qquad \beta_m^M(x) = \sum_{i=1}^{M+1} b_{m,i}^M \hat{L}_i^M(x), \qquad m = 1,2,3,4.$$

Here $L_i^M(t)$, $i = 1, \ldots, M+1$, denotes the standard hat function defined on a uniform mesh with mesh size T/M which has support in (t_{i-1}, t_{i+1}) with values $0, 1, 0$ at t_{i-1}, $t_i\, t_{i+1}$ respectively. The functions $\hat{L}_i^M(x)$ are the hat functions defined similarly on the spatial grid with uniform mesh $(x_{max} - x_{min})/M$. The above representations for q_1, q_2, q_3 and q_4 are then substituted into (16) and (17). The elements of the matrix $P(t)$ are now given by

$$\begin{aligned}
P_{ij} = &- \left(\sum_{k=1}^{M+1} a_{1k}^M L_k^M(t)\right) \sum_{l=1}^{M+1} b_{1l}^M \left\langle \frac{\partial}{\partial x} w_i^N, \frac{\partial}{\partial x}(\hat{L}_l^M(x) B_j^N)\right\rangle \\
&- \frac{1}{2}\left(\sum_{k=1}^{M+1} a_{2k}^M L_k^M(t)\right) \sum_{l=1}^{M+1} a_{2l}^M \left\langle \frac{\partial}{\partial x} w_i^N, \frac{\partial}{\partial x}(\hat{L}_l^M(x) B_j^N)\right\rangle \\
&- \left(\sum_{k=1}^{M+1} a_{1k}^M L_k^M(t)\right)\left(\sum_{k=1}^{M+1} b_{1k}^M \hat{L}_k^M(x_{min})\right) w_i^N(x_{min}) \\
&+ \left(\sum_{k=1}^{M+1} a_{1k}^M L_k^M(t)\right)\left(\sum_{k=1}^{M+1} b_{1k}^M \hat{L}_k^M(x_{max})\right) w_i^N(x_{max}) B_j^N(x_{max}) \\
&+ \left(\sum_{k=1}^{M+1} a_{4k}^M L_k^M(t)\right)\left(\sum_{l=1}^{M+1} \int_{x_{min}}^{x_{max}} b_{4l}^M L_l^M(\xi) B_j^N(\xi)\, d\xi\right) w_i^N(x_{min}) \\
&- \left(\sum_{k=1}^{M+1} a_{3k}^M L_k^M(t)\right) \sum_{l=1}^{M+1} b_{3l}^M \left\langle \hat{L}_l^M(x) B_j^N, w_i^M\right\rangle.
\end{aligned} \tag{18}$$

Therefore, the approximate parameter estimation problem is to find the vector of parameters $\tilde{q}^M = (a_{1i}^M, b_{1i}^M, a_{2i}^M, b_{2i}^M, a_{3i}^M, b_{3i}^M, a_{4i}^M, b_{4i}^M)$ that minimizes

$$J^N(q^M) = \sum_{i=1}^{n} \left|u^N(t_i, q^M) - z_i\right|^2,$$

where $u^N(t,x;q^M) = \sum_{j=1}^{N+1} B_j^N(t,x) y_j^N(t;q^M)$, and the coefficients y_j^N satisfying the system of ordinary differential equations

$$M(t)\frac{d}{dt}\,y^N = P(t)y^N + F(t), \quad \text{with initial condition} \quad u_0(z_i) = B_j^N(0,z_j)y_j^N(0).$$

We used in our numerical example as "true" parameters $\tilde{q}_1(t,x) = e^t x(1-x)$, $\tilde{q}_2(t,x) = e^t(1-x)^2$, $\tilde{q}_3(t,x) = (1-.01\sinh(7t-3))x$, $\tilde{q}_4(t,x) = 3e^t(6x-1)$, and "true" state $\tilde{u}(t,x) = e^t(1-x)$. Then, the inhomogeneous term f was calculated from (1). The interval $[0,1]$ was selected both for the spatial and time domain. The observations were generated at $n = 5$ moments of time $t_i = 0.2, 0.4, 0.6, 0.8, 1$ by $z_i = \tilde{u}_{ij} + \delta r_{ij}$, where r_{ij} are uniformly distributed random numbers with values in $[-1,1]$, $\delta = 0.01$, and $\tilde{u}_{ij} = \tilde{u}(t_i, x_j; \tilde{q})$ are the values of $\tilde{u}$ at the nodal points of the grid. For simplicity, we put $N = M = 8$. The estimated values of the parameters were plotted on a twice finer grid than the initial one by cubic interpolation. The start up values of q_i were chosen to be a flat surface $q_i^{startup} \equiv 1$, $i = 1,2,3,4$. The results of the identification procedure are depicted in Fig. 1.

The minimization problem was solved numerically with the Matlab routine *constr* by using a SQP method. The most time consuming part of the algorithm consists of the evaluation of the objective function J^N ([2], [4]); this operation requires the solutions to the semi-discrete Galerkin equations (16). Let Δt be the time discretization step, and $y^{N,n}$ approximate $y^N(n\Delta t)$. Then, given $y^{N,0} = y^N(0)$ and $y^{N,1}$, we compute $y^{N,n+1}$ from $y^{N,n}$, $y^{N,n-1}$ by the following two-step implicit scheme

$$\left(\frac{3}{2\Delta t}\,M^{n+1} - P^{n+1}\right) y^{N,n+1} - \frac{2}{\Delta t}\,M^{n+1} y^{N,n} + \frac{1}{2\Delta t} M^{n+1} y^{N,n-1} = F^{n+1}, \tag{19}$$

where $n = 1,2,\ldots$. This scheme is unconditionally stable and has a time truncation error $\mathcal{O}(|\Delta t|^2)$. To obtain $y^{N,1}$ from $y^{N,0}$ we used a Crank-Nicholson scheme which has the same accuracy (second order) and is also unconditionally stable.

At each time step $n\Delta t$, scheme (19) requires the solution to a linear system of $N+1$ equations. Efficient algebra computations are carried out at every time step due to the peculiar form of basis and test functions. To solve the ordinary differential equations (17) defining the movement of the grids we used *ode45* Matlab routine based on an explicit Runge-Kutta (4,5)-order formula.

Although, the two-step implicit scheme (19), as compared to other implicit schemes (Crank-Nicholson or backward Euler method), has enough dissipation so that it can be used for stiff problems and long range integration, we notice (see Fig. 1) a poor estimation for the wavelike part due to convective term q_1 and for the other parameters q_2, q_3 and q_4, in the neighborhood of the point $(1,1)$ of the rectangular domain $[0,1] \times [0,1]$.

In conclusion, this study demonstrates the feasibility of applying the moving finite element method to the simultaneous estimation of the parameters q_1, q_2, q_3 and q_4 involved in the size/age structured population model (1)-(3). Future work will focus on using efficient solvers for the linear system (19) and improvement of the convergence rate for the identification procedure.

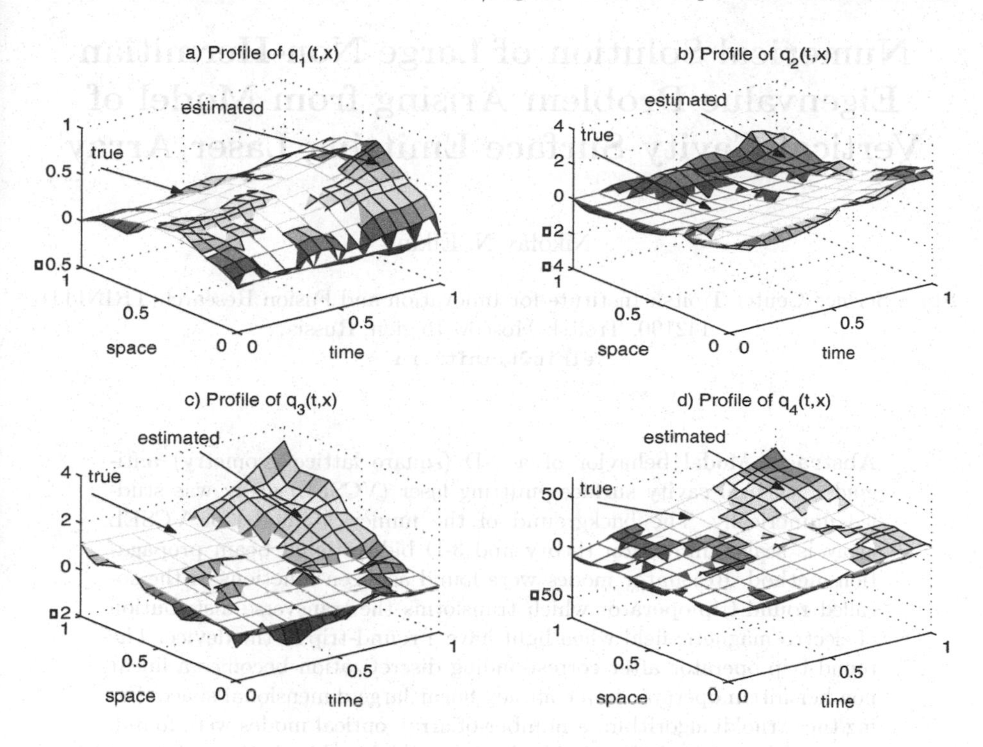

Fig. 1. True and estimated profiles of the parameters q_1, q_2, q_3 and q_4

References

1. Banks, H. T.: Computational techniques for inverse problems in size-structured stochastic population models. In: Bermudez, A. (ed.): Proc. IFIP Conf. on Optimal Control of Systems Governed by PDE (Santiago de Compostela, July 6-9, 1987). Lecture Notes in Control and Info. Sci., Vol. 114 (1989) 3–10.
2. Banks, H. T. and Kunisch, K.: Estimation Techniques for Distributed Parameter Systems. Progress in Systems and Control, 1, Birkhauser, Boston (1989)
3. Banks, H. T., Tran, H. T., and Woodward, D. E.: Estimation of variable coefficients in the Fokker-Planck equations using moving finite elements. CAMS 90-9, August (1990)
4. Dimitriu, G.: Numerical approximation of the optimal inputs for an identification problem. Intern. J. Computer Math. 70 (1998) 197–209.
5. Harrison, G. W.: Numerical solution of the Fokker-Planck equation using moving finite elements. Numerical Methods for Partial Differential Equations, 4 (1988) 219–232.
6. Lions, J. L.: Optimal Control of Systems Governed by Partial Differential Equations. Springer Verlag, New York (1971)
7. Schultz, M. H: Spline Analysis. Prentice-Hall, Englewood Cliffs (1973)

Numerical Solution of Large Non-Hermitian Eigenvalue Problem Arising from Model of Vertical Cavity Surface Emitting Laser Array

Nikolay N. Elkin

State Science Center Troitsk Institute for Innovation and Fusion Research(TRINITI), 142190, Troitsk Moscow Region, Russia,
elkin@triniti.ru

Abstract. Modal behavior of a 2-D (square lattice geometry) anti-guided vertical cavity surface emitting laser (VCSEL) array was studied numerically. The background of the numerical model of VCSEL array is scalar diffraction theory and 3-D bidirectional beam propagation method. Resonator modes were found as eigen-functions of the so-called round-trip operator which transforms the transverse distribution of electro-magnetic field when light have a round-trip in the device. The round-trip operator after corresponding discretization becomes a linear non-hermitian operator in a complex linear large dimensional space. Using the Arnoldi algorithm, a number of array optical modes were found. In calculations, both Fourier and space variable descriptions of beam propagation were combined. Calculations were made for various spacing length between elements and a size of the array. 4x4 and 10x10 laser arrays were studied numerically. Array optical modes having different symmetry properties were found. They include in-phase mode with constant phase over the array, out-of-phase mode with alternating phase between elements and modes with mixed symmetry. Conditions are found for favorable lasing of the in-phase mode providing high laser beam quality.

1 Introduction

The VCSEL array scheme is presented in Figure 1. The optical thickness of each of layers is expressed in terms of the vacuum wave-length λ_0. Looking at the scheme from bottom to top one can see the GaAs substrate which is taken to be unbounded, the Bragg reflector consisting of 31 pairs AlAs-GaAs, 1-wave cavity, containing the active layer (black strip), the index-step lattice, the Bragg reflector, consisting of 18 pairs, and the metal contact plate with lattice of windows. Furthermore, four matching layers are added to the scheme. The index-step lattice and the lattice of windows have supposedly the same geometrical arrangement. The index-step lattice is of such a small thickness that its action can be reduced to acquiring a phase step and an attenuation in their transmission through the lattice. In other words, we use the phase-screen approximation [1] in this case. VCSEL array is an open resonator of complicated configuration as a whole. The gain and index distributions of the active layers

I. Dimov et al. (Eds.): NMA 2002, LNCS 2542, pp. 430–437, 2003.

are taken to be uniform and independent of a light intensity. By this means a linear problem on eigen-oscillation is considered here.

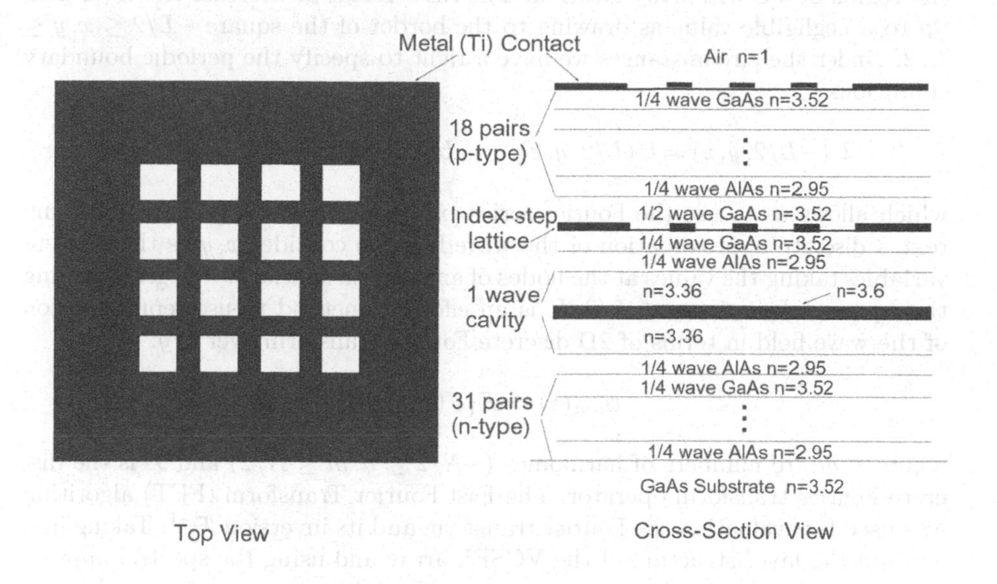

Fig. 1. Scheme of VCSEL array in two projections.

2 Formulation and Numerical Scheme

The optical field satisfies the scalar wave equation

$$\frac{\partial^2 U}{\partial x^2} + \frac{\partial^2 U}{\partial y^2} + \frac{\partial^2 U}{\partial z^2} + (k_0^2 n^2 - ik_0 g)U = 0,$$

in an assumption that the polarization effects can be neglected. Here $k_0 = 2\pi/\lambda_0$ is the wave number, n is index and g is gain, which may be negative in case of absorption.

There is a need to define the boundary condition at the interface between adjoining layers and the boundary condition at the lateral boundaries. We can specify the condition of continuity for the wave field U and its normal derivative at the interfaces.

It is not an easy task to define correctly the boundary condition at the lateral boundaries. We suppose that the leaky waves are not reflected from the lateral boundaries. Such a situation may be ensured using the absorbing boundary conditions (ABC) [2]. The ABC are realized by the artificial absorption localized near the border of the square $-L/2 \leq x, y \leq L/2$ (see below the distribution of the artificial absorption). The size L is taken so larger than the size of the design in Top view (see Figure 1) that the artificial absorption has a negligible value in the region of VCSEL array location. The ABC result in decrease the wave field up to a negligible value as drawing to the border of the square $-L/2 \leq x, y \leq L/2$. Under the circumstances we have a right to specify the periodic boundary conditions:

$$U(-L/2, y, z) = U(L/2, y, z), \qquad U(x, -L/2, z) = U(x, L/2, z)$$

which allows us to use the Fourier series expansion. Assuming in the following text a discrete approximation of the wave field we consider x, y as the discrete variables taking the values at the nodes of an uniform spatial $N \times N$ grid covering the square $-L/2 \leq x, y \leq L/2$. It is an effective method to use representation of the wave field in terms of 2D discrete Fourier transform over x, y:

$$\psi_{nm}(z) = \mathcal{F}\{U(x, y, z)\}$$

where n, m are numbers of harmonics ($-N/2 \leq n, m \leq N/2$) and $\mathcal{F}$ is the discrete Fourier transform operator. The Fast Fourier Transform (FFT) algorithm was used for evaluation the Fourier transform and its inversion $\mathcal{F}^{-1}$. Taking into account the layer structure of the VCSEL array and using the spectral approximation of the partial derivatives over x and y, the wave equation in the j-th layer can be expressed in a form:

$$\frac{d^2\psi_{nm}}{dz^2} + q_{jnm}^2 \psi_{nm} = 0, \quad q_{jnm}^2 = k_0^2 n_j^2 - ik_0 g_j - \kappa_n^2 - \nu_m^2, \quad z_j < z < z_{j+1},$$

where $\kappa_n = n\frac{2\pi}{L}$ and $\nu_m = m\frac{2\pi}{L}$ are the transverse wave vector components. Omitting indexes n, m the general solution of the last equation has a form:

$$\psi(z) = A_j \exp(iq_j z) + B_j \exp(-iq_j z), \quad \mathrm{Re}(q_j) \geq 0$$

The coefficients A_j and B_j are coupled by the translation matrix:

$$\begin{pmatrix} A_{j+1} \\ B_{J+1} \end{pmatrix} = T_{j+1} \begin{pmatrix} A_j \\ B_J \end{pmatrix}, \quad \text{where}$$

$$T_{j+1} = \frac{1}{2q_{j+1}} \begin{pmatrix} (q_{j+1} + q_j)\,\mathrm{e}^{-i(q_{j+1}-q_j)z_{j+1}} & (q_{j+1} - q_j)\,\mathrm{e}^{-i(q_{j+1}+q_j)z_{j+1}} \\ (q_{j+1} - q_j)\,\mathrm{e}^{i(q_{j+1}+q_j)z_{j+1}} & (q_{j+1} + q_j)\,\mathrm{e}^{i(q_{j+1}-q_j)z_{j+1}} \end{pmatrix}$$

Multiplying the translation matrices for the neighboring layers, T-matrices for the top Bragg reflector

$$T^t = T_l^t T_{l-1}^t \cdot \ldots \cdot T_1^t = \begin{pmatrix} t_{11}^t & t_{12}^t \\ t_{21}^t & t_{22}^t \end{pmatrix}$$

and for the bottom Bragg reflector can be found

$$T^b = T^b_{m+1} T^b_m \cdot \ldots \cdot T^b_1 = \begin{pmatrix} t^b_{11} & t^b_{12} \\ t^b_{21} & t^b_{22} \end{pmatrix},$$

where $l+1$ is the total number of layers above the index-step lattice and $m+1$ is the total number of layers below the index-step lattice. Note, that the boundary between the last top reflector layer and the metal-air lattice are not included into the matrix T^t.

We use 3-D bi-directional beam propagation method [3] for solving the model equations. The usual round-trip operator was built up in order to determine oscillating modes and their losses. Starting with the outgoing wave U_0^+ on the index-step lattice (sign '+' denotes the upward direction of propagation and sign '-' denotes the downward direction) we calculate $A_0 = \mathcal{F}(U_0^+)$. Reflected from the top Bragg reflector wave B_0 is calculated from the system

$$\begin{gathered} t^t_{11} A_0 + t^t_{12} B_0 = A_l, \qquad t^t_{21} A_0 + t^t_{22} B_0 = B_l, \\ U^+_{l+1} = \mathcal{F}^{-1}\{A_l \exp(i q_l z_{l+1})\}, \qquad U^-_{l+1} = \mathcal{F}^{-1}\{B_l \exp(-i q_l z_{l+1})\}, \\ U^-_{l+1} = \Phi(x,y) U^+_{l+1}, \text{ where } \Phi(x,y) = \left(\frac{n_0 - 1}{n_0 + 1} r(x,y) + \frac{n_0 - n_{Ti}}{n_0 + n_{Ti}} (1 - r(x,y)) \right) \end{gathered} \tag{1}$$

Here $n_0 = 3.52$ is the index of the layer adjoining the metal contact, $n_{Ti} = 3.3 - i3.26$ is the complex index of metal (Ti). The form factor $r(x,y) = 1$ at the semiconductor-air contact and is equal to 0 at the semiconductor-Ti contact. The last string of the equations (1) represents the Fresnel reflection formulas. In order to avoid a solution of a system of high dimensionality the following iterative procedure

$$U^+_{l+1} \Longrightarrow U^-_{l+1} = \Phi(x,y) U^+_{l+1} \Longrightarrow B_l = \exp(i q_l z_{l+1}) \mathcal{F}\{U^-_{l+1}\} \Longrightarrow$$

$$B_0 = -\frac{t^t_{21}}{t^t_{22}} A_0 + \frac{1}{t^t_{22}} B_l \Longrightarrow A_l = t^t_{11} A_0 + t^t_{12} B_0 \Longrightarrow U^+_{l+1} = \mathcal{F}^{-1}\{A_l \exp(i q_l z_{l+1})\}$$

was realized to solve the system (1). Then we calculate $U_0^- = \mathcal{F}^{-1}(B_0)$ and the field transmitting through the phase screen $V_0^- = U_0^- \exp(-d(x,y) + i\varphi(x,y))$, which simulates the index-step lattice. Here $\varphi(x,y)$ is a phase step and $d(x,y)$ is an attenuation. Both of these functions are structured in step-wise manner with the amplitude φ_0 for the phase and the amplitude d_0 for the attenuation. The absorbing boundary condition [2] is realized by addition the function

$$b(x,y) = b_0 \left(\frac{1}{\cosh^2\left(\frac{x+\frac{L}{2}}{h}\right)} + \frac{1}{\cosh^2\left(\frac{\frac{L}{2}-x}{h}\right)} + \frac{1}{\cosh^2\left(\frac{y+\frac{L}{2}}{h}\right)} + \frac{1}{\cosh^2\left(\frac{\frac{L}{2}-y}{h}\right)} \right) \tag{2}$$

to the primary attenuation $d(x,y)$. Reflection from the bottom Bragg reflector is computed by the following procedure:

$$C_0 = \mathcal{F}(V_0^-), \quad D_0 = -t^b_{21} C_0 / t^b_{22}, \quad V_0^+ = \mathcal{F}^{-1}(D_0)$$

in an assumption that the upward propagating wave in the substrate is absent. To complete the round-trip operator $\mathbf{P}$ it is necessary to calculate transmission through the phase screen $\mathbf{P}(U_0^+) = V_0^+ \exp(-d(x,y) + i\varphi(x,y))$. The field distribution $U_0^+(x,y)$ of the oscillating mode on the index-step lattice can be found as the solution of the eigen-problem

$$\mathbf{P}u = \gamma u \tag{3}$$

The eigenvalue γ determines the losses $\delta = 1 - |\gamma^2|$ of the mode.

The operator $\mathbf{P}$ is non-hermitian what is an inherent property of open resonators. On the other hand, the operator $\mathbf{P}$ acts in the complex linear space having high dimension $\mathcal{N} = N^2$ what is a difficult problem. Note, that we do not use matrix presentation of the round-trip operator, we have only the 'black box' allowing us to calculate the action of the operator on an arbitrary vector. These circumstances force us to use Krylov's subspace methods for solving the eigen-problem. Luckily, only a few modes having highest possible values of $|\gamma|$ are of interest for all practical purposes what allows us to reduce our task to a partial eigen-problem. We are based on the Arnoldi algorithm [4] because of its simplicity and robustness property.

In order to start Arnoldi iterations we have to determine an initial function $v(x,y)$. The appropriate way is to specify $v = \mathbf{P}^K\psi$ where $\psi = \psi(x,y)$ is an arbitrary function. All solutions of the eigen-problem (3) of interest to us are sought within the Krylov subspace

$$\mathcal{K}_M(\mathbf{P}, v) = \operatorname{span}\{v, \mathbf{P}v, \mathbf{P}^2v, \ldots, \mathbf{P}^{M-1}v\}$$

The number K of preliminary iterations and the Krylov subspace dimension M are chosen empirically. Note, that M is the number of modes to be calculated.

It is necessary to perform M steps of the Arnoldi algorithm to calculate the orthogonal basic $\{q_1, q_2, \ldots, q_M\}$ of the Krylov subspace $\mathcal{K}_M(\mathbf{P}, v)$ and the upper Hessenberg $M \times M$ matrix $H_M = Q_M^* \mathbf{P} Q_M$ where $Q_M = [q_1, q_2, \ldots, q_M]$ is the $\mathcal{N} \times M$ matrix formed by the basic vectors $q_1, q_2, \ldots, q_M$. The eigenvalues $\gamma_1, \gamma_2, \ldots, \gamma_M$ and eigenvectors $h_1, h_2, \ldots, h_M$ of the matrix H_M can be calculated using any suitable standard program. The first M eigenvalues of the round-trip operator $\mathbf{P}$ are approximated by the numbers $\gamma_1, \gamma_2, \ldots, \gamma_M$ and the corresponding eigen-functions are approximated by the expressions

$$u_1 \cong Q_M h_1,\ u_2 \cong Q_M h_2,\ \ldots,\ u_M \cong Q_M h_M.$$

All of the eigen-functions are normalized so as $\max_{x,y} |u| = 1$.

3 Error Control and Estimation

There is no exact solution of the above-stated problem in order to test the numerical scheme. The accessible error criterion for an algebraic eigen-problem solution is small value of the residuals

$$\Delta_{max} = \max_{x,y} |\mathbf{P}u - \gamma u|, \qquad \Delta_{rms} = \sqrt{\frac{1}{L^2} \int_{-L/2}^{L/2} \int_{-L/2}^{L/2} |\mathbf{P}u - \gamma u|^2 dxdy}$$

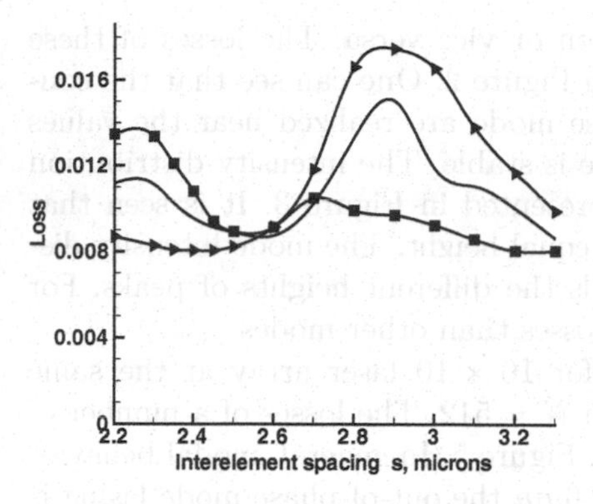

Fig. 2. Losses for the in-phase mode (curve with square markers), out-of-phase mode (triangle markers) and mode of mixed symmetry for the 4x4 array.

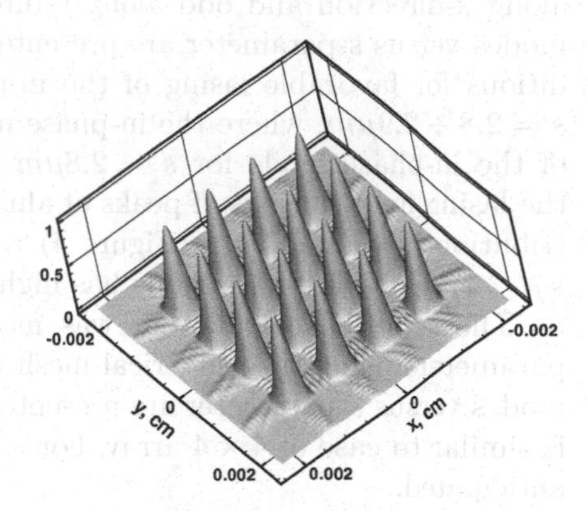

Fig. 3. Intensity distribution of the in-phase mode for s=2.8 μm.

where $\{\gamma, u\}$ is the calculated eigen-pair. The calculated mode was considered as correct if $\Delta_{max} < 0.04$ and $\Delta_{rms} < 0.005$ and was rejected otherwise. Good results was obtained for the typical values $K = 500$ and $M = 32$.

The error of a discrete approximation was estimated on physical reasons. The typical module of the eigenvalue for the lowest loss mode $|\gamma| \approx 0.995$ which corresponds the losses $\delta = 1 - |\gamma^2| \approx 0.01$. It means that a value of δ may be very sensitive to simulation error. The numerical scheme parameters described above (L, N, b_0, h) were selected on the basis of a reasonable compromise. Further variation with the purpose to decrease an error changes the value δ not more than by several percent relative to the previous one.

4 Simulation Results and Discussion

Calculations were made for 4×4 laser array with fixed size $6\,\mu m$ of the square element of the index-step lattice. The size of window in the metal contact plate was $6\,\mu m$ as well by our assumption. The size s of the inter-element spacing was a variable parameter. Numerical mesh consisted of $N \times N$ cells with $N = 256$. Wavelength was taken $\lambda_0 = 0.98\,\mu m$, integral gain of the active layer was $g_0 = 0.005$, the attenuation and phase step of the phase screen were $d_0 = 0.005$ and $\varphi_0 = 0.49$, respectively. The refraction indexes of all layers are listed in Figure 1. The parameters of absorbing boundaries (2) were selected as follows: $b_0 = 10$ and $h = 2\,\mu m$. The size $L = 57\,\mu m$ of the numerical mesh domain was defined from condition that the maximum value of edge absorption within elements nearest to the boundary is smaller by order of magnitude than the values of d_0 and g_0.

It was found that there exists a group of 3 modes with minimal losses: in-phase mode, out-of-phase mode and modes with mixed symmetry, which is even along x-direction and odd along y-direction or vice versa. The losses of these modes versus s-parameter are presented in Figure 2. One can see that the conditions for favorable lasing of the in-phase mode are realized near the values $s = 2.8 \div 2.9\mu m$, where the in-phase mode is stable. The intensity distribution of the in-phase mode for $s = 2.8\mu m$ is presented in Figure 3. It is seen that the lasing field consists of peaks of almost equal height. The mode intensity distribution for $s = 3.3\mu m$ (Figure 4) reveals the different heights of peaks. For $s < 2.7\mu m$ the in-phase mode has higher losses than other modes.

The series of calculations was made for 10×10 laser array at the same parameters using the numerical mesh with $N = 512$. The losses of a number of modes versus s-parameter are presented in Figure 5. In general, modal behavior is similar to case of 4×4 array. For $s < 2.6\mu m$ the out-of-phase mode lasing is anticipated.

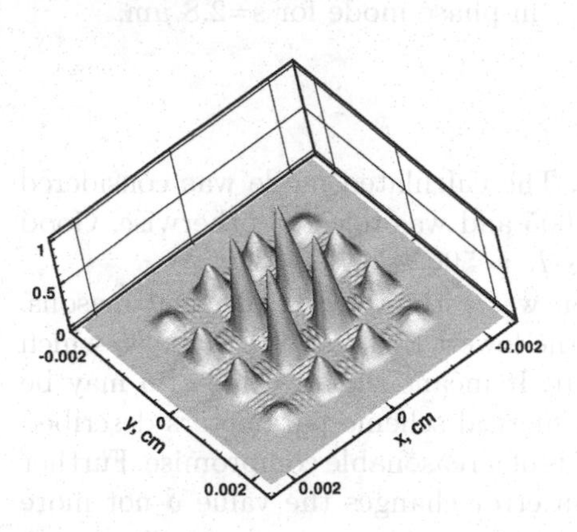

Fig. 4. Intensity distribution of the in-phase mode for s=3.3 μm.

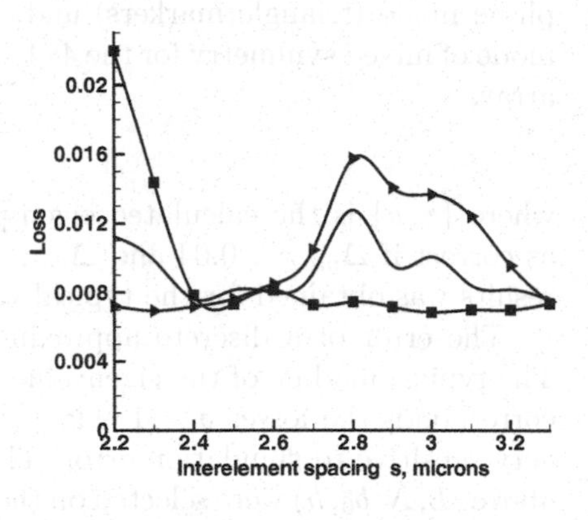

Fig. 5. Losses for the in-phase mode (curve with square markers), out-of-phase mode (triangle markers) and mode of mixed symmetry for the 10x10 array.

5 Conclusion

In summary, we can say that the proposed numerical scheme is valid for such a complicated device as the VCSEL array and allows us to receive the physically significant results. High processing speed of the numerical algorithm was achieved owing to FFT and T-matrix formalism. The Arnoldi algorithm has shown good efficiency for a partial eigen-problem for a high-dimensional non-hermitian linear operator. The computational time of the typical variant on PC of the computer family Pentium II was several minutes. The master code of developed program may be used for modeling the non-linear problem considering the influence of a light intensity on the gain and index distributions of the active layer. After appropriate modification this algorithm may be applied to modeling the VCSEL array under an external injected signal.

Acknowledgments

The author are grateful to Luke J. Mawst, the Wisconsin-Madison University who has initiated this research and to Anatoly P. Napartovich for helpful discussions. Work is partially supported by the RFBR project No. 02-02-17101.

References

1. Fox, A.G., Li, T.: Effect of gain saturation on the oscillating modes of optical masers. IEEE Journal of Quantum Electronics, QE-2 (1966) 774–783.
2. Kosloff, R. and Kosloff, D.: Absorbing Boundaries for Wave Propagation Problem. Journal of Computational Physics, 63 (1986) 363–376.
3. Scarmozzino, R., Gopinath, A., Pregla, R., and Helfert, S.: Numerical Techniques for Modeling Guided-Wave Photonic Devices. IEEE Journal of Selected Topics in Quantum Electronics, 6 (2000) 150–162.
4. Demmel, J.M.: Applied Numerical Linear Algebra. SIAM, Philadelphia, PA. (1997)

ACO Algorithm for MKP Using Various Heuristic Information

Stefka Fidanova

IRIDIA - Université Libre de Bruxelles, Av. Roosevelt 50 - Bruxelles, Belgium
fidanova@ulb.ac.be

Abstract. The ant colony optimization (ACO) algorithms are being applied successfully to diverse heavily constrained problems: traveling salesman problem, quadratic assignment problem. Early applications of ACO algorithms have been mainly concerned with solving ordering problems. In this paper, the principles of the ACO algorithm are applied to the multiple knapsack problem (MKP). In the first part of the paper we explain the basic principles of ACO algorithm. In the second part of the paper we propose different types of heuristic information and we compare the obtained results.

1 Introduction

There are many NP-hard combinatorial optimization problems for which it is impractical to find an optimal solution. Among them is the MKP. For such problems the only reasonable way is to look for heuristic algorithms that quickly produce good, although not necessarily optimal, solutions. Many researchers have focused their attention on a new class of algorithms called metaheuristic. Metaheuristic are rather general algorithmic frameworks that can be applied to several different optimization problems with few modifications. Examples of metaheuristic are simulated annealing [10], evolutionary computation [6] and tabu search [9]. Metaheuristic are often inspired by natural processes. The above-cited metaheuristic were inspired, respectively, by the physical annealing process, the Darwinian evolutionary process and the clever management of memory structures. One of the most recent nature-inspired metaheuristic is the ant colony optimization [2,3,5].

Ant algorithms were inspired by the observation of real ant colonies. Ants are social insects, that is, insects that live in colonies and whose behavior is directed more to the survival of the colony as a whole then to that of a single individual component of the colony. Social insects have captured the attention of many scientist because of the high structuration level their colonies can achieve, especially when compared to the relative simplicity of the colony's individuals. An important and interesting behavior of ant colonies is their foraging behavior, and in particular, how ants can find the shortest paths between food sources and their nest.

ACO is the recently developed, population-based approach which has been successfully applied to several NP-hard combinatorial optimization problems

I. Dimov et al. (Eds.): NMA 2002, LNCS 2542, pp. 438–444, 2003.

[4,8,13]. One of its main ideas is the indirect communication among the individuals of a colony of agents, called (artificial) ants, based on an analogy with trails of a chemical substance, called pheromone which real ants use for communication. The (artificial) pheromone trails are a kind of distributed numerical information which is modified by the ants to reflect their experience accumulated while solving a particular problem.

Early experiments with the ACO algorithms were connected with ordering problems such as the traveling salesman problem. In this paper we discuss ACO algorithm for multiple knapsack problem. We use a particular implementation of ACO algorithm, known as ant colony system (ACS). We propose different types of heuristic information and we compare obtained results.

The remainder of this paper is structured as follows. In section 2 we describe ACO algorithm. Section 3 of this paper investigates the applicability of the ACO algorithm for solving MKP. In section 4 we propose different types of heuristic information. In section 5 we show experimental results over some test problems. Finally we draw some conclusions.

2 ACO Algorithm

ACO algorithm make use of simple agents called ants which iteratively construct candidate solutions to a combinatorial optimization problem. The ants' solution construction is guided by (artificial) pheromone trail and problem dependent heuristic information. ACO algorithms can be applied to any combinatorial optimization problem by defining solution components which the ants use to iteratively construct candidate solutions and on which they may deposit a pheromone. An individual ant constructs candidate solution by starting with an empty solution and then iteratively adding solution components until a complete candidate solution is generated. We will call each point at which an ant has to decide which solution component to add to its current partial solution a choice point. After the solution construction is completed, the ants give feedback on the solutions they have constructed by depositing pheromone on solution components which they have used in their solution. After that we reinforce the pheromone on the components of the best found solution. Typically, solution components which are part of better solutions or are used by many ants will receive a higher amount of pheromone and hence will more used by the ants in future iterations of the algorithm. To avoid the search getting stuck, typically before the pheromone trails get reinforced all pheromone trails are decreased.

More precisely, partial problem solutions are seen as states: each ant moves from a state i to another state j corresponding to a more complete partial solution. At each step, each ant k computes a set of feasible expansion to its current state and moves to one of these according to a probability distribution specified as follows.

For ant k, the probability p_{ij}^k of moving from state i to a state j depends on the combination of two values:

1. The attractiveness η_{ij} of the move, as computed by some heuristic indicating the a priority desirability of that move;
2. The pheromone trail level τ_{ij} of the move, indicating how profitable it has been in the past to make that particular move: it represents therefore an a posterior indication of the desirability of that move.

3 ACO Algorithm for MKP

In the beginning ACO algorithms is made for ordering problems. They have initially been tested on the traveling salesman problem [5,3]. MKP is quite different from ordering problems. We can formulate MKP as:

$$\max \sum_{j=1}^{n} p_j x_j$$

$$\text{subject to} \quad \sum_{j=1}^{n} r_{ij} x_j \le c_i \quad i = 1, \dots, m \tag{1}$$

$$x_j \in \{0, 1\} \quad j = 1, \dots, n.$$

There are m constraints in this problem, so MKP is also called m-dimensional knapsack problem. Let $I = \{1, \dots, m\}$ and $J = \{1, \dots, n\}$, with $c_i \ge 0$ for all $i \in I$. A well-stated MKP assumes that $p_j > 0$ and $r_{ij} \le c_i \le \sum_{j=1}^{n} r_{ij}$ for all $i \in I$ and $j \in J$. Note that the $[r_{ij}]_{m \times n}$ matrix and $[c_i]_m$ vector are both non-negative.

MKP can be thought as a resource allocation problem, where we have m resources (the knapsacks) and n objects. Each resource has its own budget (knapsack capacity) and r_{ij} represents the consumption of resource j by object i. We are interested in maximizing the profit, while working with a limited budget.

MKP has received wide attention from the operation research community, because it embraces many practical problems. Applications include resource allocation in distributed systems, capital budgeting and cutting stock problems. In addition, MKP can be seen as a general model for any kind of binary problems with positive coefficients [11].

In MKP we are not interested in solutions giving a particular order. Therefore a partial solution is represented by $\tilde{S} = \{i_1, i_2, \dots, i_j\}$ and the most recent elements incorporated to $\tilde{S}$, i_j need not be involved in the process for selecting the next element. Moreover, solutions for ordering problems have a fixed length as we search for a permutation of a known number of elements. Solutions for MKP, however, do not have a fixed length.

We define the graph of the problem as follows: The nodes correspond to the items. The arcs fully connect nodes. The pheromone trail is laid on the visited arcs. For a partial solution $\tilde{S}_k = \{i_1, i_2, \dots, i_j\}$ being built by ant k, the probability $p_{i_p}^k(t)$ of selecting i_p as the next item is given as:

$$p^k_{i_p}(t) = \begin{cases} \frac{\tau_{i_j i_p}\eta_{i_p}(\tilde{S}_k(t))}{\sum_{i_q \in allowed_k(t)} \tau_{i_j i_q}\eta_{i_q}(\tilde{S}_k(t))} & \text{if } i_p \in allowed_k(t) \\ 0 & \text{otherwise} \end{cases} , \quad (2)$$

where $\tau_{i_j i_p}$ is a pheromone level on the arc (i_j, i_p), $\eta_{i_p}(\tilde{S}_k(t))$ is the heuristic and $allowed_k(t)$ is the set of remaining feasible items. Thus the higher the value of $\tau_{i_j i_p}$ and $\eta_{i_p}(\tilde{S}_k(t))$, the more profitable it is to include item i_p in the partial solution.

While building a solution, the ants visit the arcs and change their pheromone level by applying the local updating rule:

$$\tau_{i_j i_p} \leftarrow (1-\rho)\tau_{i_j i_p} + \rho\tau_0 \ , \quad (3)$$

where $0 < \rho < 1$ and τ_0 are constants.

After all ants have completed their tours, global updating is performed. The pheromone level is updated by applying the global update rule:

$$\tau_{i_j i_p} \leftarrow (1-\alpha)\tau_{i_j i_p} + \alpha\Delta\tau_{i_j i_p} \ , \quad (4)$$

where $\Delta\tau_{i_j i_p} = \begin{cases} L_{gb} & \text{if } (i_j, i_p) \in \text{global best tour} \\ 0 & \text{otherwise} \end{cases}$,

$0 < \alpha < 1$, L_{gb} is the best value of objective function from the beginning of the trial.

4 Different Types of Heuristic Information

When we calculate the selection probability we can use different types of heuristic information. We can use different problems' parameters for heuristic.

4.1 Heuristic Using Only the Profit

The first type of heuristic use only the profit of the objects. In this case $\eta_j = p_j^{d_1}$. Thus the objects with greater profit will be more desirable. We will call this heuristic *heuristic1*. We use the constraints only to verify if the solution is feasible.

4.2 Heuristic Using Constraints

The second type of heuristic use not only the profit of the objects. It use also the constraints. We propose two types of heuristics using constraints. We will call them *heuristic2* and *heuristic3* respectively.

– Heuristic2. Let $s_j = \max_i(r_{ij})$. For heuristic we use:

$$\eta_j = \begin{cases} p_j^{d_1}/s_j^{d_2} & \text{if } \max_i(r_{ij}) \neq 0 \\ p_j^{d_1} & \text{if } \max_i(r_{ij}) = 0 \end{cases}. \tag{5}$$

Hence the objects with greater profit and less maximal expenses will be more desirable.

– Heuristic3. Let $s_j = \sum_{i=1}^{m} r_{ij}$. For heuristic information we use:

$$\eta_j = \begin{cases} p_j^{d_1}/s_j^{d_2} & \text{if } s_j \neq 0 \\ p_j^{d_1} & \text{if } s_j = 0 \end{cases}. \tag{6}$$

Hence the objects with greater profit and less average expenses will be more desirable.

5 Experimental Results

We tested our algorithms on the set of 30 large MKP from "OR-Library" available within WWW access at *http://mscmga.ms.ic.ac.uk/jeb/orlib*. When we made the experiments we found the best results when $\rho = \alpha \in [0.01, 0.9]$ and $d_1, d_2 \in [1, 9]$. For other parameters we use $\tau_0 = 1$ and n ants. We stop the algorithms after 200 iterations.

We can conclude that algorithms with heuristic using constraints achieve better result than algorithm with heuristic using only profit of the objects. For all tested problems we achieve better results using heuristic with average expenses. Table 1 makes a comparison between ACO algorithms using different heuristics. It shows number of objects, number of constraints, the best value achieved by the ACO algorithms. With bold are better results.

5.1 Conclusion

In this paper we presented a version of ACO algorithms extended to handle knapsack problems. We have presented different types of heuristic informations. When we use profit of the objects and constraints we obtain better results than using only the profit of the objects. The results indicate the potential of the ACO approach for solving multiple knapsack problem. We can use similar technique for another constraint problems too. Our empirical results show that our ACO algorithm is currently among the best performing algorithms for this problem [1,12].

Table 1. Experimental results using different heuristics, n is the number of objects and m is the number of constraints

n×m	heuristic1	heuristic2	heuristic3
100×5	21640	22224	**23984**
100×5	21175	22542	**24143**
100×5	20477	22177	**23515**
100×5	21443	22390	**22874**
100×5	22169	23246	**23263**
100×5	20675	22427	**24523**
100×5	22068	24286	**25177**
100×5	21432	21771	**23204**
100×5	20501	22832	**23762**
100×5	22006	22893	**24208**
100×5	39510	41700	**42705**
100×5	40700	41804	**42445**
100×5	40329	40922	**41435**
100×5	41390	43856	**44911**
100×5	39253	40979	**42025**
100×5	40268	42130	**42671**
100×5	38138	40307	**41776**
100×5	41406	43304	**44559**
100×5	40029	42015	**43122**
100×5	41968	43135	**44364**
100×5	41968	43135	**44364**
100×5	58178	58660	**59798**
100×5	60031	60913	**61821**
100×5	58109	59117	**59694**
100×5	58592	59633	**60479**
100×5	59144	60050	**60954**
100×5	57218	58127	**58695**
100×5	59464	60193	**61406**
100×5	60180	60360	**61520**
100×5	57836	58659	**59121**
100×5	58186	59060	**59864**
Average	40117.166	41390.4	**42400.666**

Acknowledgements. Stefka Fidanova was supported by a Marie Curie Fellowship of the European Community program "Improving Human Research Potential and the Socio-Economic Knowledge Base" under contract number No HPMFCT-2000-00496. This work was supported by the "Metaheuristics Network", a Research Training Network funded by the Improving Human Potential programme of the CEC, grant HPRN-CT-1999-00106. The information provided in this paper is the sole responsibility of the authors and does not reflect the Community's opinion. The Community is not responsible for any use that might be made of data appearing in this publication.

References

1. Chu, P.C., Beasley, J.E.: A genetic algorithm for the multiple knapsack problem. Journal of Heuristics, 4 (1998) 63–86.
2. Dorigo, M., Di Caro, G.: The ant colony optimization metaheuristic. In: Corne, D., Dorigo, M., Glover, F. (eds.): New Idea in Optimization, McGrow-Hill (1999) 11–32.
3. Dorigo, M., Gambardella, L.M.: Ant colony system: A cooperative learning approach to the traveling salesman problem. IEEE Transactions on Evolutionary Computation, 1 (1999) 53–66.
4. Dorigo, M., Di Caro, G.,Gambardella, L.M.: Ant algorithms for distributed discret optimization. Artificial Life, 5 (1999) 137–172.
5. Dorigo, M., Maniezzo, V., Colorni, A.: The ant system: Optimization by a colony of cooperating agents. IEEE Transaction on Systems, Man. and Cibernetics - Part B, 26 (1996) 29–41
6. Fogel, B.: Evolutionary computation: Toward a new philosophy of machine intelligence. IEEE Press, New York (1994)
7. Gambardella, M.L., Taillard, E.D., Agazzi, G.: A multiple ant colony system for vehicle routing problems with time windows. In: Corne, D., Dorigo, M., Glover, F. (eds.): New Ideas in Optimization, McGraw-Hill (1999) 63–76.
8. Gambardella, L.M., Taillard, E.D., Dorigo, M.: Ant colonies for the QAP. J. of Oper. Res. Soc. 50 (1999) 167–176.
9. Glover, F.: Tabu search. ORSA J. of Comput. 1 (1989)
10. Kirkpatrick, S., Gelatt, C.D., Vechi, M.P.: Optimization by simulated annealing. Scienece, 220 (1983) 671–680.
11. Kochenberger, G., McCarl, G., Wymann, F.: A heuristic for general integer programming. Decision Sciences, 5 (1974) 36–44.
12. Leguizamon, G., Michalevich, Z.: A new version of the ant system for subset problems. In: Proceedings of Int. Conf. on Evolutionary Computations, Washington (1999)
13. Michel, R., Middendorf, M.: An island based ant system with lookahead for the shortest common super-sequence problem. In: Eiben, A.E., Bäck, T., Schoenauer, M., Schwefel, H-P. (eds.): Proceedings of the Fifth International Conference on Parallel Problem Solving from Nature, Lecture Notes in Computer Science, Vol. 1498 (1998) 692–708.

Discretization Methods with Discrete Minimum and Maximum Property for Convection Dominated Transport in Porous Media*

Peter Frolkovič and Jürgen Geiser

Institut of Computer Science,
Im Neuenheimer Feld 368, D-69120 Heidelberg, Germany
peter.frolkovic@iwr.uni-heidelberg.de,
juergen.geiser@iwr.uni-heidelberg.de

Abstract. Second order explicit discretization methods for convection dominated transport are studied here from the point of view of discrete minimum and maximum property of numerical solutions. These methods are based on vertex-centered finite volume methods on general unstructured computational grids. It will be shown that "standard TVD methods" [13,14] do not fulfill in general the discrete minimum and maximum property and that these methods must be modified to obtain numerical solutions with no unphysical oscillations. Finally, new methods based on our theoretical results are proposed.

1 Introduction and Mathematical Model

The main motivation for the study presented in this paper is coming from computational simulations of radioactive contaminant transport in flowing groundwater [4,6] that are based on the following mathematical equation

$$\partial_t R_\alpha(c_\alpha) + \nabla \cdot (\boldsymbol{v} c_\alpha - D \nabla c_\alpha) + \lambda_{\alpha\beta} c_\alpha = \sum_\beta \lambda_{\beta\alpha} c_\beta \,. \tag{1}$$

The unknown concentrations $c_\alpha = c_\alpha(t,x)$ are considered in $(0,T) \times \Omega \subset I\!R \times I\!R^n$ and the parameters include the retardation functions (continuous and increasing) $R_\alpha = R_\alpha(c) \geq \phi\, c$, where ϕ is the porosity of the medium, $\boldsymbol{v}$ is divergence free groundwater velocity, D is the dispersion-diffusion matrix and $\lambda_{\alpha\beta}$ are reaction constants.

The aim of this paper is to propose second order explicit upwind scheme for convective transport on vertex-centered dual mesh of finite volumes. Such method should be well defined on general unstructured computational grids, it should be based on a perfect local and global mass conservation and it should produce no unphysical oscillation for numerical solutions. Such methods of the

* This work is funded by the Federal Ministry of Economics and Technology (BMWi) under the contract number 02 E 9148 2

I. Dimov et al. (Eds.): NMA 2002, LNCS 2542, pp. 445–453, 2003.

first order with no restriction on the choice of time step were described in [7]. The numerical treatment of complex decay chains in (1) was described in [6,8].

Second order explicit upwind methods for FVM on dual meshes were considered in details in [13,14,11,12]. Particularly, the simplest "TVD method" based on piecewise defined discontinuous interpolation of numerical solution, seems to offer a best compromise between the complexity of scheme and the exactness of numerical solution.

The most important part of TVD methods is the construction of "limiters" that ensure physically acceptable numerical solutions. Unfortunately, as it will be shown later, the limiters described in [13,14,11,12] do not exhibit numerical solutions with the discrete minimum and maximum property in general.

In this paper we present theoretical results concerning the discrete minimum and maximum property of second order TVD method for vertex-centered FVM for the simplified transport equation

$$\partial_t R(c) + \nabla \cdot (\mathbf{v} c) = 0 \tag{2}$$

with details for the simplest case $R(c) \equiv \phi c$. To simplify further our presentation we consider only trivial inflow and outflow boundary conditions where $c = 0$ is considered at inflow.

Based on these results, we propose new discretization scheme that fulfills the discrete minimum and maximum property with standard CFL restriction on the choice of time step. Finally, some numerical results will be presented that confirm our theoretical results.

2 Finite Volume Discretization

The discretization methods described in this paper are related to standard (explicit in time) finite volume methods (FVM). The FVM are based on a discrete mass balance formulation, determined locally for time intervals $(t^n, t^{n+1}) \subset (0, T)$, $n = 0, 1, \ldots$ and for the computational cells $\Omega_i \subset \Omega$, $i = 1, \ldots, I$.

The finite volume mesh is constructed here as a dual mesh to a triangulation $\mathcal{T}$ [2] of the polygonal domain Ω with finite elements denoted by T^e, $e = 1, \ldots, E$. The polygonal computational cells Ω_i of the dual mesh are related to vertices x_i of the triangulation.

We introduce here only the most important notations, for details see for instance [5,7]. To do so, let $V_i = |\Omega_i|$ and let Λ_i denote the set of indices of all neighbor points x_j to x_i. The line segments Γ_{ij}, $i \neq j$ are defined by $\bar{\Omega}_i \cap \bar{\Omega}_j$.

The aim of the FVM is to construct system of algebraic equations for unknowns $c_i^n \approx c(x_i, t^n)$. The values c_i^0 are given by initial conditions. At any time point t^n, one can consider two "natural" interpolation schemes of discrete values c_i^n. First, for the primary mesh of finite elements,

$$c^n = \sum_{i=1}^{I} c_i^n \, \phi_i(x) \tag{3}$$

where ϕ_i , $i = 1, \ldots, I$ are standard globally continuous finite element basis functions [2]. Secondly, for the dual mesh of finite volumes,

$$\hat{c}^n = \sum_{i=1}^{I} c_i^n \ \varphi_i(x) \tag{4}$$

where φ_i are the piecewise constant discontinuous functions defined by $\varphi_i(x) = 1$ for $x \in \Omega_i$ and $\varphi_i(x) = 0$ otherwise.

3 Finite Volume Discretization of the Flow Equation

By integrating the flow equation $\nabla \cdot \boldsymbol{v} = 0$ over Ω_i, one obtains

$$\int_{\partial\Omega_i} \boldsymbol{n} \cdot \boldsymbol{v} \ d\gamma = 0 \ , \tag{5}$$

where $\boldsymbol{n}$ is the normal unit vector with respect to $\partial\Omega_i$. The integral equation (5) describes an analytical form of the local mass conservation for the flow. To introduce the FVM, we denote the "discrete values"

$$v_{ij} \ := \int_{\Gamma_{ij}} \boldsymbol{n} \cdot \boldsymbol{v} \ d\gamma \ \text{ with } \ v_{ji} = -v_{ij} \tag{6}$$

and the equation (5) can be transformed to the discrete form

$$v_i := \sum_{j \, \in \, out(i)} v_{ij} = \sum_{k \, \in \, in(i)} v_{ki} \ , \tag{7}$$

where $out(i) := \{j \in \Lambda_i, \ v_{ij} > 0\}$ and $in(i) := \{j \in \Lambda_i, \ v_{ij} \leq 0\}$. The values v_{ij} can be computed approximately, for instance by choosing the middle point $x_{ij} \in \Gamma_{ij}$ and $v_{ij} :\approx |\Gamma_{ij}| \ (\boldsymbol{n} \cdot \boldsymbol{v})(x_{ij})$.

4 The Discrete Minimum and Maximum Property

All numerical schemes described later, take formally the form

$$\gamma_{ii} \ R(c_i^{n+1}) \ = \ \sum_j \gamma_{ij} \ R(u_j) \ , \ i = 1, \ldots, I \ , \tag{8}$$

where j runs through indices of some finite set. If the following properties are fulfilled

$$\gamma_{ii} \ > \ 0 \, , \ \ \gamma_{ij} \ > \ 0 \, , \ \ \gamma_{ii} \ \geq \ \sum_j \gamma_{ij} \ , \tag{9}$$

then (local) discrete minimum and maximum property [9,5] is valid:

$$\min_j \ \{u_j\} \ \leq \ c_i^{n+1} \ \leq \ \max_j \ \{u_j\}. \tag{10}$$

4.1 Piecewise Constant Interpolation Case

If one considers a piecewise constant form (4) of numerical solution, the FV discretization of (2) takes the form

$$V_i R(c_i^{n+1}) = V_i R(c_i^n) - \tau^n c_i^n \sum_{j \in out(i)} v_{ij} + \tau^n \sum_{k \in in(i)} c_k^n v_{ki} \,, \tag{11}$$

where $\tau^n := t^{n+1} - t^n$. Using (7) one can rewrite (11) to the following form

$$V_i \, R(c_i^{n+1}) = \tag{12}$$

$$R(c_i^n) \, (V^i - \tau^n \sum_{k \in in(i)} r(c_i^n, c_k^n) \, v_{ki}) + \tau^n \sum_{k \in in(i)} R(c_k^n) \, r(c_i^n, c_k^n) \, v_{ki},$$

where

$$r(u,v) = \begin{cases} \frac{v-u}{R(v)-R(u)} & \text{if } u \neq v \,; \\ \frac{1}{R'(u)} & \text{if } u = v \,. \end{cases} \tag{13}$$

As $R(c) \geq \phi c$, it is clear that $r > 0$.

Next, we introduce the important definition of a "critical time step" τ_i^n

$$\tau_i^n := \frac{V_i}{\sum_{k \in in(i)} r(c_i^n, c_k^n) \, v_{ki}} \,. \tag{14}$$

It is clear that if the time step τ^n (chosen in the discretization) fulfills

$$\tau^n \leq \min_{i=1,\ldots,I} \{\tau_i^n\} \tag{15}$$

then (9) is fulfilled and the local discrete minimum and maximum property (10) for c_i^{n+1} is valid. Particularly, it takes the form

$$\min\{c_i^n, \; c_j^n, \; j \in in(i)\} \leq c_i^{n+1} \leq \max\{c_i^n, \; c_j^n, \; j \in in(i)\} \,. \tag{16}$$

The restriction (15) must be seen as the well-known CFL condition [3]. In the linear case it can be written in the form

$$\frac{\tau^n \, v_i}{\phi_i \, V_i} \leq 1 \,, \; i = 1, \ldots, I \,, \tag{17}$$

where the left hand sides of (17) can be viewed as "local grid Courant numbers" and $\phi_i := V_i^{-1} \int_{\Omega_i} \phi \, dx$.

4.2 Piecewise Linear Interpolation Case

Next we aim to develop a numerical scheme where the linear interpolation scheme (3) is used for numerical solution.

To do so, one has first to determine a piecewise linear (discontinuous) function $u^n = u^n(x)$ by using (3),

$$u^n(x_i) = c_i^n \ , \ i = 1, \ldots, I \ ,$$

$$\nabla u^n|_{V_i} = \frac{1}{V_i} \sum_{e=1}^{E} \int_{T^e \cap \Omega_i} \nabla c^n \, dx \ , \ i = 1, \ldots, I \ . \tag{18}$$

To simplify our next considerations, we describe first the linear case $R(c) \equiv \phi c$. One can derive, analogously to (11), the following discretization scheme

$$\phi_i \, V_i \, c_i^{n+1} \quad = \tag{19}$$

$$\phi_i \, V_i \, c_i^n \ - \tau^n \sum_{j \, \in out(i)} u^n|_{V_i}(x_{ij}) \, v_{ij} \ + \tau^n \sum_{k \, \in \, in(i)} u^n|_{V_k}(x_{ki}) \, v_{ki} \ .$$

It is well-known that such discretization scheme can produce oscillatory numerical solutions. The main reason for such behavior is that the "reconstructed" values $u^n(x_{ij})$ need not to respect the local extrema of origin variables $\{c_i^n, i = 1, ..., I\}$.

In several papers, see e.g. [13,14], the so called "limiters" were proposed in the framework of the so called "TVD-methods". Considering that

$$u^n|_{V_i}(x_{ij}) = c_i^n \ + \ \nabla u^n|_{V_i} \cdot (x_{ij} - x_i) \ , \tag{20}$$

the idea is to replace $u^n|_{V_i}(x_{ij})$ by

$$u_{ij}^n \ := \ c_i^n \ + \ \psi_i \, \nabla u^n|_{V_i} \cdot (x_{ij} \ - \ x_i) \ , \tag{21}$$

where $\psi_i \in < 0, 1 >$ are scalars to be determined for $i = 1, ..., I$. In [13] ψ_i were determined by requiring

$$\min\{c_i^n, \ c_k^n \ , \ k \in \Lambda_i\} \leq u_{ij}^n \leq \max\{c_i^n, \ c_k^n \ , k \in \Lambda_i\} \ . \tag{22}$$

As it will be shown later, this condition can be weakened. In fact, the following restriction for u_{ij}^n, $j \in out(i)$ will be required later,

$$\min\{c_j^n, \ \ c_k^n \ , \ k \in in(j)\} \leq u_{ij}^n \leq \max\{c_j^n, \ \ c_k^n \ , \ k \in in(j)\} \ . \tag{23}$$

Now, one can replace $u^n(x_{ij})$ in (19) by u_{ij}^n and rewrite it to the following form,

$$\phi_i \, V_i \, c_i^{n+1} \ = \ \phi_i \, V_i \, c_i^n \ - \tau^n \sum_{j \in out(i)} u_{ij}^n \, v_{ij} + \tau^n \sum_{k \in in(i)} u_{ki}^n \, v_{ki} \ . \tag{24}$$

Nevertheless, the conditions (22) or (23) are not sufficient for the local discrete minimum and maximum property of (24). To obtain such conditions, one has to determine additionally the values $\alpha_{ij}^n \in (0,1\rangle$ for each u_{ij}^n, $j \in out(i)$ such that if

$$c_i^n = \alpha_{ij}^n \, u_{ij}^n + (1 - \alpha_{ij}^n) \, u_{ij'}^n \,, \tag{25}$$

then $u_{ij'}^n$ must fulfill

$$\min\{c_i^n , \; c_k^n \, , \, k \in in(i)\} \le u_{ij'}^n \le \max\{c_i^n , \; c_k^n \, , \, k \in in(i)\} \, . \tag{26}$$

In a structured case of computation grid, e.g. in one-dimensional case with uniform discretization step $h = x_{i+1} - x_i$, one obtains $j' \in in(i)$ and $u_{ij'} = u^n|_{V_i}(x_{ij'})$.

The values α_{ij}^n can be computed, for instance if $u_{ij}^n > c_i^n$, by choosing $u_{ij'}^n = \min_{j \in in(i)}\{c_j^n\}$ and

$$\alpha_{ij}^n = \frac{c_i^n - u_{ij'}^n}{u_{ij}^n - u_{ij'}^n} \, . \tag{27}$$

Analogously, if $u_{ij}^n < c_i^n$. In fact, the requirements (26) can be used for the determination of constants ψ_i in (21).

Using (25), the scheme (24) can be written in the form,

$$V_i \, \phi_i \, c_i^{n+1} = c_i^n \left(\phi_i V_i - \tau^n \sum_{j \, \in \, out(i)} \frac{v_{ij}}{\alpha_{ij}^n} \right) \tag{28}$$

$$+ \, \tau^n \sum_{j \, \in \, out(i)} u_{ij'}^n \, v_{ij} \, \frac{1 - \alpha_{ij}^n}{\alpha_{ij}^n} + \tau^n \sum_{k \, \in \, in(i)} u_{ki}^n \, v_{ki}$$

It is clear that the discrete minimum and maximum property is fulfilled for (28), resp. (24), if

$$\frac{\tau^n}{\phi_i \, V_i} \le \left(\sum_{j \in out(i)} \frac{v_{ij}}{\alpha_{ij}^n} \right)^{-1} \le (v_i)^{-1} \min_{j \in out(i)} \alpha_{ij}^n \, . \tag{29}$$

Particularly, this means that the "standard" condition on the local grid Courant numbers (17) is not sufficient for (24) and it must be replaced by more restrictive form (29) .

For a general nonlinear form of $R = R(c)$, the scheme (24) takes the form

$$V_i R(c_i^{n+1}) = R(c_i^n)(V_i + \tau^n q) - \tau^n \sum_{j \in out(i)} R(u_{ij}^n) \, r(c_i^n , u_{ij}^n) \, v_{ij} \tag{30}$$

$$+ \tau^n \sum_{k \in in(i)} R(u_{ki}^n) \, r(c_i^n , u_{ki}^n) \, v_{ki}$$

where

$$q := \sum_{j \in out(i)} r(c_i^n , u_{ij}^n) \, v_{ij} - \sum_{k \in in(i)} r(c_i^n , u_{ki}^n) \, v_{ki} \, . \tag{31}$$

For the linear case $R(c) = \phi c$, one obtains $q \equiv 0$. A restriction on the time step, analogous to (29) can be derived, but it will be not described here.

4.3 Piecewise Linear Interpolation Case and Maximal Time Step

From previous section it is clear that for discretization schemes (23) - (26) the discrete minimum and maximum property (10) can not be obtained for the limit case of maximal local grid Courant number equals 1 (e.g. $\alpha_{ij}^n \equiv 1$). In fact, if larger time step τ^n is chosen (e.g. one obtains larger Courant numbers) then smaller values of ψ_i must be used in (21), if the discrete minimum and maximum shall be fulfilled.

To enable maximal possible time step τ^n for (24) we propose to replace u_{ij}^n in (24) (that is independent of the choice of τ^n) by the time step dependent value,

$$u_{ij}^n(\tau^n) := u^n|_{V_i}(x_{ij}) + \frac{\tau^n}{\tau_i}\left(c_i^n - u^n|_{V_i}(x_{ij})\right). \tag{32}$$

Using (32) the discretization scheme (19) can be replaced by

$$\phi_i \, V_i \, c_i^{n+1} = \phi_i \, V_i \, c_i^n - \tau^n \sum_{j \in out(i)} u_{ij}^n(\tau^n) \, v_{ij} + \tau^n \sum_{k \in in(i)} u_{ki}^n(\tau^n) \, v_{ki} \, . \tag{33}$$

Further, the limiter procedure (21) for $u_{ij}^n(\tau^n)$ to fulfill (23) and (26) must be applied. The advantages is, as $u_{ij}^n(\tau^n) \to c_i^n$ for $\tau^n \to \tau_i$, that ψ_i nearer to 1 can be chosen.

The discretization scheme (33) corresponds better to the character of piecewise linear interpolation form of u^n than (24).

5 Numerical Experiments

To illustrate the theoretical results presented in previous sections, we computed standard test example of rotating Gaussian impulse with no diffusion. The following numerical experiments were done using the software library UG [1] with aligned vertex-centered FVM [10] .

The initial unstructured grid consists of 66 elements, see the Figure 1, the left picture. This grid of the "grid level 0" was further uniformly refined up to the grid level 6 (270 336 elements). The initial function $c^0(x)$ is a circle-shaped Gaussian impulse with the maximum value 1, see the middle picture in the Figure 1, where 9 contour lines from 0.1 to 0.9 are plotted. The space dependent velocity is chosen in such a way that the initial impulse should rotate anti-clockwise and it should return to the origin position at $t = \pi/2$, see [7] for all details.

The numerical solutions for methods described in this paper were first compared for $t = \pi/8$ with respect to minimal and maximal values of numerical solutions and with the "simplified" discrete L_1 norm $E_{L_1}^l := \sum_{i=1,\dots,I} V_i \, |c_i^m - C(x_i, y_i, t^m)|$ where l denotes the grid level, $t^m = \pi/2$ and C is the exact solution. The time step $\tau = \tau_l$ was chosen constant in time, but grid level dependent, with $\tau_2 = \pi/640$ and $\tau_{l+1} = \tau_l/2$.

The first experiment was done using the (standard) TVD method described in [13] with the limiter constructed by requiring the property (22). As it was

Table 1. Comparison between the standard and modified TVD-Method

The method	l	$E^l_{L_1}$	C_{min}	C_{max}
TVD standard	2	$6.884\ 10^{-3}$	$-1.117\ 10^{-4}$	0.374
TVD standard	3	$2.283\ 10^{-3}$	$-1.33\ 10^{-6}$	0.691
TVD standard	4	$1.492\ 10^{-3}$	$-8.97\ 10^{-8}$	0.892
TVD modified	2	$6.229\ 10^{-3}$	$-1.63\ 10^{-24}$	0.406
TVD modified	3	$1.897\ 10^{-3}$	$-2.85\ 10^{-35}$	0.759
TVD modified	4	$1.380\ 10^{-3}$	0.0	0.944
TVD modified	5	$0.767\ 10^{-3}$	0.0	0.99

shown also by theoretical results, this method produce numerical solutions that can exhibit unphysical oscillations, especially for coarse (unstructured) grids, see the Table 1.

The next computations were done with the new method determined by (23) and (26), that can be viewed as a modified version of TVD method in [13] to obtain the discrete minimum and maximum property. The values of α^n_{ij} were chosen according to (27). This method clearly improves the result for this example.

Finally, we present the results for the new method (33) together with numerical convergence rates $\gamma = (\log(E^l_{L_1}) - \log(E^{l-1}_{L_1}))/\log(0.5)$. The numerical scheme (33) can be used with the maximal grid Courant number equals 1 and the numerical solution fulfills the discrete minimum and maximum property. Starting with the grid level 4 the norm of the error is the smallest one between all considered algorithms.

In the right picture of the Figure 1 the numerical solutions for the new method (33) for $t = \pi/8$, $t = \pi/4$, $t = 3\pi/8$ and $t = \pi/2$, the grid level 4, are plotted. The norm of the error was $E^4_{L_1} = 2.92\ 10^{-3}$ and the extrema were $C_{min} = -2.45\ 10^{-37}$ and $C_{max} = 0.711$.

Table 2. Convergence results for new method (33)

l	$E^l_{L_1}$	C_{min}	C_{max}	γ
2	$7.176\ 10^{-3}$	$-2.94\ 10^{-25}$	0.366	
3	$2.784\ 10^{-3}$	$-6.55\ 10^{-35}$	0.677	1.366
4	$9.708\ 10^{-4}$	0.0	0.878	1.519
5	$2.883\ 10^{-4}$	0.0	0.96	1.751
6	$1.009\ 10^{-4}$	0.0	0.986	1.513

References

1. Bastian, P., Birken, K., Johannsen, K., Lang, S., Neuss, N., Rentz-Reichert, H.: $\mathcal{UG}$ - a flexible software toolbox for solving partial differential equations. Computing and Visualization in Science **1(1)** (1997) 27–40.
2. Ciarlet, P.G.: The Finite Element Methods for Elliptic Problems. North Holland Publishing Company, Amsterdam (1978)

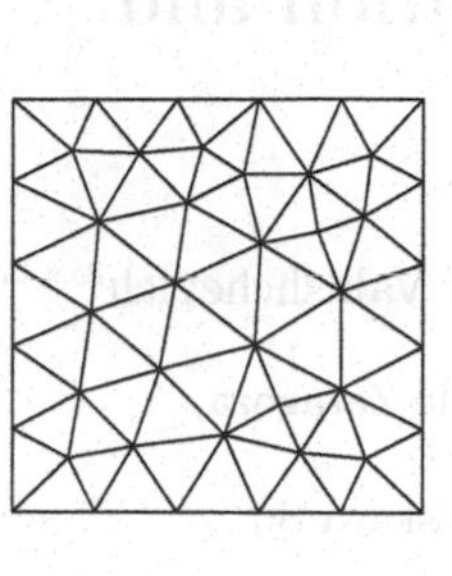
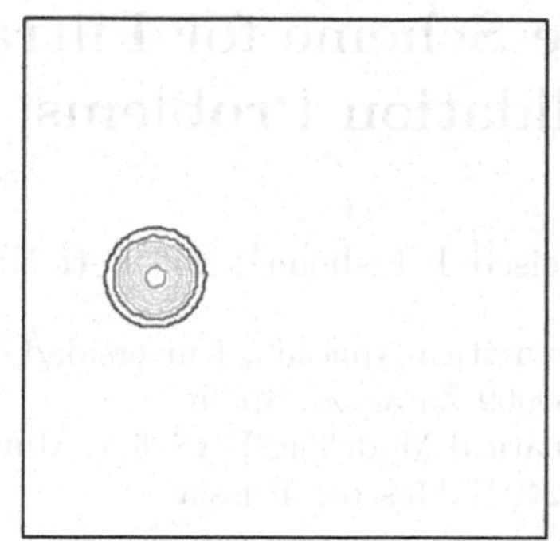
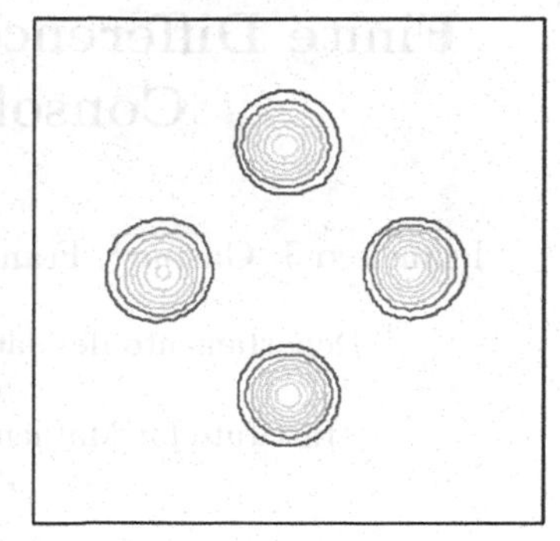

Fig. 1. The initial grid (left), the initial function $c^0(x)$ (middle) and the numerical solution for grid level 4 at $t = \pi/8$, $\pi/4$, $3\pi/4$ and $\pi/2$ (right).

3. Courant, R., Friedrics, K.O., Lewy, H.: Über die partiellen Differenzengleichungen der mathematischen Physik. Math. Annalen **100** (1928) 32–74.
4. Fein, E., Kühle, T., Noseck, U.: Entwicklung eines Programms zur dreidimensionalen Modellierung des Schadstofftransportes. Fachliches Feinkonzept, Braunschweig (2001)
5. Frolkovič, P.: Maximum principle and local mass balance for numerical solutions of transport equation coupled with variable density flow. Acta Mathematica Universitatis Comenianae **1 (67)** (1998) 137–157.
6. Frolkovič, P., Geiser, J.: Numerical Simulation of Radionuclides Transport in Double Porosity Media with Sorption. Conference of Scientific Computing, Proceedings of Algorithmy (2000) 28–36.
7. Frolkovič, P.: Flux-based method of characteristics for contaminant transport in flowing groundwater. Computing and Visualization in Science (2002) (to appear)
8. Geiser, J.: Numerical Simulation of a Model for Transport and Reaction of Radionuclides. In: Margenov, S., Wasniewski, J., and Yalamov, P. (eds.): Large-Scale Scientific Computing. Lecture Notes in Computer Science, Vol. 2179, Springer-Verlag (2001) 487–496.
9. Ikeda, T.: Maximum principle in finite element models for convection-diffusion phenomena. North-Holland-Publishing Company, Amsterdam.
10. Johannsen, K.: An Aligned 3D-Finite-Volume Method for Convection-Diffusion Problems. In: Helmig, R., Jäger, W., Kinzelbach, W., Knabner, P., Wittum, G. (eds.): Modeling and Computation in Environmental Sciences. Vieweg, Braunschweig, 59 (1997) 227–243.
11. Klöfkorn, R.: Simulation von Abbau- und Transportprozessen gelöster Schadstoffe im Grundwasser. Diplomarbeit, Institut für Angewandte Mathematik, Universität Freiburg (2001)
12. Klöfkorn, R., Kröner, D., Ohlberger, M.: Local adaptive methods for convection dominated problems. Internat. J. Numer. Methods Fluids (2002) (to appear)
13. Sonar, Th.: On the design of an upwind scheme for compressible flow on general triangulation. Numerical Analysis, 4 (1993) 135–148.
14. Sonar, Th.: Mehrdimensionale ENO-Verfahren. Advances in numerical mathematics, Teubner Stuttgart (1997)

Finite Difference Scheme for Filtration and Consolidation Problems*

Francisco J. Gaspar[1], Francisco J. Lisbona[1], and Petr N. Vabishchevich[2]

[1] Departamento de Matemática Aplicada, Universidad de Zaragoza, 50009 Zaragoza, Spain
[2] Institute for Mathematical Modelling RAS 4-A, Miusskaya Sq., 124047 Moscow, Russia

Abstract. It's well known that numerical instabilities appear in the approximation of the Biot's consolidation problem, when standard finite elements or difference methods are applied. To stabilizate this problem, we propose the use of staggered grids for the discretization. A monotone and second order finite difference scheme on this kind of grid is given. We check this properties with some numerical results.

1 Introduction

We consider the classical Biot consolidation problem for a saturated, homogeneous, isotropic, porous medium made of an incompressible solid matrix. The classical quasi–static Biot model is obtained in [2] for incompressible fluids, on the assumption that the soil is not completely saturated. The same equations can be used in the case of slightly compressible fluids in a totally saturated porous medium (see Bear [1]), with a different meaning of the γ parameter in the pressure equation. Neglecting body forces, the filtration and consolidation process is governed by the set of equations

$$-\mu\Delta\mathbf{u} - (\lambda+\mu)\operatorname{grad}\operatorname{div}\mathbf{u} + \operatorname{grad} p = 0, \tag{1}$$

$$\frac{\partial}{\partial t}(\gamma p + \operatorname{div}\mathbf{u}) - \frac{\kappa}{\eta}\Delta p = f(\mathbf{x},t), \quad x\in\Omega, \quad 0 < t \le T\,, \tag{2}$$

where λ and μ are the Lamé coefficients; $\gamma = n\beta$ where n denotes the porosity and β the compressibility coefficient of the fluid; κ is the permeability of the porous medium and η the viscosity of the fluid. The quantity $\operatorname{div}\mathbf{u}\,(\mathbf{x},t)$ represents the dilatation and then, $\gamma\, p(x,t) + \operatorname{div}\mathbf{u}\,(\mathbf{x},t)$ gives the variation in water content at time t, i.e. the increment of water volume in the soil per unit volume. A source term $f(\mathbf{x},t)$ can be used to represent a forced extraction or injection process. Typical boundary conditions for (1), (2) are

$$\mathbf{u} = 0, \quad \frac{\kappa}{\eta}\operatorname{grad} p\cdot\mathbf{n} = 0, \quad \text{on} \quad \Gamma_1,$$

$$p = 0, \quad \sigma.\mathbf{n} = \mathbf{h}, \quad \text{on} \quad \Gamma_2,$$

* This research has been partially supported by the Spanish project MCYT-FEDER BFM 2001-2521 and The Russian Foundation for Basic Research RFBR 99–01–00958

I. Dimov et al. (Eds.): NMA 2002, LNCS 2542, pp. 454–462, 2003.

where $\Gamma_1 \bigcup \Gamma_2 = \Gamma = \partial\Omega$, with Γ_1 and Γ_2 disjoint subsets of positive measure in Γ. The function $\mathbf{h}$ represents a prescribed traction on boundary Γ_2 and $\mathbf{n}$ the unit outward normal vector. The effective stress for the porous medium is defined by $\sigma = \lambda \operatorname{div} \mathbf{u} \mathbf{I} + 2\mu\epsilon(\mathbf{u})$, where $\epsilon(\mathbf{u}) = 1/2(\operatorname{grad} \mathbf{u} + \operatorname{grad} \mathbf{u}^t)$. The initial condition is taken to be

$$\gamma\, p(\mathbf{x}, 0) + \operatorname{div} \mathbf{u}\,(\mathbf{x}, 0) = 0, \quad \mathbf{x} \in \Omega,$$

because if a load is applied on Γ_2 at $t = 0$, the water moves with a finite velocity and the instantaneous change in water content is zero at this time. The initial state of the system, $u_0(x) = u(x, 0)$ and $p_0(x) = p(x, 0)$, is then described by the problem

$$\begin{aligned}
&-\mu\Delta\mathbf{u}_0 - (\lambda + \mu)\operatorname{grad}\operatorname{div}\mathbf{u}_0 + \operatorname{grad} p_0 = 0, \\
&(\gamma\, p_0 + \operatorname{div}\mathbf{u}_0) = 0, \\
&\mathbf{u}_0 = 0 \quad \text{on} \quad \Gamma_1, \quad \sigma_0\,\mathbf{n} = \mathbf{h} \quad \text{on} \quad \Gamma_2.
\end{aligned}$$

In the case $\gamma = 0$ the initial state is given by the Stokes-type problem

$$\begin{aligned}
&-\mu\Delta\mathbf{u}_0 - (\lambda + \mu)\operatorname{grad}\operatorname{div}\mathbf{u}_0 + \operatorname{grad} p_0 = 0, \\
&\operatorname{div}\mathbf{u}_0 = 0, \quad x \in \Omega, \\
&\mathbf{u}_0 = 0 \quad \text{on} \quad \Gamma_1, \quad \sigma_0\,\mathbf{n} = \mathbf{h} \quad \text{on} \quad \Gamma_2.
\end{aligned}$$

Galerkin methods have been considered for the approximation of Biot's consolidation problem in several papers [5,7]. Due to the instabilities that appear at the initial stages, when $\gamma = 0$, equal order interpolation spaces for both displacement and pressure fields must be discarded. It is well known that in classical mixed formulations, the finite element spaces must satisfy some compatibility conditions to fulfill the Brezzi–Babuska stability condition. These kinds of discretizations for the Biot's problem give in general a lower order of convergence for the pore pressure than for the displacements. To improve the rate of convergence for pressure and effective stress, Murad et al. proposed a Petrov–Galerkin post–processing technique in [6]. Also, finite difference discretization has been used for the consolidation and filtration problem in [9]. Here we follow a different approach, using a finite difference discretization on Mac type or staggered grids [4]. For the analysis, the classical methodology of Samarski [8,10] is considered. Stability and monotonicity results and second order convergence in energy and L_2–norm are obtained . For simplicity here we deal the 1D case, but this approach is applicable in high dimensional problems. The plan of the paper is as follows. In Section 2, we formulate the mathematical model for the continuous problem and we give the simplest stability estimates. In Section 3 a complete analysis of a difference scheme on staggered grids is made, giving stability and convergence results for the compressible and incompressible case. Finally, numerical test are carried out to illustrate the theoretical results.

2 Continuous Problem Formulation

After adimensionalization with respect to the length l of the porous medium domain, the Lamé stress constant $\lambda + 2\mu$, the term u_0, the permeability κ and the viscosity η, such that

$$x := \frac{x}{l}, \quad t := \frac{(\lambda + 2\mu)\kappa t}{\eta l^2}, \quad p := \frac{p}{u_0}, \quad u := \frac{(\lambda + 2\mu)u}{u_0 l},$$

the poroelastic problem (1), (2) in the one-dimensional case become

$$\begin{aligned}
&-\frac{\partial^2 u}{\partial x^2} + \frac{\partial p}{\partial x} = 0, \quad x \in (0,1) \\
&\frac{\partial}{\partial t}\left(a\,p + \frac{\partial u}{\partial x}\right) - \frac{\partial^2 p}{\partial x^2} = f(x,t), \quad x \in (0,1),\ 0 < t \leq T, \\
&\frac{\partial u}{\partial x} = -1, \quad p = 0, \quad \text{on} \quad x = 0, \\
&u = 0, \quad \frac{\partial p}{\partial x} = 0, \quad \text{on} \quad x = 1, \\
&\left(ap + \frac{\partial u}{\partial x}\right)(x) = 0, \quad \text{in} \quad (0,1), \quad t = 0,
\end{aligned} \tag{3}$$

where $f(x,t) = \alpha\, g(x,t)$ is a scaled source term and $a = \gamma\,\mu$. For convenience in the theoretical formulation we transform the previous problem into a problem with homogeneous boundary conditions and non-homogeneous initial data.

To write an operator formulation of the problem, we first introduce appropriate functional spaces and operators. Let $\mathcal{H} = L_2(0,1)$ be the set of square–integrable scalar values functions defined on $(0,1)$, with scalar product (u,v) and corresponding norm $\|u\|$. On $\mathcal{H}$ let us consider the operator $\mathcal{A}$ given by $\mathcal{A}u = -\dfrac{\partial^2 u}{\partial x^2}$, with domain

$$D(\mathcal{A}) = \{u \in \mathcal{H} \mid \frac{\partial^2 u}{\partial x^2} \in \mathcal{H},\ \frac{\partial u}{\partial x}(0) = 0, u(1) = 0\}.$$

The operator $\mathcal{A}$ is positive and self–adjoint in $\mathcal{H}$ ($\mathcal{A} \geq 0, \mathcal{A} = \mathcal{A}^*$). Similarly on $\mathcal{H}$, we define $\mathcal{B}p = -\dfrac{\partial^2 p}{\partial x^2}$, with domain

$$D(\mathcal{B}) = \{p \in \mathcal{H} \mid \frac{\partial^2 p}{\partial x^2} \in \mathcal{H},\ p(0) = 0, \frac{\partial p}{\partial x}(1) = 0\}.$$

The coupling terms in the poroelastic problem are associated with the gradient and divergence operators that we will denote by $\mathcal{G}$ and $\mathcal{D}$ respectively. Note that for all $(u,p) \in D(\mathcal{A}) \times D(\mathcal{B})$, $(\mathcal{G}p, u) = -(\mathcal{D}u, p)$. In the construction of discrete analogs for $\mathcal{A}$ and $\mathcal{B}$ we will be oriented to the fulfillment of the same properties.

Problem (3) can be written in differential operator form as the abstract initial value problem: Given s and $f : (0,T) \rightarrow \mathcal{H}$, find $(u,p) : (0,T) \rightarrow \mathcal{H} \times \mathcal{H}$ such that

$$\begin{aligned} &\mathcal{A}u + \mathcal{G}p = 0, \\ &\frac{d}{dt}(ap + \mathcal{D}u) + \mathcal{B}p = f(t), \forall t \in (0,T) \\ &ap(0) + \frac{\partial u}{\partial x}(0) = s. \end{aligned} \tag{4}$$

Associated to $\mathcal{A}$ and $\mathcal{B}$, we define the inner products

$$(u,v)_{\mathcal{A}} = (\mathcal{A}u, v), \quad u,v \in D(\mathcal{A}),$$

$$(p,q)_{\mathcal{B}} = (\mathcal{B}p, q), \quad p,q \in D(\mathcal{B}),$$

and the corresponding norms $\|u\|_{\mathcal{A}}$ and $\|p\|_{\mathcal{B}}$. The same association is made for the selfadjoint and definite positive operator $\mathcal{B}^{-1}$. For $t \leq T$, the following stability estimates can be obtained

$$\| u(t) \|_{\mathcal{A}}^2 + a \| p(t) \|^2 \leq \| u(0) \|_{\mathcal{A}}^2 + a \| p(0) \|^2 + \frac{1}{2} \int_0^t \|f(x,\theta)\|_{\mathcal{B}^{-1}}^2 d\theta, \tag{5}$$

$$\|p(t)\|_{\mathcal{B}}^2 \leq \|p(0)\|_{\mathcal{B}}^2 + \frac{1}{2(1+a)} \int_0^t \|f(x,\theta)\|^2 d\theta \quad .$$

which ensures stability with respect to the initial data and right hand side.

3 Staggered Grids

3.1 Grid and Grid-Operators

Given a positive integer N and $h = \dfrac{2H}{2N-1}$, let us define two different grids, $\overline{\omega}_p$ to discretize the pressure and $\overline{\omega}_u$ to discretize the displacement,

$$\begin{aligned} \overline{\omega}_p &= \{x_i | x_i = ih, \ i = 0, \ldots, N-1\}, \\ \overline{\omega}_u &= \left\{y_i | y_i = \tfrac{h}{2} + (i-1)h, i = 1, \ldots, N\right\}. \end{aligned}$$

The grid points for u are shown in the figure (1) by small circles, while the grid points for p are shown by filled circles.

Fig. 1. Grids for displacement ○ and for pressure •

We define the Hilbert space $H_{\overline{\omega}_p}$ of the discrete functions $p_i = p(x_i), x_i \in \overline{\omega}_p$, and the Hilbert space $H_{\overline{\omega}_u}$, of the discrete functions $u_i = u(x_i), x_i \in \overline{\omega}_u$ with inner products

$$(p,q)_{\overline{\omega}_p} = h\left(\frac{p_0 q_0}{2} + \sum_{i=1}^{N-1} p_i q_i\right), \quad (u,v)_{\overline{\omega}_u} = h\left(\sum_{i=1}^{N-1} u_i v_i + \frac{u_N v_N}{2}\right).$$

In a similar way we introduce the Hilbert spaces

$$H_{\omega_p} = \{p \in H_{\overline{\omega}_p} \,|\, p_0 = 0\}, \quad H_{\omega_u} = \{u \in H_{\overline{\omega}_u} \,|\, u_N = 0\},$$

with scalar products defined respectively by

$$(p,q)_{\omega_p} = h\sum_{i=1}^{N-1} p_i q_i, \quad (u,v)_{\omega_u} = h\sum_{i=1}^{N-1} u_i v_i,$$

and associated norms $\| \cdot \|_{\omega_p}$ and $\| \cdot \|_{\omega_u}$.

We approximate the differential operators $\mathcal{A}$ and $\mathcal{B}$ by the discrete operators A and B defined by

$$(Au)_i = \begin{cases} -\frac{1}{h}(u_x)_i, & i = 1, \\ -(u_{\bar{x}x})_i, & 2 \le i \le N-1, \\ 0 \quad i = N. \end{cases} \qquad (Bp)_i = \begin{cases} 0 \quad i = 0, \\ -(p_{\bar{x}x})_i, & 1 \le i \le N-2, \\ \frac{1}{h}(p_{\bar{x}})_i, & i = N-1, \end{cases}$$

where $u_x, u_{\bar{x}}$ are the right and left difference derivatives, using the standard indexless notation of the theory of difference schemes.

It is easy to verify that operators A and B are self–adjoints and positive in H_{ω_u} and H_{ω_p} respectively, and therefore we can define the norms associated to inner products $(u,v)_A = (Au,v)_{\overline{\omega}_u}$ and $(p,q)_B = (Bp,q)_{\overline{\omega}_p}$,

$$\| u \|_A = (u,u)_A^{1/2}, \quad \| p \|_B = (p,p)_B^{1/2}.$$

In order to approximate the gradient and divergence operators $\mathcal{G}$ and $\mathcal{D}$ we define discrete operators $G : H_{\omega_p} \to H_{\omega_u}$ and $D : H_{\omega_u} \to H_{\omega_p}$ by

$$(Gp)_i = \begin{cases} (p_{\bar{x}})_i & i = 1,\ldots,N-1, \\ 0, & i = N. \end{cases} \qquad (Du)_i = \begin{cases} 0, & i = 0, \\ (u_x)_i, & i = 1,\ldots,N-1, \end{cases}$$

which satisfy

$$(Gp,u)_{\overline{\omega}_u} = -(p,Du)_{\overline{\omega}_p}, \quad \forall (u,p) \in H_{\omega_u} \times H_{\omega_p}$$

After space approximation we have the semidiscrete problem: Given $f, s \in H_{\overline{\omega}_p}$, find $(u(t),p(t)) \in H_{\overline{\omega}_u} \times H_{\overline{\omega}_p}$, for $0 < t \le T$ such that

$$\begin{aligned} &Au(t) + Gp(t) = 0, \quad \text{on } \overline{\omega}_u \\ &\frac{d}{dt}(ap(t) + Du(t)) + Bp(t) = f, \quad \text{on } \overline{\omega}_p. \end{aligned} \tag{6}$$

with initial condition

$$ap(0) + Du(0) = s, \quad \text{on } \overline{\omega}_p,$$

which formally is similar to (4).

Problem (6) can be discretized with the weighted two level scheme

$$Au^{n+1} + Gp^{n+1} = 0, \quad n = 0, 1, \ldots, M-1, \tag{7}$$

$$a\frac{p^{n+1} - p^n}{\tau} + \frac{Du^{n+1} - Du^n}{\tau} + Bp_\sigma^{n+1} = f_\sigma^{n+1}, \quad n = 0, 1, \ldots, M-1. \tag{8}$$

For this difference scheme, we have a discrete analogue of estimate (5) for the continuous problem. That follows from the following result

Proposition 1. *For $\sigma \geq 1/2$ the solution of the difference scheme (7), (8) satisfies the a priori estimate*

$$\| u^{n+1} \|_A^2 + a \| p^{n+1} \|^2 \leq \| u^n \|_A^2 + a \| p^n \|^2 + \frac{\tau}{2} \| f_\sigma^{n+1} \|_{B^{-1}}^2 . \tag{9}$$

Proposition 2. *Let u^0 and p^0 be $O(h^2)$ approximations of $u(x,0)$ and $p(x,0)$, where $u(x,t)$ and $p(x,t)$ are smooth solutions of problem (3). Let (u^n, p^n) be the solution of the difference scheme (7),(8). For $\sigma \geq 1/2$ the convergence result*

$$\| u^{n+1} - u(.\,, t_{n+1}) \|_A + a \| p^{n+1} - p(.\,, t_{n+1}) \| = O(\tau^\nu + h^2) \tag{10}$$

where $\nu = 2$ if $\sigma = 1/2$ or $\nu = 1$ if $\sigma \neq 1/2$, holds.

Proof. Let's be the error in displacements and pressure $\delta u^n(x) = u^n(x) - u(x, t_n), x \in \omega_u$ and $\delta p^n(x) = p^n(x) - p(x, t_n), x \in \bar{\omega_p}$, respectively, that satisfy

$$A\,\delta u^{n+1} + G\delta p^{n+1} = \psi_1^{n+1}, \quad n = 0, 1, \ldots, M-1,$$

$$a\frac{\delta p^{n+1} - \delta p^n}{\tau} + \frac{D\delta u^{n+1} - D\delta u^n}{\tau} + B\delta p_\sigma^{n+1} = \psi_2^{n+1}, \quad n = 0, 1, \ldots, M-1,$$

We split the displacement error $\delta u^n = w_1^n + w_2^n$, where w_1^n is the solution of $Aw_1^n = \psi_1^n$. So, $\| w_1^n \|_A \leq \| \psi_1^n \|_{A^{-1}}$.

Now, w_2^n and δp^n are the solution to the problem

$$Aw_2^{n+1} + G\delta p^{n+1} = 0, \quad n = 0, \ldots, M-1,$$

$$a\frac{\delta p^{n+1} - \delta p^n}{\tau} + \frac{Dw_2^{n+1} - Dw_2^n}{\tau} + B\delta p_\sigma^{n+1} = \psi_2^{n+1} - \frac{Dw_1^{n+1} - Dw_1^n}{\tau},$$

Using the estimate (9), (10) follows (see [3] for details.)

From proposition (2) convergence properties for displacements in the A–norm and for pressures in the discrete L_2–norm if $a \neq 0$, follows. We now consider the convergence of pressure gradients or equivalently convergence of pressure in the B–norm.

Proposition 3. *For $\sigma \geq 1/2$ the solution of the difference scheme (7), (8) satisfies the a priori estimate*

$$\| p^{n+1} \|_B^2 \leq \| p^n \|_B^2 + \frac{\tau}{2(1+a)} \| f_\sigma^{n+1} \|_{\omega_p}^2,$$

which permits us to get the following convergence result,

Proposition 4. *Let us suppose that the same conditions as in proposition 2 hold. For $a \geq 0$ and $\sigma \geq 1/2$ we have the estimate*

$$\| p^n - p(., t_n) \|_B = O(\tau^\nu + h^2),$$

where $\nu = 2$ if $\sigma = 1/2$ or $\nu = 1$ if $\sigma \neq 1/2$.

4 Numerical Results

To validate the previous analysis, numerical experiments has been performed using a problem given by Biot in [2]. Let us consider problem (3), which corresponds to a column of soil, bounded by impermeable and rigid lateral walls and bottom, supporting a unit load on the top, which is free to drain.

The solution can be expressed, as

$$u = -\sum_{i=0}^{\infty} \frac{2}{M^2} \cos(Mx) e^{(a-M^2)t},$$

$$p = \sum_{i=0}^{\infty} \frac{2}{M} \sin(Mx) e^{(a-M^2)t}$$

where

$$M = \frac{\pi(2n+1)}{2}.$$

Figure (2) shows that for Mac type grid schemes, the pressure approximation is monotone without restrictions between step sizes.
The computational order of convergence the Mac type grid scheme is shown in the figure (3). Mac type grid schemes run well in cases with $a \geq 0$ and $\sigma \geq 1/2$, and give second order accuracy in space and time for $\sigma = 1/2$.

References

1. Bear, J., Bachmat, Y.: Introduction to Modelling Phenomena of Transport in Porous Media. Kluwer Academic, Dordrecht (1991)
2. Biot, M.: General theory of three dimensional consolidation. J. Appl. Phys. 12 (1941) 155–169.
3. Gaspar, F.J., Lisbona, F.J., Vabishchevich, P.N.: A finite difference analysis of Biot's consolidation model. *submitted*

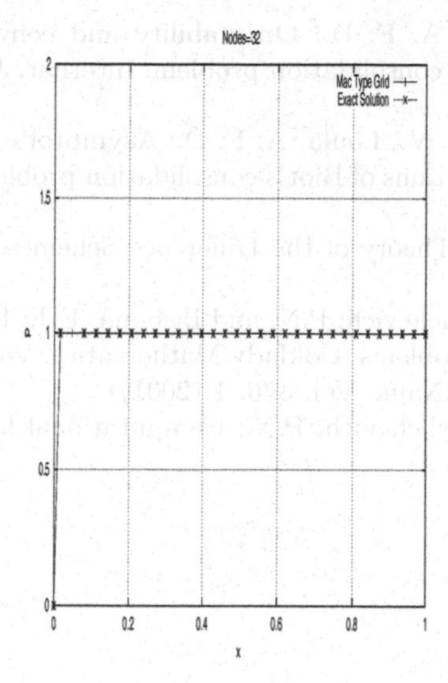

Fig. 2. Numerical solution with 32 nodes, $\sigma = 1$, $\tau = 1.d - 6$ with $a = 0$

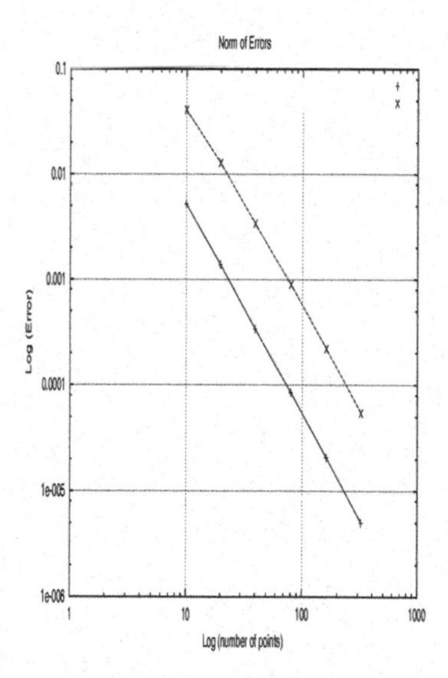

Fig. 3. Convergence, $(\| e_u \|_A^2 + a \| e_p \|^2)^{1/2}$, (+), $\| e_p \|_B$, ($\times$) for $\sigma = 1/2$ with $a = 0$

4. Harlow, F.H. and Welch, J.E.: Numerical calculation of time-dependent viscous incompressible flow of fluid with free surface. Phys. Fluids, 8 (1965)
5. Murad, M. A., Loula, A. F. D.: Improved accuracy in finite element analysis of Biot's consolidation problem. Comput. Methods Appl. Mech. Engrg., 95 (1992) 359–382.

6. Murad, M. A., Loula, A. F. D.: On stability and convergence of finite element approximations of Biot's consolidation problem. Internat. J. Numer. Methods Engrg. 37 (1994) 645–667.
7. Murad, M. A., Thomée, V., Loula, A. F. D.: Asymptotic behaviour of semidiscrete finite-element approximations of Biot's consolidation problem. SIAM J. Numer. Anal. 33 (1996) 1065–1083.
8. Samarskii, A. A.: The Theory of the Difference Schemes. Marcel Deker, New York and Basel (2001)
9. Samarskii, A.A., Vabishchevich, P.N. and Lisbona, F.J.: Difference Schemes for Filtration Consolidation Problems. Doklady Mathematics, Vol. 63, 1 (2001) (Translated from Doklady Akademii Nauk, Vol. 376, 1 (2001))
10. Samarskii, A.A., Vabishchevich, P.N.: Computational heat transfer, Vol 1 and 2, Wiley, Chichester (1995)

An Algorithm for Parallel Implementations of an Eulerian Smog Model

Krassimir Georgiev

Bulgarian Academy of Sciences,
Acad. G. Bonchev, Bl. 25-A, 1113 Sofia, Bulgaria
georgiev@parallel.bas.bg *and*
VITO - Centre for Remote Sensing and Atmospheric Processes,
Boeretang 200, B-2400 Mol, Belgium

Abstract. A parallel algorithm for an implementation of EUROS (EURopean Operational Smog) model on high-performance parallel computer platforms is presented. The atmospheric model system of partial differential equations is solved using the operator splitting technique. The parallel version of EUROS is based on an algorithm very closed to the well-known domain decomposition approach in numerical mathematics. Different decomposition of the model domain is used for the different submodels included in EUROS model. A grid refinement procedure, up to four levels in user defined subareas, leads to some difficulties in load balancing, which are discussed in the paper. The parallel algorithm reported can be used on (*i*) distributed memory parallel computers, (*ii*) parallel computers with shared memory and (*iii*) SMP computers.

Key words: air pollution modelling, parallel computing, domain decomposition
Subject classifications: 65Y05, 65Y10

1 Introduction

The EUROS model is an Eulerian atmospheric model that simulates the tropospheric ozone over Europe on a long-term basis (up to several years). It represents the various atmospheric processes responsible for ozone formation and destruction: pollutant emission, atmospheric dispersion and transport, chemical transformations and deposition. A grid refinement procedure up to four levels allows the spatial resolution in certain areas of the model domain to be reduced considerably. The vertical stratification of EUROS consists of four layers. The atmospheric model system of partial differential equations is solved using the operator splitting technique: horizontal advection, horizontal diffusion, vertical diffusion + dry deposition, wet deposition, fumigation and chemical reactions + emissions. The output results obtained by the simulation procedure with EUROS model allow to estimate the indicators currently used to evaluate the effect of ozone on vegetation and human health, such as AOTs (Accumulated exposure Over a Threshold) and NETs (Number of Exceedances of Threshold). In order to use EUROS model as an operational model for policy support the output results

I. Dimov et al. (Eds.): NMA 2002, LNCS 2542, pp. 463–470, 2003.

have to be obtained at limited computational cost (computer memory and CPU time). However, the improvements in the EUROS model can only be carried out at the price of an increased computer capacity and the improved version of the model exceed the computational limitations that were set for policy support (3 - 4 hours CPU time). An elegant solution of this problem is to implement the EUROS model on parallel computer architectures by dividing the computational load over a number of processors which are working in parallel. Moreover, the parallel tasks can be run nowadays not only on the very much expensive supercomputers but on clusters of inexpensive PC's or workstations. The parallel version of EUROS is based on an algorithm very closed to the well-known domain decomposition approach in numerical mathematics. The algorithm presented in this paper leads to a considerable reduction of the computer time and moreover it allows distribution of the input and output data, which is very huge, to the local memories of the selected processors.

The remainder of the paper is organised as follows. Short description of the EUROS model and the splitting procedure as well as the numerical techniques used can be found in Section 2. Section 3 focuses on the parallel algorithm and its properties. The final Section 4 summarises our conclusions and outlook.

2 The EUROS Model

The EUROS model ([2,7]) is used for simulating emissions input, transport processes, chemical transformation, and dry and wet deposition processes of various air polluting compounds. The model can be used to examine the time and spatial behaviour of SO_x, NO_x, O_3 and Volatile Organic Compounds (VOC's) in the lower troposphere over Europe.

In EUROS model, as usual in the atmospheric models, an operating splitting is used ([8]). The main advantage of this approach is that when each process is treated separately the most efficient numerical integration techniques can be chosen. This leads as well to limitation of the total computation time. In fact, the "*Strang splitting*" ([10]) is applied in EUROS, i.e. the model processes are treated in *forward* and *backward* order during two successive time steps. Strang operator splitting introduces a splitting error that is second order in time. Note that different time-steps can be used inside each time interval and moreover, the time-steps for the different atmospheric processes in fixed time interval can be different too.

The model domain used in EUROS is a 3-dimensional volume above the earth surface. In a horizontal direction, the model domain extends over a large part of Europe. Spherical coordinate system is used, i.e. longitudinal, latitudinal coordinates over the earth surface and a vertical coordinate is the height over the earth surface. In order to overcome some numerical problems when simulating the advection process, EUROS uses a *shifted pole coordinate system*, where the equator is shifted to 60° Northern latitude. The base grid consists of (52 x 55) grid cells with a $(0.55° \times 0.55°)$ longitude-latitude resolution on shifted pole coordinates which is about (60 km x 60 km). Local grid refinement is possible up to four levels resulting in a maximum longitude-latitude resolution of about ($0.069° \times 0.069°$) which is about (7.5 km x 7.5 km). The vertical stratification of EUROS consists of four layers: *surface layer* (50 m hight), *mixing layer* (minimum height of 300 m during the night and in the early morning), *reservoir layer* for sources emitting above the surface and mixing layers *top layer* which limits the model height to 3000 m.

The horizontal advection process is simulated by using a set of volumes (grid cells) and a discretization of the advection equation on this grid. The numerical scheme used in EUROS model is based on the **Finite Volume Method** ([6,7]). The scheme used for computation of the fluxes at the grid cell boundaries is the limited $\kappa = 1/3$ scheme, known as *van Leer scheme* or *third order upwind biased discretization* too. The computing fluxes are limited in order to prevent undershoot (negative concentrations) and overshoot. The mass conservation is guaranteed for Finite Volume Method. The changes in the concentrations in EUROS are computed by integrating a mass balance equation with Runge-Kutta methods of order three or four. The time step is automatically chosen depending on the Courant number (ratio of the mesh size and the norm of the wind speed).

The separate treatment of the advection process requires some forces to make the wind field divergence free. The procedure of *Endlich* ([5]) (an iterative process that reduces the divergence of the wind field in each step on a cell by cell basis) is used in EUROS to enforce the divergence freeness.

A simple *Forward Euler* integration method is used to obtained the new concentrations on each time step in the diffusion submodel.

Emissions of surface sources, such as traffic and domestic heating, can be found in the surface layer. The contaminants are removed due to dry deposition and vertical diffusion introduces a mass exchange between surface and mixing layers. No diffusion occurs in the upper two layers. A constant flux profile and quadratic concentration profile are used in EUROS. Following these two assumptions a system of linear differential equations for the concentrations in surface and mixing layers is derived and solved exactly.

The differential equations describing the *fumigation*, the *afternoon stratification* and *wet deposition* are solved exactly.

The chemical scheme, which is actually used in EUROS is Carbon Bond-IV (CB-IV-V scheme) mechanism which consists of 87 reactions (including 12 photochemical) and 35 species. The time evolution of the concentrations of all chemical components is governed by a production term (**P**) and a loss term (**L**). The system of ordinary differential equations, describing the productions and losses, is very stiff, because slow and very fast chemical reactions are involved. For its numerical treatment the TWOSTEP method based on the two step second order Backward Differentiation Formula ([11,12] is used.

3 The Parallel Algorithm and Its Properties

The computer simulation with EUROS is very much time consuming mainly because its model domain is very large and the size of the systems of ordinary differential equations that arise after the space discretization is enormous. As EUROS, like other air quality models (see e.g. [1,13], is used for the evaluation of emission control strategies and planning for the control of air pollution episodes, the computations should be kept tractable. The problem becomes even harder when a local refinement is used in some subareas of the model domain, which are of special interest. A simple analysis of the runs of EUROS sequential code ([3])shows that the chemistry submodel takes about 65% of the computing time and the advection submodel takes about 30% of it. The task of the parallelisation of the existing sequential codes is both extremely hard and very laborious.

The parallel computer platform, which is now available for parallel runs of EUROS consists of an operational **cluster of four Dell 410/400 workstations** containing in total **eight 400 MHz processors**. Each node is equipped with 256 Mb local memory and each processor is able to access the local memories of other nodes through the network. The processors are interconnected by means of a fast Ethernet switch. According to the different classification schemes, this architecture can be classified as *Multiple Instruction stream / Multiple Data stream* (MIMD) and *Non Uniform Memory Access* (NUMA).

Open MP Fortran Application Program Interface (OpenMP) is used for programming in a shared memory mode ([14]), and a *Message-Passing Interface Standard* (MPI) is used in a distributed memory mode ([4]). As the computer system available consists of several nodes each of them with its own memory and each node consists of several processors with shared memory, then MPI is used between nodes and OpenMP within each node (see e.g. [9]). The proposed algorithm allows to use the MPI primitives for the communications between all the processors, nevertheless the processors belong to a single node have a shared memory.

The main task in the parallel programming is the distribution of the computational work among the available processors. The arrays for the concentrations and depositions of the chemical species studied are to be split in several subarrays and each node or processor owns one of them. Three main types of decomposition are possible: *chemical, horizontal* and *vertical* ([3]. Taking into account the advanteges and disadvanteges of each of them the **horizontal decomposition**, known as *grid partitioning* or *domain decomposition* scheme, which mostly confines communication to next-neighbours interaction is chosen. It has a minimal impact on the code appearance and therefore it is the most popular strategy in the atmospheric modelling for mapping the computational work on processor network.

3.1 Parallelisation Strategy for EUROS Model by Domain Decomposition

The domain decomposition approach seems to be very useful because the EUROS model domain is regular (rectangular horizontal domain with equidistant basic grid). However, the local refinement which is applied into some specific, user defined, subareas leads to big difficulties in order to divide the domain into subdomains in such a way that a good load balance between the nodes to be achieved. In this approach the parallelism is created by partitioning the horizontal model domain into p subdomains, where p is the number of the computer nodes available ([3]). Each subdomain is assigned to different node. These subdomains are strips in south-north direction. The reason for such decomposition is that the stiffness of the system of ordinary differential equations in the chemical part (the most time consuming) depends on the location of the grid-cell in east-west direction. In this case if each subdomain covers a similar number of grid-cells then the chemistry, vertical exchange and deposition will have a similar computation cost. Communications between the neighbour nodes which manage neighbour subdomains are necessary during the advection and diffusion stage. Different decompositions are used for the different submodels: *overlapping* — for the horizontal advection and horizontal diffusion part of the model, and *non-overlapping* - for the other submodels mentioned above. In the case of an overlapping each subdomain has two cells overlap region in order to meet the requirements of the third order upwind discretization scheme used in the advection submodel. The amount of data, which have to be exchanged at each time step is determined by the number of grid-cells in φ-direction multiplied by two and by the number of the vertical layers. The domain decomposition approach leads to an algorithm where each subdomain is managed in the same way as the whole domain used to be managed in the sequential mode. The only difference is how the inner boundary grid-cells are filled.

In order to show the *load balance* in the parallel EUROS model we will consider two test examples. In the first one, there is a single area where a local refinement has to be applied. In the second example, there are five subareas where a local refinement is applied. They are on different places into the model domain, and have different sizes. A cell is refined by bisecting all its sides. It leads to an increasing of the number of the grid cells in a given θ row up to 27 times for the finest grid in the first example, and up to 26 times — for the second example, compare with the rows where there are no refined cells. An algorithm for automatic dividing of the model domain into subdomains, where the local refinement level and the coordinates of the subareas with a local refinement are parameters, is created. The results obtained for the first test example are given in Table 1. It is well seen from the fifth column of this table that in the case of one subarea where a local refinement is applied there is a good load balance in the chemical part of the algorithm (most time consuming part) and relatively good load balance in the advection-diffusion submodel (see the last column of the same table) even for the finest refinement level.

The results obtained for the second test example are given in Table 2. The load balance between the computer nodes in this case is even better than the previous one. Let us note that both examples are taken from the real-life tasks for an evaluation of emission control strategies and planning for the control of air pollution episodes in Europe. Much more results, tables and discussions can be found in [3].

Table 1. Computational load for the first test example

Local ref. level	Node	Chemistry,emissions,vertical diffusion, fumigation		Advection and diffusion	
		Lines from-to (No. of cells)	Comput. load in %	Lines from-to (No. of cells)	Comput. load in %
0	1	1-14 (728)	100	1-16 (832)	89
Total	2	15-28 (728)	100	13-30 (936)	100
2860	3	29-42 (728)	100	27-44 (936)	100
cells	4	43-55 (676)	93	41-55 (832)	89
1	1	1-17 (1082)	98	1-19 (1318)	86
Total	2	18-26 (1062)	96	16-28 (1534)	100
4312	3	27-35 (1062)	96	25-37 (1468)	96
cells	4	36-55 (1106)	100	34-55 (1342)	88
2	1	1-19 (2638)	99	1-21 (3402)	81
Total	2	20-26 (2674)	100	18-28 (4202)	100
10120	3	27-33 (2674)	100	25-35 (4202)	100
cells	4	34-55 (2134)	80	32-55 (2898)	69
3	1	1-19 (7918)	92	1-21 (10794)	75
Total	2	20-25 (8628)	100	18-27 (14380)	100
33352	3	26-31 (8628)	100	24-33 (14380)	100
cells	4	32-55 (8178)	95	30-55 (11054)	77

3.2 Input and Output Data Handling and Data Storage

The access to the local memories of the nodes is much faster than the access to the memories of the other nodes, which is organised by the message passing primitives. Therefore, it is important to store as much as possible of the input data which is for the grid-cells belong to a subdomain in the local memory of the corresponding computer node which manages it. In this way, not only each processor is working on its own subdomain but it has also direct access to all input data needed in the run. During the run, each processor prepares output data files corresponding to the grid-cells in its own subdomain. At the end of the job all these files have to be collected on one of the processors and prepared for using them in the future (scatter plots, plots with the concentrations and depositions on whole model domain, etc.). This approach leads to considerable reduction of the communications during the actual computations. Such a transfer of data is times faster than one via the MPI primitives. It can be done either inside the main program using

```
if(me.eq.0) then
   call PREP_INPUT(nproc) ! Produces input files for subdomains
endif
write(file_in,'A,i1.1,A,i1.1)')
      'cp /maindir/filename.',me,' /scratch/filename.',me
call SYSTEM(file_in)
```

Table 2. Computational load for the second test example

Local ref. level	Node	Chemistry,emissions,vertical diffusion, fumigation		Advection and diffusion	
		Lines from-to (No. of cells)	Comput. load in %	Lines from-to (No. of cells)	Comput. load in %
1	1	1-16 (1006)	95	1-18 (1176)	85
Total	2	17-27 (1022)	96	15-29 (1344)	97
4099	3	28-41 (1010)	95	26-43 (1389)	100
cells	4	42-55 (1061)	100	40-55 (1219)	88
2	1	1-19 (2353)	100	1-21 (2787)	86
Total	2	20-28 (2343)	100	18-30 (3121)	97
9055	3	29-42 (2153)	92	27-44 (3231)	100
cells	4	43-55 (2206)	94	41-55 (2580)	80
3	1	1-20 (7466)	100	1-22 (8956)	76
Total	2	21-27 (7042)	94	19-29 (9644)	82
28879	3	28-42 (7269)	97	26-44 (11824)	100
cells	4	43-55 (7102)	95	41-55 (8340)	71

or in the script for the program run by

```
extent=0
for hostname in \$LOADL_PROCESSOR_LIST
do
   global=\$globaldir_in/\$filename_in.\$extent
   local=\$hostname:\$localdir/\$filename_in.\$extent
   rcp \$global \$local
   ....................
   extent=\$((\$extent+1))
done.
```

4 Conclusions

The horizontal decomposition approach, which is the most popular strategy in the atmospheric modelling, promises an efficient parallelization due to cell or column oriented processes. The strategy for the parallelization of EUROS model is based on a computer platform consists of four computer nodes with distributed memory. Each subdomain is assigned to one of the nodes so each node manages all concentrations and depositions in it. Only advection and diffusion submodels require exchange of data during the execution of the code. The communication between nodes will be done using MPI primitives. Inside the nodes a second level of a parallelization is organized on the base of OpenMP directives. So, the existing two processors in each of the nodes perform the computations in each subdomain in parallel. In order to use efficiently the cache memory of the processors and to avoid the superlinear speedups chunks in the chemical part are used. In order to avoid some communications between nodes during the actual computations pre-processing and post-processing procedures are used. This approach leads to an algorithm where only input and output data for the subdomain which is assigned to the corresponding node is stored in its local memory and therefore the access to all data is supposed to be much faster.

Acknowledgements

The research reported in this paper was supported from the Federal Office for Scientific, Technical and cultural Affairs of Belgium under Contract No. 001058 and partly supported by the project of European Commission — BIS 21 under contract ICA1-CT-2000-70016 and by the Bulgarian Ministry of Education and Science under Grant I-901/99.

References

1. Segers, A.: Data assimilitaion in atmospheric chemistry models using Kalman filtering. PhD Thesis, Technical University of Delft, The Netherlands (2002)
2. Delobbe, L., Kinnaer, L., Mensink, C.: Optimization of chemical and advection modules in EUROS. Scientific Report on OSTC Project AS/DD/10, VITO, Belgium (2000)
3. Georgiev, K.: Optimal strategies for parallelization of the EUROS model: design and implementation. Scientific Report 2002/TAP/R/011, VITO, Belgium (2002)
4. Gropp, W., Lusk, E., Skjellum, A.: Using MPI: Portable programming with the message passing interface. MIT Press, Cambridge (1994)
5. Endlich, R. M.: An iterative method for altering the kinematic properties of wind fields. J. of Appl. Meteorology, 6 (1967) 837–844.
6. van Leer, B.: Upwind-difference methods for aerodynamic problems governed by the Euler equation. In: Engquist, B. E. et al. (eds.): Large scale computations in fluid mechanics, AMS Series (1985) 327–336.
7. van Loon, M.: Numerical methods in smog prediction. PhD Thesis. University of Amsterdam, The Netherlands (1996)
8. McRae, G. J., Goodin, W. R., Seinfeld, J. H.: Numerical solution of the atmospheric diffusion equation for chemistry reacting flows. J. Comp. Phys. 45 (1982) 1–42.
9. Owczarz, W., Zlatev, Z.: Running a large air pollution model on an IBM SMP computer. Parallel Computing, 28, 2 (2002) 355–368.
10. Strang, G.: On the construction and comparison of difference schemes. SIAM J. Numer. Anal. 5 (1968) 506–517.
11. Verver, J. G.: Gauss-Seidel iteration for stiff ODEs from chemical kinetics. SIAM J. Sci. Comput. 15 (1994) 1243–1250.
12. Verver, J. G., Simpson, D.: Explicit methods for stiff ODEs from atmospheric chemistry. Appl. Numer. Math. 18 (1995) 413–430.
13. Zlatev, Z.: Computer treatment of large air pollution models. Kluwer Academic Publ., Dordrecht-Boston-London (1995)
14. *WEB-site for OpenMP tools*, http://www.openmp.org (1999)

Generating Minimal Surfaces Subject to the Plateau Problems by Finite Element Method*

Hong Gu

Research Institute of Symbolic Computation
Johannes Kepler University, A4232, Linz, Austria
hgu@risc.uni-linz.ac.at

Abstract. There already exists avariety of softwares for generating minimal surfaces of special types. However, the convergence theories for those approximating methods are always left uncompleted. This leads to the difficulty of displaying a whole class of minimal surface in general form. In this paper, we discuss the finite element approximating methods to the minimal surfaces which are subject to the well-known Plateau probems. The differential form of the Plateau problems will be given and, for solving the associated discrete scheme, either the numerical Newton iteration method can be applied or we can try some promsing symbolic approaches. The convergence property of the numerical solutions is proved and this method will be applied to generating the minimal surface graphically on certain softwares later. The method proposed in this paper has much lower complexity and fits for inplementing the two grid and parallel algorithms to speed up the computation.

Key Words: Plateau Problem, Variational form, Convexity, Brouwer's fixed point theorem.

1 Introduction

The study of minimal surfaces is a branch of differential geometry, because the methods of differential calculus are applied to geometrical problems. One of the oldest questions is: "What is the surface of smallest area spanned by a given contour?" Such question is nontrivial despite the fact that every physical soap film appears to know the answer. The existence of the minimal surface within such a fixed contour is usually taken as the well-known Plateau Problem, which has been solved by Rado, Douglas and Osserman in the 1930's.

From the point of view of local geometry, a minimal surface can also be taken as the surface which has the zero average mean curvature on each surface point, e.g., a saddle shape.

Previously unknown and certainly unexpected minimal surfaces were found by David Hoffman and his collaborators at GANG, the Center for Geometry, Analysis, Numerics, and Graphics at the University of Massachusetts in 1985

* This work has been partially supported by the Austrian Fonds zur Förderung der wissenschaftlichen Forschung (FWF) under project nr. SFB F013/F1304

I. Dimov et al. (Eds.): NMA 2002, LNCS 2542, pp. 471–478, 2003.

(see [5] etc.). They first used their MESH computer graphics system to find these surfaces, and then later proved their existence with fully rigorous mathematics. This truly excited the minimal surface community and piqued their interest in computer graphics.

Many problems in optimal geometry require specialized software systems because there is often no explicit parametrization of a desired minimal surface. Although in the past years, lots of softwares have been produced for generating minimal surfaces graphically (e.g. [1]), there is very few numerical analysis for their approximation procedures. One possibility of using the finite element method solving minimal surfaces has been proposed in [7] but solving its discrete form is not cheap by the exact method. In this paper we will discuss the finite element solution to the Plateau problem based on another discretization form, give the proof of the convergence and then show some typical experiments.

In the next section we will first propose the differential equations to the Plateau problems and define their variational forms in the finite element space. Some error analysis for solving the linear finite element equations will be previewed in Section 3, and we will focus on the proving of the convergence property of the discrete solution to the Plateau problem mainly by Brouwer's fixed point theorem in Section 4. In Section 5 we show some experiments of generating the minimal surface on Maple software for testing this new method.

2 Preliminaries

Due to the fact that the minimal surface has the property of having zero average mean curvature at each surface point, we can define the Plateau problems in terms of a partial differential equation, which satisfying some boundary restrictions.

Let us assume that such minimal surface can be represented explicitly by $z = u(x, y)$ in a 3-d space with the boundary representation

$$u = f(x, y), \quad (x, y) \in \partial\Omega,$$

then u solves the following system of differential equations:

$$\begin{cases} -div((1 + u_y^2)u_x, (1 + u_x^2)u_y) + 6u_{xy}u_xu_y = 0 \ \text{ in } \ \Omega, \\ u = f(x, y), \quad \text{on } \partial\Omega, \end{cases} \tag{1}$$

where Ω is a strictly convex domain such that it is sufficient to assume that the solution to the previous equation is unique if $f(x, y)$, which satisfies the bounded slope condition (see [9]), is the restriction to $\partial\Omega$ of a function in the Sobolev space $W_q^2(\Omega)$, for some $q > 2$.

Set the nonlinear operator

$$A(u, v) = \int_\Omega [(1 + u_y^2)u_xv_x + (1 + u_x^2)u_yv_y + 6u_{xy}u_xu_yv]dxdy,$$

where u_x, u_{xy}... denote $\partial_x u, \partial_x(\partial_y u)$..., etc. Then the equation (2.1) weakly equivalent to the following variational form:

Looking for $w \in H_0^1(\Omega) \cap H^2(\Omega) \cap C_1(\Omega)$, such that

$$A(w + \bar{f}, v) = 0, \text{ in } \Omega, \forall v \in H_0^1(\Omega) \cap H^2(\Omega) \cap C_1(\Omega), \tag{2}$$

where $\bar{f}$ is the "regular" extension of f from $\partial\Omega$ to Ω, especially, $\bar{f} \in W_q^2(\Omega), q > 2$. (Obviously, $w + \bar{f} = u$ according to the previous uniqueness assumption.)

We now consider the discrete scheme for (2.2). Let $\Omega_h \subset \Omega$ be divided by mesh T_h with size h. $S_h(\Omega_h)$ is the finite element subspace of $H_0^1(\Omega_h) \cap H^2(\Omega_h) \cap C_1(\Omega_h)$ and I_h is the corresponding interpolation operator on $S_h(\Omega_h)$ (see [2,3] etc.). It is also assumed that the mesh partition is regular and all the grid nodes on $\partial\Omega_h$ also located on $\partial\Omega$.

Then the discrete solution $w_h \in S_h(\Omega)$ is defined by

$$A(w_h + \bar{f}, v) = 0, \text{ in } \quad \forall v \in S_h(\Omega_h). \tag{3}$$

and $w_h + \bar{f} = u_h$ will be taken as the finite element solution to the Plateau problems.

The above discrete scheme appears to be a bunch of polynomial equations so that can be practically solved either by the Newton iteration method (see [6,15] etc.), or by other promising symbolic eliminations. And furthermore, such discrtization form is cheap to be solved out by a lot of exact computations.

For the continuous solution w is unique according to the assumption, $w^h = w|_{\Omega_h} \in H_0^1(\Omega_h) \cap H^2(\Omega_h) \cap C_1(\Omega_h)$ which satisfying

$$A(w^h + \bar{f}, v) = 0, \text{ in } \quad \forall v \in H_0^1(\Omega_h) \cap H^2(\Omega_h) \cap C_1(\Omega_h)$$

also must be unique. Then we could consider the oringinal problem based on a certain convex polygon domain. For convenience we still take w^h as w, Ω_h as Ω, etc., and will prove that w_h convergence to the real solution w when the mesh size h goes to zero in the following sections.

3 Some Error Resolutions for Solving the Linear Finite Element Equations

Let $L(\cdot)$ be a linear operator defined on $S_h(\Omega)$ by

$$L(u) = -div(a_{11}u_x + a_{21}u_y, a_{12}u_x + a_{22}u_y) + b_1 u_x + b_2 u_y + cu,$$

where $a_{ij}, b_i,\ i, j = 1, 2,\ c \geq 0$ are continuous functions satisfying

$$a_{11}v_x^2 + a_{22}v_y^2 + (a_{12} + a_{21})v_x v_y \geq c_0(\nabla v)^2 \qquad \forall v \in S_h(\Omega)$$

for some $c_0 > 0$.

According to the maximum value principle, it can be proved that the following equation

$$A'(P_h u, v) = (L(P_h u), v) = (f, v) \quad \forall v \in S_h(\Omega) \tag{4}$$

is uniquely solvable. (see [10,12,13])

Namely, such mapping: $P_h : u \to P_h u$, which satisfying

$$A'(u - P_h u, v) = 0 \quad \forall v \in S_h(\Omega),$$

is the Galerkin's projection operator. And we have the following convergence theorem (see [2,11] etc.)

Theorem 1. *For $h \ll 1$, $P_h u$ admits the following estimate*

$$\|u - P_h u\|_{1,\infty} \le Ch\|u\|_{2,\infty}, \tag{5}$$

for $C > 0$.

Moreover, we introduce the discrete Green functions as, given $z \in \Omega$, the Green functions $g^z_{h,x}, g^z_{h,y} \in S_h(\Omega)$ satisfying

$$A'(v, g^z_{h,x}) = v_x(z), \quad A'(v, g^z_{h,y}) = v_y(z) \quad \forall v \in S_h(\Omega). \tag{6}$$

It has been proved that (see [14])

$$\|g^z_{h,x}\|_{1,1} \le C|\log h|, \tag{7}$$

$$\|g^z_{h,y}\|_{1,1} \le C|\log h|, \tag{8}$$

for constant $C > 0$.

4 The Approximation to the Discrete Plateau Problems

The following lemma shows the sufficient and necessary condition of being the finite element solution to (2.3).

Lemma 1. *For any w, w_h satisfies (2.2), (2.3) respectively, and $v \in S_h(\Omega)$,*

$$A(w_h + \bar{f}, v) = A(w + \bar{f}, v) + A'(w + \bar{f}; w_h - w, v) + R(w + \bar{f}, w_h + \bar{f}, v), \tag{9}$$

where $A'(w + \bar{f}; .,.)$ is the bilinear operator defined as (3.1), when we take the coefficients a_{ij}, b_i, c, $i, j = 1, 2$, respectively according to the first argument as

$$a_{11} = 1 + u_y^2, \; a_{22} = 1 + u_x^2, \quad a_{12} = a_{21} = -u_y u_x,$$

$$b_1 = 3u_y u_{xy} - 3u_{yy} u_x, \quad b_2 = 3u_x u_{xy} - 3u_{xx} u_y,$$

and $c = 0$.

Then $w_h \in S_h(\Omega)$ solves (2.3) if and only if

$$A'(w + \bar{f}; w - w_h, v) = R(w + \bar{f}, w_h + \bar{f}, v), \quad \forall v \in S_h(\Omega). \tag{10}$$

Further more, if $\|w_h\|_{2,\infty} \le K$, then the remainder R satisfying

$$R(w + \bar{f}, w_h + \bar{f}, v) \le C(K)\|w - w_h\|^2_{1,\infty}\|v\|_{1,1}. \tag{11}$$

Proof: Set $G(t) = A(w + \bar{f} + t(w_h - w), v)$. then it follows from the identity

$$G(1) = G(0) + G_t(0) + \int_0^1 G_{tt}(t)(1-t)dt.$$

It is easy to compute by local intergration that

$$G_t(0) = A'(w + \bar{f}; w_h - w, v)$$

where the coefficients of A' satisfying the given conditions from the lemma. And, by taking

$$R(w + \bar{f}, w_h + \bar{f}, v) = \int_0^1 G_{tt}(t)(1-t)dt.$$

A straightforward calculation shows that

$$\begin{aligned} &|R(w + \bar{f}, w_h + \bar{f}, v)| \\ &\le \max|G_{tt}(t)| \\ &\le C(K)\|w - w_h\|_{1,\infty}^2 \|v\|_{1,1}. \end{aligned}$$

Finally, if w_h solves (2.3), then $G(1) = 0$ and complete the proof.

Let P_h be the Galerkin projection with respect to the bilinear from $A'(w + \bar{f}; ., .)$, then we have the following theorem:

Theorem 2. *For problems of the form (2.2), there exist a constant $C > 0$, such that for mesh T_h with its size $h \ll 1$, the corresponding finite element equation (2.3) has a unique solution w_h satisfying*

$$\|w_h - P_h w\|_{1,\infty} \le Ch^2 |logh|,$$

and thus we have the following error estimation

$$\|w_h - w\|_{1,\infty} \le Ch$$

Proof: Define a nonlinear operator $\Phi : S_h(\Omega) \to S_h(\Omega)$ by

$$A'(w + \bar{f}, \Phi(v) - w, \phi) = R(w + \bar{f}, v + \bar{f}, \phi) \quad \forall \phi \in S_h(\Omega)$$

It can be proved according to the last section that, Φ is well defined and a contiunous operator, according to the result from Theorem 3.1 we can obtain that

$$\|w - P_h w\|_{1,\infty} \le Ch.$$

Then define a set $B = \{v \in S_h(\Omega) : \|v - P_h w\|_{1,\infty} \le Ch, C > 0\}$, and by the inverse estimation, there exist $C_1, C_2, C_3 > 0$, such that

$$\begin{aligned} \|v - w\|_{2,\infty} &\le \|v - I_h w\|_{2,\infty} + \|I_h w - w\|_{2,\infty} \\ &\le C_1 h^{-1}(\|v - P_h w\|_{1,\infty} + \|I_h w - P_h w\|_{1,\infty}) + C_2 \\ &\le C_3, \end{aligned}$$

then we get $\|v\|_{2,\infty}$ is uniformly bounded.

Now we can prove that $\Phi(B) \subset B$ which means Φ is a contraction operator. In fact, when we substitute ϕ by the discrete Green Function $\phi = g^z_{h,x}$ and $\phi = g^z_{h,y}$ into

$$A'(w + \bar{f}; \Phi(v) - P_h w, \phi) = R(w + \bar{f}, v + \bar{f}, \phi),$$

according to Lemma 4.1, (3.4) and (3.5), for all $v \in B$, we get

$$\begin{aligned} \|\Phi(v) - P_h w\|_{1,\infty} &\le C_0 |\text{log}h| \|w - v\|^2_{1,\infty} \\ &\le 2C_0 |\text{log}h| (\|P_h w - v\|^2_{1,\infty} + \|w - P_h w\|^2_{1,\infty}) \\ &\le 2C_0 |\text{log}h| (C^2 h^2 + C^2 h^2) \\ &\le Ch, \end{aligned}$$

where $C_0, C_1 > 0$, for $h \ll 1$.

By Brouwer's fixed point theorem, there exist a solution $w_h \in B$, such that $\Phi(w_h) = w_h$. And according to Lemma 4.1, w_h solves (2.3) and satisfies

$$\|w_h - P_h w\|_{1,\infty} \le Ch^2 |logh|.$$

For uniqueness, we consider that u_h should still maintains the minimal area property over the domain Ω as (see [4])

$$u_h = \min_{u|_{\partial\Omega} = f} \{A^*(u_h) = \int_\Omega (1 + |\nabla u_h|^2)^{1/2}\}.$$

By the fact that the operator A^* is strictly convex (see [7]) and $S_h(\Omega)$ is finite dimensional, it then follows that there exists a unique minimizing solution u_h.

5 Graphical Examples for Generating Minimal Surface

To illustrate the features of the approximation method proposed from the last sections, we now show two model examples.

Example 1.

For $\Omega = [0, 1] \times [0, 1]$, we use the piecewise bilinear finite elements with mesh size $h = 1/16$. The border curves are the restriction of smooth function $\bar{f} = 0.25 - (x - 0.5)^2$. Then the finite element result after 3 iteration steps is displayed in Figure 1.

Example 2.

This example shows the convergence property on a trival case. By given the initial border function the restriction of $xy(1 - x)(1 - y)$ on the same domain, we know the exact solution to this Plateau Problem should be a plane $z = 0$.

Figure 2 shows the maxium error between the numerical solution and the exact one will not be greater than $2 * 10^{-4}$ if we use the same grid partition as the last example. And based on the relatively coarser grid partition by 8×8 and

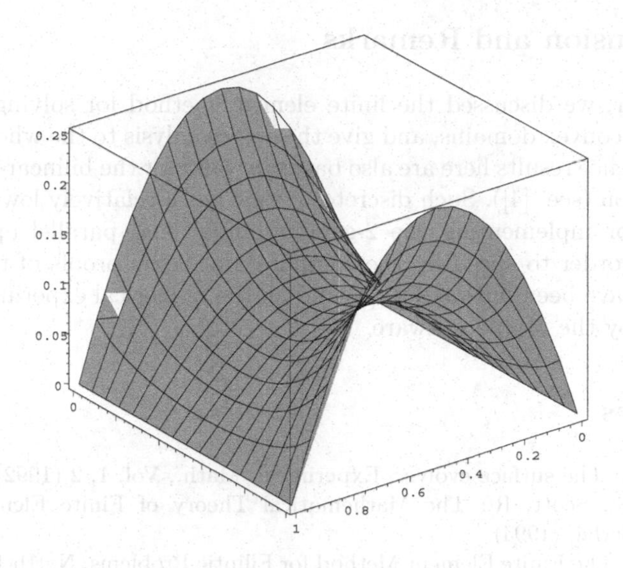

Fig. 1. The approximated minimal surface

4×4 meshes, we get the maximum errors are less than $8 * 10^{-4}$ and $4 * 10^{-3}$, respectively, which coincident with our theoretical estimations in the last section.

Those experimental datas for the computations are obtained from the Solaris operating system computer galaxy.risc.uni-linz.ac.at with 1152M memory.

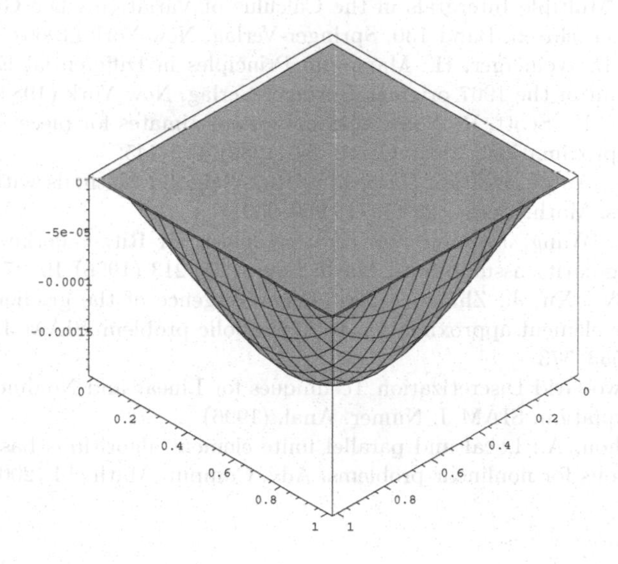

Fig. 2. The shape after computation which also indicate the error

6 Conclusion and Remarks

In this paper, we discussed the finite element method for solving the Plateau problems on convex domains, and give the error analysis to the whole approachs. The convergence results here are also basically valid for the bilinear finite element approximation (see [4]). Such discrete scheme has a relatively lower complexity and fitted for inplementing the 2-grid technique and parallel approachs (see [8,15,16]) in order to speed up the computation. Some proofs of those speedup algorithms have been finished in [4] and all the numerical experiments here are carried out by the Maple software.

References

1. Brakke, K.: The surface evolver. Experiment. Math., Vol. 1, 2 (1992) 141–165.
2. Brenner, S., Scott, R.: The Mathematical Theory of Finite Element Methods. Springer-Verlag (1994)
3. Ciarlet, P.: The Finite Element Method for Elliptic Problems. North-Holland (1978)
4. Gu, H.: The Software Generating of Minamal Surface Subject to the Plateau Problems. PhD thesis (to appear), RISC-Linz, Austria (2002)
5. Hoffman, D.: Natural minimal surfaces. Via theory and computation. Science Television, New York; American Mathematical Society, Providence, RI (1990)
6. Matthies, H., Strang, G.: The solution of nonlinear finite element equations. Internat. J. Numer. Methods Engrg., Vol. 14, 11 (1979) 1613–1626.
7. Johnson, C., Thomeé, V.: Error estimates for a finite element approximation of a minimal surface. Math. of Comp., Vol. 29, 130 (1975) 343–349.
8. Marion, M., Xu, J.: Error estimates on a new nonlinear Galerkin method based on two-grid finite elements. SIAM J. Numer. Anal. 32, 4 (1995) 1170–1184.
9. Morry, C.: Multiple Intergrals in the Calculus of Variations. Die Grundlehren der math. Wissenschften, Band 130, Springer-Verlag, New York (1966)
10. Protter, M., Weinerger, H.: Maximum Principles in Differential Equations. Corrected reprint of the 1967 original. Springer-Verlag, New York (1984)
11. Rannacher, R., Scott, R.: Some optional error estimates for piecewise linear finite element approximation. Math. Comp. 54 (1982) 437–445.
12. Schatz, A.: An observation concerning Ritz-Galerkin Methods with indefinite bilinear forms. Math. Comp. 28 (1974) 959–962.
13. Schatz, A., Wang, J.: Some new error estimates for Ritz-Galerkin methods with minimal regularity assumptions. Math. Comp. 65, 213 (1996) 19–27.
14. Thomeé, V., Xu, J., Zhang, N.: Supperconvergence of the gradient in piecewise linear finite element approximation to a parabolic problem. SIAM J. Numer. Anal. 26 (1989) 553–573.
15. Xu, J.: Two-Grid Discretization Techniques for Linear and Nonlinear Partial Differential Equation. SIAM J. Numer. Anal. (1996)
16. Xu, J., Zhou, A.: Local and parallel finite element algorithms based on two-grid discretizations for nonlinear problems. Adv. Comput. Math. 14 (2001) 393–327.

Optimal Structural Design of Biomorphic Composite Materials*

Ronald H.W. Hoppe[1] and Svetozara I. Petrova[1,2]

[1] Institute of Mathematics, University of Augsburg,
University Str.14, D-86159 Augsburg, Germany
[2] Central Laboratory for Parallel Processing, Bulgarian Academy of Sciences,
Acad. G. Bontchev Str., Block 25A, 1113 Sofia, Bulgaria

Abstract. The production of biomorphic microcellular silicon carbide ceramics from natural grown materials like wood has attracted in the last few years particular interest in the field of biomimetics. Based on the constitutive microstructural model for the inelastic behavior of the new ceramic composites the macroscale model is obtained by using the homogenization approach. The paper comments on the optimal distribution of our composite material in a suitable reference domain which can carry given loads. The structural optimization problem is solved under a set of constraints on the state variables (displacements) and design parameters (material density and angle of cell rotation). Primal–dual Newton–type interior–point method with proper optimality criteria is applied to the resulting nonconvex nonlinear optimization problem.

1 Introduction

Biotemplating is a new technology in the field of biomimetics that focuses on a material synthesis of biologically grown materials into microcellular ceramic composites by high temperature processing. The biological structures are often microstructural designed materials with a hierarchical composite morphology revealing outstanding mechanical properties such as high modulus and high tensile strength both on micro and on the macro scale. Special emphasis has been done on the production of biomorphic silicon carbide (SiC)-based ceramics from wood (cf., e.g., [4,5] and the references therein). Experiments show that their anisotropic porous microstructures are one-to-one pseudomorphous to the original wood material.

The production process of the biomorphic SiC ceramic materials comprises several processing steps ranging from the preparation of appropriate carbonized preforms by drying and high-temperature pyrolysis (800-1800°C) in inert atmosphere via chemical reactions by liquid- or gaseous-phase Si-infiltrations at 1600°C to postprocessing such as cutting and etching. In this way, the natural

* This work has been partially supported by the German National Science Foundation (DFG) under Grant No.HO877/5-1. The second author has also been supported in part by the Bulgarian NSF Grant I1001/2000.

I. Dimov et al. (Eds.): NMA 2002, LNCS 2542, pp. 479–487, 2003.

microstructure can be preserved to produce high performance ceramics with excellent structural-mechanical and thermomechanical properties offering a wide range of applications (heat insulation, particle filters, medical implants, automotive tools, etc.)

Optimal performances of the new composite materials can be obtained by tuning microstructural geometrical features that strongly influence the macrocharacteristics of the final products. Our macroscale model is obtained by using the homogenization approach which has found a lot of applications in mechanics of composite materials (see [1,3,9]). We assume a periodical distribution of the composite microstructure treated as infinitesimal square tracheidal periodicity cell consisting of two materials (carbon and SiC) and a void (no material). Detailed description of the macroscopic homogenized model obtained by asymptotic expansion of the solution of the nonhomogenized elasticity equation with a scale parameter close to zero is given in [6]. In the case of stationary microstructure we compute the effective elasticity coefficients numerically by finite element discretization of the cell. Efficient iterative solvers for the homogenized equation in a linearly elastic design macrostructure are proposed in [7].

In this paper, we focus on the development of efficient methods for the structural optimization of biomorphic microcellular SiC ceramics using homogenization modelling and model validations by experimental studies. The optimal design of microstructural materials by homogenization method is well established in structural mechanics (cf., e.g., [2,10]). The structural optimization problem applied to our macroscopic homogenized model is solved under a set of equality and inequality constraints on the state variables (displacements) and the design parameters (material density and angle of cell rotation). This task typically leads to nonconvex nonlinear minimization problem for the objective functional exhibiting a variety of local optima and saddle points. The behavior of the homogenized elasticity coefficients with respect to the design parameters is investigated and visualized in Section 2. The primal–dual Newton–type interior–point method for solving the optimization problem is applied in Section 3.

2 Design Parameters

Consider a two–dimensional bounded domain Ω of a macroscopic length L occupied by a SiC-based ceramic composite material. We assume periodically distributed constituents with a square periodicity cell Y of characteristic length l consisting of an interior part (region V) that is either void (no material) or filled with silicon surrounded by a layer of silicon carbide (region SiC) and an outer layer of carbon (region C) (see Fig. 1).

Introducing $x \in \Omega$ and $y \in Y$ as the macroscopic (slow) and microscopic (fast) variables and $\varepsilon := l/L$ as the scale parameter, the homogenized approach based on a double scale asymptotic expansion results (see [6] for details) in the homogenized elasticity tensor $\mathbf{E}^H = (E^H_{ijkl}),\ i,j,k,l = 1,2$ of the form

$$E^H_{ijkl} = \frac{1}{|Y|} \int_Y \left(E_{ijkl}(y) - E_{ijpq}(y) \frac{\partial \xi_p^{kl}}{\partial y_q} \right) dy, \tag{1}$$

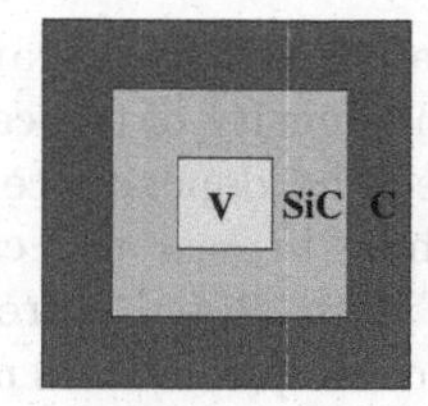

Fig. 1. The periodicity cell $Y = V \cup SiC \cup C$

where $\xi^{kl} = \xi^{kl}(y_1, y_2) \in [H^1(Y)]^2$ is a Y-periodic function which is considered as a microscopic displacement field for the following elasticity cell–problem given in a weak formulation

$$\int_Y \left(E_{ijpq}(y) \frac{\partial \xi_p^{kl}}{\partial y_q} \right) \frac{\partial \phi_i}{\partial y_j} \, dy = \int_Y E_{ijkl}(y) \frac{\partial \phi_i}{\partial y_j} \, dy, \qquad \forall \phi \in V_Y. \tag{2}$$

Here, V_Y is the set of all admissible Y-periodic virtual displacement fields.

Explicit formulas for the homogenized elasticity coefficients (1) can be found only in the case of layered materials and checkerboard structures (cf., e.g., [1,2,9]). Due to the equal solutions $\xi^{12} = \xi^{21}$ one has to solve numerically three problems (2) in the period Y. We use conforming P1 finite elements with respect to a simplicial triangulation of Y taking into account the decomposition $Y = V \cup SiC \cup C$. Note that in our numerical experiments the void is treated as a weak material which is introduced in order to avoid singularity of the stiffness matrix during computations. The values of the Young modulus and Poisson's ratio for the respective materials are given in Table 1.

Table 1. Young's modulus E and Poisson's ratio ν

material	E(GPa)	ν
weak material	0.1	0.45
silicon carbide	410	0.14
carbon	10	0.22

Efficient iterative solvers and extensive numerical experiments for the homogenized equation in a linearly elastic design macrostructure are presented in [7]. The discretized problems have been solved by the PCG method with block- or point-wise incomplete Cholesky decomposition as a preconditioner.

We consider a microstructure occupying the unit cell $Y = [0, 1]^2$. The domain of the weak material in the microcell is specified as a rectangular α by β hole. This choice allows to realize the complete solid ($\alpha = \beta = 0$) and the complete void

($\alpha = \beta = 1$). The remaining values $0 < \alpha < 1$ and $0 < \beta < 1$ characterize the generalized porous medium. The density of the cell is computed by $\mu = 1 - \alpha\beta$.

We have further investigated the dependence of the homogenized elasticity tensor $\mathbf{E}$ on the sizes of the hole in case the carbon has completely reacted with the silicon (i.e., $C = \emptyset$). Figures 2 and 3 present the relations between the effective material properties and the parameters $a = 1 - \alpha$ and $b = 1 - \beta$. Finite element discretization of the cell with 20×20 grid points varying the sizes of the rectangular hole are used to compute the homogenized elasticity coefficients.

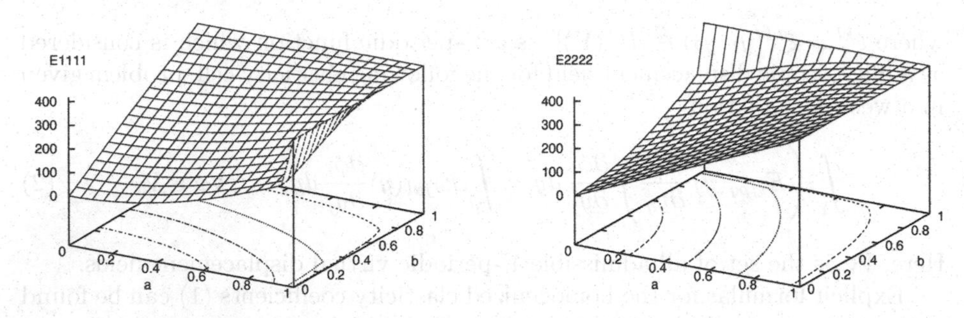

Fig. 2. Homogenized coefficients E_{1111} and E_{2222}

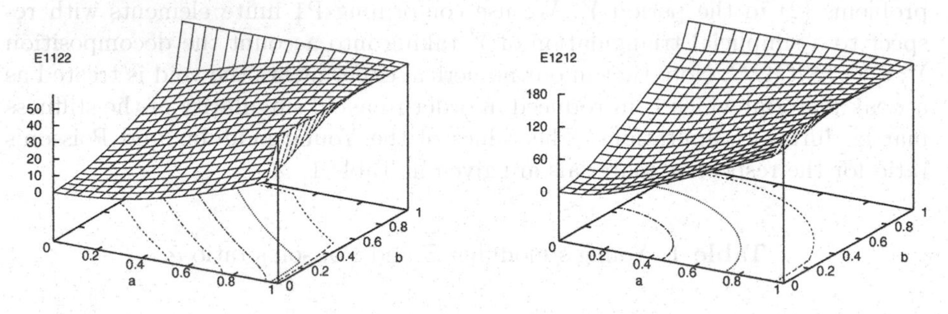

Fig. 3. Homogenized coefficients E_{1122} and E_{1212}

We also note that the homogenized anisotropic elasticity tensor in the macroscale problem strongly depends on the orientation of the microscopic unit cell. Moreover, the influence of rotation of the cell becomes very strong when the size of the microscale hole becomes large. Denote by θ the angle of cell rotation with respect to a fixed reference frame (for instance, the coordinate system). The unrotated case corresponds to $\theta = 0$. The design parameters in the structural otpimization are α, β, and the angle θ. One can compute the rotated elasticity coefficients E^R_{ijkl} for $i, j, k, l = 1, 2$ as follows

$$E^R_{ijkl} = \sum_{m,n,p,q=1}^{2} E^H_{mnpq}(\alpha, \beta)\, R_{im}(\theta)\, R_{jn}(\theta)\, R_{kp}(\theta)\, R_{lq}(\theta), \tag{3}$$

where $R_{im}(\theta)$, $1 \le i, m \le 2$, are the components of the rotation matrix.

In particular, the rotated elasticity tensor $\mathbf{E}^R$ (symmetric) has the form

$$\mathbf{E}^R = \begin{pmatrix} E^R_{1111} & E^R_{1122} & E^R_{1112} \\ E^R_{2211} & E^R_{2222} & E^R_{2212} \\ E^R_{1211} & E^R_{1222} & E^R_{1212} \end{pmatrix},$$

where the corresponding entries depend on the effective elasticity coefficients $E^H_{mnpq} = E^H_{mnpq}(\alpha, \beta)$, $m, n, p, q = 1, 2$ as follows

$$E^R_{1111} = E^H_{1111} \cos^4 \theta + E^H_{2222} \sin^4 \theta + (4E^H_{1212} + 2E^H_{1122}) \sin^2 \theta \cos^2 \theta,$$
$$E^R_{1122} = (E^H_{1111} + E^H_{2222} - 4E^H_{1212}) \sin^2 \theta \cos^2 \theta + E^H_{1122}(\cos^4 \theta + \sin^4 \theta),$$
$$E^R_{1112} = \sin \theta \cos \theta \left[(E^H_{1111} - E^H_{1122}) \cos^2 \theta - (E^H_{2222} - E^H_{1122}) \sin^2 \theta - 2E^H_{1212} \cos 2\theta\right],$$
$$E^R_{2222} = E^H_{1111} \sin^4 \theta + E^H_{2222} \cos^4 \theta + (4E^H_{1212} + 2E^H_{1122}) \sin^2 \theta \cos^2 \theta,$$
$$E^R_{2212} = \sin \theta \cos \theta \left[(E^H_{1111} - E^H_{1122}) \sin^2 \theta - (E^H_{2222} - E^H_{1122}) \cos^2 \theta + 2E^H_{1212} \cos 2\theta\right],$$
$$E^R_{1212} = (E^H_{1111} + E^H_{2222} - 2E^H_{1122}) \sin^2 \theta \cos^2 \theta + E^H_{1212} \cos^2 2\theta.$$

Figure 4 shows the behavior of the rotated elasticity coefficients for a square hole in the microcell with density $\mu = 0.8$ and $\mu = 0.26$, respectively. We vary the angle of cell rotation by values $0 \leq \theta \leq 1.5$ given in radians. Note that for a square cell the profile of the coefficient E_{1111} coincides with those of E_{2222}.

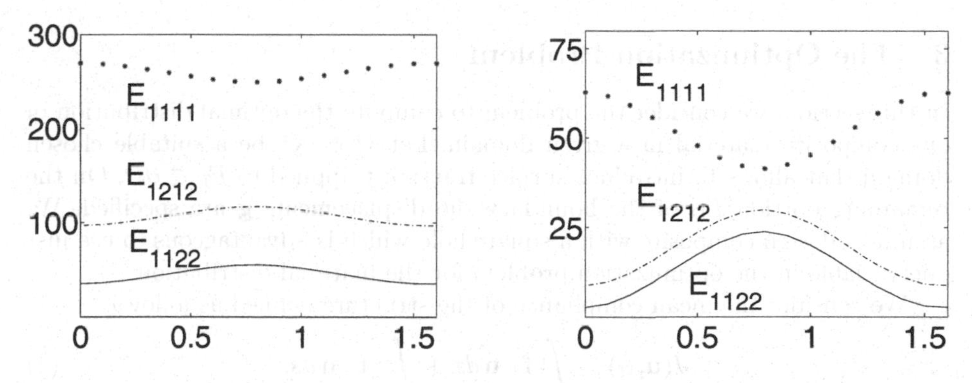

Fig. 4. Elastic coefficients w.r.t. to cell rotation (density $\mu = 0.8$ and $\mu = 0.26$)

Figure 5 displays the dependence of the homogenized elasticity coefficients on the density of the cell. We consider a square hole and compute the effective elasticity tensor for a certain number of points of the density (varying the size of the hole). In order to get a continuous variation with respect to the material density, the remaining values of the homogenized elasticity coefficients on

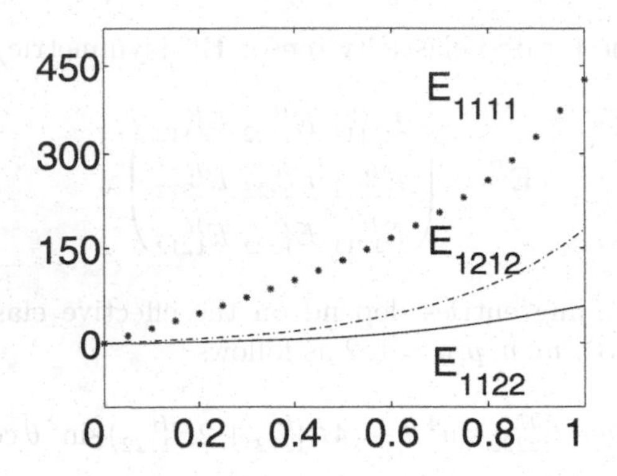

Fig. 5. Elastic coefficients w.r.t. density (square hole)

Figure 5 are interpolated by splines. One can easily observe a highly nonlinear behavior of these coefficients.

Our purpose now is to form continuous functions for the corresponding material coefficients E^H_{ijkl}. We note that E^H_{ijkl} are computed only for a fixed number of values of the cell hole sizes and interpolated by, for instance, Legendre polynomials or Bezier curves defined in the interval $[0, 1]$ to obtain a continuous functional dependence. Thus, together with the expressions (3) we compute the rotated homogenized elasticity tensor $\mathbf{E}^R$, depending on sines and cosines in θ. Concerning the optimization problem the continuous functions of the rotated material coefficients allow us to compute explicitly the gradient and Hessian of the Lagrangian function needed in the optimization loops.

3 The Optimization Problem

In this section, we consider the problem to compute the optimal distribution of our composite material in a given domain. Let $\Omega \subset \mathcal{R}^2$ be a suitable chosen domain that allows to introduce surface traction $\mathbf{t}$ applied to $\Gamma_T \subset \partial\Omega$. On the remaining portion Γ_D of the boundary the displacements $\mathbf{g}$ are specified. We assume a design composite with a square hole which is advantageous to use just one variable in the optimization problem for the material distribution.

We consider the mean compliance of the structure defined as follows

$$J(\mathbf{u}, \alpha) = \int_\Omega \mathbf{f} \cdot \mathbf{u}\, dx + \int_{\Gamma_T} \mathbf{t} \cdot \mathbf{u}\, ds, \tag{4}$$

where $\mathbf{f}$ is the external body force applied to Ω. The displacement vector $\mathbf{u} = (u_1, u_2)^T$ represents the state variables, and the vector $\alpha = (\mu, \theta)^T$ stands for the design parameters (μ - the density of the composite material and θ - the angle of rotation).

For a given volume of material M our structural optimization problem has the form

$$\inf_{\mathbf{u},\alpha} J(\mathbf{u}, \alpha),$$

subject to the following equality and inequality constraints

$$\sum_{i,j,k,l=1}^{2} \int_{\Omega} E_{ijkl}^{R} \frac{\partial u_k}{\partial x_l} \frac{\partial \phi_i}{\partial x_j}\, dx = \int_{\Omega} \mathbf{f} \cdot \phi \, dx + \int_{\Gamma_T} \mathbf{t} \cdot \phi \, ds, \qquad \forall \phi \in V_0 \quad (5)$$

$$g(\mu) := \int_{\Omega} \mu \, dx = M\,, \qquad \mu_1 \le \mu \le \mu_2, \quad (6)$$

for constants $0 \le \mu_1 < \mu_2 \le 1$. Note that (5) is the weak form of the rotated homogenized equilibrium equation. Here, $\mathbf{u} \in V_D = \{\mathbf{v} \in \mathbf{H}^1(\Omega) | \mathbf{v} = \mathbf{g} \text{ on } \Gamma_D\}$ and $V_0 = \{\mathbf{v} \in \mathbf{H}^1(\Omega) | \mathbf{v} = \mathbf{0} \text{ on } \Gamma_D\}$.

Furthermore, the state variables are discretized by conforming P1 elements with respect to a triangulation of Ω. The discretized nonlinear constrained minimization problem has the following form

$$\inf_{\mathbf{u},\mu,\theta} J(\mathbf{u}, \mu, \theta), \quad (7)$$

subject to

$$\begin{aligned} A(\mu, \theta)\, \mathbf{u} - \mathbf{b} &= 0, \\ \mu - \mu_1 &\ge 0, \\ \mu_2 - \mu &\ge 0, \end{aligned} \quad (8)$$

where $A(\mu, \theta)$ is the stiffness matrix corresponding to (5) and $\mathbf{b}$ is the discrete load vector.

The discretized constrained minimization problem is solved by primal–dual interior–point method substituting the inequality constraints in (8) by logarithmic barrier functions. Assuming that $\mu > \mu_1$ and $\mu_2 > \mu$ this substitution results in the following parametrized family of optimization subproblems

$$\inf_{\mathbf{u},\mu,\theta} \left[J(\mathbf{u}, \mu, \theta) - \rho \left(\log(\mu - \mu_1) + \log(\mu_2 - \mu) \right) \right] \quad (9)$$

subject to

$$A(\mu, \theta)\, \mathbf{u} - \mathbf{b} = 0, \quad (10)$$

where $\rho > 0$ is a suitably chosen barrier parameter. Coupling the equality constraint by Lagrangian multiplier we have the following Lagrangian function associated with the problem (9)-(10)

$$\begin{aligned} L_\rho(\mathbf{u}, \mu, \theta; \lambda) := {} & J(\mathbf{u}, \mu, \theta) - \rho \left(\log(\mu - \mu_1) + \log(\mu_2 - \mu) \right) \\ & + \lambda^T (A(\mu, \theta)\, \mathbf{u} - \mathbf{b}). \end{aligned}$$

The first-order Karush–Kuhn–Tucker (KKT) conditions are given by

$$\mathbf{F}^\rho(\mathbf{u}, \mu, \theta; \lambda) = \mathbf{0}\,, \quad (11)$$

where

$$\begin{aligned} F_1^\rho &= \nabla_{\mathbf{u}} L_\rho = \nabla_{\mathbf{u}} J + A(\mu, \theta)^T \lambda\,, \\ F_2^\rho &= \nabla_\mu L_\rho = \partial_\mu (\lambda^T A(\mu, \theta) \mathbf{u}) - \rho / d_1 + \rho / d_2\,, \\ F_3^\rho &= \nabla_\theta L_\rho = \partial_\theta (\lambda^T A(\mu, \theta) \mathbf{u})\,, \\ F_4^\rho &= \nabla_\lambda L_\rho = A(\mu, \theta)\, \mathbf{u} - \mathbf{b}\,, \end{aligned} \quad (12)$$

and $d_1 := \mu - \mu_1$ and $d_2 := \mu_2 - \mu$. Since for $\rho \to 0$ the expressions ρ/d_1 and ρ/d_2 approximate the complementarity conditions associated with (7), it is standard to introduce $z := \rho/d_1 \geq 0$ and $w = \rho/d_2 \geq 0$ serving as perturbed complementarity. Then, the primal–dual Newton-type interior–point method is applied to three sets of variables: primal feasibility $(\mathbf{u}, \mu, \theta)$, dual feasibility (λ), and perturbed complementarity conditions related to (z, w).

Denote the Lagrangian function of problem (7)-(8) by

$$
\begin{aligned}
L(\mathbf{u}, \mu, \theta; \lambda; z, w) := & \; J(\mathbf{u}, \mu, \theta) \\
& + \lambda^T (A(\mu, \theta)\,\mathbf{u} - \mathbf{b}) \\
& - z(\mu - \mu_1) - w(\mu_2 - \mu).
\end{aligned}
$$

Taking into account the KKT conditions the Newton method results in

$$
\begin{pmatrix}
0 & L_{\mathbf{u}\mu} & L_{\mathbf{u}\theta} & L_{\mathbf{u}\lambda} & 0 & 0 \\
L_{\mu\mathbf{u}} & L_{\mu\mu} & L_{\mu\theta} & L_{\mu\lambda} & -1 & 1 \\
L_{\theta\mathbf{u}} & L_{\theta\mu} & L_{\theta\theta} & L_{\theta\lambda} & 0 & 0 \\
L_{\lambda\mathbf{u}} & L_{\lambda\mu} & L_{\lambda\theta} & 0 & 0 & 0 \\
0 & z & 0 & 0 & d_1 & 0 \\
0 & -w & 0 & 0 & 0 & d_2
\end{pmatrix}
\begin{pmatrix}
\Delta\mathbf{u} \\ \Delta\mu \\ \Delta\theta \\ \Delta\lambda \\ \Delta z \\ \Delta w
\end{pmatrix}
= -
\begin{pmatrix}
\nabla_{\mathbf{u}} L \\ \nabla_{\mu} L \\ \nabla_{\theta} L \\ \nabla_{\lambda} L \\ \nabla_{z} L \\ \nabla_{w} L
\end{pmatrix}, \tag{13}
$$

where $\nabla_z L = d_1 z - \rho$ and $\nabla_w L = d_2 w - \rho$. The coefficient matrix (13) is usually referred to as the primal–dual system. It can be easily symmetrized but we do not use this approach here. Instead we eliminate the increments Δz and Δw yielding the condensed primal–dual system

$$
\begin{pmatrix}
0 & L_{\mathbf{u}\mu} & L_{\mathbf{u}\theta} & L_{\mathbf{u}\lambda} \\
L_{\mu\mathbf{u}} & \tilde{L}_{\mu\mu} & L_{\mu\theta} & L_{\mu\lambda} \\
L_{\theta\mathbf{u}} & L_{\theta\mu} & L_{\theta\theta} & L_{\theta\lambda} \\
L_{\lambda\mathbf{u}} & L_{\lambda\mu} & L_{\lambda\theta} & 0
\end{pmatrix}
\begin{pmatrix}
\Delta\mathbf{u} \\ \Delta\mu \\ \Delta\theta \\ \Delta\lambda
\end{pmatrix}
= -
\begin{pmatrix}
\nabla_{\mathbf{u}} L \\ \tilde{\nabla}_{\mu} L \\ \nabla_{\theta} L \\ \nabla_{\lambda} L
\end{pmatrix}, \tag{14}
$$

where $\tilde{L}_{\mu\mu} := L_{\mu\mu} + z/d_1 + w/d_2$ and $\tilde{\nabla}_{\mu} L := \nabla_{\mu} L - \nabla_z L/d_1 + \nabla_w L/d_2$. For details of solving the condensed primal–dual system by interior–point method using damped Newton iterations in structural optimization of electromagnetic devices we refer to [8].

References

1. Bakhvalov, N., Panasenko, G.: Averaging Processes in Periodic Media. Nauka, Moscow (1984)
2. Bendsøe, M.P.: Optimization of Structural Topology, Shape, and Material. Springer (1995)
3. Bensoussan, A., Lions, J.L., Papanicolaou, G.: Asymptotic Analysis for Periodic Structures. North–Holland, Elsevier Science Publishers, Amsterdam (1978)
4. Greil, P., Lifka, T., Kaindl A.: Biomorphic cellular silicon carbide ceramics from wood: I. Processing and microstructure. J. Europ. Ceramic Soc. 18 (1998) 1961–1973.

5. Hoffmann, C., Vogli, E., Kladny, R., Kaindl, A., Sieber, H., Greil, P.: Processing of biomorphic ceramics from wood. In: Stanzl-Tschegg, S.E. et al. (eds.): Proc. 1st Int. Symp. on Wood Mechanics, Vienna (2000) 221–229.
6. Hoppe, R. H.W., Petrova, S. I.: Structural optimization of biomorphic microcellular ceramics by homogenization approach. In: Margenov, S., Wasniewski, J., and Yalamov, P. (eds.): Large-Scale Scientific Computing. Lecture Notes in Computer Science, Vol. 2179, Springer-Verlag (2001) 353–360.
7. Hoppe, R. H.W., Petrova, S. I.: Homogenized elasticity solvers for biomorphic microcellular ceramics. In: Brezzi. F. et al. (eds.): Proc. ENUMATH 2001, July 23-28, 2001, Ischia, Italy, Springer ITALIA (to appear).
8. Hoppe, R. H.W., Petrova, S. I., Schulz, V.: Primal–dual Newton–type interior–point method for topology optimization. J. Optim. Theory Appl. 114 (2002) 545–571.
9. Jikov, V.V., Kozlov, S.M., Oleinik, O.A.: Homogenization of Differential Operators and Integral Functionals. Springer (1994)
10. Suzuki, K., Kikuchi, N.: A homogenization method for shape and topology optimization. Comput. Meth. Appl. Mech. Engrg. 93 (1991) 291–318.

A Generalized (Meshfree) Finite Difference Discretization for Elliptic Interface Problems

Oleg Iliev and Sudarshan Tiwari

Fraunhofer Institute for Industrial Mathematics (ITWM)
Gottlieb-Daimler-Strasse, Geb. 49
D-67663 Kaiserslautern, Germany
{iliev, tiwari}@itwm.fhg.de

Abstract. The aim of this paper is twofold. First, two generalized (meshfree) finite difference methods (GFDM) for the Poisson equation are discussed. These are methods due to Liszka and Orkisz (1980) [10] and to Tiwari (2001) [7]. Both methods are based on using moving least squares (MLS) approach for deriving the discretization. The relative comparison shows, that the second method is preferable because it is less sensitive to the topological restrictions on the nodes distribution. Next, an extension of the second method is presented, which allows for accounting for internal interfaces, associated with discontinuous coefficients. Results from numerical experiments illustrate the second order convergence of the proposed GFDM for interface problems.

Keywords: elliptic equation, discontinuous coefficients, interface problem, moving least squares method, meshfree method

1 Introduction

The appearance and the development of meshfree methods is motivated by the challenges in numerical simulation of processes involving significant changes of the geometry, such as multiphase flows, filling process and other free surface flows in fluid dynamics, large displacements, crack propagation in solid mechanics, etc. The traditional mesh-based method, such as finite element method (FEM) and finite difference method (FDM) often run into problems when the mesh cells (elements) deteriorate during simulation due to the geometry changes.

Meshfree methods are originally developed to simulate fluid dynamics problems, smoothed particle hydrodynamics (SPH) is referred to as the first meshfree method. In its original formulation, SPH is easy for implementation, but it provides poor accuracy/convergence [6,4]. Further, variety of meshfree methods were proposed in the last decade for solving fluid dynamics and solid mechanics problems. Detailed discussion on the development of meshfree methods and on their application, can be found, e.g., in [1,2,4,6,8,9,11]. It should be noted, that the terminology in this area is still not well established. We use the term meshfree methods, but one can find in the literature different names: particle

I. Dimov et al. (Eds.): NMA 2002, LNCS 2542, pp. 488–497, 2003.

methods, meshless, gridfree, gridless methods, cloud methods. These names are used as a notation for the group including such methods as SPH, reproducing kernel particle method (RKPM), element free Galerkin (EFG) methods, diffuse element methods (DEM), finite cloud method, generalized finite difference method (GFDM). Some of these methods are very close or equivalent to each other, and the great part of them allow a general consideration as partition of unity methods (PUMs). We refer to the above papers for detailed presentation of different classifications and of comparisons of meshfree methods. Here we will discuss only those aspects of meshfree methods, which are directly related to the particular problem we consider in this paper.

Currently, the moving least squares (MLS) method is the basis for the great part of the meshless methods, see [2,4,5,6,8] for more details. The MLS originates from the constructive approximation, where it is used in data fitting. Consider the domain $\Omega \subset R^d$, $d = 1,2,3$. For a given set (cloud, ensemble) of nodes, S, $S = \{\mathbf{x}_i | \mathbf{x}_i \in \Omega, i = 1,2,...,N\}$, and given values of a function, u_i, $i = 1,2,...,N$ in these nodes, MLS provides an algorithm for constructing a smooth function, $v(\mathbf{x}, \bar{\mathbf{x}})$, which is the best approximation to u and its derivatives in $\bar{\mathbf{x}}$ in a weighted least square sense. Further, this function v is used in different ways depending on the variant of the meshfree method. It can be used in construction of the kernel, or more general, in the partition of unity in the FEM-type meshfree methods. In generalized finite difference method (GFDM), the derivatives of u are approximated by the derivatives of v, which in turn depend on the nodal values u_i. These approximations are substituted in the partial differential equation, in order to obtain the discretization.

The following problem is considered here. Denote by $\bar{\Omega} \subset R^2$ a closed domain, and by $\partial\Omega$ its boundary. Suppose, $\bar{\Omega}$ is divided into subdomains by interfaces (which are surfaces in 3-D, or curves in 2-D), denoted by Γ. Denote by Ω the interior of $\bar{\Omega}$: $\Omega = \bar{\Omega} \setminus \{\partial\Omega \cup \Gamma\}$. For brevity, we consider the case of two subdomains, Ω_1 and Ω_2.

Consider the following second order elliptic equation

$$-\nabla(k\nabla u) = f, \quad \mathbf{x} \in \Omega,\ \mathbf{x} = (x_1, x_2), \qquad u(\mathbf{x}) = g(\mathbf{x}) \quad \mathbf{x} \in \partial\Omega, \tag{1}$$

where the diffusivity coefficient $k(\mathbf{x})$ may have jump discontinuity across Γ: $k(\mathbf{x}) = k_1(\mathbf{x})$ in Ω_1 and $k(\mathbf{x}) = k_2(\mathbf{x})$ in Ω_2. Other then Dirichlet boundary conditions can be used, as well. Here $k(\mathbf{x})$, $g(\mathbf{x})$ are given functions. Continuity of the solution $u(\mathbf{x})$, and of the normal component of the flux through the interface are required in this case:

$$[u]\,|_\Gamma = 0, \qquad \left[k\frac{\partial u}{\partial \mathbf{n}}\right]|_\Gamma = 0. \tag{2}$$

where $\mathbf{n}$ stands for the outer normal vector to Ω_1 on the interface, and $[\phi]$ denotes the jump of ϕ. Other interface conditions can be used, as well.

The paper is organized as follows. First, we present GFD methods [10,7] for the case $k(\mathbf{x}) \equiv 1$. Next, we briefly analyze stability of these methods in 1-D case. Further, we present an extension of [7] for the case of discontinuous coefficients.

The fourth section contains results from numerical experiments, and the last section is devoted to conclusions.

2 Meshfree Discretizations for the Constant Coefficients Case

We start with the case $k(\mathbf{x}) \equiv 1$, and present two GFD discretizations. Consider

$$Lu \equiv -\sum_{k=1}^{2} \frac{\partial^2 u}{\partial x_k^2} = f(\mathbf{x}), \quad \mathbf{x} \in \Omega \subset R^2 \quad + \text{proper boundary conditions on } \partial\Omega.$$

Let $u(\mathbf{x})$ be smooth enough. Denote by ω a set of N points from Ω. *We do not call ω grid or mesh, because we do not require any connections between these points.* Let $\bar{S}$ be a cloud in Ω (i.e., it is a subset of ω), associated with the node $\bar{\mathbf{x}}$. For example, these might be points from ω, which lie within a circle with radius H and center $\bar{\mathbf{x}}$. Note, this radius is usually associated with the support of weight function in the weighted least square method (to be discussed below). Thus, $\bar{S}$ consists from $\bar{\mathbf{x}}$ and nodes $\mathbf{x}_i$, $i = 1, 2, 3, ...\nu$.

Consider ν Taylor expansions

$$u(\mathbf{x}_i) = u(\bar{\mathbf{x}}) + \sum_{j=1}^{l} \frac{\partial u^{|j|}}{\partial x_1^{j_1} \partial x_2^{j_2}} \, \frac{1}{j!} \, (x_{1,i} - \bar{x}_1)^{j_1} (x_{1,i} - \bar{x}_2)^{j_2} + O(|\mathbf{x}_i - \bar{\mathbf{x}}|^{l+1}), \quad (3)$$

$i = 1, 2, ..., \nu$.

2.A. GFDM from [10]. Denote

$$\alpha_1 = \frac{\partial u}{\partial x_1}, \quad \alpha_2 = \frac{\partial u}{\partial x_2}, \quad \alpha_3 = \frac{\partial^2 u}{\partial x_1^2}, \quad \alpha_4 = \frac{\partial^2 u}{\partial x_1 \partial x_2}, \quad \alpha_5 = \frac{\partial^2 u}{\partial x_2^2},$$

Drop remainder terms and rewrite the set of ν, $\nu \geq 5$, Taylor expansions as overdetermined system with respect to unknown α-s:

$$M\alpha = \mathbf{b}, \quad \text{where } \alpha = (\alpha_1, \alpha_2, \alpha_3, \alpha_4, \alpha_5)^T, \quad (4)$$

$$M = \begin{pmatrix} h_{1,1} & h_{2,1} & \frac{1}{2}\, h_{1,1}^2 & h_{1,1}\, h_{2,1} & \frac{1}{2}\, h_{2,1} \\ h_{1,2} & h_{2,2} & \frac{1}{2}\, h_{1,2}^2 & h_{1,1}\, h_{2,2} & \frac{1}{2}\, h_{2,2} \\ \vdots & \vdots & \vdots & \vdots & \vdots \\ h_{1,\nu} & h_{2,\nu} & \frac{1}{2}\, h_{1,\nu}^2 & h_{1,\nu}\, h_{2,\nu} & \frac{1}{2}\, h_{2,\nu} \end{pmatrix},$$

$h_{1,i} = x_{1,i} - \bar{x}_1, h_{2,i} = x_{2,i} - \bar{x}_2, \mathbf{b} = ((u_1 - \bar{u}), (u_2 - \bar{u}), ..., (u_\nu - \bar{u}))^T, u_i = u(\mathbf{x}_i)$. The least square solution of the above problem (if $M^T M$ is nonsingular) is given by

$$\alpha = \left(M^T M\right)^{-1} M^T \mathbf{b},$$

where the 5×5 matrix $M^T M$ in this case is written as

$$M^T M = \sum_{i=1}^{\nu} \begin{pmatrix} h_{1,i}^2 & h_{1,i}\, h_{2,i} & \frac{1}{2} h_{1,i}^3 & h_{1,i}^2\, h_{2,i} & \frac{1}{2} h_{1,i}\, h_{2,i}^2 \\ h_{1,i}\, h_{2,i} & h_{2,i}^2 & \frac{1}{2} h_{1,i}^2\, h_{2,i} & h_{1,i}\, h_{2,i}^2 & \frac{1}{2} h_{2,i}^3 \\ \frac{1}{2} h_{1,i}^3 & \frac{1}{2} h_{1,i}^2\, h_{2,i} & \frac{1}{4} h_{1,i}^4 & \frac{1}{2} h_{1,i}^3\, h_{2,i} & \frac{1}{4} h_{1,i}^2\, h_{2,i}^2 \\ h_{1,i}^2\, h_{2,i} & h_{1,i}\, h_{2,i}^2 & \frac{1}{2} h_{1,i}^3\, h_{2,i} & h_{1,i}^2\, h_{2,i}^2 & \frac{1}{2} h_{1,i}\, h_{2,i}^3 \\ \frac{1}{2} h_{1,i}\, h_{2,i}^2 & \frac{1}{2} h_{2,i}^3 & \frac{1}{4} h_{1,i}^2\, h_{2,i}^2 & \frac{1}{2} h_{1,i}\, h_{2,i}^3 & \frac{1}{4} h_{2,i}^4 \end{pmatrix}.$$

The right hand side is

$$M^T \mathbf{b} = \left(\sum_{i=1}^{\nu} [h_{1,i}(u_i - \bar{u})], \sum_{i=1}^{\nu} [h_{2,i}(u_i - \bar{u})], \right.$$

$$\left. \sum_{i=1}^{\nu} [\tfrac{1}{2} h_{1,i}^2 (u_i - \bar{u})], \sum_{i=1}^{\nu} [h_{1,i} h_{2,i} (u_i - \bar{u})], \sum_{i=1}^{\nu} [\tfrac{1}{2} h_{2,i}^2 (u_i - \bar{u})] \right)^T .$$

In this case, there exist constants c_1, c_2, c_3, c_4, c_5 such that

$$\alpha_3 + \alpha_5 = c_1 \sum_{i=1}^{\nu} [h_{1,i}(u_i - \bar{u})] + c_2 \sum_{i=1}^{\nu} [h_{2,i}(u_i - \bar{u})] + \tag{5}$$

$$c_3 \sum_{i=1}^{\nu} [\tfrac{1}{2} h_{1,i}^2 (u_i - \bar{u})] + c_4 \sum_{i=1}^{\nu} [h_{1,i} h_{2,i} (u_i - \bar{u})] + c_5 \sum_{i=1}^{\nu} [\tfrac{1}{2} h_{2,i}^2 (u_i - \bar{u})].$$

Having in mind that α_3 and α_5 denote (approximations) to second derivatives of u, we see that the above expression gives us a discretization of the governing equation in point $\bar{\mathbf{x}} \in \omega$. Using the standard FD terminology, we can say that the stencil of discretization in $\bar{\mathbf{x}}$ consist from $\nu + 1$ nodes (in general, $\nu = \nu(\bar{\mathbf{x}})$).

Note: (i) if $\nu = 5$, i.e. number of equations in (4) is equal to number of unknown α-s, the system is not overdetermined, and we derive a standard FD scheme; (ii) LS is presented here for brevity, usually weighted LS is used. In MLS case $\alpha = \left(M^T W M\right)^{-1} \left(M^T W\right) \mathbf{b}$, where W is a weighting diagonal matrix. The weight function (i.e., the diagonal entries of W) can be quite arbitrary. We use $w(\mathbf{x}_i - \bar{\mathbf{x}}; H) = exp(-\sigma \frac{\|\mathbf{x}_i - \bar{\mathbf{x}}\|^2}{H^2})$ if $\frac{\|\mathbf{x}_i - \bar{\mathbf{x}}\|}{H} \leq 1$, and $w(\mathbf{x}_i - \bar{\mathbf{x}}; H) = 0$ elsewhere. The size of the support H defines the set of neighboring points around $\bar{\mathbf{x}}$, the positive constant σ influences the coefficients of the scheme.

In the same way, we can derive discretization for each $\mathbf{x}_j$, $j = 1, 2, ...N$ from the set ω. Rearranging terms, we can write down the generalized FD scheme as $A\mathbf{u} = \mathbf{f}$, where $\mathbf{u}$ is the vector of all N unknowns. For theoretical results concerning this GFDM see [3]. **2.B. GFDM from [7]**. Let us introduce additional notation $\alpha_0 = \bar{u}$. Further, form local matrix $\tilde{M}$: ν Taylor expansions (as in M) plus a constraint ($\alpha_3 + \alpha_5 = \bar{f}$):

$$\tilde{M} = \begin{pmatrix} 1 & h_{1,1} & h_{2,1} & \frac{1}{2}\,h_{1,1}^2 & h_{1,1}\,h_{2,1} & \frac{1}{2}\,h_{2,1} \\ 1 & h_{1,2} & h_{2,2} & \frac{1}{2}\,h_{1,2}^2 & h_{1,1}\,h_{2,2} & \frac{1}{2}\,h_{2,2} \\ \vdots & \vdots & \vdots & \vdots & \vdots & \\ 1 & h_{1,\nu} & h_{2,\nu} & \frac{1}{2}\,h_{1,\nu}^2 & h_{1,\nu}\,h_{2,\nu} & \frac{1}{2}\,h_{2,\nu} \\ 0 & 0 & 0 & 1 & 0 & 1 \end{pmatrix}. \tag{6}$$

Here $\nu \geq 6$, $\alpha = (\alpha_0, \alpha_1, \alpha_2, \alpha_3, \alpha_4, \alpha_5)^T$, $\mathbf{b} = \left(\bar{u},\ u_1,\ u_2, \ldots..,\ u_\nu,\ \bar{f}\right)^T$. The LS solution (if $\tilde{M}^T\tilde{M}$ is nonsingular) is given by $\tilde{\alpha} = \left(\tilde{M}^T\tilde{M}\right)^{-1}\tilde{M}^T\tilde{\mathbf{b}}$. In this case there exist $\tilde{c}_1, \tilde{c}_2, \ldots, \tilde{c}_\nu$ and $\tilde{c}_f$ such that

$$\bar{u} = \alpha_0 = \sum_{i=1}^{\nu} \tilde{c}_i u_i + \tilde{c}_f \bar{f}. \tag{7}$$

This relation gives us Tiwari's GFDM discretization of the governing equation in $\bar{\mathbf{x}}$.

3 Comparison of Discretizations from [10] and from [7].

The complete analysis of discretizations (5) and (7) is a difficult task. The coefficients of these discretizations, in general, are not known explicitely - as described above, they are obtained as solutions of small matrix equations. An impression about the quality of the two discretizations can be obtained by considering the 1-D case, where explicit expressions for the coefficients of the generalized FD schemes can be derived. It is known, that stability + approximation ensures convergence for the (generalized) FD discretizations. It is also known, that stability can be proved if the matrix of the FD operator is an $M - matrix$. Because we cannot formulate explicit conditions for the stability of the FD schemes, we investigate the more strong requirements - is the discretization matrix an $M - matrix$, or not. The same approach is used in [5], where they investigate coefficients of FD discretization of 1-D conservation laws, obtained with the method from [10]. Thus, in our analysis of the discretization matrices A we will check the conditions, under which the diagonal entries are negative, all the off-diagonal entries are positive, and the sum of elements in each row is non-positive.

We suppose here that the weight functions $w \equiv 1$ in both cases. Consider the discretization in a fixed node, $\bar{x}$. Let the cloud S consist from $\bar{x}$ and ν other points. Denote $h_i = x_i - \bar{x}$, $\sum = \sum_{i=1}^{\nu}$. For convenience, locate the coordinate system at x_ν. It is clear, that h_i can be positive and can be negative.

3.A. The Scheme from [10].

$$\begin{pmatrix} \sum h_i^2 & \sum \frac{1}{2} h_i^3 \\ \sum \frac{1}{2} h_i^3 & \sum \frac{1}{4} h_i^4 \end{pmatrix} \begin{pmatrix} \dfrac{\partial u}{\partial x} \\ \dfrac{\partial^2 u}{\partial x^2} \end{pmatrix} = \begin{pmatrix} \sum h_i (u_i - \bar{u}) \\ \sum \frac{1}{2} h_i^2 (u_i - \bar{u}) \end{pmatrix}$$

It is easy to see, that approximation to $\frac{\partial^2 u}{\partial x^2}$ is given by

$$C\,\frac{1}{2}\sum_i \left(h_i^2 - h_i \frac{\sum_j h_j^3}{\sum_j h_j^2} \right) (u_i - \bar{u}) \,, \quad \text{where } C > 0. \tag{8}$$

It is obvious, that the sum of off-diagonal coefficients is equal to the coefficient in front of $\bar{u}$. So, we will investigate the other condition for monotonicity, namely, the sign of the off-diagonal coefficients. This sign is determined by the sign of a_i,

$$a_i = -h_i^2 + h_i \frac{\sum_j h_j^3}{\sum_j h_j^2} = -h_i^2 \left(1 - \frac{\sum_j h_j^3}{h_i \sum_j h_j^2} \right) \cdot \tag{9}$$

3.B. The Scheme from [7].

$$\begin{pmatrix} \nu & \sum h_i & \sum \frac{1}{2} h_i^2 \\ \sum h_i & \sum h_i^2 & \sum \frac{1}{2} h_i^3 \\ \sum \frac{1}{2} h_i^2 & \sum \frac{1}{2} h_i^3 & 1 + \sum \frac{1}{4} h_i^4 \end{pmatrix} \begin{pmatrix} \bar{u} \\ \frac{\partial u}{\partial x} \\ \frac{\partial^2 u}{\partial x^2} \end{pmatrix} = \begin{pmatrix} \sum u_i \\ \sum h_i \, u_i \\ \bar{f} + \sum \frac{1}{2} h_i^2 \, u_i \end{pmatrix}$$

Note, a proper scaling can be applied to the governing equation before using it within MLS procedure, in order to improve the condition number of the local matrix equation. The signs of the discretization coefficients in this case are driven by the sign of $\tilde{a}_i$,

$$\tilde{a}_i \approx -\frac{\sum_j h_j^2 - h_i \sum_j h_j}{\nu \sum_j h_j^2 - (\sum_j h_j)^2} + O(h^4). \tag{10}$$

We investigate the signs of the off-diagonal elements in two extreme cases. First, we consider a perfectly balanced cloud, when both discretizations provide monotone matrices, and look what happen when small dissbalance of the cloud appear. More precisely, we remove one node from a pair-symmetric cloud. Second, we consider completely unbalanced stencil (all points from one side of $\bar{x}$). We add to the cloud one point from the other side of $\bar{x}$ and check when it can stabilize the respective discretization. In both cases, we check under which conditions at least one of the off-diagonal coefficients a_i or $\tilde{a}_i$ become positive. Due to the lack of space, we omit the detailed derivation, it will be presented in a forthcoming paper. Here we summarize only the results.

Behaviour in removing a point from perfectly balanced cloud. It is easy to see, that if a cloud $\bar{S}$ consist from the node $\bar{x}$ plus an even number of nodes, which are pair-symmetric with respect to $\bar{x}$, the discretization in both cases is monotone. Let us now remove a node from the above cloud. That is, consider a cloud= $\{\bar{x}$ + one or more pair nodes + an unbalanced node$\}$. It can be shown that discretization from [10] leads to appearance of positive off-diagonal elements in the matrix A for large absolute value of the unbalanced node. On the other hand, the off-diagonal coefficients $\tilde{a}_i$ from [7] are always negative.

Behaviour in adding a point to a completely unbalanced cloud. Now, consider a cloud, consisting from $\bar{x}$, one node, x_k from its left side, and $\nu - 1$ nodes from its right hand side. Cloud= $\{\bar{x} = 0 \cup$ nodes $x_i > 0 \cup$ one node $x = C < 0, \}$. In this case, off-diagonal coefficients in (9) are negative for $C \in (-\infty, C_{[10]})$, while those from (10) are negative for $C \in (-\infty, C_{[7]})$ It can be seen, that $C_{[10]} < C_{[7]}$, that is, the second approach has extended area of stability.

We can summarize, that in the considered extreme cases, the discretization from (10) is less sensitive to the dissbalance in the cloud, compare to (9). Our numerical experiments on unbalanced clouds also confirm the above considerations. Non-physical oscillations of the numerical solution obtained with discretization from (9) were observed in cases when discretization from (10) provided correct solution. Note, that the usage of weights (i.e., MLS instead of LS) smooths the above effect, but does not remove it completely. A reason for the better stability of the new scheme is that the additional equation, involved in MLS discretization, is always centered in $\bar{x}$. Although the comparison of the stability of the two discretizations is not complete, we found that the discretization (10) is preferable, and use it for discretizing interface problems.

4 Meshfree Discretization for Interface Problems

Consider now the interface problem in the case when the diffusivity coefficient is piecewise constant. Consider, for brevity, the case of two subdomains, Ω_1 and Ω_2, separated by an interface Γ. Introduce a set ω_1 of N_1 nodes in Ω_1, a set ω_2 of N_2 nodes in Ω_2, and a set ω_Γ of N_3 nodes in Γ. The problem is discretized as follows. If $\bar{x} \in \omega_1$, use only those of neighbouring nodes, which belong to $\omega_1 \bigcup \omega_\Gamma$, and apply the above described approach for the problem with constant coefficients. The matrix $\tilde{M}$ in this case will be the same as (6). In the same way, discretize in points belonging to ω_2. The most interesting is the case when $\bar{x} \in \omega_\Gamma$. In this case, we suppose that a circle with radius H contains $\nu_1 > 5$ nodes belonging to Ω_1 and $\nu_2 > 5$ nodes belonging to Ω_2. Denote by $\mathbf{n} = (n_1, n_2)$ the unit normal vector to Ω_1. Further, introduce a vector of unknowns:

$$\breve{\alpha} = (\alpha_0, \alpha_{1,-}, \alpha_{2,-}, \alpha_{3,-}, \alpha_{4,-}, \alpha_{5,-}, \alpha_{1,+}, \alpha_{2,+}, \alpha_{3,+}, \alpha_{4,+}, \alpha_{5,+})^T ,$$

$$\alpha_0 = u(\bar{x}), \quad \alpha_{1,\pm} = \frac{\partial u(\bar{x} \pm 0)}{\partial x_1}, \quad \alpha_{2,\pm} = \frac{\partial u(\bar{x} \pm 0)}{\partial x_2}, \tag{11}$$

$$\alpha_{3,\pm} = \frac{\partial^2 u(\bar{x} \pm 0)}{\partial x_1^2}, \quad \alpha_{4,\pm} = \frac{\partial^2 u(\bar{x} \pm 0)}{\partial x_1 \partial x_2}, \quad \alpha_{5,\pm} = \frac{\partial^2 u(\bar{x} \pm 0)}{\partial x_2^2}. \tag{12}$$

Here $\frac{\partial u(\bar{x}-0)}{\partial x_1}$ denotes the value of the derivative from the side of Ω_1, while $\frac{\partial u(\bar{x}+0)}{\partial x_1}$ denotes the value of the derivative from the side of Ω_2. To discretize in $\bar{x} \in \Gamma$, we form a matrix $\breve{M}$, on the base of $\nu_1 + \nu_2$ Taylor expansion, and a constraint. Instead of the governing equation, we use as a constrain the condition for continuity of the normal component of the flux at the interface. The matrix $\breve{M}$ in this case looks as follows:

$$\begin{pmatrix} 1 & \dot{h}_{1,1} & \dot{h}_{2,1} & \frac{1}{2}\dot{h}_{1,1}^2 & \dot{h}_{1,1}\dot{h}_{2,1} & \frac{1}{2}\dot{h}_{2,1} & 0 & 0 & 0 & 0 & 0 \\ 1 & \dot{h}_{1,2} & \dot{h}_{2,2} & \frac{1}{2}\dot{h}_{1,2}^2 & \dot{h}_{1,1}\dot{h}_{2,2} & \frac{1}{2}\dot{h}_{2,2} & 0 & 0 & 0 & 0 & 0 \\ \vdots & \vdots & \vdots & \vdots & \vdots & \vdots & \vdots & \vdots & \vdots & \vdots & \\ 1 & \dot{h}_{1,\nu_1} & \dot{h}_{2,\nu_1} & \frac{1}{2}\dot{h}_{1,\nu_1}^2 & \dot{h}_{1,\nu_1}\dot{h}_{2,\nu_1} & \frac{1}{2}\dot{h}_{2,\nu_1} & 0 & 0 & 0 & 0 & 0 \\ 1 & 0 & 0 & 0 & 0 & 0 & \ddot{h}_{1,1} & \ddot{h}_{2,1} & \frac{1}{2}\ddot{h}_{1,1}^2 & \ddot{h}_{1,1}\ddot{h}_{2,1} & \frac{1}{2}\ddot{h}_{2,1} \\ 1 & 0 & 0 & 0 & 0 & 0 & \ddot{h}_{1,2} & \ddot{h}_{2,2} & \frac{1}{2}\ddot{h}_{1,2}^2 & \ddot{h}_{1,1}\ddot{h}_{2,2} & \frac{1}{2}\ddot{h}_{2,2} \\ \vdots & \vdots & \vdots & \vdots & \vdots & \vdots & \vdots & \vdots & \vdots & \vdots & \\ 1 & 0 & 0 & 0 & 0 & 0 & \ddot{h}_{1,\nu_1} & \ddot{h}_{2,\nu_1} & \frac{1}{2}\ddot{h}_{1,\nu_1}^2 & \ddot{h}_{1,\nu_1}\ddot{h}_{2,\nu_1} & \frac{1}{2}\dot{h}_{2,\nu_1} \\ 0 & k_1 n_1 & k_1 n_2 & 0 & 0 & 0 & 0 & -k_2 n_1 & -k_2 n_2 & 0 & 0 \end{pmatrix}.$$

Here $\dot{h}_{1,i} = x_{1,i} - \bar{x}_1$ and $\dot{h}_{2,i} = x_{2,i} - \bar{x}_2$ for $\mathbf{x}_i = (x_{1,i}, x_{2,i}) \in \Omega_1$, while $\ddot{h}_{1,i} = x_{1,i} - \bar{x}_1$ and $\dot{h}_{2,i} = x_{2,i} - \bar{x}_2$ for $\mathbf{x}_i \in \Omega_2$.

Different iterative methods can be used for solving the large sparse system of algebraic equations, $A\mathbf{u} = \mathbf{f}$, arising after the discretization We use additive-multiplicative domain decomposition. More precisely, solution is independently updated in ω_1 and in ω_2, and after that the just calculated values are used in updating solution on ω_Γ. Gauss-Seidel or BiCG were used in each subdomain.

5 Numerical Experiments

Several numerical experiments were run in order to examine the above described discretization. Both, uniform and random distribution of nodes were used in experiments. It is clear from the above discussions, that the topology of the nodes influences the quality of the results. Trying to separately investigate the influence of the coefficients discontinuity, we present here only calculations for (almost) uniform distribution of nodes. Numerical experiments in 1-D case showed that the scheme is exact for piecewise linear functions, and it is second order accurate in other cases. That is, it shows the same accuracy as FD schemes based on harmonic averaging of the discontinuous coefficient. Of course, the discretization stencil of the meshfree discretization is larger then the discretization stencil of FDM. In 1-D and 2-D MLS derivations, we used weight function with $\sigma = 6.26$ and $H \approx 3h$, where h is some average distance between nodes. In 2-D case, we solved 3 examples in the unit square. We choose problems with known

Table 1. Example 1: straight interface. The interface Γ is given by $x = 0.5$.

$$u^{exact}(x,y) = \frac{1}{k}\sin(\frac{\pi x}{2})(x-0.5)(y-0.5)(1+x^2+y^2)$$

N	$k_1 = k_2 = 1$	$k_1 = 1$, $k_2 = 10$	$k_1 = 1$, $k_2 = 100$	$k_1 = 1$, $k_2 = 1000$	$k_1 = 10$, $k_2 = 10$
121	1.0e-3	8.8e-4	8.6e-4	8.6e-4	3.2e-4
441	2.2e-4	2.2e-4	2.2e-4	2.2e-4	8.4e-5
1681	3.6e-5	4.0e-5	5.5e-5	5.6e-5	2.0e-5

Table 2. Example 2: curvilinear interface (parabola). The interface Γ is given by $y = x^2 + 0.3$.

$$u^{exact}(x,y) = \frac{1}{k}\sin(\frac{\pi x}{2})(y - x^2 - 0.3)(y - 0.5)(1 + x^2 + y^2)$$

N	$k_1 = k_2 = 1$	$k_1 = 1,$ $k_2 = 10$	$k_1 = 1,$ $k_2 = 100$	$k_1 = 1,$ $k_2 = 1000$	$k_1 = 10,$ $k_2 = 10$
121	2.7e-3	5.4e-3	6.7e-3	6.9e-3	1.7e-3
441	6.0e-4	7.8e-4	8.7e-4	8.9e-4	3.9e-4
1681	1.7e-4	2.1e-4	2.5e-4	2.5e-4	6.1e-5

Table 3. Example 3: curvilinear interface (ellipse). Γ_S is $(x-0.6)^2 + (y-0.5)^2 = 0.09$.

$$u^{exact}(x,y) = \frac{1}{k}\sin(\frac{\pi x}{2})((x - 0.6)^2 + 4.0(y - 0.5)^2 - 0.09)(y - 0.5)(1 + x^2 + y^2)$$

N	$k_1 = k_2 = 1$	$k_1 = 1,$ $k_2 = 10$	$k_1 = 1,$ $k_2 = 100$	$k_1 = 1,$ $k_2 = 1000$	$k_1 = 10,$ $k_2 = 10$
133	7.0e-3	4.0e-3	5.1e-3	5.3e-3	1.4e-2
461	1.6e-3	6.6e-4	8.3e-4	8.1e-4	2.4e-3
1700	2.7e-4	1.7e-4	1.9e-4	2.0e-4	5.9e-4

exact solution, so that we can monitor the error between the numerical and the analytical solutions. *The maximum of this error for each of the cases is presented in tables below.* Diffusivity coefficient in Ω_1 and Ω_2 is denoted by k_1 and k_2, respectively. Computations are done for different ratios of k_1 and k_2.

It can be seen from the tables, that second order accuracy is achieved in all three cases.

6 Conclusions

Let us briefly discuss the advantages and disadvantages of the presented approach.

In general, on the negative side of the meshfree methods is the fact that they are slower compare to the grid methods. Also, the topological restrictions concerning grids are better understood, often they are formulated at an element level (minimal angle, stretching of an element, etc.) and thus it is clear how to control the quality of the grid.

On the other hand, Generalized FD methods (and other meshfree methods) are suitable for solving problems with complex or rapidly changing geometries, when grid generation or often remeshing are expensive (an aim of our research is to increase the efficiency of a recently developed meshfree projection method for multiphase and free surface incompressible flows [12]). Also, the restrictions on the nodes (points) distribution for meshfree methods is weaker, compare to the restrictions for grid-based methods. It is easy to add/remove nodes, aiming at improving the quality of the nodes distribution. However, the topological restrictions on points distribution usually have a non-local character, and therefore the control on the quality of the points distribution may be expensive. In this context, the method from [7] is better then the method from [10], because it is less sensitive to the topology of the nodes. Also, we found it to be more suitable for problems with discontinuous coefficients.

References

1. Atluri, N.R., Li, G.: Finite cloud method: a true meshless technique based on a fixed reproducing kernel approximation. Int. J. Num. Meth. Engng. 50 (2001) 2373–2410.
2. Belytschko, T., Krongauz, Y., Organ, D., Flemming, M., Krysl, P.: Meshless methods: An overview and recent developments. Comput. Methods Appl. Mech. Engng. 139 (1996) 3–47.
3. Demkowicz, L., Karafiat, A., Liszka, T.: On some convergence results for FDM with irregular mesh. Computer Meth. in Appl. Mech. and Engineering, 42 (1984) 343–355.
4. Duarte, C.A., Oden, J.T.: H-p clouds and h-p meshless methods. Num. Meth. for PDEs, Vol. 12, 6 (1996) 673–705.
5. Fuerst, J., Sonar, Th.: On meshless collocation approximations of conservation laws: Preliminary investigations on positive schemes and dissipation models. Z. Angew. Math. Mech (ZAMM), 81, 6 (2001) 403–415.
6. Griebel, M., Schweitzer, A.: A particle-partition of unity method for the solution of elliptic, parabolic, and hyperbolic PDEs. SIAM J. Dci. Comput. 22 (2000) 853–890.
7. Kuhnert, J., Tiwari, S.: Gridfree method for solving Poisson equation. Technical report of the Fraunhofer Institute for Industrial Mathematics, Kaiserslautern, Germany, No. 25 (2001)
8. Li, S., Liu, W.K.: Meshfree and particle methods and their applications. Applied Mechanics Review, 55 (2002) 1–34.
9. Liszka, T., Duarte, C., Tworzydlo, W.: H-p meshless cloud method. J. Comp. Meth. Appl. Mech. Eng. 139 (1996) 263–288.
10. Liszka, T., Orkisz, J.: The finite difference method on arbitrary irregular grid and its application in applied mechanics. Computers & Structures, 11 (1980) 83–95.
11. Park, S.-H., Youn, S-K.: The least squares meshfree method. Int. J. Num. Meth. Engng. 52 (2001) 997–1012.
12. Tiwari, S., Kuhnert, J.: Finite pointset method based on the projection method for simulations of the incompressible Navier-Stokes equations. In: Griebel, M., Schweitzer, M. (eds.): Meshfree Methods for Partial Differential Equations. Lecture notes in computer science and engineering, Springer, to appear.

Applications of Weighted Compact Scheme to Curvilinear System

Li Jiang and Chaoqun Liu

University of Texas at Arlington, Arlington TX 76019, USA

Abstract. Weighted compact scheme developed by Jiang et al has been used to capture the shock wave and small vortex with high-order in smooth area and no oscillation for the discontinuity for both rectangular and curvilinear grids.

1 Introduction

Recently compact schemes have been widely used in CFD. Standard finite difference schemes need to be at least one point wider than the desired approximation order. It is difficult to find suitable and stable boundary closure for high order schemes. However, the compact scheme can achieve higher order without increasing the stencil width. As the compact scheme has an implicit form and involves neighboring grid point derivative values, additional free parameters can be used not only to improve the accuracy but also to optimize the other properties such as resolution, stability, and conservation. A family of centered compact schemes proposed by LeLe (Lele, 1992) have been proved to have spectral-like resolution. The conservation property is also important, especially for flow with shocks.

Though the advantages of compact schemes are obvious, there are still difficulties in using them to solve problems involving shock waves or discontinuities. When they are used to differentiate a discontinuous function, the computed derivative has grid to grid oscillation. Recently the ENO (Harten, 1987; Shu, 1988, 1989) and WENO (Liu, 1994; Jiang, 1996) schemes have been widely used for shock wave capturing and have been proved very successful. These schemes check the smoothness of the candidate stencils. The ENO scheme selects the smoothest stencil, while the WENO scheme uses all the candidate stencils but with assigned weights. Inspired by the success of the WENO scheme, we have developed a new compact scheme (Jiang et al, 2001) so that the new compact scheme not only preserves the properties of compact schemes but also can be used for shock wave capturing. This new scheme preserves the characteristic of standard compact schemes achieving high order accuracy and high resolution by a compact stencil. The improvement of this new scheme over the standard compact scheme is that it can accurately capture shock waves without oscillation. The idea of the Weighted Compact Scheme is similar to the WENO scheme (Jiang and Shu 1996). In the WENO scheme, each of the candidate stencils is assigned a weight that determines the contribution of this stencil to the final approximation of the numerical flux. The weights are defined in such a

I. Dimov et al. (Eds.): NMA 2002, LNCS 2542, pp. 498–505, 2003.

way that in smooth regions it approaches certain optimal weights to achieve a higher order of accuracy, while in regions near discontinuities, the stencils that contain the discontinuities are assigned a nearly zero weight. Similarly to this idea, the Weighted Compact Scheme is constructed by the combination of the approximations of derivatives on candidate stencils. The stencil that contains the discontinuity has less contribution. In this way, the oscillation near the discontinuity can be reduced, while the characteristics of the compact scheme can still be preserved. The building blocks of the Weighted Compact Scheme are the standard compact schemes. The Weighted Compact Scheme is a hybrid of different forms of standard schemes. Compared to the WENO scheme, this Weighted Compact Scheme can achieve higher order accuracy and higher resolution with the same stencil.

2 High-Order eighted Compact Scheme

For simplicity, we consider a uniform grid. The independent variable at the node j is $x_j = h(j-1)$ for $1 \le j \le N$ and the function value at the nodes $f_j = f(x_j)$ is given. The finite difference approximation f_j' to the first derivative of the function f on the nodes can be written in the following general form while the finite difference compact scheme (Lele, 1992) is used.

$$\beta_- f_{j-2}' + \alpha_- f_{j-1}' + f_j' + \alpha_+ f_{j+1}' + \beta_+ f_{j+2}' = \frac{1}{h}(b_- f_{j-2} + a_- f_{j-1} + c f_j + a_+ f_{j+1} + b_+ f_{j+2}) \tag{1}$$

For the point j, we define three candidate stencils containing point j:

$$S_0 = (x_{j-2}, x_{j-1}, x_j), \quad S_1 = (x_{j-1}, x_j, x_{j+1}), \quad S_2 = (x_j, x_{j+1}, x_{j+2}).$$

On each of them we can get a compact scheme. By matching the Taylor series coefficients to various, the third anf fourth order compact schemes corresponding to each stencil can be derived. The coefficients are given as follows:

$$\begin{aligned}
&S_0: \beta_- = \vartheta,\ \alpha_- = 2\vartheta + 2,\ b_- = -\tfrac{5}{2}\vartheta - \tfrac{1}{2},\ a_- = 2\vartheta - 2,\ c = \tfrac{1}{2}\vartheta + \tfrac{5}{2};\\
&S_1: \qquad\qquad \alpha_- = \tfrac{1}{4},\ \alpha_+ = \tfrac{1}{4},\ a_- = -\tfrac{3}{4},\ a_+ = \tfrac{3}{4},\ c = 0;\\
&S_2: \beta_+ = \vartheta,\ \alpha_+ = 2\vartheta + 2,\ b_+ = \tfrac{5}{2}\vartheta + \tfrac{1}{2},\ a_+ = -2\vartheta + 2,\ c = -\tfrac{1}{2}\vartheta - \tfrac{5}{2},
\end{aligned} \tag{2}$$

where ϑ is a free parameter. The coefficients that are not listed are set tozero. The schemes corresponding to stencils S_0 and S_2 are third order one-sided finite difference schemes, and the scheme corresponding to S_1 is a fourth order centered scheme. With these three sets of coefficients, we get three different equations from Eq.(1). These equations are represented by $F_0,\ F_1,\ F_2$. When these equations are assigned specific weights and combined, a new scheme is obtained:

$$F = C_0 F_0 + C_1 F_1 + C_2 F_2 \tag{3}$$

where $C_0 + C_1 + C_2 = 1$. If the weights are properly chosen, the new scheme can achieve a higher order because the additional free parameters are introduced. If we set:

$$C_0 = C_2 = \frac{1}{18 - 24\vartheta}, \quad C_1 = \frac{8 - 12\vartheta}{9 - 12\vartheta} \tag{4}$$

the new scheme is at least a sixth order centered compact scheme. The process implies that the sixth order centered compact scheme can be represented by a combination of three lower order schemes.

Obviously, the scheme F is a standard finite difference compact scheme and cannot avoid the oscillation near discontinuities. Can we define the weights in such a way that the scheme has the non-oscillatory property? Then the idea of the WENO scheme is introduced to determine the new weight for each stencil. The weights are determined according to the smoothness of the function on each stencil. According to the WENO method, the new weights are defined as (Jiang *et al.*, 1996):

$$\omega_k = \frac{\gamma_k}{\sum_{i=0}^{2} \gamma_i} \quad \gamma_k = \frac{C_k}{(\epsilon + IS_k)^p} \tag{5}$$

where ϵ is a small positive number that is used to avoid the denominator becoming zero. The smoothness measurement IS_k is defined as following:

$$\begin{aligned} IS_0 &= \frac{13}{12}(f_{j-2} - 2f_{j-1} + f_j)^2 + \frac{1}{4}(f_{j-2} - 4f_{j-1} + 3f_j)^2 \\ IS_1 &= \frac{13}{12}(f_{j-1} - 2f_j + f_{j+1})^2 + \frac{1}{4}(f_{j-1} - f_{j+1})^2 \\ IS_2 &= \frac{13}{12}(f_j - 2f_{j+1} + f_{j+2})^2 + \frac{1}{4}(f_{j+2} - 4f_{j+1} + 3f_j)^2 \end{aligned} \tag{6}$$

where, the two terms on the right side can be regarded as the measurements of the curvature and the slope respectively at a certain point. Through the Taylor expansion, it can be easily proved that in smooth regions new weights ω_k satisfy:

$$\omega_k = C_k + O(h^2) \;\; and \;\; \omega_2 - \omega_0 = O(h^3). \tag{7}$$

The new scheme is then formed using these new weights:

$$F = \omega_0 F_0 + \omega_1 F_1 + \omega_2 F_2. \tag{8}$$

The leading error of F is also a combination of the errors of the original schemes F_i, which is as following:

$$\left(\frac{1}{12}\omega_0 - \frac{1}{12}\omega_2\right) f^{(4)} h^3 + \left(-\frac{1}{15}\omega_0 + \frac{1}{120}\omega_1 - \frac{1}{15}\omega_2\right) f^{(5)} h^4. \tag{9}$$

When Eq. (7) is satisfied, the leading error of the new scheme can be written as $O(h^6)$. Obviously, this new scheme is of sixth-order accuracy and has the high resolution property as the centered sixth-order compact scheme in smooth regions. But in the regions containing discontinuities, the smoothness measurement IS_k of the non-smooth stencil is large compared to that of the smooth stencil, thus the non-smooth stencil is assigned a small weight and have less contribution to the final scheme so that the non-oscillatory property is achieved.

With the new weights ω_k, the new finite difference compact scheme Eq. (8) is written in the form of Eq. (1). The coefficients of the final Weighted Compact Scheme are given as follows:

$$\beta_- = \vartheta\omega_0,\ \alpha_- = (2\vartheta+2)\omega_0 + \frac{1}{4}\omega_1,\ \alpha_+ = (2\vartheta+2)\omega_2 + \frac{1}{4}\omega_1,\ \beta_- = \vartheta\omega_2,$$

$$b_- = (-\frac{5}{2}\vartheta - \frac{1}{2})\omega_0,\ a_- = (2\vartheta-2)\omega_0 - \frac{3}{4}\omega_1,\ c = (\frac{1}{2}\vartheta + \frac{5}{2})\omega_0 - (\frac{1}{2}\vartheta + \frac{5}{2})\omega_2,$$

$$a_+ = (-2\vartheta+2)\omega_2 + \frac{3}{4}\omega_1,\ b_+ = (\frac{5}{2}\vartheta + \frac{1}{2})\omega_2.$$

If $\beta = 0$, the scheme is tridiagonal; otherwise, it is pendadiagonal. The free parameter ϑ can be used to optimize the scheme when the properties of high resolution, conservation and stability are concerned. From the above, we can find that the weights play a very important role in the Weighter Compact Scheme. They can be used to optimize the accuracy. In addition, there weights can make the new scheme non-oscillatory.

3 Computational Results for Rectangular Grids

3.1 Convection Equation

We first solve the one-dimensional convection equation.

$$\begin{aligned} &u_t + u_x = 0, &&-1 \le x \le 1 \\ &u(x,0) = u_0(x), &&\textit{periodic with a period of } 2. \end{aligned} \tag{10}$$

The initial function are:

$$u_0(x) = \begin{cases} 1, -\frac{1}{5} \le x \le \frac{1}{5}; \\ 0, \textit{otherwise}, \end{cases}$$

Figure 1 is the result of the standard compact scheme for the initial function (1). The solution is seriously damaged by wiggles generated near the discontinuities. Figures 2 illustrate the results obtained by the Weighted Compact Scheme for the given initial function There is no obvious numerical oscillation observed in the regions near the discontinuities, and good resolution has been achieved.

3.2 1D Euler Equation

We have applied the scheme to 1D Euler equation with the following initial condition:

$$U_0 = \begin{cases} (1,0,1), & x < 0; \\ (0.125,0,0.1), & x \ge 0, \end{cases}$$

The distributions of pressure, density,velocity and energy are shown in Fig. 3. The shock wave and contact discontinuity are accurately captured.

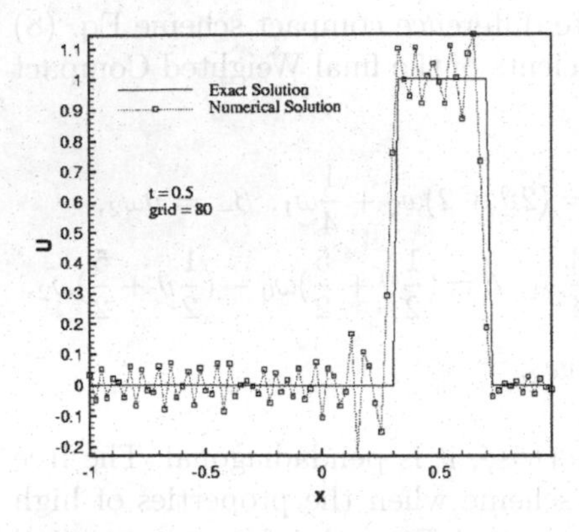

Fig. 1. Solution of standard compact scheme

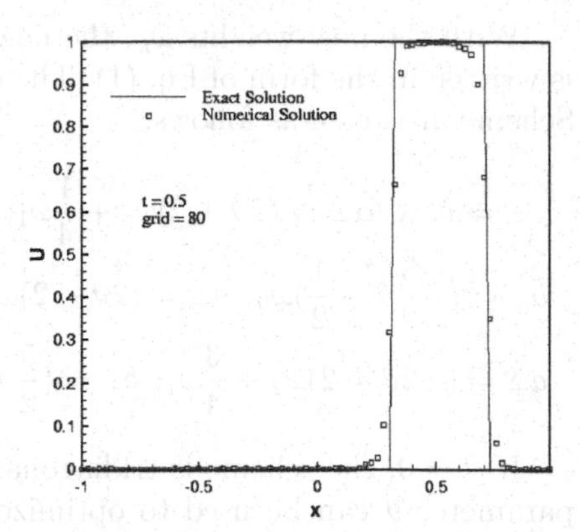

Fig. 2. The solution at $t = 0.5$

4 Computational Results for 2D Curvilinear Grids

For 2-d curvilinear coordinates, several testing cases have been conducted. The results show that the Weighted Compacted Scheme developed by Jiang (2001) can be applied to curvilinear coordinate system. The medium distortion of the grid doesn't have much effect on the solutions.

4.1 Shock-Fluctuation Interaction Problem

We apply the Weighted Compact Scheme to investigate a 2D shock-turbulence interaction problem(Adams,1996; Shu,1989). 2D Euler equations are solved. The computational domain is given by $-1.5 < x < 1.5$, $-1 < y < 1$. At time $t = 0$ a Mach 8 shock at $x = -1$ is moving in the positive x direction into a vorticity fluctuation field. The initial condition for pre-shock field is specified with vorticity fluctuation as

$$\begin{aligned} u_1 &= -c_1 sin\theta cos(xkcos\theta + yksin\theta) \\ v_1 &= c_1 cos\theta cos(xkcos\theta + yksin\theta) \\ \rho_1 &= 1 \\ p_1 &= 1 \end{aligned}$$

where c_1 is the speed of sound, $k = 2\pi$, $\theta = \frac{1}{6}$. The inital post-shock state can be derived from shock relations.

$$\begin{aligned} u_2 &= \frac{2(M^2-1)}{(\gamma+1)M^2} u_s \\ v_2 &= 0 \end{aligned}$$

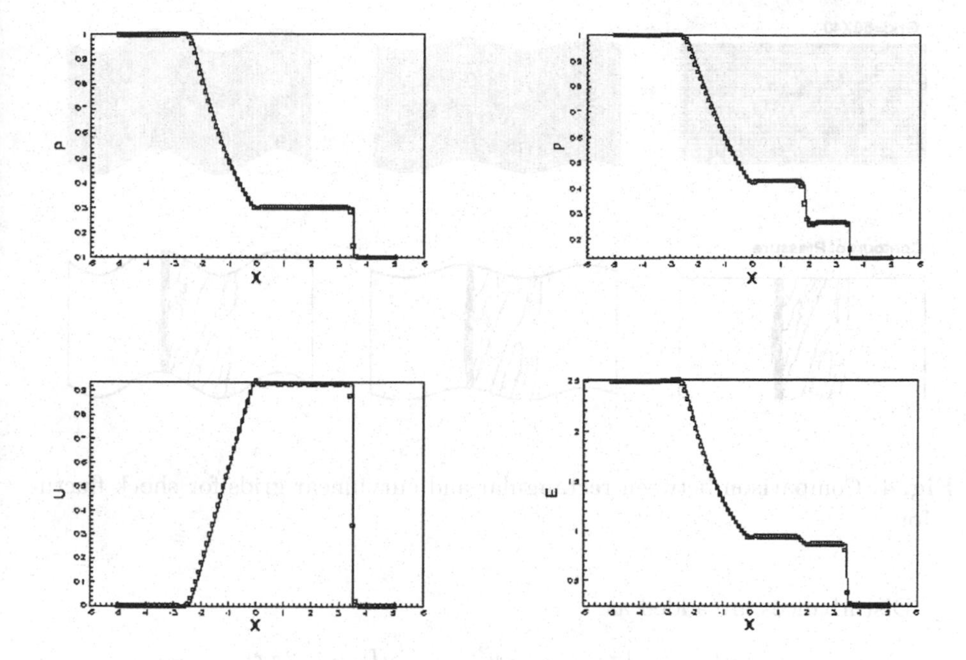

Fig. 3. The solutions to the shock-tube problem at $t = 2$, $N = 100$

$$\rho_2 = \frac{(\gamma+1)M^2}{2+(\gamma-1)M^2}\rho_1$$

$$p_2 = (1 + \frac{2\gamma(M^2-1)}{\gamma+1})p_1$$

where $u_s = Mc_1$ is the shock propagation velocity.

Fig.4 display the results obtained on different grids. The shock front is clearly shown by the pressure contour lines. After the vorticity fluctuations strike the shock, the shock front develops ripples and the vorticity fluctuations are amplified. The results are similar compared with those obtained by Shu (1989) and Adams (1996). It can be seen that the sharpness of the shock is kept very well even the grid is distorted.

4.2 Vortex Pairing in a Mixing Layer

Vortex pairing process is forced by adding velocity disturbances to the initial mean velocity profile with opposite free-stream. Mean flow and temperature fields are given by

$$u = tanh(2y)$$

$$T = 1 + \frac{\gamma-1}{2}M^2(1-u^2)$$

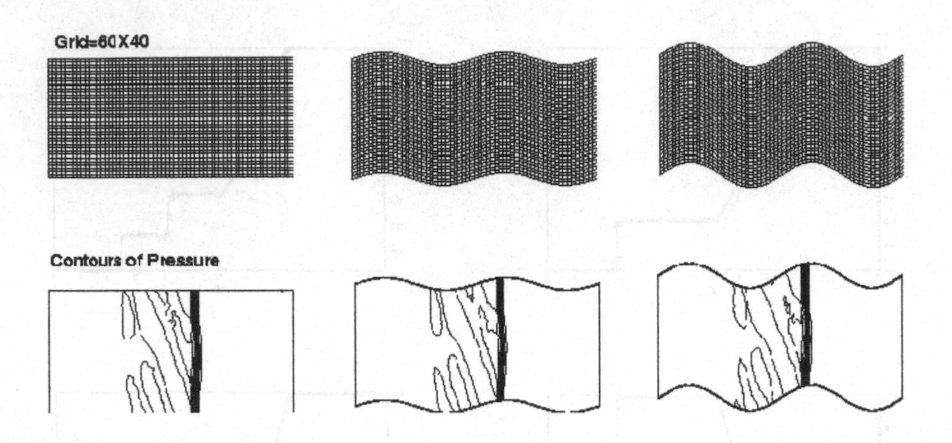

Fig. 4. Comparison between rectangular and curvilinear grids for shock fluctuation

Disturbances are added as

$$u' = -A_1 \frac{yL_x}{2\pi B} cos(\frac{4\pi x}{L_x}) exp(\frac{-y^2}{B}) - A_2 \frac{yL_x}{\pi B} cos(\frac{2\pi x}{L_x}) exp(\frac{-y^2}{B})$$

$$v' = A_1 sin(\frac{4\pi x}{L_x}) exp(\frac{-y^2}{B}) + A_2 sin(\frac{2\pi x}{L_x}) exp(\frac{-y^2}{B})$$

where $A_1 = 0.05$, $A_2 = 0.025$, $B = 10$, $L_x = 20$, $L_y = 40$, $Re = 2000$, $M = 0.8$. The following pictures show the appearance of the shock waves and their interactions with the vortices. The results show that the position and the shapes of the shock waves are not affect by the distortion of the grid.

5 Conclusion

The weighted compact scheme can preserve the high order accuracy in the smooth area while has no oscillation for discontinuity (shocks). The scheme works well for both rectangular and curvilinear grids.

References

1. Adams, N.A. and Shariff, K., (1996), A high-resolution hybrid compact-ENO scheme for shock-turbulence interaction problems, *J. Comput. Phys.*, **127**, pp.27-51.
2. Harten, A., Engquist, B., Osher, S. and Charavarthy, S.: Uniformly high order accurate essentially non-oscillatory scheme, III. J. Comput. Phys. 71 (1987) 231–303.

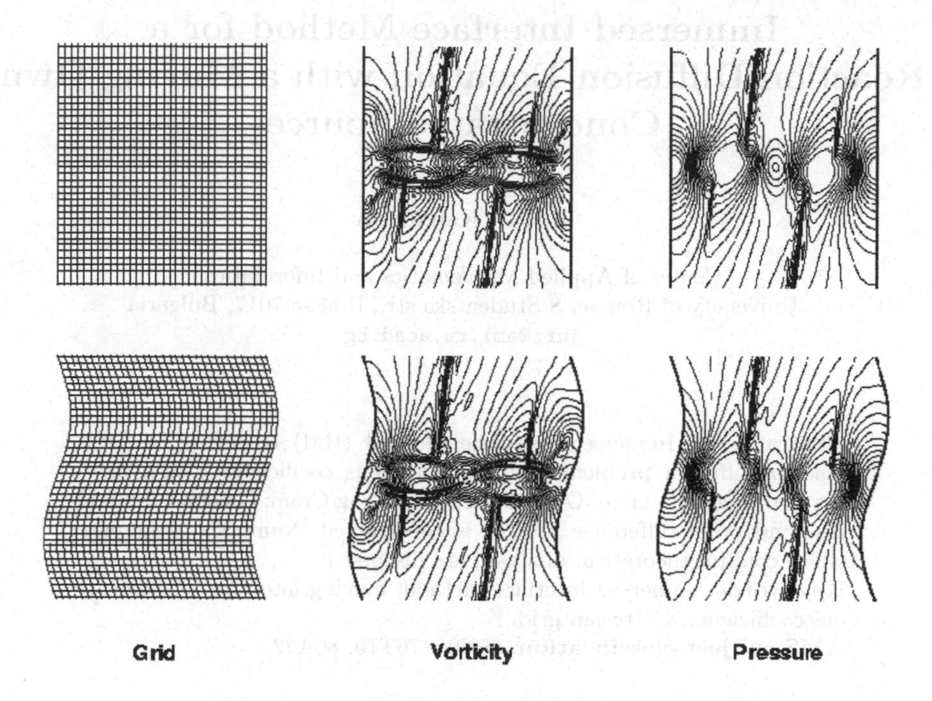

Fig. 5. Comparison between rectangular and curvilinear grids for vortex pairing

3. Jiang, G. S. and Shu, C. W.: Efficient implementation of weighted ENO scheme. J. Comput. Phys. 126 (1996) 202–228.
4. Jiang, L., Shan, H., and Liu, C.: Direct numerical simulation of boundary-layer receptivity for subsonic flow around airfoil. The Second AFOSR International Conference on DNS/LES, Rutgers, New Jersey, June 7-9 (1999)
5. Lele, S. K.: Compact finite difference schemes with spectral-like resolution. J. Comput. Phys. 103 (1992) 16–42.
6. Liu, X. D., Osher, S., and Chan, T.: Weighted essentially non-oscillatory schemes. J. Comput. Phys. 115 (1994) 200–212.
7. Jiang, L., Shan, H., and Liu, C.: Weighted Compact Scheme for Shock Capturing. International Journal of Computational Fluid Dynamics, 15 (2001) 147–155.
8. Shu, C. W. and Osher, S.: Efficient implementation of essentially non-oscillatory shock-capturing schemes. J. Comput. Phys. 77 (1988) 439–471.
9. Shu, C. W. and Osher, S.: Efficient implementation of essentially non-oscillatory shock-capturing schemes II. J. Comput. Phys. 83 (1989) 32–78.

Immersed Interface Method for a Reaction-Diffusion Equation with a Moving Own Concentrated Source

Juri D. Kandilarov

Center of Applied Mathematics and Informatics
University of Rousse, 8 Studentska str., Rousse 7017, Bulgaria
juri@ami.ru.acad.bg

Abstract. An Immersed Interface Method (IIM) is developed for a reaction-diffusion problem with discontinuous coefficients and a moving own singular source. On a regular grid using Crank-Nicolson method a second order difference scheme is constructed. Numerical examples, which confirm theoretical analysis, are discussed.

Key words: immersed interface method, moving interface, discontinuous coefficients, Cartesian grid.
AMS subject classification: 65N06, 76T10, 80A22.

1 Introduction

We consider the following problem:

$$u_t - (\beta u_x)_x + c(t)\delta(x - \xi(t))f(u) = F(x,t),\ x \in (0, \xi(t)) \cup (\xi\,(t)\,, 1), \tag{1}$$

$$u(x,0) = u_0(x),\ x \in (0,1),\ u(0,t) = u_l(t),\ u(1,t) = u_r(t),\ t > 0, \tag{2}$$

where $\beta(x,t)$, $F(x,t)$ may have discontinuity at the moving interface $\xi(t)$, $\beta\,(x,t)$ is positive and $f(u)$ is differentiable.

With some assumptions for the regularity of the solution, the problem (1), (2) is equivalent to the next one:

$$u_t - (\beta(x,t)u_x)_x = F(x,t), \quad x \in (0, \xi(t)) \cup (\xi\,(t)\,, 1), \quad t > 0, \tag{3}$$

$$[u] = 0, \qquad x = \xi(t),\ t > 0, \tag{4}$$

$$[\beta(x,t)u_x] = c(t)f(u(x,t)), \qquad x = \xi(t),\ t > 0, \tag{5}$$

and initial and boundary conditions (2).

This problem arises in biological systems, at modeling of chemically active membranes and describes chemical reaction-diffusion processes in which, due to the effect of catalyst, the reaction takes place only at some local sites. The jump of the flux typically depends on the concentration, [2, 6].

When $f = const.$, and $c(t)$ is known function, this is a heat-transfer problem in composite materials with a known local source. In [1] the immersed boundary method (Peskin's method) of second order on regular grids for the case of

I. Dimov et al. (Eds.): NMA 2002, LNCS 2542, pp. 506–513, 2003.

$\beta(x,t) = 1$ is proposed. For similar problem in [9] the authors use dynamical locally refined grids near the interface. For the case of discontinuous coefficients and $f(u) = 0$ in [8] first order of convergence of the solution of classical difference schemes is proved.

For the case of discontinuous coefficients and $[u] = c_1(t)$, $[\beta u_x] = c_2(t)$ in [5] Z. Li applies the IIM and presents schemes of second order. He also considers a Stefan problem, where on the moving interface the solution is known.

Our attention is the case when the jump of the flux depends on the solution. In [4, 10] the systems of nonlinear equations with many fixed interfaces $\xi(t) = \xi_i$, $i = 1, 2, ...$ are considered. In [3] the IIM for (2)-(5) with $\beta(x,t) = 1$ and $f(u) = c(t)u$ is discussed. In the present paper we develop the difference schemes on regular grids for the more complicated problem, when $\beta(x,t)$ is discontinuous and $f(u)$ is arbitrary differentiable function.

2 Computational Frame

We use a uniform grid $x_i = ih$, $i = 0, 1, ..., M$, $x_0 = 0$, $x_M = 1$, where h is the step size in space and $t^n = n\tau$, where τ is the step size in time. Using the Crank-Nicolson scheme [7], the semi-discrete difference scheme for (1), (2) can be written in the following general form

$$\frac{u_i^{n+1} - u_i^n}{\tau} - Q_i^{n+1/2} = \frac{1}{2}\left((\beta u_x)_{x,i}^{n+1} + (\beta u_x)_{x,i}^n\right) + \frac{1}{2}(F_i^{n+1} + F_i^n),$$

where $(\beta u_x)_{x,i}^n$ is a discrete analogy of $(\beta u_x)_x$ at (x_i, t^n) and $Q_i^{n+1/2}$ is a correction term needed when $\xi(t)$ crosses the grid line x_i at some time between t^n and t^{n+1}.

We call the point (x_i, t^n) **regular**, if the interface $\xi(t)$ does not cross the sixth point stencil of the Crank-Nicolson scheme and **irregular** in other cases. At Fig.1. with □ we denote the irregular points of kind A, when the interface crosses the stencil only in space direction and with $*$ - the points B, when the interface crosses the stencil in time direction.

In the next sections we will use the following notations to express the jump in a function $g(x,t)$ across the interface: $[g] = g(\xi^+, t) - g(\xi^-, t)$, $\quad [g]_{;t} = g(\xi, t^+) - g(\xi, t^-)$. It is easily to see that $[g] = -sign\xi'[g]_{;t}$.

The IIM for the space and time discretization involves the following steps, see [5]:

- Use the jump condition and the differential equation to get the interface relations between the quantities on each side of the interface.
- Use the interface relations to derive a modified difference scheme.
- Derive the correction term based on the difference scheme and the interface relations.

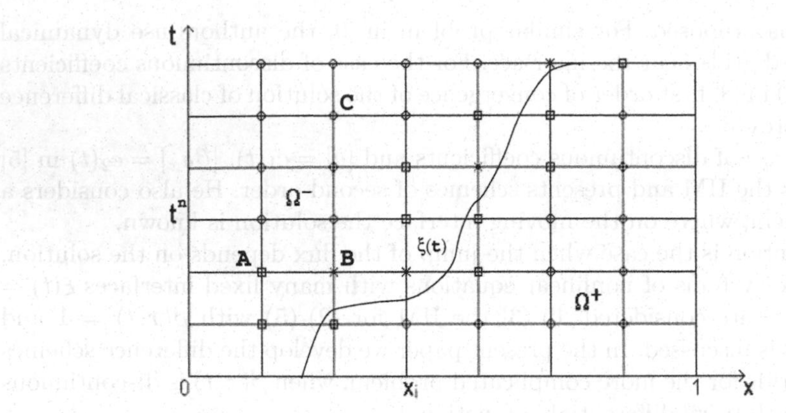

Fig. 1. Different cases of grid points: □ - the interface crosses the stencil only in space direction; ∗ - the interface crosses the stencil in time direction; ∘ - regular grid points.

3 Difference Scheme for the Linear Case

The full discretization of (2)-(5) is based on the undetermined coefficients method and has the form

$$\frac{u_i^{n+1} - u_i^n}{\tau} - Q_i^{n+1/2} = \frac{1}{2}\left((\gamma_{i,1}^{n+1} u_{i-1}^{n+1} + \gamma_{i,2}^{n+1} u_i^{n+1} + \gamma_{i,3}^{n+1} u_{i+1}^{n+1} + C_i^{n+1}) + (\gamma_{i,1}^{n} u_{i-1}^{n} + \gamma_{i,2}^{n} u_i^{n} + \gamma_{i,3}^{n} u_{i+1}^{n} + C_i^{n})\right) + \frac{1}{2}(F_i^{n+1} + F_i^n). \quad (6)$$

At regular grid points $\gamma_{i,1}^n = \beta_{i-1/2}^n/h^2$, $\gamma_{i,3}^n = \beta_{i+1/2}^n/h^2$, $\gamma_{i,2}^n = -(\gamma_{i,1}^n + \gamma_{i,3}^n)$, $Q_i^{n+1/2} = C_i^n = C_i^{n+1} = 0$. At the irregular grid points the coefficients are not standard and the additional terms need more attention.

With the process described above we have the following

Lemma 1. *If $x_I \le \xi(t) < x_{I+1}$ and $f(u) = c(t)u$, then*

$$(\beta u_x)_{x,I} = \gamma_{I,1} u_{I-1} + \gamma_{I,2} u_I + \gamma_{I,3} u_{I+1} + C_I + O(h) , \quad (7)$$

where

$$[\gamma_{I,1}, \gamma_{I,2}, \gamma_{I,3}]^T = A_I^{-1}[0, \beta_x^-, \beta^-]^T, \qquad C_I = \frac{(x_{I+1} - \xi)^2}{2\beta^+}[F(x,t)]\gamma_{I,3},$$

and A_I^{-1} is the inverse matrix of

$$A_I = \begin{pmatrix} 1 & 1 & 1 + (x_{I+1} - \xi)\frac{c(t)}{\beta^+} - \frac{(x_{I+1}-\xi)^2}{2\beta^{+2}}(\xi' + \beta_x^+)c(t) \\ (x_{I-1} - \xi) & (x_I - \xi) & (x_{I+1} - \xi)\frac{\beta^-}{\beta^+} + \frac{(x_{I+1}-\xi)^2}{2\beta^+}\left(\xi'(1 - \frac{\beta^-}{\beta^+}) - \frac{\beta_x^+\beta^-}{\beta^+} + \beta_x^-\right) \\ \frac{(x_{I-1}-\xi)^2}{2} & \frac{(x_I-\xi)^2}{2} & \frac{(x_{I+1}-\xi)^2}{2}\frac{\beta^-}{\beta^+} \end{pmatrix}$$

Proof. From the jump conditions (4), (5) we have:

$$u^+ = u^-, \; u_x^+ = \frac{c(t)u^- + \beta^- u_x^-}{\beta^+}. \tag{8}$$

Differentiating (4) with respect to t for the jump of u_t we find

$$[u_t] = -\xi'[u_x]. \tag{9}$$

From (3), (5) and (9) it follows

$$u_{xx}^+ = \frac{1}{\beta+}\left(-(\xi' + \beta_x^+)\frac{c(t)}{\beta+}u^- + \left(\xi'(1 - \frac{\beta^-}{\beta^+}) - \frac{\beta_x^+\beta^-}{\beta^+} + \beta_x^-\right)u_x^- + \beta^- u_{xx}^- - [F]\right). \tag{10}$$

We expand $u(x_{I-1}, t)$, $u(x_I, t)$ from the left hand side and $u(x_{I+1}, t)$ from the right hand side of $\xi(t)$ and put them into (7). Then we exchange the + by the – values, using (8) and (10). Vanishing the coefficients of all terms of order $O(1)$ we get to the conditions of the lemma. □

The point (x_{I+1}, t^n) is irregular too and we have the following

Lemma 2. *If $x_I \le \xi(t) < x_{I+1}$ and $f(t,u) = c(t)u$, then*

$$(\beta u_x)_{x,I+1} = \gamma_{I+1,1}u_I + \gamma_{I+1,2}u_{I+1} + \gamma_{I+1,3}u_{I+2} + C_{I+1} + O(h)\;, \tag{11}$$

where

$$[\gamma_{I+1,1}, \gamma_{I+1,2}, \gamma_{I+1,3}]^T = A_{I+1}^{-1}[0, \beta_x^+, \beta^+]^T,\; C_{I+1} = \frac{(x_I - \xi)^2}{2\beta^-}[F(x,t)]\gamma_{I+1,1},$$

and A_{I+1}^{-1} is the inverse matrix of

$$A_{I+1} = \begin{pmatrix} 1 - (x_I - \xi)\frac{c(t)}{\beta^-} + \frac{(x_I-\xi)^2}{2\beta^{-2}}(\xi' + \beta_x^-)c(t) & 1 & 1 \\ (x_I - \xi)\frac{\beta^+}{\beta^-} + \frac{(x_I-\xi)^2}{2\beta^-}\left(\xi'(1 - \frac{\beta^+}{\beta^-}) - \frac{\beta_x^-\beta^+}{\beta^-} + \beta_x^+\right) & (x_{I+1} - \xi) & (x_{I+2} - \xi) \\ \frac{(x_I-\xi)^2}{2}\frac{\beta^+}{\beta^-} & \frac{(x_{I+1}-\xi)^2}{2} & \frac{(x_{I+2}-\xi)^2}{2} \end{pmatrix}$$

What about the discretization of the u_t near the interface? If there is a grid crossing at a grid point x_i from time t^n to time t^{n+1}, meaning that $x_i \in (\xi(t^n), \xi(t^{n+1}))$, (the point of kind B, Fig.1.), then the time derivative of u is not smooth. In this case even though we can approximate the x-derivatives well at each time level, the standard Crank- Nicolson scheme needs to be corrected with a term $Q_i^{n+1/2}$ to guarantee first order of local approximation.

Lemma 3. *Suppose the equation $\xi(t) = x_i$ has a unique solution t_0^n in the interval $t^n < t_0^n < t^{n+1}$. Then*

$$\frac{u(x_i, t^{n+1}) - u(x_i, t^n)}{\tau} - Q_i^{n+1/2} = \frac{1}{2}\left(u_t(x_i, t^{n+1}) + u_t(x_i, t^n)\right) + O(\tau), \tag{12}$$

where

$$Q_i^{n+1/2} = \frac{1}{\tau}(t^{n+1/2} - t_0^n)|\xi'(t_0^n)|\,[u_x]_{t_0^n}. \tag{13}$$

Proof. We expand $u(x_i, t^n)$ and $u(x_i, t^{n+1})$ in Taylor series about time t_0^n from each side on the interface and then

$$\frac{u(x_i,t^{n+1}) - u(x_i,t^n)}{\tau} = u_t(x_i, t_0^{n-}) + \frac{[u]_{;t_0^n}}{\tau} + \frac{t_{n+1} - t_0^n}{\tau}[u_t]_{;t_0^n} + O(\tau). \quad (14)$$

On the other hand

$$u_t(x_i,t^n) = u_t(x_i,t_0^{n-}) + O(\tau);\; u_t(x_i,t^{n+1}) = u_t(x_i,t_0^{n-}) + [u_t]_{;t_0^n} + O(\tau),$$

which implies

$$u_t(x_i,t_0^{n-}) = \frac{1}{2}\left(u_t(x_i,t^{n+1}) + u_t(x_i,t^n)\right) - \frac{[u_t]_{;t_0^n}}{2} + O(\tau). \quad (15)$$

From $[u_t]_{;t_0^n} = -\xi'[u_x]_{t_0^n}$ and from (9) we have

$$[u_t]_{;t_0} = -sign\xi'(t_0^n)[u_t]_{t_0^n} = |\xi'(t_0^n)|[u_x]_{t_0^n} \quad (16)$$

The substitution of (15) and (16) into (14) gives

$$\frac{u(x_i,t^{n+1}) - u(x_i,t^n)}{\tau} = \frac{1}{2}\left(u_t(x_i,t^{n+1}) + u_t(x_i,t^n)\right)$$
$$+\frac{t^{n+1/2} - t_0^n}{\tau}|\xi'(t_0^n)|[u_x]_{t_0^n} + O(\tau),$$

which completes the proof. □

In (16) for approximation of $[u_x]_{t_0^n}$ we consider two cases. If $\xi' > 0$, then

$$[u_x]_{t_0^n} = (1 - \frac{\beta^+}{\beta^-})u_x^+(x_i,t_0^n) + \frac{c(t_0^n)}{\beta^-}u^+(x_i,t_0^n) \quad (17)$$
$$= (1 - \frac{\beta^+}{\beta^-})\frac{u(x_{i+1},t^n) - u(x_i,t^n)}{h} + \frac{c(t_0^n)}{\beta^-}u(x_i,t^n) + O(\tau + h).$$

If $\xi' < 0$, then

$$[u_x]_{t_0^n} = (\frac{\beta^-}{\beta^+} - 1)u_x^-(x_i,t_0^n) + \frac{c(t_0^n)}{\beta^+}u^-(x_i,t_0^n) \quad (18)$$
$$= (\frac{\beta^-}{\beta^+} - 1)\frac{u(x_i,t^n) - u(x_{i-1},t^n)}{h} + \frac{c(t_0^n)}{\beta^+}u(x_i,t^n) + O(\tau + h).$$

By the Lemmas 1,2,3 it follows the next

Theorem 1. *If $f(u) = u(x,t)$ and $c(t)$ is smooth, then the scheme (6) with (11)-(13) approximates the problem (1), (2) with $O(\tau + h)$ local truncation error at irregular grid points and $O(\tau^2 + h^2)$ at regular one.*

Using the maximum principle and Theorem 1 we can proof the following

Theorem 2. *If $u(x,t) \in C^2(0,T; C(\Omega) \cap C^4(\Omega^-) \cap C^4(\Omega^+))$, then the scheme (6) is stable and the numerical solution converges to the exact solution with*

$$|u(x_i,t^n) - u_i^n| = O(\tau^2 + h^2),\; i = 1,...,M-1,\; n = 1,2,...,N.$$

4 Nonlinear Problem

We will discuss only the approximation at irregular grid points. Let $f(t,u)$ is a smooth function. Then in a similar way as in the previous section we have the following semi-discretization of the derivative in space direction

$$(\beta u_x)_{x,I} = \gamma_{I,1}u_{I-1} + \gamma_{I,2}u_I + \gamma_{I,3}u_{I+1} + C_I + O(h) \ , \qquad (19)$$

where

$$[\gamma_{I,1}, \gamma_{I,2}, \gamma_{I,3}]^T = A_I^{-1}[0, \beta_x^-, \beta^-]^T \ ,$$

$$C_I = \frac{(x_{I+1}-\xi)^2}{2\beta^+}[F(x,t)]\gamma_{I,3} - \left(\frac{(x_{I+1}-\xi)}{\beta^+} - \frac{(x_{I+1}-\xi)^2}{2\beta^{+2}}(\xi' + \beta_x^+)\right)\gamma_{I,3}f(u^-)$$

and A_I^{-1} is the inverse matrix of

$$A_I = \begin{pmatrix} 1 & 1 & 1 \\ (x_{I-1}-\xi) & (x_I-\xi) & (x_{I+1}-\xi)\frac{\beta^-}{\beta^+} + \frac{(x_{I+1}-\xi)^2}{2\beta^+}\left(\xi'(1-\frac{\beta^-}{\beta^+}) - \frac{\beta_x^+\beta^-}{\beta^+} + \beta_x^-\right) \\ \frac{(x_{I-1}-\xi)^2}{2} & \frac{(x_I-\xi)^2}{2} & \frac{(x_{I+1}-\xi)^2}{2}\frac{\beta^-}{\beta^+} \end{pmatrix}$$

For $f(u^-)$ we apply the one-side approximation formula

$$f(u^-) = \frac{(\xi - x_{I-1})}{h}f(u_I) - \frac{(\xi - x_I)}{h}f(u_{I-1}) + O(h^2),$$

which preserves the stencil and guarantees the order $O(h)$ of the full discretization of $(\beta u_x)_x$. In a similar way can be obtained the approximation of (19) for $x = x_{I+1}$.

When the interface crosses the stencil in time direction, then the additional term has the form, (see (17), (18)): if $\xi' > 0$

$$Q_i^{n+1/2} = \frac{(t^{n+1/2} - t_0^n)}{\tau}|\xi'(t_0^n)|\left((1 - \frac{\beta^+}{\beta^-})\frac{u(x_{i+1},t^n) - u(x_i,t^n)}{h} + \frac{f(u_i^n)}{\beta^-}\right) + O(\tau+h),$$

and if $\xi' < 0$

$$Q_i^{n+1/2} = \frac{(t^{n+1/2} - t_0^n)}{\tau}|\xi'(t_0^n)|\left((\frac{\beta^-}{\beta^+} - 1)\frac{u(x_i,t^n) - u(x_{i-1},t^n)}{h} + \frac{f(u_i^n)}{\beta^+}\right) + O(\tau+h).$$

5 Numerical Examples

We test two numerical examples.

Example 1 Linear local own source:

$$u_t - (\beta u_x)_x = -K\delta(x - \xi(t))u, \ x \in (0, \xi(t)) \cup (\xi(t), 1), \ 0 < t \le 1, \ \xi(t) = \alpha t.$$

We choose the exact solution of the form:

$$u(x,t) = \begin{cases} 2 + exp(-\alpha(x-\alpha t)/\beta_1)), & \text{if } 0 \le x \le \xi(t); \\ 2 + 3K/\alpha + (\beta_2/\beta_1 - 3K/\alpha)exp(-\alpha(x-\alpha t)/\beta_2)) & \text{if } \xi(t) \le x \le 1. \end{cases}$$

Table 1. Grid refinement analysis for example 1 at $t = 0.1$, $\alpha = 1$, $K = 3$, and $\beta_1 = 1$, $\beta_2 = 2$ or $\beta_1 = 2$, $\beta_2 = 1$, N–number of time layers.

M	N	$\beta_1 = 1, \beta_2 = 2$		$\beta_1 = 2, \beta_2 = 1$	
		$\Vert E_M\Vert_\infty$	*ratio*	$\Vert E_M\Vert_\infty$	*ratio*
8	160	1.2160e-03	-	3.4332e-03	-
16	160	5.3424e-04	2.27	1.4445e-03	2.37
32	160	1.1640e-04	4.59	3.5414e-04	4.08
64	160	2.4842e-05	4.68	8.8148e-05	4.02
128	160	5.7882e-06	4.29	2.2849e-05	3.86
256	160	1.3736e-06	4.21	5.5180e-06	4.14
512	160	3.4492e-07	3.98	1.2993e-06	4.24

The initial and boundary conditions are found from the exact solution. In Table 1 we present the mesh refinement analysis for $\alpha = 1$, $K = 3, t = 0.1$ and two cases of diffusion coefficients: $\beta_1 = 1$, $\beta_2 = 2$ and $\beta_1 = 2$, $\beta_2 = 1$. With $\Vert E_M\Vert_\infty$ we denote the infinity norm of the error at the final time t. The ratio $\Vert E_M\Vert_\infty / \Vert E_{2M}\Vert_\infty$ (near 4) indicates second order of convergence. For the case of "wild" coefficients $\beta_1 = 120$, $\beta_2 = 1$ at time $t = 0.1$ and $M = 512$, $N = 80$ the error is $\Vert E_M\Vert_\infty = 1.5680e - 007$. For $\beta_1 = 1$, $\beta_2 = 120$ it is 1.7764e-007, which confirm the stability of the method for different choices of the coefficients

Example 2 Nonlinear local own source.

In the problem (1) let we choose $f(u) = u^{1+\theta}$, $\theta > 0$ and the exact solution to be

$$u(x,t) = \begin{cases} sin(\omega_1 x)\exp(-\beta_1\omega_1^2 t), & \text{if} \quad 0 \le x \le \xi(t), \\ sin(\omega_2(1-x))\exp(-\beta_2\omega_2^2 t), & \text{if} \quad \xi \le x \le 1. \end{cases}$$

The interface is found by the scalar equation for every t^n, $n = 1, ..., N$:

$$sin(\omega_1\xi)\exp(-\beta_1\omega_1^2 t^n) = sin(\omega_2(1-\xi))\exp(-\beta_2\omega_2^2 t^n).$$

Figure 2(a) shows how the interface moves with time. The comparison of the exact and numerical solution at $t = 0.1$ for $\beta_1 = 2$, $\beta_2 = 1$, $\omega_1 = \omega_2 = 3\pi/2, M = 64, N = 200$ is given on Figure 2(b). The solid line is the exact solution and the circles are the computed solution. The numerical results are similar to those of *Example 1* and confirm the theoretical analysis.

References

1. Beyer, R., Leveque, R.: Analysis of a one-dimensional model for the immersed boundary method. SIAM J. Numer. Anal. 29 (1992) 332–364.
2. Bimpong - Bota, K., Nizan, A., Ortoleva, P., Ross, J.: Cooperative phenomena in analysis of catalytic sites. J. Chem. Phys. 66 (1970) 3650–3683.

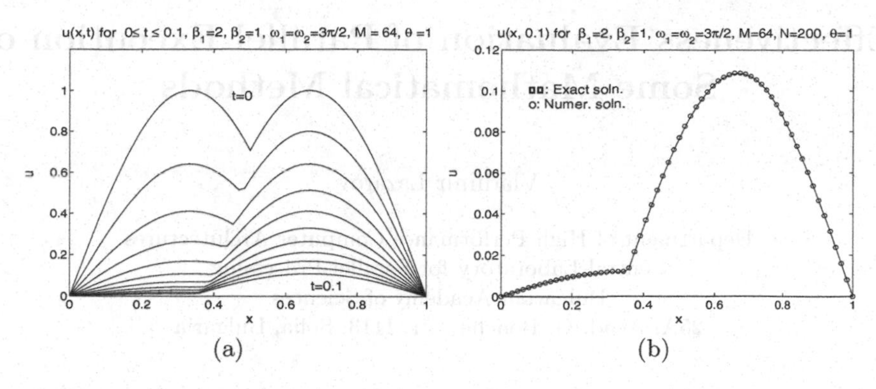

Fig. 2. (a) The profile of the computed solution $u(x,t)$ from $t = 0$ to $t = 0.1$; (b) The comparison of the exact and numerical solution at $t = 0.1$.

3. Kandilarov, J.: A second-order difference method for solution of diffusion problems with localized chemical reactions. In: Samarskii A.A., Vabishchevich P. N., Matus P. P. (eds.): Proceed. of Second Int. Conf. Finite Difference Methods: Theory and Applications. Minsk (1999) 63–67.
4. Kandilarov, J., Vulkov, L.: The immersed interface method for a nonlinear chemical diffusion equation with local sites of reactions. (submitted).
5. Li, Z.: Immersed interface method for moving interface problems. Numer. Algorithms, 14 (1997) 269–293.
6. Pierce, A., Rabitz, H.: An analysis of the effect of defect structures on catalytic surfaces by the boundary element technique. Surface Science, 202 (1988) 1–31.
7. Samarskii, A.: The Theory of Difference Schemes. Nauka, Moscow (1977) (in Russian)
8. Samarskii, A., Friazinov, I.: Convergence of a homogeneous difference schemes for heat equation with discontinuous coefficients. Comput. Math. and Math. Phys. 1 (1961) 806–824.
9. Vabishchevich, P., Matus, P., Richagov, V.: On a class of difference schemes locally refined grids. Differ. Uravn. 31, 5 (1995) 849–857.
10. Vulkov, L., Kandilarov, J.: Construction and implementation of finite-difference schemes for systems of diffusion equations with localized chemical reactions. Comp. Math. Math. Phys. 40, 5 (2000) 705–717.

Effectiveness Evaluation of Parallel Execution of Some Mathematical Methods

Vladimir Lazarov

Department of High Performance Computer Architectures
Central Laboratory for Parallel Processing
Bulgarian Academy of Sciences
25A, Acad. G. Bonchev str. 1113, Sofia, Bulgaria

1 Introduction

In many applications the essential part of the calculations in the programs are based on some well-known and defined mathematical method, such as FDM (Finite Difference Method), FEM (Finite Element Method), LU Factorization, SOR (successive over-relaxation) method, Monte Carlo method and so on. In such cases the decision how to execute effectively the application on a parallel machine should be taken according to the information how the appropriate mathematical method could be parallelized. Obviously it will be very convenient to have practical guidelines of how these methods perform in parallel versions on different parallel machines for a scale of input arguments. The following is a description of an attempt to receive representative figures for the behavior of two mathematical methods on three parallel machines.

2 Formulation of the Task

The idea is to program a core variant of a mathematical method in two versions - serial and parallel one, using the most popular parallel programming method, namely MPI (Message Passing Interface). The serial program is executed on one processing element (PE), while the parallel version is executed on increasing number of PE's. The elapsed time for execution of both versions is measured and speed-up coefficient is calculated. The procedure is repeated for number of input parameters, describing different dimensions of the input arrays and the results are presented in tables. The experiments are carried out on three parallel machines - IBM SP3, ORIGIN 3000 and CRAY T3E. On the base of the received figures some conclusions can be made and practical suggestions for parallel programmers can be formulated.

3 Finite Difference Method

The FDM method is chosen in this work because of its large implementation in many technical problems to solve partial differential equations. The existing

I. Dimov et al. (Eds.): NMA 2002, LNCS 2542, pp. 514–518, 2003.

dependences between the input variables lead to medium degree of difficulties in programming.

One-Dimensional FDM

The results of the experiments with one-dimensional FDM are presented in table 1.

Table 1. Speed-up coefficients for one-dimensional FDM with arrays from 1k to 1M elements

	SP3					O3000					T3E				
	PE2	PE3	PE4	PE5	PE6	PE2	PE3	PE4	PE5	PE6	PE2	PE3	PE4	PE5	PE6
1k	1.09	1.13	1.22	1.22	-	-	-	-	-	-	1.05	1.05	-	-	-
10k	1.2	1.26	1.28	1.28	-	1.21	1.21	-	-	-	1.28	1.35	1.37	1.39	1.40
100k	1.2	1.35	1.39	1.43	1.44	1.27	1.32	1.37	-	-	1.39	1.49	1.54	1.63	1.63
1M	1.27	1.35	1.39	1.44	1.44	1.30	1.54	1.66	1.66	-	1.38	1.51	1.57	1.62	1.66

The speed-up coefficient is calculated as

Elapsed time of the serial program (PE1) / Elapsed time of the parallel program with PEi

The elapsed time is measured between two important points, defining the start and the end of the main computational part within the programs. In all calculations the elapsed time is an average of multiple measurements in order to reduce the impact of the scheduling among the PE's during the experiments. On the other hand the experiments present results, achieved in real working conditions.

In one-dimensional FDM the computational part is not substantial one and although the program is easily parallelized the speed-up is respectively poor. The communication overhead, due to the exchange of the neighbouring variables between processes eliminates the advantage of the parallel execution.The results are similar on the three machines and even with large arrays one needs six PE's to achieve about 1,5 speed up. These figures lead to the conclusion that in applications using one-dimensional FDM the parallelization of the serial program is not justified.

Two-Dimensional FDM

In two-dimensional FDM the input arrays are two dimensional and the distribution of iterations among processes could be done in two ways - column-wise and row-wise. The choice is important, because it has a substantial impact both on cache misses and communication overhead. Here both kinds of distribution are examined with programs written in Fortran 90, where the arrays are stored in memory in column-major order. The dependences between variables are in four directions.

4 Column-wise Distribution

The results of the experiments with two-dimensional FDM column-wise distribution are presented in table 2.

Table 2. Speed-up coefficients for two-dimensional FDM column-wise distribution with arrays from 100x100 to 1k x 10k elements

	SP3					O3000					T3E				
	PE2	PE3	PE4	PE5	PE6	PE2	PE3	PE4	PE5	PE6	PE2	PE3	PE4	PE5	PE6
100x100	1.93	2.63	3.04	3.56	3.93	1.52	1.56	1.71	2.36	2.36	1.27	1.43	1.55	1.66	1.76
1k x 1k	1.92	2.82	3.72	4.65	5.54	1.8	2.72	3.45	3.48	-	1.98	2.95	3.88	4.80	5.68
1k x 4k	1.89	2.76	3.72	4.60	5.51	1.97	2.76	3.57	3.94	4.58	1.99	2.98	3.97	4.96	5.92
1k x 10k	1.86	2.76	3.69	4.52	5.38	-	-	-	-	-	2.00	2.99	3.98	4.97	5.98

Using this method the parallel execution has given very good results. The neighbouring strings that have to be exchanged between processes are contigous in memory and the computational part is decisive in the elapsed time. The speed-up is better with large arrays, because of the predominance of the computations and is close to the theoretical one. It is best achieved on Cray T3E, while on Origin 3000 the parallel efficiency decrease faster with the increase of the PE's.

5 Row-wise Distribution

The results with two-dimensional FDM row-wise distribution are presented in table 3.

Table 3. Speed-up coefficients for two-dimensional FDM column-wise distribution

	SP3					O3000					T3E				
	PE2	PE3	PE4	PE5	PE6	PE2	PE3	PE4	PE5	PE6	PE2	PE3	PE4	PE5	PE6
1k,1k	1.46	1.88	2.23	2.39	2.46	1.5	1.64	2.02	2.02	-	1.88	2.64	3.31	4.08	4.73
4k,1k	1.73	2.43	3.10	3.71	4.18	1.59	2.00	2.34	2.61	2.68	1.98	2.92	3.85	4.76	5.59
1k,4k	1.40	1.79	2.11	2.42	2.57	1.21	1.9	1.98	-	-	1.89	2.67	3.38	4.18	4.76
1k,10k	1.38	1.78	2.11	2.34	2.47	-	-	-	-	-	1.91	2.73	3.38	4.19	4.81
10k,1k	1.75	2.56	3.31	4.01	4.77	-	-	-	-	-	1.98	2.95	3.9	4.85	5.78

This method requires more complicated usage of the MPI commands, because of the necessity to create derived data types or packing and unpacking data for the boundary elements. Nevertheless, the row-wise distribution performs fairly well when the number of rows is considerably bigger than the number of columns in the array and thus decreasing the communication part in the program execution. Here the machines demonstrate the same behavior as in column-wise distribution. The speed-up on IBM SP3 is average, while Cray T3E shows values close to the theoretical ones. The coefficients on Origin 3000 are on a low level. The most possible explanation is that MPI perhaps is not the most suitable method for parallel programming on shared memory architectures.

Both ways of distribution offer good possibilities for parallel execution and the use of up to four processors is appropriate. On Cray T3E larger degree of parallelism is also justified. Comparing the elapsed time best results for both distributions were achieved on IBM SP3, twice faster than on Origin 3000. Cray T3E is inferior to IBM by 20 %.

6 Monte Carlo Method

The concrete example for implementation of the Monte Carlo method is a random walk in two-dimensions. A number of particles perform random walk and the program calculates the distribution of the distances that the particles have traveled. The program is parallelized by distributing particles among processes. The results of the experiments are shown in table 4.

Table 4. Speed-up coefficients for Monte Carlo method with particles from 100k to 10M

	SP3					O3000					T3E				
	PE2	PE3	PE4	PE5	PE6	PE2	PE3	PE4	PE5	PE6	PE2	PE3	PE4	PE5	PE6
100k	1.98	2.96	3.93	4.89	5.86	2.00	2.97	3.49	4.14	4.52	1.80	2.26	2.48	2.76	2.91
1M	2.00	3.00	4.00	5.00	6.00	2.00	3.00	3.93	4.76	5.70	1.98	2.9	3.76	4.60	5.40
10M	2.00	3.00	4.00	5.00	6.00	1.99	2.99	3.95	4.74	5.72	1.99	2.99	3.97	4.96	5.93

The efficiency of the parallelization is very high for all the three machines. The explanation lies in the fact that the time spent for communication in the parallel program is substantially smaller than that of computation. On IBM SP3 the theoretical values are reached even for smaller number of particles. In terms of elapsed time Cray T3E is twice faster than IBM SP3 with Origin 3000 between them inferior to Cray by 25 %.

We could conclude that in applications with implementation of the Monte Carlo method the use of MPI is very suitable and high speed-up coefficients could be achieved for increasing number of processors.

7 Conclusions

The results that have been received during the experiments show that it is convenient for the programmers to dispose with information regarding the behaviour of the most popular mathematical methods in parallel environment. Such information will help the choice of the programming method for the application, of the machine to be used and the suitable number of processors to achieve the desired run time specifications. The task to prepare such sort of information could be enlarged not only concerning additional mathematical methods as mentioned in the Introduction, but also to increase the dimensions of the methods. The approach can be implemented in different models of parallel programming and can be used as a practical guide.

Acknowledgments

I would like to thank all the people from CINECA for their help and hospitality and for the perfect conditions I was given in order to perform this work. I owe my special thanks to Paolo Malfetti and Christiano Calonaci from the High Performance Systems Division for the fruitful discussions and the concrete help in accessing the computing resources.

References

1. Kumar, V., Grama, A., Gupta, A., and Karypis, G.: Introduction to Parallel Programming. The Benjamin/Cummings Publishing Company, Inc. (1994)
2. Sima, D., Fontain, T., and Kacsuk, P.: Advanced Computer Architectures. Addison-Wesley Longman Inc. (1997)
3. Aoyama, Y. and Nakano, J.: RS/6000 SP: Practical MPI Programming. IBM International Technical Support Organization (1999)
4. Ellis, T., Philips, I., and Lahey, T.: Fortran 90 Programming. Addison-Wesley Publishing Company Inc. (1994)

A Numerical Approach for a Hemivariational Inequality Concerning the Dynamic Interaction between Ad acent Elastic Bodies

Asterios A. Liolios[1], Angelos A. Liolios[1], Stefan Radev[2], and Todor Angelov[2]

[1] Democritus University of Thrace, Dept. Civil Engineering,
Institute of Structural Mechanics and Earthquake Engineering,
GR-67100 Xanthi, Greece,
liolios@civil.duth.gr

[2] Institute of Mechanics, Bulgarian Academy of Sciences,
1113 Sofia, Bulgaria,
stradev@imbm.bas.bg, taa@imbm.bas.bg

Dedicated to the 70th anniversary of Academician Blagovest Sendov,
Bulgarian Academy of Sciences, Sofia, Bulgaria
and
In memoriam of Prof. Panagiotis D. Panagiotopoulos (1.1.1950-12.8.1998),
Late Professor of Steel Structures, Aristotle-University of Thessaloniki, Greece.

Abstract. A numerical treatment of an dynamic hemivariational inequality problem in structural mechanics is presented. This problem concerns the elastoplastic-fracturing unilateral contact with friction between neighboring civil engineering structures under second-order geometric effects during earthquakes. The numerical procedure is based on an incremental problem formulation and on a double discretization, in space by the finite element method and in time by the ϑ-Wilson method. The generally nonconvex constitutive contact laws are piece-wise linearized, and in each time-step a nonconvex linear complementarity problem is solved with a reduced number of unknowns.

Key words: dynamic hemivariational inequalities, computational contact mechanics, nonconvex linear complementarity problem, earthquake engineering

1 Introduction

Seismic interaction among adjacent buildings, as wellknown in earthquake engineering, is often a main cause of damages in seismically active regions, where, due to various socioeconomic reasons, the so-called continuous building system is allowed to be applied [1, 4, 10]. Thus the numerical estimation of the interaction effects to the seismic response of such buildings is significant for their earthquake resistant design, construction and repair.

Obviously the above interaction problem is very difficult from many aspects. Mathematically this problem of pounding of buildings belongs to the inequality

I. Dimov et al. (Eds.): NMA 2002, LNCS 2542, pp. 519–526, 2003.

problems of the mathematical theory of elasticity and of the structural mechanics, where the governing conditions are equalities as well as inequalities, see e.g. Panagiotopoulos [20-21], Nitsiotas [17], Maier [12,13]. These so-called unilateral problems can be treated mathematically by the variational or hemivariational inequality concept, see e.g. Panagiotopoulos [18-21]. So, the seismic response of the interacting structures system investigated here is governed by a set of equations and inequalities, which is equivalent to a dynamic hemivariational inequality in the way used by P.D. Panagiotopoulos. As wellknown, the hemivariational inequality concept has been introduced into Mechanics and Applied Mathematics by P.D. Panagiotopoulos for first time in 1983, see [18], and constitutes now the basis of the so-called Non-Smooth Mechanics.

As regards the numerical treatment of such inequality problems in earthquake engineering and multibody dynamics, some numerical approaches have already been presented, see e.g. [1, 10, 22, 24, 30].

In the present paper, a special case of seismic building interaction is treated numerically. This case concerns the unilateral elastoplastic-softening contact between adjacent structures under second-order instabilizing effects. So, the purpose here is to estimate numerically and to control actively the influence of the interaction effects on the seismic response of the adjacent structures. The latter can be obtained by suitably adjusting the gap between the buildings (if it is possible, e.g. for new constructions), and/or the contact material behaviour (hardening or softening) according to the optimal control theory in structural analysis, see e.g. [3,5,8,19,31]. Finally, the method is applied to a civil engineering example of adjacent buildings.

2 Method of Analysis

A system of only two adjacent linearly elastic structures (A) and (B) is considered here for simplicity. The extension to systems with more than two linear and/or nonlinear elastic buildings can be done in a straightforward way.

2.1 Uncoupled System Analysis

First the system of the two structures (A) and (B), considered as an uncoupled one, is discretized by the finite element method. So, assuming no interaction, the matrix equations of dynamic equilibrium are

$$\underline{M}_L \underline{\ddot{u}}_L + \underline{C}_L \underline{\dot{u}}_L + \underline{K}_L \underline{u}_L = -\underline{M}_L \underline{\ddot{u}}_g, \quad (L = A, B), \tag{1}$$

where $\underline{M}_L, \underline{C}_L, \underline{K}_L$ are the mass, damping and stiffness matrices, respectively; $\underline{u}(t)$ is the sought node displacement (relative to ground) vector corresponding to given ground earthquake excitation $\underline{u}_g(t)$ and appropriate initial conditions; and dots over symbols indicate time derivatives. Problem (1) can be solved by wellknown methods of Structural Dynamics.

2.2 Interaction Simulation

Let j_A and j_B be two associated nodes on the interface (joint), where unilateral frictional contact can take place during an earthquake. These nodes are considered (see Liolios [11]) as connected by two fictive unilateral constraints, normal to interface the first and tangential the second one. The corresponding force-reactions and retirement relative displacements are denoted by r_{jN}, z_{jN} and r_{jT}, z_{jT}, respectively. They satisfy in general nonconvex and nonmonotone constitutive relations of the following type (2), expressing mathematically the unilateral elastoplastic-softening contact with friction:

$$r_j(d_j) \in \partial R_j(d_j). \tag{2}$$

Here ∂ is the generalized gradient of Clarke, d the deformation and $R_j(\cdot)$ is the superpotential function, see e.g. Panagiotopoulos [20, 21] and [7, 15, 16, 25-27]. By definition, rel. (2) is equivalent to the following hemivariational inequality:

$$R_j^{\uparrow}(d_j, e_j - d_j) \geq r_j(d_j) \cdot (e_j - d_j), \tag{3}$$

where $R_j^{\uparrow}$ denotes subderivative and e_j virtual deformation. In engineering terminology, this inequality expresses the virtual work principle holding in inequality form for unilateral constraints.

By piecewise linearizing these relations as in [9-13] we obtain the following linear complementarity conditions:

$$r_{jN} = p_{jN}(z_{jN} - g_j + w_j) + c_j \dot{z}_{jN}, \tag{4}$$

$$w_j \geq 0, \quad r_{jN} \leq 0, \quad w_j r_{jN} = 0, \tag{5}$$

$$|r_{jT}| \leq f_j |r_{jN}|, \quad \dot{z}_{jT} \cdot r_{jT} = 0,$$

$$\dot{z}_{jT} \cdot (|r_{jT}| - f_j |r_{jN}|) = 0. \tag{6}$$

In rels. (4),(5) c_j is the damping coefficient, p_{jN} the reaction function for the normal unilateral constraint, g_j the existing normal gap and w_j a non-negative multiplier; in rels. (6) f_j is the Coulomb's friction coefficient. So, rels. (4) - (5) impose that friction phenomena (slip or adhesion) can take place only when unilateral contact occurs, i.e. when the compressive contact force r_{jN} is appeared.

2.3 Coupled System Conditions with P-delta Effects

Taking into account, now, the interaction and the second-order geometric effects (P-Delta effects), we write the incremental dynamic equilibrium conditions for the coupled system of the interacting buildings (A) and (B):

$$\underline{M}_A \Delta \underline{\ddot{u}}_A + \underline{C}_A \Delta \underline{\dot{u}}_A + (\underline{K}_A + \underline{G}_A) \Delta \underline{u}_A = -\underline{M}_A \Delta \underline{\ddot{u}}_g + \underline{T}_A \Delta \underline{r},$$

$$\underline{M}_B \Delta \ddot{u}_B + \underline{C}_B \Delta \dot{u}_B + (\underline{K}_B + \underline{G}_B) \Delta \underline{u}_B = -\underline{M}_B \Delta \underline{\ddot{u}}_g - \underline{T}_B \Delta \underline{r},$$

$$\underline{r} = \underline{r}_N + \underline{r}_T. \tag{7}$$

Here $\underline{G}_A$ and $\underline{G}_B$ are the geometric stiffness matrices, by which P-Delta effects are taken into account [2,6,12], $\underline{T}_A$ and $\underline{T}_B$ are transformation matrices, and *underliner* is the coupling vector of the normal and tangential interaction forces, satisfying (4),(5). Appropriate initial conditions are taken into account, and so the problem consists in finding the time-dependent vectors $\{\underline{u}_A, \underline{u}_B, \underline{g}, \underline{z}, \underline{r}, \underline{w}\}$ which satisfy the rels. (2)-(7) for the given earthquake excitation $\underline{u}_g(t)$.

2.4 Time Discretization and Problem Solution

Further the problem of rels. (2)-(7) is discretized in time. Because this problem is nonlinear -due to inequalities- the mode superposition method cannot be applied. Thus, as suggested in [28], direct time-integration methods have to be used. Here the ϑ-Wilson method is preferred to other implicit schemes and a suitable elimination of some unknowns is made. In each time-step we assume that the unilateral constraints remain either active or inactive by adjusting suitably the time-step. To compute what is happening, the procedure of Liolios [11] is applied. So, a nonconvex linear complementarity problem of the following form is eventually solved by available algorithms [14-15,23,29]:

$$\underline{v} \geq \underline{0}, \quad \underline{D}\underline{v} + \underline{d} \leq \underline{0}, \quad \underline{v}^T \cdot (\underline{D}\underline{v} + \underline{d}) = 0. \tag{8}$$

Due to non-convexity of the interface behaviour (frictional unilateral contact, descending branches in the relevant stress-deformation diagrammes etc., see e.g. Fig. 1d for the numerical example of the next section), the matrix $\underline{D}$ does not correspond to a strictly positive quadratic form. But the hemivariational inequality (3), interpreted from the engineering point of view for the case of a stable system (no collapse), means that the internal virtual energy of the coupled system is greater of/or equal to external virtual work. So, using at every time-step the hemivariational inequality (3) as a stability criterion , it can be proved that for most practical applications in structural mechanics, the matrix $\underline{D}$ is a P-matrix. Thus a unique solution of the nonconvex linear complementarity problem (8) can be assured [9,12].

2.5 Influence Coefficients

Further, we introduce the influence coefficients

$$c = \frac{Q^c - Q^u}{Q^u}, \tag{9}$$

where Q is the absolutely maximum value which takes a response quantity during the seismic excitation. Index (c) is for the coupled system and index (u) for the uncoupled one (i.e. without interaction). By the influence coefficients c comparison is made between the uncoupled and the coupled cases. Thus, these coefficients show whether a structural element is overstressed or understressed due to interaction.

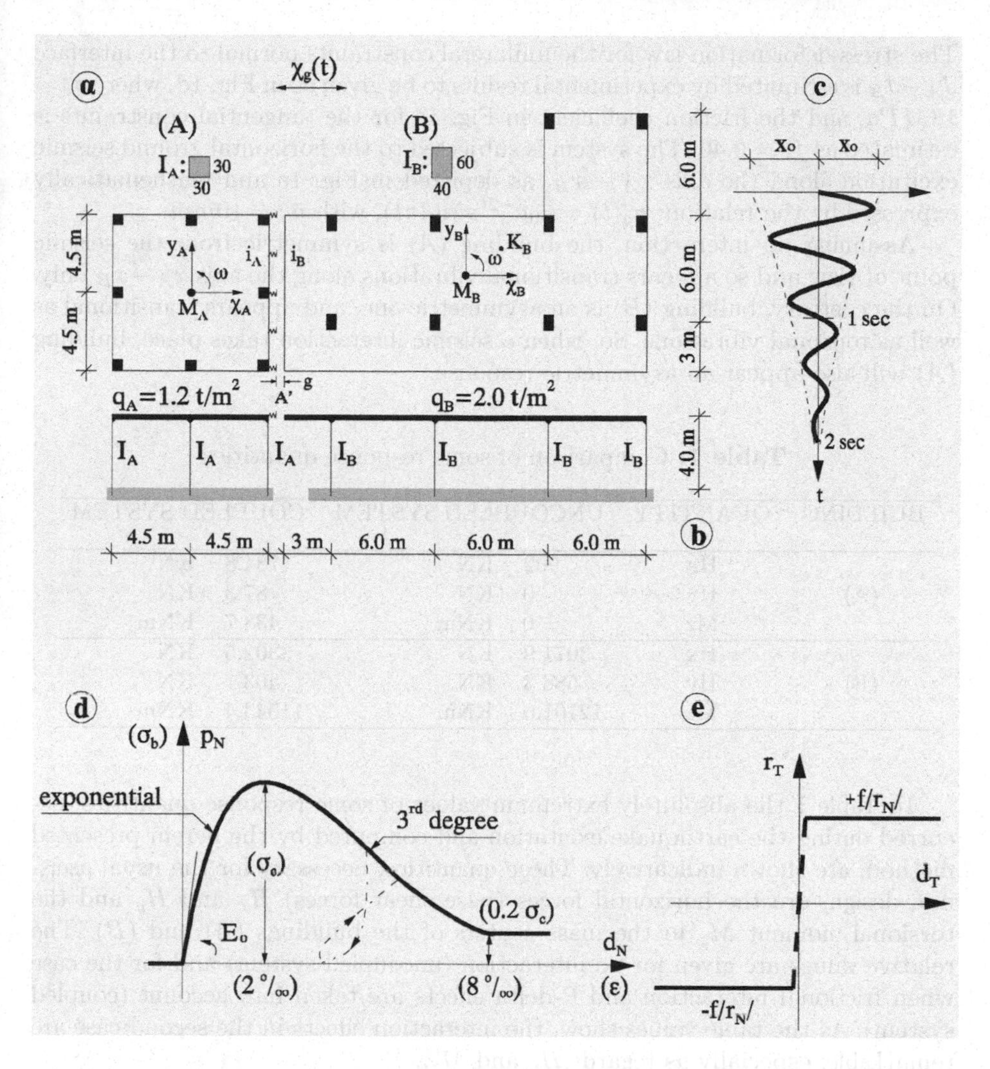

Fig. 1. Numerical example: a) Plan view of the system of buildings (A) and (B); b) Vertical section view; c) Seismic ground displacement; d) Stress-deformation law for the normal unilateral constraints; e) Stress-deformation law for the tangential unilateral constraints.

3 Numerical Example

The system of the two one-storey buildings (A) and (B) of Fig. 1a,b are of reinforced concrete with elasticity modulus $E_b = 3 \times 10^7 KN/m^2$, slab thickness $0.25m$, damping ratio 5% and beams $30/80cm$ connecting the columns tops perimetrically. The columns section is $30/30cm$ for (A) and $40/60cm$ for (B).

The stress-deformation law for the unilateral constraints normal to the interface $J_A - J_B$ is estimated by experimental results to be given as in Fig. 1d, where $\sigma_c = 18MPa$, and the friction coefficient in Fig. 1e for the tangential constraints is estimated as $f_j = 0.40$. The system is subjected to the horizontal ground seismic excitation along the axis $x_A - x_B$, as depicted in Fig. 1c and mathematically expressed by the relation: $u_g(t) = u_0 e^{-2t} sin(4\pi t)$, with $u_0 = 10mm$.

Assuming no interaction, the building (A) is symmetric from the seismic point of view and so appears transitional vibrations along the axis $x_A - x_B$ only. On the contrary, building (B) is an asymmetric one, and appears transitional as well as torsional vibrations. So, when a seismic interaction takes place, building (A) will also appear an asymmetric response.

Table 1. Comparison of some response quantities

BUILDING	QUANTITY	UNCOUPLED SYSTEM		COUPLED SYSTEM	
	Hx	902	KN	784.8	KN
(A)	Hy	0	KN	87.3	KN
	Mz	0	KNm	438.7	KNm
	Hx	3071.9	KN	3302.5	KN
(B)	Hy	588.3	KN	403.1	KN
	Mz	12161.0	KNm	11543.4	KNm

In Table 1 the absolutely extremum values of some response quantities, occurred during the earthquake excitation and computed by the herein presented method, are shown indicatively. These quantities, necessary for the usual aseismic design, are the horizontal forces (base shear forces) H_x and H_y and the torsional moment M_z in the mass centers of the buildings (A) and (B). The relative values are given for no interaction (uncoupled system) and for the case when frictional interaction and P-delta effects are taken into account (coupled system). As the table values show, the interaction effects in the second case are remarkable, especially as regards H_y and M_z.

4 Concluding Remarks

Frictional seismic interaction under second-order effects, which is often not taken into account in the usual Civil Engineering design of adjacent buildings, can change significantly the earthquake response of such structures subjected to unilateral contact. As in the numerical example has been shown, a numerical estimation of the so caused seismic interaction effects can be obtained by the herein-presented approach. Thus, the numerical procedure is realized by using available computer codes of the finite element method and the nonlinear mathematical programming (non-convex optimization algorithms).

Certainly the most complicated task, from the earthquake engineering point of view, in all the above cases is the realistic simulation of the dynamic unilateral contact behavior. To overcome this difficulty, experimental results can be used for the rational estimation of parameters involved to simulate the interface behaviour between adjacent structures. On the other hand, the herein presented numerical approach can be used effectively to estimate numerically and to control actively the influence of the interaction on the seismic response of adjacent structures. This can be obtained by using methods of the optimal control in order to adjust the gap between the buildings and/or the contact material behaviour (hardening or softening) of the structural interface elements.

Finally, it is necessary to be emphasized that the basis of the herein presented approach is the concept of the hemivariational inequalities, introduced into Mechanics and Applied Mathematics by P.D. Panagiotopoulos, see [18-21]. These inequalities constitute now the basis for the effective treatment of many unilateral problems, and especially for the Non-Smooth Mechanics. Thus, after the sudden passing away of Prof. P.D. Panagiotopoulos at 12th of August 1998, it has been proposed by the first author (Ast. Liolios) that **the hemivariational inequalities** must be called **the Panagiotopoulos Inequalities**.

References

1. Anagnostopoulos, S.A., Spiliopoulos K.V.: Analysis of Building Pounding due to Earthquakes. In: Krätzig W.B. et al (eds.): Structural Dynamics. Balkema, Rotterdam (1991) 479-484.
2. Anastasiadis, K.: Méthode simplifiée de calcul du second ordre de bâtiments à étage. Construction Metallique 4 (1986) 43-68.
3. Baniotopoulos, C.C.: Optimal control of above-ground pipelines under dynamic excitations. Int. Jnl Pressure Vessel & Piping 63 (1995) 211-222.
4. Bertero, V.V.: Observations on structural pounding. Proc. Intern. Conf. "The Mexico Earthquakes" ASCE (1987) 264-278.
5. Bisbos, C.: Aktive Steuerung erdbebenerregter Hochhäuser. Z. Angew. Math. Mech. 65 (1985) T297-T299.
6. Chen, W.F., Lui, E.M.: Structural Stability. Elsevier, New York (1981).
7. Dem'yanov, V.F., Stavroulakis, G.E., Polyakova, L.N., Panagiotopoulos, P.D.: Quasidifferentiability and nonsmooth modelling in mechanics, engineering and economics. Kluwer Academic, Dordrecht, Boston, London (1996).
8. Haslinger, J., Neittaanmäki, P. : Finite element approximation for optimal shape, material and topology design. Wiley, Chichester (2nd edition) (1996).
9. Klarbring, A.: General contact boundary conditions and the analysis of frictional systems. Int. J. Solids Structures, 22 (1986) 1377-1398.
10. Liolios, A.A.: A finite-element central-difference approach to the dynamic problem of nonconvex unilateral contact between structures. In: B. Sendov, R. Lazarov and P. Vasilevski (eds.): Numerical Methods and Applications. Bulgarian Acad. Sciences, Sofia (1984) 394-401.
11. Liolios, A.A.: A linear complementarity approach to the nonconvex dynamic problem of unilateral contact with friction between adjacent structures. Z. Angew. Math. Mech. 69 (1989) T420-T422.
12. Maier, G.: Incremental Elastoplastic Analysis in the Presence of Large Displacements and Physical Instabilizing Effects. Int. Jnl Solids and Structures 7 (1971) 345-372.

13. Maier, G.: Mathematical programming methods in structural analysis. In: Brebbia, C. & H. Tottenham (eds.): Variational methods in engineering. Proc. Int. Conf. Southampton University Press, Southampton, Vol. 2 (1973) 8/1-8/32.
14. Miettinen, M., Mäkelä M.M., Haslinger, J.: On mumerical solution of hemivariational inequalities by nonsmooth optimization methods. Journal of Global Optimization 6 (1995) 401-425.
15. Mistakidis, E.S., Stavroulakis, G.E.: Nonconvex Optimization in Mechanics: Algorithms, Heuristics and Engineering Applications. Kluwer Academic, Dordrecht, Boston, London (1998).
16. Naniewicz, Z., Panagiotopoulos, P.D.: Mathematical Theory of Hemivariational Inequalities and Applications, Marcel Dekker, New York (1995).
17. Nitsiotas, G.: Die Berechnung statisch unbestimmter Tragwerke mit einseitigen Bindungen. Ingenieur-Archiv 41 (1971) 46-60.
18. Panagiotopoulos, P.D.: Non-convex Energy Functions. Hemivariational Inequalities and Substationarity principles. Acta Mechanica 48 (1983) 111-130.
19. Panagiotopoulos, P.D.: Optimal control of structures with convex and nonconvex energy densities and variational and hemivariational inequalities. Eng. Struct. 6 (1984) 12-18.
20. Panagiotopoulos, P.D.: Inequality problems in mechanics and applications. Convex and nonconvex energy functions. Birkhäuser Verlag, Boston-Basel-Stuttgart (1985).
21. Panagiotopoulos, P.D.: Hemivariational Inequalities. Applications in Mechanics and Engineering. Springer-Verlag, Berlin, New York (1993).
22. Papadrakakis, M., Apostolopoulou, K., Bitzarakis, S., Zacharopoulos, A.: A 3D model for the analysis of building pounding during earthquakes. In: Moan, T. et al., (eds.): Structural Dynamics-Eurodyn '93. Balkema, Rotterdam (1993) 85-92.
23. Pardalos, P.: Global Optimization Algorithms for Linearly Constrained Indefinite Quadratic Problems. Computers Math. Applic. 21(6/7) (1991) 87-97.
24. Pfeiffer, F., Glocker, Ch.: Multibody Dynamics with Unilateral Contacts, Wiley, New York (1996).
25. Stavroulakis, G.E.: Quasi-variational inequalities. In: Floudas, C.A. and Pardalos, P.A. (Eds.): Encyclopedia of Optimization, Kluwer Academic, London (in press).
26. Stavroulakis, G.E., Antes, H., Panagiotopoulos, P.D.: Transient elastodynamics around cracks including contact and friction. Comp. Meth. in Appl. Mech. and Enging. Special Issue: Computational Modeling of Contact and Friction. Martins, J.A.C., Klarbring,A. (eds.): 177(3/4), (1999) 427-440.
27. Stavroulakis, G.E., Mistakidis, E.S.: Nonconvex Energy Functions: Hemivariational Inequalities. In: Floudas, C.A. and Pardalos, P.A. (Eds.): Encyclopedia of Optimization, Kluwer Academic, London (in press).
28. Talaslidis, D., Panagiotopoulos, P.D.: A linear finite element approach to the solution of the variational inequalities arising in contact problems of structural dynamics. Int. J. Num. Meth. Enging 18 (1982) 1505-1520.
29. Tzaferopoulos, M.A. and Liolios, A.A.: On a Branch and Bound Algorithm for the Solution of a Family of Softening Material Problems of Mechanics with Applications to the Analysis of Metallic Structures. Journal of Global Optimization 11 (1997) 133-149.
30. Wolf, J.P., Skrikerud, P.E.: Mutual pounding of adjacent structures during earthquakes. Nuclear Engin. Design 57 (1980) 253-275.
31. Zacharenakis, E.C.: On the Disturbance Attenuation and H-Optimization in structural analysis, Z. Angew. Math. Mech. 77 (1997) 189-195.

Parallel Performance of an MPI Solver for 3D Elasticity Problems

Ivan Lirkov

Central Laboratory for Parallel Processing
Bulgarian Academy of Sciences
Acad.G.Bonchev Str., Bl.25A, 1113 Sofia, Bulgaria
ivan@parallel.bas.bg

Abstract. The numerical solution of 3D linear elasticity equations is considered. The problem is described by a coupled system of second order elliptic partial differential equations. This system is discretized by trilinear parallelepipedal finite elements.
The Preconditioned Conjugate Gradient iterative method is used for solving of the large-scale linear algebraic systems arising after the Finite Element Method (FEM) discretization of the problem. Displacement decomposition technique is applied at the first step to construct a preconditioner using the decoupled block-diagonal part of the original matrix. Then circulant block-factorization is used for preconditioning of the obtained block-diagonal matrix. Both preconditioning techniques, displacement decomposition and circulant block-factorization, are highly parallelizable.
A parallel algorithm is invented for the proposed preconditioner. The theoretical analysis of the execution time shows that the algorithm is highly efficient for coarse-grain parallel computer systems.
A portable parallel FEM code based on MPI is developed. Numerical tests for real-life engineering problems in computational geomechanics are performed on a number of modern parallel computers: Cray T3E, Sunfire 6800, and Beowulf cluster. The reported speed-up and parallel efficiency well illustrate the parallel features of the proposed method and its implementation.
Keywords: parallel algorithms, PCG method, preconditioner, circulant matrix, elasticity problem
MSC2000: 65F10, 68W10, 74B05, 74B20, 74S05

1 Introduction

This work concerns new efficient parallel algorithms and the related program software for solving the elasticity problem in computational geomechanics. Typical application problems include the simulations of the foundation of engineering constructions (which transfer and distribute the total loading into the bed soil)

I. Dimov et al. (Eds.): NMA 2002, LNCS 2542, pp. 527–535, 2003.

and the multi-layer media with strongly varying material characteristics. Here, the spatial framework of the construction produces a composed stressed-strained state in active interaction zones. A modern design of cost-efficient construction with a sufficient guaranteed reliability requires to determine the parameters of this stressed-strained state.

The application problems are three dimensional nonlinear elasticity problems which are described mathematically by a system of partial differential equations. A finite element (or finite difference) discretization reduces the partial differential equation problem to a system of linear/nonlinear equations. To make a reliable prediction of the construction safety, which is sensitive to soil deformations, a very accurate model and a large system of sparse linear equations is required. In the real-life applications, the system can be very large containing up to several millions of unknowns. Hence, these problems have to be solved by robust and efficient parallel iterative methods on a powerful multiprocessor machine.

Note that the numerical solution of linear systems is fundamental in the elasticity problem. In fact, nonlinear equations generated from the discretization of the nonlinear elasticity problem have to be solved by an iterative procedure, in which a system of linear equations has to be solved in every step of iteration. Solving these linear systems is usually very time-consuming (costing up to 90% of the total solution time). Hence, developing fast algorithms for solving linear equations becomes the most important and fundamental issue. A highly efficient iterative method for solving linear systems significantly speed up the simulation processes of real application problems. An efficient iterative solver should not only have a fast convergence rate but also a high parallel efficiency. Moreover, the resulting program should be efficiently implemented on modern shared-memory, distributed memory, and shared-distributed memory parallel computers.

2 Elasticity Problems

For simplicity, we mainly study the 3D linear elasticity problem based on the following *two basic assumptions*: (1) the displacements are small, and (2) the material properties are isotropic.

The mathematical formulation of the 3D elasticity problem is described as follows. Let $\underline{u} = (u_1, u_2, u_3)^T$ be the displacement vector and $\underline{p}$ the volume force vector. Here T denotes the transpose of a vector or a matrix. Let us denote the matrices D, G and H by

$$D = \begin{pmatrix} \frac{\partial}{\partial x_1} & 0 & 0 & \frac{\partial}{\partial x_2} & 0 & \frac{\partial}{\partial x_3} \\ 0 & \frac{\partial}{\partial x_2} & 0 & \frac{\partial}{\partial x_1} & \frac{\partial}{\partial x_3} & 0 \\ 0 & 0 & \frac{\partial}{\partial x_3} & 0 & \frac{\partial}{\partial x_2} & \frac{\partial}{\partial x_1} \end{pmatrix}, \quad G = \begin{pmatrix} \frac{\partial}{\partial x_1} & 0 & 0 \\ 0 & \frac{\partial}{\partial x_2} & 0 \\ 0 & 0 & \frac{\partial}{\partial x_3} \\ \frac{\partial}{2\partial x_2} & \frac{\partial}{2\partial x_1} & 0 \\ 0 & \frac{\partial}{2\partial x_3} & \frac{\partial}{2\partial x_2} \\ \frac{\partial}{2\partial x_3} & 0 & \frac{\partial}{2\partial x_1} \end{pmatrix},$$

$$H = \begin{pmatrix} (1-\nu) & \nu & \nu & 0 & 0 & 0 \\ \nu & (1-\nu) & \nu & 0 & 0 & 0 \\ \nu & \nu & (1-\nu) & 0 & 0 & 0 \\ 0 & 0 & 0 & (1-2\nu) & 0 & 0 \\ 0 & 0 & 0 & 0 & (1-2\nu) & 0 \\ 0 & 0 & 0 & 0 & 0 & (1-2\nu) \end{pmatrix}.$$

Then the strain vector $\underline{\epsilon} = (\epsilon_{11}, \epsilon_{22}, \epsilon_{33}, \epsilon_{12}, \epsilon_{23}, \epsilon_{31})^T$ and the stress vector $\underline{\sigma} = (\sigma_{11}, \sigma_{22}, \sigma_{33}, \sigma_{12}, \sigma_{23}, \sigma_{31})^T$ are determined by

$$\underline{\epsilon} = G\underline{u}, \quad \text{and} \quad \underline{\sigma} = E^* H \underline{\epsilon}, \tag{1}$$

where $E^* = \frac{E}{(1+\nu)(1-2\nu)}$. Here ν and E are respectively the Poisson ratio and the deformation module.

With the above notation, the 3D elasticity problem on a computational domain Ω, can be described by a coupled system of three differential equations, which is written in the form

$$\begin{cases} D\underline{\sigma} = -\underline{p} & \text{in } \Omega \\ \underline{u} = \underline{u_D} & \text{on } \Gamma_D \\ \sum_{i=1}^{3} \sigma_{ij} n_i = \sigma_{Nj} & \text{on } \Gamma_N, \quad j = 1,2,3, \end{cases}$$

where Γ_D and Γ_N are the parts of the boundary of Ω with respectively Dirichlet and Neumann boundary conditions; and $\underline{u_D}$ and $\underline{\sigma_N}$ are respectively the given displacement and stress vectors on the boundaries Γ_D and Γ_N. Here we set $\sigma_{ji} = \sigma_{ij}$ for $i < j$.

If the Poisson ratio and the deformation module are nonlinear functions, the relations (1) represent the nonlinear nature of the generalized Hooke's law. Here the generalized Hooke's law is specified by the following additional assumption: the Poisson ratio $\nu \in (0, \frac{1}{2})$ is a constant for a given material (soil layer or constructive element). Obviously, this means that the coefficients in the boundary value problem (2) are piece-wise continuous with jumps through the inner boundaries between the different soil layers as well as between the soil and the construction elements.

With a linearization, the nonlinear equations given in (2) can be simplified to a system of three linear differential equations, which is often referred to as the Lamé equations.

Denote Sobolev spaces $\left[H_E^1(\Omega)\right]^3 = \left\{\underline{v} \in \left[H^1(\Omega)\right]^3 : \underline{v}|_{\Gamma_D} = \underline{u_D}\right\}$ and $\left[H_0^1(\Omega)\right]^3 = \left\{\underline{v} \in \left[H^1(\Omega)\right]^3 : \underline{v}|_{\Gamma_D} = 0\right\}$. The variational formulation of the Lamé equations is given below:

$$\text{find } \underline{u} \in \left[H_E^1(\Omega)\right]^3 \text{ such that} \qquad \forall \underline{v} \in \left[H_0^1(\Omega)\right]^3$$

$$\int_\Omega \left[\lambda\, div\, \underline{u}\, div\, \underline{v} + 2\mu \sum_{i=1}^{3}\sum_{j=1}^{3} \epsilon_{ij}(\underline{u})\, \epsilon_{ij}(\underline{v})\right] d\Omega = -\int_\Omega \underline{p}^T \underline{v}\, d\Omega + \int_{\Gamma_N} \underline{\sigma_N}^T \underline{v}\, d\Gamma,$$

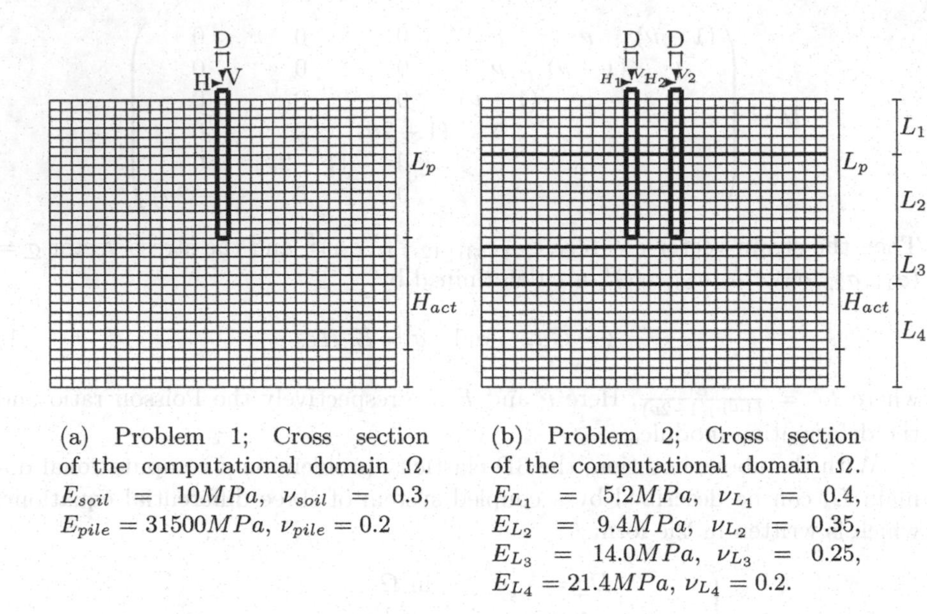

(a) Problem 1; Cross section of the computational domain Ω. $E_{soil} = 10MPa$, $\nu_{soil} = 0.3$, $E_{pile} = 31500MPa$, $\nu_{pile} = 0.2$

(b) Problem 2; Cross section of the computational domain Ω. $E_{L_1} = 5.2MPa$, $\nu_{L_1} = 0.4$, $E_{L_2} = 9.4MPa$, $\nu_{L_2} = 0.35$, $E_{L_3} = 14.0MPa$, $\nu_{L_3} = 0.25$, $E_{L_4} = 21.4MPa$, $\nu_{L_4} = 0.2$.

Fig. 1. Benchmark problems

where $\lambda > 0$ and $\mu > 0$ are the Lamé coefficients. Here $div\,\underline{u}$ is the divergence of the vector $\underline{u}$. The relations between the elasticity modulus E, ν and the material parameters λ, μ are $\lambda = \frac{\nu E}{(1+\nu)(1-2\nu)}$ and $\mu = \frac{E}{2(1+\nu)}$. We restrict our considerations to the case $\Omega = [0, x_1^{max}] \times [0, x_2^{max}] \times [0, x_3^{max}]$, where the boundary conditions on each of the sides of Ω are of a fixed type. The benchmark problems from [2] are used in the reported numerical tests. These benchmarks represent the model of a single pile in a homogeneous sandy clay soil layer (see Fig. 1(a)) and two piles in a multi-layer soil media (Fig. 1(b)). An uniform grid is used with n_1, n_2 and n_3 grid points along the coordinate directions. Then the stiffness matrix K can be written in a 3×3 block form where the blocks K_{ij} are sparse block–tridiagonal matrices of a size $n_1 n_2 n_3$.

3 DD CBF Preconditioning

The preconditioning technique used in this work is described in details in [4,3]. The theoretical analysis of the execution time of the proposed parallel algorithm is published in [4]. Here we will only sketch the construction of the preconditioner.

First, we use the approach known as *displacement decomposition* (see, e.g., [1]) to define the preconditioner M_{DD} of the matrix K. We introduce the auxiliary Laplace equation $-u_{x_1x_1} - u_{x_2x_2} - u_{x_3x_3} = f$, with boundary conditions corresponding to the considered coupled elasticity problem. This Laplace equation is discretized by the same brick finite elements as the original problem, and

K_0 is the obtained stiffness matrix. Then $M_{DD} = diag(K_0, K_0, K_0)$. The next step in our construction is to substitute in M_{DD}, K_0 by A_0, where A_0 stands for the Laplace stiffness matrix corresponding to linear finite elements. Now, let us denote by M_0 the Circulant Block-Factorization (CBF) preconditioner (see [5,4,3]) for A_0. At the last step of our construction we substitute in M_{DD}, K_0 by M_0, and get the DD CBF preconditioner defined by:

$$M_{DD\ CBF} = diag(M_0, M_0, M_0).$$

The following estimate of the condition number of the preconditioned matrix is derived in [3]

$$\kappa(M_{DD\ CBF}^{-1} K) = \mathcal{O}\left(\frac{n_{\max}}{1 - 2\nu_{\max}}\right)$$

where $n_{\max} = \max(n_1, n_2, n_3)$ and $\nu_{\max} = \max_\Omega \nu$.

Remark 1. We have observed in the performed numerical tests that a diagonal scaling of K improves the convergence rate of the iterative method in the case of problems with jumping coefficients.

4 Parallel Tests of the DD CBF Preconditioning FEM Code

In this section we report the results of the experiments executed on three parallel systems. We report here the number of iterations N_{it}, the elapsed time T_p on p processors, the speed-up $S_p = T_1/T_p$, and the parallel efficiency $E_p = S_p/p$. We have used discretizations with $n_1 = n_2 = n_3 = n$ where $n = 32, 48, 64$, and 96. The sizes of the discrete problems are $3n^3$.

The developed parallel code has been implemented in C and the parallelization has been facilitated using the MPI [6,7] and OpenMP libraries. In all cases, the optimization options of the compiler have been tuned to achieve the best performance. Times have been collected using the MPI provided timer. In all cases we report the best results from multiple runs.

In Table 1 we present results of experiments executed on Cray T3E-900 consisting of 336 Digital Alpha 450 MHz processors, with 64 or 128 MB memory on processor. The memory on one processor of Cray computer is sufficient only for the discretization with $32 \times 32 \times 32$ grid points. For larger problems we report the parallel efficiency related to the results on 6, 16, and 24 processors respectively.

Table 2 shows the results obtained on Beowulf cluster consisting of 17 PC with AMD Athlon processors, 650 MHz, 128 MB memory per computer. The memory per processor is the same and the same approach is used to compute the relative parallel efficiency.

Tables 3 and 4 shows results obtained on a Sunfire 6800 consisting of 24 UltraSPARC-III 750 MHz processors and 48 GB main memory. As expected, the parallel efficiency increases with the size of the discrete problems. The parallel

Table 1. Parallel time (in seconds), speed-up and parallel efficiency on Cray T3E-900.

		Problem 1				Problem 2			
n	p	N_{it}	T_p	S_p	E_p	N_{it}	T_p	S_p	E_p
32	1	112	124.242			471	479.214		
	2		61.371	2.02	1.012		236.371	2.03	1.014
	4		31.029	4.00	1.001		119.153	4.02	1.005
	8		15.973	7.78	0.972		60.529	7.92	0.990
	16		8.741	14.21	0.888		32.795	14.61	0.913
	32		5.104	24.34	0.761		18.991	25.23	0.789
48	6	607	462.692			1118	845.019		
	8		344.337		1.008		626.818		1.011
	12		232.377		0.996		424.325		0.996
	16		176.318		0.984		300.292		1.055
	24		121.932		0.949		206.756		1.022
	48		65.336		0.885		118.534		0.891
64	16	771	491.822			1253	793.226		
	32		245.944		1.000		405.173		0.979
	64		128.406		0.958		209.953		0.945
96	24	1164	1921.650			1941	3190.150		
	32		1424.960		1.011		2361.410		1.013
	48		960.577		1.000		1592.680		1.002

Table 2. Parallel time (in seconds), speed-up and parallel efficiency on Beowulf cluster.

		Problem 1			Problem 2		
n	p	T_p	S_p	E_p	T_p	S_p	E_p
32	1	120.962			486.297		
	2	64.125	1.89	0.943	257.513	1.89	0.944
	4	36.656	3.30	0.825	146.807	3.31	0.828
	8	18.783	6.44	0.805	74.887	6.49	0.812
	16	18.446	6.56	0.410	73.772	6.59	0.412
48	3	1006.950			1721.350		
	4	770.222		0.981	1409.420		0.916
	6	530.166		0.950	902.080		0.954
	8	537.809		0.702	984.016		0.656
	12	517.499		0.486	936.786		0.460
	16	395.148		0.478	674.949		0.478
64	8	1147.780			1909.620		
	16	1385.660		0.414	2304.120		0.414

Table 3. Parallel time (in seconds), speed-up and parallel efficiency on Sunfire 6800 using OpenMP.

		Problem 1			Problem 2		
n	p	T_p	S_p	E_p	T_p	S_p	E_p
32	1	68.271			265.898		
	2	35.388	1.93	0.965	138.818	1.92	0.958
	3	24.884	2.74	0.915	99.436	2.67	0.891
	4	18.527	3.68	0.921	71.172	3.74	0.934
	5	15.423	4.43	0.885	58.423	4.55	0.910
	6	12.594	5.42	0.904	48.040	5.53	0.922
	7	9.627	7.09	1.013	36.469	7.29	1.042
	8	6.802	10.04	1.255	25.033	10.62	1.328
	16	4.013	17.01	1.063	14.519	18.31	1.145
	24	4.401	15.51	0.646	17.889	14.86	0.619
48	1	1601.360			2726.980		
	2	803.957	1.99	0.996	1476.360	1.85	0.924
	3	551.950	2.90	0.967	1000.590	2.73	0.908
	4	407.601	3.93	0.982	739.007	3.69	0.923
	5	347.410	4.61	0.922	635.786	4.29	0.858
	6	277.599	5.77	0.961	479.173	5.69	0.949
	7	246.424	6.50	0.928	414.200	6.58	0.941
	8	208.451	7.68	0.960	376.430	7.24	0.906
	16	108.478	14.76	0.923	184.147	14.81	0.926
	24	78.358	20.44	0.852	136.314	20.01	0.834
64	1	4264.670			7087.930		
	2	2174.790	1.96	0.980	3578.770	1.98	0.990
	3	1479.230	2.88	0.961	2396.550	2.96	0.986
	4	1073.160	3.97	0.993	1751.840	4.05	1.011
	5	902.729	4.72	0.945	1455.900	4.87	0.974
	6	758.351	5.62	0.937	1219.490	5.81	0.969
	7	679.034	6.28	0.897	1111.130	6.38	0.911
	8	551.936	7.73	0.966	894.441	7.92	0.991
	16	294.812	14.47	0.904	489.250	14.49	0.905
	24	269.942	15.80	0.658	433.576	16.35	0.681
96	1	27668.500			45917.300		
	2	13881.100	1.99	0.997	23092.600	1.99	0.994
	3	9374.630	2.95	0.984	15543.400	2.95	0.985
	4	6909.890	4.00	1.001	11521.600	3.99	0.996
	5	5803.100	4.77	0.954	9647.530	4.76	0.952
	6	4740.550	5.84	0.973	7825.420	5.87	0.978
	7	4142.120	6.68	0.954	6808.080	6.74	0.964
	8	3471.910	7.97	0.996	5799.240	7.92	0.990
	16	1825.810	15.15	0.947	3017.930	15.21	0.951
	24	1504.800	18.39	0.766	2461.980	18.65	0.777

Table 4. Parallel time (in seconds), speed-up and parallel efficiency on Sunfire 6800 using MPI.

		Problem 1			Problem 2		
n	p	T_p	S_p	E_p	T_p	S_p	E_p
32	1	78.98			313.33		
	2	43.68	1.81	0.904	168.02	1.86	0.932
	4	20.21	3.91	0.977	79.13	3.96	0.990
	8	8.10	9.75	1.219	26.72	11.73	1.466
	16	3.71	21.28	1.330	12.93	24.24	1.515
48	1	1718.96			2973.82		
	2	864.19	1.99	0.995	1502.07	1.98	0.990
	3	582.46	2.95	0.984	990.31	3.00	1.001
	4	436.42	3.94	0.985	747.80	3.98	0.994
	6	291.81	5.89	0.982	534.42	5.56	0.927
	8	212.68	8.08	1.010	386.37	7.70	0.962
	16	109.80	15.65	0.978	201.43	14.76	0.923
	24	72.59	23.68	0.987	132.34	22.47	0.936
64	1	4847.60			8075.46		
	2	2351.42	2.06	1.031	3869.54	2.09	1.043
	4	1116.15	4.34	1.086	1830.77	4.41	1.103
	8	560.48	8.65	1.081	889.14	9.08	1.135
	16	284.02	17.07	1.067	465.98	17.33	1.083
96	1	30949.00			51284.61		
	2	15197.60	2.04	1.018	25321.60	2.03	1.013
	3	10145.80	3.05	1.017	16848.80	3.04	1.015
	4	7493.50	4.13	1.033	12545.60	4.09	1.022
	6	5006.49	6.18	1.030	8367.84	6.13	1.021
	8	4029.95	7.68	0.960	6644.44	7.72	0.965
	16	1862.06	16.62	1.039	3102.48	16.53	1.033
	24	1412.54	21.91	0.913	2350.58	21.82	0.909

efficiency is above 90% (except the cases where the number of processors do not divide the size of the discrete problem) which confirms our general expectations. There exist at least two reasons for the reported high efficiency: (a) the network parameters *start-up time* and *time for transferring of single word* are relatively small for the multiprocessor machines; (b) there is also some overlapping between the computations and the communications in the algorithm. Moreover, the super-linear speed-up can be seen in some of the runs. This effect has a relatively simple explanation. When the number of processors increases, the size of data per processor decreases. Thus the stronger *memory locality* increases the role of the cache memories. (The level 2 cache on the Sunfire 6800 is 8 Mbyte.)

Finally, we compare results on Cray, Sunfire, and Beowulf cluster with our previous results (see [3]) on SGI Origin 2000, SUN Ultra-Enterprise, PowerPC and Alpha clusters. Fig. 2 shows parallel speed-up for execution of one iteration on different parallel systems.

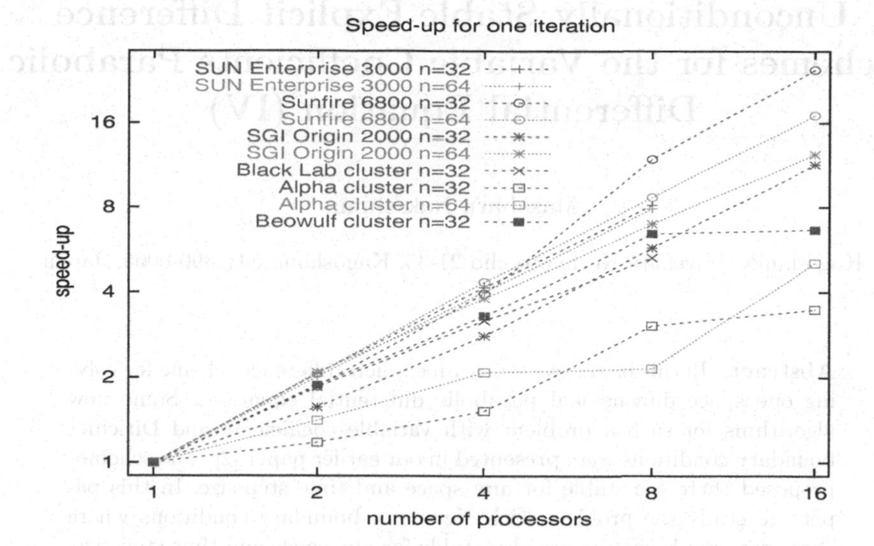

Fig. 2. Speed-up for one iteration

Acknowledgments

This research was supported by the European Commission through grant number HPRI-CT-1999-00026 (the TRACS Programme at EPCC)". The development of the code started while the author was on a TRACS research visit in EPCC, and was completed during the period of extended access to the EPCC supercomputing facilities. It was supported also by grant I-1001/2000 from the Bulgarian NSF and by Center of Excellence BIS-21 grant ICA1-2000-70016.

References

1. Blaheta, R.: Displacement decomposition-incomplete factorization preconditioning techniques for linear elasticity problems. Num. Lin. Alg. Appl., 1 (1994) 107–128.
2. Georgiev, A., Baltov, A., Margenov, S.: Hipergeos benchmark problems related to bridge engineering applications. PROJECT REPORT HG CP 94–0820–MOST–4.
3. Lirkov, I.: MPI solver for 3D elasticity problems. Mathematics and Computers in Simulation, 60, 3-6 (2003) 509–516.
4. Lirkov, I., Margenov, S.: MPI parallel implementation of CBF preconditioning for 3D elasticity problems. Mathematics and Computers in Simulation, 50 (1999) 247–254.
5. Lirkov, I., Margenov, S., Vassilevski, P.S.: Circulant block-factorization preconditioners for elliptic problems. Computing, 53 1 (1994) 59–74.
6. Snir, M., Otto, St., Huss-Lederman, St., Walker, D., Dongara, J.: MPI: The Complete Reference. Scientific and engineering computation series, The MIT Press, Cambridge, Massachusetts (1997) Second printing.
7. Walker, D., Dongara, J.: MPI: a standard Message Passing Interface. Supercomputer, 63 (1996) 56–68.

Unconditionally Stable Explicit Difference Schemes for the Variable Coefficients Parabolic Differential Equation (IV)

Masaharu Nakashima

Kagoshima University Korimoto cho 21-35, Kagoshima city 890-0065, Japan

Abstract. In this paper, we study an explicit difference scheme for solving one space dimensional parabolic differential equations. Some new algorithms for such a problem with variable coefficients and Dirichlet boundary conditions were presented in our earlier paper [2]. The schemes proposed there are stable for any space and time step-size. In this paper, we study the problem with Neumann boundary conditions where the constructed schemes are also stable for any space and time step-size. Numerical test data supporting our algorithms are presented at the end.

1 Introduction

A number of difference schemes solving partial difference equations have been proposed. The methods of lines are recognized as useful tool for numerical solution of PDEs. Du Fort, E. C., Frankel, S. P (see [1]) and some others have proposed difference schemes based on methods of lines. However, there are serious stability problems using the explicit lines methods. To overcome those problems, the author has proposed in [3] the explicit difference scheme using the idea of lines. The linear parabolic differential equation with Dirichlet boundary conditions is studied in [3] where

$$\frac{\partial u(x,t)}{\partial t} = \frac{\partial}{\partial x}\left(a(x,t)\frac{\partial u(x,t)}{\partial x}\right) + b(x,t)\frac{\partial u(x,t)}{\partial x} + c(x,t)u(x,t), \qquad (1)$$

$$(x,t) \in \Omega = \{(x,t); 0 \le x \le 1, \quad 0 \le t \le T\}$$

In this paper, we study the problem (1) with Neumann boundary conditions:

$$\frac{\partial u(x,t)}{\partial x} = \psi(t)u(x,t) + g(t), \quad (x,t) \in \partial\Omega \qquad (2)$$

where

$$\psi(t) = \begin{cases} \psi_1(t) & (>0) \qquad (x=0) \\ \psi_2(t), & (<0) \qquad (x=1) \end{cases}$$

and initial condition:

$$u(x,0) = f(x). \quad (x,t) \in \partial\Omega \cup \Omega \qquad (3)$$

I. Dimov et al. (Eds.): NMA 2002, LNCS 2542, pp. 536–544, 2003.

The explicit difference scheme which will be presented is stable without any time or space step size restriction. The outline of this paper is as follows. In § 2, replacing the second space derivatives by finite difference, we derive a set of first order ordinary differential equations. In §3, we utilize the algorithms presented in [2] to approximate the ODEs and present the explicit difference scheme, which is unconditionally stable with respect to the space and time step-sizes. In §4, we study the convergence of the difference equation derived in §3. At the end, some numerical tests justifying the results are presented.

2 Derivation of Methods for One Space Dimensional Parabolic Differential Equations

We will first transform PDE (1) into a system of ODEs. To this end, we replace the space derivatives of (1) by the following difference operators:

$$\frac{\partial}{\partial x}(a(x,t)\frac{\partial u(x,t)}{\partial x}) \cong \frac{1}{h^2}\delta(a(x,t)\delta u(x,t)),$$

$$\frac{\partial u(x,t)}{\partial x}) \cong \begin{cases} \frac{1}{h}\triangle u(x,t) & (b(x,t) \geq 0) \\ \frac{1}{h}\nabla u(x,t), & (b(x,t) \leq 0) \end{cases} \tag{4}$$

where δ stands for the central difference operator , $\triangle$ is the forward difference operator, ∇ is the backward difference operator and h is the mesh size for $x-$space respectively. Then we find that (1) reduces to the system of ordinary differential equations:

$$\frac{d}{dt}u_j(t) = \frac{1}{h^2}\delta(a_j(t)\delta(u_j(t))) + \frac{1}{2h}\{b_j(t) - |b_j(t)|\}\nabla u_j(t) + \tag{5}$$
$$\frac{1}{2h}\{|b_j(t)| + b_j(t)\}\triangle u_j(t) + c_j(t)u_j(t),)$$

$$u_j(t) = u(jh,t),\ a_j(t) = a(jh,t), b_j(t) = b(jh,t), c_j(t) = c(jh,t),$$
$$h = \frac{1}{(N+1)}.\quad j = 1,2,..,N$$

We approximate the boundary conditions by

$$\frac{u_1(t) - u_0(t)}{h} = \frac{1}{2}\psi_1(t)(u_1(t) + u_0(t)) + g(t), \quad (0,t) \in \partial\Omega \tag{6}$$

and

$$\frac{u_{N+1}(t) - u_N(t)}{h} = \frac{1}{2}\psi_2(t)(u_N(t) + u_{N+1}(t)) + g(t). \quad (1,t) \in \partial\Omega. \tag{7}$$

In order to write (5) more compactly, we use the vector notation:

$$U(t) = (u_1(t),\ u_2(t), .., \ u_N(t)),$$

where $u_i(t), (i = 1, 2, .., n)$ are defined in (5). Then, we have

$$\frac{dU(t)}{dt} = M^{(N)}(t)U(t) + G(u_0(t), u_{N+1}(t)), \tag{8}$$

with the initial condition:

$$U(0) = (f(h), f(2h), ..., f(Nh)),$$

where $M^{(N)}(t)$ is the $N \times N$ matrix

$$M^{(N)}(t) = \begin{pmatrix} \alpha_1(t) & \gamma_1(t) & 0 & 0 & 0 & 0 & 0 & 0 \\ \beta_2(t) & \alpha_2(t) & \gamma_2(t) & 0 & 0 & 0 & 0 & .. \\ 0 & \beta_3(t). & \alpha_3(t) & \gamma_3(t) & 0 & 0 & 0 & .. \\ .. & .. & .. & .. & . & .. & ... & \\ 0 & 0 & 0 & 0 & 0 & \beta_{N-1}(t) & \alpha_{N-1}(t) & \gamma_{N-1}(t) \\ 0 & 0 & 0 & 0 & 0 & 0 & \beta_N(t) & \alpha_N(t) \end{pmatrix},$$

$$G(u_0(t), u_{N+1}(t)) = \begin{pmatrix} \beta_1(t) \\ 0 \\ .. \\ 0 \\ \gamma_N(t) \end{pmatrix},$$

with elements given by

$$\begin{aligned} \beta_j(t) &= \tfrac{a_{j-\frac{1}{2}}(t)}{h^2} + \tfrac{\{|b_j(t)|-b_j(t)\}}{2h}, \quad \alpha_j(t) = -\tfrac{\{a_{j+\frac{1}{2}}(t)+a_{j-\frac{1}{2}}(t)\}}{h^2} - \tfrac{|b_j(t)|}{h} + c_j(t), \\ \gamma_j(t) &= \tfrac{a_{j+\frac{1}{2}}(t)}{h^2} + \tfrac{\{|b_j(t)|+b_j(t)\}}{2h}, \quad j = 2, .., N-1 \end{aligned} \tag{9}$$

and

$$\begin{aligned} \beta_1(t) &= \{\tfrac{-1}{h^2}a_{\frac{1}{2}}(t) + \tfrac{(b_1(t)|b_1(t)|)}{2h}\}\tfrac{2h}{2+\psi_1(t)h}g(t), \\ \alpha_1(t) &= -\tfrac{1}{h^2}\{a_{\frac{1}{2}}(t)(1 - \tfrac{2-\psi_1(t)h}{2+\psi_1(t)h}) + a_{\frac{3}{2}}(t)\} - \\ &\quad \tfrac{1}{2h}\{2|b_1(t) + (b_1(t) - |b_1(t)|)(\tfrac{2-\psi_1(t)h}{2+\psi_1(t)h})\} + c_1(t), \\ \gamma_1(t) &= \tfrac{a_{\frac{3}{2}}(t)}{h^2} + \tfrac{\{b_1(t)-|b_1(t)|\}}{2h}, \\ \beta_N(t) &= \tfrac{1}{h^2}a_{N-\frac{1}{2}}(t) - \tfrac{(b_N(t)|b_N(t)|)}{2h}, \\ \alpha_N(t) &= -\tfrac{1}{h^2}\{a_{N+\frac{1}{2}}(t)(1 - \tfrac{2+\psi_2(t)h}{2-\psi_2(t)h}) + a_{N-\frac{1}{2}}(t)\} - \\ &\quad \tfrac{1}{2h}\{2|b_N(t)| - (b_N(t) + |b_N(t)|)\,(\tfrac{2+\psi_2(t)h}{2-\psi_2(t)h})\} + c_N(t), \end{aligned}$$

and

$$\gamma_N(t) = -\{\frac{a_{N+\frac{3}{2}}(t)}{h^2} + \frac{\{b_N(t) + |b_N(t)|\}}{2h}\}\frac{2h}{2-\psi_2(t)h}g(t). \tag{10}$$

3 Difference Scheme

In this section, we present the difference scheme for (1) with (2) and (3). The following algorithm was studied in [2]:

$$Y_{n+1} = D(\xi_{ij}(t_n))\,Y_n, \tag{11}$$

where $D(\xi_{ij}(t_n))$ is the $N \times N$ matrix with elements

$$\xi_{ij}(t_n) = \begin{cases} 1 + \frac{kp_{ii}(t_n)}{1+kr} & (i = j) \\ \frac{kp_{ij}(t_n)}{1+kr}, & (i \neq j)\ (i,j = 1,2,...,N) \end{cases}$$

$(k; step\ size,\ r; constant)$

for solving the first order system of N linear ordinary differential equations:

$$\dot{Y}(t) = A(p_{i,j}(t))\, Y(t),$$

where $A(p_{i,j}(t))$ is a $N \times N$ matrix.

Utilizing the method (11) to approximate (8), we have

$$U_{l+1} = D_l^{(N)} U_l + G(u_0^l, u_{N+1}^l), (l = 0, 1, 2, ..., \frac{T}{k}) \tag{12}$$

$$U_0 = U(0),$$

where

$$U_l = (u_1^l, u_2^l, .., u_N^l),$$

and u_i^n is the numerical approximation to $u_i(nk)$, the matrix $D_l^{(N)}$ given by

$$D_l^{(N)} = \begin{pmatrix} d_{11}(l)\ d_{12}(l) & 0 & .. & 0 \\ d_{21}(l)\ d_{22}(l)\ d_{23}(l) & & .. & 0 \\ . \quad .. \quad ... & & . & .. \\ . \quad .. \quad .. & & d_{N-1N-1}(l). & d_{N-1N}(l) \\ 0.. \quad . \quad .. & & d_{NN-1}(l) & d_{NN}(l) \end{pmatrix},$$

$$G(u_0^l, u_{N+1}^l) = \begin{pmatrix} \frac{k\beta_1(lk)}{1+k|\beta_1(lk)|} \\ 0 \\ .. \\ 0 \\ \frac{k\gamma_N(lk)}{1+k|\gamma_N(lk)|} \end{pmatrix},$$

with

$$d_{ij}(l) = \begin{cases} \frac{k\beta_i(t_l)}{1+kr} & (j = i-1) \\ 1 + \frac{k\alpha_i(t_l)}{1+kr} & (j = i) \\ \frac{k\gamma_i(t_l)}{1+kr}. & (j = i+1) \qquad (i = 1, 2, ..., N) \end{cases} \tag{13}$$

Next we study the stability of (12), subject to the following definition.

Definition: The numerical processes $\{Y_n \in R^n\}$ is stable if there exists a positive constant K such that

$$\|Y_n\| \leq K,$$

where $\|.\|$ denotes some norm in R^n and the constant K depends only on the initial value.

We will use the Lemma presented below to prove the stability. First, let us consider the homogeneous equations corresponding to (11)

$$V_{l+1} = D_l^{(N)} V_l, V_0 = I, \tag{14}$$

where

$$V_l = (v_1^l, v_2^l, .., v_N^l).$$

The next stability result is obtained in the same way as in [3].

Lemma 1. *For any given step size h and k, we set the constant r of (13) to be*

$$r = \max_{0 \leq i \leq \frac{1}{h}, 0 \leq j \leq \frac{T}{k}} \{\frac{1}{h^2}\{a_{i+\frac{1}{2}}(jk) + a_{i-\frac{1}{2}}(jk)\} + \frac{1}{h}|b(ih, jk)| - c(ih, jk)\}. \tag{15}$$

Then the difference processes (14) is stable.

Using the solution of the homogeneous difference equation (14), we have the solution of the non homogeneous difference equation (12) given by

$$U_n = Z_n Z_0^{-1} U_0 + \Sigma_{i=1}^n Z_n (Z_i)^{-1} G(u_{i-1}^0, u_{i-1}^{N+1}), \tag{16}$$

where Z_n is the $n \times n$ matrix of the fundamental set whose elements are the solutions of (14). From (16) and Lemma 1, we get the following results.

Theorem 1. *If there exists a positive constant K such that*

$$|\psi_i(t_l)| \leq K, \quad (i = 1, 2), (l = 1, 12, .., \frac{T}{k})$$

and

$$|g(t_l)| \leq K, \quad (l = 1, 2, , , .\frac{T}{k})$$

where $\psi_i(t), (i = 1, 2)$ and $g(t)$ are defined in (2), and the constant r satisfies (15), then the numerical processes (12) is stable.

4 Convergence

In the section, we study the convergence of the scheme (12). The *j–th* component of U_n is given by

$$u_j^{n+1} = u_j^n + \frac{k}{1 + kr}(\eta_j^n u_j^{n-1} + \rho_j^n u_j^n + \tau_j^n u_j^{n+1}), \tag{17}$$

with

$$\rho_j^n = \alpha_j(nk), \quad \eta_j^n = \beta_j(nk), \quad \tau_j^n = \gamma_j(nk),$$

and at the boundary respectively by

$$u_2^n = u_1^n + \frac{k}{1 + kr}(\eta_1^n \tilde{u}_0^n + \rho_1^n u_1^n + \tau_1^n u_2^n), \qquad (0, nk) \in \partial\Omega \tag{18}$$

$$u_{N+1}^n = u_N^n + \frac{k}{1+kr}(\eta_N^n \tilde{u}_{N-1}^n + \rho_N^n u_N^n + \tau_N^n u_{N+1}^n), \quad (1, nk) \in \partial\Omega \tag{19}$$

with

$$\tilde{u}_0^n = (\frac{2 - \psi_1(nk)h}{2 + \psi_1(nk)h})u_1^n - \frac{g(nk)h}{2 + \psi_1(nk)h}$$

and

$$\tilde{u}_{N-1}^n = (\frac{2 + \psi_2(nk)h}{2 - \psi_2(nk)h})u_N^n - \frac{g(nk)h}{2 - \psi_2(nk)h}.$$

Utilizing the equation

$$\eta_1^n \tilde{u}_0^n + \rho_n^j u_1^n + \tau_1^n u_2^n = \eta_1^n u_0^n + \rho_1^n u_1^n + \tau_1^n u_2^n + \eta_1^n(\tilde{u}_0^n - u_0^n),$$

and the result

$$\eta_1^n(\tilde{u}_0^n - u_0^n) = O(h),$$

we have the approximation order at the boundary $(0, t) \in \partial\Omega$:

$$\eta_1^n \tilde{u}_0^n + \rho_n^j u_1^n + \tau_1^n u_2^n = \frac{\partial}{\partial x}(a(x,t)\frac{\partial u(x,t)}{\partial x}) + b(x,t)\frac{\partial u(x,t)}{\partial x} + c(x,t)u(x,t) + O(h).$$

Similarly, from the equation

$$\eta_N^n \tilde{u}_{N-1}^n + \rho_N^n u_N^n + \tau_N^n u_{N+1}^n = \eta_N^n u_{N-1}^n + \rho_N^n u_N^n + eta_N^n u_{N-1}^n + \eta_N^n(\tilde{u}_{N-1}^n - u_{N+1}^n),$$

and

$$\eta_N^n(\tilde{u}_{N-1}^n - u_{N-1}^n) = O(h),$$

we obtain the approximation order at the boundary $(1, t) \in \partial\Omega$:

$$\eta_N^n \tilde{u}_{N-1}^n + \rho_N^n u_N^n + \tau_N^n u_{N+1}^n = \frac{\partial}{\partial x}(a(x,t)\frac{\partial u(x,t)}{\partial x}) + b(x,t)\frac{\partial u(x,t)}{\partial x} + c(x,t)u(x,t) + O(h).$$

Consequently, we have the following convergence theorem.

Theorem 2. *Let us assume that the refinement path* $\{(h_i, k_i); i = 1, 2, ..; h_i, k_i \to 0 \ (i \to \infty)\}$ *in the numerical process (12) is such that*

$$\nu_i = \frac{k_i}{h_i^2}, k_i \to 0 \ (i \to \infty), \tag{20}$$

and let there exist the positive numbers $n_i(k_l)$, $m_i(h_l)$ *such that for each* h_l, k_l

$$n_i(k_l)k_l \to t \in [0, T], \quad m_i(h_l)h_l \to x \in [0, 1] \ (i \to \infty). \tag{21}$$

Then, the scheme (12) converges to the solution $u(x, t)$ *of the differential equations (1), if*

$$u_x(x,t), u_{xx}(x,t), \ u_{xxx}(x,t), \ u_{xxxx}(x,t), a(x,t), a_x(x,t), a_{xx}(x,t), b(x,t)$$

are bounded for $(x, t) \in R$,

Now, let us set the time step size k_i as

$$k_i = \{(\frac{h_i}{h_0})^{\frac{2}{(1-\alpha)}}\}k_0, \ (i = 0, 1, 2, ...), \quad (0 < \alpha < 1)$$

where h_0 and k_0 are the initial step sizes. Then,

$$\frac{k_i}{h_i^2} = k_i^{\alpha}(\frac{1}{h_0^2})k_0^{(1-\alpha)},$$

which leads to

$$\frac{k_i}{h_i^2} \to 0 \ as \ k_i \to 0.$$

Then we have the following result.

Corollary. Let us consider the refinement path L,

$$L = \{(h_i, k_i); k_i = [(h_i/h_0)^{\frac{2}{(1-\alpha)}}]k_0, \ h_i, k_i \to 0 \ (i \to \infty)\}, \quad (0 < \alpha < 1),$$

where h_0, k_0 are given initial space and time step sizes. Now, let us set the step path (h_l, k_l) on L, and let us assume that for each h_l, k_l, there exist the positive numbers $n_i(k_l), m_i(h_l)$ such that

$$n_i(k_l)k_l \to t \in [0, T], \quad m_i(h_l)h_l \to x \in [0, 1]. \ (i \to \infty).$$

Then, the scheme (12) converges to the solution $u(x, t)$ of the differential equations (1), under the same conditions for the functions $a(x, t)$, $b(x, t)$, $c(x, t)$ and $u(x, t)$, as in Theorem 2.

5 Numerical Example

Here, we study a numerical test of the differential equation (1) with (2) and (3), where the functions $a(x, t)$, $b(x, t)$ and $c(x, t)$ are given by

$$a(x,t) = \frac{1}{1 + (x - \frac{1}{2})^2}, \ b(x,t) = 250 \ sin(250\pi x), \ c(x,t) = \frac{-1}{1 + (x - \frac{1}{2})^4},$$

subject to the following initial and boundary conditions:

$$u(x,0) = \begin{cases} 1 & (0 < x < 1) \\ c, \quad (0 \le c \le 1) & (x = 0, 1) \end{cases}$$

$$\psi_i(t) = \begin{cases} \frac{100}{(1+t^3)} & (i = 1) \\ -\frac{100}{(1+t^3)} & (i = 2) \end{cases}$$

$$g(t) = \frac{1}{10+t^2}.$$

In this numerical example, the initial condition is a pulse which is continuous but $\frac{\partial u(x,t)}{\partial t}$ is discontinuous. In our numerical experience, the numerical solution

Table 1. Numerical solutions at time t = 0,5k,25k,50k,100k,1100k.

x	h	20h	50h	80h	100h
t k	0.120	0.999	0.999	0.999	0.121
t 25k	0.515	0.998	0.998	0.998	0.404
t 50k	0.599	0.997	0.997	0.997	0.584
t 100k	0.694	0.993	0.995	0.994	0.693
t 1100k	0.841	0.861	0.887	0.866	0.851

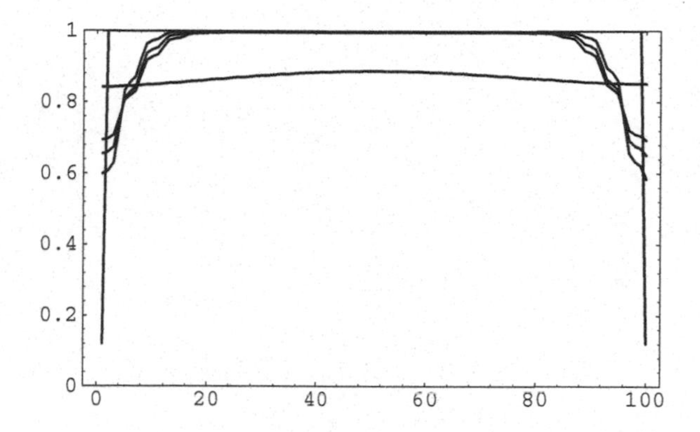

Fig. 1. Graphical representation of the result

computed by some of the well known usual schemes is oscillating and can be highly inaccurate. We present our numerical experiment in Table 1 and Fig. 1. As we see in Fig. 1, the shape of the t cross-sections of ridge at $x = 0$ and $x = 1$ make the angle almost near to the value of $\psi_1(t)$ and $\psi_2(t)$ at time t respectively. The shape becomes slowly to the constant state and once again the solution become to be steady-state, it is getting down with the same shape. We may succeed in overcoming the stability problem and assert the usefulness of our schemes.

Here, the interval [0,1] is divided into $N = 100$ equal subintervals of length $h = \frac{1}{101}$ and the number of the time steps is $k = 100$.

References

1. Du Fort, E.C., Frankel, S.P: Stability conditions in the numerical treatment on parabolic differential equations. Math. Tables and other Aids to Computation, Vol. 7 (1950)
2. Nakashima, M.: Explicit A-stable Runge-Kutta methods for linear stiff-equations. In: Mastorakis, P.D.N.E.(eds.): 3rd International Conference on: Circuits, Systems and Communications and Computers. Recent Advances in Information Science and Technology (1999) 231–236.

3. Nakashima, M.: Unconditionally stable explicit difference schemes for the variable coefficients parabolic differential equation (II). Processing Techniques and Applications. Proceedings of the International Conference on Parallel and Distributed Processing, Las Vegas, Nevada,USA (2001) 561–569.
4. Schiesser, W. E.: The Numerical Method of Lines for Integration of Partial Differential Equations, Academic Press (1991)
5. Richtmyer, R. D., Morton, D. K.: Difference Methods for Initial-value Problems, John Wiley, New York (1967)

Flexible Two-Level Parallel Implementations of a Large Air Pollution Model

Tzvetan Ostromsky[1] and Zahari Zlatev[2]

[1] Central Lab. for Parallel Processing, Bulgarian Academy of Sciences, Acad. G.Bonchev str., bl. 25-A, 1113 Sofia, Bulgaria.
ceco@parallel.bas.bg

[2] Nat. Environmental Research Institute, Frederiksborgvej 399 P. O. Box 358, DK-4000 Roskilde, Denmark.
zz@dmu.dk

Abstract. Large scale air pollution models are powerful tools, designed to meet the increasing demand in different environmental studies. The atmosphere is the most dynamic component of the environment, where the pollutants can quickly be moved in a very long distance. Therefore the advanced modeling must be done in a large computational domain. Moreover, all relevant physical, chemical and photochemical processes must be taken into account. The speed of these processes vary in a wide range. This fact implies that a small time step must be used in order to achieve both numerical stability and sufficient accuracy of the results. Thus the numerical treatment of such an air pollution model becomes in many cases a huge computational problem, a challenging task for the most powerful up-to-date supercomputers.

The Danish Eulerian Model (DEM) is used in this work. The paper focuses on the efficient parallel implementation of DEM on powerful parallel supercomputers. We present a variety of performance and scalability results, obtained on different parallel machines by using standard parallelization tools (MPI for distributed-memory parallelism and OpenMP for shared-memory parallelism). It is shown by experiments that MPI and OpenMP can both be used on separate levels of parallelism to get the best use of the clustered parallel machines. Application of the results in the environmental studies, related to high ozone concentrations in the air, is illustrated by some plots in the last section.

1 The Danish Eulerian Model

The DEM is represented mathematically by the following system of partial differential equations (PDE):

$$\frac{\partial c_s}{\partial t} = -\frac{\partial (uc_s)}{\partial x} - \frac{\partial (vc_s)}{\partial y} - \frac{\partial (wc_s)}{\partial z}$$

$$+\frac{\partial}{\partial x}\left(K_x \frac{\partial c_s}{\partial x}\right) + \frac{\partial}{\partial y}\left(K_y \frac{\partial c_s}{\partial y}\right) + \frac{\partial}{\partial z}\left(K_z \frac{\partial c_s}{\partial z}\right)$$

$$+E_s - (\kappa_{1s} + \kappa_{2s})c_s + Q_s(c_1, c_2, \ldots, c_q)\ ; \quad s = 1, \ldots, q \tag{1}$$

I. Dimov et al. (Eds.): NMA 2002, LNCS 2542, pp. 545–554, 2003.

where c_s are the concentrations of the chemical species involved in the model, u, v and w are the wind components, K_x, K_y and K_z are diffusion coefficients, E_s are the emissions, κ_{1s} and κ_{2s} are the coefficients for dry and wet deposition, and $Q_s(c_1, c_2, \dots, c_q)$ are expressions that describe the chemical reactions under consideration.

The PDE system (1) is too complex for direct numerical treatment. A splitting procedure, based on ideas in [5,6], is used to split (1) into five sub-models, representing the main physical and chemical processes ($s = 1, 2, \dots, q$): the horizontal advection (2), the horizontal diffusion (3), the chemistry and the emission (4), the deposition (5) and the vertical exchange (6):

$$\frac{\partial c_s^{(1)}}{\partial t} = -\frac{\partial (u c_s^{(1)})}{\partial x} - \frac{\partial (v c_s^{(1)})}{\partial y} \tag{2}$$

$$\frac{\partial c_s^{(2)}}{\partial t} = \frac{\partial}{\partial x}\left(K_x \frac{\partial c_s^{(2)}}{\partial x}\right) + \frac{\partial}{\partial y}\left(K_y \frac{\partial c_s^{(2)}}{\partial y}\right) \tag{3}$$

$$\frac{d c_s^{(3)}}{dt} = E_s + Q_s(c_1^{(3)}, c_2^{(3)}, \dots, c_q^{(3)}) \tag{4}$$

$$\frac{d c_s^{(4)}}{dt} = -(\kappa_{1s} + \kappa_{2s}) c_s^{(4)} \tag{5}$$

$$\frac{\partial c_s^{(5)}}{\partial t} = -\frac{\partial (w c_s^{(5)})}{\partial z} + \frac{\partial}{\partial z}\left(K_z \frac{\partial c_s^{(5)}}{\partial z}\right) \tag{6}$$

The discretization of the spatial derivatives in the right-hand-sides of the sub-models (2) – (6) results in five systems of ordinary differential equations:

$$\frac{dg^{(i)}}{dt} = f^{(i)}(t, g^{(i)}), \qquad \begin{array}{l} g^{(i)} \in R^{N_x \times N_y \times N_z \times q} \\ f^{(i)} \in R^{N_x \times N_y \times N_z \times q} \end{array}; \qquad i = 1, \dots 5$$

where N_x, N_y and N_z are the numbers of grid-points along the coordinate axes and q is the number of chemical species. The functions $f^{(i)}, i = 1, \dots 5$, depend on the particular discretization methods used in the sub-models, while the functions $g^{(i)}$, $i = 1, \dots 5$, represent approximations of the concentrations at the grid-points of the computational domain. More details about the numerical methods, used in the submodels, can be found in [1,3,4,10].

2 Space Domain Discretization

The space domain of the model is part of the Northern hemisphere (4800×4800 km.) that covers Europe, most of the Mediterian and neighboring parts of Asia and the Atlantic Ocean. Grids of different size (and, respectively, with different step) are used in the discretization of that domain. The resulting versions of the

Table 1. Versions of DEM with existing parallel codes. The 3-D version for 288 × 288 grid is not yet fully operational.

Grid size	Grid step	Number of grid cells	2-D version	3-D version
(32 × 32)	150 km	1024	Yes	Yes
(96 × 96)	50 km	9216	Yes	Yes
(288 × 288)	16.7 km	82944	Yes	Yes*
(480 × 480)	10 km	230400	Yes	No

model are given in Table 1. Some of these versions are discussed in more detail in [1,4,7,8,10,11]. In the 3-D versions the domain in vertical direction is split into several layers of different (but constant) thickness. The lower layers are thinner, while the upper are thicker. Ten layers are used in the present 3-D versions. The number of cells in the corresponding 2-D version is given in the third column of Table 1. This number should be multiplied by 10 (the number of layers) in order to get the number of cells for the corresponding 3-D version. The 2-D and 3-D versions on the medium-resolution (96 × 96) grid are mainly used in our experiments (Table 2), as well as the newly developed 3-D version for 288 × 288 grid (Table 3).

3 Parallelization Techniques

Most of the existing parallel supercomputers can be classified in one of the following three groups:

(i) Shared memory computers, like SGI Origin, SUN Enterprice, etc.;

(ii) Distributed memory computers, like CRAY T3E, IBM SP2;

(iii) Clustered (hybrid) computers – a distributed memory cluster of *nodes*, each node being a separate shared-memory parallel computer (for example, IBM SP3, Beowulf clusters, etc.).

Results on computers from all the three groups are presented in this paper.

One of the main goals in this work was to exploit efficiently the full capacity of the clustered computers (group (iii)). Nevertheless, our strategic principle in code development has been **high portability**. That is why only standard parallelization tools (MPI and OpenMP) are used in the codes under consideration. It should be emphasized that neither extensions of the standard of the above libraries, nor any special properties of the particular computers, that have been used in the experiments, are applied in these codes. With minor changes in the driver routine they can be run on any parallel computing system that supports MPI (either clustered, with shared or distributed memory).

3.1 Shared Memory Parallelization via OpenMP

It is relatively easy to obtain an efficient shared memory parallelization by exploiting the data-independent potentially parallel tasks in each submodel. This can be achieved by using only standard OpenMP [9] directives. However, it is not always possible to achieve good data locality in the large shared arrays. The small tasks are grouped in chunks where appropriate for more efficient cache utilization. A parameter CHUNKSIZE is provided in the code, which should be tuned with respect to the cache size of the target machine. The main submodels are parallelized as follows:

- **Horizontal advection and diffusion.** The horizontal advection and diffusion submodels are treated together. In fact several independent advection-diffusion subproblems with Dirichlet boundary conditions arise on each time step after the splitting procedure, one for every chemical compound on every layer (in the 3-D version). It means, there are enough parallel tasks on this stage ($N_z \times q$) and these tasks are big enough.
- **Chemistry and deposition.** The calculations of these two processes can be carried out independently for each grid-point. There are many parallel tasks ($N_x \times N_y \times N_z$), but each task is small. For the purpose of efficient cache utilization the tasks should be grouped in chunks. The parameter CHUNKSIZE is used to set their size.
- **Vertical exchange.** This stage is present only in the 3-D versions. The vertical exchange submodel splits into independent relatively simple advection-diffusion subproblems (along each vertical grid-line). The number of parallel tasks is large ($N_x \times N_y$) and they are not very big. Like on the chemistry-deposition stage, the tasks can be grouped in chunks to improve the cache utilization (see Fig. 1).

3.2 Distributed Memory Parallelization via MPI

The MPI (Message Passing Interface, [2]) was initially developed as a standard communication library for distributed memory computers. Later, proving to be efficient, portable and easy to use, it became one of the most popular parallelization tools for application programming. Now it can be used on much wider class of parallel systems, including shared-memory computers and clustered systems (each node of the cluster being a separate shared-memory computer with fixed number of processors). Thus it provides high level of portability to the codes.

Our MPI parallelization is based on the space domain partitioning. The space domain is divided into several sub-domains (the number of the sub-domains being equal to the number of MPI tasks). Each MPI task works on its own sub-domain. On each time step there is no data dependency between the MPI tasks on both the chemistry and the vertical exchange stages. This is not so with the advection-diffusion stage. Spatial grid partitioning between the MPI tasks requires overlapping of the inner boundaries and exchange of certain boundary values on the neighboring subgrids for proper treatment of the boundary conditions. This leads to two main consequences:

(i) certain computational overhead and load imbalance, leading to lower speed-up of the advection-diffusion stage in comparison with the chemistry and the vertical transport (as can be seen on Fig. 1).

(ii) communication necessity for exchanging boundary values on each time step (done in a separate **communication** stage).

In addition, two extra procedures are used for scattering the input data and gathering the results in the beginning and in the end of the run respectively.

- **Pre-processing.** In the beginning of the job the input data, stored in several large files (containing meteorological and emission data sets for the whole domain) is distributed in separate files for each of the sub-domains (to be processed in parallel). In this way, not only each MPI process will be working on its own sub-domain, but will also be accessing only the part of meteorological and emission data relevant to its sub-domain.
- **Post-processing.** During the run each MPI process prepares its own output data files. At the end of the run all the output data of a same kind from all MPI processes are collected and prepared for future use by one of the MPI processes during the post-processing procedure.

The time for pre-processing and post-processing is, in fact, overhead, introduced by the MPI partitioning strategy. Moreover, this overhead is growing up with increasing the number of MPI tasks and little can be done for its parallel processing. Thus the relative weight of these two stages grows up with increasing the number of MPI tasks, which eventually affects the total speed-up and efficiency of the MPI code.

3.3 Mixed Shared-Distributed Memory Parallelization

By mixing both parallelization techniques, described above (i.e. MPI parallelization by using domain decomposition on the top level together with OpenMP-parallelization for building second level, somewhat finer-grain parallelism), a mixed shared-distributed memory parallel version of the model is created. This code should be perfect for huge clustered supercomputers with large number of nodes. By giving each node one MPI task and an OpenMP thread to each processor of the node, optimal performance can be achieved. We should mention, however, that this code is build as an extension of the pure MPI code (by adding OpenMP directives) without destroying its structure. Although it has additional functionality on machines with OpenMP, on machines without OpenMP it is virtually the same MPI code (the OpenMP directives are treated as comments). Thus it looses neither portability, nor efficiency with respect to the the pure MPI code. Some results of experiments with such codes for grid size 96×96 and 288×288 are given in the next section, together with the results of the pure MPI codes.

4 Numerical Results

In Table 2 results from experiments with the 2-D and 3-D versions of DEM (96×96 grid) on three supercomputers of different classes are presented.

Table 2. Total computing time (measured in seconds), speed-up and efficiency of the 2-D (upper table) and 3-D (lower table) versions of DEM. Three supercomputers of different type, all in CINECA – Bologna, are used in the experiments: (i) IBM SP Power 3 (clustered machine, 8 nodes, 16 PE/node, 375 MHz); (ii) SGI Origin 3800 (shared-memory machine, 32 PE R14000/500 MHz); (iii) CRAY T3E - 1200E (distributed-memory machine, 256 PE). There are no experiments on 1 and 2 processors on the CRAY T3E for the 3-D version (the places marked with *) due to the time limit of 6 hours per job.

2-D version of DEM on the CINECA supercomputers (96 × 96 × 1) grid, CHUNKSIZE=48									
	IBM sp3			SGI Origin 3800			CRAY T3E		
Proc.	Time [sec]	Speed -up	**E** %	Time [sec]	Speed -up	**E** %	Time [sec]	Speed -up	**E** %
1	2148			2261			6042		
2	1047	2.1	103	1080	2.0	100	3050	2.0	99
4	528	4.1	102	537	4.2	105	1520	4.0	99
6	356	6.0	101	349	6.5	108	1033	5.8	97
8	273	7.9	98	268	8.4	105	780	7.7	97
12	192	11.2	93	183	12.4	103	530	11.4	95
16	157	13.7	86	142	15.9	99	425	14.2	89
24	124	17.3	72	117	19.3	81	315	19.2	80
32	113	19.9	62	85	26.6	83	257	23.5	73
48	95	22.6	47				198	30.5	64

3-D version of DEM on the CINECA supercomputers (96 × 96 × 10) grid, CHUNKSIZE=48									
	IBM sp3			SGI Origin 3800			CRAY T3E		
Proc.	Time [sec]	Speed -up	**E** %	Time [sec]	Speed -up	**E** %	Time [sec]	Speed -up	**E** %
1	21516			21653			*	*	*
2	10147	2.1	106	10487	2.1	103	*	*	*
3	6863	3.1	105	7057	3.1	102	19532	3.0	100
4	5110	4.2	105	5412	4.0	100	14653	4.0	100
6	3516	6.1	102	3480	6.2	104	9813	6.0	100
8	2586	8.3	104	2658	8.1	102	7258	8.1	101
12	1759	12.2	102	1797	12.0	100	4867	12.0	100
16	1431	15.0	94	1376	15.7	98	3770	15.5	97
24	940	22.9	95	964	22.5	94	2566	22.8	95
32	764	28.2	88	723	29.9	94	2020	29.0	91
48	607	35.4	74				1444	40.6	85
24x2	**561**	**38.4**	**80**						
12x4	**543**	**39.6**	**83**						

The scalability of the main computational stages of the 3-D MPI code on the T3E is shown in Fig. 1. While the chemistry and the vertical transport stages scale nearly perfect, this is not the case with the advection. The main reasons are

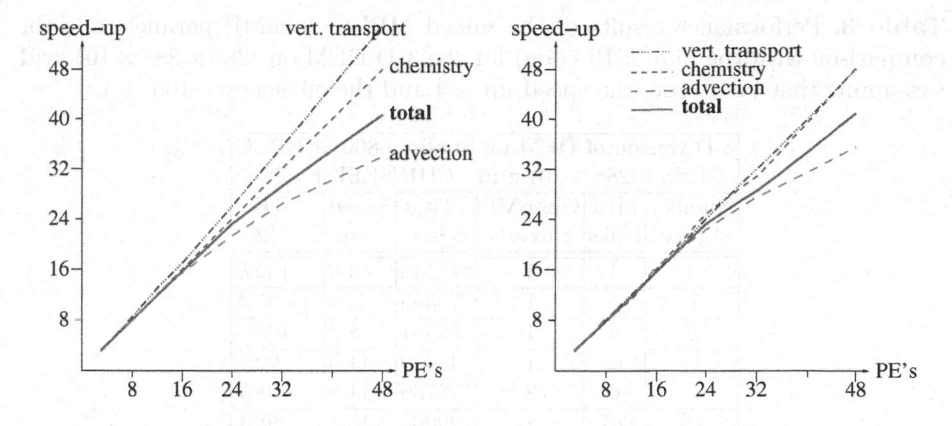

Fig. 1. Scalability of the main stages of the 3-D MPI code on the T3E. Experiments with two versions of the vertical transport stage are presented in the two plots. Results of the simpler version (without using chunks on the vertical transport stage) are given in the left plot. The version in the right plot uses chunks in order to improve the reuse of data in the cache. The superlinear speed-up of the vertical transport in the left plot indicates the cache-size effect on the performance (avoided by using chunks). Both versions use chunks on the chemistry stage.

the overhead due to subdomain overlapping and the non-optimal load-balance on that stage.

Some results of the new 3-D version for 288 × 288 grid on Sunfire 6800 computer (24 PE UltraSPARC 3, 750 MHz) at the Edinburgh Parallel Computer Centre (EPCC) are presented in Table 3.

5 Using the Model for Ozone Pollution Estimation with Respect to the Human Health

One of the most dangerous pollutants with strong influence in many areas like crops production, ecology and human health is the tropospheric ozone. Not only its concentrations are evaluated by DEM, but also several functions, related to one of the above areas [12,13,14].

Let us call *"bad" days* the days, in which at least one of the 8-hour averages of the ozone concentration exceeds the critical value of 60 ppb.. In these days people suffering from asthmatic deceases have extra difficulties. There is a discussion in the European Union about this problem. The main suggestion is to reduce the emissions so that the number of "bad" days in one year do not exceed 20.

The plots in Fig. 2 represent in different colors the areas with given number of "bad" days in 1997. The 2-D version of the model with the fine-resolution (10 × 10 km.) grid is used to create these plots. Fig. 2 (a) represents the overall situation in Europe. The maximal value, 74 "bad" days, exceeds by a factor

Table 3. Performance results of the mixed MPI - OpenMP parallel code (in comparison with the pure MPI code) for the 3-D DEM on $(288 \times 288 \times 10)$ grid (assuming that on 4 proc. the speed-up is 4 and the efficiency – 100 %).

3-D version of DEM on Sunfire 6800 at EPCC $(288 \times 288 \times 10)$ grid, CHUNKSIZE=48					
Number of proc.	MPI tasks	OpenMP threads	Time [sec]	Speed -up	E [%]
4	4	1	114140	4.0	100%
6	6	1	79355	5.8	96%
8	8	1	55481	8.2	103%
12	12	1	41430	11.0	92%
12	6	2	38769	11.8	98%
16	16	1	29004	15.7	98%
16	8	2	29003	15.7	98%
24	24	1	19805	23.1	96%
24	12	2	22747	20.1	84%
24	6	4	22280	20.5	85%

of nearly four the accepted norm 20. The situation in Italy (Fig. 2 (b)) is not better. In some places in the Northern part of the country the global maximum of 74 "bad" days is reached.

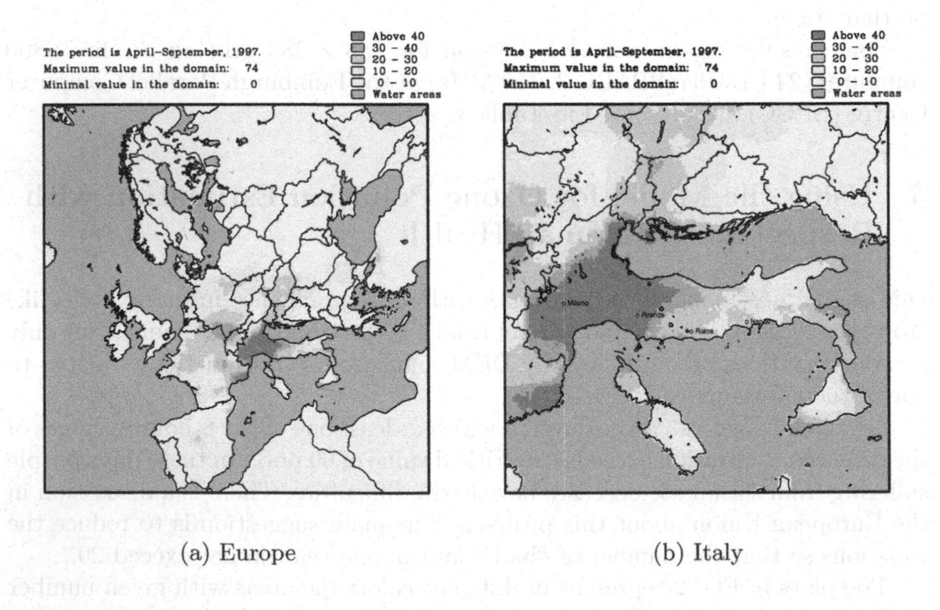

(a) Europe (b) Italy

Fig. 2. Number of days with an 8-hour average ozone concentration more than the critical value (60 ppb.) in 1997.

6 Conclusions and Plans for Future ork

The standard parallelization tools (MPI for distributed-memory parallelism and OpenMP for shared-memory parallelism) can both be used on separate levels of parallelism to get the best use of the clustered parallel machines. The new hybrid code is highly portable and shows good efficiency and scalability on a variety of parallel machines.

An important and challenging task is development of a refined grid 3-D version of the model, in which the spatial domain is discretized on a $(480 \times 480 \times 10)$ grid. This is a huge computational task. Its solution is a big challenge to the power of the existing now supercomputers. Such a code is under development and positive results are expected in the near future.

If a long sequence of scenarios has to be run, then the two-dimensional versions are usually used. The 3-D versions are more accurate, but about 10 times more expensive (on both time and storage resources) with respect to the corresponding 2-D versions. This fact illustrates the need for further improvements (faster numerical algorithms, better exploitation of the potential power of the modern supercomputers, faster and bigger supercomputers, etc.).

Acknowledgments

This research is supported in part by grant I-901/99 from the Bulgarian NSF. Most of the results, presented in the paper, are obtained by Tz. Ostromsky during his visit to CINECA – Bologna, Italy in November 2001. This visit was supported by the EC via Access to Research Infrastructure action of the Improving Human Potential Programme via the MINOS project. We would like to thank all the people from CINECA for their hospitality and perfect conditions provided to the MINOS visitors. Special thanks to Prof. Giovani Erbacci and to Sigismondo Boschi for the technical support in order to get the best use of the supercomputing facilities at CINECA.

References

1. Alexandrov, V., Sameh, A., Siddique, Y., and Zlatev, Z.: Numerical integration of chemical ODE problems arising in air pollution models. Env. Modeling and Assessment, 2 (1997) 365–377.
2. Gropp, W., Lusk, E., and Skjellum, A.: Using MPI: Portable programming with the message passing interface. MIT Press, Cambridge, Massachusetts (1994)
3. Hesstvedt, E., Hov, Ø., and Isaksen, I. A.: Quasi-steady-state approximations in air pollution modeling: comparison of two numerical schemes for oxidant prediction. Int. Journal of Chemical Kinetics, 10 (1978) 971–994.
4. Hov, Ø., Zlatev, Z., Berkowicz, R., Eliassen, A., and Prahm, L. P.: Comparison of numerical techniques for use in air pollution models with non-linear chemical reactions. Atmospheric Environment, 23 (1988) 967–983.

5. Marchuk, G. I.: Mathematical modeling for the problem of the environment. Studies in Mathematics and Applications, 16, North-Holland, Amsterdam (1985)
6. McRae, G. J., Goodin, W. R., and Seinfeld, J. H.: Numerical solution of the atmospheric diffusion equations for chemically reacting flows. J. Comp. Physics, 45 (1984) 1–42.
7. Ostromsky, Tz., Owczarz, W., Zlatev, Z.: Computational Challenges in Large-scale Air Pollution Modelling. Proc. 2001 International Conference on Supercomputing in Sorrento, ACM Press (2001) 407–418.
8. Ostromsky, Tz., Zlatev, Z.: Parallel Implementation of a Large-scale 3-D Air Pollution Model. In: Margenov, S., Waśniewski, J., Yalamov, P. (eds.): Large-Scale Scientific Computing. Lecture Notes in Computer Science, Vol. 2179, Springer-Verlag (2001) 309–316.
9. WEB-site for OPEN MP tools, http://www.openmp.org
10. Zlatev, Z.: Computer treatment of large air pollution models, Kluwer (1995)
11. Zlatev, Z., Dimov, I., Georgiev, K.: Three-dimensional version of the Danish Eulerian Model. Zeitschrift für Angewandte Mathematik und Mechanik, 76 (1996) 473–476.
12. Zlatev, Z., Dimov, I., Ostromsky, Tz., Geernaert, G., Tzvetanov, I., and Bastrup-Birk, A.: Calculating Losses of Crops in Denmark Caused by High Ozone Levels. Env. Modeling and Assessment, 6 (2001) 35–55.
13. Zlatev, Z., Fenger, J., and Mortensen, L.: Relationships between emission sources and excess ozone concentrations. Computers and Math. with Appl. 32 (1996) 101–123.
14. Zlatev, Z., Geernaert, G., and Skov, H.: A Study of ozone critical levels in Denmark. EUROSAP Newsletter 36 (1999) 1–9.

High Accuracy Difference Schemes on Unstructured Triangle Grids

Igor Popov, Sergey Polyakov, and Yuri Karamzin

Institute for Mathematical Modelling,
Moscow 125047, Miusskaya sq. 4a, Russia,
popov@imamod.ru, http://www.imamod.ru

Abstract. The problem of constructing of higher order difference schemes for parabolic equations in domains of arbitrary form on triangle grids is invetsigated. We proposed the second order schemes oriented to arbitrary grids which can not satisfy the Delaunay criteria. We suggested also the forth order schemes on triangle grids similar to uniform ones.

1 Introduction

The last decade was marked by high productivity and operating memory growth in modern computation systems, so that the opportunity of generation of unstructured triangle grids for technological problems appeared. In this connection the requirement of efficient difference methods construction for mathematical physics problems in complicated domains on unstructured grids arised.

In this report the construction and investigation of difference schemes for 2D-heat equation is considered using unstructured triangle grids. The selection of such a model is motivated by the presence in heat equation of two main differential operators: divergence and gradient. The investigation of characteristics for these operator difference analogues on unstructured grids gives the opportunity to extend the theoretical and practical results for another types of partial differential equations including elliptic and hyperbolic ones.

In this report we consider second and high order difference schemes for technological calculations on defected grids which do not satisfy the Delaunay criteria. The high order difference schemes are oriented to triangle grids similar to regular ones.

2 Differential Problem

Let us consider the process of heat diffusion in closed two dimentional domain D with boundary ∂D. In dimensionless variables it is described by differential equation

$$\frac{\partial u}{\partial t} = \operatorname{div} \mathbf{w} - qu + f, \quad (x, y) \in D, \quad t > 0, \tag{1}$$

where $u = u(x, y, t)$ is dimentionless temperature variation in the domain D relative to surround media temperature, $\mathbf{w} = \mathbf{K} \operatorname{grad} u$ is heat flux within the

I. Dimov et al. (Eds.): NMA 2002, LNCS 2542, pp. 555–562, 2003.

accuracy of sign, $\mathbf{K}$ is diagonal matrix with the components $k_{ii}(x,y,t) \geq k_0 > 0$, $q = q(x,y,t) \geq 0$ is heating or cooling coefficient of media, $f = f(x,y,t)$ is intensity of volumetric heat sources. Equation (1) is complemented with boundary and initial conditions

$$(\mathbf{w},\mathbf{n}) = -\eta u, \quad (x,y) \in \partial D, \tag{2}$$

$$u|_{t=0} = u_0(x,y), \quad (x,y) \in D. \tag{3}$$

Here $\mathbf{n}$ is external normal vector to the boundary ∂D, $\eta \geq 0$ is boundary heat transfer coefficient, $u_0(x,y)$ — initial distribution of u.

Later we will suppose that the solution of Eq. (1)–(3) exists and is a unique one. We suppose also that the functions k_{ii}, q and f are piece wise continuous. The following integral identity takes place

$$\frac{\partial}{\partial t}\iint\limits_D u\,dxdy = -\oint\limits_{\partial D} \eta u\,dl + \iint\limits_D (-qu+f)\,dxdy. \tag{4}$$

3 Construction of Difference Schemes

We assume that the unstructured grid $\omega_P = \{P_i = (x_i,y_i),\ i = 1,\dots,N\}$ is given in domain D and it includes the boundary grid points. The nondegenerate triangulation $T = \{T_m = \Delta(P_{i_m},P_{j_m},P_{k_m}),\ P_{i_m},P_{j_m},P_{k_m} \in \omega_P,\ m = 1,\dots,M,\ M = N-2\}$ is constructed on ω_P. We state that

1) T includes all the grid points of ω_P;
2) the area of triangle $S(T_m) > 0$ for all $T_m \in T$;
3) $S(T_m \cap T_n) = 0$ for $m \neq n$;
4) $D_h = \bigcup\limits_{m=1}^{M} T_m$ has the same connectedness as domain D;[1]
5) the ratio $\xi = S(D_h)/S(D) = 1 - \varepsilon$ $(0 < \varepsilon < 1)$ and $\lim\limits_{h\to 0} \xi = 1$.

If T does not satisfy the Delaunay criteria then we will suppose that

$$\max(\alpha_m,\beta_m,\gamma_m) \leq \pi - \delta, \quad \delta \in (0, 2\pi/3], \quad m = 1,\dots,M, \tag{5}$$

where α_m, β_m, γ_m are angles of triangle T_m.

Let the problem (1)–(3) be solved at the time segment $[0, t_{max}]$ and the uniform grid $\omega_t = \{t_n = n\cdot\tau, n = 0,\dots,N_t, \tau = t_{max}/N_t\}$ is given.

To construct the difference schemes we will use the finite volume method [1,2] combined with the flux method. The initial equation is divided in two with the help of flux vector. This approach was used early for difference schemes on orthogonal meshes [3]. In this case the temperature u was given on quadrangle

[1] Here the index h characterizes the size of grid. We assume that the D_h is grid analog of domain D and $\lim_{h\to 0} D_h = D$.

cell and considered constant in it, and fluxes were given at the cell faces. In our case the temperature is assigned to grid points of ω_P, and the fluxes are approximated in a special way so that the boundary approximation became natural for all types of boundary conditions (Dirichlet, Neumann, mixed and inhomogeneous). The similar approach is used in [4].

Let us introduce the concept of control volume. For this purpose at each point $P_i \in \omega_P$ we determine the set of all triangles having the top P_i: $H_i = \{T_m \in T, P_i \in T_m\}$ (this set is named as template). Let the quantity of such triangles be N_i. We enumerate them counter-clockwise: $T_{m_1}, \ldots, T_{N_i}$ so that each triangle T_{m_j} is defined by points P_i, P_{i_j}, $P_{i_{j+1}}$ (see Fig. 1a). In each triangle we define the point of median intersection M_{m_j} ($j = 1, \ldots, N_i$). Using these points we construct closed broken line L_i. The domain bounded by line L_i will be the control volume V_i at point P_i with area S_i (Fig. 1a). If point P_i belongs to the boundary then line L_i can be closed using the projection of points M_{m_j} at boundary edges plus point P_i (Fig. 1b).

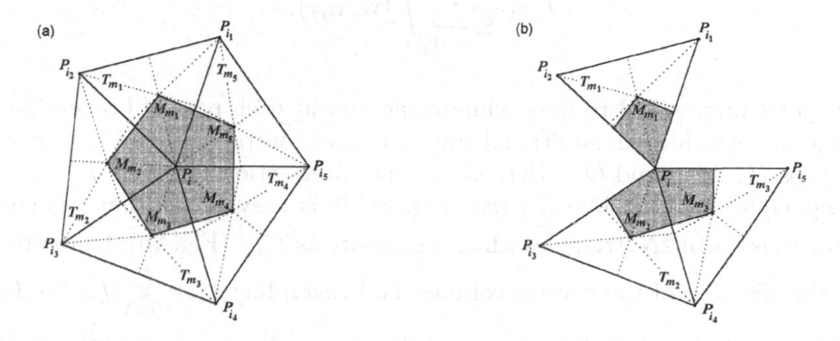

Fig. 1. Definition of control volume for internal (a) and boundary (b) points

Now we integrate the initial Eq. (1) over the control volume and time segment $[t_n, t_{n+1}]$. Using elementary transformation we get

$$\frac{1}{S_i} \iint\limits_{V_i} \frac{u(x,y,t_{n+1}) - u(x,y,t_n)}{\tau}\, dxdy = \frac{1}{S_i \tau} \iint\limits_{V_i} \int\limits_{t_n}^{t_{n+1}} (\operatorname{div} \mathrm{w} - qu + f)\, dt\, dxdy.$$

Let us construct for example the implicit difference scheme. For this purpose we approximate the time integral by the value of function in point t_{n+1} multiplied by step τ. The space integrals we replace by the function value at point P_i multiplied by the area of control volume S_i. As a result we obtain the next approximate relation:

$$\frac{u_i^{n+1} - u_i^n}{\tau} \approx \Lambda_i^{n+1} u_i^{n+1} - q_i^{n+1} u_i^{n+1} + f_i^{n+1}, \qquad (6)$$

where we used the standard notations: $\varphi_i^n = \varphi(x_i, y_i, t_n)$. Term $\Lambda_i^{n+1} u_i^{n+1}$ is some approximation of flux integral $I_i^{n+1} = \frac{1}{S_i} \iint\limits_{V_i} \operatorname{div} \mathrm{w}^{n+1}\, dxdy$.

Let us construct the approximation of flux integral. For this purpose we consider first the case when point P_i is the internal one. Then we transform the flux integral with the help of divergence theorem:

$$I_i \equiv \frac{1}{S_i} \iint\limits_{V_i} \operatorname{div} \mathbf{w}\, dxdy = \frac{1}{S_i} \int\limits_{\partial V_i} (\mathbf{w}, \mathbf{n})\, dl.$$

Here $\partial V_i \equiv L_i$ is the boundary of control volume, $\mathbf{n}$ is external normal vector to boundary ∂V_i, dl is the element of ∂V_i. Taking into account that ∂V_i is closed broken line containing N_i parts ($L_i = \bigcup\limits_{j=1}^{N_i} L_{ij}$) with normal vectors $\mathbf{n} = \mathbf{n}_{ij}$, the last formula can be converted to

$$I_i = \frac{1}{S_i} \sum_{j=1}^{N_i} \int\limits_{L_{ij}} (\mathbf{w}, \mathbf{n}_{ij}). \tag{7}$$

As a result we need to approximate the flux at each part of broken line. To do this, we introduce an additional subdivision of control volume V_i on triangles with tops P_i, M_{m_j} and Q_{i_j}. Here Q_{i_j} is the intersection of edges $[P_i, P_{i_j}]$ with corresponding parts of line L_{ij} (see Fig. 2). It is easy to see that the control volume consists of $2N_i$ triangles which we denote as $\tilde{T}_{m_j}^{(\pm)}$ (Fig. 2b). In accordance with this division of the control volume the broken line $L_i = \bigcup\limits_{j=1}^{N_i} (L_{m_j}^{(+)} + L_{m_j}^{(-)})$. Introducing the new notations for normal vector $\mathbf{n}_{m_j}^{(\pm)}$ on corresponding parts of broken line we get from (7)

$$I_i = \frac{1}{S_i} \sum_{j=1}^{N_i} \Big(\int\limits_{L_{m_j}^{(+)}} (\mathbf{w}, \mathbf{n}_{m_j}^{(+)}) + \int\limits_{L_{m_j}^{(-)}} (\mathbf{w}, \mathbf{n}_{m_j}^{(-)}) \Big). \tag{8}$$

At each part of broken line we can pass from flux and normal vectors to its ortogonal ones $\mathbf{W} = (W_x, W_y) = (-w_y, w_x)$ and $\boldsymbol{\nu} = (\nu_x, \nu_y) = (-n_y, n_x)$. The relation (8) will take the form

$$I_i = \frac{1}{S_i} \sum_{j=1}^{N_i} \Big(\int\limits_{L_{m_j}^{(+)}} (\mathbf{W}, \boldsymbol{\nu}_{m_j}^{(+)}) + \int\limits_{L_{m_j}^{(-)}} (\mathbf{W}, \boldsymbol{\nu}_{m_j}^{(-)}) \Big). \tag{9}$$

Note that for co-normal vectors $\boldsymbol{\nu}_{m_j}^{(\pm)}$ the next formulas are valid:

$$\boldsymbol{\nu}_{m_j}^{(\pm)} = \mathbf{l}_{m_j}^{(\pm)} / |\mathbf{l}_{m_j}^{(\pm)}|, \quad \mathbf{l}_{m_j}^{(+)} = \overrightarrow{M_{m_j} Q_{m_j}^{(+)}}, \quad \mathbf{l}_{m_j}^{(-)} = \overrightarrow{Q_{m_j}^{(-)} M_{m_j}}. \tag{10}$$

Here the new notations for Q_{ij} are used. Now we approximate integrals (9) by carrying out from integral the value of scalar product at some point. Then we arrive to:

$$I_i = \frac{1}{S_i}\sum_{j=1}^{N_i}\Big[(\mathbf{W}_{m_j}^{(+)}, \mathbf{l}_{m_j}^{(+)}) + (\mathbf{W}_{m_j}^{(-)}, \mathbf{l}_{m_j}^{(-)})\Big]. \tag{11}$$

Here $\mathbf{W}_{m_j}^{(\pm)}$ are unknown modified heat fluxes.

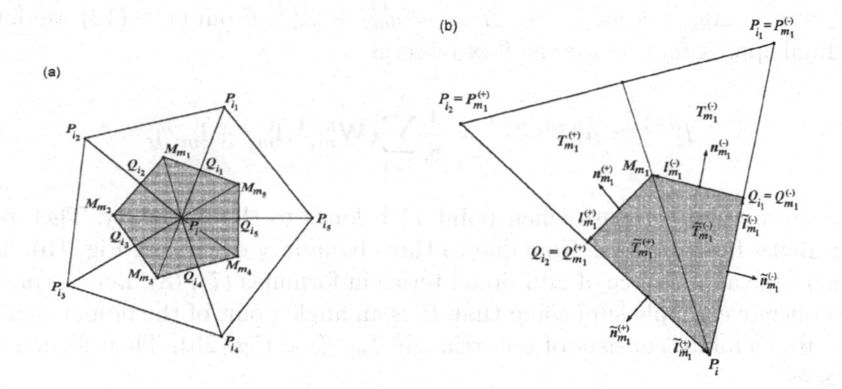

Fig. 2. Division of control volume onto parts (a). Distribution of normals and co-normals at one part of control volume (b)

To construct the second order scheme, it is enough to assume that the fluxes $\mathbf{W}_{m_j}^{(\pm)}$ are constant inside triangles T_{m_j}. Then $\mathbf{W}_{m_j}^{(+)} = \mathbf{W}_{m_j}^{(-)} = \mathbf{W}_{m_j}$. To define vector $\mathbf{W}_{m_j}$, one can use different approaches. The most general of them is the following [1]:

$$W_x = -k_{22}\,\frac{\partial u}{\partial y},\; W_y = k_{11}\,\frac{\partial u}{\partial x}, \quad \text{or} \quad \frac{W_x}{k_{22}} = -\frac{\partial u}{\partial y},\; \frac{W_y}{k_{11}} = \frac{\partial u}{\partial x}.$$

Let us integrate last two equations over triangle T_{m_j}. Taking into account the Green's formula we obtain

$$\int\limits_{T_{m_j}} \frac{W_x}{k_{22}}\,dxdy = \oint\limits_{\partial T_{m_j}} udx, \qquad \int\limits_{T_{m_j}} \frac{W_y}{k_{11}}\,dxdy = \oint\limits_{\partial T_{m_j}} udy.$$

We divide these equations by the area S_{m_j} and approximate the right hand integrals using the trapezoid formula through the values u_i, $u_{m_j}^{(\pm)}$ taken in the tops of triangle T_{m_j}: $P_i = (x_i, y_i)$, $P_{m_j}^{(\pm)} = (x_{m_j}^{(\pm)}, y_{m_j}^{(\pm)})$. As a result we get the following approximation for $\mathbf{W}_{m_j}$:

$$W_x \approx b_{m_j} \left[\bar{u}^{(-)}_{m_j} \Delta x^{(-)}_{m_j} + \bar{u}_{m_j} \Delta x_{m_j} - \bar{u}^{(+)}_{m_j} \Delta x^{(+)}_{m_j} \right],$$

$$W_y \approx a_{m_j} \left[\bar{u}^{(-)}_{m_j} \Delta y^{(-)}_{m_j} + \bar{u}_{m_j} \Delta y_{m_j} - \bar{u}^{(+)}_{m_j} \Delta y^{(+)}_{m_j} \right], \qquad (12)$$

$$b^{-1}_{m_j} = \frac{1}{S_{m_j}} \int\limits_{T_{m_j}} \frac{dxdy}{k_{22}}, \quad a^{-1}_{m_j} = \frac{1}{S_{m_j}} \int\limits_{T_{m_j}} \frac{dxdy}{k_{11}},$$

where $\bar{u}^{(\pm)}_{m_j} = 0.5(u^{(\pm)}_{m_j} + u_i)$, $\bar{u}_{m_j} = 0.5(u^{(+)}_{m_j} + u^{(-)}_{m_j})$, $\Delta x^{(\pm)}_{m_j} = x^{(\pm)}_{m_j} - x_i$, $\Delta x_{m_j} = x^{(+)}_{m_j} - x^{(-)}_{m_j}$, $\Delta y^{(\pm)}_{m_j} = y^{(\pm)}_{m_j} - y_i$, $\Delta y_{m_j} = y^{(+)}_{m_j} - y^{(-)}_{m_j}$. From (11)–(12) we obtain the final approximation for the flux integral

$$I^{n+1}_i \approx \Lambda^{n+1}_i u^{n+1}_i = \frac{1}{S_i} \sum_{j=1}^{N_i} (\mathbf{W}^{n+1}_{m_j}, \mathbf{l}^{(+)}_{m_j} + \mathbf{l}^{(-)}_{m_j}). \qquad (13)$$

Now we consider the case when point P_i belongs to the boundary. Then some of triangles from H_i have from one to three boundary edges (see Fig. 1b). That leads to the appearance of additional terms in formulas (7), (8). Let us consider one concrete example supposing that P_i is an angle point of the domain. In this case, its template consists of one triangle T_{m_1} (see Fig. 2b). Then formula (8) looks as

$$I_i = \frac{1}{S_i} \Big(\int\limits_{L^{(+)}_{m_1}} (\mathbf{w}, \mathbf{n}^{(+)}_{m_1}) + \int\limits_{L^{(-)}_{m_1}} (\mathbf{w}, \mathbf{n}^{(-)}_{m_1}) + \int\limits_{\tilde{L}^{(+)}_{m_1}} (\mathbf{w}, \mathbf{n}^{(+)}_{m_1}) + \int\limits_{\tilde{L}^{(-)}_{m_1}} (\mathbf{w}, \mathbf{n}^{(-)}_{m_1}) \Big). \qquad (14)$$

If parts of $\tilde{L}^{(\pm)}_{m_1}$ are located exactly on boundary of the domain then condition (2) takes place on $\tilde{L}^{(\pm)}_{m_1}$ and it can be used in (14):

$$I_i = \frac{1}{S_i} \Big(\int\limits_{L^{(+)}_{m_1}} (\mathbf{w}, \mathbf{n}^{(+)}_{m_1}) + \int\limits_{L^{(-)}_{m_1}} (\mathbf{w}, \mathbf{n}^{(-)}_{m_1}) - \int\limits_{\tilde{L}^{(+)}_{m_1}} \eta u \, dl - \int\limits_{\tilde{L}^{(-)}_{m_1}} \eta u \, dl \Big). \qquad (15)$$

If the parts of $\tilde{L}^{(\pm)}_{m_1}$ approximate the corresponding parts of ∂D with some accuracy, then formula (15) will be also approximate. In this case an error will be proportional to area of triangle S_{m_1} if we lay some conditions on initial triangulation T. This corresponds to second order of accuracy for space variables.

Now we obtain analog of (13) from formula (15). For this purpose we will approximate the first two integrals in right part of (15) as it was done above. The last two integrals are approximated using the formulas $u \approx (1-l)u_i + l u^{(\pm)}_{m_1}$ valid on $\tilde{L}^{(\pm)}_{m_1}$. Then

$$\int\limits_{\tilde{L}^{(\pm)}_{m_1}} \eta u \, dl \approx 0.5 \eta \, l \left((2-l) u_i + l u^{(\pm)}_{m_1} \right), \qquad (16)$$

where $l = |\tilde{\mathbf{l}}_{m_1}^{(\pm)}|$, $\tilde{\mathbf{l}}_{m_1}^{(+)} = \overrightarrow{Q_{m_1}^{(+)}P_i}$, $\tilde{\mathbf{l}}_{m_1}^{(-)} = \overrightarrow{P_iQ_{m_1}^{(-)}}$. If we introduce operators $\eta_{m_j}^{(\pm)}$:

$$\eta_{m_j}^{(\pm)} u_i = \begin{cases} 0.5\eta\, l\left((2-l)u_i + l u_{m_j}^{(\pm)}\right), & P_i \in \partial D \text{ and } P_{m_j}^{(\pm)} \in \partial D, \\ 0, \; P_i \notin \partial D \text{ or } P_{m_j}^{(\pm)} \notin \partial D, \end{cases}$$

then we can join results of (13) and (15)–(16) and can obtain a general approximation for the flux integral at any point of grid:

$$I_i^{n+1} \approx \Lambda_i^{n+1} u_i^{n+1} = \frac{1}{S_i}\sum_{j=1}^{N_i}\left[(\mathbf{W}_{m_j}^{n+1}, \mathbf{l}_{m_j}^{(+)} + \mathbf{l}_{m_j}^{(-)}) - \eta_{m_j}^{(+)} u_i - \eta_{m_j}^{(-)} u_i\right]. \tag{17}$$

As a result we can write the finite difference scheme for grid function U on $\Omega = \omega_P \times \omega_t$:

$$\frac{U_i^{n+1} - U_i^n}{\tau} = \Lambda_i^{n+1} U_i^{n+1} - q_i^{n+1} U_i^{n+1} + f_i^{n+1}, \quad U_i^0 = u_0(P_i). \tag{18}$$

This scheme has the second order of accuracy in spatial variables, it is conservative (analog of Eq. (4) takes place on grid) and satisfies difference maximum principle. One can prove that the difference solution of scheme converges in grid norm $L_2(\omega_P) \times C(\omega_t)$ to smooth solution of (1)–(3) with rate $O(\overline{\Delta l}^2 + \tau)$, where $\overline{\Delta l}$ is average size of triangle's edges from T.

Let us do some remarks about scheme construction and its accuracy in some special cases. First, if we use grid with obtuse-angled triangles near the boundary, then control volume must be determined as it is shown in Fig. 3. This modification of scheme does not decrease its accuracy order.

Second, to raise order of accuracy we offer to enlarge template for flux term using additional triangles closely related to H_i. In this case flux is approximated in each triangle T_{m_j} with the help of six values of function u. Moreover, three of them are taken from neighbour triangles. Then we get the fourth order of accuracy inside and second order near the boundary. However the rate of convergence will be fourth order.

4 Conclusion

Recently we developed original approach for generation of unstructured triangle grids for nonconvex multiply connected domains with adaptive small adjustment of grid to the boundary [5]. Our grid generator has the opportunity to make quasi-uniformly triangulation. In this work we proposed new difference schemes for parabolic boundary problems in domains of complex forms. These schemes allow to solve effectively both linear and quasi-linear problems on triangle grids, and to obtain high accuracy of computations.

Acknowledgments

The present work has been supported by the Russian Fund for Basic Research (Grant No 02-01-00699).

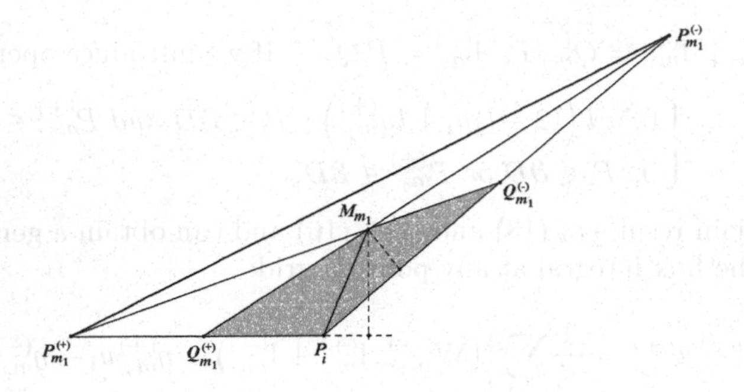

Fig. 3. Definition of control volume in the case of obtuse-angled boundary triangle

References

1. Samarskii, A.A.: Theory of Difference Schemes, Nauka, Moscow (1983) (in Russ.).
2. Fryazinov, I.V.: Balance method and variation-difference schemes. Differential equations, Vol. 16, 7 (1980) 1332–1341 (in Russ.)
3. Samarskii, A.A., Koldoba, A.V., Poveshchenko, Yu.A., Tishkin, V.F., Favorskii, A.P.: Difference schemes on irregular grids. Nauka, Minsk (1996) (in Russ.)
4. Atluri, S.N., Gallagher, R.H., Zienkiewitz, O.C.: Hybrid and Mixed Finite Element Methods. John Wiley & Sons, New York (1983)
5. Popov, I.V., Polyakov, S.V.: Construction of adaptive irregular triangle grids for 2D multiconnected nonconvex domains. Matematicheskoe modelirovanie, Vol. 14, 6 (2002) 25–35 (in Russ.)

The PCD Method

Ahmed Tahiri*

Service de Métrologie Nucléaire, ULB, (CP 165/84),
50 av. F. D. Roosevelt, B-1050, Brussels, Belgium.

Abstract. We propose in this contribution a new BVP discretization method which represents the unknown distribution as well as its derivatives by piecewise constant distributions (PCD) but on distinct meshes. Once the meshes are chosen, it is relatively straightforward to define an approximate variational formulation of the BVP on the associated PCD spaces and hence to derive the discrete equations. We end with the same scheme as the corner mesh box method and we display a precise relation between the exact solution and our approximation, which holds in the absence of absorption. We also compare these results with other approaches using a mixed variational formulation of the same BVP and sharing with our method the use of several meshes. We show that PCD approximations can also be used in this context leading, on rectangular meshes, to the same scheme as the centered box method.

1 Introduction

The purpose of the present contribution is to introduce a new boundary value problems (BVP) discretization technique based on the use of piecewise constant distributions (PCD) to represent the unknown distribution as well as its derivatives. The use of piecewise constant approximations has been considered in numerous classical contributions [1,2,4,11,12] and raises persistent interest today as illustrated by many recent works [3,5,6,7,8]. Their simultaneous use for the unknown distribution and its derivatives has however not been covered. This feature requires the introduction of distinct meshes to represent each distribution and new mathematical tools for the convergence analysis of the resulting discrete scheme. For simplicity, we restrict the present contribution to the analysis of the 2D diffusion equation on a nonuniform rectangular mesh and we borrow some technical results from our forthcoming PhD thesis [10].

We consider solving the following BVP on a rectangular domain Ω:

$$-\text{div}(\, p(x)\, \nabla u(x)) + q(x)u(x) = s(x) \qquad x \in \Omega \tag{1}$$

$$u(x) = 0 \qquad x \in \Gamma_0 \tag{2}$$

$$n \cdot \nabla\, u(x) = 0 \qquad x \in \Gamma_1 \tag{3}$$

* We would like to gratefully thank Professor R. Beauwens for his fruitful suggestions and helpful advices.

I. Dimov et al. (Eds.): NMA 2002, LNCS 2542, pp. 563–571, 2003.

where n denotes the unit normal to $\Gamma = \partial\Omega$ and $\Gamma = \Gamma_0 \cup \Gamma_1$. We assume that p and q are in $L^\infty(\Omega)$, that $p(x)$ is strictly positive on $\overline{\Omega}$, that $q(x)$ is nonnegative on Ω and that we have a well posed problem.
The discrete version of this problem will be based on its variational formulation:

$$\text{find} \quad u \in H \quad \text{such that} \quad \forall\, v \in H \quad a(u, v) = (s, v) \tag{4}$$

where $H = H^1_{\Gamma_0}(\Omega) = \{v \in H^1(\Omega),\ v = 0 \text{ on } \Gamma_0\}$,

$$a(u,v) = \int_\Omega p(x)\,\nabla u(x) \cdot \nabla v(x)\,dx + \int_\Omega q(x) u(x)\, v(x)\,dx\,, \tag{5}$$

and (s, v) denotes the $L^2(\Omega)$ scalar product.

2 PCD Discretization

2.1 PCD Spaces

The PCD discretization splits the open domain Ω under investigation into elements Ω_ℓ, open subsets of Ω, such that

$$\overline{\Omega} = \bigcup_{\ell=1}^{N} \overline{\Omega}_\ell \qquad \Omega_k \cap \Omega_\ell = \emptyset \qquad \text{if} \quad k \neq \ell \quad \forall\, k, \ell \leq N$$

and defines several submeshes on each element Ω_ℓ for the representation of $v \in H^1(\Omega)$ and of its derivatives $\partial_i v$ $(i = 1, 2)$. These representations, denoted v_h and $\partial_{hi}\, v_h$ $(i = 1, 2)$ respectively, are piecewise constant on one of these submeshes (a specific one for each) with the additional requirement for v_h that it must be continuous across the element boundaries (i.e. along the normal to the element boundary). Here, we consider rectangular meshes and the operators ∂_{hi} $(i = 1, 2)$ are finite difference quotients taken along the element edges. The submeshes used to define $v_h|_{\Omega_\ell}$, $\partial_{h1}\, v_h|_{\Omega_\ell}$ and $\partial_{h2}\, v_h|_{\Omega_\ell}$ on a rectangular element

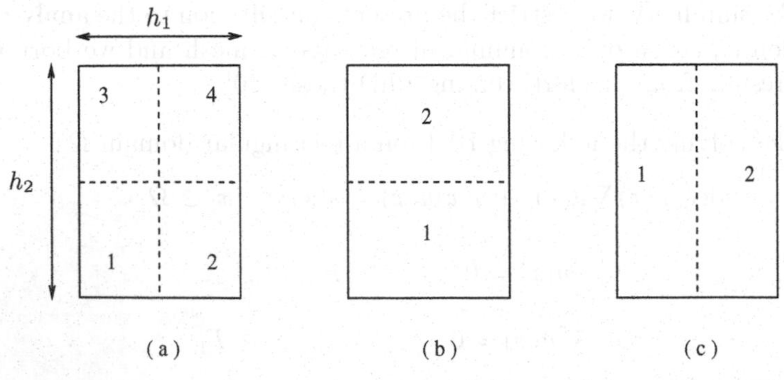

Fig. 1. Submeshes used to represent v_h (a), $\partial_{h1}\, v_h$ (b) and $\partial_{h2}\, v_h$ (c) on Ω_ℓ

Ω_ℓ are represented on Fig. 1; $v_h|_{\Omega_\ell}$ is the piecewise constant distribution with 4 values v_{hi} on the regions denoted i with $i = 1, ..., 4$ on Fig. 1 (a); $\partial_{h1} v_h|_{\Omega_\ell}$ is the piecewise constant distribution with constant values:

$$(\partial_{h1} v_h)_1 = \frac{v_{h2} - v_{h1}}{h_1} \quad , \quad (\partial_{h1} v_h)_2 = \frac{v_{h4} - v_{h3}}{h_1}$$

on the regions denoted 1, 2 on Fig. 1 (b) and $\partial_{h2} v_h|_{\Omega_\ell}$ is similarly the piecewise constant distribution with constant values:

$$(\partial_{h2} v_h)_1 = \frac{v_{h3} - v_{h1}}{h_2} \quad , \quad (\partial_{h2} v_h)_2 = \frac{v_{h4} - v_{h2}}{h_2}$$

on the regions denoted 1, 2 on Fig. 1 (c). In addition, v_h must be continuous across element boundaries. Thus for example if the bottom boundary of Ω_ℓ is common with the top boundary of Ω_k, one must have that $v_{h1}(\Omega_\ell) = v_{h3}(\Omega_k)$ and $v_{h2}(\Omega_\ell) = v_{h4}(\Omega_k)$.

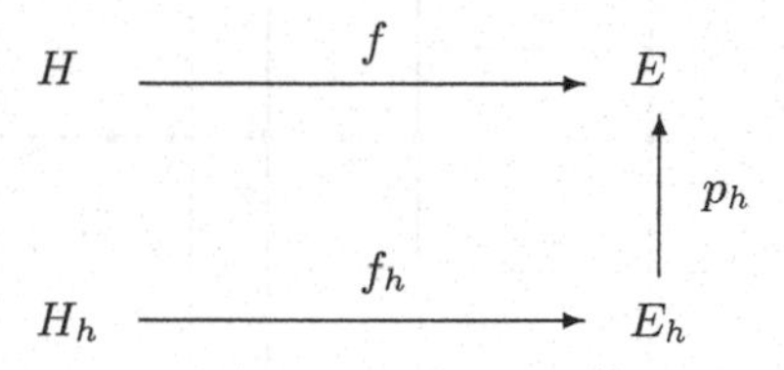

Fig. 2. Structure of the discretization analysis

The spaces associated with this discretization are defined as follow. E denotes $(L^2(\Omega))^3$ with norm $\|(u, v, w)\|^2 = \|u\|^2 + \|v\|^2 + \|w\|^2$. F denotes the subspace of E of the elements of the form $(v, \partial_1 v, \partial_2 v)$.
H_{h0} and H_{hi} $(i = 1, 2)$ denote the spaces of piecewise constant distributions used to define v_h and $\partial_{hi} v_h$ $(i = 1, 2)$, equipped with the $L^2(\Omega)$ scalar product. $E_h = H_{h0} \times H_{h1} \times H_{h2}$ with norm $\|(u_h, v_h, w_h)\|^2 = \|u_h\|^2 + \|v_h\|^2 + \|w_h\|^2$ and F_h is the subspace of E_h of the elements of the form $(v_h, \partial_{h1} v_h, \partial_{h2} v_h)$.
We further denote by H_h the space H_{h0} equipped with the inner product

$$(v_h, w_h)_h = (v_h, w_h) + \sum_{i=1}^{2} (\partial_{hi} v_h, \partial_{hi} w_h) \tag{6}$$

Clearly H and H_h are isomorphic to F and F_h respectively and we let f and f_h denote the bijections of H and H_h into E and E_h ($F = f(H)$ and $F_h = f_h(H_h)$). We further denote by p_h the canonical injection of E_h into E. These operators are represented on Fig. 2. The motivation for using this space structure is that, while we cannot directly compare the elements of H and H_h, we can use the norm of E to measure the distance between elements $f(v) = (v, \partial_1 v, \partial_2 v)$ of F and $f_h(v_h) = (v_h, \partial_{h1} v_h, \partial_{h2} v_h)$ of F_h.

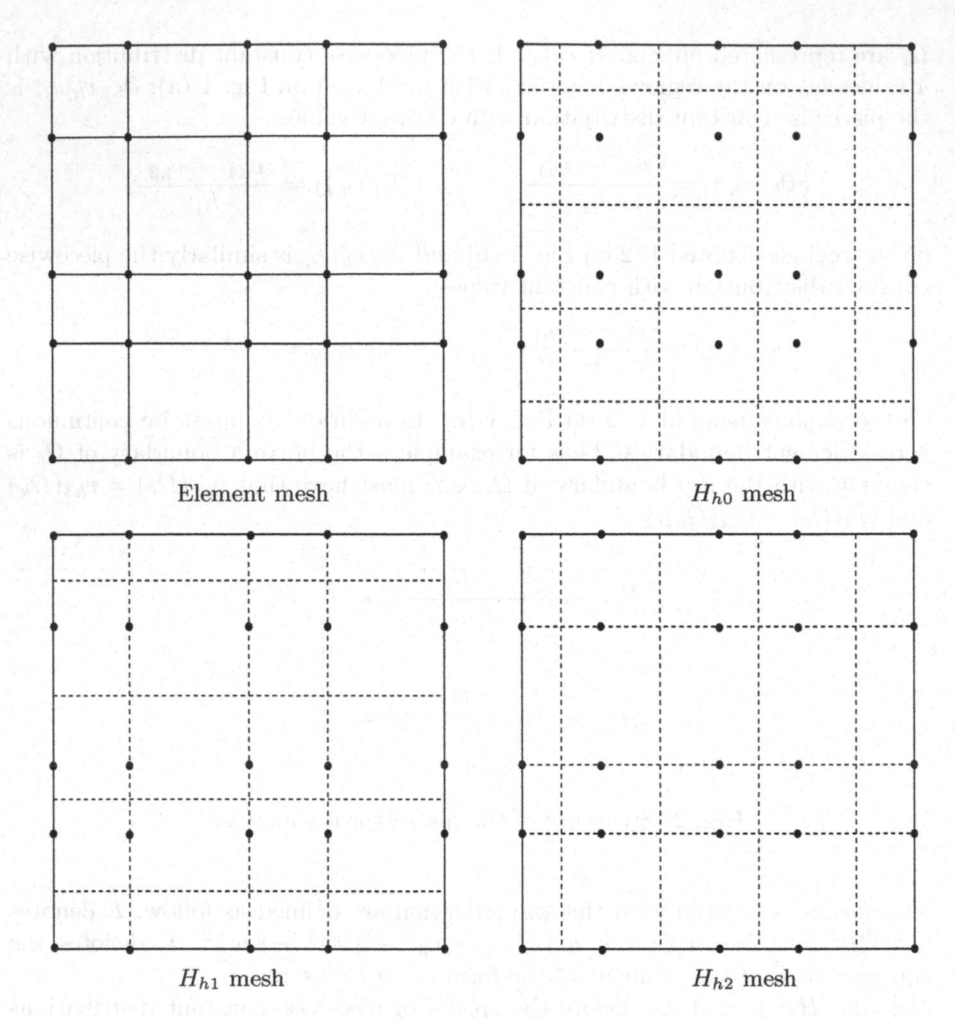

Fig. 3. Discrete meshes

Figure 3 (top–left) provides an example of a rectangular element mesh. On the same figure we also represent the meshes H_{h0} and H_{hi} $(i = 1, 2)$ used to define the piecewise constant distributions v_h and $\partial_{hi}\, v_h$, $i = 1, 2$. Each of these meshes defines cells which are useful for distinct purposes. The elements are denoted by Ω_ℓ , $\ell \in L$; we similarly denote the cells of the other meshes by $\Omega_{\ell i}$, $\ell \in L_i$, $i =$ 0, 1, 2 respectively. It is of interest to note that each node of the mesh may be uniquely associated with a cell of H_{h0}; we therefore denote them by N_ℓ , $\ell \in L_0$.

2.2 PCD Equations

We now define the discrete problem to be solved in H_h by

$$\text{find} \quad u_h \in H_h \qquad \text{such that} \qquad \forall\, v_h \in H_h \quad a_h(\,u_h,\, v_h\,) = (s,\, v_h) \tag{7}$$

where

$$a_h(u_h, v_h) = \sum_{i=1}^{2} (p(x)\,\partial_{hi} u_h, \partial_{hi} v_h)_\Omega + (q(x)\, u_h, v_h)_\Omega \tag{8}$$

The discrete matrix is obtained as usual by introducing a basis $(\phi_i)_{i\in I}$ of the space H_h, expanding the unknown u_h in this basis $u_h = \sum_{i\in I} \xi_i\, \phi_i$ and expressing the variational condition (7) by

$$a_h(u_h, \phi_j) = \sum_{i\in I} a_h(\phi_j, \phi_i)\, \xi_i = (s, \phi_j)_\Omega$$

for all $j \in I$, whence the linear system $\mathcal{A}\,\xi = b$ with stiffness matrix $\mathcal{A} = (a_{ij}) = (a_h(\phi_i, \phi_j))$, right hand side b with components $b_j = (s, \phi_j)_\Omega$, $j \in I$ and unknown vector ξ with components ξ_i, $i \in I$.
Since $p(x)$ and $q(x)$ are in $L^\infty(\Omega)$, it is clear that $a_h(u_h, v_h)$ is uniformly continuous on H_h. On the other hand, we have shown in [10] that, provided Γ_0 is of positive measure, $a_h(u_h, v_h)$ is also uniformly coercive on H_h.
Therefore, the associated norm

$$\| v_h \|_{a_h} = (a_h(v_h, v_h))^{1/2}$$

is uniformly equivalent to the norm of H_h, i.e. there exist positive constants α, β independent of the mesh size h (defined as $h = \max(h_\ell)$, $\ell \in L_0$ with $h_\ell = \mathrm{diam}(\Omega_\ell)$) such that: $\forall\, v_h \in H_h \quad \alpha \| v_h \|_h \le \| v_h \|_{a_h} \le \beta \| v_h \|_h$.

3 Convergence Analysis

As mentioned in §2, the error between the solution $u \in H$ of (1) and the discrete solution $u_h \in H_h$ of (7) will be measured by the distance

$$\| f(u) - f_h(u_h) \|_E \tag{9}$$

between their representations in E.
As usual, the bounds that can be obtained depend on the regularity of u. Here, we assume that $u \in H^2(\Omega)$ with some comments on other issues. In that case, u is continuous on $\overline{\Omega}$ and we can then define its interpolant u_I in H_h through

$$u_I(N_\ell) = u(N_\ell) \quad \text{for all nodes } N_\ell,\ \ell \in L_0 \tag{10}$$

and rely on interpolation theory in H_h.
More specifically, we can then use the following result proved in [10].

Lemma 1. *Under the general assumptions and notation defined above, there exists a positive constant C independent of the mesh size h such that $\forall v \in H^2(\Omega)$*

$$\| f(v) - f_h(v_I) \|_E \le C\, h \,\| v \|_{2,\Omega} \tag{11}$$

where v_I denotes the interpolant of v in H_h.

By this result, the problem of bounding (9) is reduced to that of bounding $\| f_h(u_I) - f_h(u_h) \|_E = \| u_I - u_h \|_h$. Since the a_h-norm is uniformly equivalent to the H_h-norm, we may equivalently try to bound $\| u_I - u_h \|_{a_h}$.
First we consider bounding $\| u_d - u_h \|_{a_h}$ where u_d is an arbitrary element of H_h and u_h the solution of (7). we have

$$\| u_d - u_h \|_{a_h} = \sup_{\substack{v_h \in H_h \\ v_h \neq 0}} \frac{| a_h(u_d, v_h) - a_h(u_h, v_h) |}{\| v_h \|_h} \tag{12}$$

$$= \sup_{\substack{v_h \in H_h \\ v_h \neq 0}} \frac{| a_h(u_d, v_h) - a(u, v_h) |}{\| v_h \|_h}$$

because,

$$\forall v_h \in H_h \qquad a_h(u_h, v_h) = (s, v_h) = a(u, v_h)$$

where, in the last term, the derivatives of v_h are understood in distribution sense. Explicitly

$$\begin{aligned} & a_h(u_d, v_h) - a(u, v_h) \\ & = (p(x)\,\partial_{h1} u_d, \partial_{h1} v_h) + (p(x)\,\partial_{h2} u_d, \partial_{h2} v_h) + (q(x)\,u_d, v_h) \\ & - (p(x)\,\partial_1 u, \partial_1 v_h) - (p(x)\,\partial_2 u, \partial_2 v_h) - (q(x)\,u, v_h) \end{aligned} \tag{13}$$

In this expression of the error, we see that $\partial_i v_h$ appear with Dirac behaviours across specific lines and, because this is clearly incompatible with coefficients $p(x)$ that would be discontinuous across the same lines, we must introduce some restriction on the choice of the mesh, namely that material discontinuities (i. e. discontinuities of $p(x)$) should never match grid lines of the H_{h0} mesh. The best practical way to ensure this restriction is to require that material discontinuities be always grid lines of the element mesh.
Returning now to our concern of bounding (13), we first note that the $q(x)$ contribution is readily bounded by

$$\frac{| (q(x)(u_d - u), v_h) |}{\| v_h \|_h} \leq \| q \|_{\infty,\Omega} \frac{| (u_d - u, v_h) |}{\| v_h \|_h} \leq \| q \|_{\infty,\Omega} \| u_d - u \|$$

The other terms are most easily analysed on the cells $\Omega_{\ell i}$ of the H_{hi} meshes $(i = 1, 2)$. The contribution of an arbitrary cell $\Omega_{\ell i}$ is

$$\begin{aligned} & (p(x)\,\partial_{hi} u_d, \partial_{hi} v_h)_{\Omega_{\ell i}} - (p(x)\,\partial_i u, \partial_i v_h)_{\Omega_{\ell i}} \\ & = \left(\langle p(x) \rangle_{\Omega_{\ell i}} \partial_{hi} u_d \, \partial_{hi} v_h - \langle p(x)\,\partial_i u \rangle_{E_{\ell i}} \partial_{hi} v_h \right) \cdot | \Omega_{\ell i} | \\ & = \left(\langle p(x) \rangle_{\Omega_{\ell i}} \partial_{hi} u_d - \langle p(x)\,\partial_i u \rangle_{E_{\ell i}} \right) \cdot \partial_{hi} v_h \cdot | \Omega_{\ell i} | \end{aligned}$$

where $E_{\ell 1}$ ($E_{\ell 2}$) denotes the vertical (horizontal) median line of the cell $\Omega_{\ell 1}$ ($\Omega_{\ell 2}$) and $\langle\ \rangle_Q$ denotes averages on Q defined by

$$\langle f(x) \rangle_Q = \frac{1}{|Q|} \int_Q f(x)\, dx$$

To bound this local $p(x)$ contribution, it is sufficient to bound

$$(\langle p(x) \rangle_{\Omega_{\ell i}}\, \partial_{hi}\, u_d \; - \; \langle p(x)\, \partial_i\, u \rangle_{E_{\ell i}})$$

which can be done through

$$\left| \langle p(x) \rangle_{\Omega_{\ell i}}\, \partial_{hi}\, u_d \; - \; \langle p(x)\, \partial_i\, u \rangle_{E_{\ell i}} \right| \; \leq \; C \,\| p \|_{\infty,\Omega} \left| \partial_{hi}\, u_d \; - \; \langle \partial_i\, u \rangle_{E_{\ell i}} \right|$$

In the case where $u_I = u_d$, this brings us back to interpolation theory in H_h space, whence

Theorem 1. *Assume that the solution u of (1) belongs to $H^2(\Omega)$. Then, there exists a constant $C > 0$ (independent of h), such that:*

$$\| u_I \; - \; u_h \|_h \; = \; \| f_h(u_I) \; - \; f_h(u_h) \|_E \; \leq \; C\, h \,\| u \|_{2,\Omega}$$

and

$$\| f(u) \; - \; f_h(u_h) \|_E \; \leq \; C\, h \,\| u \|_{2,\Omega}$$

where u_h is the solution of the problem (7).

Under higher regularity assumptions an error bound of order $O(h^2)$ can be obtained. If the solution u is only in $H^1(\Omega)$, we can still prove the convergence of u_h to u.
It is also of interest to consider the case where $u_d = u_h$ since in that case, the error must be zero. When $q(x) = 0$, the $p(x)$ contribution of each cell $\Omega_{\ell i}$, $i = 1, 2$ must then be zero, whence

Theorem 2. *If $q(x) = 0$, the solution u_h of (7) satisfies*

$$\partial_{hi}\, u_h \,|\, \Omega_{\ell i} \; = \; \frac{1}{\langle p(x) \rangle_{\Omega_{\ell i}}} \,\langle p(x)\, \partial_i\, u \rangle_{E_{\ell i}} \qquad \ell \in L_i\,, \; i = 1, 2 \qquad (14)$$

where u is the solution of (1)

Beyond giving us a relation between the exact and approximate solutions, (14) shows that in this case, the method reduces to the control volume method under its corner mesh version. In more general cases, we may still consider our approach as an interpretation of this box method.

4 Mixed Formulation

4.1 Dual Mixed Variational Principle

The variational formulation (4)–(5) used as basic framework in our developments so far is more precisely called primal variational formulation. One can also use a dual variational formulation with unknown distribution $\mathbf{J}$

$$\mathbf{J} = -p(x)\,\nabla u(x) \tag{15}$$

as well as mixed formulation that use both u and $\mathbf{J}$.
The so-called dual mixed formulation of the problem (1)–(3) writes

$$\text{find} \quad (\, u\,,\, \mathbf{J}\,) \in L^2(\Omega) \times \mathbf{H}_{\Gamma_1}(\text{div}, \Omega) \tag{16}$$

such that $\forall\,(\, v\,,\, \mathbf{K}\,) \in L^2(\Omega) \times \mathbf{H}_{\Gamma_1}(\text{div}, \Omega)$

$$(\, \operatorname{div} \mathbf{J}\,,\, v\,) + (\, q(x)u\,,\, v\,) = (\, s\,,\, v\,) \tag{17}$$

$$(\, (p(x))^{-1}\,\mathbf{J}\,,\, \mathbf{K}\,) - (\, u\,,\, \operatorname{div} \mathbf{K}\,) = 0 \tag{18}$$

where

$$\mathbf{H}_{\Gamma_1}(\text{div}, \Omega) = \{\, \mathbf{K} \in (L^2(\Omega))^2\,,\; \operatorname{div} \mathbf{K} \in L^2(\Omega)\,,\; \mathbf{n} \cdot \mathbf{K} = 0 \;\text{ on }\; \Gamma_1\,\}$$

with scalar product

$$(\, \mathbf{J}\,,\, \mathbf{K}\,)_{H(div,\Omega)} = (\, \mathbf{J}\,,\, \mathbf{K}\,) + (\, \operatorname{div} \mathbf{J}\,,\, \operatorname{div} \mathbf{K}\,) \tag{19}$$

Using this formulation with zeroth order Raviart–Thomas finite elements [9] and special numerical quadrature formulas, an interpretation of the centered mesh box scheme has been obtained (cf. [12]). We briefly sketch below how the same stencil may also be interpreted as a PCD discretization of (16)–(18).

4.2 PCD Discretization

For this purpose, we now choose as elements the cells $\Omega_{\ell 0}\,,\ \ell \in L_0$ of the mesh H_{h0} with u_h constant on each element $\Omega_{\ell 0}$ and we use 2-value submeshes (determined by the vertical and horizontal median lines of $\Omega_{\ell 0}$) to represent the components $J_{hi}\,,\ i = 1,\, 2$ of $\mathbf{J}$ on each cell, with the additional requirement that J_{hi} must be continuous across cell boundaries orthogonal to direction i (J_{hi} is thus constant on the $\Omega_{\ell i}$ cells of the mesh $H_{hi}\,,\ i = 1,\, 2$). Finally, the distributions $\partial_{hi}\, J_{hi}$ are chosen as constant on each element $\Omega_{\ell 0}$ and determined by the edge-to-edge difference quotient of J_{hi} in direction i on each cell $\Omega_{\ell 0}$.
Note that in the present case, contrarily to the former one, the material discontinuities (i. e. the discontinuities of $p(x)$) should always match those of the H_{h0} mesh. We then define the PCD discretization of (16)–(18) by:

$$\text{find} \quad (\, u_h\,,\, J_{h1}\,,\, J_{h2}\,) \in H_{h0} \times H_{h1} \times H_{h2}$$

such that $\forall\,(v_h\,,\,K_{h1}\,,\,K_{h2}\,) \in H_{h0} \times H_{h1} \times H_{h2}$

$$(\,\partial_{h1}\,J_{h1} + \partial_{h2}\,J_{h2}\,,\,v_h\,) + (\,q(x)\,u_h\,,\,v_h\,) = (\,s\,,\,v_h\,)$$

$$\big(\,(p(x))^{-1}\,J_{h1}\,,\,K_{h1}\,\big) + \big(\,(p(x))^{-1}\,J_{h2}\,,\,K_{h2}\,\big) - \big(\,u_h\,,\,\partial_{h1}\,K_{h1} + \partial_{h2}\,K_{h2}\,\big) = 0$$

which leads to the centered mesh box scheme.

5 Concluding Remarks

The main isssue of this contribution is the introduction of a new BVP discretization method. It is based on the use of piecewise constant distributions to represent the unknown distribution as well as its derivatives on distinct meshes. The PCD method leads to a very sparse discrete system, and has the standard first order estimate under the H^2–regularity of the exact solution.
We note that triangular elements may also be introduced in order to be able to follow any shape (on the domain Ω) by combining the use of mesh refinement and triangular elements, see [10].

References

1. Aubin,J. P.: Approximation des espaces de distributions et ses opérateurs différentiels. Bull. Soc. Math. France, 12 (1967) 1–139.
2. Bank, R. E., Rose, D. J.: Some error estimates for the box method. SIAM J. Num. Anal., 24 (1987) 777–787.
3. Cai, Z., Mandel, J., McCormick, S.: The finite volume element method for diffusion equations on general triangulations. SIAM J. Numer. Anal., 28 (1991) 392–402.
4. Cea, J.: Approximation variationnelle des problèmes aux limites. Ann. Inst. Fourier, 14 (1964) 345–444.
5. Chou, S. H., Kwak, D. Y.: Mixed covolume methods on rectangular grids for elliptic problems. SIAM J. Num. Anal., 37 (2000) 758–771.
6. Chou, S. H., Kwak, D. Y., Vassilevski, P. S.: Mixed covolume methods for elliptic problems on triangular grids. SIAM J. Num. Anal., 35 (1998) 1850–1861.
7. Ewing, R., Iliev, O., Lazarov, R.: A modified finite volume approximation of second-order elliptic equations with discontinuous coefficients. SIAM J. Sci. Comput., 23 (2001) 1334–1350.
8. Lazarov, R. D, Mishev, I. D, Vassilevski, P. S.: Finite volume methods for convection diffusion problems. SIAM J. Num. Anal., 33 (1996) 31–55.
9. Raviart, P. A., Thomas, J. M.: A mixed finite element method for second-order elliptic problems. Lecture Notes in Mathematics, Springer-Verlag, Heidelberg, 606 (1977) 292–315.
10. Tahiri, A.: A compact discretization method with local mesh refinement. PhD thesis, Service de Métrologie Nucléaire, ULB, Brussels, Belgium (09/2002)
11. Temam, R.: Analyse Numérique. Presses Univ. de France, Paris (1970)
12. Weiser, A., Wheeler, M. F.: On convergence of block-centered finite differences for elliptic problems. SIAM J. Num. Anal., 25 (1988) 351–375.

Author Index

Lecture Notes in Computer Science

For information about Vols. 1–2498

please contact your bookseller or Springer-Verlag

Vol. 2499: S.D. Richardson (Ed.), Machine Translation: From Research to Real Users. Proceedings, 2002. XXI, 254 pages. 2002. (Subseries LNAI).

Vol. 2500: E. Grädel, W. Thomas, T. Wilke (Eds.), Automata Logics, and Infinite Games. VIII, 385 pages. 2002.

Vol. 2501: D. Zheng (Ed.), Advances in Cryptology – ASIACRYPT 2002. Proceedings, 2002. XIII, 578 pages. 2002.

Vol. 2502: D. Gollmann, G. Karjoth, M. Waidner (Eds.), Computer Security – ESORICS 2002. Proceedings, 2002. X, 281 pages. 2002.

Vol. 2503: S. Spaccapietra, S.T. March, Y. Kambayashi (Eds.), Conceptual Modeling – ER 2002. Proceedings, 2002. XX, 480 pages. 2002.

Vol. 2504: M.T. Escrig, F. Toledo, E. Golobardes (Eds.), Topics in Artificial Intelligence. Proceedings, 2002. XI, 432 pages. 2002. (Subseries LNAI).

Vol. 2506: M. Feridun, P. Kropf, G. Babin (Eds.), Management Technologies for E-Commerce and E-Business Applications. Proceedings, 2002. IX, 209 pages. 2002.

Vol. 2507: G. Bittencourt, G.L. Ramalho (Eds.), Advances in Artificial Intelligence. Proceedings, 2002. XIII, 418 pages. 2002. (Subseries LNAI).

Vol. 2508: D. Malkhi (Ed.), Distributed Computing. Proceedings, 2002. X, 371 pages. 2002.

Vol. 2509: C.S. Calude, M.J. Dinneen, F. Peper (Eds.), Unconventional Models in Computation. Proceedings, 2002. VIII, 331 pages. 2002.

Vol. 2510: H. Shafazand, A Min Tjoa (Eds.), EurAsia-ICT 2002: Information and Communication Technology. Proceedings, 2002. XXIII, 1020 pages. 2002.

Vol. 2511: B. Stiller, M. Smirnow, M. Karsten, P. Reichl (Eds.), From QoS Provisioning to QoS Charging. Proceedings, 2002. XIV, 348 pages. 2002.

Vol. 2512: C. Bussler, R. Hull, S. McIlraith, M.E. Orlowska, B. Pernici, J. Yang (Eds.), Web Services, E-Business, and the Semantic Web. Proceedings, 2002. XI, 277 pages. 2002.

Vol. 2513: R. Deng, S. Qing, F. Bao, J. Zhou (Eds.), Information and Communications Security. Proceedings, 2002. XII, 496 pages. 2002.

Vol. 2514: M. Baaz, A. Voronkov (Eds.), Logic for Programming, Artificial Intelligence, and Reasoning. Proceedings, 2002. XIII, 465 pages. 2002. (Subseries LNAI).

Vol. 2515: F. Boavida, E. Monteiro, J. Orvalho (Eds.), Protocols and Systems for Interactive Distributed Multimedia. Proceedings, 2002. XIV, 372 pages. 2002.

Vol. 2516: A. Wespi, G. Vigna, L. Deri (Eds.), Recent Advances in Intrusion Detection. Proceedings, 2002. X, 327 pages. 2002.

Vol. 2517: M.D. Aagaard, J.W. O'Leary (Eds.), Formal Methods in Computer-Aided Design. Proceedings, 2002. XI, 399 pages. 2002.

Vol. 2518: P. Bose, P. Morin (Eds.), Algorithms and Computation. Proceedings, 2002. XIII, 656 pages. 2002.

Vol. 2519: R. Meersman, Z. Tari, et al. (Eds.), On the Move to Meaningful Internet Systems 2002: CoopIS, DOA, and ODBASE. Proceedings, 2002. XXIII, 1367 pages. 2002.

Vol. 2521: A. Karmouch, T. Magedanz, J. Delgado (Eds.), Mobile Agents for Telecommunication Applications. Proceedings, 2002. XII, 317 pages. 2002.

Vol. 2522: T. Andreasen, A. Motro, H. Christiansen, H. Legind Larsen (Eds.), Flexible Query Answering. Proceedings, 2002. XI, 386 pages. 2002. (Subseries LNAI).

Vol. 2523: B.S. Kaliski Jr., Ç.K. Koç, C. Paar (Eds.), Cryptographic Hardware and Embedded Systems – CHES 2002. Proceedings, 2002. XIV, 612 pages. 2002.

Vol. 2525: H.H. Bülthoff, S.-Whan Lee, T.A. Poggio, C. Wallraven (Eds.), Biologically Motivated Computer Vision. Proceedings, 2002. XIV, 662 pages. 2002.

Vol. 2526: A. Colosimo, A. Giuliani, P. Sirabella (Eds.), Medical Data Analysis. Proceedings, 2002. IX, 222 pages. 2002.

Vol. 2527: F.J. Garijo, J.C. Riquelme, M. Toro (Eds.), Advances in Artificial Intelligence – IBERAMIA 2002. Proceedings, 2002. XVIII, 955 pages. 2002. (Subseries LNAI).

Vol. 2528: M.T. Goodrich, S.G. Kobourov (Eds.), Graph Drawing. Proceedings, 2002. XIII, 384 pages. 2002.

Vol. 2529: D.A. Peled, M.Y. Vardi (Eds.), Formal Techniques for Networked and Distributed Sytems – FORTE 2002. Proceedings, 2002. XI, 371 pages. 2002.

Vol. 2531: J. Padget, O. Shehory, D. Parkes, N. Sadeh, W.E. Walsh (Eds.), Agent-Mediated Electronic Commerce IV. Proceedings, 2002. XVII, 341 pages. 2002. (Subseries LNAI).

Vol. 2532: Y.-C. Chen, L.-W. Chang, C.-T. Hsu (Eds.), Advances in Multimedia Information Processing – PCM 2002. Proceedings, 2002. XXI, 1255 pages. 2002.

Vol. 2533: N. Cesa-Bianchi, M. Numao, R. Reischuk (Eds.), Algorithmic Learning Theory. Proceedings, 2002. XI, 415 pages. 2002. (Subseries LNAI).

Vol. 2534: S. Lange, K. Satoh, C.H. Smith (Ed.), Discovery Science. Proceedings, 2002. XIII, 464 pages. 2002.

Vol. 2535: N. Suri (Ed.), Mobile Agents. Proceedings, 2002. X, 203 pages. 2002.

Vol. 2536: M. Parashar (Ed.), Grid Computing – GRID 2002. Proceedings, 2002. XI, 318 pages. 2002.

Vol. 2537: D.G. Feitelson, L. Rudolph, U. Schwiegelshohn (Eds.), Job Scheduling Strategies for Parallel Processing. Proceedings, 2002. VII, 237 pages. 2002.

Vol. 2538: B. König-Ries, K. Makki, S.A.M. Makki, N. Pissinou, P. Scheuermann (Eds.), Developing an Infrastructure for Mobile and Wireless Systems. Proceedings 2001. X, 183 pages. 2002.

Vol. 2539: K. Börner, C. Chen (Eds.), Visual Interfaces to Digital Libraries. X, 233 pages. 2002.

Vol. 2540: W.I. Grosky, F. Plášil (Eds.), SOFSEM 2002: Theory and Practice of Informatics. Proceedings, 2002. X, 289 pages. 2002.

Vol. 2541: T. Barkowsky, Mental Representation and Processing of Geographic Knowledge. X, 174 pages. 2002. (Subseries LNAI).

Vol. 2542: I. Dimov, I. Lirkov, S. Margenov, Z. Zlatev (Eds.), Numerical Methods and Applications. Proceedings, 2002. X, 174 pages. 2003.

Vol. 2544: S. Bhalla (Ed.), Databases in Networked Information Systems. Proceedings 2002. X, 285 pages. 2002.

Vol. 2545: P. Forbrig, Q. Limbourg, B. Urban, J. Vanderdonckt (Eds.), Interactive Systems. Proceedings 2002. XII, 574 pages. 2002.

Vol. 2546: J. Sterbenz, O. Takada, C. Tschudin, B. Plattner (Eds.), Active Networks. Proceedings, 2002. XIV, 267 pages. 2002.

Vol. 2547: R. Fleischer, B. Moret, E. Meineche Schmidt (Eds.), Experimental Algorithmics. XVII, 279 pages. 2002.

Vol. 2548: J. Hernández, Ana Moreira (Eds.), Object-Oriented Technology. Proceedings, 2002. VIII, 223 pages. 2002.

Vol. 2549: J. Cortadella, A. Yakovlev, G. Rozenberg (Eds.), Concurrency and Hardware Design. XI, 345 pages. 2002.

Vol. 2550: A. Jean-Marie (Ed.), Advances in Computing Science – ASIAN 2002. Proceedings, 2002. X, 233 pages. 2002.

Vol. 2551: A. Menezes, P. Sarkar (Eds.), Progress in Cryptology – INDOCRYPT 2002. Proceedings, 2002. XI, 437 pages. 2002.

Vol. 2552: S. Sahni, V.K. Prasanna, U. Shukla (Eds.), High Performance Computing – HiPC 2002. Proceedings, 2002. XXI, 735 pages. 2002.

Vol. 2553: B. Andersson, M. Bergholtz, P. Johannesson (Eds.), Natural Language Processing and Information Systems. Proceedings, 2002. X, 241 pages. 2002.

Vol. 2554: M. Beetz, Plan-Based Control of Robotic Agents. XI, 191 pages. 2002. (Subseries LNAI).

Vol. 2555: E.-P. Lim, S. Foo, C. Khoo, H. Chen, E. Fox, S. Urs, T. Costantino (Eds.), Digital Libraries: People, Knowledge, and Technology. Proceedings, 2002. XVII, 535 pages. 2002.

Vol. 2556: M. Agrawal, A. Seth (Eds.), FST TCS 2002: Foundations of Software Technology and Theoretical Computer Science. Proceedings, 2002. XI, 361 pages. 2002.

Vol. 2557: B. McKay, J. Slaney (Eds.), AI 2002: Advances in Artificial Intelligence. Proceedings, 2002. XV, 730 pages. 2002. (Subseries LNAI).

Vol. 2558: P. Perner, Data Mining on Multimedia Data. X, 131 pages. 2002.

Vol. 2559: M. Oivo, S. Komi-Sirviö (Eds.), Product Focused Software Process Improvement. Proceedings, 2002. XV, 646 pages. 2002.

Vol. 2560: S. Goronzy, Robust Adaptation to Non-Native Accents in Automatic Speech Recognition. Proceedings, 2002. XI, 144 pages. 2002. (Subseries LNAI).

Vol. 2561: H.C.M. de Swart (Ed.), Relational Methods in Computer Science. Proceedings, 2001. X, 315 pages. 2002.

Vol. 2562: V. Dahl, P. Wadler (Eds.), Practical Aspects of Declarative Languages. Proceedings, 2003. X, 315 pages. 2002.

Vol. 2566: T.Æ. Mogensen, D.A. Schmidt, I.H. Sudborough (Eds.), The Essence of Computation. XIV, 473 pages. 2002.

Vol. 2567: Y.G. Desmedt (Ed.), Public Key Cryptography – PKC 2003. Proceedings, 2003. XI, 365 pages. 2002.

Vol. 2568: M. Hagiya, A. Ohuchi (Eds.), DNA Computing. Proceedings, 2002. XI, 338 pages. 2003.

Vol. 2569: D. Gollmann, G. Karjoth, M. Waidner (Eds.), Computer Security – ESORICS 2002. Proceedings, 2002. XIII, 648 pages. 2002. (Subseries LNAI).

Vol. 2570: M. Jünger, G. Reinelt, G. Rinaldi (Eds.), Combinatorial Optimization – Eureka, You Shrink!. Proceedings, 2001. X, 209 pages. 2003.

Vol. 2571: S.K. Das, S. Bhattacharya (Eds.), Distributed Computing. Proceedings, 2002. XIV, 354 pages. 2002.

Vol. 2572: D. Calvanese, M. Lenzerini, R. Motwani (Eds.), Database Theory – ICDT 2003. Proceedings, 2003. XI, 455 pages. 2002.

Vol. 2574: M.-S. Chen, P.K. Chrysanthis, M. Sloman, A. Zaslavsky (Eds.), Mobile Data Management. Proceedings, 2003. XII, 414 pages. 2003.

Vol. 2575: L.D. Zuck, P.C. Attie, A. Cortesi, S. Mukhopadhyay (Eds.), Verification, Model Checking, and Abstract Interpretation. Proceedings, 2003. XI, 325 pages. 2003.

Vol. 2576: S. Cimato, C. Galdi, G. Persiano (Eds.), Security in Communication Networks. Proceedings, 2002. IX, 365 pages. 2003.

Vol. 2578: F.A.P. Petitcolas (Ed.), Information Hiding. Proceedings, 2002. IX, 427 pages. 2003.

Vol. 2580: H. Erdogmus, T. Weng (Eds.), COTS-Based Software Systems. Proceedings, 2003. XVIII, 261 pages. 2003.

Vol. 2581: J.S. Sichman, F. Bousquet, P. Davidsson (Eds.), Multi-Agent-Based Simulation II. Proceedings, 2002. X, 195 pages. 2003. (Subseries LNAI).

Vol. 2583: S. Matwin, C. Sammut (Eds.), Inductive Logic Programming. Proceedings, 2002. X, 351 pages. 2003. (Subseries LNAI).

Vol. 2588: A. Gelbukh (Ed.), Computational Linguistics and Intelligent Text Processing. Proceedings, 2003. XV, 648 pages. 2003.

Vol. 2594: A. Asperti, B. Buchberger, J.H. Davenport (Eds.), Mathematical Knowledge Management. Proceedings, 2003. X, 225 pages. 2003.

Vol. 2598: R. Klein, H.-W. Six, L. Wegner (Eds.), Computer Science in Perspective. X, 357 pages. 2003.

Vol. 2600: S. Mendelson, A.J. Smola, Advanced Lectures on Machine Learning. Proceedings, 2002. IX, 259 pages. 2003. (Subseries LNAI).